DISCRETE-TIME CONTROL SYSTEMS

DISCRETE-TIME CONTROL SYSTEMS

Second Edition

Katsuhiko Ogata
University of Minnesota

PRENTICE HALL, Upper Saddle River, New Jersey 07458

Library of Congress Cataloging-in-Publication Data

Ogata, Katsuhiko.
   Discrete-time control systems / Katsuhiko Ogata. — 2nd ed.
     p.   cm.
   Includes bibliographical references and index.
   ISBN 0-13-034281-5
    1. Discrete-time systems. 2. Control theory. I. Title.
QA402.04 1994   94-19896
629.8′3—dc20   CIP

Editorial/production supervision: Lynda Griffiths/TKM Productions
Cover design: Karen Salzbach
Production coordinator: David Dickey/Bill Scazzero

 © 1995, 1987 by Prentice-Hall, Inc.
Simon & Schuster / A Viacom Company
Upper Saddle River, New Jersey 07458

Printed in the United States of America
10 9 8 7 6

ISBN 0-13-034281-5

Prentice-Hall International (UK) Limited, *London*
Prentice-Hall of Australia Pty. Limited, *Sydney*
Prentice-Hall Canada Inc., *Toronto*
Prentice-Hall Hispanoamericana, S.A., *Mexico*
Prentice-Hall of India Private Limited, *New Delhi*
Prentice-Hall of Japan, Inc., *Tokyo*
Simon & Schuster Asia Pte. Ltd., *Singapore*
Editora Prentice-Hall do Brasil, Ltda., *Rio de Janeiro*

Contents

Chapter 3

z-Plane Analysis of Discrete-Time Control Systems   74

Chapter 4

Design of Discrete-Time Control Systems by Conventional Methods   173

Chapter 5

State-Space Analysis   293

Chapter 6

Pole Placement and Observer Design   377

Chapter 7

Polynomial Equations Approach to Control Systems Design   517

Chapter 8

Quadratic Optimal Control Systems   566

Appendix A

Vector-Matrix Analysis   633

Appendix B

z Transform Theory   681

Appendix C

Pole Placement Design with Vector Control   704

References   730

Index   735

Preface

This book presents a comprehensive treatment of the analysis and design of discrete-time control systems. It is written as a textbook for courses on discrete-time control systems or digital control systems for senior and first-year graduate level engineering students.

In this second edition, some of the older material has been deleted and new material has been added throughout the book. The most significant feature of this edition is a greatly expanded treatment of the pole-placement design with minimum-order observer by means of the state-space approach (Chapter 6) and the polynomial-equations approach (Chapter 7).

In this book all materials are presented in such a way that the reader can follow the discussions easily. All materials necessary for understanding the subject matter presented (such as proofs of theorems and steps for deriving important equations for pole placement and observer design) are included to ease understanding of the subject matter presented.

The theoretical background materials for designing control systems are discussed in detail. Once the theoretical aspects are understood, the reader can use MATLAB with advantage to obtain numerical solutions that involve various types of vector-matrix operations. It is assumed that the reader is familiar with the material presented in my book *Solving Control Engineering Problems with MATLAB* (Prentice Hall) or its equivalent.

The prerequisites for the reader are a course on introductory control systems, a course on ordinary differential equations, and familiarity with MATLAB computations. (If the reader is not familiar with MATLAB, it may be studied concurrently.)

Since this book is written from the engineer's point of view, the basic concepts involved are emphasized and highly mathematical arguments are carefully avoided in the presentation. The entire text has been organized toward a gradual development of discrete-time control theory.

The text is organized into eight chapters and three appendixes. The outline of the book is as follows: Chapter 1 gives an introduction to discrete-time control systems. Chapter 2 presents the z transform theory necessary for the study of discrete-time control systems. Chapter 3 discusses the z plane analysis of discrete-time systems, including impulse sampling, data hold, sampling theorem, pulse transfer function, and digital filters. Chapter 4 treats the design of discrete-time control systems by conventional methods. This chapter includes stability analysis of closed-loop systems in the z plane, transient and steady-state response analyses, and design based on the root-locus method, frequency-response method, and analytical method.

Chapter 5 presents state-space analysis, including state-space representations of discrete-time systems, pulse transfer function matrix, discretization method, and Liapunov stability analysis. Chapter 6 discusses pole-placement and observer design. This chapter contains discussions on controllability, observability, pole placement, state observers, and servo systems. Chapter 7 treats the polynomial equations approach to control systems design. This chapter first discusses the Diophantine equation and then presents the polynomial equations approach to control systems design. Finally, model matching control systems are designed using the polynomial equations approach. Chapter 8 presents quadratic optimal control. Both finite-stage and infinite-stage quadratic optimal control problems are discussed. This chapter concludes with a design problem based on quadratic optimal control solved with MATLAB.

Appendix A presents a summary of vector-matrix analysis. Appendix B gives useful theorems of the z transform theory that were not presented in Chapter 2, the inversion integral method, and the modified z transform method. Appendix C discusses the pole-placement design problem when the control signal is a vector quantity.

Examples are presented at strategic points throughout the book so that the reader will have a better understanding of the subject matter discussed. In addition, a number of solved problems (A problems) are provided at the end of each chapter, except Chapter 1. These problems represent an integral part of the text. It is suggested that the reader study all these problems carefully to obtain a deeper understanding of the topics discussed. In addition, many unsolved problems (B problems) are provided for use as homework or quiz problems.

Most of the materials presented in this book have been class-tested in senior and first-year graduate level courses on control systems at the University of Minnesota.

All the materials in this book may be covered in two quarters. In a semester course, the instructor will have some flexibility in choosing the subjects to be covered. In a quarter course, a good part of the first six chapters may be covered. An instructor using this text can obtain a complete solutions manual from the publisher. This book can also serve as a self-study book for practicing engineers who wish to study discrete-time control theory by themselves.

Appreciation is due to my former students who solved all the solved problems (A problems) and unsolved problems (B problems) and made numerous constructive comments about the material in this book.

Katsuhiko Ogata

DISCRETE-TIME
CONTROL
SYSTEMS

1

Introduction to Discrete-Time Control Systems

1–1 INTRODUCTION

In recent years there has been a rapid increase in the use of digital controllers in control systems. Digital controls are used for achieving optimal performance—for example, in the form of maximum productivity, maximum profit, minimum cost, or minimum energy use.

Most recently, the application of computer control has made possible "intelligent" motion in industrial robots, the optimization of fuel economy in automobiles, and refinements in the operation of household appliances and machines such as microwave ovens and sewing machines, among others. Decision-making capability and flexibility in the control program are major advantages of digital control systems.

The current trend toward digital rather than analog control of dynamic systems is mainly due to the availability of low-cost digital computers and the advantages found in working with digital signals rather than continuous-time signals.

Types of Signals. A continuous-time signal is a signal defined over a continuous range of time. The amplitude may assume a continuous range of values or may assume only a finite number of distinct values. The process of representing a variable by a set of distinct values is called *quantization*, and the resulting distinct values are called *quantized* values. The quantized variable changes only by a set of distinct steps.

An analog signal is a signal defined over a continuous range of time whose amplitude can assume a continuous range of values. Figure 1–1(a) shows a continuous-time analog signal, and Figure 1–1(b) shows a continuous-time quantized signal (quantized in amplitude only).

1

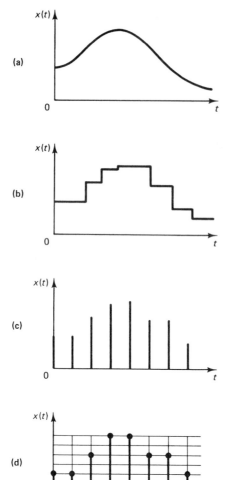

Figure 1–1 (a) Continuous-time analog signal; (b) continuous-time quantized signal; (c) sampled-data signal; (d) digital signal.

Notice that the analog signal is a special case of the continuous-time signal. In practice, however, we frequently use the terminology "continuous-time" in lieu of "analog." Thus in the literature, including this book, the terms "continuous-time signal" and "analog signal" are frequently interchanged, although strictly speaking they are not quite synonymous.

A discrete-time signal is a signal defined only at discrete instants of time (that is, one in which the independent variable t is quantized). In a discrete-time signal, if the amplitude can assume a continuous range of values, then the signal is called a *sampled-data signal*. A sampled-data signal can be generated by sampling an analog signal at discrete instants of time. It is an amplitude-modulated pulse signal. Figure 1–1(c) shows a sampled-data signal.

A digital signal is a discrete-time signal with quantized amplitude. Such a signal can be represented by a sequence of numbers, for example, in the form of binary numbers. (In practice, many digital signals are obtained by sampling analog signals and then quantizing them; it is the quantization that allows these analog signals to be read as finite binary words.) Figure 1–1(d) depicts a digital signal. Clearly, it is a signal quantized both in amplitude and in time. The use of the digital controller requires quantization of signals both in amplitude and in time.

The term "discrete-time signal" is broader than the term "digital signal" or the term "sampled-data signal." In fact, a discrete-time signal can refer either to a digital signal or to a sampled-data signal. In practical usage, the terms "discrete time" and "digital" are often interchanged. However, the term "discrete time" is frequently used in theoretical study, while the term "digital" is used in connection with hardware or software realizations.

In control engineering, the controlled object is a plant or process. It may be a physical plant or process or a nonphysical process such as an economic process. Most plants and processes involve continuous-time signals; therefore, if digital controllers are involved in the control systems, signal conversions (analog to digital and digital to analog) become necessary. Standard techniques are available for such signal conversions; we shall discuss them in Section 1–4.

Loosely speaking, terminologies such as discrete-time control systems, sampled-data control systems, and digital control systems imply the same type or very similar types of control systems. Precisely speaking, there are, of course, differences in these systems. For example, in a sampled-data control system both continuous-time and discrete-time signals exist in the system; the discrete-time signals are amplitude-modulated pulse signals. Digital control systems may include both continuous-time and discrete-time signals; here, the latter are in a numerically coded form. Both sampled-data control systems and digital control systems are discrete-time control systems.

Many industrial control systems include continuous-time signals, sampled-data signals, and digital signals. Therefore, in this book we use the term "discrete-time control systems" to describe the control systems that include some forms of sampled-data signals (amplitude-modulated pulse signals) and/or digital signals (signals in numerically coded form).

Systems Dealt With in This Book.   The discrete-time control systems considered in this book are mostly linear and time invariant, although nonlinear and/or time-varying systems are occasionally included in discussions. A linear system is one in which the principle of superposition applies. Thus, if y_1 is the response of the system to input x_1 and y_2 the response to input x_2, then the system is linear if and only if, for every scalar α and β, the response to input $\alpha x_1 + \beta x_2$ is $\alpha y_1 + \beta y_2$.

A linear system may be described by linear differential or linear difference equations. A time-invariant linear system is one in which the coefficients in the differential equation or difference equation do not vary with time, that is, one in which the properties of the system do not change with time.

Discrete-Time   Control   Systems   and   Continuous-Time   Control   Systems.
Discrete-time control systems are control systems in which one or more variables can change only at discrete instants of time. These instants, which we shall denote by kT or t_k ($k = 0, 1, 2, \ldots$), may specify the times at which some physical measurement

is performed or the times at which the memory of a digital computer is read out. The time interval between two discrete instants is taken to be sufficiently short that the data for the time between them can be approximated by simple interpolation.

Discrete-time control systems differ from continuous-time control systems in that signals for a discrete-time control system are in sampled-data form or in digital form. If a digital computer is involved in a control system as a digital controller, any sampled data must be converted into digital data.

Continuous-time systems, whose signals are continuous in time, may be described by differential equations. Discrete-time systems, which involve sampled-data signals or digital signals and possibly continuous-time signals as well, may be described by difference equations after the appropriate discretization of continuous-time signals.

Sampling Processes. The sampling of a continuous-time signal replaces the original continuous-time signal by a sequence of values at discrete time points. A sampling process is used whenever a control system involves a digital controller, since a sampling operation and quantization are necessary to enter data into such a controller. Also, a sampling process occurs whenever measurements necessary for control are obtained in an intermittent fashion. For example, in a radar tracking system, as the radar antenna rotates, information about azimuth and elevation is obtained once for each revolution of the antenna. Thus, the scanning operation of the radar produces sampled data. In another example, a sampling process is needed whenever a large-scale controller or computer is time-shared by several plants in order to save cost. Then a control signal is sent out to each plant only periodically and thus the signal becomes a sampled-data signal.

The sampling process is usually followed by a quantization process. In the quantization process the sampled analog amplitude is replaced by a digital amplitude (represented by a binary number). Then the digital signal is processed by the computer. The output of the computer is sampled and fed to a hold circuit. The output of the hold circuit is a continuous-time signal and is fed to the actuator. We shall present details of such signal-processing methods in the digital controller in Section 1–4.

The term "discretization," rather than "sampling," is frequently used in the analysis of multiple-input–multiple-output systems, although both mean basically the same thing.

It is important to note that occasionally the sampling operation or discretization is entirely fictitious and has been introduced only to simplify the analysis of control systems that actually contain only continuous-time signals. In fact, we often use a suitable discrete-time model for a continuous-time system. An example is a digital-computer simulation of a continuous-time system. Such a digital-computer-simulated system can be analyzed to yield parameters that will optimize a given performance index.

Most of the material presented in this book deals with control systems that can be modeled as linear time-invariant discrete-time systems. It is important to mention that many digital control systems are based on continuous-time design techniques. Since a wealth of experience has been accumulated in the design of continuous-time controllers, a thorough knowledge of them is highly valuable in designing discrete-time control systems.

1–2 DIGITAL CONTROL SYSTEMS

Figure 1–2 depicts a block diagram of a digital control system showing a configuration of the basic control scheme. The system includes the feedback control and the feedforward control. In designing such a control system, it should be noted that the "goodness" of the control system depends on individual circumstances. We need to choose an appropriate performance index for a given case and design a controller so as to optimize the chosen performance index.

Signal Forms in a Digital Control System.   Figure 1–3 shows a block diagram of a digital control system. The basic elements of the system are shown by the blocks. The controller operation is controlled by the clock. In such a digital control system, some points of the system pass signals of varying amplitude in either continuous time or discrete time, while other points pass signals in numerical code, as depicted in the figure.

The output of the plant is a continuous-time signal. The error signal is converted into digital form by the sample-and-hold circuit and the analog-to-digital converter. The conversion is done at the sampling time. The digital computer

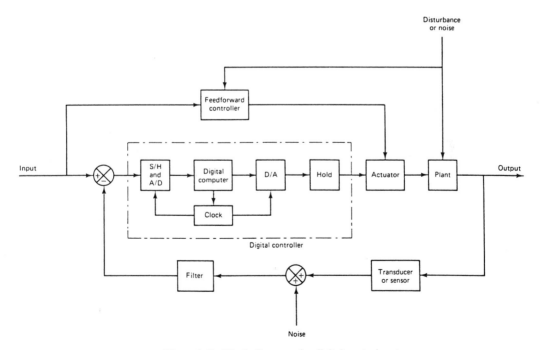

Figure 1–2   Block diagram of a digital control system.

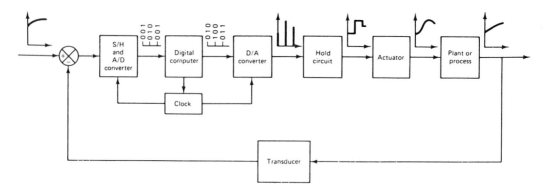

Figure 1–3   Block diagram of a digital control system showing signals in binary or graphic form.

processes the sequences of numbers by means of an algorithm and produces new sequences of numbers. At every sampling instant a coded number (usually a binary number consisting of eight or more binary digits) must be converted to a physical control signal, which is usually a continuous-time or analog signal. The digital-to-analog converter and the hold circuit convert the sequence of numbers in numerical code into a piecewise continuous-time signal. The real-time clock in the computer synchronizes the events. The output of the hold circuit, a continuous-time signal, is fed to the plant, either directly or through the actuator, to control its dynamics.

The operation that transforms continuous-time signals into discrete-time data is called *sampling* or *discretization*. The reverse operation, the operation that transforms discrete-time data into a continuous-time signal, is called *data-hold*; it amounts to a reconstruction of a continuous-time signal from the sequence of discrete-time data. It is usually done using one of the many extrapolation techniques. In many cases it is done by keeping the signal constant between the successive sampling instants. (We shall discuss such extrapolation techniques in Section 1–4.)

The sample-and-hold (S/H) circuit and analog-to-digital (A/D) converter convert the continuous-time signal into a sequence of numerically coded binary words. Such an A/D conversion process is called *coding* or *encoding*. The combination of the S/H circuit and analog-to-digital converter may be visualized as a switch that closes instantaneously at every time interval T and generates a sequence of numbers in numerical code. The digital computer operates on such numbers in numerical code and generates a desired sequence of numbers in numerical code. The digital-to-analog (D/A) conversion process is called *decoding*.

Definitions of Terms.   Before we discuss digital control systems in detail, we need to define some of the terms that appear in the block diagram of Figure 1–3.

Sample-and-Hold (S/H).   "Sample-and-hold" is a general term used for a sample-and-hold amplifier. It describes a circuit that receives an analog input signal and holds this signal at a constant value for a specified period of time. Usually the signal is electrical, but other forms are possible, such as optical and mechanical.

Analog-to-Digital Converter (A/D).   An analog-to-digital converter, also called an encoder, is a device that converts an analog signal into a digital signal, usually a numerically coded signal. Such a converter is needed as an interface between an analog component and a digital component. A sample-and-hold circuit is often an integral part of a commercially available A/D converter. The conversion of an analog signal into the corresponding digital signal (binary number) is an approximation, because the analog signal can take on an infinite number of values, whereas the variety of different numbers that can be formed by a finite set of digits is limited. This approximation process is called *quantization*. (More on quantization is presented in Section 1–3.)

Digital-to-Analog Converter (D/A).   A digital-to-analog converter, also called a decoder, is a device that converts a digital signal (numerically coded data) into an analog signal. Such a converter is needed as an interface between a digital component and an analog component.

Plant or Process.   A plant is any physical object to be controlled. Examples are a furnace, a chemical reactor, and a set of machine parts functioning together to perform a particular operation, such as a servo system or a spacecraft.

A process is generally defined as a progressive operation or development marked by a series of gradual changes that succeed one another in a relatively fixed way and lead toward a particular result or end. In this book we call any operation to be controlled a process. Examples are chemical, economic, and biological processes.

The most difficult part in the design of control systems may lie in the accurate modeling of a physical plant or process. There are many approaches to the plant or process model, but, even so, a difficulty may exist, mainly because of the absence of precise process dynamics and the presence of poorly defined random parameters in many physical plants or processes. Thus, in designing a digital controller, it is necessary to recognize the fact that the mathematical model of a plant or process in many cases is only an approximation of the physical one. Exceptions are found in the modeling of electromechanical systems and hydraulic-mechanical systems, since these may be modeled accurately. For example, the modeling of a robot arm system may be accomplished with great accuracy.

Transducer.   A transducer is a device that converts an input signal into an output signal of another form, such as a device that converts a pressure signal into a voltage output. The output signal, in general, depends on the past history of the input.

Transducers may be classified as analog transducers, sampled-data transducers, or digital transducers. An analog transducer is a transducer in which the input and output signals are continuous functions of time. The magnitudes of these signals may be any values within the physical limitations of the system. A sampled-data transducer is one in which the input and output signals occur only at discrete instants of time (usually periodic), but the magnitudes of the signals, as in the case of the analog transducer, are unquantized. A digital transducer is one in which the input and output signals occur only at discrete instants of time and the signal magnitudes are quantized (that is, they can assume only certain discrete levels).

Types of Sampling Operations.   As stated earlier, a signal whose independent variable t is discrete is called a discrete-time signal. A sampling operation is basic in transforming a continuous-time signal into a discrete-time signal.

There are several different types of sampling operations of practical importance:

1. *Periodic sampling.* In this case, the sampling instants are equally spaced, or $t_k = kT$ $(k = 0, 1, 2, \dots)$. Periodic sampling is the most conventional type of sampling operation.
2. *Multiple-order sampling.* The pattern of the t_k's is repeated periodically; that is, $t_{k+r} - t_k$ is constant for all k.
3. *Multiple-rate sampling.* In a control system having multiple loops, the largest time constant involved in one loop may be quite different from that in other loops. Hence, it may be advisable to sample slowly in a loop involving a large time constant, while in a loop involving only small time constants the sampling rate must be fast. Thus, a digital control system may have different sampling periods in different feedback paths or may have multiple sampling rates.
4. *Random sampling.* In this case, the sampling instants are random, or t_k is a random variable.

In this book we shall treat only the case where the sampling is periodic.

1–3 QUANTIZING AND QUANTIZATION ERROR

The main functions involved in analog-to-digital conversion are sampling, amplitude quantizing, and coding. When the value of any sample falls between two adjacent "permitted" output states, it must be read as the permitted output state nearest the actual value of the signal. The process of representing a continuous or analog signal by a finite number of discrete states is called *amplitude quantization.* That is, "quantizing" means transforming a continuous or analog signal into a set of discrete states. (Note that quantizing occurs whenever a physical quantity is represented numerically.)

The output state of each quantized sample is then described by a numerical code. The process of representing a sample value by a numerical code (such as a binary code) is called *encoding* or *coding.* Thus, encoding is a process of assigning a digital word or code to each discrete state. The sampling period and quantizing levels affect the performance of digital control systems. So they must be determined carefully.

Quantizing.   The standard number system used for processing digital signals is the binary number system. In this system the code group consists of n pulses each indicating either "on" (1) or "off" (0). In the case of quantizing, n "on–off" pulses can represent 2^n amplitude levels or output states.

The quantization level Q is defined as the range between two adjacent decision points and is given by

$$Q = \frac{\text{FSR}}{2^n}$$

where the FSR is the full-scale range. Note that the leftmost bit of the natural binary code has the most weight (one-half of the full scale) and is called the most significant bit (MSB). The rightmost bit has the least weight ($1/2^n$ times the full scale) and is called the least significant bit (LSB). Thus,

$$\text{LSB} = \frac{\text{FSR}}{2^n}$$

The least significant bit is the quantization level Q.

Quantization Error.   Since the number of bits in the digital word is finite, A/D conversion results in a finite resolution. That is, the digital output can assume only a finite number of levels, and therefore an analog number must be rounded off to the nearest digital level. Hence, any A/D conversion involves quantization error. Such quantization error varies between 0 and $\pm\frac{1}{2}Q$. This error depends on the fineness of the quantization level and can be made as small as desired by making the quantization level smaller (that is, by increasing the number of bits n). In practice, there is a maximum for the number of bits n, and so there is always some error due to quantization. The uncertainty present in the quantization process is called *quantization noise*.

To determine the desired size of the quantization level (or the number of output states) in a given digital control system, the engineer must have a good understanding of the relationship between the size of the quantization level and the resulting error. The variance of the quantization noise is an important measure of quantization error, since the variance is proportional to the average power associated with the noise.

Figure 1–4(a) shows a block diagram of a quantizer together with its input–output characteristics. For an analog input $x(t)$, the output $y(t)$ takes on only a finite number of levels, which are integral multiples of the quantization level Q.

In numerical analysis the error resulting from neglecting the remaining digits is called the *round-off error*. Since the quantizing process is an approximating process in that the analog quantity is approximated by a finite digital number, the quantization error is a round-off error. Clearly, the finer the quantization level is, the smaller the round-off error.

Figure 1–4(b) shows an analog input $x(t)$ and the discrete output $y(t)$, which is in the form of a staircase function. The quantization error $e(t)$ is the difference between the input signal and the quantized output, or

$$e(t) = x(t) - y(t)$$

Note that the magnitude of the quantized error is

$$0 \le |e(t)| \le \tfrac{1}{2}Q$$

For a small quantization level Q, the nature of the quantization error is similar to that of random noise. And, in effect, the quantization process acts as a source of random noise. In what follows we shall obtain the variance of the quantization noise. Such variance can be obtained in terms of the quantization level Q.

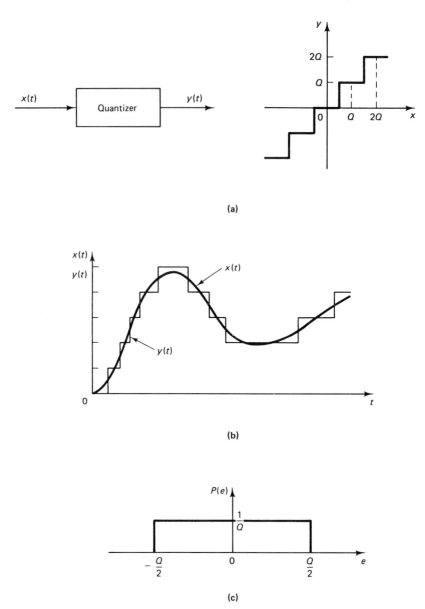

Figure 1–4 (a) Block diagram of a quantizer and its input–output characteristics; (b) analog input $x(t)$ and discrete output $y(t)$; (c) probability distribution $P(e)$ of quantization error $e(t)$.

Suppose that the quantization level Q is small and we assume that the quantization error $e(t)$ is distributed uniformly between $-\frac{1}{2}Q$ and $\frac{1}{2}Q$ and that this error acts as a white noise. [This is obviously a rather rough assumption. However, since the quantization error signal $e(t)$ is of a small amplitude, such an assumption may be acceptable as a first-order approximation.] The probability distribution $P(e)$ of signal $e(t)$ may be plotted as shown in Figure 1–4(c). The average value of $e(t)$ is zero, or $e(t) = 0$. Then the variance σ^2 of the quantization noise is

$$\sigma^2 = E[e(t) - \overline{e(t)}]^2 = \frac{1}{Q}\int_{-Q/2}^{Q/2} \xi^2 \, d\xi = \frac{Q^2}{12}$$

Thus, if the quantization level Q is small compared with the average amplitude of the input signal, then the variance of the quantization noise is seen to be one-twelfth of the square of the quantization level.

1-4 DATA ACQUISITION, CONVERSION, AND DISTRIBUTION SYSTEMS

With the rapid growth in the use of digital computers to perform digital control actions, both the data-acquisition system and the distribution system have become an important part of the entire control system.

The signal conversion that takes place in the digital control system involves the following operations:

1. Multiplexing and demultiplexing
2. Sample and hold
3. Analog-to-digital conversion (quantizing and encoding)
4. Digital-to-analog conversion (decoding)

Figure 1–5(a) shows a block diagram of a data-acquisition system, and Figure 1–5(b) shows a block diagram of a data-distribution system.

In the data-acquisition system the input to the system is a physical variable such as position, velocity, acceleration, temperature, or pressure. Such a physical variable is first converted into an electrical signal (a voltage or current signal) by a suitable

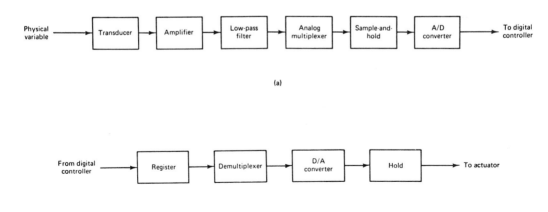

Figure 1–5 (a) Block diagram of a data-acquisition system; (b) block diagram of a data-distribution system.

transducer. Once the physical variable is converted into a voltage or current signal, the rest of the data-acquisition process is done by electronic means.

In Figure 1–5(a) the amplifier (frequently an operational amplifier) that follows the transducer performs one or more of the following functions: It amplifies the voltage output of the transducer; it converts a current signal into a voltage signal; or it buffers the signal. The low-pass filter that follows the amplifier attenuates the high-frequency signal components, such as noise signals. (Note that electronic noises are random in nature and may be reduced by low-pass filters. However, such common electrical noises as power-line interference are generally periodic and may be reduced by means of notch filters.) The output of the low-pass filter is an analog signal. This signal is fed to the analog multiplexer. The output of the multiplexer is fed to the sample-and-hold circuit, whose output is, in turn, fed to the analog-to-digital converter. The output of the converter is the signal in digital form; it is fed to the digital controller.

The reverse of the data-acquisition process is the data-distribution process. As shown in Figure 1–5(b), a data-distribution system consists of registers, a demultiplexer, digital-to-analog converters, and hold circuits. It converts the signal in digital form (binary numbers) into analog form. The output of the D/A converter is fed to the hold circuit. The output of the hold circuit is fed to the analog actuator, which, in turn, directly controls the plant under consideration.

In the following, we shall discuss each individual component involved in the signal-processing system.

Analog Multiplexer. An analog-to-digital converter is the most expensive component in a data-acquisition system. The analog multiplexer is a device that performs the function of time-sharing an A/D converter among many analog channels. The processing of a number of channels with a digital controller is possible because the width of each pulse representing the input signal is very narrow, so the empty space during each sampling period may be used for other signals. If many signals are to be processed by a single digital controller, then these input signals must be fed to the controller through a multiplexer.

Figure 1–6 shows a schematic diagram of an analog multiplexer. The analog

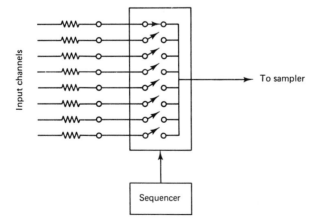

Figure 1–6 Schematic diagram of an analog multiplexer.

multiplexer is a multiple switch (usually an electronic switch) that sequentially switches among many analog input channels in some prescribed fashion. The number of channels, in many instances, is 4, 8, or 16. At a given instant of time, only one switch is in the "on" position. When the switch is on in a given input channel, the input signal is connected to the output of the multiplexer for a specified period of time. During the connection time the sample-and-hold circuit samples the signal voltage (analog signal) and holds its value, while the analog-to-digital converter converts the analog value into digital data (binary numbers). Each channel is read in a sequential order, and the corresponding values are converted into digital data in the same sequence.

Demultiplexer. The demultiplexer, which is synchronized with the input sampling signal, separates the composite output digital data from the digital controller into the original channels. Each channel is connected to a D/A converter to produce the output analog signal for that channel.

Sample-and-Hold Circuits. A sampler in a digital system converts an analog signal into a train of amplitude-modulated pulses. The hold circuit holds the value of the sampled pulse signal over a specified period of time. The sample-and-hold is necessary in the A/D converter to produce a number that accurately represents the input signal at the sampling instant. Commercially, sample-and-hold circuits are available in a single unit, known as a sample-and-hold (S/H). Mathematically, however, the sampling operation and the holding operation are modeled separately (see Section 3–2). It is common practice to use a single analog-to-digital converter and multiplex many sampled analog inputs into it.

In practice, sampling duration is very short compared with the sampling period T. When the sampling duration is negligible, the sampler may be considered an "ideal sampler." An ideal sampler enables us to obtain a relatively simple mathematical model for a sample-and-hold. (Such a mathematical model will be discussed in detail in Section 3–2).

Figure 1–7 shows a simplified diagram for the sample-and-hold. The S/H circuit is an analog circuit (simply a voltage memory device) in which an input voltage is acquired and then stored on a high-quality capacitor with low leakage and low dielectric absorption characteristics.

In Figure 1–7 the electronic switch is connected to the hold capacitor. Operational amplifier 1 is an input buffer amplifier with a high input impedance. Operational amplifier 2 is the output amplifier; it buffers the voltage on the hold capacitor.

There are two modes of operation for a sample-and-hold circuit: the tracking mode and the hold mode. When the switch is closed (that is, when the input signal is connected), the operating mode is the tracking mode. The charge on the capacitor in the circuit tracks the input voltage. When the switch is open (the input signal is disconnected), the operating mode is the hold mode and the capacitor voltage holds constant for a specified time period. Figure 1–8 shows the tracking mode and the hold mode.

Note that, practically speaking, switching from the tracking mode to the hold mode is not instantaneous. If the hold command is given while the circuit is in the

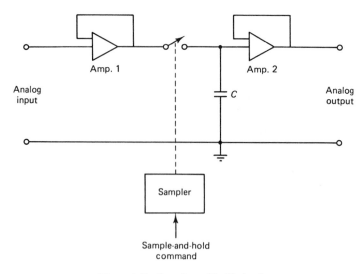

Figure 1–7   Sample-and-hold circuit.

tracking mode, then the circuit will stay in the tracking mode for a short while before reacting to the hold command. The time interval during which the switching takes place (that is, the time interval when the measured amplitude is uncertain) is called the *aperture time*.

The output voltage during the hold mode may decrease slightly. The hold mode droop may be reduced by using a high-input-impedance output buffer amplifier. Such an output buffer amplifier must have very low bias current.

The sample-and-hold operation is controlled by a periodic clock.

Types of Analog-to-Digital (A/D) Converters.   As stated earlier, the process by which a sampled analog signal is quantized and converted to a binary number is called *analog-to-digital conversion*. Thus, an A/D converter transforms an analog

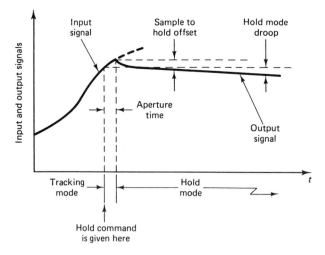

Figure 1–8   Tracking mode and hold mode.

signal (usually in the form of a voltage or current) into a digital signal or numerically coded word. In practice, the logic is based on binary digits composed of 0s and 1s, and the representation has only a finite number of digits. The A/D converter performs the operations of sample-and-hold, quantizing, and encoding. Note that in the digital system a pulse is supplied every sampling period T by a clock. The A/D converter sends a digital signal (binary number) to the digital controller each time a pulse arrives.

Among the many A/D circuits available, the following types are used most frequently:

1. Successive-approximation type
2. Integrating type
3. Counter type
4. Parallel type

Each of these four types has its own advantages and disadvantages. In any particular application, the conversion speed, accuracy, size, and cost are the main factors to be considered in choosing the type of A/D converter. (If greater accuracy is needed, for example, the number of bits in the output signal must be increased.)

As will be seen, analog-to-digital converters use as part of their feedback loops digital-to-analog converters. The simplest type of A/D converter is the counter type. The basic principle on which it works is that clock pulses are applied to the digital counter in such a way that the output voltage of the D/A converter (that is, part of the feedback loop in the A/D converter) is stepped up one least significant bit (LSB) at a time, and the output voltage is compared with the analog input voltage once for each pulse. When the output voltage has reached the magnitude of the input voltage, the clock pulses are stopped. The counter output voltage is then the digital output.

The successive-approximation type of A/D converter is much faster than the counter type and is the one most frequently used. Figure 1–9 shows a schematic diagram of the successive-approximation type of A/D converter.

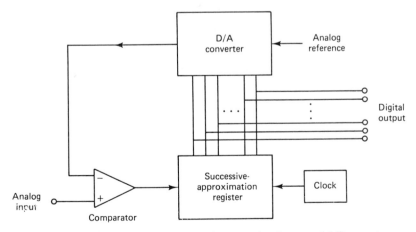

Figure 1–9   Schematic diagram of a successive-approximation-type of A/D converter.

The principle of operation of this type of A/D converter is as follows. The successive-approximation register (SAR) first turns on the most significant bit (half the maximum) and compares it with the analog input. The comparator decides whether to leave the bit on or turn it off. If the analog input voltage is larger, the most significant bit is set on. The next step is to turn on bit 2 and then compare the analog input voltage with three-fourths of the maximum. After n comparisons are completed, the digital output of the successive-approximation register indicates all those bits that remain on and produces the desired digital code. Thus, this type of A/D converter sets 1 bit each clock cycle, and so it requires only n clock cycles to generate n bits, where n is the resolution of the converter in bits. (The number n of bits employed determines the accuracy of conversion.) The time required for the conversion is approximately 2 μsec or less for a 12-bit conversion.

Errors in A/D Converters.   Actual analog-to-digital signal converters differ from the ideal signal converter in that the former always have some errors, such as offset error, linearity error, and gain error, the characteristics of which are shown in Figure 1–10. Also, it is important to note that the input–output characteristics change with time and temperature.

Finally, it is noted that commercial converters are specified for three basic temperature ranges: commercial (0°C to 70°C), industrial (−25°C to 85°C), and military (−55°C to 125°C).

Digital-to-Analog (D/A) Converters.   At the output of the digital controller the digital signal must be converted to an analog signal by the process called *digital-to-analog conversion*. A D/A converter is a device that transforms a digital input (binary numbers) to an analog output. The output, in most cases, is the voltage signal.

For the full range of the digital input, there are 2^n corresponding different analog values, including 0. For the digital-to-analog conversion there is a one-to-one correspondence between the digital input and the analog output.

Two methods are commonly used for digital-to-analog conversion: the method using weighted resistors, and the one using the R–$2R$ ladder network. The former is simple in circuit configuration, but its accuracy may not be very good. The latter is a little more complicated in configuration, but is more accurate.

Figure 1–11 shows a schematic diagram of a D/A converter using weighted resistors. The input resistors of the operational amplifier have their resistance values weighted in a binary fashion. When the logic circuit receives binary 1, the switch (actually an electronic gate) connects the resistor to the reference voltage. When the logic circuit receives binary 0, the switch connects the resistor to ground. The digital-to-analog converters used in common practice are of the parallel type: all bits act simultaneously upon application of a digital input (binary numbers).

The D/A converter thus generates the analog output voltage corresponding to the given digital voltage. For the D/A converter shown in Figure 1–11, if the binary number is $b_3 b_2 b_1 b_0$, where each of the b's can be either a 0 or a 1, then the output is

$$V_o = \frac{R_o}{R}\left(b_3 + \frac{b_2}{2} + \frac{b_1}{4} + \frac{b_0}{8}\right)V_{\text{ref}}$$

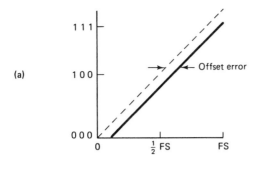

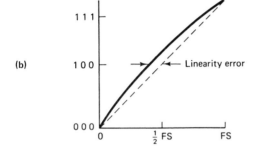

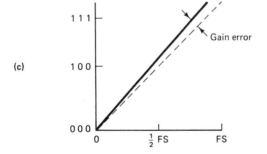

Figure 1–10 Errors in A/D converters: (a) offset error; (b) linearity error; (c) gain error.

Notice that as the number of bits is increased the range of resistor values becomes large and consequently the accuracy becomes poor.

Figure 1–12 shows a schematic diagram of an n-bit D/A converter using an R–$2R$ ladder circuit. Note that with the exception of the feedback resistor (which is $3R$) all resistors involved are either R or $2R$. This means that a high level of accuracy can be achieved. The output voltage in this case can be given by

$$V_o = \frac{1}{2}\left(b_{n-1} + \frac{1}{2}b_{n-2} + \cdots + \frac{1}{2^{n-1}}b_0 \right)V_{\text{ref}}$$

Reconstructing the Input Signal by Hold Circuits.   The sampling operation produces an amplitude-modulated pulse signal. The function of the hold operation is

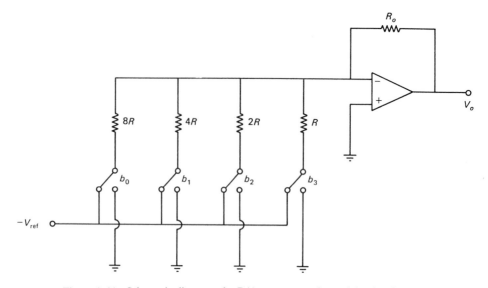

Figure 1–11 Schematic diagram of a D/A converter using weighted resistors.

to reconstruct the analog signal that has been transmitted as a train of pulse samples. That is, the purpose of the hold operation is to fill in the spaces between sampling periods and thus roughly reconstruct the original analog input signal.

The hold circuit is designed to extrapolate the output signal between successive points according to some prescribed manner. The staircase waveform of the output shown in Figure 1–13 is the simplest way to reconstruct the original input signal. The hold circuit that produces such a staircase waveform is called a *zero-order hold*. Because of its simplicity, the zero-order hold is commonly used in digital control systems.

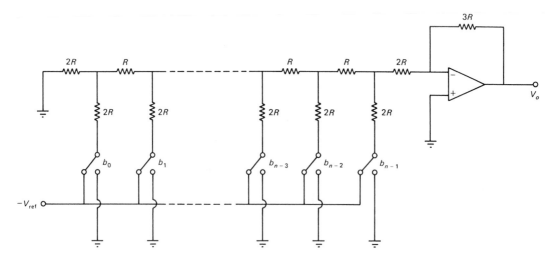

Figure 1–12   *n*-Bit D/A converter using an *R*–2*R* ladder circuit.

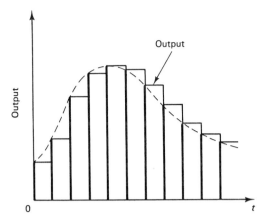

Figure 1–13 Output from a zero-order hold.

More sophisticated hold circuits are available than the zero-order hold. These are called higher-order hold circuits and include the first-order hold and the second-order hold. Higher-order hold circuits will generally reconstruct a signal more accurately than a zero-order hold, but with some disadvantages, as explained next.

The first-order hold retains the value of the previous sample, as well as the present one, and predicts, by extrapolation, the next sample value. This is done by generating an output slope equal to the slope of a line segment connecting previous and present samples and projecting it from the value of the present sample, as shown in Figure 1–14.

As can easily be seen from the figure, if the slope of the original signal does not change much, the prediction is good. If, however, the original signal reverses its slope, then the prediction is wrong and the output goes in the wrong direction, thus causing a large error for the sampling period considered.

An interpolative first-order hold, also called a *polygonal* hold, reconstructs the original signal much more accurately. This hold circuit also generates a straight-line output whose slope is equal to that joining the previous sample value and the present sample value, but this time the projection is made from the current sample point with

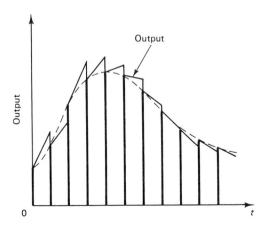

Figure 1–14 Output from a first-order hold.

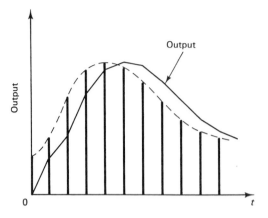

Figure 1–15 Output from an interpolative first-order hold (polygonal hold).

the amplitude of the previous sample. Hence, the accuracy in reconstructing the original signal is better than for other hold circuits, but there is a one-sampling-period delay, as shown in Figure 1–15. In effect, the better accuracy is achieved at the expense of a delay of one sampling period. From the viewpoint of the stability of closed-loop systems, such a delay is not desirable, and so the interpolative first-order hold (polygonal hold) is not used in control system applications.

1–5 CONCLUDING COMMENTS

In concluding this chapter we shall compare digital controllers and analog controllers used in industrial control systems and review digital control of processes. Then we shall present an outline of the book.

Digital Controllers and Analog Controllers. Digital controllers operate only on numbers. Decision making is one of their important functions. They are often used to solve problems involved in the optimal overall operation of industrial plants.

Digital controllers are extremely versatile. They can handle nonlinear control equations involving complicated computations or logic operations. A very much wider class of control laws can be used in digital controllers than in analog controllers. Also, in the digital controller, by merely issuing a new program the operations being performed can be changed completely. This feature is particularly important if the control system is to receive operating information or instructions from some computing center where economic analysis and optimization studies are made.

Digital controllers are capable of performing complex computations with constant accuracy at high speed and can have almost any desired degree of computational accuracy at relatively little increase in cost.

Originally, digital controllers were used as components only in large-scale control systems. At present, however, thanks to the availability of inexpensive microcomputers, digital controllers are being used in many large- and small-scale control systems. In fact, digital controllers are replacing the analog controllers that have been used in many small-scale control systems. Digital controllers are often superior in performance and lower in price than their analog counterparts.

Analog controllers represent the variables in an equation by continuous physical quantities. They can easily be designed to serve satisfactorily as non-decision-making controllers. But the cost of analog computers or analog controllers increases rapidly as the complexity of the computations increases, if constant accuracy is to be maintained.

There are additional advantages of digital controllers over analog controllers. Digital components, such as sample-and-hold circuits, A/D and D/A converters, and digital transducers, are rugged in construction, highly reliable, and often compact and lightweight. Moreover, digital components have high sensitivity, are often cheaper than their analog counterparts, and are less sensitive to noise signals. And, as mentioned earlier, digital controllers are flexible in allowing programming changes.

Digital Control of Processes.   In industrial process control systems, it is generally not practical to operate for a very long time at steady state, because certain changes may occur in production requirements, raw materials, economic factors, and processing equipments and techniques. Thus, the transient behavior of industrial processes must always be taken into consideration. Since there are interactions among process variables, using only one process variable for each control agent is not suitable for really complete control. By the use of a digital controller, it is possible to take into account all process variables, together with economic factors, production requirements, equipment performance, and all other needs, and thereby to accomplish optimal control of industrial processes.

Note that a system capable of controlling a process as completely as possible will have to solve complex equations. The more complete the control, the more important it is that the correct relations between operating variables be known and used. The system must be capable of accepting instructions from such varied sources as computers and human operators and must also be capable of changing its control subsystem completely in a short time. Digital controllers are most suitable in such situations. In fact, an advantage of the digital controller is flexibility, that is, ease of changing control schemes by reprogramming.

In the digital control of a complex process, the designer must have a good knowledge of the process to be controlled and must be able to obtain its mathematical model. (The mathematical model may be obtained in terms of differential equations or difference equations, or in some other form.) The designer must be familiar with the measurement technology associated with the output of the process and other variables involved in the process. He or she must have a good working knowledge of digital computers as well as modern control theory. If the process is complicated, the designer must investigate several different approaches to the design of the control system. In this respect, a good knowledge of simulation techniques is helpful.

Outline of the Book.   The objective of this book is to present a detailed account of the control theory that is relevant to the analysis and design of discrete-time control systems. Our emphasis is on understanding the basic concepts involved.

In this book, digital controllers are often designed in the form of pulse transfer functions or equivalent difference equations, which can be easily implemented in the form of computer programs.

The outline of the book is as follows. Chapter 1 has presented introductory material. Chapter 2 presents the z transform theory. This chapter includes z transforms of elementary functions, important properties and theorems of the z transform, the inverse z transform, and the solution of difference equations by the z transform method. Chapter 3 treats background materials for the z plane analysis of control systems. This chapter includes discussions of impulse sampling and reconstruction of original signals from sampled signals, pulse transfer functions, and realization of digital controllers and digital filters.

Chapter 4 first presents mapping between the s plane and the z plane and then discusses stability analysis of closed-loop systems in the z plane, followed by transient and steady-state response analyses, design by the root-locus and frequency-response methods, and an analytical design method. Chapter 5 gives state-space representation of discrete-time systems, the solution of discrete-time state-space equations, and the pulse transfer function matrix. Then, discretization of continuous-time state-space equations and Liapunov stability analysis are treated.

Chapter 6 presents control systems design in the state space. We begin the chapter with a detailed presentation of controllability and observability. We then present design techniques based on pole placement, followed by discussion of full-order state observers and minimum-order state observers. We conclude this chapter with the design of servo systems. Chapter 7 treats the polynomial-equations approach to the design of control systems. We begin the chapter with discussions of Diophantine equations. Then we present the design of regulator systems and control systems using the solution of Diophantine equations. The approach here is an alternative to the pole-placement approach combined with minimum-order observers. The design of model-matching control systems is included in this chapter. Finally, Chapter 8 treats quadratic optimal control problems in detail.

The state-space analysis and design of discrete-time control systems, presented in Chapters 5, 6, and 8, make extensive use of vectors and matrices. In studying these chapters the reader may, as need arises, refer to Appendix A, which summarizes the basic materials of vector-matrix analysis. Appendix B presents materials in z transform theory not included in Chapter 2. Appendix C treats pole-placement design problems when the control is a vector quantity.

In each chapter, except Chapter 1, the main text is followed by solved problems and unsolved problems. The reader should study all solved problems carefully. Solved problems are an integral part of the text. Appendixes A, B, and C are followed by solved problems. The reader who studies these solved problems will have an increased understanding of the material presented.

2

The z Transform

2–1 INTRODUCTION

A mathematical tool commonly used for the analysis and synthesis of discrete-time control systems is the z transform. The role of the z transform in discrete-time systems is similar to that of the Laplace transform in continuous-time systems.

In a linear discrete-time control system, a linear difference equation characterizes the dynamics of the system. To determine the system's response to a given input, such a difference equation must be solved. With the z transform method, the solutions to linear difference equations become algebraic in nature. (Just as the Laplace transformation transforms linear time-invariant differential equations into algebraic equations in s, the z transformation transforms linear time-invariant difference equations into algebraic equations in z.)

The main objective of this chapter is to present definitions of the z transform, basic theorems associated with the z transform, and methods for finding the inverse z transform. Solving difference equations by the z transform method is also discussed.

Discrete-Time Signals. Discrete-time signals arise if the system involves a sampling operation of continuous-time signals. The sampled signal is $x(0), x(T)$, $x(2T), \ldots$, where T is the sampling period. Such a sequence of values arising from the sampling operation is usually written as $x(kT)$. If the system involves an iterative process carried out by a digital computer, the signal involved is a number sequence $x(0), x(1), x(2) \ldots$. The sequence of numbers is usually written as $x(k)$, where the argument k indicates the order in which the number occurs in the sequence, for example, $x(0), x(1), x(2) \ldots$. Although $x(k)$ is a number sequence, it can be considered as a sampled signal of $x(t)$ when the sampling period T is 1 sec.

The *z* transform applies to the continuous-time signal $x(t)$, sampled signal $x(kT)$, and the number sequence $x(k)$. In dealing with the *z* transform, if no confusion occurs in the discussion, we occasionally use $x(kT)$ and $x(k)$ interchangeably. [That is, to simplify the presentation, we occasionally drop the explicit appearance of T and write $x(kT)$ as $x(k)$.]

Outline of the Chapter. Section 2–1 has presented introductory remarks. Section 2–2 presents the definition of the *z* transform and associated subjects. Section 2–3 gives *z* transforms of elementary functions. Important properties and theorems of the *z* transform are presented in Section 2–4. Both analytical and computational methods for finding the inverse *z* transform are discussed in Section 2–5. Section 2–6 presents the solution of difference equations by the *z* transform method. Finally, Section 2–7 gives concluding comments.

2–2 THE z TRANSFORM

The *z* transform method is an operational method that is very powerful when working with discrete-time systems. In what follows we shall define the *z* transform of a time function or a number sequence.

In considering the *z* transform of a time function $x(t)$, we consider only the sampled values of $x(t)$, that is, $x(0), x(T), x(2T), \ldots$, where T is the sampling period.

The *z* transform of a time function $x(t)$, where t is nonnegative, or of a sequence of values $x(kT)$, where k takes zero or positive integers and T is the sampling period, is defined by the following equation:

$$X(z) = \mathcal{Z}[x(t)] = \mathcal{Z}[x(kT)] = \sum_{k=0}^{\infty} x(kT)z^{-k} \qquad (2\text{--}1)$$

For a sequence of numbers $x(k)$, the *z* transform is defined by

$$X(z) = \mathcal{Z}[x(k)] = \sum_{k=0}^{\infty} x(k)z^{-k} \qquad (2\text{--}2)$$

The *z* transform defined by Equation (2–1) or (2–2) is referred to as the *one-sided z transform*.

The symbol $\mathcal{Z}$ denotes "the *z* transform of." In the one-sided *z* transform, we assume $x(t) = 0$ for $t < 0$ or $x(k) = 0$ for $k < 0$. Note that z is a complex variable.

Note that, when dealing with a time sequence $x(kT)$ obtained by sampling a time signal $x(t)$, the *z* transform $X(z)$ involves T explicitly. However, for a number sequence $x(k)$, the *z* transform $X(z)$ does not involve T explicitly.

The *z* transform of $x(t)$, where $-\infty < t < \infty$, or of $x(k)$, where k takes integer values ($k = 0, \pm 1, \pm 2, \cdots$), is defined by

$$X(z) = \mathcal{Z}[x(t)] = \mathcal{Z}[x(kT)] = \sum_{k=-\infty}^{\infty} x(kT)z^{-k} \qquad (2\text{--}3)$$

or

$$X(z) = \mathcal{Z}[x(k)] = \sum_{k=-\infty}^{\infty} x(k)z^{-k} \qquad (2\text{--}4)$$

The *z* transform defined by Equation (2-3) or (2-4) is referred to as the *two-sided z transform*. In the two-sided *z* transform, the time function $x(t)$ is assumed to be nonzero for $t < 0$ and the sequence $x(k)$ is considered to have nonzero values for $k < 0$. Both the one-sided and two-sided *z* transforms are series in powers of z^{-1}. (The latter involves both positive and negative powers of z^{-1}.) In this book, only the one-sided *z* transform is considered in detail.

For most engineering applications the one-sided *z* transform will have a convenient closed-form solution in its region of convergence. Note that whenever $X(z)$, an infinite series in z^{-1}, converges outside the circle $|z| = R$, where R is called the *radius of absolute convergence*, in using the *z* transform method for solving discrete-time problems it is not necessary each time to specify the values of *z* over which $X(z)$ is convergent.

Notice that expansion of the right-hand side of Equation (2-1) gives

$$X(z) = x(0) + x(T)z^{-1} + x(2T)z^{-2} + \cdots + x(kT)z^{-k} + \cdots \qquad (2\text{-}5)$$

Equation (2-5) implies that the *z* transform of any continuous-time function $x(t)$ may be written in the series form by inspection. The z^{-k} in this series indicates the position in time at which the amplitude $x(kT)$ occurs. Conversely, if $X(z)$ is given in the series form as above, the inverse *z* transform can be obtained by inspection as a sequence of the function $x(kT)$ that corresponds to the values of $x(t)$ at the respective instants of time.

If the *z* transform is given as a ratio of two polynomials in *z*, then the inverse *z* transform may be obtained by several different methods, such as the direct division method, the computational method, the partial-fraction-expansion method, and the inversion integral method (see Section 2-5 for details.)

2-3 *z* TRANSFORMS OF ELEMENTARY FUNCTIONS

In the following we shall present *z* transforms of several elementary functions. It is noted that in one-sided *z* transform theory, in sampling a discontinuous function $x(t)$, we assume that the function is continuous from the right; that is, if discontinuity occurs at $t = 0$, then we assume that $x(0)$ is equal to $x(0+)$ rather than to the average at the discontinuity, $[x(0-) + x(0+)]/2$.

Unit-Step Function.   Let us find the *z* transform of the unit-step function

$$x(t) = \begin{cases} 1(t), & 0 \le t \quad t \geq 0 \\ 0, & t < 0 \end{cases}$$

As just noted, in sampling a unit-step function we assume that this function is continuous from the right; that is, $1(0) = 1$. Then, referring to Equation (2-1), we have

$$X(z) = \mathcal{Z}[1(t)] = \sum_{k=0}^{\infty} 1 z^{-k} = \sum_{k=0}^{\infty} z^{-k}$$

$$= 1 + z^{-1} + z^{-2} + z^{-3} + \cdots$$

$$= \frac{1}{1 - z^{-1}}$$

$$= \frac{z}{z - 1}$$

Notice that the series converges if $|z| > 1$. In finding the z transform, the variable z acts as a dummy operator. It is not necessary to specify the region of z over which $X(z)$ is convergent. It suffices to know that such a region exists. The z transform $X(z)$ of a time function $x(t)$ obtained in this way is valid throughout the z plane except at poles of $X(z)$.

It is noted that $1(k)$ as defined by

$$1(k) = \begin{cases} 1, & k = 0, 1, 2, \ldots \\ 0, & k < 0 \end{cases}$$

is commonly called a *unit-step sequence*.

Unit-Ramp Function.   Consider the unit-ramp function

$$x(t) = \begin{cases} t, & 0 \le t \quad t \gtrless 0 \\ 0, & t < 0 \end{cases}$$

Notice that

$$x(kT) = kT, \quad k = 0, 1, 2, \ldots$$

Figure 2–1 depicts the sampled unit-ramp signal. The magnitudes of the sampled values are proportional to the sampling period T. The z transform of the unit-ramp function can be written as

$$X(z) = \mathcal{Z}[t] = \sum_{k=0}^{\infty} x(kT)z^{-k} = \sum_{k=0}^{\infty} kTz^{-k} = T\sum_{k=0}^{\infty} kz^{-k}$$

$$= T(z^{-1} + 2z^{-2} + 3z^{-3} + \cdots)$$

$$= T\frac{z^{-1}}{(1 - z^{-1})^2}$$

$$= \frac{Tz}{(z - 1)^2}$$

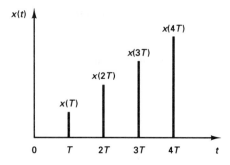

Figure 2–1   Sampled unit-ramp signal.

Note that it is a function of the sampling period T.

Polynomial Function a^k.   Let us obtain the z transform of $x(k)$ as defined by

not linked to a sample period, not

$$x(k) = \begin{cases} a^k, & k = 0, 1, 2, \ldots \\ 0, & k < 0 \end{cases}$$

where a is a constant. Referring to the definition of the z transform given by Equation (2–2), we obtain

$$X(z) = \mathscr{Z}[a^k] = \sum_{k=0}^{\infty} x(k)z^{-k} = \sum_{k=0}^{\infty} a^k z^{-k}$$

$$= 1 + az^{-1} + a^2 z^{-2} + a^3 z^{-3} + \cdots$$

$$= \frac{1}{1 - az^{-1}}$$

$$= \frac{z}{z - a}$$

Exponential Function.   Let us find the z transform of

$$x(t) = \begin{cases} e^{-at}, & 0 \le t \\ 0, & t < 0 \end{cases}$$

replace t with kT

Since

$$x(kT) = e^{-akT}, \qquad k = 0, 1, 2, \ldots$$

we have

$$X(z) = \mathscr{Z}[e^{-at}] = \sum_{k=0}^{\infty} x(kT)z^{-k} = \sum_{k=0}^{\infty} e^{-akT} z^{-k}$$

$$= 1 + e^{-aT} z^{-1} + e^{-2aT} z^{-2} + e^{-3aT} z^{-3} + \cdots$$

$$= \frac{1}{1 - e^{-aT} z^{-1}}$$

$$= \frac{z}{z - e^{-aT}}$$

Sinusoidal Function.   Consider the sinusoidal function

$$x(t) = \begin{cases} \sin \omega t, & 0 \le t \\ 0, & t < 0 \end{cases}$$

$t \ge 0$

Noting that

$$e^{j\omega t} = \cos \omega t + j \sin \omega t$$

$$e^{-j\omega t} = \cos \omega t - j \sin \omega t$$

we have

$$\sin \omega t = \frac{1}{2j}(e^{j\omega t} - e^{-j\omega t})$$

Since the z transform of the exponential function is

$$\mathcal{Z}[e^{-at}] = \frac{1}{1 - e^{-aT}z^{-1}}$$

we have

$$X(z) = \mathcal{Z}[\sin \omega t] = \mathcal{Z}\left[\frac{1}{2j}(e^{j\omega t} - e^{-j\omega t})\right]$$

$$= \frac{1}{2j}\left(\frac{1}{1 - e^{j\omega T}z^{-1}} - \frac{1}{1 - e^{-j\omega T}z^{-1}}\right)$$

$$= \frac{1}{2j}\frac{(e^{j\omega T} - e^{-j\omega T})z^{-1}}{1 - (e^{j\omega T} + e^{-j\omega T})z^{-1} + z^{-2}}$$

$$= \frac{z^{-1}\sin \omega T}{1 - 2z^{-1}\cos \omega T + z^{-2}}$$

$$= \frac{z \sin \omega T}{z^2 - 2z \cos \omega T + 1}$$

Example 2–1

Obtain the z transform of the cosine function

$$x(t) = \begin{cases} \cos \omega t, & 0 \le t \qquad t \gg 0 \\ 0, & t < 0 \end{cases}$$

If we proceed in a manner similar to the way we treated the z transform of the sine function, we have

$$X(z) = \mathcal{Z}[\cos \omega t] = \tfrac{1}{2}\mathcal{Z}[e^{j\omega t} + e^{-j\omega t}]$$

$$= \frac{1}{2}\left(\frac{1}{1 - e^{j\omega T}z^{-1}} + \frac{1}{1 - e^{-j\omega T}z^{-1}}\right)$$

$$= \frac{1}{2}\frac{2 - (e^{-j\omega T} + e^{j\omega T})z^{-1}}{1 - (e^{j\omega T} + e^{-j\omega T})z^{-1} + z^{-2}}$$

$$= \frac{1 - z^{-1}\cos \omega T}{1 - 2z^{-1}\cos \omega T + z^{-2}}$$

$$= \frac{z^2 - z \cos \omega T}{z^2 - 2z \cos \omega T + 1}$$

Example 2–2

Obtain the z transform of

$$X(s) = \frac{1}{s(s + 1)}$$

Whenever a function in s is given, one approach for finding the corresponding z transform is to convert $X(s)$ into $x(t)$ and then find the z transform of $x(t)$. Another approach is to expand $X(s)$ into partial fractions and use a z transform table to find the z transforms of the expanded terms. Still other approaches will be discussed in Section 3–3.

The inverse Laplace transform of $X(s)$ is

$$x(t) = 1 - e^{-t}, \quad 0 \leq t \qquad t \geq 0$$

Hence,

$$X(z) = \mathcal{Z}[1 - e^{-t}] = \frac{1}{1 - z^{-1}} - \frac{1}{1 - e^{-T}z^{-1}} \qquad (3\,\&\,4)$$

$$= \frac{(1 - e^{-T})z^{-1}}{(1 - z^{-1})(1 - e^{-T}z^{-1})}$$

$$= \frac{(1 - e^{-T})z}{(z - 1)(z - e^{-T})}$$

Comments.   Just as in working with the Laplace transformation, a table of z transforms of commonly encountered functions is very useful for solving problems in the field of discrete-time systems. Table 2-1 is such a table.

TABLE 2-1   TABLE OF z TRANSFORMS

	$X(s)$	$x(t)$	$x(kT)$ or $x(k)$	$X(z)$
1.	—	—	Kronecker delta $\delta_0(k)$ $1, \quad k = 0$ $0, \quad k \neq 0$	1
2.	—	—	$\delta_0(n - k)$ $1, \quad n = k$ $0, \quad n \neq k$	z^{-k}
3.	$\dfrac{1}{s}$	$1(t)$	$1(k)$	$\dfrac{1}{1 - z^{-1}}$
4.	$\dfrac{1}{s + a}$	e^{-at}	e^{-akT}	$\dfrac{1}{1 - e^{-aT}z^{-1}}$
5.	$\dfrac{1}{s^2}$	t	kT	$\dfrac{Tz^{-1}}{(1 - z^{-1})^2}$
6.	$\dfrac{2}{s^3}$	t^2	$(kT)^2$	$\dfrac{T^2 z^{-1}(1 + z^{-1})}{(1 - z^{-1})^3}$
7.	$\dfrac{6}{s^4}$	t^3	$(kT)^3$	$\dfrac{T^3 z^{-1}(1 + 4z^{-1} + z^{-2})}{(1 - z^{-1})^4}$
8.	$\dfrac{a}{s(s + a)}$	$1 - e^{-at}$	$1 - e^{-akT}$	$\dfrac{(1 - e^{-aT})z^{-1}}{(1 - z^{-1})(1 - e^{-aT}z^{-1})}$
9.	$\dfrac{b - a}{(s + a)(s + b)}$	$e^{-at} - e^{-bt}$	$e^{-akT} - e^{-bkT}$	$\dfrac{(e^{-aT} - e^{-bT})z^{-1}}{(1 - e^{-aT}z^{-1})(1 - e^{-bT}z^{-1})}$
10.	$\dfrac{1}{(s + a)^2}$	te^{-at}	kTe^{-akT}	$\dfrac{Te^{-aT}z^{-1}}{(1 - e^{-aT}z^{-1})^2}$
11.	$\dfrac{s}{(s + a)^2}$	$(1 - at)e^{-at}$	$(1 - akT)e^{-akT}$	$\dfrac{1 - (1 + aT)e^{-aT}z^{-1}}{(1 - e^{-aT}z^{-1})^2}$

TABLE 2–1 (continued)

	$X(s)$	$x(t)$	$x(kT)$ or $x(k)$	$X(z)$
12.	$\dfrac{2}{(s+a)^3}$	$t^2 e^{-at}$	$(kT)^2 e^{-akT}$	$\dfrac{T^2 e^{-aT}(1 + e^{-aT}z^{-1})z^{-1}}{(1 - e^{-aT}z^{-1})^3}$
13.	$\dfrac{a^2}{s^2(s+a)}$	$at - 1 + e^{-at}$	$akT - 1 + e^{-akT}$	$\dfrac{[(aT - 1 + e^{-aT}) + (1 - e^{-aT} - aTe^{-aT})z^{-1}]z^{-1}}{(1 - z^{-1})^2(1 - e^{-aT}z^{-1})}$
14.	$\dfrac{\omega}{s^2 + \omega^2}$	$\sin \omega t$	$\sin \omega kT$	$\dfrac{z^{-1}\sin \omega T}{1 - 2z^{-1}\cos \omega T + z^{-2}}$
15.	$\dfrac{s}{s^2 + \omega^2}$	$\cos \omega t$	$\cos \omega kT$	$\dfrac{1 - z^{-1}\cos \omega T}{1 - 2z^{-1}\cos \omega T + z^{-2}}$
16.	$\dfrac{\omega}{(s+a)^2 + \omega^2}$	$e^{-at}\sin \omega t$	$e^{-akT}\sin \omega kT$	$\dfrac{e^{-aT}z^{-1}\sin \omega T}{1 - 2e^{-aT}z^{-1}\cos \omega T + e^{-2aT}z^{-2}}$
17.	$\dfrac{s+a}{(s+a)^2 + \omega^2}$	$e^{-at}\cos \omega t$	$e^{-akT}\cos \omega kT$	$\dfrac{1 - e^{-aT}z^{-1}\cos \omega T}{1 - 2e^{-aT}z^{-1}\cos \omega T + e^{-2aT}z^{-2}}$
18.			a^k	$\dfrac{1}{1 - az^{-1}}$
19.			a^{k-1} $k = 1, 2, 3, \ldots$	$\dfrac{z^{-1}}{1 - az^{-1}}$
20.			ka^{k-1}	$\dfrac{z^{-1}}{(1 - az^{-1})^2}$
21.			$k^2 a^{k-1}$	$\dfrac{z^{-1}(1 + az^{-1})}{(1 - az^{-1})^3}$
22.			$k^3 a^{k-1}$	$\dfrac{z^{-1}(1 + 4az^{-1} + a^2 z^{-2})}{(1 - az^{-1})^4}$
23.			$k^4 a^{k-1}$	$\dfrac{z^{-1}(1 + 11az^{-1} + 11a^2 z^{-2} + a^3 z^{-3})}{(1 - az^{-1})^5}$
24.			$a^k \cos k\pi$	$\dfrac{1}{1 + az^{-1}}$
25.			$\dfrac{k(k-1)}{2!}$	$\dfrac{z^{-2}}{(1 - z^{-1})^3}$
26.			$\dfrac{k(k-1)\cdots(k-m+2)}{(m-1)!}$	$\dfrac{z^{-m+1}}{(1 - z^{-1})^m}$
27.			$\dfrac{k(k-1)}{2!}a^{k-2}$	$\dfrac{z^{-2}}{(1 - az^{-1})^3}$
28.		$\dfrac{k(k-1)\cdots(k-m+2)}{(m-1)!}a^{k-m+1}$		$\dfrac{z^{-m+1}}{(1 - az^{-1})^m}$

$x(t) = 0$, for $t < 0$.

$x(kT) = x(k) = 0$, for $k < 0$.

Unless otherwise noted, $k = 0, 1, 2, 3, \ldots$.

2–4 IMPORTANT PROPERTIES AND THEOREMS OF THE z TRANSFORM

The use of the z transform method in the analysis of discrete-time control systems may be facilitated if theorems of the z transform are referred to. In this section we present important properties and useful theorems of the z transform. We assume that the time function $x(t)$ is z-transformable and that $x(t)$ is zero for $t < 0$.

Multiplication by a Constant.   If $X(z)$ is the z transform of $x(t)$, then

$$\mathscr{Z}[ax(t)] = a\,\mathscr{Z}[x(t)] = aX(z)$$

where a is a constant.

To prove this, note that by definition

$$\mathscr{Z}[ax(t)] = \sum_{k=0}^{\infty} ax(kT)z^{-k} = a\sum_{k=0}^{\infty} x(kT)z^{-k} = aX(z)$$

Linearity of the z Transform.   The z transform possesses an important property: linearity. This means that, if $f(k)$ and $g(k)$ are z-transformable and α and β are scalars, then $x(k)$ formed by a linear combination

$$x(k) = \alpha f(k) + \beta g(k)$$

has the z transform

$$X(z) = \alpha F(z) + \beta G(z)$$

where $F(z)$ and $G(z)$ are the z transforms of $f(k)$ and $g(k)$, respectively.

The linearity property can be proved by referring to Equation (2–2) as follows:

$$
\begin{aligned}
X(z) = \mathscr{Z}[x(k)] &= \mathscr{Z}[\alpha f(k) + \beta g(k)] \\
&= \sum_{k=0}^{\infty} [\alpha f(k) + \beta g(k)]z^{-k} \\
&= \alpha \sum_{k=0}^{\infty} f(k)z^{-k} + \beta \sum_{k=0}^{\infty} g(k)z^{-k} \\
&= \alpha \mathscr{Z}[f(k)] + \beta \mathscr{Z}[g(k)] \\
&= \alpha F(z) + \beta G(z)
\end{aligned}
$$

Multiplication by a^k.   If $X(z)$ is the z transform of $x(k)$, then the z transform of $a^k x(k)$ can be given by $X(a^{-1}z)$:

$$\mathscr{Z}[a^k x(k)] = X(a^{-1}z) \tag{2–6}$$

This can be proved as follows:

$$
\begin{aligned}
\mathscr{Z}[a^k x(k)] &= \sum_{k=0}^{\infty} a^k x(k)z^{-k} = \sum_{k=0}^{\infty} x(k)(a^{-1}z)^{-k} \\
&= X(a^{-1}z)
\end{aligned}
$$

Shifting Theorem.   The shifting theorem presented here is also referred to as the real translation theorem. If $x(t) = 0$ for $t < 0$ and $x(t)$ has the z transform $X(z)$, then

$$\mathscr{Z}[x(t - nT)] = z^{-n}X(z) \tag{2-7}$$

and

$$\mathscr{Z}[x(t + nT)] = z^n\left[X(z) - \sum_{k=0}^{n-1} x(kT)z^{-k}\right] \tag{2-8}$$

where n is zero or a positive integer.

To prove Equation (2–7), note that

$$\mathscr{Z}[x(t - nT)] = \sum_{k=0}^{\infty} x(kT - nT)z^{-k}$$

$$= z^{-n}\sum_{k=0}^{\infty} x(kT - nT)z^{-(k-n)} \tag{2-9}$$

By defining $m = k - n$, Equation (2–9) can be written as follows:

$$\mathscr{Z}[x(t - nT)] = z^{-n}\sum_{m=-n}^{\infty} x(mT)z^{-m}$$

Since $x(mT) = 0$ for $m < 0$, we may change the lower limit of the summation from $m = -n$ to $m = 0$. Hence,

$$\mathscr{Z}[x(t - nT)] = z^{-n}\sum_{m=0}^{\infty} x(mT)z^{-m} = z^{-n}X(z) \tag{2-10}$$

Thus, multiplication of a z transform by z^{-n} has the effect of delaying the time function $x(t)$ by time nT. (That is, move the function to the right by time nT.)

To prove Equation (2–8), we note that

$$\mathscr{Z}[x(t + nT)] = \sum_{k=0}^{\infty} x(kT + nT)z^{-k}$$

$$= z^n\sum_{k=0}^{\infty} x(kT + nT)z^{-(k+n)}$$

$$= z^n\left[\sum_{k=0}^{\infty} x(kT + nT)z^{-(k+n)} + \sum_{k=0}^{n-1} x(kT)z^{-k} - \sum_{k=0}^{n-1} x(kT)z^{-k}\right]$$

$$= z^n\left[\sum_{k=0}^{\infty} x(kT)z^{-k} - \sum_{k=0}^{n-1} x(kT)z^{-k}\right]$$

$$= z^n\left[X(z) - \sum_{k=0}^{n-1} x(kT)z^{-k}\right]$$

For the number sequence $x(k)$, Equation (2–8) can be written as follows:

$$\mathscr{Z}[x(k + n)] = z^n\left[X(z) - \sum_{k=0}^{n-1} x(k)z^{-k}\right]$$

From this last equation, we obtain

$$\mathscr{Z}[x(k + 1)] = zX(z) - zx(0) \tag{2-11}$$

$$\mathscr{Z}[x(k + 2)] = z\mathscr{Z}[x(k + 1)] - zx(1) = z^2X(z) - z^2x(0) - zx(1) \tag{2-12}$$

Similarly,

$$\mathcal{Z}[x(k + n)]$$
$$= z^n X(z) - z^n x(0) - z^{n-1} x(1) - z^{n-2} x(2) - \cdots - zx(n - 1) \quad (2\text{-}13)$$

where *n* is a positive integer.

Remember that multiplication of the *z* transform $X(z)$ by z has the effect of advancing the signal $x(kT)$ by one step (1 sampling period) and that multiplication of the *z* transform $X(z)$ by z^{-1} has the effect of delaying the signal $x(kT)$ by one step (1 sampling period).

Example 2-3

Find the *z* transforms of unit-step functions that are delayed by 1 sampling period and 4 sampling periods, respectively, as shown in Figure 2-2(a) and (b).

Using the shifting theorem given by Equation (2-7), we have

$$\mathcal{Z}[1(t - T)] = z^{-1}\mathcal{Z}[1(t)] = z^{-1}\frac{1}{1 - z^{-1}} = \frac{z^{-1}}{1 - z^{-1}}$$

Also,

$$\mathcal{Z}[1(t - 4T)] = z^{-4}\mathcal{Z}[1(t)] = z^{-4}\frac{1}{1 - z^{-1}} = \frac{z^{-4}}{1 - z^{-1}}$$

(Note that z^{-1} represents a delay of 1 sampling period T, regardless of the value of T.)

Example 2-4

Obtain the *z* transform of

$$f(a) = \begin{cases} a^{k-1}, & k = 1, 2, 3, \ldots \\ 0, & k \leq 0 \end{cases}$$

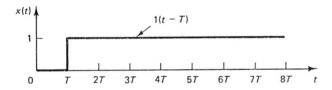

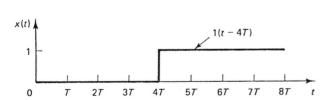

Figure 2-2 (a) Unit-step function delayed by 1 sampling period; (b) unit-step function delayed by 4 sampling periods.

Referring to Equation (2–7), we have

$$\mathcal{Z}[x(k-1)] = z^{-1}X(z)$$

The z transform of a^k is

$$\mathcal{Z}[a^k] = \frac{1}{1-az^{-1}}$$

and so

$$\mathcal{Z}[f(a)] = \mathcal{Z}[a^{k-1}] = z^{-1}\frac{1}{1-az^{-1}} = \frac{z^{-1}}{1-az^{-1}}$$

where $k = 1, 2, 3, \ldots$.

Example 2–5

Consider the function $y(k)$, which is a sum of functions $x(h)$, where $h = 0, 1, 2, \ldots, k$, such that

$$y(k) = \sum_{h=0}^{k} x(h), \qquad k = 0, 1, 2, \ldots$$

where $y(k) = 0$ for $k < 0$. Obtain the z transform of $y(k)$.

First note that

$$y(k) = x(0) + x(1) + \cdots + x(k-1) + x(k)$$
$$y(k-1) = x(0) + x(1) + \cdots + x(k-1)$$

Hence,

$$y(k) - y(k-1) = x(k), \qquad k = 0, 1, 2, \ldots$$

Therefore,

$$\mathcal{Z}[y(k) - y(k-1)] = \mathcal{Z}[x(k)]$$

or

$$Y(z) - z^{-1}Y(z) = X(z)$$

which yields

$$Y(z) = \frac{1}{1-z^{-1}}X(z)$$

where $X(z) = \mathcal{Z}[x(k)]$.

Complex Translation Theorem. If $x(t)$ has the z transform $X(z)$, then the z transform of $e^{-at}x(t)$ can be given by $X(ze^{aT})$. This is known as the *complex translation theorem*.

To prove this theorem, note that

$$\mathcal{Z}[e^{-at}x(t)] = \sum_{k=0}^{\infty} x(kT)e^{-akT}z^{-k} = \sum_{k=0}^{\infty} x(kT)(ze^{aT})^{-k} = X(ze^{aT}) \qquad (2\text{–}14)$$

Thus, we see that replacing z in $X(z)$ by ze^{aT} gives the z transform of $e^{-at}x(t)$.

Example 2–6

Given the z transforms of $\sin \omega t$ and $\cos \omega t$, obtain the z transforms of $e^{-at}\sin \omega t$ and $e^{-at}\cos \omega t$, respectively, by using the complex translation theorem.

Noting that

$$Z[\sin \omega t] = \frac{z^{-1} \sin \omega T}{1 - 2z^{-1} \cos \omega T + z^{-2}}$$

we substitute ze^{aT} for z to obtain the z transform of $e^{-at} \sin \omega t$, as follows:

$$Z[e^{-at} \sin \omega t] = \frac{e^{-aT} z^{-1} \sin \omega T}{1 - 2e^{-aT} z^{-1} \cos \omega T + e^{-2aT} z^{-2}}$$

Similarly, for the cosine function, we have

$$Z[\cos \omega t] = \frac{1 - z^{-1} \cos \omega T}{1 - 2z^{-1} \cos \omega T + z^{-2}}$$

By substituting ze^{aT} for z in the z transform of $\cos \omega t$, we obtain

$$Z[e^{-at} \cos \omega t] = \frac{1 - e^{-aT} z^{-1} \cos \omega T}{1 - 2e^{-aT} z^{-1} \cos \omega T + e^{-2aT} z^{-2}}$$

Example 2–7

Obtain the z transform of te^{-at}.
    Notice that

$$Z[t] = \frac{Tz^{-1}}{(1 - z^{-1})^2} = X(z)$$

Thus,

$$Z[te^{-at}] = X(ze^{aT}) = \frac{Te^{-aT} z^{-1}}{(1 - e^{-aT} z^{-1})^2}$$

Initial Value Theorem.   If $x(t)$ has the z transform $X(z)$ and if $\lim_{z \to \infty} X(z)$ exists, then the initial value $x(0)$ of $x(t)$ or $x(k)$ is given by

$$x(0) = \lim_{z \to \infty} X(z) \tag{2–15}$$

To prove this theorem, note that

$$X(z) = \sum_{k=0}^{\infty} x(k)z^{-k} = x(0) + x(1)z^{-1} + x(2)z^{-2} + \cdots$$

Letting $z \to \infty$ in this last equation, we obtain Equation (2–15). The behavior of the signal in the neighborhood of $t = 0$ or $k = 0$ can thus be determined by the behavior of $X(z)$ at $z = \infty$.
    The initial value theorem is convenient for checking z transform calculations for possible errors. Since $x(0)$ is usually known, a check of the initial value by $\lim_{z \to \infty} X(z)$ can easily spot errors in $X(z)$, if any exist.

Example 2–8

Determine the initial value $x(0)$ if the z transform of $x(t)$ is given by

$$X(z) = \frac{(1 - e^{-T})z^{-1}}{(1 - z^{-1})(1 - e^{-T}z^{-1})}$$

By using the initial value theorem, we find

$$x(0) = \lim_{z \to \infty} \frac{(1 - e^{-T})z^{-1}}{(1 - z^{-1})(1 - e^{-T}z^{-1})} = 0$$

Referring to Example 2–2, notice that this $X(z)$ was the z transform of

$$x(t) = 1 - e^{-t}$$

and thus $x(0) = 0$, which agrees with the result obtained earlier.

Final Value Theorem.   Suppose that $x(k)$, where $x(k) = 0$ for $k < 0$, has the z transform $X(z)$ and that all the poles of $X(z)$ lie inside the unit circle, with the possible exception of a simple pole at $z = 1$. [This is the condition for the stability of $X(z)$, or the condition for $x(k)\,(k = 0, 1, 2, \ldots)$ to remain finite.] Then the final value of $x(k)$, that is, the value of $x(k)$ as k approaches infinity, can be given by

$$\lim_{k \to \infty} x(k) = \lim_{z \to 1} [(1 - z^{-1})X(z)] \qquad (2\text{–}16)$$

To prove the final value theorem, note that

$$\mathscr{Z}[x(k)] = X(z) = \sum_{k=0}^{\infty} x(k)z^{-k}$$

$$\mathscr{Z}[x(k - 1)] = z^{-1}X(z) = \sum_{k=0}^{\infty} x(k - 1)z^{-k}$$

Hence,

$$\sum_{k=0}^{\infty} x(k)z^{-k} - \sum_{k=0}^{\infty} x(k - 1)z^{-k} = X(z) - z^{-1}X(z)$$

Taking the limit as z approaches unity, we have

$$\lim_{z \to 1} \left[\sum_{k=0}^{\infty} x(k)z^{-k} - \sum_{k=0}^{\infty} x(k - 1)z^{-k} \right] = \lim_{z \to 1} [(1 - z^{-1})X(z)]$$

Because of the assumed stability condition and the condition that $x(k) = 0$ for $k < 0$, the left-hand side of this last equation becomes

$$\sum_{k=0}^{\infty} [x(k) - x(k - 1)] = [x(0) - x(-1)] + [x(1) - x(0)]$$

$$+ [x(2) - x(1)] + \cdots = x(\infty) = \lim_{k \to \infty} x(k)$$

Hence,

$$\lim_{k \to \infty} x(k) = \lim_{z \to 1} [(1 - z^{-1})X(z)]$$

which is Equation (2–16). The final value theorem is very useful in determining the behavior of $x(k)$ as $k \to \infty$ from its z transform $X(z)$.

Example 2–9

Determine the final value $x(\infty)$ of

$$X(z) = \frac{1}{1 - z^{-1}} - \frac{1}{1 - e^{-aT}z^{-1}}, \qquad a > 0$$

by using the final value theorem.

By applying the final value theorem to the given $X(z)$, we obtain

$$x(\infty) = \lim_{z \to 1} [(1 - z^{-1})X(z)]$$

$$= \lim_{z \to 1} \left[(1 - z^{-1}) \left(\frac{1}{1 - z^{-1}} - \frac{1}{1 - e^{-aT} z^{-1}} \right) \right]$$

$$= \lim_{z \to 1} \left(1 - \frac{1 - z^{-1}}{1 - e^{-aT} z^{-1}} \right) = 1$$

It is noted that the given $X(z)$ is actually the z transform of

$$x(t) = 1 - e^{-at}$$

By substituting $t = \infty$ in this equation, we have

$$x(\infty) = \lim_{t \to \infty} (1 - e^{-at}) = 1$$

As a matter of course, the two results agree.

Summary.   In this section we have presented important properties and theorems of the z transform that will prove to be useful in solving many z transform problems. For the purpose of convenient reference, these important properties and theorems are summarized in Table 2-2. (Many of the theorems presented in this table are discussed in this section. Those not discussed here but included in the table are derived or proved in Appendix B.)

2-5 THE INVERSE z TRANSFORM

The z transformation serves the same role for discrete-time control systems that the Laplace transformation serves for continuous-time control systems. For the z transform to be useful, we must be familiar with methods for finding the inverse z transform.

The notation for the inverse z transform is $\mathcal{Z}^{-1}$. The inverse z transform of $X(z)$ yields the corresponding time sequence $x(k)$.

It should be noted that only the time sequence at the sampling instants is obtained from the inverse z transform. Thus, the inverse z transform of $X(z)$ yields a unique $x(k)$, but does not yield a unique $x(t)$. This means that the inverse z transform yields a time sequence that specifies the values of $x(t)$ only at discrete instants of time, $t = 0, T, 2T, \ldots$, and says nothing about the values of $x(t)$ at all other times. That is, many different time functions $x(t)$ can have the same $x(kT)$. See Figure 2-3.

When $X(z)$, the z transform of $x(kT)$ or $x(k)$, is given, the operation that determines the corresponding $x(kT)$ or $x(k)$ is called the *inverse z transformation*. An obvious method for finding the inverse z transform is to refer to a z transform table. However, unless we refer to an extensive z transform table, we may not be able to find the inverse z transform of a complicated function of z. (If we use a less extensive table of z transforms, it is necessary to express a complex z transform as a sum of simpler z transforms. Refer to the partial-fraction-expansion method presented in this section.)

Other than referring to z transform tables, four methods for obtaining the inverse z transform are commonly available:

TABLE 2–2 IMPORTANT PROPERTIES AND THEOREMS OF THE z TRANSFORM

	$x(t)$ or $x(k)$	$\mathscr{Z}[x(t)]$ or $\mathscr{Z}[x(k)]$
1.	$ax(t)$	$aX(z)$
2.	$ax_1(t) + bx_2(t)$	$aX_1(z) + bX_2(z)$
3.	$x(t + T)$ or $x(k + 1)$	$zX(z) - zx(0)$
4.	$x(t + 2T)$	$z^2 X(z) - z^2 x(0) - zx(T)$
5.	$x(k + 2)$	$z^2 X(z) - z^2 x(0) - zx(1)$
6.	$x(t + kT)$	$z^k X(z) - z^k x(0) - z^{k-1} x(T) - \cdots - zx(kT - T)$
7.	$x(t - kT)$	$z^{-k} X(z)$
8.	$x(n + k)$	$z^k X(z) - z^k x(0) - z^{k-1} x(1) - \cdots - zx(k - 1)$
9.	$x(n - k)$	$z^{-k} X(z)$
10.	$tx(t)$	$-Tz \dfrac{d}{dz} X(z)$
11.	$kx(k)$	$-z \dfrac{d}{dz} X(z)$
12.	$e^{-at} x(t)$	$X(ze^{aT})$
13.	$e^{-ak} x(k)$	$X(ze^{a})$
14.	$a^k x(k)$	$X\left(\dfrac{z}{a}\right)$
15.	$ka^k x(k)$	$-z \dfrac{d}{dz} X\left(\dfrac{z}{a}\right)$
16.	$x(0)$	$\lim\limits_{z \to \infty} X(z)$ if the limit exists
17.	$x(\infty)$	$\lim\limits_{z \to 1} [(1 - z^{-1})X(z)]$ if $(1 - z^{-1})X(z)$ is analytic on and outside the unit circle
18.	$\nabla x(k) = x(k) - x(k - 1)$	$(1 - z^{-1})X(z)$
19.	$\Delta x(k) = x(k + 1) - x(k)$	$(z - 1)X(z) - zx(0)$
20.	$\displaystyle\sum_{k=0}^{n} x(k)$	$\dfrac{1}{1 - z^{-1}} X(z)$
21.	$\dfrac{\partial}{\partial a} x(t, a)$	$\dfrac{\partial}{\partial a} X(z, a)$
22.	$k^m x(k)$	$\left(-z \dfrac{d}{dz}\right)^m X(z)$
23.	$\displaystyle\sum_{k=0}^{n} x(kT)y(nT - kT)$	$X(z)Y(z)$
24.	$\displaystyle\sum_{k=0}^{\infty} x(k)$	$X(1)$

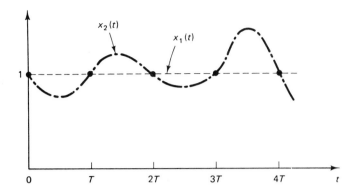

Figure 2–3   Two different continuous-time functions, $x_1(t)$ and $x_2(t)$, that have the same values at $t = 0, T, 2T, \ldots$.

1. Direct division method
2. Computational method
3. Partial-fraction-expansion method
4. Inversion integral method

In obtaining the inverse z transform, we assume, as usual, that the time sequence $x(kT)$ or $x(k)$ is zero for $k < 0$.

Before we present the four methods, however, a few comments on poles and zeros of the pulse transfer function are in order.

Poles and Zeros in the z Plane.   In engineering applications of the z transform method, $X(z)$ may have the form

$$X(z) = \frac{b_0 z^m + b_1 z^{m-1} + \cdots + b_m}{z^n + a_1 z^{n-1} + \cdots + a_n} \qquad (m \le n) \qquad (2\text{–}17)$$

or

$$X(z) = \frac{b_0(z - z_1)(z - z_2) \cdots (z - z_m)}{(z - p_1)(z - p_2) \cdots (z - p_n)}$$

where the p_i's $(i = 1, 2, \ldots, n)$ are the poles of $X(z)$ and the z_j's $(j = 1, 2, \ldots, m)$ the zeros of $X(z)$.

The locations of the poles and zeros of $X(z)$ determine the characteristics of $x(k)$, the sequence of values or numbers. As in the case of the s plane analysis of linear continuous-time control systems, we often use a graphical display in the z plane of the locations of the poles and zeros of $X(z)$.

Note that in control engineering and signal processing $X(z)$ is frequently expressed as a ratio of polynomials in z^{-1}, as follows:

$$X(z) = \frac{b_0 z^{-(n-m)} + b_1 z^{-(n-m+1)} + \cdots + b_m z^{-n}}{1 + a_1 z^{-1} + a_2 z^{-2} + \cdots + a_n z^{-n}} \qquad (2\text{–}18)$$

where z^{-1} is interpreted as the unit delay operator. In this chapter, where the basic properties and theorems of the z transform method are presented, $X(z)$ may be expressed in terms of powers of z, as given by Equation (2–17), or in terms of powers of z^{-1}, as given by Equation (2–18), depending on the circumstances.

In finding the poles and zeros of $X(z)$, it is convenient to express $X(z)$ as a ratio of polynomials in z. For example,

$$X(z) = \frac{z^2 + 0.5z}{z^2 + 3z + 2} = \frac{z(z + 0.5)}{(z + 1)(z + 2)}$$

Clearly, $X(z)$ has poles at $z = -1$ and $z = -2$ and zeros at $z = 0$ and $z = -0.5$. If $X(z)$ is written as a ratio of polynomials in z^{-1}, however, the preceding $X(z)$ can be written as

$$X(z) = \frac{1 + 0.5z^{-1}}{1 + 3z^{-1} + 2z^{-2}} = \frac{1 + 0.5z^{-1}}{(1 + z^{-1})(1 + 2z^{-1})}$$

Although poles at $z = -1$ and $z = -2$ and a zero at $z = -0.5$ are clearly seen from the expression, a zero at $z = 0$ is not explicitly shown, and so the beginner may fail to see the existence of a zero at $z = 0$. Therefore, in dealing with the poles and zeros of $X(z)$, it is preferable to express $X(z)$ as a ratio of polynomials in z, rather than polynomials in z^{-1}. In addition, in obtaining the inverse z transform by use of the inversion integral method, it is desirable to express $X(z)$ as a ratio of polynomials in z, rather than polynomials in z^{-1}, to avoid any possible errors in determining the number of poles at the origin of function $X(z)z^{k-1}$.

Direct Division Method.   In the direct division method we obtain the inverse z transform by expanding $X(z)$ into an infinite power series in z^{-1}. This method is useful when it is difficult to obtain the closed-form expression for the inverse z transform or it is desired to find only the first several terms of $x(k)$.

The direct division method stems from the fact that if $X(z)$ is expanded into a power series in z^{-1}, that is, if

$$X(z) = \sum_{k=0}^{\infty} x(kT)z^{-k}$$

$$= x(0) + x(T)z^{-1} + x(2T)z^{-2} + \cdots + x(kT)z^{-k} + \cdots$$

or

$$X(z) = \sum_{k=0}^{\infty} x(k)z^{-k}$$

$$= x(0) + x(1)z^{-1} + x(2)z^{-2} + \cdots + x(k)z^{-k} + \cdots$$

then $x(kT)$ or $x(k)$ is the coefficient of the z^{-k} term. Hence, the values of $x(kT)$ or $x(k)$ for $k = 0, 1, 2, \ldots$ can be determined by inspection.

If $X(z)$ is given in the form of a rational function, the expansion into an infinite power series in increasing powers of z^{-1} can be accomplished by simply dividing the numerator by the denominator, where both the numerator and denominator of $X(z)$ are written in increasing powers of z^{-1}. If the resulting series is convergent, the coefficients of the z^{-k} term in the series are the values $x(kT)$ of the time sequence or the values of $x(k)$ of the number sequence.

Although the present method gives the values of $x(0), x(T), x(2T), \ldots$ or the values of $x(0), x(1), x(2), \ldots$ in a sequential manner, it is usually difficult to obtain an expression for the general term from a set of values of $x(kT)$ or $x(k)$.

Example 2–10

Find $x(k)$ for $k = 0, 1, 2, 3, 4$ when $X(z)$ is given by

$$X(z) = \frac{10z + 5}{(z - 1)(z - 0.2)}$$

First, rewrite $X(z)$ as a ratio of polynomials in z^{-1}, as follows:

$$X(z) = \frac{10z^{-1} + 5z^{-2}}{1 - 1.2z^{-1} + 0.2z^{-2}}$$

Dividing the numerator by the denominator, we have

$$
\begin{array}{r}
10z^{-1} + 17z^{-2} + 18.4z^{-3} + 18.68z^{-4} + \cdots \\
\hline
1 - 1.2z^{-1} + 0.2z^{-2} \overline{)\, 10z^{-1} + 5z^{-2}} \\
\underline{10z^{-1} - 12z^{-2} + 2z^{-3}} \\
17z^{-2} - 2z^{-3} \\
\underline{17z^{-2} - 20.4z^{-3} + 3.4z^{-4}} \\
18.4z^{-3} - 3.4z^{-4} \\
\underline{18.4z^{-3} - 22.08z^{-4} + 3.68z^{-5}} \\
18.68z^{-4} - 3.68z^{-5} \\
\underline{18.68z^{-4} - 22.416z^{-5} + 3.736z^{-6}}
\end{array}
$$

Thus,

$$X(z) = 10z^{-1} + 17z^{-2} + 18.4z^{-3} + 18.68z^{-4} + \cdots$$

By comparing this infinite series expansion of $X(z)$ with $X(z) = \sum_{k=0}^{\infty} x(k)z^{-k}$, we obtain

$$x(0) = 0$$
$$x(1) = 10$$
$$x(2) = 17$$
$$x(3) = 18.4$$
$$x(4) = 18.68$$

As seen from this example, the direct division method may be carried out by hand calculations if only the first several terms of the sequence are desired. In general, the method does not yield a closed-form expression for $x(k)$, except in special cases.

Example 2–11

Find $x(k)$ when $X(z)$ is given by

$$X(z) = \frac{1}{z + 1} = \frac{z^{-1}}{1 + z^{-1}}$$

By dividing the numerator by the denominator, we obtain

$$X(z) = \frac{z^{-1}}{1 + z^{-1}} = z^{-1} - z^{-2} + z^{-3} - z^{-4} + \cdots$$

By comparing this infinite series expansion of $X(z)$ with $X(z) = \sum_{k=0}^{\infty} x(k)z^{-k}$, we obtain

$$x(0) = 0$$
$$x(1) = 1$$
$$x(2) = -1$$
$$x(3) = 1$$
$$x(4) = -1$$
$$\vdots$$

This is an alternating signal of 1 and -1, which starts from $k = 1$. Figure 2–4 shows a plot of this signal.

Example 2–12

Obtain the inverse z transform of

$$X(z) = 1 + 2z^{-1} + 3z^{-2} + 4z^{-3}$$

The transform $X(z)$ is already in the form of a power series in z^{-1}. Since $X(z)$ has a finite number of terms, it corresponds to a signal of finite length. By inspection, we find

$$x(0) = 1$$
$$x(1) = 2$$
$$x(2) = 3$$
$$x(3) = 4$$

All other $x(k)$ values are zero.

Computational Method.   In what follows, we present two computational approaches to obtain the inverse z transform.

1. MATLAB approach
2. Difference equation approach

Consider a system $G(z)$ defined by

$$G(z) = \frac{0.4673z^{-1} - 0.3393z^{-2}}{1 - 1.5327z^{-1} + 0.6607z^{-2}} \tag{2-19}$$

In finding the inverse z transform, we utilize the Kronecker delta function $\delta_0(kT)$, where

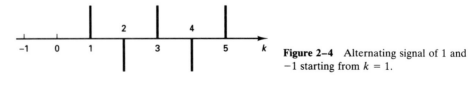

Figure 2–4   Alternating signal of 1 and -1 starting from $k = 1$.

$$\delta_0(kT) = 1, \qquad \text{for } k = 0$$
$$= 0, \qquad \text{for } k \neq 0$$

Assume that $x(k)$, the input to the system $G(z)$, is the Kronecker delta input, or

$$x(k) = 1, \qquad \text{for } k = 0$$
$$= 0, \qquad \text{for } k \neq 0$$

The z transform of the Kronecker delta input is

$$X(z) = 1$$

Using the Kronecker delta input, Equation (2–19) can be rewritten as

$$G(z) = \frac{Y(z)}{X(z)} = \frac{0.4673z^{-1} - 0.3393z^{-2}}{1 - 1.5327z^{-1} + 0.6607z^{-2}}$$

$$= \frac{0.4673z - 0.3393}{z^2 - 1.5327z + 0.6607} \tag{2–20}$$

MATLAB Approach.   MATLAB can be used for finding the inverse z transform. Referring to Equation (2–20), the input $X(z)$ is the z transform of the Kronecker delta input. In MATLAB the Kronecker delta input is given by

$$x = [1 \quad \text{zeros}(1,N)]$$

where N corresponds to the end of the discrete-time duration of the process considered.

Since the z transform of the Kronecker delta input $X(z)$ is equal to unity, the response of the system to this input is

$$Y(z) = G(z) = \frac{0.4673z^{-1} - 0.3393z^{-2}}{1 - 1.5327z^{-1} + 0.6607z^{-2}} = \frac{0.4673z - 0.3393}{z^2 - 1.5327z + 0.6607}$$

Hence the inverse z transform of $G(z)$ is given by $y(0), y(1), y(2), \dots$. Let us obtain $y(k)$ up to $k = 40$.

To obtain the inverse z transform of $G(z)$ with MATLAB, we proceed as follows: Enter the numerator and denominator as follows:

$$\text{num} = [0 \quad 0.4673 \quad -0.3393]$$
$$\text{den} = [1 \quad -1.5327 \quad 0.6607]$$

Enter the Kronecker delta input.

$$x = [1 \quad \text{zeros}(1,40)]$$

Then enter the command

$$y = \text{filter(num,den,x)}$$

to obtain the response $y(k)$ from $k = 0$ to $k = 40$.

Summarizing, the MATLAB program to obtain the inverse z transform or the response to the Kronecker delta input is as shown in MATLAB Program 2-1.

MATLAB Program 2–1

```
% ---------- Finding inverse z transform ----------

% ***** Finding the inverse z transform of G(z) is the same as
% finding the response of the system Y(z)/X(z) = G(z) to the
% Kronecker delta input *****

% ***** Enter the numerator and denominator of G(z) *****

num = [0   0.4673   -0.3393];
den = [1   -1.5327   0.6607];

% ***** Enter the Kronecker delta input x and filter command
% y = filter(num,den,x) *****

x = [1   zeros(1,40)];
y = filter(num,den,x)
```

If this program is executed, the screen will show the output $y(k)$ from $k = 0$ to 40 as follows:

```
y =
  Columns 1 through 7
          0     0.4673     0.3769     0.2690     0.1632     0.0725     0.0032
  Columns 8 through 14
    -0.0429   -0.0679    -0.0758    -0.0712    -0.0591    -0.0436   -0.0277
  Columns 15 through 21
    -0.0137   -0.0027     0.0050     0.0094     0.0111     0.0108     0.0092
  Columns 22 through 28
     0.0070    0.0046     0.0025     0.0007    -0.0005    -0.0013   -0.0016
  Columns 29 through 35
    -0.0016   -0.0014    -0.0011    -0.0008    -0.0004    -0.0002     0.0000
  Columns 36 through 41
     0.0002    0.0002     0.0002     0.0002     0.0002     0.0001
```

(Note that MATLAB computations begin from column 1 and end at column 41, rather than from column 0 to column 40.) These values give the inverse z transform of $G(z)$. That is,

$$y(0) = 0$$

$$y(1) = 0.4673$$

$$y(2) = 0.3769$$

$$y(3) = 0.2690$$

$$\vdots$$

$$y(40) = 0.0001$$

To plot the values of the inverse z transform of $G(z)$, follow the procedure given in the following.

Plotting Response to the Kronecker Delta Input.   Consider the system given by Equation (2–20). A possible MATLAB program to obtain the response of this system to the Kronecker delta input is shown in MATLAB Program 2–2. The corresponding plot is shown in Figure 2–5.

```
MATLAB Program 2-2

% --------- Response to Kronecker delta input ---------

num = [0   0.4673   -0.3393];
den = [1   -1.5327   0.6607];
x = [1   zeros(1,40)];
k = 0:40;
y = filter(num,den,x);
plot(k,y,'o')
v = [0   40   -1   1];
axis(v);
grid
title('Response to Kronecker Delta Input')
xlabel('k')
ylabel('y(k)')
```

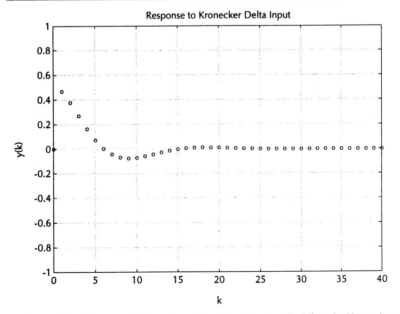

Figure 2–5   Response of the system defined by Equation (2–20) to the Kronecker delta input.

If we wish to connect consecutive points (open circles, o) by straight lines, we need to modify the plot command from plot(k,y,'o') to plot(k,y,'o',k,y,'–').

Difference Equation Approach.   Noting that Equation (2–20) can be written as

$$(z^2 - 1.5327z + 0.6607)Y(z) = (0.4673z - 0.3393)X(z)$$

we can convert this equation into the difference equation as follows:

$$y(k + 2) - 1.5327y(k + 1) + 0.6607y(k)$$

$$= 0.4673x(k + 1) - 0.3393x(k)   \qquad (2\text{–}21)$$

where $x(0) = 1$ and $x(k) = 0$ for $k \neq 0$, and $y(k) = 0$ for $k < 0$. [$x(k)$ is the Kronecker delta input.]

The initial data $y(0)$ and $y(1)$ can be determined as follows: By substituting $k = -2$ into Equation (2–21), we find

$$y(0) - 1.5327y(-1) + 0.6607y(-2) = 0.4673x(-1) - 0.3393x(-2)$$

from which we get

$$y(0) = 0$$

Next, by substituting $k = -1$ into Equation (2–21), we obtain

$$y(1) - 1.5327y(0) + 0.6607y(-1) = 0.4673x(0) - 0.3393x(-1)$$

from which we get

$$y(1) = 0.4673$$

Finding the inverse *z* transform of $Y(z)$ now becomes a matter of solving the following difference equation for $y(k)$:

$$y(k + 2) - 1.5327y(k + 1) + 0.6607y(k)$$

$$= 0.4673x(k + 1) - 0.3393x(k)   \qquad (2\text{–}22)$$

with the initial data $y(0) = 0$, $y(1) = 0.4673$, $x(0) = 1$, and $x(k) = 0$ for $k \neq 0$. Equation (2–22) can be solved easily by hand, or by use of BASIC, FORTRAN, or other.

Partial-Fraction-Expansion Method.   The partial-fraction expansion method presented here, which is parallel to the partial-fraction-expansion method used in Laplace transformation, is widely used in routine problems involving *z* transforms. The method requires that all terms in the partial fraction expansion be easily recognizable in the table of *z* transform pairs.

To find the inverse *z* transform, if $X(z)$ has one or more zeros at the origin ($z = 0$), then $X(z)/z$ or $X(z)$ is expanded into a sum of simple first- or second-order terms by partial fraction expansion, and a *z* transform table is used to find the corresponding time function of each expanded term. It is noted that the only reason that we expand $X(z)/z$ into partial fractions is that each expanded term has a form that may easily be found from commonly available *z* transform tables.

Example 2–13

Before we discuss the partial-fraction-expansion method, we shall review the shifting theorem. Consider the following $X(z)$:

$$X(z) = \frac{z^{-1}}{1 - az^{-1}}$$

By writing $zX(z)$ as $Y(z)$, we obtain

$$zX(z) = Y(z) = \frac{1}{1 - az^{-1}}$$

Referring to Table 2–1, the inverse z transform of $Y(z)$ can be obtained as follows:

$$\mathscr{Z}^{-1}[Y(z)] = y(k) = a^k$$

Hence, the inverse z transform of $X(z) = z^{-1} Y(z)$ is given by

$$\mathscr{Z}^{-1}[X(z)] = x(k) = y(k - 1)$$

Since $y(k)$ is assumed to be zero for all $k < 0$, we have

$$x(k) = \begin{cases} y(k - 1) = a^{k-1}, & k = 1, 2, 3, \ldots \\ 0, & k \le 0 \end{cases}$$

Consider $X(z)$ as given by

$$X(z) = \frac{b_0 z^m + b_1 z^{m-1} + \cdots + b_{m-1} z + b_m}{z^n + a_1 z^{n-1} + \cdots + a_{n-1} z + a_n}, \qquad m \le n$$

To expand $X(z)$ into partial fractions, we first factor the denominator polynomial of $X(z)$ and find the poles of $X(z)$:

$$X(z) = \frac{b_0 z^m + b_1 z^{m-1} + \cdots + b_{m-1} z + b_m}{(z - p_1)(z - p_2) \cdots (z - p_n)}$$

We then expand $X(z)/z$ into partial fractions so that each term is easily recognizable in a table of z transforms. If the shifting theorem is utilized in taking inverse z transforms, however, $X(z)$, instead of $X(z)/z$, may be expanded into partial fractions. The inverse z transform of $X(z)$ is obtained as the sum of the inverse z transforms of the partial fractions.

A commonly used procedure for the case where all the poles are of simple order and there is at least one zero at the origin (that is, $b_m = 0$) is to divide both sides of $X(z)$ by z and then expand $X(z)/z$ into partial fractions. Once $X(z)/z$ is expanded, it will be of the form

$$\frac{X(z)}{z} = \frac{a_1}{z - p_1} + \frac{a_2}{z - p_2} + \cdots + \frac{a_n}{z - p_n}$$

The coefficient a_i can be determined by multiplying both sides of this last equation by $z - p_i$ and setting $z = p_i$. This will result in zero for all the terms on the right-hand side except the a_i term, in which the multiplicative factor $z - p_i$ has been canceled by the denominator. Hence, we have

$$a_i = \left[(z - p_i) \frac{X(z)}{z} \right]_{z=p_i}$$

Note that such determination of a_i is valid only for simple poles.

If $X(z)/z$ involves a multiple pole, for example, a double pole at $z = p_1$ and no other poles, then $X(z)/z$ will have the form

$$\frac{X(z)}{z} = \frac{c_1}{(z - p_1)^2} + \frac{c_2}{z - p_1}$$

The coefficients c_1 and c_2 are determined from

$$c_1 = \left[(z - p_1)^2 \frac{X(z)}{z} \right]_{z=p_1}$$

$$c_2 = \left\{ \frac{d}{dz} \left[(z - p_1)^2 \frac{X(z)}{z} \right] \right\}_{z=p_1}$$

It is noted that if $X(z)/z$ involves a triple pole at $z = p_1$, then the partial fractions must include a term $(z + p_1)/(z - p_1)^3$. (See Problem A–2–8.)

Example 2–14

Given the z transform

$$X(z) = \frac{(1 - e^{-aT})z}{(z - 1)(z - e^{-aT})}$$

where a is a constant and T is the sampling period, determine the inverse z transform $x(kT)$ by use of the partial-fraction-expansion method.

The partial fraction expansion of $X(z)/z$ is found to be

$$\frac{X(z)}{z} = \frac{1}{z - 1} - \frac{1}{z - e^{-aT}}$$

Thus,

$$X(z) = \frac{1}{1 - z^{-1}} - \frac{1}{1 - e^{-aT}z^{-1}}$$

From Table 2–1 we find

$$\mathscr{Z}^{-1}\left[\frac{1}{1 - z^{-1}} \right] = 1$$

$$\mathscr{Z}^{-1}\left[\frac{1}{1 - e^{-aT}z^{-1}} \right] = e^{-akT}$$

Hence, the inverse z transform of $X(z)$ is

$$x(kT) = 1 - e^{-akT}, \qquad k = 0, 1, 2, \ldots$$

Example 2–15

Let us obtain the inverse z transform of

$$X(z) = \frac{z^2 + z + 2}{(z - 1)(z^2 - z + 1)}$$

by use of the partial-fraction-expansion method.

We may expand $X(z)$ into partial fractions as follows:

$$X(z) = \frac{4}{z - 1} + \frac{-3z + 2}{z^2 - z + 1} = \frac{4z^{-1}}{1 - z^{-1}} + \frac{-3z^{-1} + 2z^{-2}}{1 - z^{-1} + z^{-2}}$$

Noting that the two poles involved in the quadratic term of this last equation are complex conjugates, we rewrite $X(z)$ as follows:

$$X(z) = \frac{4z^{-1}}{1 - z^{-1}} - 3\left(\frac{z^{-1} - 0.5z^{-2}}{1 - z^{-1} + z^{-2}}\right) + \frac{0.5z^{-2}}{1 - z^{-1} + z^{-2}}$$

$$= 4z^{-1}\frac{1}{1 - z^{-1}} - 3z^{-1}\frac{1 - 0.5z^{-1}}{1 - z^{-1} + z^{-2}} + z^{-1}\frac{0.5z^{-1}}{1 - z^{-1} + z^{-2}}$$

Since

$$\mathscr{Z}[e^{-akT}\cos\omega kT] = \frac{1 - e^{-aT}z^{-1}\cos\omega T}{1 - 2e^{-aT}z^{-1}\cos\omega T + e^{-2aT}z^{-2}}$$

$$\mathscr{Z}[e^{-akT}\sin\omega kT] = \frac{e^{-aT}z^{-1}\sin\omega T}{1 - 2e^{-aT}z^{-1}\cos\omega T + e^{-2aT}z^{-2}}$$

by identifying $e^{-2aT} = 1$ and $\cos\omega T = \frac{1}{2}$ in this case, we have $\omega T = \pi/3$ and $\sin\omega T = \sqrt{3}/2$. Hence, we obtain

$$\mathscr{Z}^{-1}\left[\frac{1 - 0.5z^{-1}}{1 - z^{-1} + z^{-2}}\right] = 1^k\cos\frac{k\pi}{3}$$

and

$$\mathscr{Z}^{-1}\left[\frac{0.5z^{-1}}{1 - z^{-1} + z^{-2}}\right] = \mathscr{Z}^{-1}\left[\frac{1}{\sqrt{3}}\frac{(\sqrt{3}/2)z^{-1}}{1 - z^{-1} + z^{-2}}\right] = \frac{1}{\sqrt{3}}1^k\sin\frac{k\pi}{3}$$

Thus, we have

$$x(k) = 4(1^{k-1}) - 3(1^{k-1})\cos\frac{(k - 1)\pi}{3} + \frac{1}{\sqrt{3}}(1^{k-1})\sin\frac{(k - 1)\pi}{3}$$

Rewriting, we have

$$x(k) = \begin{cases} 4 - 3\cos\dfrac{(k - 1)\pi}{3} + \dfrac{1}{\sqrt{3}}\sin\dfrac{(k - 1)\pi}{3}, & k = 1, 2, 3, \ldots \\ 0, & k \le 0 \end{cases}$$

The first several values of $x(k)$ are given by

$$x(0) = 0$$
$$x(1) = 1$$
$$x(2) = 3$$
$$x(3) = 6$$
$$x(4) = 7$$
$$x(5) = 5$$
$$\vdots$$

Note that the inverse z transform of $X(z)$ can also be obtained as follows:

$$X(z) = 4z^{-1}\frac{1}{1 - z^{-1}} - 3\left(\frac{z^{-1}}{1 - z^{-1} + z^{-2}}\right) + 2z^{-1}\frac{z^{-1}}{1 - z^{-1} + z^{-2}}$$

Since

$$\mathscr{Z}^{-1}\left[\frac{z^{-1}}{1 - z^{-1}}\right] = \begin{cases} 1, & k = 1, 2, 3, \ldots \\ 0, & k \le 0 \end{cases}$$

and

$$\mathcal{Z}^{-1}\left[\frac{z^{-1}}{1 - z^{-1} + z^{-2}}\right] = \frac{2}{\sqrt{3}}(1^k)\sin\frac{k\pi}{3}$$

we have

$$x(k) = \begin{cases} 4 - 2\sqrt{3}\sin\dfrac{k\pi}{3} + \dfrac{4}{\sqrt{3}}\sin\dfrac{(k-1)\pi}{3}, & k = 1, 2, 3, \ldots \\ 0, & k \le 0 \end{cases}$$

Although this solution may look different from the one obtained earlier, both solutions are correct and yield the same values for $x(k)$.

Inversion Integral Method.   This is a useful technique for obtaining the inverse z transform. The inversion integral for the z transform $X(z)$ is given by

$$\mathcal{Z}^{-1}[X(z)] = x(kT) = x(k) = \frac{1}{2\pi j}\oint_C X(z)z^{k-1}\,dz \tag{2-23}$$

where C is a circle with its center at the origin of the z plane such that all poles of $X(z)z^{k-1}$ are inside it. [For the derivation of Equation (2–23), see Appendix B.]

The equation for giving the inverse z transform in terms of residues can be derived by using theory of complex variables. It can be obtained as follows:

$$x(kT) = x(k) = K_1 + K_2 + \cdots + K_m$$

$$= \sum_{i=1}^{m}[\text{residue of } X(z)z^{k-1} \text{ at pole } z = z_i \text{ of } X(z)z^{k-1}] \tag{2-24}$$

where $K_1, K_2, \ldots, K_m$ denote the residues of $X(z)z^{k-1}$ at poles $z_1, z_2, \ldots, z_m$, respectively. (For the derivation of this equation, see Appendix B.) In evaluating residues, note that if the denominator of $X(z)z^{k-1}$ contains a simple pole $z = z_i$ then the corresponding residue K is given by

$$K = \lim_{z \to z_i}[(z - z_i)X(z)z^{k-1}] \tag{2-25}$$

If $X(z)z^{k-1}$ contains a multiple pole z_j of order q, then the residue K is given by

$$K = \frac{1}{(q-1)!}\lim_{z \to z_j}\frac{d^{q-1}}{dz^{q-1}}[(z - z_j)^q X(z)z^{k-1}] \tag{2-26}$$

Note that the values of k in Equations (2–24), (2–25), and (2–26) are nonnegative integer values.

If $X(z)$ has a zero of order r at the origin, then $X(z)z^{k-1}$ in Equation (2–24) will involve a zero of order $r + k - 1$ at the origin. If $r \ge 1$, then $r + k - 1 \ge 0$ for $k \ge 0$, and there is no pole at $z = 0$ in $X(z)z^{k-1}$. However, if $r \le 0$, then there will be a pole at $z = 0$ for one or more nonnegative values of k. In such a case, separate inversion of Equation (2–24) is necessary for each such value of k. (See Problem A–2–9.)

It should be noted that the inversion integral method, when evaluated by residues, is a very simple technique for obtaining the inverse z transform, provided that $X(z)z^{k-1}$ has no poles at the origin, $z = 0$. If, however, $X(z)z^{k-1}$ has a simple pole or a multiple pole at $z = 0$, then calculations may become cumbersome and the partial-fraction-expansion method may prove to be simpler to apply. On the other hand, in certain problems the partial-fraction-expansion approach may become laborious. Then, the inversion integral method proves to be very convenient.

Example 2-16

Obtain $x(kT)$ by using the inversion integral method when $X(z)$ is given by

$$X(z) = \frac{z(1 - e^{-aT})}{(z - 1)(z - e^{-aT})}$$

Note that

$$X(z)z^{k-1} = \frac{(1 - e^{-aT})z^k}{(z - 1)(z - e^{-aT})}$$

For $k = 0, 1, 2, \ldots, X(z)z^{k-1}$ has two simple poles, $z = z_1 = 1$ and $z = z_2 = e^{-aT}$. Hence, from Equation (2-24), we have

$$x(k) = \sum_{i=1}^{2}\left[\text{residue of } \frac{(1 - e^{-aT})z^k}{(z - 1)(z - e^{-aT})} \text{ at pole } z = z_i\right]$$

$$= K_1 + K_2$$

where

$$K_1 = [\text{residue at simple pole } z = 1]$$

$$= \lim_{z\to 1}\left[(z - 1)\frac{(1 - e^{-aT})z^k}{(z - 1)(z - e^{-aT})}\right] = 1$$

$$K_2 = [\text{residue at simple pole } z = e^{-aT}]$$

$$= \lim_{z\to e^{-aT}}\left[(z - e^{-aT})\frac{(1 - e^{-aT})z^k}{(z - 1)(z - e^{-aT})}\right] = -e^{-akT}$$

Hence,

$$x(kT) = K_1 + K_2 = 1 - e^{-akT}, \qquad k = 0, 1, 2, \ldots$$

Example 2-17

Obtain the inverse z transform of

$$X(z) = \frac{z^2}{(z - 1)^2(z - e^{-aT})}$$

by using the inversion integral method.

Notice that

$$X(z)z^{k-1} = \frac{z^{k+1}}{(z - 1)^2(z - e^{-aT})}$$

For $k = 0, 1, 2, \ldots, X(z)z^{k-1}$ has a simple pole at $z = z_1 = e^{-aT}$ and a double pole at $z = z_2 = 1$. Hence, from Equation (2-24), we obtain

$$x(k) = \sum_{i=1}^{2}\left[\text{residue of } \frac{z^{k+1}}{(z - 1)^2(z - e^{-aT})} \text{ at pole } z = z_i\right]$$

$$= K_1 + K_2$$

where

$$K_1 = [\text{residue at simple pole } z = e^{-aT}]$$

$$= \lim_{z \to e^{-aT}} \left[(z - e^{-aT}) \frac{z^{k+1}}{(z-1)^2(z-e^{-aT})} \right] = \frac{e^{-a(k+1)T}}{(1-e^{-aT})^2}$$

$$K_2 = [\text{residue at double pole } z = 1]$$

$$= \frac{1}{(2-1)!} \lim_{z \to 1} \frac{d}{dz} \left[(z-1)^2 \frac{z^{k+1}}{(z-1)^2(z-e^{-aT})} \right]$$

$$= \lim_{z \to 1} \frac{d}{dz} \left(\frac{z^{k+1}}{z-e^{-aT}} \right)$$

$$= \lim_{z \to 1} \frac{(k+1)z^k(z-e^{-aT}) - z^{k+1}}{(z-e^{-aT})^2}$$

$$= \frac{k}{1-e^{-aT}} - \frac{e^{-aT}}{(1-e^{-aT})^2}$$

Hence,

$$x(kT) = K_1 + K_2 = \frac{e^{-aT}e^{-akT}}{(1-e^{-aT})^2} + \frac{k}{1-e^{-aT}} - \frac{e^{-aT}}{(1-e^{-aT})^2}$$

$$= \frac{kT}{T(1-e^{-aT})} - \frac{e^{-aT}(1-e^{-akT})}{(1-e^{-aT})^2}, \qquad k = 0, 1, 2, \ldots$$

2-6 z TRANSFORM METHOD FOR SOLVING DIFFERENCE EQUATIONS

Difference equations can be solved easily by use of a digital computer, provided the numerical values of all coefficients and parameters are given. However, closed-form expressions for $x(k)$ cannot be obtained from the computer solution, except for very special cases. The usefulness of the z transform method is that it enables us to obtain the closed-form expression for $x(k)$.

Consider the linear time-invariant discrete-time system characterized by the following linear difference equation:

$$x(k) + a_1 x(k-1) + \cdots + a_n x(k-n)$$

$$= b_0 u(k) + b_1 u(k-1) + \cdots + b_n u(k-n) \qquad (2\text{--}27)$$

where $u(k)$ and $x(k)$ are the system's input and output, respectively, at the kth iteration. In describing such a difference equation in the z plane, we take the z transform of each term in the equation.

Let us define

$$\mathcal{Z}[x(k)] = X(z)$$

Then $x(k+1), x(k+2), x(k+3), \ldots$ and $x(k-1), x(k-2), x(k-3), \ldots$ can be expressed in terms of $X(z)$ and the initial conditions. Their exact z transforms were derived in Section 2–4 and are summarized in Table 2–3 for convenient reference.

Next we present two example problems for solving difference equations by the z transform method.

TABLE 2–3   z TRANSFORMS OF $x(k + m)$ AND $x(k - m)$

Discrete function	z Transform
$x(k + 4)$	$z^4 X(z) - z^4 x(0) - z^3 x(1) - z^2 x(2) - zx(3)$
$x(k + 3)$	$z^3 X(z) - z^3 x(0) - z^2 x(1) - zx(2)$
$x(k + 2)$	$z^2 X(z) - z^2 x(0) - zx(1)$
$x(k + 1)$	$zX(z) - zx(0)$
$x(k)$	$X(z)$
$x(k - 1)$	$z^{-1} X(z)$
$x(k - 2)$	$z^{-2} X(z)$
$x(k - 3)$	$z^{-3} X(z)$
$x(k - 4)$	$z^{-4} X(z)$

Example 2–18

Solve the following difference equation by use of the z transform method:

$$x(k + 2) + 3x(k + 1) + 2x(k) = 0, \qquad x(0) = 0, \qquad x(1) = 1$$

First note that the z transforms of $x(k + 2)$, $x(k + 1)$, and $x(k)$ are given, respectively, by

$$\mathscr{Z}[x(k + 2)] = z^2 X(z) - z^2 x(0) - zx(1)$$
$$\mathscr{Z}[x(k + 1)] = zX(z) - zx(0)$$
$$\mathscr{Z}[x(k)] = X(z)$$

Taking the z transforms of both sides of the given difference equation, we obtain

$$z^2 X(z) - z^2 x(0) - zx(1) + 3zX(z) - 3zx(0) + 2X(z) = 0$$

Substituting the initial data and simplifying gives

$$X(z) = \frac{z}{z^2 + 3z + 2} = \frac{z}{(z + 1)(z + 2)} = \frac{z}{z + 1} - \frac{z}{z + 2}$$

$$= \frac{1}{1 + z^{-1}} - \frac{1}{1 + 2z^{-1}}$$

Noting that

$$\mathscr{Z}^{-1}\left[\frac{1}{1 + z^{-1}}\right] = (-1)^k, \qquad \mathscr{Z}^{-1}\left[\frac{1}{1 + 2z^{-1}}\right] = (-2)^k$$

we have

$$x(k) = (-1)^k - (-2)^k, \qquad k = 0, 1, 2, \ldots$$

Example 2–19

Obtain the solution of the following difference equation in terms of $x(0)$ and $x(1)$:

$$x(k + 2) + (a + b)x(k + 1) + abx(k) = 0$$

where a and b are constants and $k = 0, 1, 2, \ldots$.

The z transform of this difference equation can be given by

$$[z^2 X(z) - z^2 x(0) - zx(1)] + (a + b)[zX(z) - zx(0)] + abX(z) = 0$$

or

$$[z^2 + (a + b)z + ab]X(z) = [z^2 + (a + b)z]x(0) + zx(1)$$

Solving this last equation for $X(z)$ gives

$$X(z) = \frac{[z^2 + (a + b)z]x(0) + zx(1)}{z^2 + (a + b)z + ab}$$

Notice that constants a and b are the negatives of the two roots of the characteristic equation. We shall now consider separately two cases: (a) $a \neq b$ and (b) $a = b$.

(a) For the case where $a \neq b$, expanding $X(z)/z$ into partial fractions, we obtain

$$\frac{X(z)}{z} = \frac{bx(0) + x(1)}{b - a} \frac{1}{z + a} + \frac{ax(0) + x(1)}{a - b} \frac{1}{z + b}, \qquad a \neq b$$

from which we get

$$X(z) = \frac{bx(0) + x(1)}{b - a} \frac{1}{1 + az^{-1}} + \frac{ax(0) + x(1)}{a - b} \frac{1}{1 + bz^{-1}}$$

The inverse z transform of $X(z)$ gives

$$x(k) = \frac{bx(0) + x(1)}{b - a}(-a)^k + \frac{ax(0) + x(1)}{a - b}(-b)^k, \qquad a \neq b$$

where $k = 0, 1, 2, \ldots$.

(b) For the case where $a = b$, the z transform $X(z)$ becomes

$$X(z) = \frac{(z^2 + 2az)x(0) + zx(1)}{z^2 + 2az + a^2}$$

$$= \frac{zx(0)}{z + a} + \frac{z[ax(0) + x(1)]}{(z + a)^2}$$

$$= \frac{x(0)}{1 + az^{-1}} + \frac{[ax(0) + x(1)]z^{-1}}{(1 + az^{-1})^2}$$

The inverse z transform of $X(z)$ gives

$$x(k) = x(0)(-a)^k + [ax(0) + x(1)]k(-a)^{k-1}, \qquad a = b$$

where $k = 0, 1, 2, \ldots$.

2-7 CONCLUDING COMMENTS

In this chapter the basic theory of the z transform method has been presented. The z transform serves the same purpose for linear time-invariant discrete-time systems as the Laplace transform provides for linear time-invariant continuous-time systems.

The computer method of analyzing data in discrete time results in difference equations. With the z transform method, linear time-invariant difference equations can be transformed into algebraic equations. This facilitates the transient response analysis of the digital control system. Also, the z transform method allows us to use conventional analysis and design techniques available to analog (continuous-time) control systems, such as the root-locus technique. Frequency-response analysis and design can be carried out by converting the z plane into the w plane. Also, the z-transformed characteristic equation allows us to apply a simple stability test, such as the Jury stability criterion. These subjects will be discussed in detail in Chapters 3 and 4.

EXAMPLE PROBLEMS AND SOLUTIONS

Problem A–2–1

Obtain the z transform of $\mathbf{G}^k$, where $\mathbf{G}$ is an $n \times n$ constant matrix.

Solution By definition, the z transform of $\mathbf{G}^k$ is

$$\mathcal{Z}[\mathbf{G}^k] = \sum_{k=0}^{\infty} \mathbf{G}^k z^{-k}$$

$$= \mathbf{I} + \mathbf{G}z^{-1} + \mathbf{G}^2 z^{-2} + \mathbf{G}^3 z^{-3} + \cdots$$

$$= (\mathbf{I} - \mathbf{G}z^{-1})^{-1}$$

$$= (z\mathbf{I} - \mathbf{G})^{-1} z$$

Note that $\mathbf{G}^k$ can be obtained by taking the inverse z transform of $(\mathbf{I} - \mathbf{G}z^{-1})^{-1}$ or $(z\mathbf{I} - \mathbf{G})^{-1} z$. That is,

$$\mathbf{G}^k = \mathcal{Z}^{-1}[(\mathbf{I} - \mathbf{G}z^{-1})^{-1}] = \mathcal{Z}^{-1}[(z\mathbf{I} - \mathbf{G})^{-1} z]$$

Problem A–2–2

Obtain the z transform of k^2.

Solution By definition, the z transform of k^2 is

$$\mathcal{Z}[k^2] = \sum_{k=0}^{\infty} k^2 z^{-k} = z^{-1} + 4z^{-2} + 9z^{-3} + 16z^{-4} + \cdots$$

$$= z^{-1}(1 + z^{-1})(1 + 3z^{-1} + 6z^{-2} + 10z^{-3} + 15z^{-4} + \cdots)$$

$$= \frac{z^{-1}(1 + z^{-1})}{(1 - z^{-1})^3}$$

Here we have used the closed-form expression $(1 - z^{-1})^{-3}$ for the infinite series involved in the problem. (See Appendix B.)

Problem A–2–3

Obtain the z transform of ka^{k-1} by two methods.

Solution

Method 1. By definition, the z transform of ka^{k-1} is given by

$$\mathcal{Z}[ka^{k-1}] = \sum_{k=0}^{\infty} ka^{k-1} z^{-k}$$

$$= z^{-1} + 2az^{-2} + 3a^2 z^{-3} + 4a^3 z^{-4} + \cdots$$

$$= z^{-1}(1 + 2az^{-1} + 3a^2 z^{-2} + 4a^3 z^{-3} + \cdots)$$

$$= \frac{z^{-1}}{(1 - az^{-1})^2}$$

Method 2.   The summation expression for the z transform of ka^{k-1} can also be written as follows:

$$\mathscr{Z}[ka^{k-1}] = \sum_{k=0}^{\infty} ka^{k-1}z^{-k} = a^{-1}\sum_{k=0}^{\infty} ka^{k}z^{-k} = \frac{1}{a}\sum_{k=0}^{\infty} k\left(\frac{z}{a}\right)^{-k}$$

$$= \frac{1}{a}\frac{(z/a)^{-1}}{[1-(z/a)^{-1}]^2} = \frac{z^{-1}}{(1-az^{-1})^2}$$

Problem A–2–4

Show that

$$\mathscr{Z}\left[\sum_{h=0}^{k} x(h)\right] = \frac{1}{1-z^{-1}}X(z)$$

$$\mathscr{Z}\left[\sum_{h=0}^{k-1} x(h)\right] = \frac{z^{-1}}{1-z^{-1}}X(z)$$

and

$$\sum_{k=0}^{\infty} x(k) = \lim_{z\to 1} X(z) \tag{2–28}$$

Also show that

$$\mathscr{Z}\left[\sum_{h=i}^{k} x(h)\right] = \frac{1}{1-z^{-1}}\left[X(z) - \sum_{h=0}^{i-1} x(h)z^{-h}\right] \tag{2–29}$$

where $1 \le i \le k - 1$.

Solution   Define

$$y(k) = \sum_{h=0}^{k} x(h), \qquad k = 0, 1, 2, \ldots$$

so that

$$y(0) = x(0)$$
$$y(1) = x(0) + x(1)$$
$$y(2) = x(0) + x(1) + x(2)$$
$$\vdots$$
$$y(k) = x(0) + x(1) + x(2) + \cdots + x(k)$$

Then, clearly

$$y(k) - y(k-1) = x(k)$$

By writing the z transforms of $x(k)$ and $y(k)$ as $X(z)$ and $Y(z)$, respectively, and by taking the z transform of this last equation, we have

$$Y(z) - z^{-1}Y(z) = X(z)$$

Hence,

$$Y(z) = \frac{1}{1-z^{-1}}X(z)$$

or

$$\mathcal{Z}\left[\sum_{h=0}^{k} x(h)\right] = \mathcal{Z}[y(k)] = Y(z) = \frac{1}{1 - z^{-1}}X(z)$$

and

$$\mathcal{Z}\left[\sum_{h=0}^{k-1} x(h)\right] = \mathcal{Z}[y(k - 1)] = z^{-1}Y(z) = \frac{z^{-1}}{1 - z^{-1}}X(z)$$

By using the final value theorem, we find

$$\lim_{k \to \infty} y(k) = \lim_{k \to \infty}\left[\sum_{h=0}^{k} x(h)\right] = \lim_{z \to 1}\left[(1 - z^{-1})\frac{1}{1 - z^{-1}}X(z)\right]$$

or

$$\sum_{h=0}^{\infty} x(h) = \sum_{k=0}^{\infty} x(k) = \lim_{z \to 1} X(z)$$

Next, to prove Equation (2–29), first define

$$\bar{y}(k) = \sum_{h=i}^{k} x(h) = x(i) + x(i + 1) + \cdots + x(k)$$

where $1 \le i \le k - 1$. Define also

$$\tilde{X}(z) = x(i)z^{-i} + x(i + 1)z^{-(i+1)} + \cdots + x(k)z^{-k} + \cdots$$

Then, noting that

$$X(z) = \mathcal{Z}[x(k)] = \sum_{k=0}^{\infty} x(k)z^{-k} = x(0) + x(1)z^{-1} + x(2)z^{-2} + \cdots$$

we obtain

$$\tilde{X}(z) = X(z) - \sum_{h=0}^{i-1} x(h)z^{-h}$$

Since

$$\bar{y}(k) - \bar{y}(k - 1) = x(k), \qquad k = i, i + 1, i + 2, \ldots$$

the z transform of this last equation becomes

$$\tilde{Y}(z) - z^{-1}\tilde{Y}(z) = \tilde{X}(z)$$

[Note that the z transform of $x(k)$, which begins with $k = i$, is $\tilde{X}(z)$, not $X(z)$.] Thus,

$$\mathcal{Z}\left[\sum_{h=i}^{k} x(h)\right] = \tilde{Y}(z) = \frac{1}{1 - z^{-1}}\tilde{X}(z) = \frac{1}{1 - z^{-1}}\left[X(z) - \sum_{h=0}^{i-1} x(h)z^{-h}\right]$$

Problem A–2–5

Obtain the z transform of the curve $x(t)$ shown in Figure 2–6. Assume that the sampling period T is 1 sec.

Solution   From Figure 2–6 we obtain

$$x(0) = 0$$

$$x(1) = 0.25$$

$$x(2) = 0.50$$

$$x(3) = 0.75$$

$$x(k) = 1, \qquad k = 4, 5, 6, \ldots$$

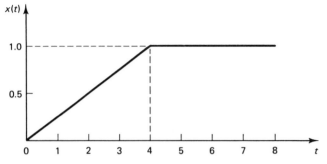

Figure 2–6   Curve $x(t)$.

Then the z transform of $x(k)$ can be given by

$$X(z) = \sum_{k=0}^{\infty} x(k)z^{-k}$$

$$= 0.25z^{-1} + 0.50z^{-2} + 0.75z^{-3} + z^{-4} + z^{-5} + z^{-6} + \cdots$$

$$= 0.25(z^{-1} + 2z^{-2} + 3z^{-3}) + z^{-4}\frac{1}{1 - z^{-1}}$$

$$= \frac{z^{-1} + z^{-2} + z^{-3} + z^{-4}}{4(1 - z^{-1})}$$

$$= \frac{1}{4}\frac{z^{-1}(1 + z^{-1} + z^{-2} + z^{-3})(1 - z^{-1})}{(1 - z^{-1})^2}$$

$$= \frac{1}{4}\frac{z^{-1}(1 - z^{-4})}{(1 - z^{-1})^2}$$

Notice that the curve $x(t)$ can be written as

$$x(t) = \tfrac{1}{4}t - \tfrac{1}{4}(t - 4)1(t - 4)$$

where $1(t - 4)$ is the unit-step function occurring at $t = 4$. Since the sampling period $T = 1$ sec, the z transform of $x(t)$ can also be obtained as follows:

$$X(z) = \mathscr{Z}[x(t)] = \mathscr{Z}[\tfrac{1}{4}t] - \mathscr{Z}[\tfrac{1}{4}(t - 4)1(t - 4)]$$

$$= \frac{1}{4}\frac{z^{-1}}{(1 - z^{-1})^2} - \frac{1}{4}\frac{z^{-4}z^{-1}}{(1 - z^{-1})^2}$$

$$= \frac{1}{4}\frac{z^{-1}(1 - z^{-4})}{(1 - z^{-1})^2}$$

Problem A–2–6

Consider $X(z)$, where

$$X(z) = \frac{2z^3 + z}{(z - 2)^2(z - 1)}$$

Obtain the inverse z transform of $X(z)$.

Solution   We shall expand $X(z)/z$ into partial fractions as follows:

$$\frac{X(z)}{z} = \frac{2z^2 + 1}{(z - 2)^2(z - 1)} = \frac{9}{(z - 2)^2} - \frac{1}{z - 2} + \frac{3}{z - 1}$$

Then

$$X(z) = \frac{9z^{-1}}{(1 - 2z^{-1})^2} - \frac{1}{1 - 2z^{-1}} + \frac{3}{1 - z^{-1}}$$

The inverse z transforms of the individual terms give

$$\mathcal{Z}^{-1}\left[\frac{z^{-1}}{(1 - 2z^{-1})^2}\right] = k(2^{k-1}), \qquad k = 0, 1, 2, \ldots$$

$$\mathcal{Z}^{-1}\left[\frac{1}{1 - 2z^{-1}}\right] = 2^k, \qquad k = 0, 1, 2, \ldots$$

$$\mathcal{Z}^{-1}\left[\frac{1}{1 - z^{-1}}\right] = 1$$

and therefore

$$x(k) = 9k(2^{k-1}) - 2^k + 3, \qquad k = 0, 1, 2, \ldots$$

Problem A–2–7

Obtain the inverse z transform of

$$X(z) = \frac{z + 2}{(z - 2)z^2}$$

Solution   Expanding $X(z)$ into partial fractions, we obtain

$$X(z) = \frac{1}{z - 2} - \frac{1}{z^2} - \frac{1}{z} = \frac{z^{-1}}{1 - 2z^{-1}} - z^{-2} - z^{-1}$$

[Note that in this example $X(z)$ involves a double pole at $z = 0$. Hence the partial fraction expansion must include the terms $1/(z^2)$ and $1/z$.] By referring to Table 2–1, we find the inverse z transform of each term of this last equation. That is,

$$\mathcal{Z}^{-1}\left[\frac{z^{-1}}{1 - 2z^{-1}}\right] = \begin{cases} 2^{k-1}, & k = 1, 2, 3, \ldots \\ 0, & k \leq 0 \end{cases}$$

$$\mathcal{Z}^{-1}[z^{-2}] = \begin{cases} 1, & k = 2 \\ 0, & k \neq 2 \end{cases}$$

$$\mathcal{Z}^{-1}[z^{-1}] = \begin{cases} 1, & k = 1 \\ 0, & k \neq 1 \end{cases}$$

Hence, the inverse z transform of $X(z)$ can be given by

$$x(k) = \begin{cases} 0 - 0 - 0 = 0, & k = 0 \\ 1 - 0 - 1 = 0, & k = 1 \\ 2 - 1 - 0 = 1, & k = 2 \\ 2^{k-1} - 0 - 0 = 2^{k-1}, & k = 3, 4, 5, \ldots \end{cases}$$

Rewriting, we have

$$x(k) = \begin{cases} 0, & k = 0, 1 \\ 1, & k = 2 \\ 2^{k-1}, & k = 3, 4, 5, \ldots \end{cases}$$

To verify this result, the direct division method may be applied to this problem. Noting that

$$X(z) = \frac{z + 2}{(z - 2)z^2} = \frac{z^{-2} + 2z^{-3}}{1 - 2z^{-1}}$$

$$= z^{-2} + 4z^{-3} + 8z^{-4} + 16z^{-5} + 32z^{-6} + \cdots$$

$$= z^{-2} + (2^{3-1})z^{-3} + (2^{4-1})z^{-4} + (2^{5-1})z^{-5} + (2^{6-1})z^{-6} + \cdots$$

we find

$$x(k) = \begin{cases} 0, & k = 0, 1 \\ 1, & k = 2 \\ 2^{k-1}, & k = 3, 4, 5, \ldots \end{cases}$$

Problem A–2–8

Obtain the inverse z transform of

$$X(z) = \frac{z^{-2}}{(1 - z^{-1})^3}$$

Solution   The inverse z transform of $z^{-2}/(1 - z^{-1})^3$ is not available from most z transform tables. It is possible, however, to write the given $X(z)$ as a sum of z transforms that are commonly available in z transform tables. Since the denominator of $X(z)$ is $(1 - z^{-1})^3$ and the z transform of k^2 is $z^{-1}(1 + z^{-1})/(1 - z^{-1})^3$, let us rewrite $X(z)$ as

$$X(z) = \frac{z^{-2}}{(1 - z^{-1})^3} = \frac{z^{-1}(1 + z^{-1})}{(1 - z^{-1})^3} - \frac{z^{-1}}{(1 - z^{-1})^3}$$

$$= \frac{z^{-1}(1 + z^{-1})}{(1 - z^{-1})^3} - \frac{z^{-1} - z^{-2} + z^{-2}}{(1 - z^{-1})^3}$$

or

$$\frac{z^{-2}}{(1 - z^{-1})^3} = \frac{z^{-1}(1 + z^{-1})}{(1 - z^{-1})^3} - \frac{z^{-1}}{(1 - z^{-1})^2} - \frac{z^{-2}}{(1 - z^{-1})^3}$$

from which we obtain the following partial fraction expansion:

$$\frac{z^{-2}}{(1 - z^{-1})^3} = \frac{1}{2}\left[\frac{z^{-1}(1 + z^{-1})}{(1 - z^{-1})^3} - \frac{z^{-1}}{(1 - z^{-1})^2} \right]$$

The z transforms of the two terms on the right-hand side of this last equation can be found from Table 2–1. Thus,

$$x(k) = \mathscr{Z}^{-1}\left[\frac{z^{-2}}{(1 - z^{-1})^3} \right] = \frac{1}{2}(k^2 - k) = \frac{1}{2}k(k - 1), \qquad k = 0, 1, 2, \ldots$$

It is noted that if the given $X(z)$ is expanded into other partial fractions then the inverse z transform may not be obtained.

As an alternative approach, the inverse z transform of $X(z)$ may be obtained by use of the inversion integral method. First, note that

$$X(z)z^{k-1} = \frac{z^k}{(z - 1)^3}$$

Hence, for $k = 0, 1, 2, \ldots, X(z)z^{k-1}$ has a triple pole at $z = 1$. Referring to Equation (2–24), we have

$$x(k) = \left[\text{residue of } \frac{z^k}{(z - 1)^3} \text{ at triple pole } z = 1 \right]$$

$$= \frac{1}{(3-1)!} \lim_{z \to 1} \frac{d^2}{dz^2} \left[(z-1)^3 \frac{z^k}{(z-1)^3} \right]$$

$$= \frac{1}{2!} \lim_{z \to 1} \frac{d^2}{dz^2} (z^k)$$

$$= \frac{1}{2} \lim_{z \to 1} \frac{d}{dz} (kz^{k-1})$$

$$= \frac{1}{2} \lim_{z \to 1} [k(k-1)z^{k-2}]$$

$$= \frac{1}{2} k(k-1), \qquad k = 0, 1, 2, \ldots$$

Problem A–2–9

Using the inversion integral method, obtain the inverse z transform of

$$X(z) = \frac{10}{(z-1)(z-2)}$$

Solution   Note that

$$X(z)z^{k-1} = \frac{10z^{k-1}}{(z-1)(z-2)}$$

For $k = 0$, notice that $X(z)z^{k-1}$ becomes

$$X(z)z^{k-1} = \frac{10}{(z-1)(z-2)z}, \qquad k = 0$$

Hence, for $k = 0$, $X(z)z^{k-1}$ has three simple poles, $z = z_1 = 1$, $z = z_2 = 2$, and $z = z_3 = 0$. For $k = 1, 2, 3, \ldots$, however, $X(z)z^{k-1}$ has only two simple poles, $z = z_1 = 1$ and $z = z_2 = 2$. Therefore, we must consider $x(0)$ and $x(k)$ (where $k = 1, 2, 3, \ldots$) separately.

For $k = 0$.   For this case, referring to Equation (2–24), we have

$$x(0) = \sum_{i=1}^{3} \left[\text{residue of } \frac{10}{(z-1)(z-2)z} \text{ at pole } z = z_i \right]$$

$$= K_1 + K_2 + K_3$$

where

$$K_1 = [\text{residue at simple pole } z = 1]$$

$$= \lim_{z \to 1} \left[(z-1) \frac{10}{(z-1)(z-2)z} \right] = -10$$

$$K_2 = [\text{residue at simple pole } z = 2]$$

$$= \lim_{z \to 2} \left[(z-2) \frac{10}{(z-1)(z-2)z} \right] = 5$$

$$K_3 = [\text{residue at simple pole } z = 0]$$

$$= \lim_{z \to 0} \left[z \frac{10}{(z-1)(z-2)z} \right] = 5$$

Hence,

$$x(0) = K_1 + K_2 + K_3 = -10 + 5 + 5 = 0$$

For k = 1, 2, 3, ...   For this case, Equation (2–24) becomes

$$x(k) = \sum_{i=1}^{2} \left[\text{residue of } \frac{10z^{k-1}}{(z-1)(z-2)} \text{ at pole } z = z_i \right]$$

$$= K_1 + K_2$$

where

$$K_1 = [\text{residue at simple pole } z = 1]$$

$$= \lim_{z \to 1} \left[(z-1) \frac{10z^{k-1}}{(z-1)(z-2)} \right] = -10$$

$$K_2 = [\text{residue at simple pole } z = 2]$$

$$= \lim_{z \to 2} \left[(z-2) \frac{10z^{k-1}}{(z-1)(z-2)} \right] = 10(2^{k-1})$$

Thus,

$$x(k) = K_1 + K_2 = -10 + 10(2^{k-1}) = 10(2^{k-1} - 1), \qquad k = 1, 2, 3, \ldots$$

Hence, the inverse z transform of the given $X(z)$ can be written

$$x(k) = \begin{cases} 0, & k = 0 \\ 10(2^{k-1} - 1), & k = 1, 2, 3, \ldots \end{cases}$$

An alternative way to write $x(k)$ for $k \geq 0$ is

$$x(k) = 5\delta_0(k) + 10(2^{k-1} - 1), \qquad k = 0, 1, 2, \ldots$$

where $\delta_0(k)$ is the Kronecker delta function and is given by

$$\delta_0(k) = \begin{cases} 1, & \text{for } k = 0 \\ 0, & \text{for } k \neq 0 \end{cases}$$

Problem A–2–10

Obtain the inverse z transform of

$$X(z) = \frac{z(z+2)}{(z-1)^2} \tag{2–30}$$

by use of the four methods presented in Section 2–5.

Solution

Method 1: Direct division method.   We first rewrite $X(z)$ as a ratio of two polynomials in z^{-1}:

$$X(z) = \frac{1 + 2z^{-1}}{(1 - z^{-1})^2} = \frac{1 + 2z^{-1}}{1 - 2z^{-1} + z^{-2}}$$

Dividing the numerator by the denominator, we get

$$X(z) = 1 + 4z^{-1} + 7z^{-2} + 10z^{-3} + \cdots$$

Hence,

$$x(0) = 1$$

$$x(1) = 4$$

$$x(2) = 7$$

$$x(3) = 10$$

$$\vdots$$

Method 2: Computational method (MATLAB approach). $X(z)$ can be written as

$$X(z) = \frac{z^2 + 2z}{z^2 - 2z + 1}$$

Hence, the inverse z transform of $X(z)$ can be obtained with MATLAB as follows: Define

$$\text{num} = \begin{bmatrix} 1 & 2 & 0 \end{bmatrix}$$
$$\text{den} = \begin{bmatrix} 1 & -2 & 1 \end{bmatrix}$$

If the values of $x(k)$ for $k = 0, 1, 2, \dots, 30$ are desired, then enter the Kronecker delta input as follows:

$$u = \begin{bmatrix} 1 & \text{zeros}(1,30) \end{bmatrix}$$

Then enter the command

$$x = \text{filter(num,den,u)}$$

See MATLAB Program 2–3. [The screen will show the output $x(k)$ from $k = 0$ to $k = 30$.] (MATLAB computations begin from column 1 and end at column 31, rather

---

MATLAB Program 2–3

```
num = [1  2  0];
den = [1  -2  1];
u = [1   zeros(1,30)];
x = filter(num,den,u)


x =

 Columns 1 through 12

         1      4      7     10     13     16     19     22     25     28     31     34

 Columns 13 through 24

        37     40     43     46     49     52     55     58     61     64     67     70

 Columns 25 through 31

        73     76     79     82     85     88     91
```

than from column 0 to column 30.) The values $x(k)$ give the inverse *z* transform of $X(z)$. That is,

$$x(0) = 1$$
$$x(1) = 4$$
$$x(2) = 7$$
$$\vdots$$
$$x(30) = 91$$

Method 3: Partial-fraction-expansion method. We expand $X(z)$ into the following partial fractions:

$$X(z) = \frac{z(z + 2)}{(z - 1)^2} = 1 + \frac{3z}{(z - 1)^2} + \frac{1}{z - 1} = 1 + \frac{3z^{-1}}{(1 - z^{-1})^2} + \frac{z^{-1}}{1 - z^{-1}}$$

Then, noting that

$$\mathcal{Z}^{-1}[1] = \begin{cases} 1, & k = 0 \\ 0, & k = 1, 2, 3, \ldots \end{cases}$$

$$\mathcal{Z}^{-1}\left[\frac{z^{-1}}{(1 - z^{-1})^2}\right] = k, \qquad k = 0, 1, 2, \ldots$$

$$\mathcal{Z}^{-1}\left[\frac{z^{-1}}{1 - z^{-1}}\right] = \begin{cases} 1, & k = 1, 2, 3, \ldots \\ 0, & k \le 0 \end{cases}$$

we obtain

$$x(0) = 1$$
$$x(k) = 3k + 1, \qquad k = 1, 2, 3, \ldots$$

which can be combined into one equation as follows:

$$x(k) = 3k + 1, \qquad k = 0, 1, 2, \ldots$$

Note that if we expand $X(z)$ into the following partial fractions

$$X(z) = 1 + \frac{4}{z - 1} + \frac{3}{(z - 1)^2} = 1 + \frac{4z^{-1}}{1 - z^{-1}} + \frac{3z^{-2}}{(1 - z^{-1})^2}$$

then the inverse *z* transform of $X(z)$ becomes

$$x(0) = 1$$
$$x(k) = 4 + 3(k - 1) = 3k + 1, \qquad k = 1, 2, 3, \ldots$$

or

$$x(k) = 3k + 1, \qquad k = 0, 1, 2, \ldots$$

which is the same as the result obtained by expanding $X(z)$ into the other partial fractions. [Remember that $X(z)$ can be expanded into different partial fractions, but the final result for the inverse *z* transform is the same.]

Method 4: Inversion integral method. First, note that

$$X(z)z^{k-1} = \frac{(z + 2)z^k}{(z - 1)^2}$$

For $k = 0, 1, 2, \ldots, X(z)z^{k-1}$ has a double pole at $z = 1$. Hence, referring to Equation (2–24), we have

$$x(k) = \left[\text{residue of } \frac{(z + 2)z^k}{(z - 1)^2} \text{ at double pole } z = 1 \right]$$

Thus,

$$x(k) = \frac{1}{(2 - 1)!} \lim_{z \to 1} \frac{d}{dz} \left[(z - 1)^2 \frac{(z + 2)z^k}{(z - 1)^2} \right]$$

$$= \lim_{z \to 1} \frac{d}{dz} [(z + 2)z^k]$$

$$= 3k + 1, \qquad k = 0, 1, 2, \ldots$$

Problem A–2–11

Solve the following difference equation:

$$2x(k) - 2x(k - 1) + x(k - 2) = u(k)$$

where $x(k) = 0$ for $k < 0$ and

$$u(k) = \begin{cases} 1, & k = 0, 1, 2, \ldots \\ 0, & k < 0 \end{cases}$$

Solution   By taking the z transform of the given difference equation,

$$2X(z) - 2z^{-1}X(z) + z^{-2}X(z) = \frac{1}{1 - z^{-1}}$$

Solving this last equation for $X(z)$, we obtain

$$X(z) = \frac{1}{1 - z^{-1}} \frac{1}{2 - 2z^{-1} + z^{-2}} = \frac{z^3}{(z - 1)(2z^2 - 2z + 1)}$$

Expanding $X(z)$ into partial fractions, we get

$$X(z) = \frac{z}{z - 1} + \frac{-z^2 + z}{2z^2 - 2z + 1} = \frac{1}{1 - z^{-1}} + \frac{-1 + z^{-1}}{2 - 2z^{-1} + z^{-2}}$$

Notice that the two poles involved in the quadratic term in this last equation are complex conjugates. Hence, we rewrite $X(z)$ as follows:

$$X(z) = \frac{1}{1 - z^{-1}} - \frac{1}{2} \frac{1 - 0.5z^{-1}}{1 - z^{-1} + 0.5z^{-2}} + \frac{1}{2} \frac{0.5z^{-1}}{1 - z^{-1} + 0.5z^{-2}}$$

By referring to the formulas for the z transforms of damped cosine and damped sine functions, we identify $e^{-2aT} = 0.5$ and $\cos \omega T = 1/\sqrt{2}$ for this problem. Hence, we get $\omega T = \pi/4$, $\sin \omega T = 1/\sqrt{2}$, and $e^{-aT} = 1/\sqrt{2}$. Then the inverse z transform of $X(z)$ can be written as

$$x(k) = 1 - \tfrac{1}{2}e^{-akT} \cos \omega kT + \tfrac{1}{2}e^{-akT} \sin \omega kT$$

$$= 1 - \frac{1}{2}\left(\frac{1}{\sqrt{2}}\right)^k \cos \frac{k\pi}{4} + \frac{1}{2}\left(\frac{1}{\sqrt{2}}\right)^k \sin \frac{k\pi}{4}, \qquad k = 0, 1, 2, \ldots$$

from which we obtain

$$x(0) = 0.5$$

$$x(1) = 1$$

$$x(2) = 1.25$$
$$x(3) = 1.25$$
$$x(4) = 1.125$$
$$\vdots$$

Problem A–2–12

Consider the difference equation

$$x(k + 2) - 1.3679x(k + 1) + 0.3679x(k) = 0.3679u(k + 1) + 0.2642u(k)$$

where $x(k)$ is the output and $x(k) = 0$ for $k \leq 0$ and where $u(k)$ is the input and is given by

$$u(k) = 0, \quad k < 0$$
$$u(0) = 1$$
$$u(1) = 0.2142$$
$$u(2) = -0.2142$$
$$u(k) = 0, \quad k = 3, 4, 5, \ldots$$

Determine the output $x(k)$.

Solution   Taking the z transform of the given difference equation, we obtain

$$[z^2 X(z) - z^2 x(0) - zx(1)] - 1.3679[zX(z) - zx(0)] + 0.3679X(z)$$
$$= 0.3679[zU(z) - zu(0)] + 0.2642U(z) \quad (2\text{–}31)$$

By substituting $k = -1$ into the given difference equation, we find

$$x(1) - 1.3679x(0) + 0.3679x(-1) = 0.3679u(0) + 0.2642u(-1)$$

Since $x(0) = x(-1) = 0$ and since $u(-1) = 0$ and $u(0) = 1$, we obtain

$$x(1) = 0.3679u(0) = 0.3679$$

By substituting the initial data

$$x(0) = 0, \quad x(1) = 0.3679, \quad u(0) = 1$$

into Equation (2–31), we get

$$z^2 X(z) - 0.3679z - 1.3679zX(z) + 0.3679X(z)$$
$$= 0.3679zU(z) - 0.3679z + 0.2642U(z)$$

Solving for $X(z)$, we find

$$X(z) = \frac{0.3679z + 0.2642}{z^2 - 1.3679z + 0.3679} U(z)$$

The z transform of the input $u(k)$ is

$$U(z) = \mathscr{Z}[u(k)] = 1 + 0.2142z^{-1} - 0.2142z^{-2}$$

Hence,

$$X(z) = \frac{0.3679z + 0.2642}{z^2 - 1.3679z + 0.3679}(1 + 0.2142z^{-1} - 0.2142z^{-2})$$

$$= \frac{0.3679z^{-1} + 0.3430z^{-2} - 0.02221z^{-3} - 0.05659z^{-4}}{1 - 1.3679z^{-1} + 0.3679z^{-2}}$$

$$= 0.3679z^{-1} + 0.8463z^{-2} + z^{-3} + z^{-4} + z^{-5} + \cdots$$

Thus, the inverse z transform of $X(z)$ gives

$$x(0) = 0$$

$$x(1) = 0.3679$$

$$x(2) = 0.8463$$

$$x(k) = 1, \quad k = 3, 4, 5, \ldots$$

Problem A–2–13

Consider the difference equation

$$x(k + 2) = x(k + 1) + x(k)$$

where $x(0) = 0$ and $x(1) = 1$. Note that $x(2) = 1, x(3) = 2, x(4) = 3, \ldots$. The series $0, 1, 1, 2, 3, 5, 8, 13, \ldots$ is known as the Fibonacci series. Obtain the general solution $x(k)$ in a closed form. Show that the limiting value of $x(k + 1)/x(k)$ as k approaches infinity is $(1 + \sqrt{5})/2$, or approximately 1.6180.

Solution   By taking the z transform of this difference equation, we obtain

$$z^2 X(z) - z^2 x(0) - zx(1) = zX(z) - zx(0) + X(z)$$

Solving for $X(z)$ gives

$$X(z) = \frac{z^2 x(0) + zx(1) - zx(0)}{z^2 - z - 1}$$

By substituting the initial data $x(0) = 0$ and $x(1) = 1$ into this last equation, we have

$$X(z) = \frac{z}{z^2 - z - 1}$$

$$= \frac{1}{\sqrt{5}} \left(\frac{z}{z - \dfrac{1 + \sqrt{5}}{2}} - \frac{z}{z - \dfrac{1 - \sqrt{5}}{2}} \right)$$

$$= \frac{1}{\sqrt{5}} \left(\frac{1}{1 - \dfrac{1 + \sqrt{5}}{2}z^{-1}} - \frac{1}{1 - \dfrac{1 - \sqrt{5}}{2}z^{-1}} \right)$$

The inverse z transform of $X(z)$ is

$$x(k) = \frac{1}{\sqrt{5}} \left[\left(\frac{1 + \sqrt{5}}{2} \right)^k - \left(\frac{1 - \sqrt{5}}{2} \right)^k \right], \quad k = 0, 1, 2, \ldots$$

Note that although this last equation involves $\sqrt{5}$ the square roots in the right-hand side of this last equation cancel out, and the values of $x(k)$ for $k = 0, 1, 2, \ldots$ turn out to be positive integers.

The limiting value of $x(k + 1)/x(k)$ as k approaches infinity is obtained as follows:

$$\lim_{k \to \infty} \frac{x(k + 1)}{x(k)} = \lim_{k \to \infty} \frac{\left(\dfrac{1 + \sqrt{5}}{2}\right)^{k+1} - \left(\dfrac{1 - \sqrt{5}}{2}\right)^{k+1}}{\left(\dfrac{1 + \sqrt{5}}{2}\right)^{k} - \left(\dfrac{1 - \sqrt{5}}{2}\right)^{k}}$$

Since $|(1 - \sqrt{5})/2| < 1$,

$$\lim_{k \to \infty} \left(\frac{1 - \sqrt{5}}{2}\right)^{k} \to 0$$

Hence,

$$\lim_{k \to \infty} \frac{x(k + 1)}{x(k)} = \lim_{k \to \infty} \frac{\left(\dfrac{1 + \sqrt{5}}{2}\right)^{k+1}}{\left(\dfrac{1 + \sqrt{5}}{2}\right)^{k}} = \frac{1 + \sqrt{5}}{2} = 1.6180$$

Problem A–2–14

Referring to Problem A–2–13, write a MATLAB program to generate the Fibonacci series. Carry out the Fibonacci series to $k = 30$.

Solution   The z transform of the difference equation

$$x(k + 2) = x(k + 1) + x(k)$$

is given by

$$z^2 X(z) - z^2 x(0) - zx(1) = zX(z) - zx(0) + X(z)$$

Solving this equation for $X(z)$ and substituting the initial data $x(0) = 0$ and $x(1) = 1$, we get

$$X(z) = \frac{z}{z^2 - z - 1}$$

The inverse z transform of $X(z)$ will give the Fibonacci series.

   To get the inverse z transform of $X(z)$, obtain the response of this system to the Kronecker delta input. MATLAB Program 2–4 will yield the Fibonacci series.

MATLAB Program 2–4

```
% ---------- Fibonacci series ----------

% ***** The Fibonacci series can be generated as the
% response of X(z) to the Kronecker delta input, where
% X(z) = z/(z^2 - z - 1) *****

num = [0  1  0];
den = [1  -1  -1];
u = [1  zeros(1,30)];
x = filter(num,den,u)
```

The filtered output y shown next gives the Fibonacci series.

---

x =

Columns 1 through 6

0	1	1	2	3	5

Columns 7 through 12

8	13	21	34	55	89

Columns 13 through 18

144	233	377	610	987	1597

Columns 19 through 24

2584	4181	6765	10946	17711	28657

Columns 25 through 30

46368	75025	121393	196418	317811	514229

Column 31

832040

Note that column 1 corresponds to $k = 0$ and column 31 corresponds to $k = 30$. The Fibonacci series is given by

$$x(0) = 0$$
$$x(1) = 1$$
$$x(2) = 1$$
$$x(3) = 2$$
$$x(4) = 3$$
$$x(5) = 5$$
$$\vdots$$
$$x(29) = 514{,}229$$
$$x(30) = 832{,}040$$

Problem A–2–15

Consider the difference equation

$$x(k + 2) + \alpha x(k + 1) + \beta x(k) = 0 \tag{2–32}$$

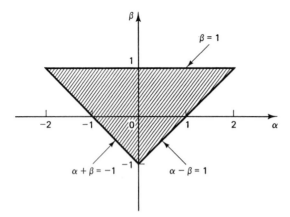

Figure 2-7 Region in the $\alpha\beta$ plane in which the solution series of Equation (2–32), subjected to initial conditions, is finite.

Find the conditions on α and β for which the solution series $x(k)$ for $k = 0, 1, 2, \ldots$, subjected to initial conditions, is finite.

Solution   Let us define

$$\alpha = a + b, \qquad \beta = ab$$

Then, referring to Example 2–19, the solution $x(k)$ for $k = 0, 1, 2, \ldots$ can be given by

$$x(k) = \begin{cases} \dfrac{bx(0) + x(1)}{b - a}(-a)^k + \dfrac{ax(0) + x(1)}{a - b}(-b)^k, & a \neq b \\ x(0)(-a)^k + [ax(0) + x(1)]k(-a)^{k-1}, & a = b \end{cases}$$

The solution series $x(k)$ for $k = 0, 1, 2, \ldots$, subjected to initial conditions $x(0)$ and $x(1)$, is finite if the absolute values of a and b are less than unity. Thus, on the $\alpha\beta$ plane, three critical points can be located:

$$\alpha = 2, \qquad \beta = 1$$
$$\alpha = -2, \qquad \beta = 1$$
$$\alpha = 0, \qquad \beta = -1$$

The interior of the region bounded by lines connecting these points satisfies the condition $|a| < 1, |b| < 1$. The boundary lines can be given by $\beta = 1$, $\alpha - \beta = 1$, and $\alpha + \beta = -1$. See Figure 2–7. If point (α, β) lies inside the shaded triangular region, then the solution series $x(k)$ for $k = 0, 1, 2, \ldots$, subjected to initial conditions $x(0)$ and $x(1)$, is finite.

PROBLEMS

Problem B–2–1

Obtain the z transform of

$$x(t) = \frac{1}{a}(1 - e^{-at})$$

where a is a constant.

Problem B–2–2

Obtain the z transform of k^3.

Problem B–2–3

Obtain the z transform of $t^2 e^{-at}$.

Problem B–2–4

Obtain the z transform of the following $x(k)$:

$$x(k) = 9k(2^{k-1}) - 2^k + 3, \qquad k = 0, 1, 2, \ldots$$

Assume that $x(k) = 0$ for $k < 0$.

Problem B–2–5

Find the z transform of

$$x(k) = \sum_{h=0}^{k} a^h$$

where a is a constant.

Problem B–2–6

Show that

$$\mathcal{Z}[k(k-1)a^{k-2}] = \frac{(2!)z}{(z-a)^3}$$

$$\mathcal{Z}[k(k-1)\cdots(k-h+1)a^{k-h}] = \frac{(h!)z}{(z-a)^{h+1}}$$

Problem B–2–7

Obtain the z transform of the curve $x(t)$ shown in Figure 2–8.

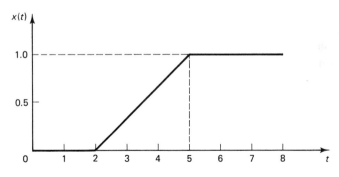

Figure 2–8   Curve $x(t)$.

Problem B–2–8

Obtain the inverse z transform of

$$X(z) = \frac{1 + 2z + 3z^2 + 4z^3 + 5z^4}{z^4}$$

Problem B–2–9

Find the inverse z transform of

$$X(z) = \frac{z^{-1}(0.5 - z^{-1})}{(1 - 0.5z^{-1})(1 - 0.8z^{-1})^2}$$

Use (1) the partial-fraction-expansion method and (2) the MATLAB method. Write a MATLAB program for finding $x(k)$, the inverse z transform of $X(z)$.

Problem B–2–10

Given the z transform

$$X(z) = \frac{z^{-1}}{(1 - z^{-1})(1 + 1.3z^{-1} + 0.4z^{-2})}$$

determine the initial and final values of $x(k)$. Also find $x(k)$, the inverse z transform of $X(z)$, in a closed form.

Problem B–2–11

Obtain the inverse z transform of

$$X(z) = \frac{1 + z^{-1} - z^{-2}}{1 - z^{-1}}$$

Use (1) the inversion integral method and (2) the MATLAB method.

Problem B–2–12

Obtain the inverse z transform of

$$X(z) = \frac{z^{-3}}{(1 - z^{-1})(1 - 0.2z^{-1})}$$

in a closed form.

Problem B–2–13

By using the inversion integral method, obtain the inverse z transform of

$$X(z) = \frac{1 + 6z^{-2} + z^{-3}}{(1 - z^{-1})(1 - 0.2z^{-1})}$$

Problem B–2–14

Find the inverse z transform of

$$X(z) = \frac{z^{-1}(1 - z^{-2})}{(1 + z^{-2})^2}$$

Use (1) the direct division method and (2) the MATLAB method.

Problem B–2–15

Obtain the inverse z transform of

$$X(z) = \frac{0.368z^2 + 0.478z + 0.154}{(z - 1)z^2}$$

by use of the inversion integral method.

Problem B–2–16

Find the solution of the following difference equation:

$$x(k + 2) - 1.3x(k + 1) + 0.4x(k) = u(k)$$

where $x(0) = x(1) = 0$ and $x(k) = 0$ for $k < 0$. For the input function $u(k)$, consider the following two cases:

$$u(k) = \begin{cases} 1, & k = 0, 1, 2, \ldots \\ 0, & k < 0 \end{cases}$$

and

$$u(0) = 1$$
$$u(k) = 0, \qquad k \neq 0$$

Solve this problem both analytically and computationally with MATLAB.

Problem B–2–17

Solve the following difference equation:

$$x(k + 2) - x(k + 1) + 0.25x(k) = u(k + 2)$$

where $x(0) = 1$ and $x(1) = 2$. The input function $u(k)$ is given by

$$u(k) = 1, \qquad k = 0, 1, 2, \ldots$$

Solve this problem both analytically and computationally with MATLAB.

Problem B–2–18

Consider the difference equation:

$$x(k + 2) - 1.3679x(k + 1) + 0.3679x(k) = 0.3679u(k + 1) + 0.2642u(k)$$

where $x(k) = 0$ for $k \leq 0$. The input $u(k)$ is given by

$$u(k) = 0, \qquad k < 0$$
$$u(0) = 1.5820$$
$$u(1) = -0.5820$$
$$u(k) = 0, \qquad k = 2, 3, 4, \ldots$$

Determine the output $x(k)$. Solve this problem both analytically and computationally with MATLAB.

3

z-Plane Analysis of Discrete-Time Control Systems

3–1 INTRODUCTION

The z transform method is particularly useful for analyzing and designing single-input–single-output linear time-invariant discrete-time control systems. This chapter presents background material necessary for the analysis and design of discrete-time control systems in the z plane. The main advantage of the z transform method is that it enables the engineer to apply conventional continuous-time design methods to discrete-time systems that may be partly discrete time and partly continuous time.

Throughout this book we assume that the sampling operation is uniform; that is, only one sampling rate exists in the system and the sampling period is constant. If a discrete-time control system involves two or more samplers in the system, we assume that all samplers are synchronized and have the same sampling rate or sampling frequency.

Outline of the Chapter. The outline of this chapter is as follows. Section 3–1 gives introductory remarks. Section 3–2 presents a method to treat the sampling operation as a mathematical representation of the operation of taking samples $x(kT)$ from a continuous-time signal $x(t)$ by impulse modulation. This section includes derivations of the transfer functions of the zero-order hold and first-order hold.

Section 3–3 deals with the convolution integral method for obtaining the z transform. Reconstructing the original continuous-time signal from the sampled signal is the main subject matter of Section 3–4. Based on the fact that the Laplace transform of the sampled signal is periodic, we present the sampling theorem. Section 3–5 discusses the pulse transfer function. Mathematical modeling of digital controllers in terms of pulse transfer functions is discussed. Section 3–6 treats the realization of digital controllers and digital filters.

3-2 IMPULSE SAMPLING AND DATA HOLD

Discrete-time control systems may operate partly in discrete time and partly in continuous time. Thus, in such control systems some signals appear as discrete-time functions (often in the form of a sequence of numbers or a numerical code) and other signals as continuous-time functions. In analyzing discrete-time control systems, the z transform theory plays an important role. To see why the z transform method is useful in the analysis of discrete-time control systems, we first introduce the concept of impulse sampling and then discuss data hold.

Impulse Sampling.   We shall consider a fictitious sampler commonly called an *impulse sampler*. The output of this sampler is considered to be a train of impulses that begins with $t = 0$, with the sampling period equal to T and the strength of each impulse equal to the sampled value of the continuous-time signal at the corresponding sampling instant. A pictorial diagram of the impulse sampler is shown in Figure 3–1. [We assume $x(t) = 0$ for $t < 0$.] (Since, mathematically, an impulse is defined as having an infinite amplitude with zero width, it is graphically represented by an arrow with an amplitude representing the strength of impulse.)

The impulse-sampled output is a sequence of impulses, with the strength of each impulse equal to the magnitude of $x(t)$ at the corresponding instant of time. [That is, at time $t = kT$, the impulse is $x(kT)\delta(t - kT)$. Note that $\delta(t - kT) = 0$ unless $t = kT$.] We shall use the notation $x^*(t)$ to represent the impulse-sampled output. The sampled signal $x^*(t)$, a train of impulses, can thus be represented by the infinite summation

$$x^*(t) = \sum_{k=0}^{\infty} x(kT)\delta(t - kT)$$

or

$$x^*(t) = x(0)\delta(t) + x(T)\delta(t - T) + \cdots + x(kT)\delta(t - kT) + \cdots \qquad (3\text{--}1)$$

We shall define a train of unit impulses as $\delta_T(t)$, or

$$\delta_T(t) = \sum_{k=0}^{\infty} \delta(t - kT)$$

The sampler output is equal to the product of the continuous-time input $x(t)$ and the train of unit impulses $\delta_T(t)$. Consequently, the sampler may be considered a modulator with the input $x(t)$ as the modulating signal and the train of unit impulses $\delta_T(t)$ as the carrier, as shown in Figure 3–2.

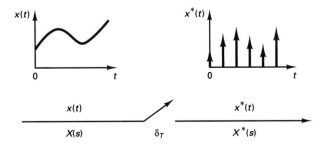

Figure 3–1 Impulse sampler.

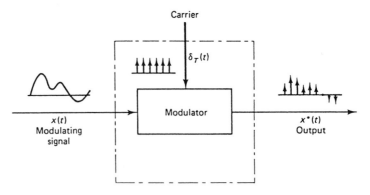

Figure 3–2   Impulse sampler as a modulator.

Next, consider the Laplace transform of Equation (3–1):

$$X^*(s) = \mathcal{L}[x^*(t)] = x(0)\mathcal{L}[\delta(t)] + x(T)\mathcal{L}[\delta(t - T)]$$
$$+ x(2T)\mathcal{L}[\delta(t - 2T)] + \cdots$$
$$= x(0) + x(T)e^{-Ts} + x(2T)e^{-2Ts} + \cdots$$
$$= \sum_{k=0}^{\infty} x(kT)e^{-kTs} \tag{3-2}$$

Notice that if we define

$$e^{Ts} = z$$

or

$$s = \frac{1}{T}\ln z$$

then Equation (3–2) becomes

$$X^*(s)\Big|_{s=(1/T)\ln z} = \sum_{k=0}^{\infty} x(kT)z^{-k} \tag{3-3}$$

The right-hand side of Equation (3–3) is exactly the same as the right-hand side of Equation (2–1): It is the z transform of the sequence $x(0), x(T), x(2T), \ldots$, generated from $x(t)$ at $t = kT$, where $k = 0, 1, 2, \ldots$. Hence we may write

$$X^*(s)\Big|_{s=(1/T)\ln z} = X(z)$$

and Equation (3–3) becomes

$$X^*(s)\Big|_{s=(1/T)\ln z} = X^*\left(\frac{1}{T}\ln z\right) = X(z) = \sum_{k=0}^{\infty} x(kT)z^{-k} \tag{3-4}$$

Note that the variable z is a complex variable and T is the sampling period. [It should be stressed that the notation $X(z)$ does not signify $X(s)$ with s replaced by z, but rather $X^*(s = T^{-1}\ln z)$.]

Summary.   Let us summarize what we have just stated. If the continuous-time signal $x(t)$ is impulse sampled in a periodic manner, mathematically the sampled signal may be represented by

$$x^*(t) = \sum_{k=0}^{\infty} x(t)\delta(t - kT)$$

In the impulse sampler the switch may be thought of as closing instantaneously every sampling period T and generating impulses $x(kT)\delta(t - kT)$. Such a sampling process is called *impulse sampling*. The impulse sampler is introduced for mathematical convenience; it is a fictitious sampler and it does not exist in the real world.

The Laplace transform of the impulse-sampled signal $x^*(t)$ has been shown to be the same as the z transform of signal $x(t)$ if e^{Ts} is defined as z, or $e^{Ts} = z$.

Data-Hold Circuits.   In a conventional sampler, a switch closes to admit an input signal every sampling period T. In practice, the sampling duration is very short in comparison with the most significant time constant of the plant. A sampler converts a continuous-time signal into a train of pulses occurring at the sampling instants $t = 0, T, 2T, \ldots$, where T is the sampling period. (Note that between any two consecutive sampling instants the sampler transmits no information. Two signals whose respective values at the sampling instants are equal will give rise to the same sampled signal.)

Data-hold is a process of generating a continuous-time signal $h(t)$ from a discrete-time sequence $x(kT)$. A hold circuit converts the sampled signal into a continuous-time signal, which approximately reproduces the signal applied to the sampler. The signal $h(t)$ during the time interval $kT \le t < (k + 1)T$ may be approximated by a polynomial in τ as follows:

$$h(kT + \tau) = a_n \tau^n + a_{n-1} \tau^{n-1} + \cdots + a_1 \tau + a_0 \qquad (3\text{-}5)$$

where $0 \le \tau < T$. Note that signal $h(kT)$ must equal $x(kT)$, or

$$h(kT) = x(kT)$$

Hence, Equation (3-5) can be written as follows:

$$h(kT + \tau) = a_n \tau^n + a_{n-1} \tau^{n-1} + \cdots + a_1 \tau + x(kT) \qquad (3\text{-}6)$$

If the data-hold circuit is an nth-order polynomial extrapolator, it is called an nth-order hold. Thus, if $n = 1$, it is called a first-order hold. [The nth-order hold uses the past $n + 1$ discrete data $x((k - n)T), x((k - n + 1)T), \ldots, x(kT)$ to generate a signal $h(kT + \tau)$.]

Because a higher-order hold uses past samples to extrapolate a continuous-time signal between the present sampling instant and the next sampling instant, the accuracy of approximating the continuous-time signal improves as the number of past samples used is increased. However, this better accuracy is obtained at the cost of a greater time delay. In closed-loop control systems, any added time delay in the loop will decrease the stability of the system and in some cases may even cause system instability.

The simplest data-hold is obtained when $n = 0$ in Equation (3-6), that is, when

$$h(kT + \tau) = x(kT) \qquad (3\text{-}7)$$

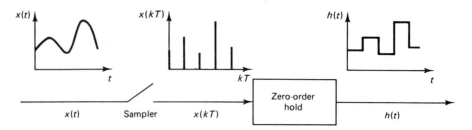

Figure 3–3   Sampler and zero-order hold.

where $0 \leq \tau < T$ and $k = 0, 1, 2, \ldots$. Equation (3–7) implies that the circuit holds the amplitude of the sample from one sampling instant to the next. Such a data-hold is called a *zero-order hold*, or clamper, or staircase generator. The output of the zero-order hold is a staircase function. In this book, unless otherwise stated, we assume that the hold circuit is of zero order.

It will be seen later that the transfer function G_h of the zero-order hold may be given by

$$G_h = \frac{1 - e^{-Ts}}{s}$$

Zero-Order Hold.   Figure 3–3 shows a sampler and a zero-order hold. The input signal $x(t)$ is sampled at discrete instants and the sampled signal is passed through the zero-order hold. The zero-order hold circuit smoothes the sampled signal to produce the signal $h(t)$, which is constant from the last sampled value until the next sample is available. That is,

$$h(kT + t) = x(kT), \qquad \text{for } 0 \leq t < T \tag{3–8}$$

We shall obtain a mathematical model of the combination of a real sampler and zero-order circuit, as shown in Figure 3–4(a). Utilizing the fact that the integral of an impulse function is a constant, we may assume that the zero-order hold is an integrator, and the input to the zero-order hold circuit is a train of impulses. Then a mathematical model for the real sampler and zero-order hold may be constructed as shown in Figure 3–4(b), where $G_{h0}(s)$ is the transfer function of the zero-order hold and $x^*(t)$ is the impulse sampled signal of $x(t)$.

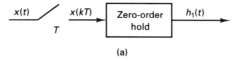

(a)

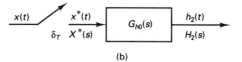

(b)

Figure 3–4   (a) A real sampler and zero-order hold; (b) mathematical model that consists of an impulse sampler and transfer function $G_{h0}(s)$.

Consider the sampler and zero-order hold shown in Figure 3–4(a). Assume that the signal $x(t)$ is zero for $t < 0$. Then the output $h_1(t)$ is related to $x(t)$ as follows:

$$h_1(t) = x(0)[1(t) - 1(t - T)] + x(T)[1(t - T) - 1(t - 2T)]$$
$$+ x(2T)[1(t - 2T) - 1(t - 3T)] + \cdots$$
$$= \sum_{k=0}^{\infty} x(kT)[1(t - kT) - 1(t - (k + 1)T)] \qquad (3\text{–}9)$$

Since

$$\mathscr{L}[1(t - kT)] = \frac{e^{-kTs}}{s}$$

the Laplace transform of Equation (3–9) becomes

$$\mathscr{L}[h_1(t)] = H_1(s) = \sum_{k=0}^{\infty} x(kT) \frac{e^{-kTs} - e^{-(k+1)Ts}}{s}$$
$$= \frac{1 - e^{-Ts}}{s} \sum_{k=0}^{\infty} x(kT)e^{-kTs} \qquad (3\text{–}10)$$

Next, consider the mathematical model shown in Figure 3–4(b). The output of this model must be the same as that of the real zero-order hold, or

$$\mathscr{L}[h_2(t)] = H_2(s) = H_1(s)$$

Thus,

$$H_2(s) = \frac{1 - e^{-Ts}}{s} \sum_{k=0}^{\infty} x(kT)e^{-kTs} \qquad (3\text{–}11)$$

From Figure 3–4(b), we have

$$H_2(s) = G_{h0}(s)X^*(s) \qquad (3\text{–}12)$$

Since

$$X^*(s) = \sum_{k=0}^{\infty} x(kT)e^{-kTs}$$

Equation (3–11) may be written as

$$H_2(s) = \frac{1 - e^{-Ts}}{s} X^*(s) \qquad (3\text{–}13)$$

By comparing Equations (3–12) and (3–13), we see that the transfer function of the zero-order hold may be given by

$$G_{h0}(s) = \frac{1 - e^{-Ts}}{s}$$

Note that, mathematically, the system shown in Figure 3–4(a) is equivalent to the system shown in Figure 3–4(b) from the viewpoint of the input–output relationship. That is, a real sampler and zero-order hold can be replaced by a mathematically equivalent continuous-time system that consists of an impulse sampler and a transfer function $(1 - e^{-Ts})/s$. The two sampling processes will be distinguished (as they are in Figure 3–4) by the manner in which the sampling switches are drawn.

Transfer Function of First-Order Hold.   Although we do not use first-order holds in control systems, it is worthwhile to see what the transfer function of first-order holds may look like. We shall show that the transfer function of the first-order hold may be given by

$$G_{h1}(s) = \left(\frac{1 - e^{-Ts}}{s}\right)^2 \frac{Ts + 1}{T} \tag{3–14}$$

Next we shall derive Equation (3–14).

We have stated that Equation (3–6) describes the output of an nth-order hold circuit. For the first-order hold, $n = 1$. Let us substitute $n = 1$ into Equation (3–6). Then we have

$$h(kT + \tau) = a_1 \tau + x(kT) \tag{3–15}$$

where $0 \le \tau < T$ and $k = 0, 1, 2, \ldots$. By applying the condition that

$$h((k - 1)T) = x((k - 1)T)$$

the constant a_1 can be determined as follows:

$$h((k - 1)T) = -a_1 T + x(kT) = x((k - 1)T)$$

or

$$a_1 = \frac{x(kT) - x((k - 1)T)}{T}$$

Hence, Equation (3–15) becomes

$$h(kT + \tau) = x(kT) + \frac{x(kT) - x((k - 1)T)}{T} \tau \tag{3–16}$$

where $0 \le \tau < T$. The extrapolation process of the first-order hold is based on Equation (3–16). The continuous-time output signal $h(t)$ obtained by use of the first-order hold is a piecewise-linear signal, as shown in Figure 3–5.

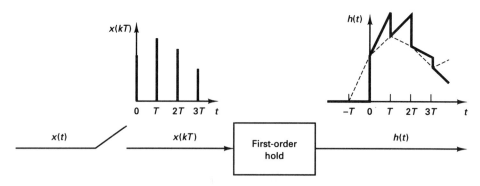

Figure 3–5   Input and output of a first-order hold.

To derive the transfer function of the first-order hold circuit, it is convenient to assume a simple function for $x(t)$. For example, a unit-step function, a unit-impulse function, and a unit-ramp function are good choices for $x(t)$.

Suppose that we choose a unit-step function as $x(t)$. Then, for the real sampler and first-order hold shown in Figure 3–6(a), the output $h(t)$ of the first-order hold consists of straight lines that are extrapolations of the two preceding sampled values. The output $h(t)$ is shown in the diagram. The output curve $h(t)$ may be written as follows:

$$h(t) = \left(1 + \frac{t}{T}\right)1(t) - \frac{t - T}{T}1(t - T) - 1(t - T)$$

The Laplace transform of this last equation becomes

$$H(s) = \left(\frac{1}{s} + \frac{1}{Ts^2}\right) - \frac{1}{Ts^2}e^{-Ts} - \frac{1}{s}e^{-Ts}$$

$$= \frac{1 - e^{-Ts}}{s} + \frac{1 - e^{-Ts}}{Ts^2}$$

$$= (1 - e^{-Ts})\frac{Ts + 1}{Ts^2} \qquad (3\text{--}17)$$

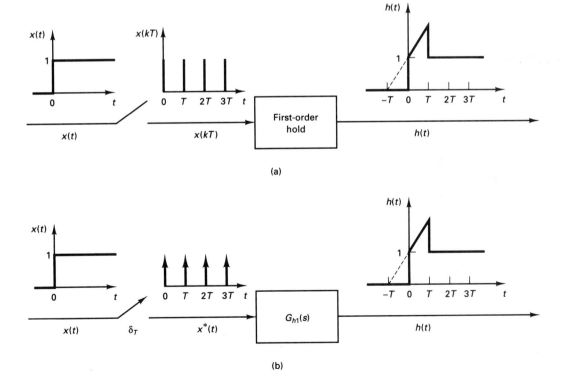

(a)

(b)

Figure 3–6 (a) Real sampler cascaded with first-order hold; (b) mathematical model consisting of impulse sampler and $G_{h1}(s)$.

Figure 3–6(b) shows a mathematical model of the real sampler cascaded with the first-order hold shown in Figure 3–6(a). The mathematical model consists of the impulse sampler and $G_{h1}(s)$, the transfer function of the first-order hold. The output signal of this model is the same as the output of the real system. Hence, the output $H(s)$ is also given by Equation (3–17).

The Laplace transform of the input $x^*(t)$ to the first-order hold $G_{h1}(s)$ is

$$X^*(s) = \sum_{k=0}^{\infty} 1(kT)e^{-kTs} = \frac{1}{1 - e^{-Ts}}$$

Hence, the transfer function $G_{h1}(s)$ of the first-order hold is given by

$$G_{h1}(s) = \frac{H(s)}{X^*(s)} = (1 - e^{-Ts})^2 \frac{Ts + 1}{Ts^2}$$

$$= \left(\frac{1 - e^{-Ts}}{s}\right)^2 \frac{Ts + 1}{T}$$

Note that a real sampler combined with a first-order hold is equivalent to an impulse sampler combined with a transfer function $(1 - e^{-Ts})^2(Ts + 1)/(Ts^2)$.

Similarly, transfer functions of higher-order hold circuits may be derived by following the procedure presented. However, since higher-order hold circuits ($n \geq 2$) are not practical from the viewpoint of delay (which may cause system instability) and noise effects, we shall not derive their transfer functions here. (The zero-order hold is the simplest and is used most frequently in practice.)

Summary.   Let us summarize what we have presented so far about impulse sampling.

1. A real sampler samples the input signal periodically and produces a sequence of pulses as the output. While the sampling duration (pulse width) of the real sampler is very small (but will never become zero), the assumption of zero width, which implies that a sequence of pulses becomes a sequence of impulses whose strengths are equal to the continuous-time signal at the sampling instants, simplifies the analysis of discrete-time systems. Such an assumption is valid if the sampling duration is very small compared with the significant time constant of the system and if a hold circuit is connected to the output of the sampler.

2. When we transform e^{Ts} to z, the concept of impulse sampling (which is purely a mathematical process) enables us to analyze by the z transform method discrete-time control systems that involve samplers and hold circuits. This means that by use of the complex variable z the techniques developed for the Laplace transform methods can be readily applied to analyze discrete-time systems involving sampling operations.

3. As pointed out earlier, once the real sampler and zero-order hold are mathematically replaced by the impulse sampler and transfer function $(1 - e^{-Ts})/s$, the system becomes a continuous-time system. This simplifies the analysis of the discrete-time control system, since we may apply the techniques available to continuous-time control systems.

4. It is repeated that the impulse sampler is a fictitious sampler introduced purely for the purpose of mathematical analysis. It is not possible to physically implement such a sampler that generates impulses.

3-3 OBTAINING THE z TRANSFORM BY THE CONVOLUTION INTEGRAL METHOD

In this section we shall obtain the z transform of $x(t)$ by using the convolution integral method.

Consider the impulse sampler shown in Figure 3-7. The output of the impulse sampler is

$$x^*(t) = \sum_{k=0}^{\infty} x(t)\delta(t - kT) = x(t)\sum_{k=0}^{\infty} \delta(t - kT) \tag{3-18}$$

Noting that

$$\mathcal{L}[\delta(t - kT)] = e^{-kTs}$$

we have

$$\mathcal{L}\left[\sum_{k=0}^{\infty} \delta(t - kT)\right] = 1 + e^{-Ts} + e^{-2Ts} + e^{-3Ts} + \cdots = \frac{1}{1 - e^{-Ts}}$$

Since

$$X^*(s) = \mathcal{L}[x^*(t)] = \mathcal{L}\left[x(t)\sum_{k=0}^{\infty} \delta(t - kT)\right]$$

we see that $X^*(s)$ is the Laplace transform of the product of two time functions, $x(t)$ and $\sum_{k=0}^{\infty} \delta(t - kT)$. Note that it is not equal to the product of the two corresponding Laplace transforms.

The Laplace transform of the product of two Laplace-transformable functions $f(t)$ and $g(t)$ can be given by

$$\mathcal{L}[f(t)g(t)] = \int_0^{\infty} f(t)g(t)e^{-st}\,dt$$

$$= \frac{1}{2\pi j}\int_{c-j\infty}^{c+j\infty} F(p)G(s - p)\,dp \tag{3-19}$$

[For the derivation of Equation (3-19), see Problem A-3-4.]

Let us substitute $x(t)$ and $\sum_{k=0}^{\infty} \delta(t - kT)$ for $f(t)$ and $g(t)$, respectively. Then the Laplace transform of $X^*(s)$, where

$$X^*(s) = \mathcal{L}\left[x(t)\sum_{k=0}^{\infty} \delta(t - kT)\right]$$

Figure 3-7 Impulse sampler.

can be given by

$$X^*(s) = \mathcal{L}\left[x(t)\sum_{k=0}^{\infty}\delta(t - kT)\right]$$

$$= \frac{1}{2\pi j}\int_{c-j\infty}^{c+j\infty} X(p)\frac{1}{1 - e^{-T(s-p)}}\,dp \tag{3-20}$$

where the integration is along the line from $c - j\infty$ to $c + j\infty$ and this line is parallel to the imaginary axis in the p plane and separates the poles of $X(p)$ from those of $1/[1 - e^{-T(s-p)}]$. Equation (3–20) is the convolution integral. It is a well-known fact that such an integral can be evaluated in terms of residues by forming a closed contour consisting of the line from $c - j\infty$ to $c + j\infty$ and a semicircle of infinite radius in the left or right half-plane, provided that the integral along the added semicircle is a constant (zero or a nonzero constant). That is, we rewrite Equation (3–20) as follows:

$$X^*(s) = \frac{1}{2\pi j}\int_{c-j\infty}^{c+j\infty} X(p)\frac{1}{1 - e^{-T(s-p)}}\,dp$$

$$= \frac{1}{2\pi j}\oint \frac{X(p)}{1 - e^{-T(s-p)}}\,dp - \frac{1}{2\pi j}\int_{\Gamma} \frac{X(p)}{1 - e^{-T(s-p)}}\,dp \tag{3-21}$$

where Γ is a semicircle of infinite radius in the left or right half p plane.

  There are two ways to evaluate this integral (one using an infinite semicircle in the left-half plane and the other an infinite semicircle in the right half-plane); we shall describe the results obtained by the two cases separately.

  In our analysis here, we assume that the poles of $X(s)$ lie in the left half-plane and that $X(s)$ can be expressed as a ratio of polynomials in s, or

$$X(s) = \frac{q(s)}{p(s)}$$

where $q(s)$ and $p(s)$ are polynomials in s. We also assume that $p(s)$ is of a higher degree in s than $q(s)$, which means that

$$\lim_{s\to\infty} X(s) = 0$$

Evaluation of the Convolution Integral in the Left Half-Plane. We shall evaluate the convolution integral given by Equation (3–21) using a closed contour in the left half of the p plane as shown in Figure 3–8. Using this closed contour, Equation (3–21) may be evaluated as follows: Noting that the denominator of $X(s)$ is of higher degree in s than the numerator, the integral along Γ_L vanishes. Hence,

$$X^*(s) = \frac{1}{2\pi j}\oint \frac{X(p)}{1 - e^{-T(s-p)}}\,dp$$

This integral is equal to the sum of the residues of $X(p)$ in the closed contour.

$$X^*(s) = \sum\left[\text{residue of }\frac{X(p)}{1 - e^{-T(s-p)}}\text{ at pole of }X(p)\right] \tag{3-22}$$

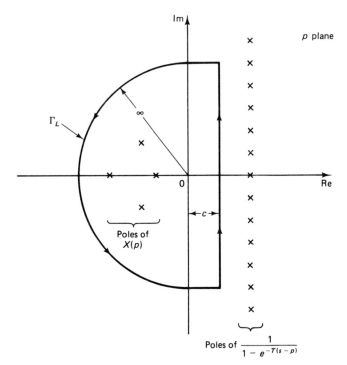

Poles of $\dfrac{1}{1 - e^{-T(s - p)}}$

Figure 3–8 Closed contour in the left half of the p plane.

[Refer to Problem A–3–6 for the derivation of Equation (3–22).] By substituting z for e^{Ts} in Equation (3–22), we have

$$X(z) = \sum \left[\text{residue of } \frac{X(p)z}{z - e^{Tp}} \text{ at pole of } X(p) \right]$$

By changing the complex variable notation from p to s, we obtain

$$X(z) = \sum \left[\text{residue of } \frac{X(s)z}{z - e^{Ts}} \text{ at pole of } X(s) \right] \qquad (3\text{–}23)$$

Assume that $X(s)$ has poles $s_1, s_2, \ldots, s_m$. If a pole at $s = s_j$ is a simple pole, then the corresponding residue K_j is

$$K_j = \lim_{s \to s_j} \left[(s - s_j) \frac{X(s)z}{z - e^{Ts}} \right] \qquad (3\text{–}24)$$

If a pole at $s = s_i$ is a multiple pole of order n_i, then the residue K_i is

$$K_i = \frac{1}{(n_i - 1)!} \lim_{s \to s_i} \frac{d^{n_i - 1}}{ds^{n_i - 1}} \left[(s - s_i)^{n_i} \frac{X(s)z}{z - e^{Ts}} \right] \qquad (3\text{–}25)$$

Example 3–1

Given

$$X(s) = \frac{1}{s^2(s + 1)}$$

obtain $X(z)$ by use of the convolution integral in the left half-plane.

Note that $X(s)$ has a double pole at $s = 0$ and a simple pole at $s = -1$. Hence, Equation (3–23) becomes

$$X(z) = \sum \left[\text{residue of } \frac{X(s)z}{z - e^{Ts}} \text{ at pole of } X(s) \right]$$

$$= \frac{1}{(2 - 1)!} \lim_{s \to 0} \frac{d}{ds} \left[s^2 \frac{1}{s^2(s + 1)} \frac{z}{z - e^{Ts}} \right] + \lim_{s \to -1} \left[(s + 1) \frac{1}{s^2(s + 1)} \frac{z}{z - e^{Ts}} \right]$$

$$= \lim_{s \to 0} \frac{-z[z - e^{Ts} + (s + 1)(-T)e^{Ts}]}{(s + 1)^2(z - e^{Ts})^2} + \frac{1}{(-1)^2} \frac{z}{z - e^{-T}}$$

$$= \frac{-z(z - 1 - T)}{(z - 1)^2} + \frac{z}{z - e^{-T}} = \frac{z^2(T - 1 + e^{-T}) + z(1 - e^{-T} - Te^{-T})}{(z - 1)^2(z - e^{-T})}$$

$$= \frac{(T - 1 + e^{-T})z^{-1} + (1 - e^{-T} - Te^{-T})z^{-2}}{(1 - z^{-1})^2(1 - e^{-T}z^{-1})}$$

Evaluation of the Convolution Integral in the Right Half-Plane. Let us next evaluate the convolution integral given by Equation (3–21) in the right half of the p plane. Let us choose the closed contour shown in Figure 3–9, which consists of the line from $c - j\infty$ to $c + j\infty$ and Γ_R, the portion of the semicircle of infinite radius in the right half of the p plane that lies to the right of this line. The closed contour encloses all poles of $1/[1 - e^{-T(s-p)}]$, but it does not enclose any poles of $X(p)$. Now $X^*(s)$ can be written as

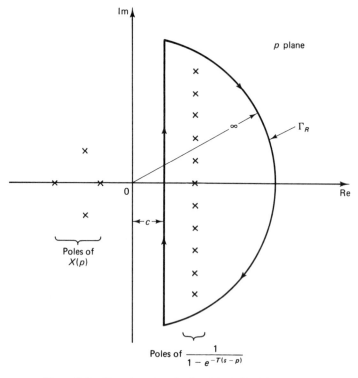

Figure 3–9   Closed contour in the right half of the p plane.

$$X^*(s) = \frac{1}{2\pi j} \int_{c-j\infty}^{c+j\infty} \frac{X(p)}{1 - e^{-T(s-p)}} dp$$

$$= \frac{1}{2\pi j} \oint \frac{X(p)}{1 - e^{-T(s-p)}} dp - \frac{1}{2\pi j} \int_{\Gamma_R} \frac{X(p)}{1 - e^{-T(s-p)}} dp \qquad (3\text{-}26)$$

In evaluating the integrals on the right-hand side of Equation (3–26), we need to consider two cases separately: one case where the denominator of $X(s)$ is two or more degrees higher in s than the numerator, and another case where the denominator of $X(s)$ is only one degree higher in s than the numerator.

Case 1: $X(s)$ has a denominator two or more degrees higher in s than the numerator. For this case, since $X(s)$ possesses at least two more poles than zeros, we have

$$\lim_{s \to \infty} sX(s) = x(0+) = 0$$

Then the integral along Γ_R is zero. Thus, in the present case

$$\frac{1}{2\pi j} \int_{\Gamma_R} \frac{X(p)}{1 - e^{-T(s-p)}} dp = 0$$

and $X^*(s)$ can be obtained as follows:

$$X^*(s) = \frac{1}{T} \sum_{k=-\infty}^{\infty} X(s + j\omega_s k) \qquad (3\text{-}27)$$

[For the derivation of Equation (3–27), see Problem A–3–7.] Thus

$$X(z) = \frac{1}{T} \sum_{k=-\infty}^{\infty} X(s + j\omega_s k) \Big|_{s=(1/T) \ln z} \qquad (3\text{-}28)$$

Note that this expression of the z transform is useful in proving the sampling theorem (see Section 3–4). However, it is very tedious to obtain z transform expressions of commonly encountered functions by this method.

Case 2: $X(s)$ has a denominator one degree higher in s than the numerator. For this case, $\lim_{s \to \infty} sX(s) = x(0+) \neq 0 < \infty$ and the integral along Γ_R is not zero. [The nonzero value is associated with the initial value $x(0+)$ of $x(t)$.] It can be shown that the contribution of the integral along Γ_R in Equation (3–26) is $-\frac{1}{2}x(0+)$. That is,

$$\frac{1}{2\pi j} \int_{\Gamma_R} \frac{x(p)}{1 - e^{-T(s-p)}} dp = -\frac{1}{2}x(0+)$$

Then the integral term on the right-hand side of Equation (3–26) becomes

$$X^*(s) = \frac{1}{T} \sum_{k=-\infty}^{\infty} X(s + j\omega_s k) + \frac{1}{2}x(0+) \qquad (3\text{-}29)$$

Example 3–2

Show that $X^*(s)$ is periodic with period $2\pi/\omega_s$.
Referring to Equation (3–29),

$$X^*(s) = \frac{1}{T} \sum_{h=-\infty}^{\infty} X(s + j\omega_s h) + \frac{1}{2}x(0+)$$

Hence,

$$X^*(s + j\omega_s k) = \frac{1}{T} \sum_{h=-\infty}^{\infty} X(s + j\omega_s k + j\omega_s h) + \frac{1}{2}x(0+)$$

Let $k + h = m$. Then this last equation becomes

$$X^*(s + j\omega_s k) = \frac{1}{T} \sum_{m=-\infty}^{\infty} X(s + j\omega_s m) + \frac{1}{2}x(0+) = X^*(s)$$

Therefore, we have

$$X^*(s) = X^*(s \pm j\omega_s k), \qquad k = 0, 1, 2, \ldots$$

Thus, $X^*(s)$ is periodic, with period $2\pi/\omega_s$. This means that, if a function $X(s)$ has a pole at $s = s_1$ in the s plane, then $X^*(s)$ has poles at $s = s_1 \pm j\omega_s k$ $(k = 0, 1, 2, \ldots)$.

Obtaining z Transforms of Functions Involving the Term $(1 - e^{-Ts})/s$. We shall here consider the function $X(s)$ involving $(1 - e^{-Ts})/s$. Suppose the transfer function $G(s)$ follows the zero-order hold. Then the product of the transfer function of the zero-order hold and $G(s)$ becomes

$$X(s) = \frac{1 - e^{-Ts}}{s} G(s)$$

In what follows we shall obtain the z transform of such an $X(s)$.
Note that $X(s)$ can be written as follows:

$$X(s) = (1 - e^{-Ts})\frac{G(s)}{s} = (1 - e^{-Ts})G_1(s) \qquad (3\text{-}30)$$

where

$$G_1(s) = \frac{G(s)}{s}$$

Consider the function

$$X_1(s) = e^{-Ts} G_1(s) \qquad (3\text{-}31)$$

Since $X_1(s)$ is a product of two Laplace-transformed functions, the inverse Laplace transform of Equation (3–31) can be given by the following convolution integral:

$$x_1(t) = \int_0^t g_0(t - \tau)g_1(\tau)\, d\tau$$

where

$$g_0(t) = \mathcal{L}^{-1}[e^{-Ts}] = \delta(t - T)$$
$$g_1(t) = \mathcal{L}^{-1}[G_1(s)]$$

Thus,

$$x_1(t) = \int_0^t \delta(t - T - \tau)g_1(\tau)\, d\tau$$
$$= g_1(t - T)$$

Hence, by writing

$$\mathcal{Z}[g_1(t)] = G_1(z)$$

the z transform of $x_1(t)$ becomes

$$\mathcal{Z}[x_1(t)] = \mathcal{Z}[g_1(t - T)] = z^{-1}G_1(z)$$

Referring to Equations (3–30) and (3–31), we have

$$X(z) = \mathcal{Z}[G_1(s) - e^{-Ts}G_1(s)]$$
$$= \mathcal{Z}[g_1(t)] - \mathcal{Z}[x_1(t)]$$
$$= G_1(z) - z^{-1}G_1(z)$$
$$= (1 - z^{-1})G_1(z)$$

or

$$X(z) = \mathcal{Z}[X(s)] = (1 - z^{-1})\mathcal{Z}\left[\frac{G(s)}{s}\right] \tag{3-32}$$

We have thus shown that if $X(s)$ involves a factor $(1 - e^{-Ts})$ then, in obtaining the z transform of $X(s)$, the term $1 - e^{-Ts} = 1 - z^{-1}$ may be factored out so that $X(z)$ becomes the product of $(1 - z^{-1})$ and the z transform of the remaining term.

Similarly, if the transfer function $G(s)$ follows the first-order hold $G_{h1}(s)$, where

$$G_{h1}(s) = \left(\frac{1 - e^{-Ts}}{s}\right)^2 \frac{Ts + 1}{T}$$

then the z transform of the function

$$X(s) = \left(\frac{1 - e^{-Ts}}{s}\right)^2 \frac{Ts + 1}{T} G(s)$$

can be obtained as follows. Since

$$X(s) = (1 - e^{-Ts})^2 \frac{Ts + 1}{Ts^2} G(s)$$

by employing the same approach as used in obtaining Equation (3–32), we have

$$X(z) = \mathcal{Z}\left[(1 - e^{-Ts})^2 \frac{Ts + 1}{Ts^2} G(s)\right]$$
$$= (1 - z^{-1})^2 \mathcal{Z}\left[\frac{Ts + 1}{Ts^2} G(s)\right] \tag{3-33}$$

Equation (3–33) can be used for obtaining the z transform of the function involving the first-order hold circuit.

Example 3–3

Obtain the z transform of

$$X(s) = \frac{1 - e^{-Ts}}{s} \frac{1}{s + 1}$$

Referring to Equation (3–32), we have

$$X(z) = \mathcal{Z}\left[\frac{1 - e^{-Ts}}{s}\frac{1}{s + 1}\right]$$

$$= (1 - z^{-1})\mathcal{Z}\left[\frac{1}{s(s + 1)}\right]$$

$$= (1 - z^{-1})\mathcal{Z}\left[\frac{1}{s} - \frac{1}{s + 1}\right]$$

$$= (1 - z^{-1})\left(\frac{1}{1 - z^{-1}} - \frac{1}{1 - e^{-T}z^{-1}}\right)$$

$$= \frac{(1 - e^{-T})z^{-1}}{1 - e^{-T}z^{-1}}$$

3–4 RECONSTRUCTING ORIGINAL SIGNALS FROM SAMPLED SIGNALS

Sampling Theorem.   If the sampling frequency is sufficiently high compared with the highest-frequency component involved in the continuous-time signal, the amplitude characteristics of the continuous-time signal may be preserved in the envelope of the sampled signal.

To reconstruct the original signal from a sampled signal, there is a certain minimum frequency that the sampling operation must satisfy. Such a minimum frequency is specified in the sampling theorem. We shall assume that a continuous-time signal $x(t)$ has a frequency spectrum as shown in Figure 3–10. This signal $x(t)$ does not contain any frequency components above ω_1 radians per second.

Sampling Theorem.   If ω_s, defined as $2\pi/T$, where T is the sampling period, is greater than $2\omega_1$, or

$$\omega_s > 2\omega_1$$

where ω_1 is the highest-frequency component present in the continuous-time signal $x(t)$, then the signal $x(t)$ can be reconstructed completely from the sampled signal $x^*(t)$.

The theorem implies that if $\omega_s > 2\omega_1$ then, from the knowledge of the sampled signal, it is theoretically possible to reconstruct exactly the original continuous-time signal. In what follows, we shall use an intuitive graphical approach to explain the sampling theorem. For an analytical approach, see Problem A–3–10.

To show the validity of this sampling theorem, we need to find the frequency

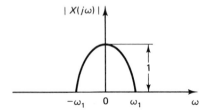

Figure 3–10   A frequency spectrum.

spectrum of the sampled signal $x^*(t)$. The Laplace transform of $x^*(t)$ has been obtained in Section 3–3 and is given by Equation (3–27) or (3–29), depending on whether $x(0+) = 0$ or not. To obtain the frequency spectrum, we substitute $j\omega$ for s in Equation (3–27). [In discussing frequency spectra, we need not be concerned with the value of $x(0+)$.] Thus,

$$X^*(j\omega) = \frac{1}{T} \sum_{k=-\infty}^{\infty} X(j\omega + j\omega_s k)$$

$$= \cdots + \frac{1}{T}X(j(\omega - \omega_s)) + \frac{1}{T}X(j\omega) + \frac{1}{T}X(j(\omega + \omega_s)) + \cdots \qquad (3\text{–}34)$$

Equation (3–34) gives the frequency spectrum of the sampled signal $x^*(t)$. We see that the frequency spectrum of the impulse-sampled signal is reproduced an infinite number of times and is attenuated by the factor $1/T$. Thus, the process of impulse modulation of the continuous-time signal produces a series of sidebands. Since $X^*(s)$ is periodic with period $2\pi/\omega_s$, as shown in Example 3–2, or

$$X^*(s) = X^*(s \pm j\omega_s k), \qquad k = 0, 1, 2, \ldots$$

if a function $X(s)$ has a pole at $s = s_1$, then $X^*(s)$ has poles at $s = s_1 \pm j\omega_s k$ ($k = 0, 1, 2, \ldots$).

Figures 3–11(a) and (b) show plots of the frequency spectra $X^*(j\omega)$ versus ω

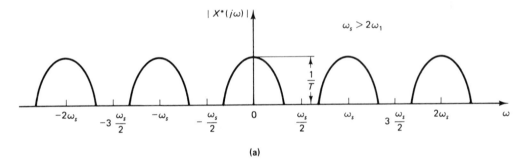

(a)

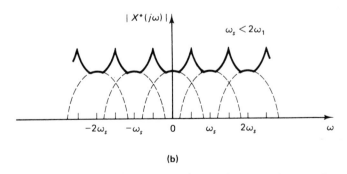

(b)

Figure 3–11 Plots of the frequency spectra $|X^*(j\omega)|$ versus ω for two values of sampling frequency ω_s: (a) $\omega_s > 2\omega_1$; (b) $\omega_s < 2\omega_1$.

for two values of the sampling frequency ω_s. Figure 3–11(a) corresponds to $\omega_s > 2\omega_1$, while Figure 3–11(b) corresponds to $\omega_s < 2\omega_1$. Each plot of $|X^*(j\omega)|$ versus ω consists of $|X(j\omega)|/T$ repeated every $\omega_s = 2\pi/T$ rad/sec. In the frequency spectrum of $|X^*(j\omega)|$ the component $|X(j\omega)|/T$ is called the *primary component*, and the other components, $|X(j(\omega \pm \omega_s k))|/T$, are called *complementary components*.

If $\omega_s > 2\omega_1$, no two components of $|X^*(j\omega)|$ will overlap, and the sampled frequency spectrum will be repeated every ω_s rad/sec.

If $\omega_s < 2\omega_1$, the original shape of $|X(j\omega)|$ no longer appears in the plot of $|X^*(j\omega)|$ versus ω because of the superposition of the spectra. Therefore, we see that the continuous-time signal $x(t)$ can be reconstructed from the impulse-sampled signal $x^*(t)$ by filtering if and only if $\omega_s > 2\omega_1$.

It is noted that although the requirement on the minimum sampling frequency is specified by the sampling theorem as $\omega_s > 2\omega_1$, where ω_1 is the highest-frequency component present in the signal, practical considerations on the stability of the closed-loop system and other design considerations may make it necessary to sample at a frequency much higher than this theoretical minimum. (Frequently, ω_s is chosen to be $10\omega_1$ to $20\omega_1$.)

Ideal Low-Pass Filter. The amplitude frequency spectrum of the ideal low-pass filter $G_l(j\omega)$ is shown in Figure 3–12. The magnitude of the ideal filter is unity over the frequency range $-\frac{1}{2}\omega_s \leq \omega \leq \frac{1}{2}\omega_s$ and is zero outside this frequency range.

The sampling process introduces an infinite number of complementary components (sideband components) in addition to the primary component. The ideal filter will attenuate all such complementary components to zero and will pass only the primary component, provided the sampling frequency ω_s is greater than twice the highest-frequency component of the continuous-time signal. Such an ideal filter reconstructs the continuous-time signal represented by the samples. Figure 3–13 shows the frequency spectra of the signals before and after ideal filtering.

The frequency spectrum at the output of the ideal filter is $1/T$ times the frequency spectrum of the original continuous-time signal $x(t)$. Since the ideal filter has constant-magnitude characteristics for the frequency region $-\frac{1}{2}\omega_s \leq \omega \leq \frac{1}{2}\omega_s$, there is no distortion at any frequency within this frequency range. That is, there is no phase shift in the frequency spectrum of the ideal filter. (The phase shift of the ideal filter is zero.)

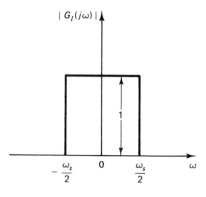

Figure 3–12 Amplitude frequency spectrum of the ideal low-pass filter.

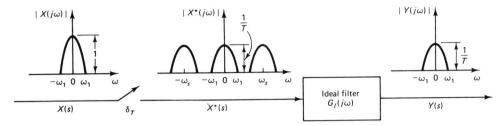

Figure 3–13 Frequency spectra of the signals before and after ideal filtering.

It is noted that if the sampling frequency is less than twice the highest-frequency component of the original continuous-time signal, then because of the frequency spectrum overlap of the primary component and complementary components, even the ideal filter cannot reconstruct the original continuous-time signal. (In practice, the frequency spectrum of the continuous-time signal in a control system may extend beyond $\pm \frac{1}{2}\omega_s$, even though the amplitudes at the higher frequencies are small.)

Ideal Low-Pass Filter Is Not Physically Realizable.   Let us find the impulse response function of the ideal filter. It will be shown that for the ideal filter an output is required prior to the application of the input to the filter. Thus, it is not physically realizable.

Since the frequency spectrum of the ideal filter is given by

$$G_I(j\omega) = \begin{cases} 1, & -\frac{1}{2}\omega_s \le \omega \le \frac{1}{2}\omega_s \\ 0, & \text{elsewhere} \end{cases}$$

the inverse Fourier transform of the frequency spectrum gives

$$g_I(t) = \frac{1}{2\pi} \int_{-\infty}^{\infty} G_I(j\omega) e^{j\omega t} \, d\omega$$

$$= \frac{1}{2\pi} \int_{-\omega_s/2}^{\omega_s/2} e^{j\omega t} \, d\omega$$

$$= \frac{1}{2\pi jt} (e^{(1/2)j\omega_s t} - e^{-(1/2)j\omega_s t})$$

$$= \frac{1}{\pi t} \sin \frac{\omega_s t}{2}$$

or

$$g_I(t) = \frac{1}{T} \frac{\sin (\omega_s t/2)}{\omega_s t/2} \tag{3-35}$$

Equation (3–35) gives the unit-impulse response of the ideal filter. Figure 3–14 shows a plot of $g_I(t)$ versus t. Notice that the response extends from $t = -\infty$ to $t = \infty$. This implies that there is a response for $t < 0$ to a unit impulse applied at $t = 0$. (That is, the time response begins before an input is applied.) This cannot be true in the physical world. Hence, such an ideal filter is physically unrealizable. [In many

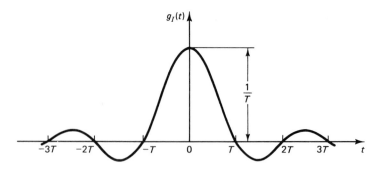

Figure 3-14 Impulse response $g_I(t)$ of ideal filter.

communications systems, however, it is possible to approximate $g_I(t)$ closely by adding a phase lag, which means adding a delay to the filter. In feedback control systems, increasing phase lag is not desirable from the viewpoint of stability. Therefore, we avoid adding a phase lag to approximate the ideal filter.]

Because the ideal filter is unrealizable and because signals in practical control systems generally have higher-frequency components and are not ideally band limited, it is not possible, in practice, to exactly reconstruct a continuous-time signal from the sampled signal, no matter what sampling frequency is chosen. (In other words, practically speaking, it is not possible to reconstruct exactly a continuous-time signal in a practical control system once it is sampled.)

Frequency-Response Characteristics of the Zero-Order Hold.   The transfer function of a zero-order hold is

$$G_{h0}(s) = \frac{1 - e^{-Ts}}{s} \tag{3-36}$$

To compare the zero-order hold with the ideal filter, we shall obtain the frequency-response characteristics of the transfer function of the zero-order hold. By substituting $j\omega$ for s in Equation (3–36), we obtain

$$G_{h0}(j\omega) = \frac{1 - e^{-Tj\omega}}{j\omega}$$

$$= \frac{2e^{-(1/2)Tj\omega}(e^{(1/2)Tj\omega} - e^{-(1/2)Tj\omega})}{2j\omega}$$

$$= T\frac{\sin(\omega T/2)}{\omega T/2}e^{-(1/2)Tj\omega}$$

The amplitude of the frequency spectrum of $G_{h0}(j\omega)$ is

$$|G_{h0}(j\omega)| = T\left|\frac{\sin(\omega T/2)}{\omega T/2}\right| \tag{3-37}$$

The magnitude becomes zero at the frequency equal to the sampling frequency and at integral multiples of the sampling frequency.

Figure 3–15(a) shows the frequency-response characteristics of the zero-order hold. As can be seen from Figure 3–15, there are undesirable gain peaks at frequencies of $3\omega_s/2$, $5\omega_s/2$, and so on. Notice that the magnitude is more than 3 dB down ($0.637 = -3.92$ dB) at frequency $\frac{1}{2}\omega_s$. Because the magnitude decreases gradually as the frequency increases, the complementary components gradually attenuate to zero. Since the magnitude characteristics of the zero-order hold are not constant, if a system is connected to the sampler and zero-order hold, distortion of the frequency spectra occurs in the system.

The phase-shift characteristics of the zero-order hold can be obtained as follows. Note that $\sin(\omega T/2)$ alternates positive and negative values as ω increases from 0 to ω_s, ω_s to $2\omega_s$, $2\omega_s$ to $3\omega_s$, and so on. Thus, the phase curve [Figure 3–15(a), bottom] is discontinuous at $\omega = k\omega_s = 2\pi k/T$, where $k = 1, 2, 3, \ldots$. Such a discontinuity or a switch from a positive value to a negative value, or vice versa, may be considered to be a phase shift of $\pm 180°$. In Figure 3–15(a), phase shift is assumed to be $-180°$. (It could be assumed to be $+180°$ as well.) Thus,

$$\underline{/G_{h0}(j\omega)} = \underline{/T\frac{\sin(\omega T/2)}{\omega T/2}} \; \underline{/e^{-(1/2)j\omega T}}$$

$$= \underline{/\sin\frac{\omega T}{2}} + \underline{/e^{-(1/2)j\omega T}} = \underline{/\sin\frac{\omega T}{2} - \frac{\omega T}{2}}$$

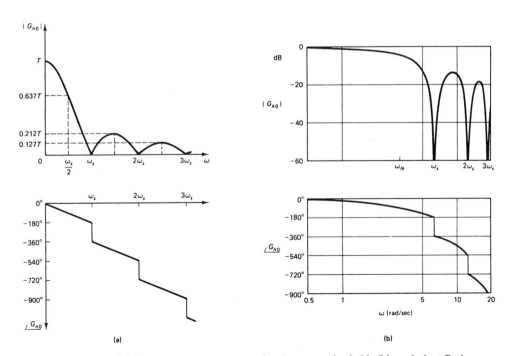

Figure 3–15 (a) Frequency-response curves for the zero-order hold; (b) equivalent Bode diagram when $T = 1$ sec.

where

$$\angle e^{-(1/2)j\omega T} = -\frac{\omega T}{2}$$

$$\angle \sin\frac{\omega T}{2} = 0° \quad \text{or} \quad \pm 180°$$

A modification of the presentation of the frequency-response diagram of Figure 3–15(a) is shown in Figure 3–15(b). The diagram shown in Figure 3–15(b) is the Bode diagram of the zero-order hold. The sampling period T is assumed to be 1 sec, or $T = 1$. Notice that the magnitude curve approaches $-\infty$ decibels at frequency points that are integral multiples of the sampling frequency $\omega_s = 2\pi/T = 6.28$ rad/sec. Discontinuities of the phase curve [Figure 3–15(b), bottom] occur at these frequency points.

To summarize what we have stated, the frequency spectrum of the output of the zero-order hold includes complementary components, since the magnitude characteristics show that the magnitude of $G_{h0}(j\omega)$ is not zero for $|\omega| > \frac{1}{2}\omega_s$, except at points where $\omega = \pm\omega_s, \omega = \pm 2\omega_s, \omega = \pm 3\omega_s, \ldots$. In the phase curve there are phase discontinuities of $\pm 180°$ at frequency points that are multiples of ω_s. Except for these phase discontinuities, the phase characteristic is linear in ω.

Figure 3–16 shows the comparison of the ideal filter and the zero-order hold. For the sake of comparison, the magnitudes $|G(j\omega)|$ are normalized. We see that the zero-order hold is a low-pass filter, although its function is not quite good. Often, additional low-pass filtering of the signal before sampling is necessary to effectively remove frequency components higher than $\frac{1}{2}\omega_s$.

The accuracy of the zero-order hold as an extrapolator depends on the sampling frequency ω_s. That is, the output of the hold may be made as close to the original continuous-time signal as possible by letting the sampling period T become as small as practically possible.

Folding.   The phenomenon of the overlap in the frequency spectra is known as *folding*. Figure 3–17 shows the regions where folding error occurs. The frequency $\frac{1}{2}\omega_s$ is called the *folding frequency* or *Nyquist frequency* ω_N. That is,

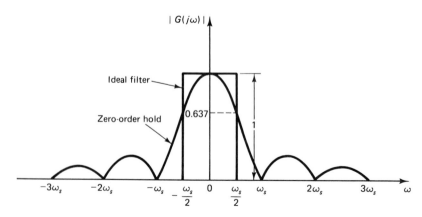

Figure 3–16   Comparison of the ideal filter and the zero-order hold.

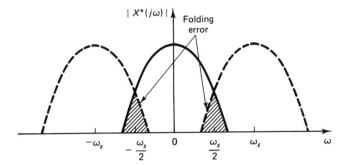

Figure 3–17 Diagram showing the regions where folding error occurs.

$$\omega_N = \frac{1}{2}\omega_s = \frac{\pi}{T}$$

In practice, signals in control systems have high-frequency components, and some folding effect will almost always exist. For example, in an electromechanical system some signal may be contaminated by noises. The frequency spectrum of the signal, therefore, may include low-frequency components as well as high-frequency noise components (that is, noises at 60 or 400 Hz). Since sampling at frequencies higher than 400 Hz is not practical, the high frequency will be folded in and will appear as a low frequency. Remember that all signals with frequencies higher than $\frac{1}{2}\omega_s$ appear as signals of frequencies between 0 and $\frac{1}{2}\omega_s$. In fact, in certain cases, a signal of zero frequency may appear in the output.

Aliasing.   In the frequency spectra of an impulse-sampled signal $x^*(t)$, where $\omega_s < 2\omega_1$, as shown in Figure 3–18, consider an arbitrary frequency point ω_2 that falls in the region of the overlap of the frequency spectra. The frequency spectrum at $\omega = \omega_2$ comprises two components, $|X^*(j\omega_2)|$ and $|X^*(j(\omega_s - \omega_2))|$. The latter component comes from the frequency spectrum centered at $\omega = \omega_s$. Thus, the frequency

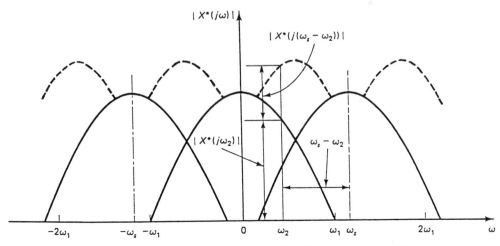

Figure 3–18 Frequency spectra of an impulse-sampled signal $x^*(t)$.

spectrum of the sampled signal at $\omega = \omega_2$ includes components not only at frequency ω_2 but also at frequency $\omega_s - \omega_2$ (in general, at $n\omega_s \pm \omega_2$, where n is an integer). When the composite spectrum is filtered by a low-pass filter, such as a zero-order hold, some higher harmonics will still be present in the output. The frequency component at $\omega = n\omega_s \pm \omega_2$ (where n is an integer) will appear in the output as if it were a frequency component at $\omega = \omega_2$. It is not possible to distinguish the frequency spectrum at $\omega = \omega_2$ from that at $\omega = n\omega_s \pm \omega_2$.

As shown in Figure 3–18, the phenomenon that the frequency component $\omega_s - \omega_2$ (in general, $n\omega_s \pm \omega_2$, where n is an integer) shows up at frequency ω_2 when the signal $x(t)$ is sampled is called *aliasing*. This frequency $\omega_s - \omega_2$ (in general, $n\omega_s \pm \omega_2$) is called an *alias* of ω_2.

It is important to remember that the sampled signals are the same if the two frequencies differ by an integral multiple of the sampling frequency ω_s. If a signal is sampled at a slow frequency such that the sampling theorem is not satisfied, then high frequencies are "folded in" and appear as low frequencies.

To avoid aliasing, we must either choose the sampling frequency high enough ($\omega_s > 2\omega_1$, where ω_1 is the highest-frequency component present in the signal) or use a prefilter ahead of the sampler to reshape the frequency spectrum of the signal (so that the frequency spectrum for $\omega > \frac{1}{2}\omega_s$ is negligible) before the signal is sampled.

Hidden Oscillation.   It is noted that, if the continuous-time signal $x(t)$ involves a frequency component equal to n times the sampling frequency ω_s (where n is an integer), then that component may not appear in the sampled signal. For example, if the signal

$$x(t) = x_1(t) + x_2(t) = \sin t + \sin 3t$$

where $x_1(t) = \sin t$ and $x_2(t) = \sin 3t$, is sampled at $t = 0, 2\pi/3, 4\pi/3, \ldots$ (the sampling frequency ω_s is 3 rad/sec), then the sampled signal will not show the frequency component with $\omega = 3$ rad/sec, the frequency equal to ω_s. (See Figure 3–19.)

Even though the signal $x(t)$ involves an oscillation with $\omega = 3$ rad/sec [that is, the component $x_2(t) = \sin 3t$], the sampled signal does not show this oscillation. Such an oscillation existing in $x(t)$ between the sampling periods is called a *hidden oscillation*.

3–5 THE PULSE TRANSFER FUNCTION

The transfer function for the continuous-time system relates the Laplace transform of the continuous-time output to that of the continuous-time input, while the pulse transfer function relates the z transform of the output at the sampling instants to that of the sampled input.

Before we discuss the pulse transfer function, it is appropriate to discuss convolution summation.

Convolution Summation.   Consider the response of a continuous-time system driven by an impulse-sampled signal (a train of impulses) as shown in Figure 3–20. Suppose that $x(t) = 0$ for $t < 0$. The impulse-sampled signal $x^*(t)$ is the input to the continuous-time system whose transfer function is $G(s)$. The output of the system

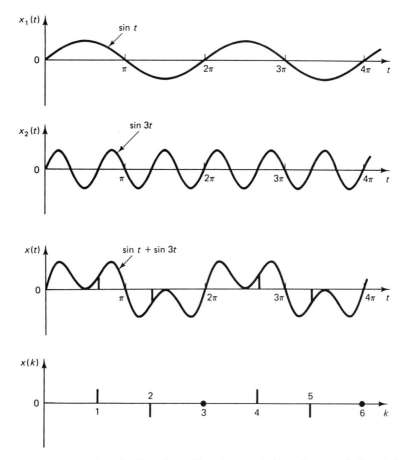

Figure 3–19   Plots of $x_1(t) = \sin t$, $x_2(t) = \sin 3t$, and $x(t) = \sin t + \sin 3t$. Sampled signal $x(k)$, where sampling frequency $\omega_s = 3$ rad/sec, does not show oscillation with frequency $\omega = 3$ rad/sec.

is assumed to be a continuous-time signal $y(t)$. If at the output there is another sampler, which is synchronized in phase with the input sampler and operates at the same sampling period, then the output is a train of impulses. We assume that $y(t) = 0$ for $t < 0$.

The z transform of $y(t)$ is

$$\mathcal{Z}[y(t)] = Y(z) = \sum_{k=0}^{\infty} y(kT)z^{-k} \tag{3–38}$$

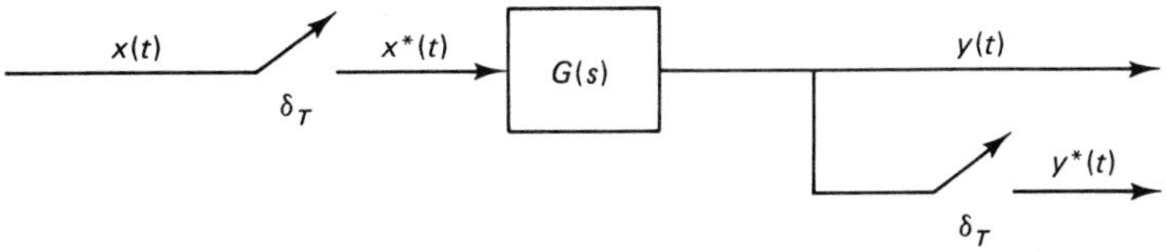

Figure 3–20   Continuous-time system $G(s)$ driven by an impulse-sampled signal.

In the absence of the output sampler, if we consider a fictitious sampler (which is synchronized in phase with the input sampler and operates at the same sampling period) at the output and observe the sequence of values taken by $y(t)$ only at instants $t = kT$, then the z transform of the output $y^*(t)$ can also be given by Equation (3–38).

For the continuous-time system, it is a well-known fact that the output $y(t)$ of the system is related to the input $x(t)$ by the convolution integral, or

$$y(t) = \int_0^t g(t - \tau)x(\tau)\, d\tau = \int_0^t x(t - \tau)g(\tau)\, d\tau$$

where $g(t)$ is the weighting function of the system or the impulse-response function of the system. For discrete-time systems we have a convolution summation, which is similar to the convolution integral. Since

$$x^*(t) = \sum_{k=0}^{\infty} x(t)\delta(t - kT) = \sum_{k=0}^{\infty} x(kT)\delta(t - kT)$$

is a train of impulses, the response $y(t)$ of the system to the input $x^*(t)$ is the sum of the individual impulse responses. Hence,

$$y(t) = \begin{cases} g(t)x(0), & 0 \le t < T \\ g(t)x(0) + g(t - T)x(T), & T \le t < 2T \\ g(t)x(0) + g(t - T)x(T) + g(t - 2T)x(2T), & 2T \le t < 3T \\ \quad\vdots \\ g(t)x(0) + g(t - T)x(T) + \cdots + g(t - kT)x(kT), & kT \le t < (k + 1)T \end{cases}$$

Noting that for a physical system a response cannot precede the input, we have $g(t) = 0$ for $t < 0$ or $g(t - kT) = 0$ for $t < kT$. Consequently, the preceding equations may be combined into one equation:

$$y(t) = g(t)x(0) + g(t - T)x(T) + g(t - 2T)x(2T) + \cdots + g(t - kT)x(kT)$$

$$= \sum_{h=0}^{k} g(t - hT)x(hT) \qquad 0 \le t \le kT$$

The values of the output $y(t)$ at the sampling instants $t = kT$ $(k = 0, 1, 2, \ldots)$ are given by

$$y(kT) = \sum_{h=0}^{k} g(kT - hT)x(hT) \qquad\qquad (3\text{–}39)$$

$$= \sum_{h=0}^{k} x(kT - hT)g(hT) \qquad\qquad (3\text{–}40)$$

where $g(kT)$ is the system's weighting sequence. [The inverse z transform of $G(z)$ is called the *weighting sequence*.] The summation in Equation (3–39) or (3–40) is called the *convolution summation*. Note that the simplified notation

$$y(kT) = x(kT) * g(kT)$$

is often used for the convolution summation.

Since we assumed that $x(t) = 0$ for $t < 0$, we have $x(kT - hT) = 0$ for $h > k$. Also, since $g(kT - hT) = 0$ for $h > k$, we may assume that the values of h in

Equations (3–39) and (3–40) can be taken from 0 to ∞ rather than from 0 to k without changing the value of the summation. Therefore, Equations (3–39) and (3–40) can be rewritten as follows:

$$y(kT) = \sum_{h=0}^{\infty} g(kT - hT)x(hT) \tag{3-41}$$

$$= \sum_{h=0}^{\infty} x(kT - hT)g(hT) \tag{3-42}$$

It is noted that if $G(s)$ is a ratio of polynomials in s and if the degree of the denominator polynomial exceeds the degree of the numerator polynomial only by 1 the output $y(t)$ is discontinuous, as shown in Figure 3–21(a). When $y(t)$ is discontinuous, Equations (3–41) and (3–42) yield the values immediately after the sampling instants, that is, $y(0+), y(T+), \ldots, y(kT+)$. Such values do not portray the actual response curve.

If the degree of the denominator polynomial exceeds that of the numerator polynomial by 2 or more, however, the output $y(t)$ is continuous, as shown in Figure 3–21(b). When $y(t)$ is continuous, Equations (3–41) and (3–42) yield the values at

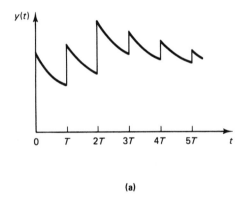

(a)

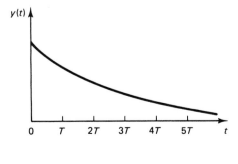

(b)

Figure 3–21   (a) Plot of output $y(t)$ (impulse response) versus t when the degree of the denominator polynomial of $G(s)$ is higher by 1 than that of the numerator polynomial; (b) plot of output $y(t)$ versus t when the degree of the denominator polynomial of $G(s)$ is higher by 2 or more than that of the numerator polynomial.

the sampling instants. The values $y(k)$ in such a case portray the actual response curve.

In analyzing discrete-time control systems it is important to remember that the system response to the impulse-sampled signal may not portray the correct time-response behavior of the actual system unless the transfer function $G(s)$ of the continuous-time part of the system has at least two more poles than zeros, so that $\lim_{s \to \infty} sG(s) = 0$.

Pulse Transfer Function.   From Equation (3–41) we have

$$y(kT) = \sum_{h=0}^{\infty} g(kT - hT)x(hT), \qquad k = 0, 1, 2, \ldots$$

where $g(kT - hT) = 0$ for $h > k$. Hence, the z transform of $y(kT)$ becomes

$$Y(z) = \sum_{k=0}^{\infty} y(kT)z^{-k}$$

$$= \sum_{k=0}^{\infty} \sum_{h=0}^{\infty} g(kT - hT)x(hT)z^{-k}$$

$$= \sum_{m=0}^{\infty} \sum_{h=0}^{\infty} g(mT)x(hT)z^{-(m+h)}$$

$$= \sum_{m=0}^{\infty} g(mT)z^{-m} \sum_{h=0}^{\infty} x(hT)z^{-h}$$

$$= G(z)X(z) \qquad\qquad (3\text{–}43)$$

where $m = k - h$ and

$$G(z) = \sum_{m=0}^{\infty} g(mT)z^{-m} = z \text{ transform of } g(t)$$

Equation (3–43) relates the pulsed output $Y(z)$ of the system to the pulse input $X(z)$. It provides a means for determining the z transform of the output sequence for any input sequence. Dividing both sides of Equation (3–43) by $X(z)$ gives

$$G(z) = \frac{Y(z)}{X(z)} \qquad\qquad (3\text{–}44)$$

$G(z)$ given by Equation (3–44), the ratio of the output $Y(z)$ and the input $X(z)$, is called the *pulse transfer function* of the discrete-time system. It is the z transform of the weighting sequence. Figure 3–22 shows a block diagram for a pulse transfer function $G(z)$, together with the input $X(z)$ and the output $Y(z)$. As seen in Equation (3–43), the z transform of the output signal can be obtained as the product of the system's pulse transfer function and the z transform of the input signal.

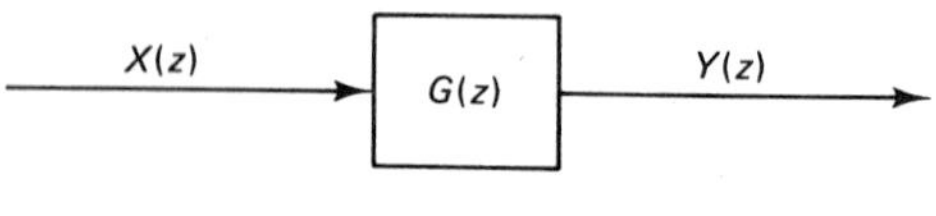

Figure 3–22 Block diagram for a pulse-transfer-function system.

Note that $G(z)$ is also the z transform of the system's response to the Kronecker delta input:

$$x(kT) = \delta_0(kT) = \begin{cases} 1, & \text{for } k = 0 \\ 0, & \text{for } k \neq 0 \end{cases}$$

Since the z transform of the Kronecker delta input is

$$X(z) = \sum_{k=0}^{\infty} x(kT)z^{-k} = 1$$

then, referring to Equation (3–44), the response $Y(z)$ to the Kronecker delta input is

$$Y(z) = G(z)$$

Thus, the system's response to the Kronecker delta input is $G(z)$, the z transform of the weighting sequence. This fact is parallel to the fact that $G(s)$ is the Laplace transform of the system's weighting function, which is the system's response to the unit-impulse function.

Starred Laplace Transform of the Signal Involving Both Ordinary and Starred Laplace Transforms.   In analyzing discrete-time control systems, we often find that some signals in the system are starred (meaning that signals are impulse sampled) and others are not. To obtain pulse transfer functions and to analyze discrete-time control systems, therefore, we must be able to obtain the transforms of output signals of systems that contain sampling operations in various places in the loops.

Suppose the impulse sampler is followed by a linear continuous-time element whose transfer function is $G(s)$, as shown in Figure 3–23. In the following analysis we assume that all initial conditions are zero in the system. Then the output $Y(s)$ is

$$Y(s) = G(s)X^*(s) \qquad (3-45)$$

Notice that $Y(s)$ is a product of $X^*(s)$, which is periodic with period $2\pi/\omega_s$, and $G(s)$, which is not periodic. The fact that the impulse-sampled signals are periodic can be seen from the fact that

$$X^*(s) = X^*(s \pm j\omega_s k), \qquad k = 0, 1, 2, \ldots \qquad (3-46)$$

(See Example 3–2.)

In the following we shall show that in taking the starred Laplace transform of Equation (3–45) we may factor out $X^*(s)$ so that

$$Y^*(s) = [G(s)X^*(s)]^* = [G(s)]^*X^*(s) = G^*(s)X^*(s) \qquad (3-47)$$

This fact is very important in deriving the pulse transfer function and also in simplifying the block diagram of the discrete-time control system.

To derive Equation (3–47), note that the inverse Laplace transform of $Y(s)$ given by Equation (3–45) can be written as follows:

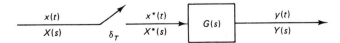

Figure 3–23   Impulse-sampled system.

$$y(t) = \mathcal{L}^{-1}[G(s)X^*(s)]$$

$$= \int_0^t g(t - \tau)x^*(\tau)\,d\tau$$

$$= \int_0^t g(t - \tau) \sum_{k=0}^{\infty} x(\tau)\delta(\tau - kT)\,d\tau$$

$$= \sum_{k=0}^{\infty} \int_0^t g(t - \tau)x(\tau)\delta(\tau - kT)\,d\tau$$

$$= \sum_{k=0}^{\infty} g(t - kT)x(kT)$$

Then the z transform of $y(t)$ becomes

$$Y(z) = \mathcal{Z}[y(t)] = \sum_{n=0}^{\infty}\left[\sum_{k=0}^{\infty} g(nT - kT)x(kT)\right]z^{-n}$$

$$= \sum_{m=0}^{\infty}\sum_{k=0}^{\infty} g(mT)x(kT)z^{-(k+m)}$$

where $m = n - k$. Thus,

$$Y(z) = \sum_{m=0}^{\infty} g(mT)z^{-m}\sum_{k=0}^{\infty} x(kT)z^{-k}$$

$$= G(z)X(z) \tag{3-48}$$

Since the z transform can be understood as the starred Laplace transform with e^{Ts} replaced by z, the z transform may be considered to be a shorthand notation for the starred Laplace transform. Thus, Equation (3–48) may be expressed as

$$Y^*(s) = G^*(s)X^*(s)$$

which is Equation (3–47). We have thus shown that by taking the starred Laplace transform of both sides of Equation (3–45) we obtain Equation (3–47).

To summarize what we have obtained, note that Equations (3–45) and (3–47) state that in taking the starred Laplace transform of a product of transforms, where some are ordinary Laplace transforms and others are starred Laplace transforms, the functions already in starred transforms can be factored out of the starred Laplace transform operation.

It is noted that systems become periodic under starred Laplace transform operations. Such periodic systems are generally more complicated to analyze than the original nonperiodic ones, but the former may be analyzed without difficulty if carried out in the z plane (that is, by use of the pulse-transfer-function approach).

General Procedures for Obtaining Pulse Transfer Functions.   Here we shall present general procedures for obtaining the pulse transfer function of a system that has an impulse sampler at the input to the system, as shown in Figure 3–24(a).

The pulse transfer function $G(z)$ of the system shown in Figure 3–24(a) is

$$\frac{Y(z)}{X(z)} = G(z) = \mathcal{Z}[G(s)]$$

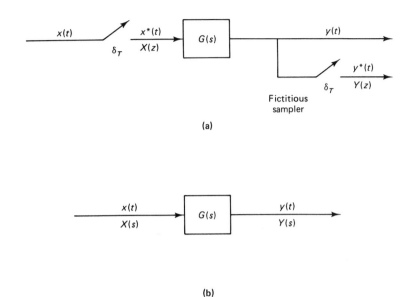

Figure 3–24   (a) Continuous-time system with an impulse sampler at the input; (b) continuous-time system.

Next, consider the system shown in Figure 3–24(b). The transfer function $G(s)$ is given by

$$\frac{Y(s)}{X(s)} = G(s)$$

The important fact to remember is that the pulse transfer function for this system is not $\mathcal{Z}[G(s)]$, because of the absence of the input sampler.

The presence or absence of the input sampler is crucial in determining the pulse transfer function of a system, because, for example, for the system shown in Figure 3–24(a), the Laplace transform of the output $y(t)$ is

$$Y(s) = G(s)X^*(s)$$

Hence, by taking the starred Laplace transform of $Y(s)$, we have

$$Y^*(s) = G^*(s)X^*(s)$$

or, in terms of the z transform,

$$Y(z) = G(z)X(z)$$

while, for the system shown in Figure 3–24(b), the Laplace transform of the output $y(t)$ is

$$Y(s) = G(s)X(s)$$

which yields

$$Y^*(s) = [G(s)X(s)]^* = [GX(s)]^*$$

or, in terms of the z transform,

$$Y(z) = \mathcal{Z}[Y(s)] = \mathcal{Z}[G(s)X(s)] = \mathcal{Z}[GX(s)] = GX(z) \neq G(z)X(z)$$

The fact that the z transform of $G(s)X(s)$ is not equal to $G(z)X(z)$ will be discussed in detail later in this section.

In discussing the pulse transfer function, we assume that there is a sampler at the input of the element in consideration. The presence or absence of a sampler at the output of the element (or the system) does not affect the pulse transfer function, because, if the sampler is not physically present at the output side of the system, it is always possible to assume that a fictitious sampler is present at the output. This means that, although the output signal is continuous, we can consider the values of the output only at $t = kT$ ($k = 0, 1, 2, \ldots$) and thus get sequence $y(kT)$.

Note that only for the case where the input to the system $G(s)$ is an impulse-sampled signal is the pulse transfer function given by

$$G(z) = \mathcal{Z}[G(s)]$$

Examples 3–4 and 3–5 demonstrate the methods for obtaining the pulse transfer function.

Example 3–4

Obtain the pulse transfer function $G(z)$ of the system shown in Figure 3–24(a), where $G(s)$ is given by

$$G(s) = \frac{1}{s + a}$$

Note that there is a sampler at the input of $G(s)$ and therefore the pulse transfer function is $G(z) = \mathcal{Z}[G(s)]$.

Method 1. By referring to Table 2–1, we have

$$\mathcal{Z}\left[\frac{1}{s + a}\right] = \frac{1}{1 - e^{-aT}z^{-1}}$$

Hence,

$$G(z) = \frac{1}{1 - e^{-aT}z^{-1}}$$

Method 2. The impulse response function for the system is obtained as follows:

$$g(t) = \mathcal{L}^{-1}[G(s)] = e^{-at}$$

Hence,

$$g(kT) = e^{-akT}, \qquad k = 0, 1, 2, \ldots$$

Therefore,

$$G(z) = \sum_{k=0}^{\infty} g(kT)z^{-k} = \sum_{k=0}^{\infty} e^{-akT}z^{-k} = \sum_{k=0}^{\infty} (e^{aT}z)^{-k}$$

$$= \frac{1}{1 - e^{-aT}z^{-1}}$$

Example 3–5

Obtain the pulse transfer function of the system shown in Figure 3–24(a), where $G(s)$ is given by

$$G(s) = \frac{1 - e^{-Ts}}{s} \frac{1}{s(s + 1)}$$

Note that there is a sampler at the input of $G(s)$.

Method 1. $G(s)$ involves the term $(1 - e^{-Ts})$; therefore, referring to Equation (3–32), we obtain the pulse transfer function as follows:

$$G(z) = \mathcal{Z}[G(s)] = \mathcal{Z}\left[(1 - e^{-Ts})\frac{1}{s^2(s + 1)}\right]$$

$$= (1 - z^{-1})\mathcal{Z}\left[\frac{1}{s^2(s + 1)}\right]$$

$$= (1 - z^{-1})\mathcal{Z}\left[\frac{1}{s^2} - \frac{1}{s} + \frac{1}{s + 1}\right]$$

From Table 2–1, the z transform of each of the partial-fraction-expansion terms can be found. Thus,

$$G(z) = (1 - z^{-1})\left[\frac{Tz^{-1}}{(1 - z^{-1})^2} - \frac{1}{1 - z^{-1}} + \frac{1}{1 - e^{-T}z^{-1}}\right]$$

$$= \frac{(T - 1 + e^{-T})z^{-1} + (1 - e^{-T} - Te^{-T})z^{-2}}{(1 - z^{-1})(1 - e^{-T}z^{-1})} \qquad (3\text{–}49)$$

Method 2. The given transfer function $G(s)$ can be written as follows:

$$G(s) = (1 - e^{-Ts})\left(\frac{1}{s^2} - \frac{1}{s} + \frac{1}{s + 1}\right)$$

Therefore, by taking the inverse Laplace transform, we have the following impulse response function:

$$g(t) = (t - 1 + e^{-t})1(t) - [t - T - 1 + e^{-(t-T)}]1(t - T)$$

or

$$g(kT) = (kT - 1 + e^{-kT}) - [kT - T - 1 + e^{-(kT-T)}]1((k - 1)T)$$

$$= \begin{cases} e^{-kT} + T - e^{-(kT-T)}, & k = 1, 2, 3, \ldots \\ 0, & k = 0 \end{cases}$$

Then the pulse transfer function $G(z)$ can be obtained as follows:

$$G(z) = \sum_{k=0}^{\infty} g(kT)z^{-k}$$

$$= \sum_{k=0}^{\infty} [e^{-kT} + T - e^{-(kT-T)}]z^{-k} + e^{T} - 1 - T$$

$$= (1 - e^{T})\sum_{k=0}^{\infty} e^{-kT}z^{-k} + T\sum_{k=0}^{\infty} z^{-k} + e^{T} - 1 - T$$

$$= (1 - e^{T})\frac{1}{1 - e^{-T}z^{-1}} + \frac{T}{1 - z^{-1}} + e^{T} - 1 - T$$

$$= \frac{(T - 1 + e^{-T})z^{-1} + (1 - e^{-T} - Te^{-T})z^{-2}}{(1 - z^{-1})(1 - e^{-T}z^{-1})}$$

Pulse Transfer Function of Cascaded Elements.   Consider the systems shown in Figures 3–25(a) and (b). Here we assume that the samplers are synchronized and have the same sampling period. We shall show that the pulse transfer function of the system shown in Figure 3–25(a) is $G(z)H(z)$, while that shown in Figure 3–25(b) is $\mathcal{Z}[G(s)H(s)] = \mathcal{Z}[GH(s)] = GH(z)$, which is different from $G(z)H(z)$.

Consider the system shown in Figure 3–25(a). From the diagram we obtain

$$U(s) = G(s)X^*(s)$$

$$Y(s) = H(s)U^*(s)$$

Hence, by taking the starred Laplace transform of each of these two equations, we get

$$U^*(s) = G^*(s)X^*(s)$$

$$Y^*(s) = H^*(s)U^*(s)$$

Consequently,

$$Y^*(s) = H^*(s)U^*(s) = H^*(s)G^*(s)X^*(s)$$

or

$$Y^*(s) = G^*(s)H^*(s)X^*(s)$$

In terms of the z transform notation,

$$Y(z) = G(z)H(z)X(z)$$

The pulse transfer function between the output $y^*(t)$ and input $x^*(t)$ is therefore given by

$$\frac{Y(z)}{X(z)} = G(z)H(z)$$

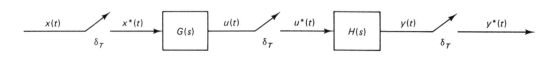

(a)

(b)

Figure 3–25   (a) Sampled system with a sampler between cascaded elements $G(s)$ and $H(s)$; (b) sampled system with no sampler between cascaded elements $G(s)$ and $H(s)$.

Next, consider the system shown in Figure 3–25(b). From the diagram we find

$$Y(s) = G(s)H(s)X^*(s) = GH(s)X^*(s)$$

where

$$GH(s) = G(s)H(s)$$

Taking the starred Laplace transform of $Y(s)$, we have

$$Y^*(s) = [GH(s)]^*X^*(s)$$

In terms of the z transform notation,

$$Y(z) = GH(z)X(z)$$

and the pulse transfer function between the output $y^*(t)$ and input $x^*(t)$ is

$$\frac{Y(z)}{X(z)} = GH(z) = \mathscr{Z}[GH(s)]$$

Note that

$$G(z)H(z) \neq GH(z) = \mathscr{Z}[GH(s)]$$

Hence, the pulse transfer functions of the systems shown in Figures 3–25(a) and (b) are different. We will now verify this statement in Example 3–6.

Example 3–6

Consider the systems shown in Figures 3–26(a) and (b). Obtain the pulse transfer function $Y(z)/X(z)$ for each of these two systems.

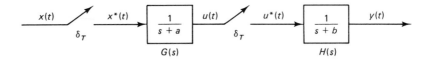

(a)

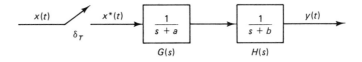

(b)

Figure 3–26   (a) Sampled system with a sampler between elements $G(s) = 1/(s + a)$ and $H(s) = 1/(s + b)$; (b) sampled system with no sampler between elements $G(s)$ and $H(s)$.

For the system of Figure 3–26(a), the two transfer functions $G(s)$ and $H(s)$ are separated by a sampler. We assume that the two samplers shown are synchronized and have the same sampling period. The pulse transfer function for this system is

$$\frac{Y(z)}{X(z)} = \frac{Y(z)}{U(z)}\frac{U(z)}{X(z)} = H(z)G(z) = G(z)H(z)$$

Hence,

$$\frac{Y(z)}{X(z)} = G(z)H(z) = \mathscr{Z}\left[\frac{1}{s+a}\right]\mathscr{Z}\left[\frac{1}{s+b}\right] = \frac{1}{1-e^{-aT}z^{-1}}\frac{1}{1-e^{-bT}z^{-1}}$$

For the system shown in Figure 3–26(b), the pulse transfer function $Y(z)/X(z)$ is obtained as follows:

$$\frac{Y(z)}{X(z)} = \mathscr{Z}[G(s)H(s)] = \mathscr{Z}\left[\frac{1}{s+a}\frac{1}{s+b}\right]$$

$$= \mathscr{Z}\left[\frac{1}{b-a}\left(\frac{1}{s+a} - \frac{1}{s+b}\right)\right]$$

$$= \frac{1}{b-a}\left(\frac{1}{1-e^{-aT}z^{-1}} - \frac{1}{1-e^{-bT}z^{-1}}\right)$$

Hence,

$$\frac{Y(z)}{X(z)} = GH(z) = \frac{1}{b-a}\left[\frac{(e^{-aT}-e^{-bT})z^{-1}}{(1-e^{-aT}z^{-1})(1-e^{-bT}z^{-1})}\right]$$

Clearly, we see that the pulse transfer functions of the two systems are different; that is,

$$G(z)H(z) \neq GH(z)$$

Therefore, we must be careful to observe whether or not there is a sampler between cascaded elements.

Pulse Transfer Function of Closed-Loop Systems. In a closed-loop system the existence or nonexistence of an output sampler within the loop makes a difference in the behavior of the system. (If there is an output sampler outside the loop, it will make no difference in the closed-loop operation.)

Consider the closed-loop control system shown in Figure 3–27. In this system, the actuating error is sampled. From the block diagram,

$$E(s) = R(s) - H(s)C(s)$$

$$C(s) = G(s)E^*(s)$$

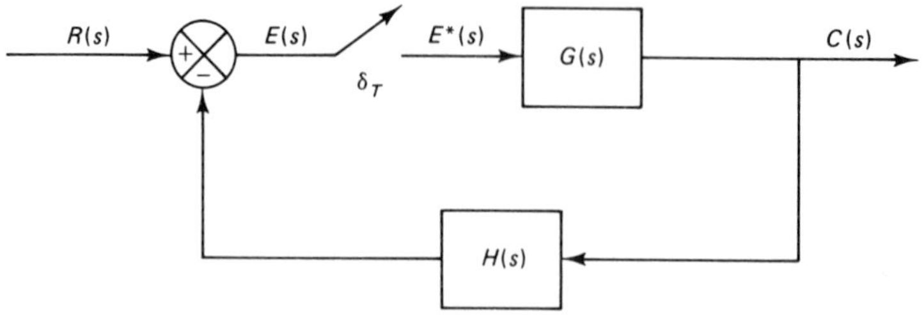

Figure 3–27   Closed-loop control system.

Hence,

$$E(s) = R(s) - H(s)G(s)E^*(s)$$

Then, by taking the starred Laplace transform, we obtain

$$E^*(s) = R^*(s) - GH^*(s)E^*(s)$$

or

$$E^*(s) = \frac{R^*(s)}{1 + GH^*(s)}$$

Since

$$C^*(s) = G^*(s)E^*(s)$$

we obtain

$$C^*(s) = \frac{G^*(s)R^*(s)}{1 + GH^*(s)}$$

In terms of the z transform notation, the output can be given by

$$C(z) = \frac{G(z)R(z)}{1 + GH(z)}$$

The inverse z transform of this last equation gives the values of the output at the sampling instants. [Note that the actual output $c(t)$ of the system is a continuous-time signal. The inverse z transform of $C(z)$ will not give the continuous-time output $c(t)$.] The pulse transfer function for the present closed-loop system is

$$\frac{C(z)}{R(z)} = \frac{G(z)}{1 + GH(z)} \tag{3–50}$$

Table 3–1 shows five typical configurations for closed-loop discrete-time control systems. Here, the samplers are synchronized and have the same sampling period. For each configuration, the corresponding output $C(z)$ is shown. Notice that some discrete-time closed-loop control systems cannot be represented by $C(z)/R(z)$ (that is, they do not have pulse transfer functions) because the input signal $R(s)$ cannot be separated from the system dynamics. Although the pulse transfer function may not exist for certain system configurations, the same techniques discussed in this chapter can still be applied for analyzing them.

Pulse Transfer Function of a Digital Controller.   The pulse transfer function of a digital controller may be obtained from the required input–output characteristics of the digital controller.

Suppose the input to the digital controller is $e(k)$ and the output is $m(k)$. In general, the output $m(k)$ may be given by the following type of difference equation:

$$m(k) + a_1 m(k - 1) + a_2 m(k - 2) + \cdots + a_n m(k - n)$$
$$= b_0 e(k) + b_1 e(k - 1) + \cdots + b_n e(k - n) \tag{3–51}$$

TABLE 3–1 FIVE TYPICAL CONFIGURATIONS FOR CLOSED-LOOP DISCRETE-TIME CONTROL SYSTEMS

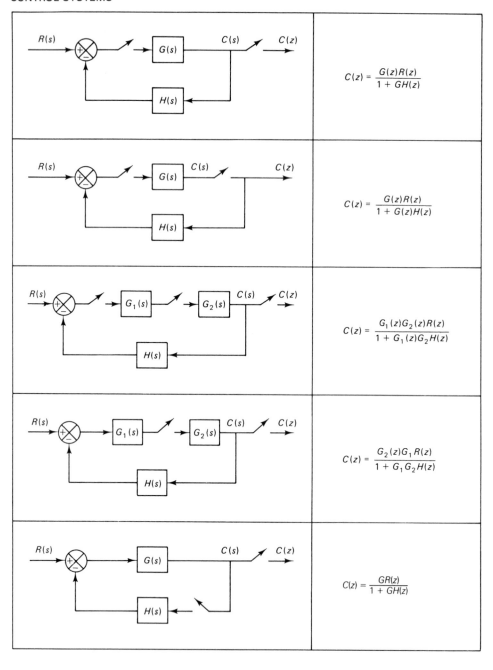

The z transform of Equation (3–51) gives

$$M(z) + a_1 z^{-1} M(z) + a_2 z^{-2} M(z) + \cdots + a_n z^{-n} M(z)$$
$$= b_0 E(z) + b_1 z^{-1} E(z) + \cdots + b_n z^{-n} E(z)$$

or

$$(1 + a_1 z^{-1} + a_2 z^{-2} + \cdots + a_n z^{-n})M(z) = (b_0 + b_1 z^{-1} + \cdots + b_n z^{-n})E(z)$$

The pulse transfer function $G_D(z)$ of the digital controller may then be given by

$$G_D(z) = \frac{M(z)}{E(z)} = \frac{b_0 + b_1 z^{-1} + \cdots + b_n z^{-n}}{1 + a_1 z^{-1} + \cdots + a_n z^{-n}} \qquad (3\text{–}52)$$

The use of the pulse transfer function $G_D(z)$ in the form of Equation (3–52) enables us to analyze digital control systems in the z plane.

Closed-Loop Pulse Transfer Function of a Digital Control System.   Figure 3–28(a) shows a block diagram of a digital control system. Here, the sampler, A/D converter, digital controller, zero-order hold, and D/A converter produce a continuous-time (piecewise-constant) control signal $u(t)$ to be fed to the plant. Figure 3–28(b) shows the transfer functions of blocks involved in the system.

The transfer function of the digital controller is shown as $G_D^*(s)$. In the actual

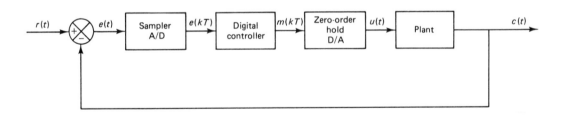

(a)

(b)

Figure 3–28   (a) Block diagram of a digital control system; (b) equivalent block diagram showing transfer functions of blocks.

system the computer (digital controller) solves a difference equation whose input–output relationship is given by the pulse transfer function $G_D(z)$.

In the present system the output signal $c(t)$ is fed back for comparison with the input signal $r(t)$. The error signal $e(t) = r(t) - c(t)$ is sampled, and the analog signal is converted to a digital signal through the A/D device. The digital signal $e(kT)$ is fed to the digital controller, which operates on the sampled sequence $e(kT)$ in some desirable manner to produce the signal $m(kT)$.

This desirable relationship between the sequences $m(kT)$ and $e(kT)$ is specified by the pulse transfer function $G_D(z)$ of the digital controller. [By properly selecting the poles and zeros of $G_D(z)$, a number of input–output characteristics can be generated.]

Referring to Figure 3–28(b), let us define

$$\frac{1 - e^{-Ts}}{s} G_p(s) = G(s)$$

From Figure 3–28(b), notice that

$$C(s) = G(s)G_D^*(s)E^*(s)$$

or

$$C^*(s) = G^*(s)G_D^*(s)E^*(s)$$

In terms of the z transform notation,

$$C(z) = G(z)G_D(z)E(z)$$

Since

$$E(z) = R(z) - C(z)$$

we have

$$C(z) = G_D(z)G(z)[R(z) - C(z)]$$

and, therefore,

$$\frac{C(z)}{R(z)} = \frac{G_D(z)G(z)}{1 + G_D(z)G(z)} \tag{3–53}$$

Equation (3–53) gives the closed-loop pulse transfer function of the digital control system shown in Figure 3–28(b). The performance of such a closed-loop system can be improved by the proper choice of $G_D(z)$, the pulse transfer function of the digital controller. We shall later discuss a variety of forms for $G_D(z)$ to be used in obtaining optimal performance for various given performance indexes.

In the following, we shall consider only a simple case, where the pulse transfer function $G_D(z)$ is of the PID (proportional plus integral plus derivative) type.

Pulse Transfer Function of a Digital PID Controller.   The analog PID control scheme has been used successfully in many industrial control systems for over half a century. The basic principle of the PID control scheme is to act on the variable to be manipulated through a proper combination of three control actions: proportional control action (where the control action is proportional to the actuating error signal, which is the difference between the input and the feedback signal), integral control action (where the control action is proportional to the integral of the actuating error signal), and derivative control action (where the control action is proportional to the derivative of the actuating error signal).

Where many plants are controlled directly by a single digital computer (as in a control scheme in which from several loops to several hundred loops are controlled by a single digital controller), the majority of the control loops may be handled by PID control schemes.

The PID control action in analog controllers is given by

$$m(t) = K\left[e(t) + \frac{1}{T_i}\int_0^t e(t)\, dt + T_d \frac{de(t)}{dt}\right] \tag{3-54}$$

where $e(t)$ is the input to the controller (the actuating error signal), $m(t)$ is the output of the controller (the manipulating signal), K is the proportional gain, T_i is the integral time (or reset time), and T_d is the derivative time (or rate time).

To obtain the pulse transfer function for the digital PID controller, we may discretize Equation (3-54). By approximating the integral term by the trapezoidal summation and the derivative term by a two-point difference form, we obtain

$$m(kT) = K\left\{e(kT) + \frac{T}{T_i}\left[\frac{e(0) + e(T)}{2} + \frac{e(T) + e(2T)}{2} + \cdots\right.\right.$$

$$\left.\left. + \frac{e((k-1)T) + e(kT)}{2}\right] + T_d\frac{e(kT) - e((k-1)T)}{T}\right\}$$

or

$$m(kT) = K\left\{e(kT) + \frac{T}{T_i}\sum_{h=1}^{k}\frac{e((h-1)T) + e(hT)}{2}\right.$$

$$\left. + \frac{T_d}{T}[e(kT) - e((k-1)T)]\right\} \tag{3-55}$$

Define

$$\frac{e((h-1)T) + e(hT)}{2} = f(hT), \qquad f(0) = 0$$

Figure 3-29 shows the function $f(hT)$. Then

$$\sum_{h=1}^{k}\frac{e((h-1)T) + e(hT)}{2} = \sum_{h=1}^{k} f(hT)$$

Taking the z transform of this last equation, we obtain

$$\mathscr{Z}\left[\sum_{h=1}^{k}\frac{e((h-1)T) + e(hT)}{2}\right] = \mathscr{Z}\left[\sum_{h=1}^{k} f(hT)\right] = \frac{1}{1 - z^{-1}}[F(z) - f(0)]$$

$$= \frac{1}{1 - z^{-1}} F(z)$$

(For the derivation of this last equation, refer to Problem A-2-4.) Notice that

$$F(z) = \mathscr{Z}[f(hT)] = \frac{1 + z^{-1}}{2} E(z)$$

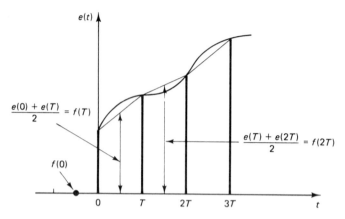

Figure 3–29   Diagram depicting function $f(hT)$.

Hence,

$$\mathcal{Z}\left[\sum_{h=1}^{k}\frac{e((h-1)T)+e(hT)}{2}\right]=\frac{1+z^{-1}}{2(1-z^{-1})}E(z)$$

Then the z transform of Equation (3–55) gives

$$M(z)=K\left[1+\frac{T}{2T_i}\frac{1+z^{-1}}{1-z^{-1}}+\frac{T_d}{T}(1-z^{-1})\right]E(z)$$

This last equation may be rewritten as follows:

$$M(z)=K\left[1-\frac{T}{2T_i}+\frac{T}{T_i}\frac{1}{1-z^{-1}}+\frac{T_d}{T}(1-z^{-1})\right]E(z)$$

$$=\left[K_P+\frac{K_I}{1-z^{-1}}+K_D(1-z^{-1})\right]E(z)$$

where

$$K_P=K-\frac{KT}{2T_i}=K-\frac{K_I}{2}=\text{proportional gain}$$

$$K_I=\frac{KT}{T_i}=\text{integral gain}$$

$$K_D=\frac{KT_d}{T}=\text{derivative gain}$$

Notice that the proportional gain K_P for the digital PID controller is smaller than the proportional gain K for the analog PID controller by $K_I/2$.

The pulse transfer function for the digital PID controller becomes

$$G_D(z)=\frac{M(z)}{E(z)}=K_P+\frac{K_I}{1-z^{-1}}+K_D(1-z^{-1}) \qquad (3\text{–}56)$$

The pulse transfer function of the digital PID controller given by Equation (3–56) is commonly referred to as the *positional form* of the PID control scheme.

The other form commonly used in the digital PID control scheme is referred to as the *velocity form*. To derive the velocity-form PID control equation, we consider the backward difference in $m(kT)$, that is, the difference between $m(kT)$ and $m((k-1)T)$. With some assumptions and manipulations, we obtain

$$M(z) = -K_P C(z) + K_I \frac{R(z) - C(z)}{1 - z^{-1}} - K_D(1 - z^{-1})C(z) \qquad (3\text{--}57)$$

[For the derivation of Equation (3–57), see Problem A–3–17.] Equation (3–57) gives the velocity-form PID control scheme. Figure 3–30 shows the block diagram realization of the velocity-form digital PID control scheme. Notice that in Equation (3–57) only the integral control term involves the input $R(z)$. Hence, the integral term cannot be excluded from the digital controller if the velocity form is used.

An advantage of the velocity-form PID control scheme is that initialization is not necessary when the operation is switched from manual to automatic. Thus, if there are sudden large changes in the set point or at the start of the process operation, the velocity-form PID control scheme exhibits better response characteristics than the positional-form PID control scheme. Another advantage of the velocity-form PID control scheme is that it is useful in suppressing excessive corrections in process control systems.

Linear control laws in the form of PID control actions, in both positional form and velocity form, are basic in digital controls because they frequently give satisfactory solutions to many practical control problems, in particular, process control problems. Note that, in digital controllers, control laws can be implemented by software, and therefore the hardware restrictions of analog PID controllers can be completely ignored. (For a comparison of the frequency-response characteristics of analog and digital PID controllers, see Problem A–3–16.)

Example 3–7

Consider the control system with a digital PID controller shown in Figure 3–31(a). (The PID controller here is in the positional form.) The transfer function of the plant is assumed to be

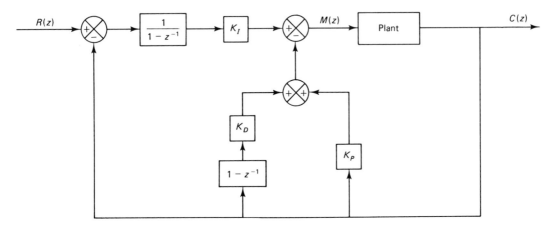

Figure 3–30 Block diagram realization of the velocity-form digital PID control scheme.

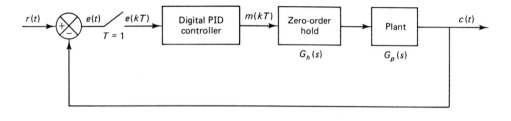

(a)

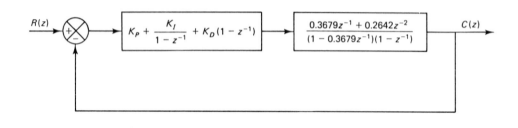

(b)

Figure 3–31 (a) Block diagram of a control system; (b) equivalent block diagram.

$$G_p(s) = \frac{1}{s(s + 1)}$$

and the sampling period T is assumed to be 1 sec. Then the transfer function of the zero-order hold becomes

$$G_h(s) = \frac{1 - e^{-s}}{s}$$

Since

$$\mathcal{Z}\left[\frac{1 - e^{-s}}{s}\frac{1}{s(s + 1)}\right] = G(z) = \frac{0.3679z^{-1} + 0.2642z^{-2}}{(1 - 0.3679z^{-1})(1 - z^{-1})} \qquad (3\text{–}58)$$

we may redraw the block diagram of Figure 3–31(a) as shown in Figure 3–31(b).

Let us obtain the unit-step response of this system when the digital controller is a PID controller with $K_P = 1$, $K_I = 0.2$, and $K_D = 0.2$. The pulse transfer function of the digital controller is given by

$$G_D(z) = \frac{1.4 - 1.4z^{-1} + 0.2z^{-2}}{1 - z^{-1}}$$

Then the closed-loop pulse transfer function becomes

$$\frac{C(z)}{R(z)} = \frac{G_D(z)G(z)}{1 + G_D(z)G(z)}$$

$$= \frac{0.5151z^{-1} - 0.1452z^{-2} - 0.2963z^{-3} + 0.0528z^{-4}}{1 - 1.8528z^{-1} + 1.5906z^{-2} - 0.6642z^{-3} + 0.0528z^{-4}}$$

We shall use the MATLAB approach to obtain the unit-step response.

Obtaining Transient Response with MATLAB.   Let us assume that we want the unit-step response up to $k = 40$. Then the input $r(t)$ may be written as

$$r = \text{ones}(1,41)$$

A MATLAB program for obtaining the unit-step response for this system is shown in MATLAB Program 3–1. The resulting output $c(k)$ versus k is plotted in Figure 3–32.

```
MATLAB Program 3–1

% ---------- Unit-step response ----------

num = [0   0.5151   -0.1452   -0.2963   0.0528];
den = [1   -1.8528   1.5906   -0.6642   0.0528];
r = ones(1,41);
k = 0:40;
c = filter(num,den,r );
plot(k,c,'o',k,c,'-')
v = [0   40   0   2];
axis(v);
grid
title('Unit-Step Response')
xlabel('k')
ylabel('c(k)')
```

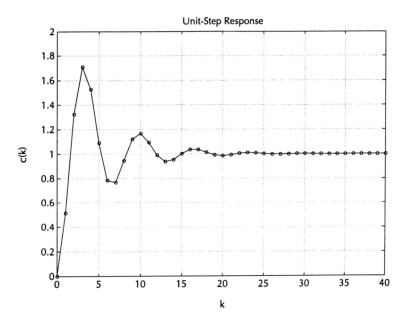

Figure 3–32 Unit-step response.

The response of this system to a unit-ramp input can be obtained by entering MATLAB Program 3–2 into the computer. The resulting plot is shown in Figure 3–33. Note that this system exhibits no steady-state error in the ramp response.

```
MATLAB Program 3–2

% ---------- Unit-ramp response ----------

num = [0   0.5151   -0.1452   -0.2963   0.0528];
den = [1   -1.8528   1.5906   -0.6642   0.0528];
k = 0:40;
r = [k];
c = filter(num,den,r);
plot(k,c,'o',k,c,'-',k,k,'--')
v = [0   16   0   16];
axis(v);
grid
title('Unit-Ramp Response')
xlabel('k')
ylabel('c(k)')
```

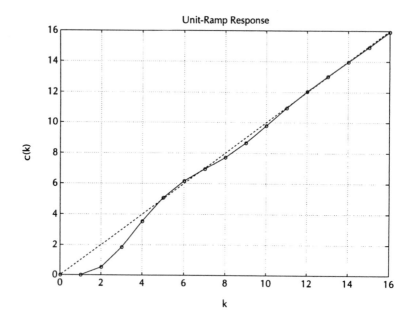

Comments.   PID controllers for such process control systems as temperature systems, pressure systems, and liquid-level systems are usually tuned experimentally. In fact, in the PID control of any industrial plant, where its dynamics are not well known or defined, the controller variables (K_P, K_I, and K_D) must be determined experimentally. Such determination or tuning may be made using step changes in the reference or disturbance signal. A few established procedures are available for such a purpose. Basically, tuning (determining the constants K_P, K_I, and K_D) is accomplished by systematically varying the values until reasonably good response characteristics are obtained.

For digital PID controllers used for process control systems, the sampling period must be chosen properly. Many process control systems have fairly large time constants. A rule of thumb in the selection of the sampling period (sampling period $T = 2\pi/\omega_s$, where ω_s is the sampling frequency) in process control systems is that for temperature control systems the sampling period should be 10 to 30 sec, for pressure control systems 1 to 5 sec, and for liquid-level control systems 1 to 10 sec.

Obtaining Response Between Consecutive Sampling Instants.   The z transform analysis will not give information on the response between two consecutive sampling instants. In ordinary cases this is not serious, because if the sampling theorem is satisfied, then the output will not vary very much between any two consecutive sampling instants. In certain cases, however, we may need to find the response between consecutive sampling instants.

Three methods for providing a response between two consecutive sampling instants are commonly available:

1. Laplace transform method
2. Modified z transform method
3. State-space method

Here we shall briefly discuss the Laplace transform method. The modified z transform method is presented in Appendix B. (Those readers interested in the modified z transform should read Section B–4.) The state-space method will be discussed in Section 5–5.

Laplace Transform Method.   Consider, for example, the system shown in Figure 3–27. The output $C(s)$ can be given by

$$C(s) = G(s)E^*(s) = G(s)\frac{R^*(s)}{1 + GH^*(s)}$$

Thus,

$$c(t) = \mathcal{L}^{-1}[C(s)] = \mathcal{L}^{-1}\left[G(s)\frac{R^*(s)}{1 + GH^*(s)}\right] \qquad (3\text{–}59)$$

Equation (3–59) will give the continuous-time response $c(t)$. Hence, the response at any time between two consecutive sampling instants can be calculated by the use of Equation (3–59). [See Problem A–3–18 for sample calculations of the right-hand side of Equation (3–59).]

3-6 REALIZATION OF DIGITAL CONTROLLERS AND DIGITAL FILTERS

In this section we discuss realization methods for pulse transfer functions that represent digital controllers and digital filters. Realization of digital controllers and digital filters may involve either software or hardware or both. In general, "realization" of a pulse transfer function means determining the physical layout for the appropriate combination of arithmetic and storage operations.

In a software realization we obtain computer programs for the digital computer involved. In a hardware realization we build a special-purpose processor using such circuitry as digital adders, multipliers, and delay elements (shift registers with a sampling period T as a unit time delay).

In the field of digital signal processing, a *digital filter* is a computational algorithm that converts an input sequence of numbers into an output sequence in such a way that the characteristics of the signal are changed in some prescribed fashion. That is, a digital filter processes a digital signal by passing desirable frequency components of the digital input signal and rejecting undesirable ones. In general terms, a digital controller is a form of digital filter.

Note that there are important differences between the digital signal processing used in communications and that used in control. In digital control the processing of signals must be done in real time. In communications, signal processing need not be done in real time, and therefore delays can be tolerated in the processing to improve accuracy.

This section deals with the block diagram realization of digital filters using delay elements, adders, and multipliers. Here several different structures of block diagram realizations will be discussed. Such block diagram realizations can be used as a basis for a software or hardware design. In fact, once the block diagram realization is completed, the physical realization in hardware or software is straightforward. Note that in a block diagram realization a pulse transfer function of z^{-1} represents a delay of one time unit (see Figure 3–34.) (Note also that in the *s* plane z^{-1} corresponds to a pure delay e^{-Ts}.)

In what follows we shall deal with the digital filters that are used for filtering and control purposes. The general form of the pulse transfer function between the output $Y(z)$ and input $X(z)$ is given by

$$G(z) = \frac{Y(z)}{X(z)} = \frac{b_0 + b_1 z^{-1} + b_2 z^{-2} + \cdots + b_m z^{-m}}{1 + a_1 z^{-1} + a_2 z^{-2} + \cdots + a_n z^{-n}}, \qquad n \geq m \qquad (3\text{–}60)$$

where the a_i's and b_i's are real coefficients (some of them may be zero). The pulse transfer function is in this form for many digital controllers. For example, the pulse transfer function of the PID controller given by Equation (3–56) can be expressed in the form of Equation (3–60), as follows:

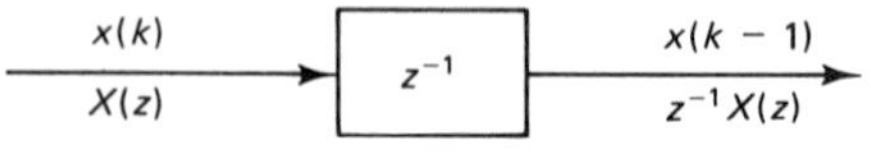

Figure 3–34 Pulse transfer function showing a delay of one time unit.

$$G_D(z) = \frac{(K_P + K_I + K_D) - (K_P + 2K_D)z^{-1} + K_D z^{-2}}{1 - z^{-1}}$$

$$= \frac{b_0 + b_1 z^{-1} + b_2 z^{-2}}{1 + a_1 z^{-1} + a_2 z^{-2}}$$

where

$$a_1 = -1$$

$$a_2 = 0$$

$$b_0 = K_P + K_I + K_D$$

$$b_1 = -(K_P + 2K_D)$$

$$b_2 = K_D$$

We shall now discuss the direct programming and the standard programming of digital filters. In these programmings, coefficients a_i and b_i (which are real quantities) appear as multipliers in the block diagram realization. Those block diagram schemes where the coefficients a_i and b_i appear directly as multipliers are called *direct structures*.

Direct Programming. Consider the digital filter given by Equation (3–60). Notice that the pulse transfer function has n poles and m zeros. Figure 3–35 shows a block diagram realization of the filter. The fact that this block diagram represents Equation (3–60) can be seen easily, since from the diagram we have

$$Y(z) = -a_1 z^{-1} Y(z) - a_2 z^{-2} Y(z) - \cdots - a_n z^{-n} Y(z) + b_0 X(z)$$
$$+ b_1 z^{-1} X(z) + \cdots + b_m z^{-m} X(z)$$

Rearranging this last equation yields Equation (3–60).

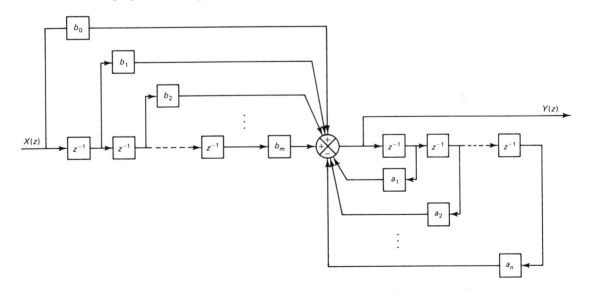

Figure 3–35 Block diagram realization of a filter showing direct programming.

The type of realization here is called *direct programming*. Direct programming means that we realize the numerator and denominator of the pulse transfer function using separate sets of delay elements. The numerator uses a set of m delay elements and the denominator uses a different set of n delay elements. Thus, the total number of delay elements used in direct programming is $m + n$.

The number of delay elements used in direct programming can be reduced. In fact, the number of delay elements can be reduced from $n + m$ to n (where $n \geq m$). The programming method that uses a minimum possible number of delay elements is called *standard programming*.

In practice, we try to use the minimum number of delay elements in realizing a given pulse transfer function. Therefore, the direct programming that requires more than the minimum number of delay elements is more or less of academic value rather than of practical value.

Standard Programming.   As previously stated, the number of delay elements required in direct programming can be reduced. In fact, the number of delay elements used in realizing the pulse transfer function given by Equation (3–60) can be reduced from $n + m$ to n (where $n \geq m$) by rearranging the block diagram, as will be discussed here.

First, rewrite the pulse transfer function $Y(z)/X(z)$ given by Equation (3–60) as follows:

$$\frac{Y(z)}{X(z)} = \frac{Y(z)}{H(z)} \frac{H(z)}{X(z)}$$

$$= (b_0 + b_1 z^{-1} + b_2 z^{-2} + \cdots + b_m z^{-m}) \frac{1}{1 + a_1 z^{-1} + a_2 z^{-2} + \cdots + a_n z^{-n}}$$

where

$$\frac{Y(z)}{H(z)} = b_0 + b_1 z^{-1} + b_2 z^{-2} + \cdots + b_m z^{-m} \tag{3–61}$$

and

$$\frac{H(z)}{X(z)} = \frac{1}{1 + a_1 z^{-1} + a_2 z^{-2} + \cdots + a_n z^{-n}} \tag{3–62}$$

Then, draw block diagrams for the systems given by Equations (3–61) and (3–62), respectively. To draw the block diagrams, we may rewrite Equation (3–61) as

$$Y(z) = b_0 H(z) + b_1 z^{-1} H(z) + \cdots + b_m z^{-m} H(z) \tag{3–63}$$

and Equation (3–62) as

$$H(z) = X(z) - a_1 z^{-1} H(z) - a_2 z^{-2} H(z) - \cdots - a_n z^{-n} H(z) \tag{3–64}$$

Then from Equation (3–63) we obtain Figure 3–36(a). Similarly, we get Figure 3–36(b) from Equation (3–64). The combination of these two block diagrams gives the block diagram for the digital filter $G(z)$, as shown in Figure 3–36(c). The block diagram realization as presented here is based on the standard programming. Notice that we use only n delay elements. The coefficients $a_1, a_2, \ldots, a_n$ appear as feedback elements, and the coefficients $b_0, b_1, \ldots, b_m$ appear as feedforward elements.

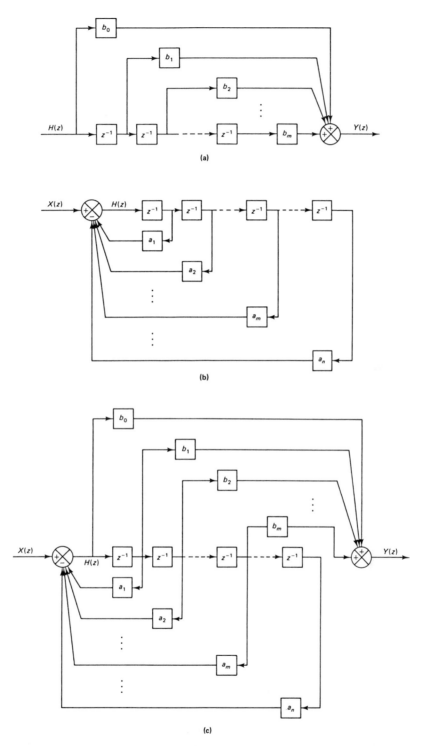

Figure 3–36 (a) Block diagram realization of Equation (3–63); (b) block diagram realization of Equation (3–64); (c) block diagram realization of the digital filter given by Equation (3–60) by standard programming.

125

The block diagrams in Figures 3–35 and 3–36(c) are equivalent, but the latter uses n delay elements, while the former uses $n + m$ delay elements. Obviously, the latter, which uses a smaller number of delay elements, is preferred.

Comments. Note first that the use of a minimal number of delay elements saves memory space in digital controllers. Also, the use of a minimal number of summing points is desirable.

In realizing digital controllers or digital filters, it is important to have a good level of accuracy. Basically, three sources of errors affect the accuracy:

1. The error due to the quantization of the input signal into a finite number of discrete levels. (In Chapter 1 we discussed this type of error, which may be considered an additive source of noise, called *quantization noise*. The quantization noise may be considered white noise; the variance of the noise is $\sigma^2 = Q^2/12$.)
2. The error due to the accumulation of round-off errors in the arithmetic operations in the digital system.
3. The error due to quantization of the coefficients a_i and b_i of the pulse transfer function. This error may become large as the order of the pulse transfer function is increased. That is, in a higher-order digital filter in direct structure, small errors in the coefficients a_i and b_i cause large errors in the locations of the poles and zeros of the filter.

These three errors arise because of the practical limitations of the number of bits that represent various signal samples and coefficients. Note that the third type of error listed may be reduced by mathematically decomposing a higher-order pulse transfer function into a combination of lower-order pulse transfer functions. In this way, the system may be made less sensitive to coefficient inaccuracies.

For decomposing higher-order pulse transfer functions in order to avoid the coefficient sensitivity problem, the following three approaches are commonly used.

1. Series programming
2. Parallel programming
3. Ladder programming

We shall discuss these three programmings next.

Series Programming. The first approach used to avoid the sensitivity problem is to implement the pulse transfer function $G(z)$ as a series connection of first-order and/or second-order pulse transfer functions. If $G(z)$ can be written as a product of pulse transfer functions $G_1(z), G_2(z), \ldots, G_p(z)$, or

$$G(z) = G_1(z)G_2(z) \cdots G_p(z)$$

then the digital filter for $G(z)$ may be given as a series connection of the component digital filters $G_1(z), G_2(z), \ldots, G_p(z)$, as shown in Figure 3–37.

In most cases the $G_i(z)$ $(i = 1, 2, \ldots, p)$ are chosen to be either first- or second-order functions. If the poles and zeros of $G(z)$ are known, $G_1(z), G_2(z), \ldots, G_p(z)$ can be obtained by grouping a pair of conjugate complex poles and a pair of

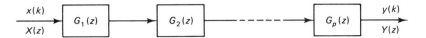

Figure 3-37 Digital filter $G(z)$ decomposed into a series connection of $G_1(z)$, $G_2(z), \ldots, G_p(z)$.

conjugate complex zeros to produce a second-order function or by grouping real poles and real zeros to produce either first- or second-order functions. It is, of course, possible to group two real zeros with a pair of conjugate complex poles, or vice versa. The grouping is, in a sense, arbitrary. It is desirable to group several different ways to see which is best with respect to the number of arithmetic operations required, the range of coefficients, and so forth.

To summarize, $G(z)$ may be decomposed as follows:

$$G(z) = G_1(z)G_2(z) \cdots G_p(z)$$

$$= \prod_{i=1}^{j} \frac{1 + b_i z^{-1}}{1 + a_i z^{-1}} \prod_{i=j+1}^{p} \frac{1 + e_i z^{-1} + f_i z^{-2}}{1 + c_i z^{-1} + d_i z^{-2}}$$

The block diagram for

$$\frac{Y(z)}{X(z)} = \frac{1 + b_i z^{-1}}{1 + a_i z^{-1}} \tag{3-65}$$

and that for

$$\frac{Y(z)}{X(z)} = \frac{1 + e_i z^{-1} + f_i z^{-2}}{1 + c_i z^{-1} + d_i z^{-2}} \tag{3-66}$$

are shown in Figures 3-38(a) and (b), respectively. The block diagram for the digital filter $G(z)$ is a series connection of p component digital filters such as shown in Figures 3-38(a) and (b).

Parallel Programming.   The second approach to avoiding the coefficient sensitivity problem is to expand the pulse transfer function $G(z)$ into partial fractions. If $G(z)$ is expanded as a sum of A, $G_1(z), G_2(z), \ldots, G_q(z)$, or so that

$$G(z) = A + G_1(z) + G_2(z) + \cdots + G_q(z)$$

where A is simply a constant, then the block diagram for the digital filter $G(z)$ can be obtained as a parallel connection of $q + 1$ digital filters, as shown in Figure 3-39.

Because of the presence of the constant term A, the first- and second-order functions can be chosen in simpler forms. That is, $G(z)$ may be expressed as

$$G(z) = A + G_1(z) + G_2(z) + \cdots + G_q(z)$$

$$= A + \sum_{i=1}^{j} G_i(z) + \sum_{i=j+1}^{q} G_i(z)$$

$$= A + \sum_{i=1}^{j} \frac{b_i}{1 + a_i z^{-1}} + \sum_{i=j+1}^{q} \frac{e_i + f_i z^{-1}}{1 + c_i z^{-1} + d_i z^{-2}}$$

The block diagram for

$$\frac{Y(z)}{X(z)} = \frac{b_i}{1 + a_i z^{-1}} \tag{3-67}$$

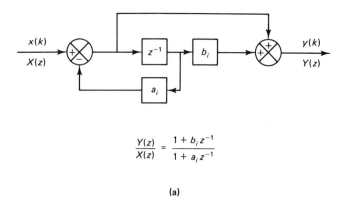

$$\frac{Y(z)}{X(z)} = \frac{1 + b_i z^{-1}}{1 + a_i z^{-1}}$$

(a)

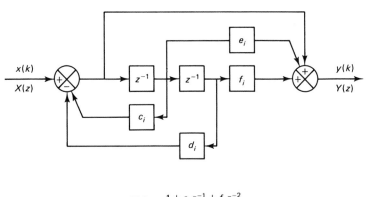

$$\frac{Y(z)}{X(z)} = \frac{1 + e_i z^{-1} + f_i z^{-2}}{1 + c_i z^{-1} + d_i z^{-2}}$$

(b)

Figure 3–38 (a) Block diagram representation of Equation (3–65); (b) block diagram representation of Equation (3–66).

and that for

$$\frac{Y(z)}{X(z)} = \frac{e_i + f_i z^{-1}}{1 + c_i z^{-1} + d_i z^{-2}} \tag{3–68}$$

are shown in Figures 3–40(a) and (b), respectively. The parallel connection of $q + 1$ component digital filters as shown in Figure 3–40 will produce the block diagram for the digital filter $G(z)$.

Ladder Programming. The third approach to avoiding the coefficient sensitivity problem is to implement a ladder structure, that is, to expand the pulse transfer

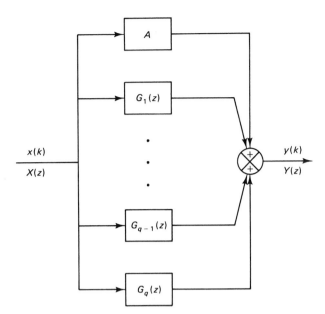

Figure 3-39 Digital filter $G(z)$ decomposed as a parallel connection of A, $G_1(z)$, $G_2(z)$, ..., $G_q(z)$.

function $G(z)$ into the following continued-fraction form and to program according to this equation:

$$G(z) = A_0 + \cfrac{1}{B_1 z + \cfrac{1}{A_1 + \cfrac{1}{B_2 z + \cfrac{1}{\ddots \atop A_{n-1} + \cfrac{1}{B_n z + \cfrac{1}{A_n}}}}}} \qquad (3\text{-}69)$$

The programming method based on this scheme is called *ladder programming*. Let us define

$$G_i^{(B)}(z) = \frac{1}{B_i z + G_i^{(A)}(z)}, \qquad i = 1, 2, \ldots, n-1$$

$$G_i^{(A)}(z) = \frac{1}{A_i + G_{i+1}^{(B)}(z)}, \qquad i = 1, 2, \ldots, n-1$$

$$G_n^{(B)}(z) = \frac{1}{B_n z + \cfrac{1}{A_n}}$$

Then $G(z)$ may be written as

$$G(z) = A_0 + G_1^{(B)}(z)$$

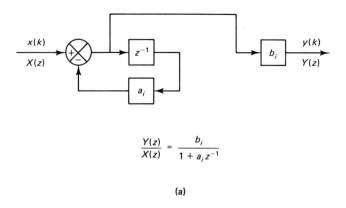

$$\frac{Y(z)}{X(z)} = \frac{b_i}{1 + a_i z^{-1}}$$

(a)

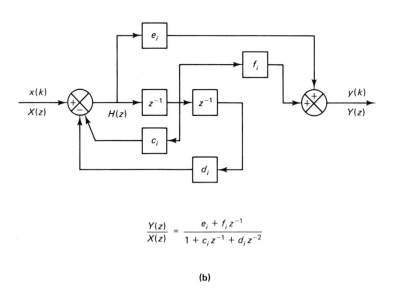

$$\frac{Y(z)}{X(z)} = \frac{e_i + f_i z^{-1}}{1 + c_i z^{-1} + d_i z^{-2}}$$

(b)

Figure 3–40 (a) Block diagram representation of Equation (3–67); (b) block diagram representation of Equation (3–68).

We shall explain this programming method by using a simple example where $n = 2$. That is,

$$G(z) = A_0 + \cfrac{1}{B_1 z + \cfrac{1}{A_1 + \cfrac{1}{B_2 z + \cfrac{1}{A_2}}}}$$

By the use of the functions $G_1^{(A)}(z)$, $G_1^{(B)}(z)$, and $G_2^{(B)}(z)$, the transfer function $G(z)$ may be written as follows:

$$G(z) = A_0 + \cfrac{1}{B_1 z + \cfrac{1}{A_1 + G_2^{(B)}(z)}}$$

$$= A_0 + \cfrac{1}{B_1 z + G_1^{(A)}(z)}$$

$$= A_0 + G_1^{(B)}(z)$$

Notice that $G_i^{(B)}(z)$ may be written as

$$G_i^{(B)}(z) = \frac{Y_i(z)}{X_i(z)} = \frac{1}{B_i z + G_i^{(A)}(z)} \tag{3-70}$$

or

$$X_i(z) - G_i^{(A)}(z)Y_i(z) = B_i z Y_i(z)$$

The block diagram for $G_i^{(B)}(z)$ given by Equation (3–70) is shown in Figure 3–41(a). Similarly, the block diagram for $G_i^{(A)}(z)$, which may be given by

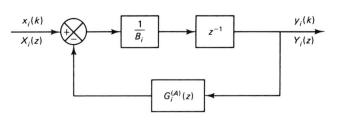

(a)

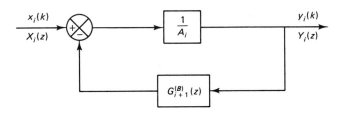

(b)

Figure 3–41   (a) Block diagram for $G_i^{(B)}(z)$ given by Equation (3–70); (b) block diagram for $G_i^{(A)}(z)$ given by Equation (3–71).

$$G_i^{(A)}(z) = \frac{Y_i(z)}{X_i(z)} = \frac{1}{A_i + G_{i+1}^{(B)}(z)} \tag{3-71}$$

or

$$X_i(z) - G_{i+1}^{(B)}(z)Y_i(z) = A_i Y_i(z)$$

may be drawn as shown in Figure 3–41(b). Note that

$$G_n^{(A)}(z) = \frac{1}{A_n}$$

By combining component digital filters as shown in Figure 3–42(a), it is possible to draw the block diagram of the digital filter $G(z)$ as shown in Figure 3–42(b). [Note that Figures 3–42(a) and (b) correspond to the case where $n = 2$.]

Comments. Digital filters based on ladder programming have advantages with respect to coefficient sensitivity and accuracy. Realization of the ladder structure is achieved by expanding $G(z)$ into continued fractions around the origin.

It is noted that the continued-fraction expansion given by Equation (3–69) is not the only way possible. There are a few different ways to construct the ladder

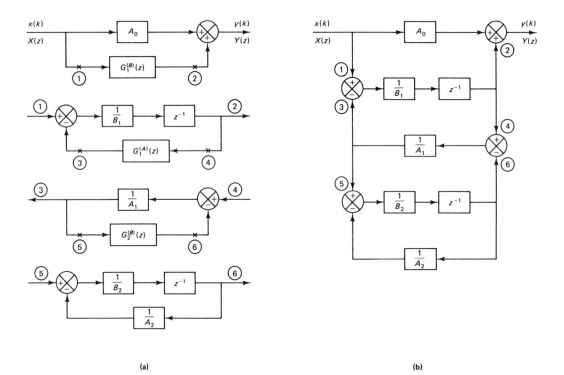

(a)   (b)

Figure 3–42 (a) Component block diagrams for ladder programming of $G(z)$ given by Equation (3–69) when $n = 2$; (b) combination of component block diagrams showing ladder programming of $G(z)$.

structure. For example, a digital filter $G(z)$ may be structured as a continued-fraction expansion form around the origin in terms of z^{-1}, as follows:

$$G(z) = \hat{A}_0 + \cfrac{1}{\hat{B}_1 z^{-1} + \cfrac{1}{\hat{A}_1 + \cfrac{1}{\hat{B}_2 z^{-1} + \cfrac{1}{\ddots \cfrac{}{\hat{A}_{n-1} + \cfrac{1}{\hat{B}_n z^{-1} + \cfrac{1}{\hat{A}_n}}}}}}}$$

Also, instead of $G(z)$, its inverse $1/G(z)$ may be expanded into continued-fraction forms in terms of z or z^{-1} in order to carry out the ladder programming.

Example 3–8

Obtain the block diagrams for the following pulse-transfer-function system (a digital filter) by (1) direct programming, (2) standard programming, and (3) ladder programming:

$$\frac{Y(z)}{X(z)} = G(z) = \frac{2 - 0.6z^{-1}}{1 + 0.5z^{-1}}$$

1. *Direct programming*. Since the given pulse transfer function can be written as

$$Y(z) = -0.5z^{-1}Y(z) + 2X(z) - 0.6z^{-1}X(z)$$

direct programming yields the block diagram shown in Figure 3–43. Notice that we need two delay elements.

2. *Standard programming*. We shall first rewrite the pulse transfer function as follows:

$$\frac{Y(z)}{X(z)} = \frac{Y(z)}{H(z)}\frac{H(z)}{X(z)} = (1 - 0.3z^{-1})\frac{2}{1 + 0.5z^{-1}}$$

where

$$\frac{Y(z)}{H(z)} = 1 - 0.3z^{-1}$$

and

$$\frac{H(z)}{X(z)} = \frac{2}{1 + 0.5z^{-1}}$$

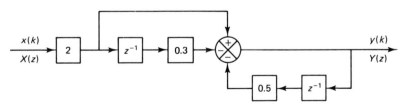

Figure 3–43 Block diagram realization of $Y(z)/X(z) = (2 - 0.6z^{-1})/(1 + 0.5z^{-1})$ (direct programming).

Block diagram realizations of these last two equations are shown in Figure 3–44(a) and (b), respectively. If we combine these two diagrams, we obtain the block diagram for the digital filter $Y(z)/X(z)$, as shown in Figure 3–44(c). Notice that the number of delay elements required has been reduced to 1 by the standard programming.

3. *Ladder programming*. We shall first rewrite the given $Y(z)/X(z)$ in the ladder form as follows:

$$\frac{Y(z)}{X(z)} = G(z) = \frac{2z - 0.6}{z + 0.5} = 2 + \frac{-1.6}{z + 0.5} = 2 + \frac{1}{-0.625z + \dfrac{1}{-3.2}}$$

Thus, $A_0 = 2$ and

$$G_1^{(B)}(z) = \frac{1}{-0.625z + \dfrac{1}{-3.2}} = \frac{1}{-\dfrac{1}{1.6}z - 0.3125}$$

(a)

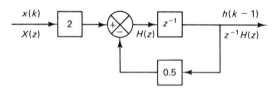

(b)

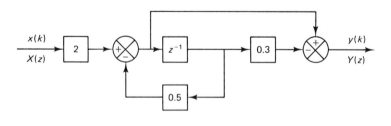

(c)

Figure 3–44 (a) Block diagram realization of $Y(z)/H(z) = 1 - 0.3z^{-1}$; (b) block diagram realization of $H(z)/X(z) = 2/(1 + 0.5z^{-1})$; (c) combination of block diagrams in parts (a) and (b) (standard programming).

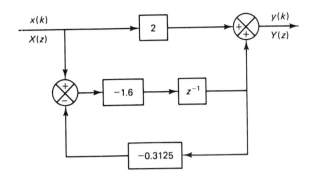

Figure 3–45 Block diagram realization of $Y(z)/X(z) = (2 - 0.6z^{-1})/(1 + 0.5z^{-1})$ (ladder programming).

Hence, we obtain

$$Y(z) = 2X(z) + G_1^{(B)}(z)X(z)$$

Referring to Figure 3–41(a) for the block diagram of $G_1^{(B)}(z)$, we obtain the block diagram of the digital filter $Y(z)/X(z)$ as shown in Figure 3–45. Notice that we need only one delay element.

Infinite-Impulse Response Filter and Finite-Impulse Response Filter.   Digital filters may be classified according to the duration of the impulse response. Consider a digital filter defined by the following pulse transfer function:

$$\frac{Y(z)}{X(z)} = \frac{b_0 + b_1 z^{-1} + \cdots + b_m z^{-m}}{1 + a_1 z^{-1} + a_2 z^{-2} + \cdots + a_n z^{-n}} \qquad (3\text{–}72)$$

where $n \geq m$. In terms of the difference equation,

$$y(k) = -a_1 y(k - 1) - a_2 y(k - 2) - \cdots - a_n y(k - n)$$
$$+ b_0 x(k) + b_1 x(k - 1) + \cdots + b_m x(k - m)$$

The impulse response of the digital filter defined by Equation (3–72), where we assume not all a_i's are zero, has an infinite number of nonzero samples, although their magnitudes may become negligibly small as k increases. This type of digital filter is called an *infinite-impulse response filter*. Such a digital filter is also called a *recursive filter*, because the previous values of the output together with the present and past values of the input are used in processing the signal to obtain the current output $y(k)$. Because of the recursive nature, errors in previous outputs may accumulate. A recursive filter may be recognized by the presence of both a_i and b_i in the block diagram realization.

Next, consider a digital filter where the coefficients a_i are all zero, or where

$$\frac{Y(z)}{X(z)} = b_0 + b_1 z^{-1} + b_2 z^{-2} + \cdots + b_m z^{-m} \qquad (3\text{–}73)$$

In terms of the difference equation,

$$y(k) = b_0 x(k) + b_1 x(k - 1) + \cdots + b_m x(k - m)$$

The impulse response of the digital filter defined by Equation (3–73) is limited to a finite number of samples defined over a finite range of time intervals; that is, the impulse response sequence is finite. This type of digital filter is called a *finite-impulse response filter*. It is also called a *nonrecursive filter* or a *moving-average filter*.

In a nonrecursive realization, the present value of the output depends only on the present and past values of the input. The finite-impulse response filter can be recognized by the absence of the a_i's in the block diagram realization.

Realization of a Finite-Impulse Response Filter. We shall next consider the realization of the finite-impulse response filter.

Let us define the finite-impulse response sequence (weighting sequence) of the digital filter as $g(kT)$. If the input $x(kT)$ is applied to this filter, then the output $y(kT)$ can be given by

$$y(kT) = \sum_{h=0}^{k} g(hT)x(kT - hT)$$

$$= g(0)x(kT) + g(T)x((k-1)T) + \cdots + g(kT)x(0) \qquad (3\text{-}74)$$

The output $y(kT)$ is a convolution summation of the input signal and the impulse response sequence. The right-hand side of Equation (3–74) consists of $k + 1$ terms. Thus, the output $y(kT)$ is given in terms of the past k inputs $x(0), x(T), \ldots, x((k-1)T)$ and the current input $x(kT)$. Notice that as k increases it is physically not possible to process all past values of input to produce the current output. We need to limit the number of the past values of the input to process.

Suppose we decide to employ the N immediate past values of the input $x((k-1)T), x((k-2)T), \ldots, x((k-N)T)$ and the current input $x(kT)$. This is equivalent to approximating the right-hand side of Equation (3–74) by the $N + 1$ most recent input values including the current one, or

$$y(kT) = g(0)x(kT) + g(T)x((k-1)T) + \cdots + g(NT)x((k-N)T) \qquad (3\text{-}75)$$

Since Equation (3–75) is a difference equation, the corresponding digital filter in the z plane can be obtained as follows. By taking the z transform of Equation (3–75), we have

$$Y(z) = g(0)X(z) + g(T)z^{-1}X(z) + \cdots + g(NT)z^{-N}X(z) \qquad (3\text{-}76)$$

Figure 3–46 shows the block diagram realization of this filter.

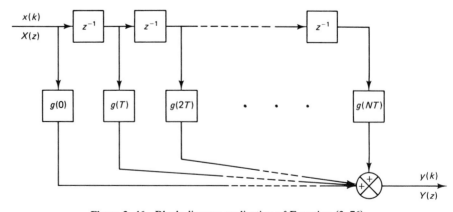

Figure 3–46 Block diagram realization of Equation (3–76).

The characteristics of the finite-impulse response filter can be summarized as follows:

1. The finite-impulse response filter is nonrecursive. Thus, because of the lack of feedback, the accumulation of errors in past outputs can be avoided in the processing of the signal.
2. Implementation of the finite-impulse response filter does not require feedback, so the direct programming and standard programming are identical. Also, implementation may be achieved by high-speed convolution using the fast Fourier transform.
3. The poles of the pulse transfer function of the finite-impulse response filter are at the origin, and therefore it is always stable.
4. If the input signal involves high-frequency components, then the number of delay elements needed in the finite-impulse response filter increases and the amount of time delay becomes large. (This is a disadvantage of the finite-impulse response filter compared with the infinite-impulse response filter.)

Example 3–9

The digital filter discussed in Example 3–8 is a recursive filter. Modify this digital filter and realize it as a nonrecursive filter. Then obtain the response of this nonrecursive filter to a Kronecker delta input.

By dividing the numerator of the recursive filter $G(z)$ by the denominator, we obtain

$$G(z) = \frac{2 - 0.6z^{-1}}{1 + 0.5z^{-1}}$$

$$= 2 - 1.6z^{-1} + 0.8z^{-2} - 0.4z^{-3} + 0.2z^{-4} - 0.1z^{-5} + 0.05z^{-6} - 0.025z^{-7} + \cdots$$

By arbitrarily truncating this series at z^{-7}, we obtain the desired nonrecursive filter, as follows:

$$\frac{Y(z)}{X(z)} = 2 - 1.6z^{-1} + 0.8z^{-2} - 0.4z^{-3} + 0.2z^{-4} - 0.1z^{-5}$$

$$+ 0.05z^{-6} - 0.025z^{-7} \qquad (3\text{–}77)$$

Figure 3–47 shows the block diagram for this nonrecursive digital filter. Notice that we need a large number of delay elements to obtain a good level of accuracy.

Noting that the digital filter is the z transform of the impulse response sequence, the inverse z transform of the digital filter gives the impulse response sequence. By taking the inverse z transform of the nonrecursive filter given by Equation (3–77), we obtain

$$y(kT) = 2x(kT) - 1.6x((k - 1)T) + 0.8x((k - 2)T) - 0.4x((k - 3)T)$$

$$+ 0.2x((k - 4)T) - 0.1x((k - 5)T) + 0.05x((k - 6)T) - 0.025x((k - 7)T)$$

For the Kronecker delta input, where $x(0) = 1$ and $x(kT) = 0$ for $k \neq 0$, this last equation gives

$$y(0) = 2$$

$$y(T) = -1.6$$

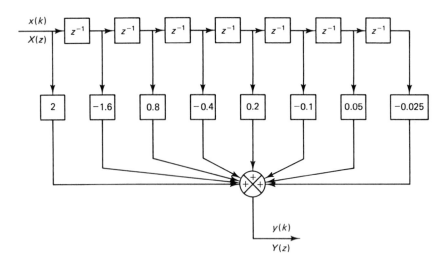

Figure 3–47 Block diagram for the digital filter given by Equation (3–77) (non-recursive form).

$$y(2T) = 0.8$$
$$y(3T) = -0.4$$
$$y(4T) = 0.2$$
$$y(5T) = -0.1$$
$$y(6T) = 0.05$$
$$y(7T) = -0.025$$

The impulse response sequence for this digital filter is shown in Figure 3–48.

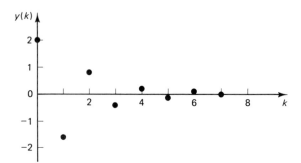

Figure 3–48 Impulse response sequence for the digital filter given by Equation (3–77).

EXAMPLE PROBLEMS AND SOLUTIONS

Problem A–3–1

Consider a zero-order hold preceded by a sampler. Figure 3–49 shows the input $x(t)$ to the sampler and the output $y(t)$ of the zero-order hold. In the zero-order hold the value of the last sample is retained until the next sample is taken.

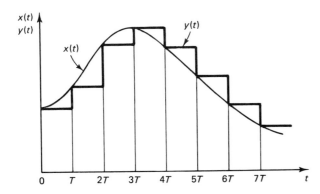

Figure 3–49 Input and output curves for a zero-order hold.

Obtain the expression for $y(t)$. Then find $Y(s)$ and obtain the transfer function of the zero-order hold.

Solution   From Figure 3–49 we obtain

$$y(t) = x(0)[1(t) - 1(t - T)] + x(T)[1(t - T) - 1(t - 2T)]$$
$$+ x(2T)[1(t - 2T) - 1(t - 3T)] + \cdots$$

The Laplace transform of $y(t)$ is

$$Y(s) = x(0)\left(\frac{1}{s} - \frac{e^{-Ts}}{s}\right) + x(T)\left(\frac{e^{-Ts}}{s} - \frac{e^{-2Ts}}{s}\right)$$
$$+ x(2T)\left(\frac{e^{-2Ts}}{s} - \frac{e^{-3Ts}}{s}\right) + \cdots$$
$$= \frac{1 - e^{-Ts}}{s}[x(0) + x(T)e^{-Ts} + x(2T)e^{-2Ts} + \cdots]$$
$$= \frac{1 - e^{-Ts}}{s}X^*(s)$$

where

$$X^*(s) = \sum_{k=0}^{\infty} x(kT)e^{-kTs} = \mathcal{L}\left[\sum_{k=0}^{\infty} x(kT)\delta(t - kT)\right]$$

The transfer function of the zero-order hold is thus

$$G_{h0} = \frac{Y(s)}{X^*(s)} = \frac{1 - e^{-Ts}}{s}$$

Problem A–3–2

Consider a first-order hold preceded by a sampler. The input to the sampler is $x(t)$ and the output of the first-order hold is $y(t)$. In the first-order hold the output $y(t)$ for $kT \leq t < (k + 1)T$ is the straight line that is the extrapolation of the two preceding sampled values, $x((k - 1)T)$ and $x(kT)$, as shown in Figure 3–50. The equation for the output $y(t)$ is

$$y(t) = \frac{t - kT}{T}[x(kT) - x((k - 1)T)] + x(kT), \qquad kT \leq t < (k + 1)T \qquad (3\text{–}78)$$

Obtain the transfer function of the first-order hold, assuming a simple function such as an impulse function at $t = 0$ as the input $x(t)$.

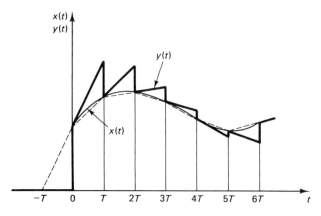

Figure 3–50 Input and output curves for a first-order hold.

Solution. For an impulse input of magnitude $x(0)$ such that $x^*(t) = x(0)\delta(t)$, the output $y(t)$ given by Equation (3–78) becomes as shown in Figure 3–51. The mathematical expression for $y(t)$ is

$$y(t) = x(0)\left(1 + \frac{t}{T}\right)1(t) - \left[2x(0) + 2x(0)\frac{t-T}{T}\right]1(t-T)$$

$$+ \left[x(0) + x(0)\frac{t-2T}{T}\right]1(t-2T)$$

Hence,

$$Y(s) = x(0)\left(\frac{1}{s} + \frac{1}{Ts^2}\right) - 2x(0)\left(\frac{e^{-Ts}}{s} + \frac{e^{-Ts}}{Ts^2}\right) + x(0)\left(\frac{e^{-2Ts}}{s} + \frac{e^{-2Ts}}{Ts^2}\right)$$

$$= x(0)(1 - 2e^{-Ts} + e^{-2Ts})\left(\frac{1}{s} + \frac{1}{Ts^2}\right)$$

$$= x(0)\frac{Ts + 1}{Ts^2}(1 - e^{-Ts})^2$$

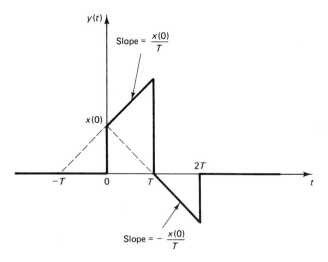

Figure 3–51 Output curve of the first-order hold when the input is a unit-impulse function.

Since

$$X^*(s) = \mathcal{L}[x^*(t)] = \mathcal{L}[x(0)\delta(t)] = x(0)$$

the transfer function of the first-order hold is obtained as follows:

$$G_{h1}(s) = \frac{Y(s)}{X^*(s)} = \frac{Ts + 1}{T}\left(\frac{1 - e^{-Ts}}{s}\right)^2$$

Problem A–3–3

Consider the function

$$X(s) = \frac{1 - e^{-Ts}}{s}$$

Show that $s = 0$ is not a pole of $X(s)$. Show also that

$$Y(s) = \frac{1 - e^{-Ts}}{s^2}$$

has a simple pole at $s = 0$.

Solution   If a transfer function involves a transcendental term e^{-Ts}, then it may be replaced by a series valid in the vicinity of the pole in question.
    For the function

$$X(s) = \frac{1 - e^{-Ts}}{s} \tag{3–79}$$

let us obtain the Laurent series expansion about the pole at the origin. Since, in the vicinity of the origin, e^{-Ts} may be replaced by

$$e^{-Ts} = 1 - Ts + \frac{(Ts)^2}{2!} - \frac{(Ts)^3}{3!} + \cdots \tag{3–80}$$

substitution of Equation (3–80) into Equation (3–79) gives

$$X(s) = \frac{1}{s}\left[Ts - \frac{(Ts)^2}{2!} + \frac{(Ts)^3}{3!} - \cdots\right]$$

$$= T - \frac{T^2 s}{2!} + \frac{T^3 s^2}{3!} - \cdots$$

which is the Laurent series expansion of $X(s)$. From this last equation we see that $s = 0$ is not a pole of $X(s)$.
    Next, consider $Y(s)$. Since

$$Y(s) = \frac{1 - e^{-Ts}}{s^2}$$

it may be expanded into the Laurent series as

$$Y(s) = \frac{T}{s} - \frac{T^2}{2!} + \frac{T^3 s}{3!} - \cdots$$

We see that the pole at the origin ($s = 0$) is of order 1, or is a simple pole.

Problem A–3–4

Show that the Laplace transform of the product of two Laplace transformable functions $f(t)$ and $g(t)$ can be given by

$$\mathcal{L}[f(t)g(t)] = \frac{1}{2\pi j} \int_{c-j\infty}^{c+j\infty} F(p)G(s - p) \, dp \tag{3-81}$$

Solution   The Laplace transform of the product of $f(t)$ and $g(t)$ is given by

$$\mathcal{L}[f(t)g(t)] = \int_0^\infty f(t)g(t)e^{-st} \, dt \tag{3-82}$$

Note that the inversion integral is

$$f(t) = \frac{1}{2\pi j} \int_{c-j\infty}^{c+j\infty} F(s)e^{st} \, ds, \qquad t > 0$$

where c is the abscissa of convergence for $F(s)$. Thus,

$$\mathcal{L}[f(t)g(t)] = \frac{1}{2\pi j} \int_0^\infty \int_{c-j\infty}^{c+j\infty} F(p)e^{pt} \, dp \, g(t)e^{-st} \, dt$$

Because of the uniform convergence of the integrals considered, we may invert the order of integration:

$$\mathcal{L}[f(t)g(t)] = \frac{1}{2\pi j} \int_{c-j\infty}^{c+j\infty} F(p) \, dp \int_0^\infty g(t)e^{-(s-p)t} \, dt$$

Noting that

$$\int_0^\infty g(t)e^{-(s-p)t} \, dt = G(s - p)$$

we obtain

$$\mathcal{L}[f(t)g(t)] = \frac{1}{2\pi j} \int_{c-j\infty}^{c+j\infty} F(p)G(s - p) \, dp \tag{3-83}$$

Problem A–3–5

Show that the Laplace transform of

$$x^*(t) = \sum_{k=0}^{\infty} x(t)\delta(t - kT) = x(t) \sum_{k=0}^{\infty} \delta(t - kT) \tag{3-84}$$

can be given by

$$X^*(s) = \mathcal{L}\left[x(t) \sum_{k=0}^{\infty} \delta(t - kT)\right]$$

$$= \frac{1}{2\pi j} \int_{c-j\infty}^{c+j\infty} X(p) \frac{1}{1 - e^{-T(s-p)}} \, dp \tag{3-85}$$

Solution   Referring to Equation (3–83), rewritten as

$$\mathcal{L}[f(t)g(t)] = \frac{1}{2\pi j} \int_{c-j\infty}^{c+j\infty} F(p)G(s - p) \, dp$$

where

$$f(t) = x(t) \qquad \text{and} \qquad g(t) = \sum_{k=0}^{\infty} \delta(t - kT)$$

and noting that

$$\mathcal{L}[\delta(t - kT)] = e^{-kTs}$$

we have

$$\mathcal{L}\left[\sum_{k=0}^{\infty} \delta(t - kT)\right] = 1 + e^{-Ts} + e^{-2Ts} + e^{-3Ts} + \cdots = \frac{1}{1 - e^{-Ts}}$$

Since

$$G(s) = \mathcal{L}\left[\sum_{k=0}^{\infty} \delta(t - kT)\right] = \frac{1}{1 - e^{-Ts}}$$

we have

$$G(s - p) = \frac{1}{1 - e^{-T(s-p)}}$$

Notice that the poles of $1/[1 - e^{-T(s-p)}]$ may be obtained by solving the equation

$$1 - e^{-T(s-p)} = 0$$

or

$$-T(s - p) = \pm j2\pi k, \qquad k = 0, 1, 2, \ldots$$

so that the poles are

$$p = s \pm j\frac{2\pi}{T}k = s \pm j\omega_s k, \qquad k = 0, 1, 2, \ldots$$

where $\omega_s = 2\pi/T$. Thus, there are infinitely many simple poles along a line parallel to the $j\omega$ axis.

The Laplace transform of $x^*(t)$ can now be written as

$$X^*(s) = \mathcal{L}\left[x(t)\sum_{k=0}^{\infty} \delta(t - kT)\right]$$

$$= \frac{1}{2\pi j}\int_{c-j\infty}^{c+j\infty} X(p)\frac{1}{1 - e^{-T(s - p)}}dp \qquad (3\text{–}86)$$

where the integration is along the line from $c - j\infty$ to $c + j\infty$, and this line is parallel to the imaginary axis in the p plane and separates the poles of $X(p)$ from those of $1/[1 - e^{-T(s-p)}]$. Equation (3–86) is the convolution integral. It is a well-known fact that such an integral can be evaluated in terms of residues by forming a closed contour consisting of the line from $c - j\infty$ to $c + j\infty$ and a semicircle of infinite radius in the left or right half-plane, provided that the integral along the added semicircle is a constant (zero or a nonzero constant). There are two ways to evaluate this integral (one using an infinite semicircle in the left half-plane and the other an infinite semicircle in the right half-plane); we shall consider these two cases separately in Problems A–3–6 and A–3–7.

Problem A–3–6

Referring to Equation (3–86), rewritten as

$$X^*(s) = \frac{1}{2\pi j}\int_{c-j\infty}^{c+j\infty} X(p)\frac{1}{1 - e^{-T(s-p)}}dp$$

show that, by performing the integration in the left half-plane, $X^*(s)$ may be given by

$$X^*(s) = \sum\left[\text{residue of } \frac{X(p)}{1 - e^{-T(s-p)}} \text{ at pole of } X(p)\right] \qquad (3\text{–}87)$$

By substituting z for e^{Ts} in Equation (3–87), we have

$$X(z) = \sum \left[\text{residue of } \frac{X(p)z}{z - e^{Tp}} \text{ at pole of } X(p) \right]$$

By changing the complex variable notation from p to s, we obtain

$$X(z) = \sum \left[\text{residue of } \frac{X(s)z}{z - e^{Ts}} \text{ at pole of } X(s) \right]$$

$$= \sum_{i=1}^{h} \frac{1}{(n_i - 1)!} \lim_{s \to s_i} \frac{d^{n_i-1}}{ds^{n_i-1}} \left[(s - s_i)^{n_i} \frac{X(s)z}{z - e^{Ts}} \right]$$

$$+ \sum_{j=h+1}^{m} \lim_{s \to s_j} \left[(s - s_j) \frac{X(s)z}{z - e^{Ts}} \right]$$

where we assumed that $X(z)$ has h different multiple poles and $m - h$ simple poles ($m \geq h$). We assume that the poles of $X(s)$ lie in the left half-plane and that $X(s)$ can be expressed as a ratio of polynomials in s, or

$$X(s) = \frac{q(s)}{p(s)}$$

where $q(s)$ and $p(s)$ are polynomials in s. We also assume that $p(s)$ is of a higher degree in s than $q(s)$, which means that

$$\lim_{s \to \infty} X(s) = 0$$

Solution   We shall evaluate the convolution integral given by Equation (3–86) using a closed contour in the left half of the p plane as shown in Figure 3–52. Using this closed contour, Equation (3–86) may be written as

$$X^*(s) = \frac{1}{2\pi j} \int_{c-j\infty}^{c+j\infty} X(p) \frac{1}{1 - e^{-T(s-p)}} dp$$

$$= \frac{1}{2\pi j} \oint \frac{X(p)}{1 - e^{-T(s-p)}} dp - \frac{1}{2\pi j} \int_{\Gamma_L} \frac{X(p)}{1 - e^{-T(s-p)}} dp \qquad (3–88)$$

where the closed contour consists of the line from $c - j\infty$ to $c + j\infty$ and Γ_L, which in turn consists of a semicircle of infinite radius and the horizontal lines at $j\infty$ and $-j\infty$, which connect the line from $c - j\infty$ to $c + j\infty$ with the semicircle in the left half of the p plane. We choose a value of c such that all the poles of $X(p)$ lie to the left of the line from $c - j\infty$ to $c + j\infty$ and all the poles of $1/[1 - e^{-T(s-p)}]$ lie to the right of this line. The closed contour encloses all poles of $X(p)$, while the poles of $1/[1 - e^{-T(s-p)}]$ are outside the closed contour.

Because we have assumed that the denominator of $X(s)$ is of a higher degree in s than the numerator, the integral along Γ_L (the infinite semicircle in the left half-plane plus the horizontal lines at $j\infty$ and $-j\infty$, which connect the line from $c - j\infty$ to $c + j\infty$ with the semicircle) vanishes. Hence,

$$X^*(s) = \frac{1}{2\pi j} \oint \frac{X(p)}{1 - e^{-T(s-p)}} dp$$

This integral is equal to the sum of the residues of $X(p)$ in the closed contour. (Refer to Appendix B for the residue theorem.) Therefore,

$$X^*(s) = \sum \left[\text{residue of } \frac{X(p)}{1 - e^{-T(s-p)}} \text{ at pole of } X(p) \right] \qquad (3–89)$$

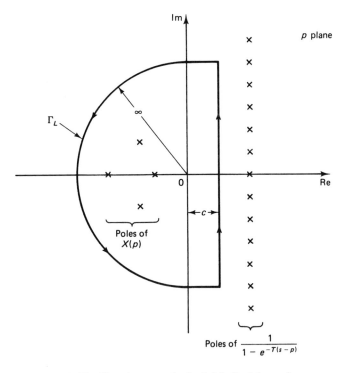

Figure 3–52 Closed contour in the left half of the p plane.

By substituting z for e^{Ts} in Equation (3–89), we have

$$X(z) = \sum \left[\text{residue of } \frac{X(p)z}{z - e^{Tp}} \text{ at pole of } X(p) \right]$$

By changing the complex variable notation from p to s, we obtain

$$X(z) = \sum \left[\text{residue of } \frac{X(s)z}{z - e^{Ts}} \text{ at pole of } X(s) \right] \qquad (3\text{–}90)$$

Let us assume that $X(s)$ has poles $s_1, s_2, \ldots, s_m$. If a pole at $s = s_j$ is a simple pole, then the corresponding residue K_j is

$$K_j = \lim_{s \to s_j} \left[(s - s_j) \frac{X(s)z}{z - e^{Ts}} \right] \qquad (3\text{–}91)$$

If a pole at $s = s_i$ is a multiple pole of order n_i, then the residue K_i is

$$K_i = \frac{1}{(n_i - 1)!} \lim_{s \to s_i} \frac{d^{n_i - 1}}{ds^{n_i - 1}} \left[(s - s_i)^{n_i} \frac{X(s)z}{z - e^{Ts}} \right] \qquad (3\text{–}92)$$

Therefore, if $X(s)$ has a multiple pole s_1 of order n_1, a multiple pole s_2 of order $n_2, \ldots,$ a multiple pole s_h of order n_h, and simple poles $s_{h+1}, s_{h+2}, \ldots, s_m$, then $X(z)$ given by Equation (3–90) can be written as

$$X(z) = \sum \left[\text{residue of } \frac{X(s)z}{z - e^{Ts}} \text{ at pole of } X(s) \right]$$

$$= \sum_{i=1}^{h} \frac{1}{(n_i - 1)!} \lim_{s \to s_i} \frac{d^{n_i-1}}{ds^{n_i-1}} \left[(s - s_i)^{n_i} \frac{X(s)z}{z - e^{Ts}} \right]$$

$$+ \sum_{j=h+1}^{m} \lim_{s \to s_j} \left[(s - s_j) \frac{X(s)z}{z - e^{Ts}} \right] \qquad (3\text{-}93)$$

where n_i is the order of the multiple pole at $s = s_i$.

Problem A–3–7

Referring to Equation (3–86), rewritten as

$$X^*(s) = \frac{1}{2\pi j} \int_{c-j\infty}^{c+j\infty} \frac{X(p)}{1 - e^{-T(s-p)}} dp$$

show that by performing this integration in the right half p plane, $X^*(s)$ may be given by

$$X^*(s) = \frac{1}{T} \sum_{k=-\infty}^{\infty} X(s + j\omega_s k) \qquad (3\text{-}94)$$

provided that the denominator of $X(s)$ is two or more degrees higher in s than the numerator. Show that if the denominator of $X(s)$ is only one degree higher in s than the numerator then

$$X^*(s) = \frac{1}{T} \sum_{k=-\infty}^{\infty} X(s + j\omega_s k) + \frac{1}{2}x(0+) \qquad (3\text{-}95)$$

Solution   Let us evaluate the convolution integral given by Equation (3–86) in the right half of the p plane. Let us choose the closed contour shown in Figure 3–53, which consists of the line from $c - j\infty$ to $c + j\infty$ and Γ_R, the portion of the semicircle of infinite radius in the right half of the p plane that lies to the right of this line. The closed contour encloses all poles of $1/[1 - e^{-T(s-p)}]$, but it does not enclose any poles of $X(p)$. Now $X^*(s)$ can be written as

$$X^*(s) = \frac{1}{2\pi j} \int_{c-j\infty}^{c+j\infty} \frac{X(p)}{1 - e^{-T(s-p)}} dp$$

$$= \frac{1}{2\pi j} \oint \frac{X(p)}{1 - e^{-T(s-p)}} dp - \frac{1}{2\pi j} \int_{\Gamma_R} \frac{X(p)}{1 - e^{-T(s-p)}} dp \qquad (3\text{-}96)$$

Let us investigate the integral along Γ_R, the portion of the infinite semicircle to the right of the line from $c - j\infty$ to $c + j\infty$. Since infinitely many poles of $1/[1 - e^{-T(s-p)}]$ lie on a line parallel to the $j\omega$ axis, the evaluation of the integral along Γ_R is not as simple as in the previous case, where the closed contour enclosed a finite number of poles of $X(p)$ in the left half of the p plane.

In almost all physical control systems, as s becomes large, $X(s)$ tends to zero at least as fast as $1/s$. Hence, in what follows, we consider two cases, one where the denominator of $X(s)$ is two or more degrees higher in s than the numerator and another where the denominator of $X(s)$ is only one degree higher in s than the numerator.

Case 1: $X(s)$ Possesses at Least Two More Poles Than Zeros.   Referring to the theory of complex variables, it can be shown that the integral along Γ_R is zero if the degree of the denominator $p(s)$ of $X(s)$ is greater by at least 2 than the degree of the numerator $q(s)$; that is, if $X(s)$ possesses at least two more poles than zeros, which implies that

$$\lim_{s \to \infty} sX(s) = x(0+) = 0$$

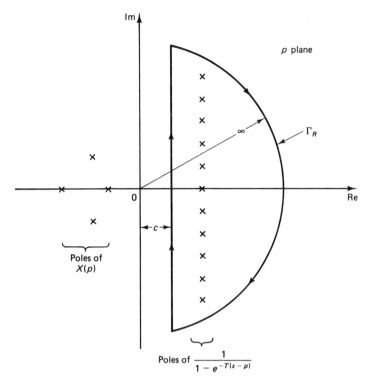

Figure 3–53   Closed contour in the right half of the p plane.

then the integral along Γ_R is zero. Thus, in the present case

$$\frac{1}{2\pi j}\int_{\Gamma_R}\frac{X(p)}{1-e^{-T(s-p)}}\,dp = 0$$

Therefore, Equation (3–96) simplifies to

$$X^*(s) = \frac{1}{2\pi j}\oint\frac{X(p)}{1-e^{-T(s-p)}}\,dp \qquad\qquad (3\text{–}97)$$

The integral along the closed contour given by Equation (3–97) can be obtained by evaluating the residues at the infinite number of poles at $p = s \pm j\omega_s k$. Thus,

$$X^*(s) = -\sum_{k=-\infty}^{\infty}\left[\lim_{p\to s+j\omega_s k}\left\{[p - (s + j\omega_s k)]\frac{X(p)}{1-e^{-T(s-p)}}\right\}\right]$$

The minus sign in front of the right-hand side of this last equation comes from the fact that the contour integration along the path Γ_R is taken in the clockwise direction. Using L'Hôpital's rule, we obtain

$$X^*(s) = -\sum_{k=-\infty}^{\infty}\left.\frac{X(p)}{\dfrac{d}{dp}[1-e^{-T(s-p)}]}\right|_{p=s+j\omega_s k}$$

Noting that

$$\frac{d}{dp}[1 - e^{-T(s-p)}]_{p=s+j\omega_s k} = -Te^{-T(s-p)}\Big|_{p=s+j\omega_s k} = -Te^{jT\omega_s k} = -Te^{j2\pi k} = -T$$

we have

$$X^*(s) = -\sum_{k=-\infty}^{\infty} \frac{X(p)}{-T}\Big|_{p=s+j\omega_s k}$$

or

$$X^*(s) = \frac{1}{T}\sum_{k=-\infty}^{\infty} X(s + j\omega_s k) \tag{3-98}$$

Thus,

$$X(z) = \frac{1}{T}\sum_{k=-\infty}^{\infty} X(s + j\omega_s k)\Big|_{s=(1/T)\ln z} \tag{3-99}$$

Note that this expression of the *z* transform is useful in proving the sampling theorem (see Section 3–4). However, it is very tedious to obtain *z* transform expressions of commonly encountered functions by this method.

Case 2: X(s) Has a Denominator One Degree Higher in s Than the Numerator. For this case $\lim_{s\to\infty} sX(s) = x(0+) \neq 0 < \infty$ and the integral along Γ_R is not zero. [The nonzero value is associated with the initial value $x(0+)$ of $x(t)$.] It can be shown that the contribution of the integral along Γ_R in Equation (3–96) is $-\frac{1}{2}x(0+)$. That is,

$$\frac{1}{2\pi j}\int_{\Gamma_R} \frac{x(p)}{1 - e^{-T(s-p)}} dp = -\frac{1}{2}x(0+)$$

Then the integral term on the right-hand side of Equation (3–96) becomes

$$X^*(s) = \frac{1}{T}\sum_{k=-\infty}^{\infty} X(s + j\omega_s k) + \frac{1}{2}x(0+) \tag{3-100}$$

Problem A–3–8

Consider the function

$$x(t) = \begin{cases} e^{-at}, & t \geq 0 \\ 0, & t < 0 \end{cases}$$

Obtain $X(z)$ by using the convolution integral in the right half-plane.

Solution   The Laplace transform of $x(t)$ is

$$X(s) = \frac{1}{s + a}$$

Clearly, $\lim_{s\to\infty} sX(s) = x(0+) = 1$, or the function has a jump discontinuity at $t = 0$. Hence we must use Equation (3–95). Referring to this equation, we have

$$X^*(s) = \frac{1}{T}\sum_{k=-\infty}^{\infty} X(s + j\omega_s k) + \frac{1}{2}x(0+)$$

$$= \frac{1}{T}\left\{\sum_{k=1}^{\infty} [X(s + j\omega_s k) + X(s - j\omega_s k)] + X(s)\right\} + \frac{1}{2}$$

$$= \frac{1}{T} \left[\sum_{k=1}^{\infty} \left(\frac{1}{s + j\omega_s k + a} + \frac{1}{s - j\omega_s k + a} \right) + \frac{1}{s + a} \right] + \frac{1}{2}$$

$$= \frac{1}{T} \left[\sum_{k=1}^{\infty} \frac{2(s + a)}{(s + a)^2 + (\omega_s k)^2} + \frac{1}{s + a} \right] + \frac{1}{2}$$

$$= \frac{1}{2\pi} \left[\sum_{k=1}^{\infty} \frac{2(s + a)/\omega_s}{\left(\dfrac{s + a}{\omega_s} \right)^2 + k^2} + \frac{\omega_s}{s + a} \right] + \frac{1}{2} \qquad (3\text{--}101)$$

Referring to a formula available in mathematical tables,

$$\sum_{k=1}^{\infty} \frac{2x}{x^2 + k^2} + \frac{1}{x} = \pi \frac{1 + e^{-2\pi x}}{1 - e^{-2\pi x}}$$

and noting that

$$2\pi \frac{s + a}{\omega_s} = T(s + a)$$

we can rewrite Equation (3–101) in the form

$$X^*(s) = \frac{\pi}{2\pi} \frac{1 + e^{-T(s+a)}}{1 - e^{-T(s-a)}} + \frac{1}{2}$$

$$= \frac{1}{2} \frac{1 + e^{-T(s+a)} + 1 - e^{-T(s+a)}}{1 - e^{-T(s+a)}}$$

$$= \frac{1}{2} \frac{2}{1 - e^{-T(s+a)}}$$

$$= \frac{1}{1 - e^{-aT} e^{-Ts}}$$

or

$$X(z) = \frac{1}{1 - e^{-aT} z^{-1}}$$

Thus, we have obtained $X(z)$ by using the convolution integral in the right half-plane. [This process of obtaining the z transform is very tedious because an infinite series of $X(s + j\omega_s k)$ is involved. The example here is presented for demonstration purposes only. One should use other methods for obtaining the z transform.]

Problem A–3–9

Obtain the z transform of

$$X(s) = \frac{s}{(s + 1)^2(s + 2)}$$

by using (1) the partial-fraction-expansion method and (2) the residue method.

Solution

1. *Partial-fraction-expansion method.* Since $X(s)$ can be expanded into the form

$$X(s) = \frac{2}{s + 1} - \frac{1}{(s + 1)^2} - \frac{2}{s + 2}$$

we have

$$X(z) = 2\left(\frac{1}{1 - e^{-T}z^{-1}}\right) - \frac{Te^{-T}z^{-1}}{(1 - e^{-T}z^{-1})^2} - 2\left(\frac{1}{1 - e^{-2T}z^{-1}}\right)$$

$$= \frac{2 - 2e^{-T}z^{-1} - Te^{-T}z^{-1}}{(1 - e^{-T}z^{-1})^2} - \frac{2}{1 - e^{-2T}z^{-1}}$$

2. Residue method. Referring to Equation (3–93) and noting that $X(s)$ has a double pole at $s = -1$ and a simple pole at $s = -2$, we have

$$X(z) = \frac{1}{(2 - 1)!} \lim_{s \to -1} \frac{d}{ds}\left[(s + 1)^2 \frac{s}{(s + 1)^2(s + 2)} \frac{z}{z - e^{Ts}}\right]$$

$$+ \lim_{s \to -2}\left[(s + 2)\frac{s}{(s + 1)^2(s + 2)} \frac{z}{z - e^{Ts}}\right]$$

$$= \frac{2z^2 - 2ze^{-T} - Tze^{-T}}{(z - e^{-T})^2} - \frac{2z}{z - e^{-2T}}$$

$$= \frac{2 - 2e^{-T}z^{-1} - Te^{-T}z^{-1}}{(1 - e^{-T}z^{-1})^2} - \frac{2}{1 - e^{-2T}z^{-1}}$$

Problem A–3–10

Consider a continuous-time signal $x(t)$ with frequency spectrum limited to between $-\omega_1$ and ω_1. That is,

$$X(j\omega) = 0, \qquad \text{for } \omega < -\omega_1 \text{ and } \omega_1 < \omega$$

Prove that if this signal is sampled with frequency $\omega_s > 2\omega_1$ then the Fourier transform of $x(t)$ is uniquely determined by $x(kT), k = \ldots, -2, -1, 0, 1, 2, \ldots$, and the original continuous-time signal $x(t)$ can be given by a sum of an infinite series of weighted sampled values $x(kT)$ as follows:

$$x(t) = \sum_{k=-\infty}^{\infty} x(kT)\frac{\sin[\omega_s(t - kT)/2]}{\omega_s(t - kT)/2}$$

(This is Shannon's sampling theorem.)

Solution   The Fourier transform of $x(t)$ is given by

$$X(j\omega) = \int_{-\infty}^{\infty} e^{-j\omega t}x(t)\, dt$$

and the inverse Fourier transform is given by

$$x(t) = \frac{1}{2\pi}\int_{-\infty}^{\infty} e^{j\omega t} X(j\omega)\, d\omega$$

Define the sampled version of $x(t)$ as $x^*(t)$. Then $x^*(t)$ can be given by

$$x^*(t) = \cdots + x(-T)\delta(t + T) + x(0)\delta(t) + x(T)\delta(t - T) + \cdots$$

$$= \sum_{k=-\infty}^{\infty} x(kT)\delta(t - kT)$$

The Fourier transform of $x^*(t)$ is

$$X^*(j\omega) = \int_{-\infty}^{\infty} e^{-j\omega t}x^*(t)\, dt = \int_{-\infty}^{\infty} e^{-j\omega t}\left[\sum_{k=-\infty}^{\infty} x(kT)\delta(t - kT)\right] dt$$

$$= \sum_{k=-\infty}^{\infty} x(kT)e^{-j\omega kT}$$

Thus, $X^*(j\omega)$ is uniquely determined by $x(kT), k = \dots, -2, -1, 0, 1, 2, \dots$.

Referring to Equation (3–27), the Fourier transform of $x^*(t)$ can be given by

$$X^*(j\omega) = \frac{1}{T} \sum_{k=-\infty}^{\infty} X(j\omega + j\omega_s k)$$

Since the frequency spectrum of the original continuous-time signal $x(t)$ is limited to between $-\omega_1$ and ω_1, we have

$$X(j\omega) = 0, \qquad \text{for } \omega < -\omega_1 \text{ and } \omega_1 < \omega$$

Since the sampling frequency ω_s is greater than $2\omega_1$, we have

$$X(j\omega) = 0, \qquad \text{for } \omega < -\tfrac{1}{2}\omega_s \text{ and } \tfrac{1}{2}\omega_s < \omega$$

Hence,

$$X^*(j\omega) = \frac{1}{T}[\cdots + X(j\omega + j\omega_s) + X(j\omega) + X(j\omega - j\omega_s) + \cdots]$$

$$= \frac{1}{T}X(j\omega)$$

Thus, we obtain

$$X(j\omega) = \begin{cases} TX^*(j\omega), & -\tfrac{1}{2}\omega_s \leq \omega \leq \tfrac{1}{2}\omega_s \\ 0, & \omega < -\tfrac{1}{2}\omega_s, \tfrac{1}{2}\omega_s < \omega \end{cases}$$

The inverse Fourier transform of $X(j\omega)$ gives

$$x(t) = \frac{1}{2\pi}\int_{-\infty}^{\infty} e^{j\omega t} X(j\omega)\, d\omega$$

$$= \frac{T}{2\pi}\int_{-\omega_s/2}^{\omega_s/2} e^{j\omega t} X^*(j\omega)\, d\omega$$

$$= \frac{1}{\omega_s}\int_{-\omega_s/2}^{\omega_s/2} e^{j\omega t}\left[\sum_{k=-\infty}^{\infty} x(kT)e^{-j\omega kT}\right] d\omega$$

$$= \frac{1}{\omega_s}\sum_{k=-\infty}^{\infty} x(kT)\int_{-\omega_s/2}^{\omega_s/2} e^{j\omega(t-kT)}\, d\omega$$

$$= \frac{1}{\omega_s}\sum_{k=-\infty}^{\infty} x(kT)\frac{e^{j\omega(t-kT)}}{j(t-kT)}\bigg|_{-\omega_s/2}^{\omega_s/2}$$

$$= \sum_{k=-\infty}^{\infty} x(kT)\frac{\sin[\omega_s(t-kT)/2]}{\omega_s(t-kT)/2}$$

Hence, we have shown that the original continuous-time signal $x(t)$ can be reconstructed from the sampled data $x(kT)$. [Note that unless $X(j\omega) = 0$ for $\omega < -\omega_1$ and $\omega_1 < \omega$ the continuous-time signal $x(t)$ cannot be determined by sampled data $x(kT)$, $k = \dots, -2, -1, 0, 1, 2, \dots$.]

Problem A–3–11

Draw the magnitude and phase curves of the first-order hold. Then compare the magnitude and phase characteristics of the first-order hold with those of the zero-order hold.

Solution   The transfer function of the first-order hold is

$$G_{h1}(s) = \frac{Ts + 1}{T}\left(\frac{1 - e^{-Ts}}{s}\right)^2$$

By substituting $j\omega$ for s in $G_{h1}(s)$, we obtain

$$G_{h1}(j\omega) = \frac{Tj\omega + 1}{T}\left(\frac{1 - e^{-Tj\omega}}{j\omega}\right)^2$$

$$= \frac{Tj\omega + 1}{T}\left[e^{-j(1/2)T\omega}\frac{e^{j(1/2)T\omega} - e^{-j(1/2)T\omega}}{j\omega}\right]^2$$

$$= \frac{Tj\omega + 1}{T}e^{-jT\omega}\left[\frac{2j\sin(T\omega/2)}{j\omega}\right]^2$$

$$= \frac{Tj\omega + 1}{T}e^{-jT\omega}\frac{4\sin^2(T\omega/2)}{\omega^2}$$

Hence,

$$|G_{h1}(j\omega)| = T\sqrt{1 + T^2\omega^2}\left[\frac{\sin(T\omega/2)}{T\omega/2}\right]^2$$

$$\angle G_{h1}(j\omega) = \angle Tj\omega + 1 + \angle e^{-jT\omega}$$

$$= \tan^{-1}T\omega - T\omega$$

$$= \tan^{-1}\frac{2\pi\omega}{\omega_s} - \frac{2\pi\omega}{\omega_s}$$

where we have used the relationship $T = 2\pi/\omega_s$.

At a few selected values of ω, we have

$$|G_{h1}(j0)| = T \qquad\qquad \angle G_{h1}(j0) = 0°$$

$$\left|G_{h1}\left(j\frac{\pi}{T}\right)\right| = 1.336T \qquad \angle G_{h1}\left(j\frac{\pi}{T}\right) = -107.7°$$

$$\left|G_{h1}\left(j\frac{2\pi}{T}\right)\right| = 0 \qquad\qquad \angle G_{h1}\left(j\frac{2\pi}{T}\right) = -279.0°$$

Figure 3–54 shows plots of the magnitude and phase characteristics of the first-order hold and those of the zero-order hold. From Figure 3–54 it is seen that both the zero-order hold and the first-order hold are not quite satisfactory low-pass filters. They allow significant transmission above the Nyquist frequency, $\omega_N = \pi/T$. It is important, therefore, that the signal be low-pass-filtered before the sampling operation so that the frequency components above the Nyquist frequency are negligible.

Problem A–3–12

Consider the zero-order hold shown in Figure 3–55. From the diagram we have

$$Y(s) = G(s)X^*(s) = \frac{1 - e^{-Ts}}{s}X^*(s) \qquad\qquad (3\text{--}102)$$

Show that

$$Y^*(s) = X^*(s)$$

Solution   By taking the starred Laplace transform of Equation (3–102), we have

$$Y^*(s) = \left(\frac{1 - e^{-Ts}}{s}\right)^* X^*(s)$$

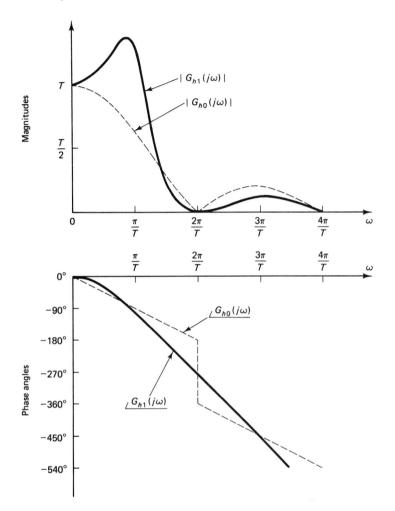

Figure 3–54 Magnitude and phase characteristics of the first-order hold and those of the zero-order hold.

In terms of the z transform notation, we have

$$Y(z) = \mathcal{Z}\left[\frac{1 - e^{-Ts}}{s}\right] X(z)$$

where

$$\mathcal{Z}\left[\frac{1 - e^{-Ts}}{s}\right] = (1 - z^{-1})\mathcal{Z}\left[\frac{1}{s}\right] = (1 - z^{-1})\frac{1}{1 - z^{-1}} = 1$$

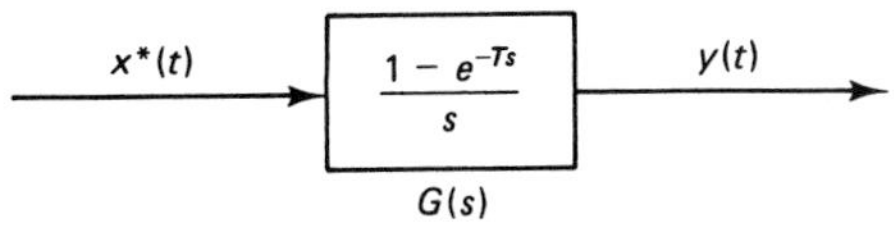

$$x^*(t) \longrightarrow \boxed{\dfrac{1 - e^{-Ts}}{s}} \longrightarrow y(t)$$

$$G(s)$$

Figure 3–55 Zero-order hold.

Hence,

$$Y(z) = X(z)$$

In terms of the starred Laplace transform notation, this last equation can be written as

$$Y^*(s) = X^*(s)$$

Problem A–3–13

Obtain the weighting sequence of the system defined by

$$G_n(z) = \frac{1}{(1 + az^{-1})^n}$$

for $n = 1, 2$, and 3, respectively.

Solution   For $n = 1$, we have

$$G_1(z) = \frac{1}{1 + az^{-1}} = 1 - az^{-1} + a^2 z^{-2} - a^3 z^{-3} + \cdots$$

Hence, the weighting sequence $g_1(k)$ is found to be

$$g_1(k) = (-a)^k$$

For $n = 2$, we obtain

$$G_2(z) = \frac{1}{(1 + az^{-1})^2} = \frac{1 - az^{-1} + a^2 z^{-2} - a^3 z^{-3} + \cdots}{1 + az^{-1}}$$

$$= 1 - 2az^{-1} + 3a^2 z^{-2} - 4a^3 z^{-3} + \cdots$$

Hence, the weighting sequence $g_2(k)$ is

$$g_2(k) = (k + 1)(-a)^k$$

For $n = 3$, we get

$$G_3(z) = \frac{1}{(1 + az^{-1})^3} = \frac{1 - 2az^{-1} + 3a^2 z^{-2} - 4a^3 z^{-3} + \cdots}{1 + az^{-1}}$$

$$= 1 - 3az^{-1} + 6a^2 z^{-2} - 10a^3 z^{-3} + \cdots$$

Hence, the weighting sequence $g_3(k)$ is

$$g_3(k) = \frac{(k + 2)(k + 1)}{2}(-a)^k$$

Problem A–3–14

Obtain the discrete-time output $C(z)$ of the closed-loop control system shown in Figure 3–56. Also, obtain the continuous-time output $C(s)$.

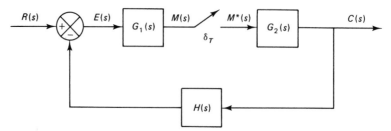

Figure 3–56   Discrete-time control system.

Solution   From the diagram we have

$$C(s) = G_2(s)M^*(s)$$
$$M(s) = G_1(s)E(s)$$
$$E(s) = R(s) - H(s)C(s)$$

Hence,

$$M(s) = G_1(s)[R(s) - H(s)C(s)]$$
$$= G_1(s)R(s) - G_1(s)H(s)G_2(s)M^*(s)$$

Taking the starred Laplace transform of this last equation, we obtain

$$M^*(s) = [G_1 R(s)]^* - [G_1 G_2 H(s)]^* M^*(s)$$

or

$$M^*(s) = \frac{[G_1 R(s)]^*}{1 + [G_1 G_2 H(s)]^*}$$

Since $C(s) = G_2(s)M^*(s)$, we have

$$C^*(s) = G_2^*(s)M^*(s) = \frac{G_2^*(s)[G_1 R(s)]^*}{1 + [G_1 G_2 H(s)]^*}$$

In terms of the z transform notation,

$$C(z) = \frac{G_2(z)G_1 R(z)}{1 + G_1 G_2 H(z)}$$

This last equation gives the discrete-time output $C(z)$.

The continuous-time output $C(s)$ can be obtained from the following equation:

$$C(s) = G_2(s)M^*(s) = G_2(s)\frac{[G_1 R(s)]^*}{1 + [G_1 G_2 H(s)]^*}$$

Notice that $[G_1 R(s)]^*/\{1 + [G_1 G_2 H(s)]^*\}$ is a series of impulses. The continuous-time output $C(s)$ is the response of $G_2(s)$ to the sequence of such impulses. [See Problem A–3–18 for details of determining the continuous-time output $c(t)$, the inverse Laplace transform of $C(s)$.]

Problem A–3–15

Consider the system shown in Figure 3–57. Obtain the closed-loop pulse transfer function $C(z)/R(z)$. Also, obtain the expression for $C(s)$.

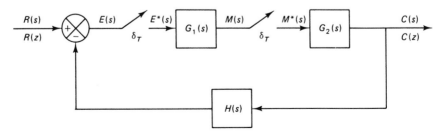

Figure 3–57   Discrete-time control system.

Solution   From the diagram we have

$$C(s) = G_2(s)M^*(s)$$

$$M(s) = G_1(s)E^*(s)$$

$$E(s) = R(s) - H(s)C(s) = R(s) - H(s)G_2(s)M^*(s)$$

Taking the starred Laplace transforms of both sides of the last three equations gives

$$C^*(s) = G_2^*(s)M^*(s)$$

$$M^*(s) = G_1^*(s)E^*(s)$$

$$E^*(s) = R^*(s) - HG_2^*(s)M^*(s)$$

Solving for $C^*(s)$ gives

$$C^*(s) = G_2^*(s)G_1^*(s)[R^*(s) - HG_2^*(s)M^*(s)]$$

or

$$C^*(s) = G_1^*(s)G_2^*(s)R^*(s) - G_1^*(s)G_2^*(s)HG_2^*(s)M^*(s)$$

$$= G_1^*(s)G_2^*(s)R^*(s) - G_1^*(s)HG_2^*(s)C^*(s)$$

Thus,

$$C^*(s)[1 + G_1^*(s)HG_2^*(s)] = G_1^*(s)G_2^*(s)R^*(s)$$

or

$$\frac{C^*(s)}{R^*(s)} = \frac{G_1^*(s)G_2^*(s)}{1 + G_1^*(s)HG_2^*(s)}$$

In terms of the z transform notation, we have

$$\frac{C(z)}{R(z)} = \frac{G_1(z)G_2(z)}{1 + G_1(z)HG_2(z)}$$

The continuous-time output $C(s)$ can be obtained from the following equation:

$$C(s) = G_2(s)M^*(s) = G_2(s)\frac{G_1^*(s)R^*(s)}{1 + G_1^*(s)HG_2^*(s)}$$

Problem A–3–16

Consider the analog PID controller and the digital PID controller. The equation for the analog PID controller is

$$m(t) = K\left[e(t) + \frac{1}{T_i}\int_0^t e(t)\,dt + T_d\frac{de(t)}{dt}\right]$$

where $e(t)$ is the input to the controller and $m(t)$ is the output of the controller. The transfer function of the analog PID controller is

$$G(s) = \frac{M(s)}{E(s)} = K\left(1 + \frac{1}{T_i s} + T_d s\right)$$

The pulse transfer function of the digital PID controller in the positional form is as given by Equation (3–56):

$$G_D(z) = \frac{M(z)}{E(z)} = K_P + \frac{K_I}{1 - z^{-1}} + K_D(1 - z^{-1})$$

where $K_P = K - \frac{1}{2}K_I$.

Compare the polar plots (frequency-response characteristics) of the analog PID controller with those of the digital PID controller.

Solution   For the analog PID controller, the frequency-response characteristics can be obtained by substituting $j\omega$ for s in $G(s)$. Thus,

$$G(j\omega) = K\left(1 + \frac{1}{T_i j\omega} + T_d j\omega\right)$$

$$= K\left(1 - j\frac{1}{T_i \omega} + T_d j\omega\right) \tag{3-103}$$

For the digital PID controller, the frequency-response characteristics can be obtained by substituting $z = e^{j\omega T}$ into $G_D(z)$:

$$G_D(e^{j\omega T}) = K_P + \frac{K_I}{1 - e^{-j\omega T}} + K_D(1 - e^{-j\omega T})$$

$$= K_P + \frac{K_I}{1 - \cos\omega T + j\sin\omega T} + K_D(1 - \cos\omega T + j\sin\omega T)$$

$$= K_P + \frac{K_I}{2}\left(1 - j\frac{\sin\omega T}{1 - \cos\omega T}\right) + K_D(1 - \cos\omega T + j\sin\omega T) \tag{3-104}$$

We shall first compare separately the P action, the I action, and the D action of the analog controller with their counterparts in the digital controller. Notice that in the proportional action (P action) the digital controller has a gain $K_I/2$ less than the corresponding gain in the analog controller, since $K_P = K - \frac{1}{2}K_I$. See Figure 3–58(a).

For the integral action (I action) the real parts of the polar plots of the analog controller and digital controller differ by $K_I/2$, as shown in Figure 3–58(b).

When the proportional action and integral action are combined, then the real parts of the polar plots for the analog PI action and the digital PI action become the same, as shown in Figure 3–58(c).

The polar plots of the derivative action (D action) for the analog controller and the digital controller differ very much, as shown in Figure 3–58(d). Hence, there are considerable differences in the analog D action and the digital D action.

The qualitative polar plot of the analog PID controller can be obtained from Equation (3–103) by varying ω from 0 to ∞, as shown in Figure 3–59(a). Similarly, the qualitative polar plot of the digital PID controller can be obtained from Equation (3–104) by varying ω from 0 to π/T, as shown in Figure 3–59(b).

Note that, although the polar plots of the analog PI controller and the digital PI controller are similar, there are significant differences between the polar plots of the analog PID controller and the digital PID controller.

Problem A–3–17

In Section 3–5 we derived the pulse transfer function for the PID controller in *positional form*. Referring to Figure 3–28, the pulse transfer function for the digital PID controller was derived as

$$G_D(z) = \frac{M(z)}{E(z)} = K_P + \frac{K_I}{1 - z^{-1}} + K_D(1 - z^{-1})$$

Using $\nabla m(kT) = m(kT) - m((k-1)T)$ derive the *velocity-form* PID control equation.

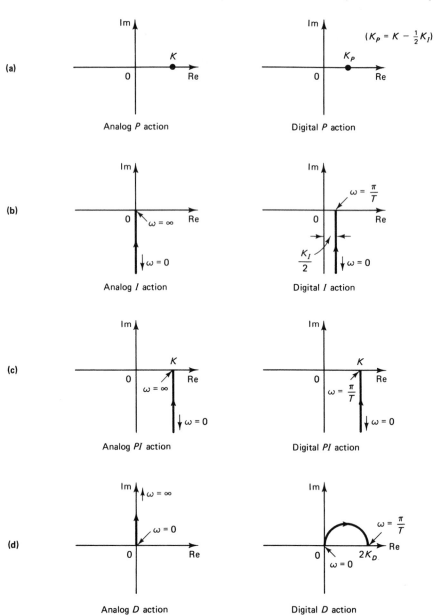

Figure 3–58 Polar plots of analog and digital controllers with (a) proportional action, (b) integral action, (c) proportional plus integral action, and (d) derivative action.

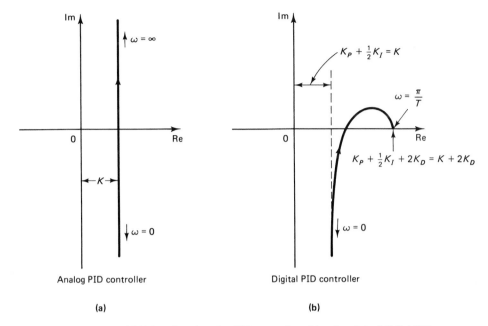

Figure 3–59   (a) Polar plot of analog PID controller; (b) polar plot of digital PID controller.

Solution   Note that

$$\nabla m(kT) = m(kT) - m((k-1)T)$$

$$= K\left\{ e(kT) - e((k-1)T) + \frac{T}{2T_i}[e(kT) + e((k-1)T)] \right.$$

$$\left. + \frac{T_d}{T}[e(kT) - 2e((k-1)T) + e((k-2)T)] \right\}$$

$$= K_P[e(kT) - e((k-1)T)] + K_I e(kT)$$

$$+ K_D[e(kT) - 2e((k-1)T) + e((k-2)T)] \qquad (3\text{-}105)$$

where we have used the relationships $K_P = K - \frac{1}{2}K_I$, $K_I = KT/T_i$, and $K_D = KT_d/T$. (For these relationships, refer to the derivations of the positional form of the digital PID control equation.) Equation (3–105) takes into consideration the variation of the positional form in one sampling period.

Suppose the actuating error $e(kT)$ is the difference between the input $r(kT)$ and the output $c(kT)$, or

$$e(kT) = r(kT) - c(kT)$$

By substituting this last equation into Equation (3–105), we obtain

$$\nabla m(kT) = K_P[r(kT) - r((k-1)T) - c(kT) + c((k-1)T)]$$

$$+ K_I[r(kT) - c(kT)] + K_D[r(kT) - 2r((k-1)T)$$

$$+ r((k-2)T) - c(kT) + 2c((k-1)T) - c((k-2)T)] \qquad (3\text{-}106)$$

The velocity-form PID control scheme given by Equation (3–106) may be modified into a somewhat different form to cope with sudden large changes in the set point. Since the proportional and derivative control actions produce a large change in the controller output when the signal entering the controller makes a sudden large change, to suppress such a large change in the controller output, the digital proportional and derivative terms may be modified as discussed next.

If changes in the set point [input $r(kT)$] are a series of step changes, then immediately after a step change takes place, the input $r(kT)$ stays constant for a while until the next step change takes place. Hence, in Equation (3–106) we assume that

$$r(kT) = r((k - 1)T) = r((k - 2)T)$$

(Note that this is true if the input stays constant. But we assume that this holds true even if a step change takes place.) Then Equation (3–106) may be modified to

$$\nabla m(kT) = -K_P[c(kT) - c((k - 1)T)] + K_I[r(kT) - c(kT)]$$
$$- K_D[c(kT) - 2c((k - 1)T) + c((k - 2)T)] \qquad (3\text{–}107)$$

The z transform of Equation (3–107) gives

$$(1 - z^{-1})M(z) = -K_P(1 - z^{-1})C(z) + K_I[R(z) - C(z)]$$
$$- K_D(1 - 2z^{-1} + z^{-2})C(z)$$

Simplifying, we obtain

$$M(z) = -K_P C(z) + K_I \frac{R(z) - C(z)}{1 - z^{-1}} - K_D(1 - z^{-1})C(z) \qquad (3\text{–}108)$$

Equation (3–108) gives the velocity-form PID control scheme. The block diagram realization of the velocity-form digital PID control scheme was shown in Figure 3–30.

Problem A–3–18

Consider the system shown in Figure 3–60(a). Obtain the continuous-time output $c(t)$ so that the output between any two consecutive sampling instants can be determined. Find the expression for the continuous-time output $c(t)$. The sampling period T is 1 sec.

Solution   For the system shown in Figure 3–60(a), we have

$$C(s) = G(s)E^*(s)$$
$$E(s) = R(s) - C(s)$$

Hence,

$$E^*(s) = R^*(s) - C^*(s) = R^*(s) - G^*(s)E^*(s)$$

or

$$E^*(s) = \frac{R^*(s)}{1 + G^*(s)}$$

Thus,

$$C(s) = G(s)\frac{R^*(s)}{1 + G^*(s)}$$

The continuous-time output $c(t)$ can therefore be obtained as the inverse Laplace transform of $C(s)$:

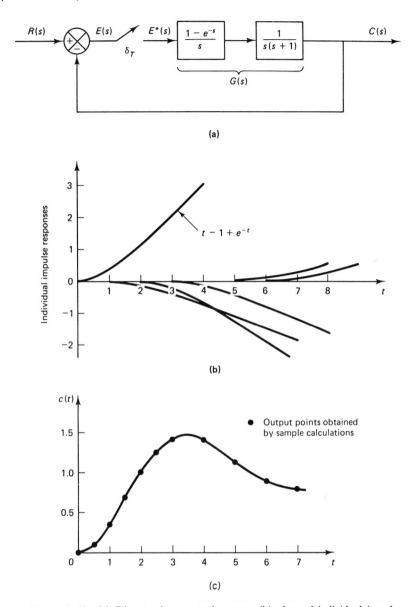

Figure 3–60 (a) Discrete-time control system; (b) plots of individual impulse responses; (c) plot of continuous-time output $c(t)$ versus t.

$$c(t) = \mathcal{L}^{-1}[C(s)] = \mathcal{L}^{-1}\left[G(s)\frac{R^*(s)}{1 + G^*(s)}\right]$$

For the present system,

$$G(s) = \frac{1 - e^{-s}}{s}\frac{1}{s(s + 1)}$$

Hence,

$$c(t) = \mathscr{L}^{-1}\left[\frac{1 - e^{-s}}{s} \frac{1}{s(s + 1)} \frac{R^*(s)}{1 + G^*(s)}\right]$$

Let us define

$$X^*(s) = (1 - e^{-s})\frac{R^*(s)}{1 + G^*(s)}$$

Then the z transform expression for this last equation is

$$X(z) = (1 - z^{-1})\frac{R(z)}{1 + G(z)}$$

Referring to Equation (3–58) for the z transform of $G(s)$, we obtain

$$X(z) = (1 - z^{-1})\frac{\dfrac{1}{1 - z^{-1}}}{1 + \dfrac{0.3679z^{-1} + 0.2642z^{-2}}{(1 - 0.3679z^{-1})(1 - z^{-1})}}$$

$$= \frac{1 - 1.3679z^{-1} + 0.3679z^{-2}}{1 - z^{-1} + 0.6321z^{-2}}$$

Hence, noting that the sampling period T is 1 sec or $T = 1$, we have

$$X^*(s) = \frac{1 - 1.3679e^{-s} + 0.3679e^{-2s}}{1 - e^{-s} + 0.6321e^{-2s}}$$

Therefore,

$$c(t) = \mathscr{L}^{-1}\left[\frac{1}{s^2(s + 1)} \frac{1 - 1.3679e^{-s} + 0.3679e^{-2s}}{1 - e^{-s} + 0.6321e^{-2s}}\right]$$

$$= \mathscr{L}^{-1}\left[\frac{1}{s^2(s + 1)}(1 - 0.3679e^{-s} - 0.6321e^{-2s} - 0.3996e^{-3s}\right.$$

$$\left. + 0e^{-4s} + 0.2526e^{-5s} + 0.2526e^{-6s} + \cdots)\right]$$

Since

$$\frac{1}{s^2(s + 1)} = \frac{1}{s^2} - \frac{1}{s} + \frac{1}{s + 1}$$

the inverse Laplace transform of this last equation is

$$\mathscr{L}^{-1}\left[\frac{1}{s^2(s + 1)}\right] = t - 1 + e^{-t}$$

Hence, we obtain

$$c(t) = (t - 1 + e^{-t}) - 0.3679[(t - 1) - 1 + e^{-(t-1)}]1(t - 1)$$

$$- 0.6321[(t - 2) - 1 + e^{-(t-2)}]1(t - 2)$$

$$- 0.3996[(t - 3) - 1 + e^{-(t-3)}]1(t - 3)$$

$$+ 0.0000[(t - 4) - 1 + e^{-(t-4)}]1(t - 4)$$

$$+ 0.2526[(t - 5) - 1 + e^{-(t-5)}]1(t - 5)$$

$$+ 0.2526[(t - 6) - 1 + e^{-(t-6)}]1(t - 6)$$

$$+ \cdots \tag{3–109}$$

Figure 3–60(b) shows plots of individual impulse responses given by Equation (3–109). [Observe that $c(t)$ consists of the sum of impulse responses that occur at $t = 0, t = 1$, $t = 2, \ldots$ with weighting factors $1, -0.3679, -0.6321, \ldots$.]

From Equation (3–109) we see that for time intervals $0 \le t < 1, 1 \le t < 2$, $2 \le t < 3, \ldots$ the output $c(t)$ is the sum of impulse responses as follows:

$$c(t) = \begin{cases} t - 1 + e^{-t}, & 0 \le t < 1 \\ (t - 1 + e^{-t}) - 0.3679[(t - 1) - 1 + e^{-(t-1)}]1(t - 1), & 1 \le t < 2 \\ (t - 1 + e^{-t}) - 0.3679[(t - 1) - 1 + e^{-(t-1)}]1(t - 1) \\ \quad -0.6321[(t - 2) - 1 + e^{-(t-2)}]1(t - 2), & 2 \le t < 3 \\ \vdots \end{cases}$$

Sample calculations for several values of t follow:

$c(0) = 0 - 1 + 1 = 0$

$c(0.5) = 0.5 - 1 + 0.6065 = 0.1065$

$c(1.0) = 0.3679 - 0.3679 \times 0 = 0.3679$

$c(1.5) = 0.7231 - 0.3679 \times 0.1065 = 0.6839$

$c(2.0) = 1.1353 - 0.3679 \times 0.3679 = 1.0000$

$c(2.5) = 1.5821 - 0.3679 \times 0.7231 - 0.6321 \times 0.1065 = 1.2487$

$c(3.0) = 2.0498 - 0.3679 \times 1.1353 - 0.6321 \times 0.3679 = 1.3996$

$c(4.0) = 3.0183 - 0.3679 \times 2.0498 - 0.6321 \times 1.1353 - 0.3996 \times 0.3679 = 1.3996$

$c(5.0) = 4.0067 - 0.3679 \times 3.0183 - 0.6321 \times 2.0498 - 0.3996 \times 1.1353$

$\qquad + 0 \times 0.3679 = 1.1469$

$c(6.0) = 5.0025 - 0.3679 \times 4.0067 - 0.6321 \times 3.0183 - 0.3996 \times 2.0498$

$\qquad + 0 \times 1.1353 + 0.2526 \times 0.3679 = 0.8944$

$\vdots$

The continuous-time output $c(t)$ thus obtained is plotted in Figure 3–60(c).

Problem A–3–19

Consider the digital filter defined by

$$G(z) = \frac{Y(z)}{X(z)} = \frac{4(z - 1)(z^2 + 1.2z + 1)}{(z + 0.1)(z^2 - 0.3z + 0.8)}$$

Draw a series realization diagram and a parallel realization diagram. (Use one first-order section and one second-order section.)

Solution   We shall first consider the series realization scheme. To limit the coefficients to real quantities, we group the second-order term in the numerator (which has complex zeros) and the second-order term in the denominator (which has complex poles). Hence, we group $G(z)$ as follows:

$$G(z) = 4 \frac{z - 1}{z + 0.1} \frac{z^2 + 1.2z + 1}{z^2 - 0.3z + 0.8}$$

$$= 4 \frac{1 - z^{-1}}{1 + 0.1z^{-1}} \frac{1 + 1.2z^{-1} + z^{-2}}{1 - 0.3z^{-1} + 0.8z^{-2}}$$

Figure 3–61(a) shows a series realization diagram.

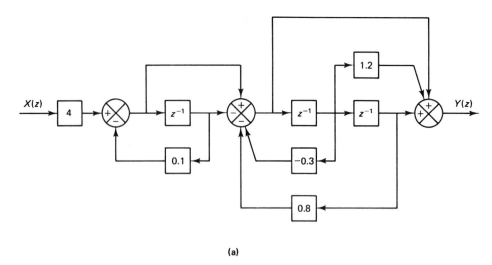

(a)

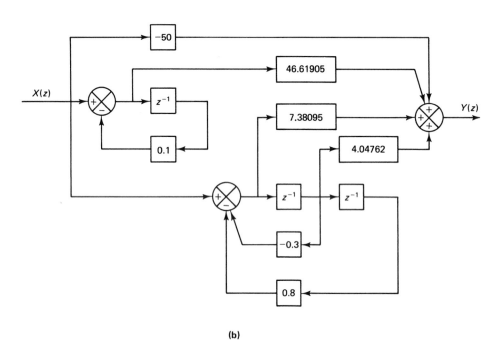

(b)

Figure 3–61   Block diagram realizations of the digital filter considered in Problem A–3–19. (a) Series realization; (b) parallel realization.

Next, we shall consider the parallel realization scheme. Expansion of $G(z)/z$ into partial fractions gives

$$\frac{G(z)}{z} = \frac{4(z-1)(z^2 + 1.2z + 1)}{z(z+0.1)(z^2 - 0.3z + 0.8)}$$

$$= -\frac{50}{z} + \frac{\frac{979}{21}}{z + 0.1} + \frac{\frac{155}{21}z + \frac{85}{21}}{z^2 - 0.3z + 0.8}$$

Then $G(z)$ can be written as follows:

$$G(z) = -50 + \frac{46.61905}{1 + 0.1z^{-1}} + \frac{7.38095 + 4.04762z^{-1}}{1 - 0.3z^{-1} + 0.8z^{-2}}$$

Figure 3–61(b) shows a parallel realization diagram.

Problem A–3–20

A slowly changing continuous-time signal $x(t)$ is sampled every T sec. Assume that changes in signal $x(t)$ are very slow compared to the sampling frequency. Show that in the z plane $(1 - z^{-1})/T$ corresponds to "differentiation," just as s corresponds to "differentiation" in the s plane.

Solution   For a slowly changing signal $x(t)$, the derivative of $x(t)$ can be approximated by

$$v(t) = \frac{dx(t)}{dt} = \frac{1}{T}[x(kT) - x((k-1)T)]$$

The z transform of this equation gives

$$V(z) = \frac{1}{T}[X(z) - z^{-1}X(z)] = \frac{1}{T}(1 - z^{-1})X(z)$$

from which we obtain the block diagram shown in Figure 3–62(a). This diagram corresponds to the s plane differentiation shown in Figure 3–62(b). Note that the block diagram shown in Figure 3–62(a) can be modified to that shown in Figure 3–63. (For approximation of "integration" in the z plane, see Problems B–3–25 through B–3–27.)

(a)   (b)

Figure 3–62   (a) Block diagram for "differentiation" in the z plane; (b) block diagram for "differentiation" in the s plane.

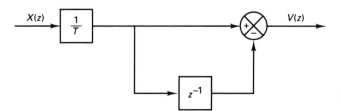

Figure 3–63   Approximate "differentiator" in the z plane.

PROBLEMS

Problem B–3–1

Show that the circuit shown in Figure 3–64 acts as a zero-order hold.

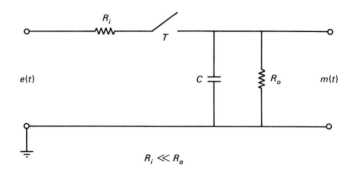

Figure 3–64   Circuit approximating a zero-order hold.

Problem B–3–2

Consider the circuit shown in Figure 3–65. Derive a difference equation describing the system dynamics when the input voltage applied is piecewise constant, or

$$e(t) = e(kT), \qquad kT \le t < (k + 1)T$$

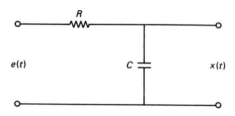

Figure 3–65   *RC* circuit.

(Derive first a differential equation and then discretize it to obtain a difference equation.)

Problem B–3–3

Consider the impulse sampler and first-order hold shown in Figure 3–66. Derive the transfer function of the first-order hold, assuming a unit-ramp function as the input $x(t)$ to the sampler.

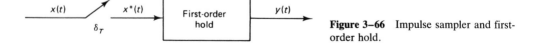

Figure 3–66   Impulse sampler and first-order hold.

Problem B–3–4

Consider a transfer function system

$$X(s) = \frac{s + 3}{(s + 1)(s + 2)}$$

Obtain the pulse transfer function by two different methods.

Problem B–3–5

Obtain the z transform of

$$X(s) = \frac{K}{(s + a)(s + b)}$$

Use the residue method and the method based on the impulse response function.

Problem B–3–6

Obtain the z transform of

$$X(s) = \frac{1 - e^{-Ts}}{s} \frac{1}{(s + a)^2}$$

Problem B–3–7

Consider the difference equation system

$$y(k + 1) + 0.5y(k) = x(k)$$

where $y(0) = 0$. Obtain the response $y(k)$ when the input $x(k)$ is a unit-step sequence. Also, obtain the MATLAB solution.

Problem B–3–8

Consider the difference equation system

$$y(k + 2) + y(k) = x(k)$$

where $y(k) = 0$ for $k < 0$. Obtain the response $y(k)$ when the input $x(k)$ is a unit-step sequence. Also, obtain the MATLAB solution.

Problem B–3–9

Obtain the weighting sequence $g(k)$ of the system described by the difference equation

$$y(k) - ay(k - 1) = x(k), \qquad -1 < a < 1$$

If two systems described by this last equation are connected in series, what is the weighting sequence of the resulting system?

Problem B–3–10

Consider the system described by

$$y(k) - y(k - 1) + 0.24y(k - 2) = x(k) + x(k - 1)$$

where $x(k)$ is the input and $y(k)$ is the output of the system.

Determine the weighting sequence of the system. Assuming that $y(k) = 0$ for $k < 0$, determine the response $y(k)$ when the input $x(k)$ is a unit-step sequence. Also, obtain the MATLAB solution.

Problem B–3–11

Consider the system

$$G(z) = \frac{1 - 0.5z^{-1}}{(1 - 0.3z^{-1})(1 + 0.7z^{-1})}$$

Obtain the response of this system to a unit-step sequence input. Also, obtain the MATLAB solution.

Problem B–3–12

Obtain the response $y(kT)$ of the following system:

$$\frac{Y(s)}{X^*(s)} = \frac{1}{(s + 1)(s + 2)}$$

where $x(t)$ is the unit-step function and $x^*(t)$ is its impulse-sampled version. Assume that the sampling period T is 0.1 sec.

Problem B–3–13

Consider the system defined by

$$\frac{Y(z)}{U(z)} = H(z) = \frac{0.5z^3 + 0.4127z^2 + 0.1747z - 0.0874}{z^3}$$

Using the convolution equation

$$y(k) = \sum_{j=0}^{k} h(k - j)u(j)$$

obtain the response $y(k)$ to a unit-step sequence input $u(k)$.

Problem B–3–14

Assume that a sampled signal $X^*(s)$ is applied to a system $G(s)$. Assume also that the output of $G(s)$ is $Y(s)$ and $y(0+) = 0$.

$$Y(s) = G(s)X^*(s)$$

Using the relationship

$$Y^*(s) = \frac{1}{T} \sum_{k=-\infty}^{\infty} Y(s + j\omega_s k)$$

show that

$$Y^*(s) = G^*(s)X^*(s)$$

Problem B–3–15

Obtain the closed-loop pulse transfer function of the system shown in Figure 3–67.

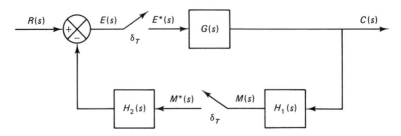

Figure 3–67   Discrete-time control system.

Problem B–3–16

Obtain the closed-loop pulse transfer function of the system shown in Figure 3–68.

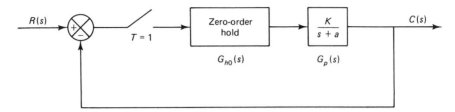

Figure 3–68   Discrete-time control system.

Problem B–3–17

Consider the discrete-time control system shown in Figure 3–69. Obtain the discrete-time output $C(z)$ and the continuous-time output $C(s)$ in terms of the input and the transfer functions of the blocks.

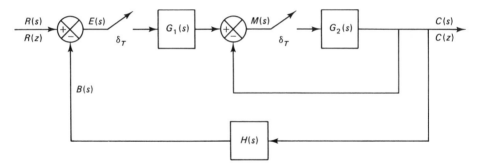

Figure 3–69   Discrete-time control system.

Problem B–3–18

Consider the discrete-time control system shown in Figure 3–70. Obtain the output sequence $c(kT)$ of the system when it is subjected to a unit-step input. Assume that the sampling period T is 1 sec. Also, obtain the continuous-time output $c(t)$.

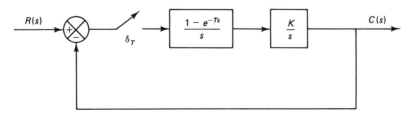

Figure 3–70   Discrete-time control system.

Problem B–3–19

Obtain in a closed form the response sequence $c(kT)$ of the system shown in Figure 3–71 when it is subjected to a Kronecker delta input $r(k)$. Assume that the sampling period T is 1 sec.

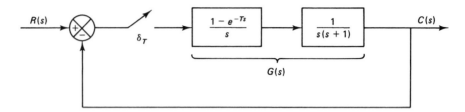

Figure 3–71   Discrete-time control system.

Problem B–3–20

Consider the system shown in Figure 3–72. Assuming that the sampling period T is 0.2 sec and the gain constant K is unity, determine the response $c(kT)$ for $k = 0, 1, 2, 3,$ and 4 when the input $r(t)$ is a unit-step function. Also, determine the final value $c(\infty)$.

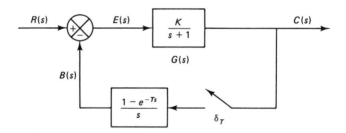

Figure 3–72   Discrete-time control system.

Problem B–3–21

Obtain the closed-loop pulse transfer function $C(z)/R(z)$ of the digital control system shown in Figure 3–30. Assume that the pulse transfer function of the plant is $G(z)$. (Note that the system shown in Figure 3–30 is a velocity-form PID control of the plant.)

Problem B–3–22

Assume that a digital filter is given by the following difference equation:

$$y(k) + a_1 y(k - 1) + a_2 y(k - 2) = b_1 x(k) + b_2 x(k - 1)$$

Draw block diagrams for the filter using (1) direct programming, (2) standard programming, and (3) ladder programming.

Problem B–3–23

Consider the digital filter defined by

$$G(z) = \frac{2 + 2.2z^{-1} + 0.2z^{-2}}{1 + 0.4z^{-1} - 0.12z^{-2}}$$

Realize this digital filter in the series scheme, the parallel scheme, and the ladder scheme.

Problem B–3–24

Referring to the approximate differentiator shown in Figure 3–63, draw a graph of the output $v(k)$ versus k when the input $x(k)$ is a unit-step sequence.

Problem B–3–25

Consider the system shown in Figure 3–73. Show that the pulse transfer function $Y(z)/X(z)$ is given by

$$\frac{Y(z)}{X(z)} = T\!\left(\frac{1}{1 - z^{-1}}\right)$$

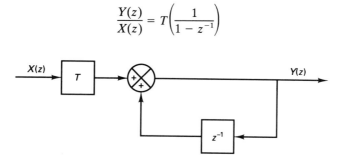

Figure 3–73   Digital integrator without delay.

Assuming that $y(kT) = 0$ for $k < 0$, show that

$$y(kT) = T[x(0) + x(T) + \cdots + x(kT)]$$

Thus, the output $y(kT)$ approximates the area made by the input. Hence, the system acts as an integrator. Because $y(0) = Tx(0)$, the output appears as soon as $x(0)$ enters the system. This integrator is commonly called a digital integrator without delay.

Draw a graph of the output $y(kT)$ when the input $x(kT)$ is a unit-step sequence.

Problem B–3–26

Consider the system shown in Figure 3–74. Show that the pulse transfer function $Y(z)/X(z)$ is given by

$$\frac{Y(z)}{X(z)} = T\!\left(\frac{z^{-1}}{1 - z^{-1}}\right)$$

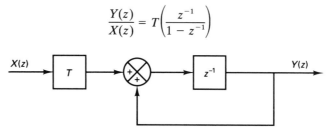

Figure 3–74   Digital integrator with delay.

Assuming that $y(kT) = 0$ for $k < 0$, show that

$$y(kT) = T[x(0) + x(T) + \cdots + x((k - 1)T)]$$

The output $y(kT)$ approximates the area made by the input. Since $y(0) = 0$ and $y(T) = Tx(0)$, the output starts to appear at $t = T$. This integrator is called a digital integrator with delay.

Draw a graph of the output $y(kT)$ when the input $x(kT)$ is a unit-step sequence.

Problem B–3–27

Consider the system shown in Figure 3–75. Show that the pulse transfer function $Y(z)/X(z)$ is given by

$$\frac{Y(z)}{X(z)} = \frac{T}{2}\left(\frac{1}{1-z^{-1}} + \frac{z^{-1}}{1-z^{-1}}\right)$$

This system is a combination of the digital integrators without delay and with delay, as presented in Problems B–3–25 and B–3–26, respectively.

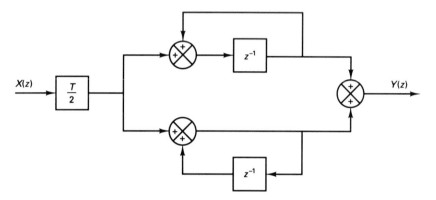

Figure 3–75   Digital bilinear integrator.

Assuming that $y(kT) = 0$ for $k < 0$, obtain $y(kT)$ in terms of $x(0), x(T), \ldots,$ $x(kT)$. This integrator is called a digital bilinear integrator. Draw a graph of the output $y(kT)$ when the input $x(kT)$ is a unit-step sequence.

4

Design of Discrete-Time Control Systems by Conventional Methods

4-1 INTRODUCTION

In this chapter we first present mapping from the s plane to the z plane and then discuss stability of closed-loop control systems in the z plane. Next we treat three different design methods for single-input–single-output discrete-time or digital control systems. The first method is based on the root-locus technique using pole–zero configurations in the z plane. The second method is based on the frequency-response method in the w plane. The third method is an analytical method in which we attempt to obtain a desired behavior of the closed-loop system by manipulating the pulse transfer function of the digital controller.

Design techniques for continuous-time control systems based on conventional transform methods (the root-locus and frequency-response methods) have become well established since the 1950s. Conventional transform methods are especially useful for designing industrial control systems. In fact, in the past, many industrial digital control systems were successfully designed on the basis of conventional transform methods. Both familiarity with the root-locus and frequency-response techniques and experiences gained in the design of analog controllers are immensely valuable in designing discrete-time control systems.

Outline of the Chapter. Section 4–1 has presented introductory material. Section 4–2 treats mapping from the s plane to the z plane. Section 4–3 discusses the Jury stability criterion for closed-loop control systems in the z plane. Section 4–4 summarizes transient and steady-state response characteristics of discrete-time control systems. The design technique based on the root-locus method is presented in Section 4–5. Section 4–6 first reviews the frequency-response method and then presents frequency-response techniques using the w transformation for designing discrete-time control systems. Section 4–7 treats an analytical design method.

4–2 MAPPING BETWEEN THE s PLANE AND THE z PLANE

The absolute stability and relative stability of the linear time-invariant continuous-time closed-loop control system are determined by the locations of the closed-loop poles in the s plane. For example, complex closed-loop poles in the left half of the s plane near the $j\omega$ axis will exhibit oscillatory behavior, and closed-loop poles on the negative real axis will exhibit exponential decay.

Since the complex variables z and s are related by $z = e^{Ts}$, the pole and zero locations in the z plane are related to the pole and zero locations in the s plane. Therefore, the stability of the linear time-invariant discrete-time closed-loop system can be determined in terms of the locations of the poles of the closed-loop pulse transfer function. It is noted that the dynamic behavior of the discrete-time control system depends on the sampling period T. In terms of poles and zeros in the z plane, their locations depend on the sampling period T. In other words, a change in the sampling period T modifies the pole and zero locations in the z plane and causes the response behavior to change.

Mapping of the Left Half of the s Plane into the z Plane. In the design of a continuous-time control system, the locations of the poles and zeros in the s plane are very important in predicting the dynamic behavior of the system. Similarly, in designing discrete-time control systems, the locations of the poles and zeros in the z plane are very important. In the following paragraphs we shall investigate how the locations of the poles and zeros in the s plane compare with the locations of the poles and zeros in the z plane.

When impulse sampling is incorporated into the process, the complex variables z and s are related by the equation

$$z = e^{Ts}$$

This means that a pole in the s plane can be located in the z plane through the transformation $z = e^{Ts}$. Since the complex variable s has real part σ and imaginary part ω, we have

$$s = \sigma + j\omega$$

and

$$z = e^{T(\sigma + j\omega)} = e^{T\sigma} e^{jT\omega} = e^{T\sigma} e^{j(T\omega + 2\pi k)}$$

From this last equation we see that poles and zeros in the s plane, where frequencies differ in integral multiples of the sampling frequency $2\pi/T$, are mapped into the same locations in the z plane. This means that there are infinitely many values of s for each value of z.

Since σ is negative in the left half of the s plane, the left half of the s plane corresponds to

$$|z| = e^{T\sigma} < 1$$

The $j\omega$ axis in the s plane corresponds to $|z| = 1$. That is, the imaginary axis in the s plane (the line $\sigma = 0$) corresponds to the unit circle in the z plane, and the interior of the unit circle corresponds to the left half of the s plane.

Primary Strip and Complementary Strips.   Note that since $\underline{/z} = \omega T$ the angle of *z* varies from $-\infty$ to ∞ as ω varies from $-\infty$ to ∞. Consider a representative point on the $j\omega$ axis in the *s* plane. As this point moves from $-j\frac{1}{2}\omega_s$ to $j\frac{1}{2}\omega_s$ on the $j\omega$ axis, where ω_s is the sampling frequency, we have $|z| = 1$, and $\underline{/z}$ varies from $-\pi$ to π in the counterclockwise direction in the *z* plane. As the representative point moves from $j\frac{1}{2}\omega_s$ to $j\frac{3}{2}\omega_s$ on the $j\omega$ axis, the corresponding point in the *z* plane traces out the unit circle once in the counterclockwise direction. Thus, as the point in the *s* plane moves from $-\infty$ to ∞ on the $j\omega$ axis, we trace the unit circle in the *z* plane an infinite number of times. From this analysis, it is clear that each strip of width ω_s in the left half of the *s* plane maps into the inside of the unit circle in the *z* plane. This implies that the left half of the *s* plane may be divided into an infinite number of periodic strips as shown in Figure 4–1. The primary strip extends from $j\omega = -j\frac{1}{2}\omega_s$ to $j\frac{1}{2}\omega_s$. The complementary strips extend from $j\frac{1}{2}\omega_s$ to $j\frac{3}{2}\omega_s$, $j\frac{3}{2}\omega_s$ to $j\frac{5}{2}\omega_s$, ..., and from $-j\frac{1}{2}\omega_s$ to $-j\frac{3}{2}\omega_s$, $-j\frac{3}{2}\omega_s$ to $-j\frac{5}{2}\omega_s$, ...

In the primary strip, if we trace the sequence of points 1–2–3–4–5–1 in the *s* plane as shown by the circled numbers in Figure 4–2(a), then this path is mapped into the unit circle centered at the origin of the *z* plane, as shown in Figure 4–2(b). The corresponding points 1, 2, 3, 4, and 5 in the *z* plane are shown by the circled numbers in Figure 4–2(b).

The area enclosed by any of the complementary strips is mapped into the same unit circle in the *z* plane. This means that the correspondence between the *z* plane

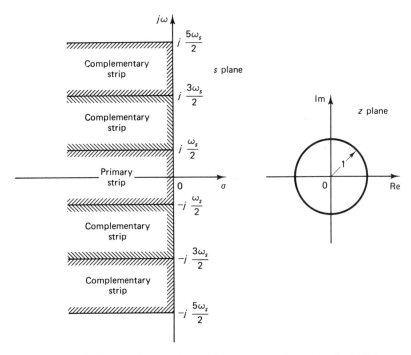

Figure 4–1   Periodic strips in the *s* plane and the corresponding region (unit circle centered at the origin) in the *z* plane.

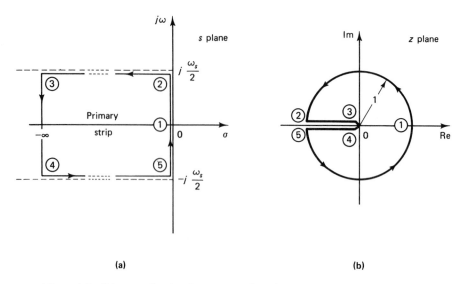

(a)   (b)

Figure 4–2 Diagrams showing the correspondence between the primary strip in the s plane and the unit circle in the z plane: (a) a path in the s plane; (b) the corresponding path in the z plane.

and the s plane is not unique. A point in the z plane corresponds to an infinite number of points in the s plane, although a point in the s plane corresponds to a single point in the z plane.

Since the entire left half of the s plane is mapped into the interior of the unit circle in the z plane, the entire right half of the s plane is mapped into the exterior of the unit circle in the z plane. As mentioned earlier, the $j\omega$ axis in the s plane maps into the unit circle in the z plane. Note that, if the sampling frequency is at least twice as fast as the highest-frequency component involved in the system, then every point in the unit circle in the z plane represents frequencies between $-\frac{1}{2}\omega_s$ and $\frac{1}{2}\omega_s$.

In what follows we shall investigate the mapping of some of the commonly used contours in the s plane into the z plane. Specifically, we shall map constant-attenuation loci, constant-frequency loci, and constant-damping-ratio loci.

Constant-Attenuation Loci.   A constant-attenuation line (a line plotted as σ = constant) in the s plane maps into a circle of radius $z = e^{T\sigma}$ centered at the origin in the z plane, as shown in Figure 4–3.

Settling Time t_s.   The settling time is determined by the value of attenuation σ of the dominant closed-loop poles. If the settling time is specified, it is possible to draw a line $\sigma = -\sigma_1$ in the s plane corresponding to a given settling time. The region to the left of the line $\sigma = -\sigma_1$ in the s plane corresponds to the inside of a circle with radius $e^{-\sigma_1 T}$ in the z plane, as shown in Figure 4–4.

Constant-Frequency Loci.   A constant-frequency locus $\omega = \omega_1$ in the s plane is mapped into a radial line of constant angle $T\omega_1$ (in radians) in the z plane, as shown in Figure 4–5. Note that constant-frequency lines at $\omega = \pm\frac{1}{2}\omega_s$ in the left half of the s plane correspond to the negative real axis in the z plane between 0 and -1, since

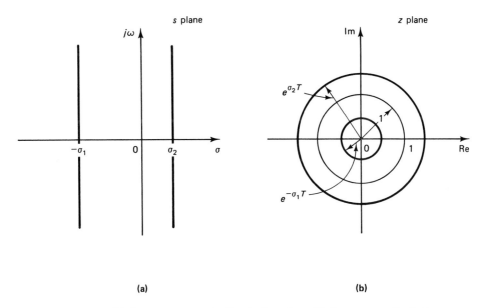

Figure 4-3   (a) Constant-attenuation lines in the *s* plane; (b) the corresponding loci in the *z* plane.

$T(\pm\frac{1}{2}\omega_s) = \pm\pi$. Constant-frequency lines at $\omega = \pm\frac{1}{2}\omega_s$ in the right half of the *s* plane correspond to the negative real axis in the *z* plane between -1 and $-\infty$. The negative real axis in the *s* plane corresponds to the positive real axis in the *z* plane between 0 and 1. And constant frequency lines at $\omega = \pm n\omega_s$ ($n = 0, 1, 2, \ldots$) in the right half of the *s* plane map into the positive real axis in the *z* plane between 1 and ∞.

The region bounded by constant-frequency lines $\omega = \omega_1$ and $\omega = -\omega_2$ (where both ω_1 and ω_2 lie between $-\frac{1}{2}\omega_s$ and $\frac{1}{2}\omega_s$) and constant-attenuation lines $\sigma = -\sigma_1$ and $\sigma = -\sigma_2$, as shown in Figure 4-6(a), is mapped into a region bounded by two radial lines and two circular arcs, as shown in Figure 4-6(b).

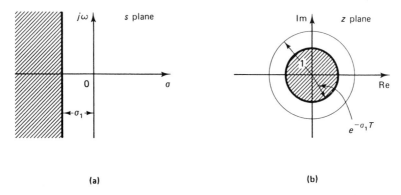

Figure 4-4   (a) Region for settling time T_s less than $4/\sigma_1$ in the *s* plane; (b) region for settling time T_s less than $4/\sigma_1$ in the *z* plane.

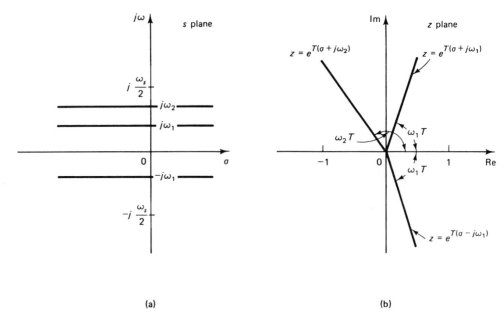

Figure 4–5   (a) Constant-frequency loci in the s plane; (b) the corresponding loci in the z plane.

Constant-Damping-Ratio Loci.   A constant-damping-ratio line (a radial line) in the s plane is mapped into a spiral in the z plane. This can be seen as follows. In the s plane a constant-damping-ratio line can be given by

$$s = -\zeta\omega_n + j\omega_n\sqrt{1 - \zeta^2} = -\zeta\omega_n + j\omega_d$$

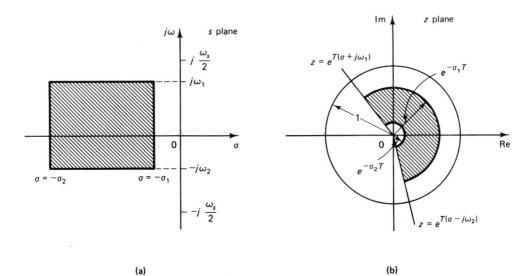

Figure 4–6   (a) Region bounded by lines $\omega = \omega_1$, $\omega = -\omega_2$, $\sigma = -\sigma_1$, and $\sigma = -\sigma_2$ in the s plane; (b) the corresponding region in the z plane.

where $\omega_d = \omega_n \sqrt{1 - \zeta^2}$ [see Figure 4–7(a)]. In the z plane this line becomes

$$z = e^{Ts} = \exp(-\zeta\omega_n T + j\omega_d T)$$

$$= \exp\left(-\frac{2\pi\zeta}{\sqrt{1 - \zeta^2}}\frac{\omega_d}{\omega_s} + j2\pi\frac{\omega_d}{\omega_s}\right)$$

Hence,

$$|z| = \exp\left(-\frac{2\pi\zeta}{\sqrt{1 - \zeta^2}}\frac{\omega_d}{\omega_s}\right) \tag{4–1}$$

and

$$\underline{/z} = 2\pi\frac{\omega_d}{\omega_s} \tag{4–2}$$

Thus, the magnitude of z decreases and the angle of z increases linearly as ω_d increases, and the locus in the z plane becomes a logarithmic spiral, as shown in Figure 4–7(b).

Notice that for a given ratio of ω_d/ω_s the magnitude $|z|$ becomes a function only of ζ, and the angle of z becomes a constant. For example, if the damping ratio is specified as 0.3, or $\zeta = 0.3$, then for $\omega_d = 0.25\omega_s$ we have

$$|z| = \exp\left(-\frac{2\pi \times 0.3}{\sqrt{1 - 0.3^2}} \times 0.25\right) = 0.610$$

$$\underline{/z} = 2\pi \times 0.25 = 0.5\pi = 90°$$

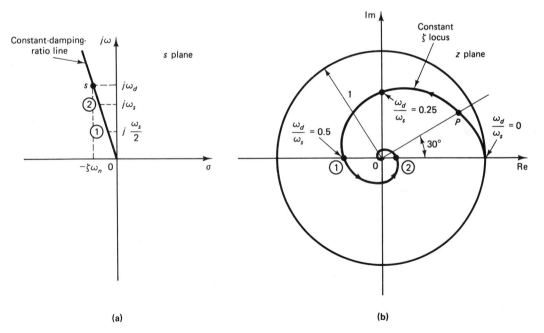

(a)   (b)

Figure 4–7   (a) Constant-damping-ratio line in the s plane; (b) the corresponding locus in the z plane.

For $\omega_d = 0.5\omega_s$,

$$|z| = \exp\left(-\frac{2\pi \times 0.3}{\sqrt{1 - 0.3^2}} \times 0.5\right) = 0.3725$$

$$\underline{/z} = 2\pi \times 0.5 = \pi = 180°$$

Thus, the spiral can be graduated in terms of a normalized frequency ω_d/ω_s [see Figure 4–7(b)]. Once the sampling frequency ω_s is specified, the numerical value of ω_d at any point on the spiral can be determined. For example, at point P in Figure 4–7(b), ω_d can be determined as follows. If, for example, the sampling frequency is specified as $\omega_s = 10\pi$ rad/sec, then at point P

$$\underline{/z} = \frac{\pi}{6} = 2\pi\frac{\omega_d}{\omega_s}$$

Hence, ω_d at point P is

$$\omega_d = \tfrac{1}{12}\omega_s = \tfrac{5}{6}\pi \text{ rad/sec}$$

Note that if a constant-damping-ratio line is in the second or third quadrant in the s plane then the spiral decays within the unit circle in the z plane. However, if a constant-damping-ratio line is in the first or fourth quadrant in the s plane (which corresponds to negative damping), then the spiral grows outside the unit circle. Figure 4–8 shows constant-damping-ratio loci for $\zeta = 0$, $\zeta = 0.2$, $\zeta = 0.4$, $\zeta = 0.6$, $\zeta = 0.8$, and $\zeta = 1$. The $\zeta = 1$ locus is a horizontal line between points $z = 0$ and $z = 1$. (Note that Figure 4–8 shows only the loci in the upper half of the z plane, which correspond to $0 \le \omega \le \tfrac{1}{2}\omega_s$. The loci corresponding to $-\tfrac{1}{2}\omega_s \le \omega \le 0$ are the mirror images of the loci in the upper half of the z plane about the horizontal axis.)

Notice that the constant ζ loci are normal to the constant ω_n loci in the s plane, as shown in Figure 4–9(a). In the z plane mapping, constant ω_n loci intersect constant ζ spirals at right angles, as shown in Figure 4–9(b). A mapping such as this, which preserves both the size and the sense of angles, is called a *conformal mapping*.

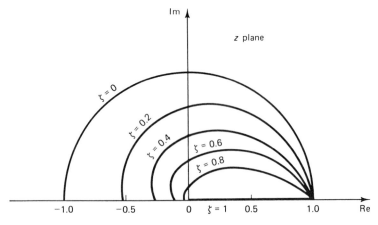

Figure 4–8   Constant-damping-ratio loci in the z plane.

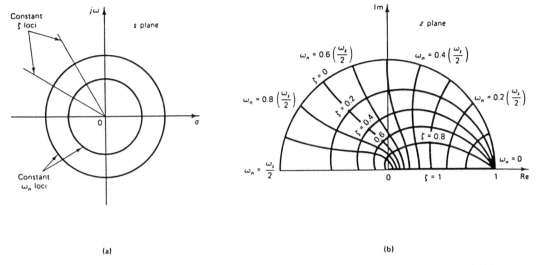

Figure 4-9   (a) Diagram showing orthogonality of the constant ζ loci and constant ω_n loci in the s plane; (b) the corresponding diagram in the z plane.

s Plane and z Plane Regions for $\zeta > \zeta_1$.   Figure 4-10 shows constant ζ loci ($\zeta = \zeta_1$) in both the s plane and the z plane. Note that the logarithmic spirals shown correspond to the primary strip in the s plane. (If the sampling theorem is satisfied, then we need to consider only the primary strip in the s plane.)

If all the poles in the s plane are specified as having a damping ratio not less than a specified value ζ_1, then the poles must lie to the left of the constant-damping ratio line in the s plane (the shaded region). In the z plane, the poles must lie in the region bounded by logarithmic spirals corresponding to $\zeta = \zeta_1$ (the shaded region).

Example 4-1

Specify the region in the z plane that corresponds to a desirable region (shaded region) in the s plane bounded by lines $\omega = \pm\omega_1$, lines $\zeta = \zeta_1$, and a line $\sigma = -\sigma_1$, as shown in Figure 4-11(a).

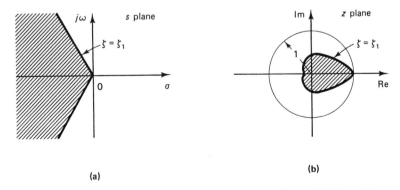

Figure 4-10   (a) Region for $\zeta > \zeta_1$ in the s plane; (b) region for $\zeta > \zeta_1$ in the z plane.

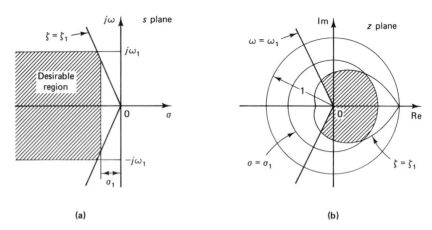

Figure 4–11   (a) A desirable region in the s plane for closed-loop pole locations; (b) corresponding region in the z plane.

On the basis of the preceding discussions on mapping from the s plane to the z plane, the desirable region can be mapped to the z plane as in Figure 4–11(b).

Note that if the dominant closed-loop poles of the continuous-time control system are required to be in the desirable region specified in the s plane, then the dominant closed-loop poles of the equivalent discrete-time control system must lie inside the region in the z plane that corresponds to the desirable region in the s plane. Once the discrete-time control system is designed, the system response characteristics must be checked by experiments or simulation. If the response characteristics are not satisfactory, then closed-loop pole and zero locations must be modified until satisfactory results are obtained.

Comments.   For discrete-time control systems, it is necessary to pay particular attention to the sampling period T. This is because, if the sampling period is too long and the sampling theorem is not satisfied, then frequency folding occurs and the effective pole and zero locations will be changed.

Suppose a continuous-time control system has closed-loop poles at $s = -\sigma_1 \pm j\omega_1$ in the s plane. If the sampling operation is involved in this system and if $\omega_1 > \frac{1}{2}\omega_s$, where ω_s is the sampling frequency, then frequency folding occurs and the system behaves as if it had poles at $s = -\sigma_1 \pm j(\omega_1 \pm n\omega_s)$, where $n = 1, 2, 3, \ldots$. This means that the sampling operation folds the poles outside the primary strip back into the primary strip, and the poles will appear at $s = -\sigma_1 \pm j(\omega_1 - \omega_s)$; see Figure 4–12(a). On the z plane those poles are mapped into one pair of conjugate complex poles, as shown in Figure 4–12(b). When frequency folding occurs, oscillations with frequency $\omega_s - \omega_1$, rather than frequency ω_1, are observed.

4–3 STABILITY ANALYSIS OF CLOSED-LOOP SYSTEMS IN THE z PLANE

Stability Analysis of a Closed-Loop System.   In what follows we shall discuss the stability of linear time-invariant single-input–single-output discrete-time control systems. Consider the following closed-loop pulse-transfer function system:

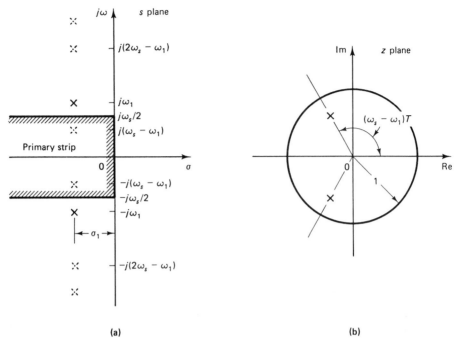

(a)   (b)

Figure 4–12 (a) Diagram showing s plane poles at $-\sigma_1 \pm j\omega_1$ and folded poles appearing at $-\sigma_1 \pm j(\omega_1 \pm \omega_s)$, $-\sigma_1 \pm j(\omega_1 \pm 2\omega_s)$, ...; (b) z plane mapping of s plane poles at $-\sigma_1 \pm j\omega_1$, $-\sigma_1 \pm j(\omega_1 \pm \omega_s)$, $-\sigma_1 \pm j(\omega_1 \pm 2\omega_s)$, ...

$$\frac{C(z)}{R(z)} = \frac{G(z)}{1 + GH(z)} \tag{4–3}$$

The stability of the system defined by Equation (4–3), as well as of other types of discrete-time control systems, may be determined from the locations of the closed-loop poles in the z plane, or the roots of the characteristic equation

$$P(z) = 1 + GH(z) = 0$$

as follows:

1. For the system to be stable, the closed-loop poles or the roots of the characteristic equation must lie within the unit circle in the z plane. Any closed-loop pole outside the unit circle makes the system unstable.

2. If a simple pole lies at $z = 1$, then the system becomes critically stable. Also, the system becomes critically stable if a single pair of conjugate complex poles lies on the unit circle in the z plane. Any multiple closed-loop pole on the unit circle makes the system unstable.

3. Closed-loop zeros do not affect the absolute stability and therefore may be located anywhere in the z plane.

Thus, a linear time-invariant single-input–single-output discrete-time closed-loop control system becomes unstable if any of the closed-loop poles lies outside the unit circle and/or any multiple closed-loop pole lies on the unit circle in the z plane.

Example 4–2

Consider the closed-loop control system shown in Figure 4–13. Determine the stability of the system when $K = 1$. The open-loop transfer function $G(s)$ of the system is

$$G(s) = \frac{1 - e^{-s}}{s} \cdot \frac{1}{s(s + 1)}$$

Referring to Equation (3–58), the z transform of $G(s)$ is

$$G(z) = \mathcal{Z}\left[\frac{1 - e^{-s}}{s} \cdot \frac{1}{s(s + 1)}\right] = \frac{0.3679z + 0.2642}{(z - 0.3679)(z - 1)} \tag{4–4}$$

Since the closed-loop pulse transfer function for the system is

$$\frac{C(z)}{R(z)} = \frac{G(z)}{1 + G(z)}$$

the characteristic equation is

$$1 + G(z) = 0$$

which becomes

$$(z - 0.3679)(z - 1) + 0.3679z + 0.2642 = 0$$

or

$$z^2 - z + 0.6321 = 0$$

The roots of the characteristic equation are found to be

$$z_1 = 0.5 + j0.6181, \qquad z_2 = 0.5 - j0.6181$$

Since

$$|z_1| = |z_2| < 1$$

the system is stable.

It is important to note that in the absence of the sampler a second-order system is always stable. In the presence of the sampler, however, a second-order system such as this can become unstable for large values of gain. In fact, it can be shown that the second-order system shown in Figure 4–13 will become unstable if $K > 2.3925$. (See Example 4–7.)

Methods for Testing Absolute Stability.   Three stability tests can be applied directly to the characteristic equation $P(z) = 0$ without solving for the roots. Two of them are the Schur–Cohn stability test and the Jury stability test. These two tests

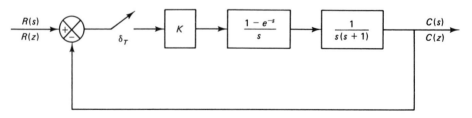

Figure 4–13   Closed-loop control system of Example 4–2.

reveal the existence of any unstable roots (the roots that lie outside the unit circle in the z plane). However, these tests neither give the locations of unstable roots nor indicate the effects of parameter changes on the system stability, except for the simple case of low-order systems. (See Example 4-7.) The third method is based on the bilinear transformation coupled with the Routh stability criterion, which will be outlined later in this section. (In Chapter 5 we discuss Liapunov stability analysis, which is applicable to control systems defined in state space.)

Both the Schur–Cohn stability test and the Jury stability test may be applied to polynomial equations with real or complex coefficients. The computations required in the Jury test, when the polynomial equation involves only real coefficients, are much simpler than those required in the Schur–Cohn test. Since the coefficients of the characteristic equations corresponding to physically realizable systems are always real, the Jury test is preferred to the Schur–Cohn test.

The Jury Stability Test.   In applying the Jury stability test to a given characteristic equation $P(z) = 0$, we construct a table whose elements are based on the coefficients of $P(z)$. Assume that the characteristic equation $P(z)$ is a polynomial in z as follows:

$$P(z) = a_0 z^n + a_1 z^{n-1} + \cdots + a_{n-1} z + a_n \tag{4-5}$$

where $a_0 > 0$. Then the Jury table becomes as given in Table 4-1.

Notice that the elements in the first row consist of the coefficients in $P(z)$ arranged in the ascending order of powers of z. The elements in the second row consist of the coefficients of $P(z)$ arranged in the descending order of powers of z. The elements for rows 3 through $2n - 3$ are given by the following determinants:

TABLE 4-1 GENERAL FORM OF THE JURY STABILITY TABLE

Row	z^0	z^1	z^2	z^3	$\cdots$	z^{n-2}	z^{n-1}	z^n
1	a_n	a_{n-1}	a_{n-2}	a_{n-3}	$\cdots$	a_2	a_1	a_0
2	a_0	a_1	a_2	a_3	$\cdots$	a_{n-2}	a_{n-1}	a_n
3	b_{n-1}	b_{n-2}	b_{n-3}	b_{n-4}	$\cdots$	b_1	b_0	
4	b_0	b_1	b_2	b_3	$\cdots$	b_{n-2}	b_{n-1}	
5	c_{n-2}	c_{n-3}	c_{n-4}	c_{n-5}	$\cdots$	c_0		
6	c_0	c_1	c_2	c_3	$\cdots$	c_{n-2}		
$\vdots$	$\vdots$							
$2n - 5$	p_3	p_2	p_1	p_0				
$2n - 4$	p_0	p_1	p_2	p_3				
$2n - 3$	q_2	q_1	q_0					

$$b_k = \begin{vmatrix} a_n & a_{n-1-k} \\ a_0 & a_{k+1} \end{vmatrix}, \qquad k = 0, 1, 2, \ldots, n-1$$

$$c_k = \begin{vmatrix} b_{n-1} & b_{n-2-k} \\ b_0 & b_{k+1} \end{vmatrix}, \qquad k = 0, 1, 2, \ldots, n-2$$

$$\vdots$$

$$q_k = \begin{vmatrix} p_3 & p_{2-k} \\ p_0 & p_{k+1} \end{vmatrix}, \qquad k = 0, 1, 2$$

Note that the last row in the table consists of three elements. (For second-order systems, $2n - 3 = 1$ and the Jury table consists only of one row containing three elements.) Notice that the elements in any even-numbered row are simply the reverse of the immediately preceding odd-numbered row.

Stability Criterion by the Jury Test. A system with the characteristic equation $P(z) = 0$ given by Equation (4–5), rewritten as

$$P(z) = a_0 z^n + a_1 z^{n-1} + \cdots + a_{n-1} z + a_n$$

where $a_0 > 0$, is stable if the following conditions are all satisfied:

1. $|a_n| < a_0$
2. $P(z)|_{z=1} > 0$
3. $P(z)|_{z=-1} \begin{cases} >0 \text{ for } n \text{ even} \\ <0 \text{ for } n \text{ odd} \end{cases}$
4. $|b_{n-1}| > |b_0|$
   $|c_{n-2}| > |c_0|$
   $\vdots$
   $|q_2| > |q_0|$

Example 4–3

Construct the Jury stability table for the following characteristic equation:

$$P(z) = a_0 z^4 + a_1 z^3 + a_2 z^2 + a_3 z + a_4$$

where $a_0 > 0$. Write the stability conditions.

Referring to the general case of the Jury stability table given by Table 4–1, a Jury stability table for the fourth-order system may be constructed as shown in Table 4–2. This table is slightly modified from the standard form and is convenient for the computations of the b's and c's. The determinant given in middle of each row gives the value of b or c written on the right-hand side of the same row.

The stability conditions are as follows:

1. $|a_4| < a_0$
2. $P(1) = a_0 + a_1 + a_2 + a_3 + a_4 > 0$
3. $P(-1) = a_0 - a_1 + a_2 - a_3 + a_4 > 0, \qquad n = 4 = \text{even}$
4. $|b_3| > |b_0|$
   $|c_2| > |c_0|$

TABLE 4–2 JURY STABILITY TABLE FOR THE FOURTH-ORDER SYSTEM

Row	z^0	z^1	z^2	z^3	z^4	
	$\begin{vmatrix} a_4 \\ a_0 \end{vmatrix}$			$\begin{vmatrix} a_0 \\ a_4 \end{vmatrix}$		$=b_3$
	$\begin{vmatrix} a_4 \\ a_0 \end{vmatrix}$			$\begin{vmatrix} a_1 \\ a_3 \end{vmatrix}$		$=b_2$
	$\begin{vmatrix} a_4 \\ a_0 \end{vmatrix}$		$\begin{vmatrix} a_2 \\ a_2 \end{vmatrix}$			$=b_1$
1 2	$\begin{vmatrix} a_4 & a_3 \\ a_0 & a_1 \end{vmatrix}$					$=b_0$
	$\begin{vmatrix} b_3 \\ b_0 \end{vmatrix}$			$\begin{vmatrix} b_0 \\ b_3 \end{vmatrix}$		$=c_2$
	$\begin{vmatrix} b_3 \\ b_0 \end{vmatrix}$		$\begin{vmatrix} b_1 \\ b_2 \end{vmatrix}$			$=c_1$
3 4	$\begin{vmatrix} b_3 & b_2 \\ b_0 & b_1 \end{vmatrix}$					$=c_0$
5	c_2	c_1	c_0			

It is noted that the value of c_1 (or, in the case of the nth-order system, the value of q_1) is not used in the stability test, and therefore the computation of c_1 (or q_1) may be omitted.

Example 4–4

Examine the stability of the following characteristic equation:

$$P(z) = z^4 - 1.2z^3 + 0.07z^2 + 0.3z - 0.08 = 0$$

Notice that for this characteristic equation

$$a_0 = 1$$
$$a_1 = -1.2$$
$$a_2 = 0.07$$
$$a_3 = 0.3$$
$$a_4 = -0.08$$

Clearly, the first condition, $|a_4| < a_0$, is satisfied. Let us examine the second condition for stability:

$$P(1) = 1 - 1.2 + 0.07 + 0.3 - 0.08 = 0.09 > 0$$

The second condition is satisfied. The third condition for stability becomes

$$P(-1) = 1 + 1.2 + 0.07 - 0.3 - 0.08 = 1.89 > 0, \qquad n = 4 = \text{even}$$

Hence the third condition is satisfied.

We now construct the Jury stability table. Referring to Example 4–3, we compute the values of b_3, b_2, b_1, and b_0 and c_2 and c_0. The result is shown in Table 4–3. (Although the value of c_1 is shown in the table, c_1 is not needed in the stability test and therefore need not be computed.) From this table, we get

$$|b_3| = 0.994 > 0.204 = |b_0|$$
$$|c_2| = 0.946 > 0.315 = |c_0|$$

Thus both parts of the fourth condition given in Example 4–3 are satisfied. Since all conditions for stability are satisfied, the given characteristic equation is stable, or all roots lie inside the unit circle in the z plane.

As a matter of fact, the given characteristic equation $P(z)$ can be factored as follows:

$$P(z) = (z - 0.8)(z + 0.5)(z - 0.5)(z - 0.4)$$

As a matter of course, the result obtained above agrees with the fact that all roots are within the unit circle in the z plane.

TABLE 4–3   JURY STABILITY TABLE FOR THE SYSTEM OF EXAMPLE 4–4

Row	z^0	z^1	z^2	z^3	z^4	
	$\begin{vmatrix} -0.08 & & & & 1 \\ 1 & & & & -0.08 \end{vmatrix}$					$= b_3 = -0.994$
	$\begin{vmatrix} -0.08 & & & -1.2 \\ 1 & & & 0.3 \end{vmatrix}$					$= b_2 = 1.176$
	$\begin{vmatrix} -0.08 & & 0.07 \\ 1 & & 0.07 \end{vmatrix}$					$= b_1 = -0.0756$
1	$\begin{vmatrix} -0.08 & 0.3 \\ 1 & -1.2 \end{vmatrix}$					$= b_0 = -0.204$
2						
	$\begin{vmatrix} -0.994 & & & -0.204 \\ -0.204 & & & -0.994 \end{vmatrix}$					$= c_2 = 0.946$
	$\begin{vmatrix} -0.994 & & -0.0756 \\ -0.204 & & 1.176 \end{vmatrix}$					$= c_1 = -1.184$
3	$\begin{vmatrix} -0.994 & 1.176 \\ -0.204 & -0.0756 \end{vmatrix}$					$= c_0 = 0.315$
4						
5	0.946	-1.184	0.315			

Example 4–5

Examine the stability of the characteristic equation given by

$$P(z) = z^3 - 1.1z^2 - 0.1z + 0.2 = 0$$

First we identify the coefficients:

$$a_0 = 1$$
$$a_1 = -1.1$$
$$a_2 = -0.1$$
$$a_3 = 0.2$$

The conditions for stability in the Jury test for the third-order system are as follows:

1. $|a_3| < a_0$
2. $P(1) > 0$
3. $P(-1) < 0, \quad n = 3 = \text{odd}$
4. $|b_2| > |b_0|$

The first condition, $|a_3| < a_0$, is clearly satisfied. Now we examine the second condition of the Jury stability test:

$$P(1) = 1 - 1.1 - 0.1 + 0.2 = 0$$

This indicates that at least one root is at $z = 1$. Therefore, the system is at best critically stable. The remaining tests determine whether the system is critically stable or unstable. (If the given characteristic equation represents a control system, critical stability will not be desired. The stability test may be stopped at this point.)

The third condition of the Jury test gives

$$P(-1) = -1 - 1.1 + 0.1 + 0.2 = -1.8 < 0, \quad n = 3 = \text{odd}$$

The third condition is satisfied. Now we examine the fourth condition of the Jury test. Simple computations give $b_2 = -0.96$ and $b_0 = -0.12$. Hence,

$$|b_2| > |b_0|$$

The fourth condition of the Jury test is satisfied.

From the above analysis we conclude that the given characteristic equation has one root on the unit circle ($z = 1$) and its other two roots within the unit circle in the z plane. Hence, the system is critically stable.

Example 4–6

A control system has the following characteristic equation:

$$P(z) = z^3 - 1.3z^2 - 0.08z + 0.24 = 0$$

Determine the stability of the system.

We first identify the coefficients:

$$a_0 = 1$$
$$a_1 = -1.3$$
$$a_2 = -0.08$$
$$a_3 = 0.24$$

Clearly, the first condition for stability, $|a_3| < a_0$, is satisfied. Next, we examine the second condition for stability:

$$P(1) = 1 - 1.3 - 0.08 + 0.24 = -0.14 < 0$$

The test indicates that the second condition for stability is violated. The system is therefore unstable. We may stop the test here.

Example 4–7

Consider the discrete-time unity-feedback control system (with sampling period $T = 1$ sec) whose open-loop pulse transfer function is given by

$$G(z) = \frac{K(0.3679z + 0.2642)}{(z - 0.3679)(z - 1)}$$

Determine the range of gain K for stability by use of the Jury stability test.
    The closed-loop pulse transfer function becomes

$$\frac{C(z)}{R(z)} = \frac{K(0.3679z + 0.2642)}{z^2 + (0.3679K - 1.3679)z + 0.3679 + 0.2642K}$$

Thus, the characteristic equation for the system is

$$P(z) = z^2 + (0.3679K - 1.3679)z + 0.3679 + 0.2642K = 0$$

Since this is a second-order system, the Jury stability conditions may be written as follows:

    1. $|a_2| < a_0$
    2. $P(1) > 0$
    3. $P(-1) > 0$, $n = 2 =$ even

    We shall now apply the first condition for stability. Since $a_2 = 0.3679 + 0.2642K$ and $a_0 = 1$, the first condition for stability becomes

$$|0.3679 + 0.2642K| < 1$$

or

$$2.3925 > K > -5.1775 \qquad\qquad (4\text{–}6)$$

The second condition for stability becomes

$$P(1) = 1 + (0.3679K - 1.3679) + 0.3679 + 0.2642K = 0.6321K > 0$$

which gives

$$K > 0 \qquad\qquad (4\text{–}7)$$

The third condition for stability gives

$$P(-1) = 1 - (0.3679K - 1.3679) + 0.3679 + 0.2642K = 2.7358 - 0.1037K > 0$$

which yields

$$26.382 > K \qquad\qquad (4\text{–}8)$$

For stability, gain constant K must satisfy inequalities (4–6), (4–7), and (4–8). Hence,

$$2.3925 > K > 0$$

The range of gain constant K for stability is between 0 and 2.3925.

If gain K is set equal to 2.3925, then the system becomes critically stable (meaning that sustained oscillations exist at the output). The frequency of the sustained oscillations can be determined if 2.3925 is substituted for K in the characteristic equation and the resulting equation is investigated. With $K = 2.3925$, the characteristic equation becomes

$$z^2 - 0.4877z + 1 = 0$$

The characteristic roots are at $z = 0.2439 \pm j0.9698$. Noting that the sampling period T is equal to 1 sec, from Equation (4–2) we have

$$\omega_d = \frac{\omega_s}{2\pi} \underline{/z} = \frac{2\pi}{2\pi} \underline{/z} = \tan^{-1} \frac{0.9698}{0.2439} = 1.3244 \text{ rad/sec}$$

The frequency of the sustained oscillations is 1.3244 rad/sec.

Stability Analysis by Use of the Bilinear Transformation and Routh Stability Criterion.   Another method frequently used in the stability analysis of discrete-time control systems is to use the bilinear transformation coupled with the Routh stability criterion. The method requires transformation from the z plane to another complex plane, the w plane. Those who are familiar with the Routh–Hurwitz stability criterion will find the method simple and straightforward. However, the amount of computation required is much more than that required in the Jury stability criterion.

The bilinear transformation defined by

$$z = \frac{w + 1}{w - 1}$$

which, when solved for w, gives

$$w = \frac{z + 1}{z - 1}$$

maps the inside of the unit circle in the z plane into the left half of the w plane. This can be seen as follows. Let the real part of w be called σ and the imaginary part ω, so that

$$w = \sigma + j\omega$$

Since the inside of the unit circle in the z plane is

$$|z| = \left| \frac{w + 1}{w - 1} \right| = \left| \frac{\sigma + j\omega + 1}{\sigma + j\omega - 1} \right| < 1$$

or

$$\frac{(\sigma + 1)^2 + \omega^2}{(\sigma - 1)^2 + \omega^2} < 1$$

we get

$$(\sigma + 1)^2 + \omega^2 < (\sigma - 1)^2 + \omega^2$$

which yields

$$\sigma < 0$$

Thus, the inside of the unit circle in the z plane ($|z| < 1$) corresponds to the left half of the w plane. The unit circle in the z plane is mapped into the imaginary axis in the w plane, and the outside of the unit circle in the z plane is mapped into the right half of the w plane. (It is pointed out that, although the w plane is similar to the s plane in that it maps the inside of the unit circle to the left half-plane, it is by no means quantitatively equivalent to the s plane. Therefore, estimating the relative stability of the system from the pole locations in the w plane is difficult.)

In the stability analysis using the bilinear transformation coupled with the Routh stability criterion, we first substitute $(w + 1)/(w - 1)$ for z in the characteristic equation

$$P(z) = a_0 z^n + a_1 z^{n-1} + \cdots + a_{n-1} z + a_n = 0$$

as follows:

$$a_0 \left(\frac{w + 1}{w - 1} \right)^n + a_1 \left(\frac{w + 1}{w - 1} \right)^{n-1} + \cdots + a_{n-1} \frac{w + 1}{w - 1} + a_n = 0$$

Then, clearing the fractions by multiplying both sides of this last equation by $(w - 1)^n$, we obtain

$$Q(w) = b_0 w^n + b_1 w^{n-1} + \cdots + b_{n-1} w + b_n = 0$$

Once we transform $P(z) = 0$ into $Q(w) = 0$, it is possible to apply the Routh stability criterion in the same manner as in continuous-time systems.

It is noted that the bilinear transformation coupled with the Routh stability criterion will indicate exactly how many roots of the characteristic equation lie in the right half of the w plane and how many lie on the imaginary axis. However, such information about the exact number of unstable poles is usually not needed in control systems design, because unstable or critically stable control systems are not desired. As mentioned earlier, the amount of computation required in this approach is much more than that required in the Jury stability test. Therefore, we shall not go any further on this subject here. We refer the reader to Problem A–4–3, where the present method is used for stability analysis.

A Few Comments on the Stability of Closed-Loop Control Systems

1. If we are interested in the effect of a system parameter on the stability of a closed-loop control system, a root-locus diagram may prove to be useful. MATLAB may be used to compute and plot a root-locus diagram.
2. It is noted that in testing the stability of a characteristic equation it may be simpler, in some cases, to find the roots of the characteristic equation directly by use of MATLAB.
3. It is important to point out that stability has nothing to do with the system's ability to follow a particular input. The error signal in a closed-loop control system may increase without bound, even if the system is stable. (Refer to Section 4–4 for a discussion of error constants.)

4–4 TRANSIENT AND STEADY-STATE RESPONSE ANALYSIS

Absolute stability is a basic requirement of all control systems. In addition, good relative stability and steady-state accuracy are also required of any control system, whether continuous time or discrete time.

In this section we shall discuss transient response and steady-state response characteristics of closed-loop control systems. The transient response refers to that portion of the response due to the closed-loop poles of the system, and the steady-state response refers to that portion of the response due to the poles of the input or forcing function.

Discrete-time control systems are very frequently analyzed with "standard" inputs such as step inputs, ramp inputs, or sinusoidal inputs. This is because the system's response to any arbitrary input may be estimated from its response to such standard inputs. In this section, we shall consider the response of the discrete-time control system to time-domain inputs such as step inputs.

Transient Response Specifications.   In many practical cases, the desired performance characteristics of control systems, whether they are continuous time or discrete time, are specified in terms of time-domain quantities. This is because systems with energy storage cannot respond instantaneously and will always exhibit transient response whenever they are subjected to inputs or disturbances.

Frequently, the performance characteristics of a control system are specified in terms of the transient response to a unit-step input, since the unit-step input is easy to generate and is sufficiently drastic to provide useful information on both the transient response and the steady-state response characteristics of the system.

The transient response of a system to a unit-step input depends on the initial conditions. For convenience in comparing transient responses of various systems, it is a common practice to use the standard initial condition: the system is at rest initially and the output and all its time-derivatives are zero. The response characteristics can then be easily compared.

The transient response of a practical control system, where the output signal is continuous time, often exhibits damped oscillations before reaching the steady state. (This is true for the majority of discrete-time or digital control systems because the plants to be controlled are in most cases continuous time and, therefore, the output signals are continuous time.)

Consider, for example, the digital control system shown in Figure 4–14. The

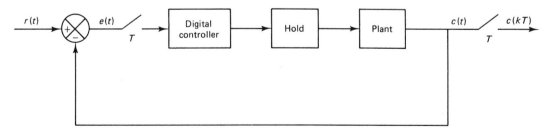

Figure 4–14   A digital control system.

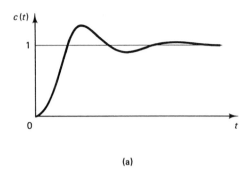

(a)

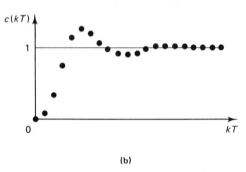

(b)

Figure 4–15 (a) Unit-step response of the system shown in Figure 4–14; (b) discrete-time output in the unit-step response.

output $c(t)$ of such a system to a unit-step input may exhibit damped oscillations as shown in Figure 4–15(a). Figure 4–15(b) shows the discrete-time output $c(kT)$.

Just as in the case of continuous-time control systems, the transient response of a digital control system may be characterized not only by the damping ratio and damped natural frequency, but also by the rise time, maximum overshoot, settling time, and so forth, in response to a step input. In fact, in specifying such transient response characteristics, it is common to specify the following quantities:

TRANSIENT RESPONSE SPECIFICATIONS

1. Delay time t_d
2. Rise time t_r
3. Peak time t_p
4. Maximum overshoot M_p
5. Settling time t_s

The aforementioned transient response specifications in the unit-step response are defined in what follows and are shown graphically in Figure 4–16.

1. *Delay time t_d.* The delay time is the time required for the response to reach half the final value the very first time.

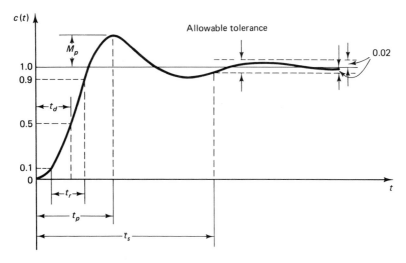

Figure 4–16   Unit-step response curve showing transient response specifications t_d, t_r, t_p, M_p, and t_s.

2. *Rise time t_r*. The rise time is the time required for the response to rise from 10% to 90%, or 5% to 95%, or 0% to 100% of its final value, depending on the situation. For underdamped second-order systems, the 0% to 100% rise time is commonly used. For overdamped systems and systems with transportation lags, the 10% to 90% rise time is commonly used.

3. *Peak time t_p*. The peak time is the time required for the response to reach the first peak of the overshoot.

4. *Maximum overshoot M_p*. The maximum overshoot is the maximum peak value of the response curve measured from unity. If the final steady-state value of the response differs from unity, then it is common to use the maximum percent overshoot. It is defined by the relation

$$\text{Maximum percent overshoot} = \frac{c(t_p) - c(\infty)}{c(\infty)} \times 100\%$$

The amount of the maximum (percent) overshoot directly indicates the relative stability of the system.

5. *Settling time t_s*. The settling time is the time required for the response curve to reach and stay within a range about the final value of a size specified as an absolute percentage of the final value, usually 2%. The settling time is related to the largest time constant of the control system.

The time-domain specifications just given are quite important since most control systems are time-domain systems; that is, they must exhibit acceptable time responses. (This means that the control system being designed must be modified until the transient response is satisfactory.)

Not all the specifications we have just defined necessarily apply to any given case. For example, for an overdamped system, the peak time and maximum overshoot terms do not apply. On the other hand, other specifications may be involved: for systems that yield steady-state errors for step inputs, the error must be kept within a specified percentage level. (Detailed discussions of steady-state errors will be given later in this section.)

Let us assume that the sampling theorem is satisfied and no frequency folding occurs. The nature of the transient response of a discrete-time control system to a given input depends on the actual locations of the closed-loop poles and zeros in the z plane. Consider the discrete-time control system defined by

$$\frac{C(z)}{R(z)} = \frac{b_0 z^n + b_1 z^{n-1} + \cdots + b_n}{z^n + a_1 z^{n-1} + \cdots + a_n} \tag{4-9}$$

where $R(z)$ is the z transform of the input and $C(z)$ is the z transform of the output. The transient response of such a system to the Kronecker delta input, step input, ramp input, and so on, can be obtained easily by use of MATLAB. See, for example, Example 4–8.

Example 4–8

Consider the discrete-time control system defined by

$$\frac{C(z)}{R(z)} = \frac{0.4673z - 0.3393}{z^2 - 1.5327z + 0.6607} \tag{4-10}$$

Obtain the unit-step response of this system.

A MATLAB program for obtaining the unit-step response is shown in MATLAB Program 4–1. The resulting plot of $c(k)$ versus k is shown in Figure 4–17.

```
MATLAB Program 4–1

% ---------- Unit-step response ----------

num = [0  0.4673  -0.3393];
den = [1  -1.5327  0.6607];
r = ones(1,41);
k = 0: 40;
c = filter(num,den,r );
plot(k,c,'o')
v = [0  40  0  1.6];
axis(v);
grid
title('Unit-Step Response')
xlabel('k')
ylabel('c(k)')
```

Steady-State Error Analysis.   An important feature associated with transient response is steady-state error. The steady-state performance of a stable control system is generally judged by the steady-state error due to step, ramp, and acceleration inputs. In what follows we shall investigate a type of steady-state error that is caused by the inability of a system to follow particular types of inputs. (It should be noted that, besides this type of steady-state error, there are errors that can be attributed to other causes, such as imperfections in system components, static

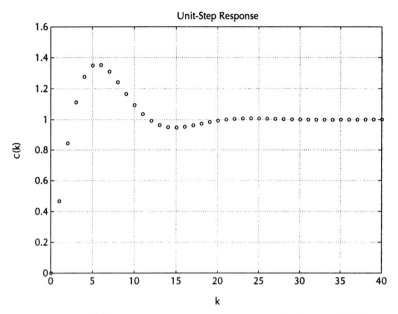

Figure 4-17   Unit-step response of the system defined by Equation (4-10).

friction, backlash, or deterioration or aging of components. In this section, however, we shall not discuss steady-state error due to such causes.)

Any physical control system inherently suffers steady-state error in response to certain types of inputs. That is, a system may have no steady-state error with step inputs, but the same system may exhibit nonzero steady-state error in response to ramp inputs. Whether or not a given system will exhibit steady-state error in response to a given type of input depends on the type of open-loop transfer function of the system.

Consider the continuous-time control system whose open-loop transfer function $G(s)H(s)$ is given by

$$G(s)H(s) = \frac{K(T_a s + 1)(T_b s + 1) \cdots (T_m s + 1)}{s^N(T_1 s + 1)(T_2 s + 1) \cdots (T_p s + 1)}$$

The term s^N in the denominator represents a pole of multiplicity N at the origin. It is customary to classify the system according to the number of integrators in the open-loop transfer function.

A system is said to be of type 0, type 1, type 2,..., if $N = 0$, $N = 1$, $N = 2, \ldots$, respectively. Type 0 systems will exhibit finite steady-state errors in response to step inputs and infinite errors in response to ramp and higher-order inputs. Type 1 systems will exhibit no steady-state error in response to step inputs, finite steady-state errors in response to ramp inputs, and infinite steady-state errors in response to acceleration and higher-order inputs. As the type number is increased, accuracy is improved. However, increasing the type number aggravates the stability problem. A compromise between steady-state accuracy and relative stability (transient response characteristics) is always necessary.

The concepts of static error constants can be extended to the discrete-time control system, as discussed in what follows.

Discrete-time control systems can be classified according to the number of open-loop poles at $z = 1$. (An open-loop pole at $z = 1$ corresponds to an integrator in the loop.) Suppose the open-loop pulse transfer function is given by the equation

$$\text{Open-loop pulse transfer function} = \frac{1}{(z - 1)^N} \frac{B(z)}{A(z)}$$

where $B(z)/A(z)$ contains neither a pole nor a zero at $z = 1$. Then the system can be classified as a type 0 system, a type 1 system, or a type 2 system according to whether $N = 0$, $N = 1$, or $N = 2$, respectively. The system type specifies the steady-state characteristics or steady-state accuracy.

The physical meaning of the static error constants for discrete-time control systems is the same as that for continuous-time control systems, except that the former transmit information only at the sampling instants.

Consider the discrete-time control system shown in Figure 4–18. We assume that the system is stable so that the final value theorem can be applied to find the steady-state values. From the diagram we have the actuating error

$$e(t) = r(t) - b(t)$$

We shall consider the steady-state actuating error at the sampling instants. Note that from the final value theorem we have

$$\lim_{k \to \infty} e(kT) = \lim_{z \to 1} [(1 - z^{-1})E(z)] \tag{4-11}$$

For the system shown in Figure 4–18, define

$$G(z) = (1 - z^{-1})\mathcal{Z}\left[\frac{G_p(s)}{s}\right]$$

and

$$GH(z) = (1 - z^{-1})\mathcal{Z}\left[\frac{G_p(s)H(s)}{s}\right]$$

Then we have

$$\frac{C(z)}{R(z)} = \frac{G(z)}{1 + GH(z)}$$

and

$$E(z) = R(z) - B(z) = R(z) - GH(z)E(z)$$

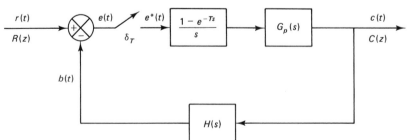

Figure 4–18   Discrete-time control system.

or

$$E(z) = \frac{1}{1 + GH(z)} R(z) \qquad (4\text{-}12)$$

By substituting Equation (4-12) into Equation (4-11), we obtain

$$e_{ss} = \lim_{z \to 1} \left[(1 - z^{-1}) \frac{1}{1 + GH(z)} R(z) \right] \qquad (4\text{-}13)$$

As in the case of the continuous-time control system, we consider three types of inputs: unit-step, unit-ramp, and unit-acceleration inputs.

Static Position Error Constant.   For a unit-step input $r(t) = 1(t)$, we have

$$R(z) = \frac{1}{1 - z^{-1}}$$

By substituting this last equation into Equation (4-13) the steady-state actuating error in response to a unit-step input can be obtained as follows:

$$e_{ss} = \lim_{z \to 1} \left[(1 - z^{-1}) \frac{1}{1 + GH(z)} \frac{1}{1 - z^{-1}} \right] = \lim_{z \to 1} \frac{1}{1 + GH(z)}$$

We define the static position error constant K_p as follows:

$$K_p = \lim_{z \to 1} GH(z) \qquad (4\text{-}14)$$

Then the steady-state actuating error in response to a unit-step input can be obtained from the equation

$$e_{ss} = \frac{1}{1 + K_p} \qquad (4\text{-}15)$$

The steady-state actuating error in response to a unit-step input becomes zero if $K_p = \infty$, which requires that $GH(z)$ have at least one pole at $z = 1$.

Static Velocity Error Constant.   For a unit-ramp input $r(t) = t1(t)$, we have

$$R(z) = \frac{Tz^{-1}}{(1 - z^{-1})^2}$$

By substituting this last equation into Equation (4-13), we have

$$e_{ss} = \lim_{z \to 1} \left[(1 - z^{-1}) \frac{1}{1 + GH(z)} \frac{Tz^{-1}}{(1 - z^{-1})^2} \right] = \lim_{z \to 1} \frac{T}{(1 - z^{-1})GH(z)}$$

Now we define the static velocity error constant K_v as follows:

$$K_v = \lim_{z \to 1} \frac{(1 - z^{-1})GH(z)}{T} \qquad (4\text{-}16)$$

Then the steady-state actuating error in response to a unit-ramp input can be given by

$$e_{ss} = \frac{1}{K_v} \qquad (4\text{-}17)$$

If $K_v = \infty$, then the steady-state actuating error in response to a unit-ramp input is zero. This requires $GH(z)$ to possess a double pole at $z = 1$.

Static Acceleration Error Constant. For a unit acceleration input $r(t) = \frac{1}{2}t^2 1(t)$, we have

$$R(z) = \frac{T^2(1 + z^{-1})z^{-1}}{2(1 - z^{-1})^3}$$

By substituting this last equation into Equation (4–13), we obtain

$$e_{ss} = \lim_{z \to 1}\left[(1 - z^{-1})\frac{1}{1 + GH(z)}\frac{T^2(1 + z^{-1})z^{-1}}{2(1 - z^{-1})^3}\right] = \lim_{z \to 1}\frac{T^2}{(1 - z^{-1})^2 GH(z)}$$

We define the static acceleration error constant K_a as follows:

$$K_a = \lim_{z \to 1}\frac{(1 - z^{-1})^2 GH(z)}{T^2} \qquad (4\text{–}18)$$

Then the steady-state actuating error becomes

$$e_{ss} = \frac{1}{K_a} \qquad (4\text{–}19)$$

The steady-state actuating error in response to a unit-acceleration input becomes zero if $K_a = \infty$. This requires $GH(z)$ to possess a triple pole at $z = 1$.

Equations (4–15), (4–17), and (4–19) give the expressions for steady-state actuating errors of the discrete-time control system shown in Figure 4-18 at the sampling instants for a unit-step, unit-ramp, and unit-acceleration input, respectively.

Summary. It is important to emphasize that the actuating error is the difference between the reference input and the feedback signal, not the difference between the reference input and the output. From the foregoing analysis we see that a type 0 system will exhibit a constant steady-state actuating error in response to a step input and an infinite actuating error in response to ramp, acceleration, or higher-order inputs. A type 1 system will exhibit a zero steady-state actuating error in response to a step input, a constant steady-state error in response to a ramp input, and an infinite steady-state actuating error in response to acceleration or higher-order inputs.

TABLE 4–4 SYSTEM TYPES AND THE CORRESPONDING
STEADY-STATE ERRORS IN RESPONSE TO STEP, RAMP, AND
ACCELERATION INPUTS FOR THE DISCRETE-TIME CONTROL
SYSTEM SHOWN IN FIGURE 4–18

	Steady-state errors in response to		
System	Step input $r(t) = 1$	Ramp input $r(t) = t$	Acceleration input $r(t) = \frac{1}{2}t^2$
Type 0 system	$\dfrac{1}{1 + K_p}$	∞	∞
Type 1 system	0	$\dfrac{1}{K_v}$	∞
Type 2 system	0	0	$\dfrac{1}{K_a}$

Table 4–4 lists system types and the corresponding steady-state errors in response to step, ramp, and acceleration inputs for the discrete-time control system of the configuration shown in Figure 4–18.

The steady-state error analysis just presented applies to the closed-loop discrete-time control system shown in Figure 4–18. For a different closed-loop configuration, it is noted that if the closed-loop discrete-time control system has a closed-loop pulse transfer function, then the static error constants can be determined by an analysis similar to the one just presented. Table 4–5 lists the static error constants for typical closed-loop configurations of discrete-time control systems. If the closed-loop discrete-time control system does not have a closed-loop pulse transfer function, however, the static error constants cannot be defined, because the input signal cannot be separated from the system dynamics.

It is important to note that the terms "position error," "velocity error," and "acceleration error" mean steady-state deviations in the output position. A finite velocity error implies that after transients have died out the input and output move at the same velocity, but have a finite position difference.

TABLE 4–5 STATIC ERROR CONSTANTS FOR TYPICAL CLOSED-LOOP CONFIGURATIONS OF DISCRETE-TIME CONTROL SYSTEMS

Closed-loop configuration	Values of K_p, K_v, and K_a
	$$K_p = \lim_{z \to 1} GH(z)$$ $$K_v = \lim_{z \to 1} \frac{(1 - z^{-1})GH(z)}{T}$$ $$K_a = \lim_{z \to 1} \frac{(1 - z^{-1})^2 GH(z)}{T^2}$$
	$$K_p = \lim_{z \to 1} G(z)H(z)$$ $$K_v = \lim_{z \to 1} \frac{(1 - z^{-1})G(z)H(z)}{T}$$ $$K_a = \lim_{z \to 1} \frac{(1 - z^{-1})^2 G(z)H(z)}{T^2}$$
	$$K_p = \lim_{z \to 1} G_1(z)HG_2(z)$$ $$K_v = \lim_{z \to 1} \frac{(1 - z^{-1})G_1(z)HG_2(z)}{T}$$ $$K_a = \lim_{z \to 1} \frac{(1 - z^{-1})^2 G_1(z)HG_2(z)}{T^2}$$
	$$K_p = \lim_{z \to 1} G_1(z)G_2(z)H(z)$$ $$K_v = \lim_{z \to 1} \frac{(1 - z^{-1})G_1(z)G_2(z)H(z)}{T}$$ $$K_a = \lim_{z \to 1} \frac{(1 - z^{-1})^2 G_1(z)G_2(z)H(z)}{T^2}$$

Response to Disturbances.   In examining transient response characteristics and steady-state errors, it is important to note that the effects of disturbances, in addition to those of reference inputs, must be explored.

For the system shown in Figure 4–19(a), let us assume that the reference input is zero, or $R(z) = 0$, but the system is subjected to disturbance $N(z)$. For this case the system block diagram can be redrawn as shown in Figure 4–19(b). Then the response $C(z)$ to disturbance $N(z)$ can be found from the closed-loop pulse transfer function

$$\frac{C(z)}{N(z)} = \frac{G(z)}{1 + G_D(z)G(z)}$$

If $|G_D(z)G(z)| \gg 1$, then we find

$$\frac{C(z)}{N(z)} \cong \frac{1}{G_D(z)}$$

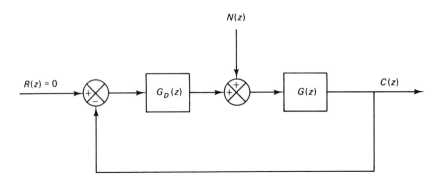

(a)

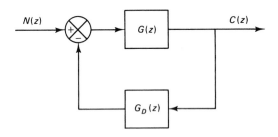

(b)

Figure 4–19   (a) Digital closed-loop control system subjected to reference input and disturbance input; (b) modified block diagram where the disturbance input is considered the input to the system.

Since the system error is

$$E(z) = R(z) - C(z) = -C(z)$$

we find the error $E(z)$ due to the disturbance $N(z)$ to be

$$E(z) = -\frac{1}{G_D(z)}N(z)$$

Thus, the larger the gain of $G_D(z)$ is, the smaller the error $E(z)$. If $G_D(z)$ includes an integrator [which means that $G_D(z)$ has a pole at $z = 1$], then the steady-state error due to a constant disturbance is zero. This may be seen as follows. Since for a constant disturbance of magnitude N we have

$$N(z) = \frac{N}{1 - z^{-1}}$$

if $G_D(z)$ involves a pole at $z = 1$, then it may be written as

$$G_D(z) = \frac{\hat{G}_D(z)}{z - 1} = \frac{\hat{G}_D(z)z^{-1}}{1 - z^{-1}}$$

where $\hat{G}_D(z)$ does not involve any zeros at $z = 1$. Then the steady-state error can be given by

$$e_{ss} = \lim_{z \to 1}\left[(1 - z^{-1})E(z)\right] = \lim_{z \to 1}\left[(1 - z^{-1})\frac{-N(z)}{G_D(z)}\right]$$

$$= -\lim_{z \to 1}\left[(1 - z^{-1})\frac{N}{1 - z^{-1}}\frac{1}{G_D(z)}\right] = -\lim_{z \to 1}\frac{(1 - z^{-1})N}{\hat{G}_D(z)z^{-1}} = 0$$

If a linear system is subjected to both the reference input and a disturbance input, then the resulting error is the sum of the errors due to the reference input and the disturbance input. The total error must be kept within acceptable limits.

Note that the point where the disturbance enters the system is very important in adjusting the gain of $G_D(z)G(z)$. For example, consider the system shown in Figure 4–20(a). The closed-loop pulse transfer function for the disturbance is

$$\frac{C(z)}{N(z)} = -\frac{E(z)}{N(z)} = \frac{1}{1 + G_D(z)G(z)}$$

To minimize the effects of disturbance $N(z)$ on the system error $E(z)$, the gain of $G_D(z)G(z)$ must be made as large as possible. However, for the system shown in Figure 4–20(b), the closed-loop pulse transfer function for the disturbance is

$$\frac{C(z)}{N(z)} = -\frac{E(z)}{N(z)} = -\frac{G_D(z)G(z)}{1 + G_D(z)G(z)}$$

and to minimize the effects of disturbance $N(z)$ on the system error $E(z)$, the gain of $G_D(z)G(z)$ must be made as small as possible.

Therefore, it is advantageous to obtain the expression for $E(z)/N(z)$ before concluding whether the gain of $G_D(z)G(z)$ should be large or small to minimize the error due to disturbances. It is important to remember, however, that the magnitude of the gain cannot be determined solely from the disturbance considerations. It must be determined by considering the responses to both reference and disturbance inputs. If the frequency regions for the reference input and disturbance input are sufficiently apart, a suitable filter may be inserted in the system. If the frequency

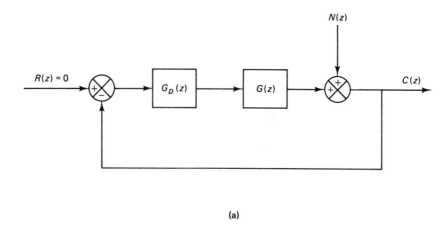

(a)

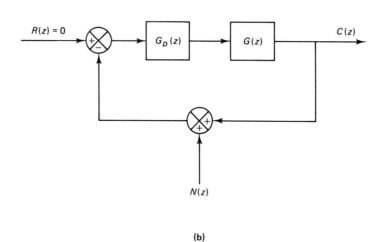

(b)

Figure 4–20   (a) Digital closed-loop control system subjected to reference input and disturbance input; (b) digital closed-loop control system where the disturbance enters the feedback loop.

regions overlap, then modification of the block diagram configuration may become necessary to get acceptable responses to both reference and disturbance inputs.

4–5 DESIGN BASED ON THE ROOT-LOCUS METHOD

As discussed in Section 4–4, the relative stability of the discrete-time control system may be investigated with respect to the unit circle in the z plane. For example, if the closed-loop poles are complex conjugates and lie inside the unit circle, the unit-step response will be oscillatory.

In addition to the transient response characteristics of a given system, it is often necessary to investigate the effects of the system gain and/or sampling period on the absolute and relative stability of the closed-loop system. For such purposes the root-locus method proves to be very useful.

The root-locus method developed for continuous-time systems can be extended to discrete-time systems without modifications, except that the stability boundary is changed from the $j\omega$ axis in the s plane to the unit circle in the z plane. The reason the root-locus method can be extended to discrete-time systems is that the characteristic equation for the discrete-time system is of the same form as that for the continuous-time system in the s plane. For example, for the system shown in Figure 4–21 the characteristic equation is

$$1 + G(z)H(z) = 0$$

which is of exactly the same form as the equation for root-locus analysis in the s plane. However, the pole locations for closed-loop systems in the z plane must be interpreted differently from those in the s plane.

In this section we shall demonstrate the application of the root-locus method to the design of discrete-time or digital control systems.

Computer programs for calculating and tracing root loci are available for most computer systems. In particular, MATLAB provides a convenient means to plot root loci for both continuous-time and discrete-time closed-loop systems. Exact plotting of the root loci can be done on the computer and, therefore, we may not need graphical plotting procedures. However, skill in plotting root loci is an advantage, since it will enable the control engineer to make quick graphical plots for given problems to speed up preliminary stages of system design. In fact, the experienced control engineer frequently uses the root-locus approach to a preliminary design to locate the dominant closed-loop poles at desired positions in the z plane and then uses a digital simulation to improve the closed-loop performance.

Angle and Magnitude Conditions.   In many linear time-invariant discrete-time control systems, the characteristic equation may have either of the following two forms:

$$1 + G(z)H(z) = 0$$

and

$$1 + GH(z) = 0$$

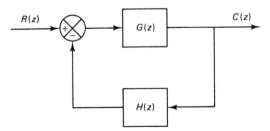

Figure 4–21   Closed-loop control system.

To combine these two forms into one, let us define the characteristic equation as

$$1 + F(z) = 0 \qquad\qquad (4\text{--}20)$$

where

$$F(z) = G(z)H(z) \qquad \text{or} \qquad F(z) = GH(z)$$

Note that $F(z)$ is the open-loop pulse transfer function. The characteristic equation given by Equation (4–20) can then be written as

$$F(z) = -1$$

Since $F(z)$ is a complex quantity, this last equation can be split into two equations by equating first the angles and then the magnitudes of the two sides to obtain

ANGLE CONDITION:

$$\underline{/F(z)} = \pm 180°(2k + 1), \qquad k = 0, 1, 2, \ldots$$

MAGNITUDE CONDITION:

$$|F(z)| = 1$$

The values of z that fulfill both the angle and the magnitude conditions are the roots of the characteristic equation, or the closed-loop poles.

A plot of the points in the complex plane satisfying the angle condition alone is the root locus. The roots of the characteristic equation (the closed-loop poles) corresponding to a given value of the gain can be located on the root loci by use of the magnitude condition. The details of applying the angle and magnitude conditions to obtain the closed-loop poles are presented in the following.

General Procedure for Constructing Root Loci.   For a complicated system with many open-loop poles and zeros, constructing a root-locus plot may seem complicated, but actually it is not difficult if the established rules for constructing root loci are applied.

By locating particular points and asymptotes and by computing angles of departure from complex poles and angles of arrival at complex zeros, it is possible to construct root loci without difficulty. Note that while root loci may be conveniently drawn with a digital computer, if manual construction of the root locus plot is attempted, we essentially proceed on a trial-and-error basis. But the number of trials required can be greatly reduced if the established rules are used.

Because the open-loop conjugate complex poles and conjugate complex zeros, if any, are always located symmetrically about the real axis, the root loci are always symmetric with respect to the real axis. Hence, we need only construct the upper half of the root loci and draw the mirror image of the upper half in the lower half of the z plane. Remember that the angles of the complex quantities originating from the open-loop poles and open-loop zeros and drawn to a test point z are measured in the counterclockwise direction. We shall now present the general rules and procedures for constructing root loci.

General Rules for Constructing Root Loci

1. Obtain the characteristic equation

$$1 + F(z) = 0$$

and then rearrange this equation so that the parameter of interest, such as gain K, appears as the multiplying factor in the form

$$1 + \frac{K(z + z_1)(z + z_2) \cdots (z + z_m)}{(z + p_1)(z + p_2) \cdots (z + p_n)} = 0$$

In the present discussion, we assume that the parameter of interest is gain K, where $K > 0$. From the factored form of the open-loop pulse transfer function, locate the open-loop poles and zeros in the z plane. [Note that if $F(z) = G(z)H(z)$ then the open-loop zeros are zeros of $G(z)H(z)$, while the closed-loop zeros consist of the zeros of $G(z)$ and the poles of $H(z)$.]

2. Find the starting points and terminating points of the root loci. Find also the number of separate branches of the root loci. The points on the root loci corresponding to $K = 0$ are open-loop poles and those corresponding to $K = \infty$ are open-loop zeros. Hence, as K is increased from zero to infinity, a root locus starts from an open-loop pole and terminates at a finite open-loop zero or an open-loop zero at infinity. This means that a root-locus plot will have just as many branches as there are roots of the characteristic equation. [If the zeros at infinity are included in the count, $F(z)$ has the same number of zeros as poles.]

If the number n of closed-loop poles is the same as the number of open-loop poles, then the number of individual root locus branches terminating at finite open-loop zeros is equal to the number m of the open-loop zeros. The remaining $n - m$ branches terminate at infinity (at $n - m$ implicit zeros at infinity) along asymptotes.

3. Determine the root loci on the real axis. Root loci on the real axis are determined by open-loop poles and zeros lying on it. The conjugate complex poles and zeros of the open-loop pulse transfer function have no effect on the location of the root loci on the real axis because the angle contribution of a pair of conjugate complex poles or zeros is 360° on the real axis. Each portion of the root locus on the real axis extends over a range from a pole or zero to another pole or zero.

In constructing the root loci on the real axis, choose a test point on it. If the total number of real poles and real zeros to the right of this test point is odd, then this point lies on a root locus. The root locus and its complement form alternate segments along the real axis.

4. Determine the asymptotes of the root loci. If the test point z is located far from the origin, then the angles of all the complex quantities may be considered the same. One open-loop zero and one open-loop pole then each cancel the effects of the other.

Therefore, the root loci for very large values of z must be asymptotic to straight lines whose angles are given as follows:

$$\text{Angle of asymptote} = \frac{\pm 180°(2N + 1)}{n - m}, \qquad N = 0, 1, 2, \ldots$$

where

$$n = \text{number of finite poles of } F(z)$$

$$m = \text{number of finite zeros of } F(z)$$

Here, $N = 0$ corresponds to the asymptote with the smallest angle with the real axis. Although N assumes an infinite number of values, the angle repeats itself, as N is increased, and the number of distinct asymptotes is $n - m$.

All the asymptotes intersect on the real axis. The point at which they do so is obtained as follows. Since

$$F(z) = \frac{K[z^m + (z_1 + z_2 + \cdots + z_m)z^{m-1} + \cdots + z_1 z_2 \cdots z_m]}{z^n + (p_1 + p_2 + \cdots + p_n)z^{n-1} + \cdots + p_1 p_2 \cdots p_n}$$

$$= \frac{K}{z^{n-m} + [(p_1 + p_2 + \cdots + p_n) - (z_1 + z_2 + \cdots + z_m)]z^{n-m-1} + \cdots}$$

for a large value of z this last equation may be approximated as follows:

$$F(z) \doteq \frac{K}{\left[z + \dfrac{(p_1 + p_2 + \cdots + p_n) - (z_1 + z_2 + \cdots + z_m)}{n - m}\right]^{n-m}}$$

If the abscissa of the intersection of the asymptotes and the real axis is denoted by $-\sigma_a$, then

$$-\sigma_a = -\frac{(p_1 + p_2 + \cdots + p_n) - (z_1 + z_2 + \cdots + z_m)}{n - m} \tag{4-21}$$

Because all the complex poles and zeros occur in conjugate pairs, $-\sigma_a$ given by Equation (4–21) is always a real quantity.

Once the intersection of the asymptotes and the real axis is found, the asymptotes can be readily drawn in the complex z plane.

5. Find the breakaway and break-in points. Because of the conjugate symmetry of the root loci, the breakaway points and break-in points either lie on the real axis or occur in conjugate complex pairs.

If a root locus lies between two adjacent open-loop poles on the real axis, then there exists at least one breakaway point between the two poles. Similarly, if the root locus lies between two adjacent zeros (one zero may be located at $-\infty$) on the real axis, then there always exists at least one break-in point between the two zeros.

If the root locus lies between an open-loop pole and a zero (finite or infinite) on the real axis, then there may exist no breakaway or break-in points or there may exist both breakaway and break-in points.

If the characteristic equation

$$1 + F(z) = 0$$

is written as

$$1 + \frac{KB(z)}{A(z)} = 0$$

where $KB(z)/A(z) = F(z)$, then

$$K = -\frac{A(z)}{B(z)} \tag{4–22}$$

and the breakaway and break-in points (which correspond to multiple roots) can be determined from the roots of

$$\frac{dK}{dz} = -\frac{A'(z)B(z) - A(z)B'(z)}{B^2(z)} = 0 \tag{4–23}$$

where the prime indicates differentiation with respect to z. (See Problem A–4–5 for a proof.)

If the value of K corresponding to a root $z = z_0$ of $dK/dz = 0$ is positive, point $z = z_0$ is an actual breakaway or break-in point. Since K is assumed to be non-negative, if the value of K thus obtained is negative, then point $z = z_0$ is neither a breakaway nor a break-in point.

Note that this approach can be used when there are complex poles and/or complex zeros.

6. Determine the angle of departure (or angle of arrival) of the root loci from the complex poles (or at the complex zeros). To sketch the root loci with reasonable accuracy, we must find the direction of the root loci near the complex poles and zeros. The angle of departure (or angle of arrival) of the root locus from a complex pole (or at a complex zero) can be found by subtracting from 180° the sum of all the angles of lines (complex quantities) from all other poles and zeros to the complex pole (or complex zero) in question, with appropriate signs included. The angle of departure is shown in Figure 4–22.

7. Find the points where the root loci cross the imaginary axis. The points where the root loci intersect the imaginary axis can be found by setting $z = j\nu$ in the characteristic equation (which involves undetermined gain K), equating both the real part and the imaginary part to zero, and solving for ν and K. The values of ν and

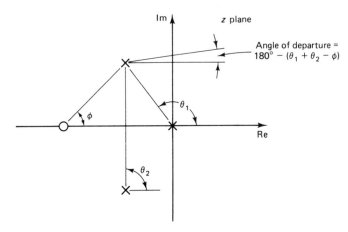

Figure 4–22   Diagram showing angle of departure.

K thus found give the location at which the root loci cross the imaginary axis and the value of the corresponding gain K, respectively.

8. Any point on the root loci is a possible closed-loop pole. A particular point will be a closed-loop pole when the value of gain K satisfies the magnitude condition. Conversely, the magnitude condition enables us to determine the value of gain K at any specific root location on the locus. The magnitude condition is

$$|F(z)| = 1$$

or

$$\left| \frac{(z + z_1)(z + z_2) \cdots (z + z_m)}{(z + p_1)(z + p_2) \cdots (z + p_n)} \right| = \frac{1}{K} \tag{4-24}$$

If gain K of the open-loop pulse transfer function is given in the problem, then by applying the magnitude condition, Equation (4–24), it is possible to locate the closed-loop poles for a given K on each branch of the root loci by a trial-and-error method.

Cancellation of Poles of G(z) With Zeros of H(z).   It is important to note that if $F(z) = G(z)H(z)$ and the denominator of $G(z)$ and the numerator of $H(z)$ involve common factors then the corresponding open-loop poles and zeros will cancel each other, reducing the degree of the characteristic equation by one or more. The root locus plot of $G(z)H(z)$ will not show all the roots of the characteristic equation, but only the roots of the reduced equation.

To obtain the complete set of closed-loop poles, we must add the canceled pole or poles of $G(z)H(z)$ to those closed-loop poles obtained from the root locus plot of $G(z)H(z)$. The important thing to remember is that a canceled pole of $G(z)H(z)$ is a closed-loop pole of the system.

As an example, consider the case where $G(z)$ and $H(z)$ of the system shown in Figure 4–21 are given by

$$G(z) = \frac{z + c}{(z + a)(z + b)}$$

and

$$H(z) = \frac{z + a}{z + d}$$

Then, clearly, the pole $z = -a$ of $G(z)$ and the zero $z = -a$ of $H(z)$ cancel each other, resulting in

$$G(z)H(z) = \frac{z + c}{(z + a)(z + b)} \frac{z + a}{z + d} = \frac{z + c}{(z + b)(z + d)}$$

However, the closed-loop pulse transfer function of the system is

$$\frac{C(z)}{R(z)} = \frac{G(z)}{1 + G(z)H(z)} = \frac{(z + c)(z + d)}{(z + a)[(z + b)(z + d) + z + c]}$$

and we see that $z = -a$, the canceled pole of $G(z)H(z)$, is a closed-loop pole of the closed-loop system.

Note, however, that if pole–zero cancellation occurs in the feed-forward pulse transfer function, then the same pole–zero cancellation occurs in the closed-loop pulse transfer function. Consider again the system shown in Figure 4–21, where we assume

$$G(z) = G_D(z)G_1(z), \qquad H(z) = 1$$

Suppose pole–zero cancellation occurs in $G_D(z)G_1(z)$. For example, suppose

$$G_D(z)G_1(z) = \frac{z + b}{z + a} \frac{z + d}{(z + b)(z + c)} = \frac{z + d}{(z + a)(z + c)}$$

Then the closed-loop pulse transfer function becomes

$$\frac{C(z)}{R(z)} = \frac{G_D(z)G_1(z)}{1 + G_D(z)G_1(z)} = \frac{(z + b)(z + d)}{(z + b)[(z + a)(z + c) + z + d]}$$

$$= \frac{z + d}{(z + a)(z + c) + z + d}$$

Because of the pole–zero cancellation, the third-order system becomes one of second order.

It is important to summarize that the effect of pole–zero cancellation in $G(z)$ and $H(z)$ is different from that of pole–zero cancellation in the feed-forward pulse transfer function (such as pole–zero cancellation in the digital controller and the plant). In the former, the canceled pole is still a pole of the closed-loop system, whereas in the latter the canceled pole does not appear as a pole of the closed-loop system (in the latter the order of the system is reduced by the number of canceled poles).

Root-Locus Diagrams of Digital Control Systems.   In what follows we shall investigate the effects of gain K and sampling period T on the relative stability of the closed-loop control system. Consider the system shown in Figure 4–23. Assume that the digital controller is of the integral type, or that

$$G_D(z) = \frac{K}{1 - z^{-1}} = K\frac{z}{z - 1}$$

Let us draw root locus diagrams for the system for three values of the sampling period T: 0.5 sec, 1 sec, and 2 sec. Let us also determine the critical value of K for each case. And finally let us locate the closed-loop poles corresponding to $K = 2$ for each of the three cases.

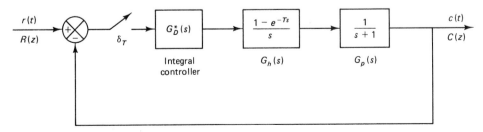

Figure 4–23   Digital control system.

We shall first obtain the z transform of $G_h(s)G_p(s)$:

$$\mathcal{Z}[G_h(s)G_p(s)] = \mathcal{Z}\left[\frac{1 - e^{-Ts}}{s} \frac{1}{s + 1}\right]$$

$$= (1 - z^{-1})\mathcal{Z}\left[\frac{1}{s(s + 1)}\right]$$

$$= (1 - z^{-1})\mathcal{Z}\left[\frac{1}{s} - \frac{1}{s + 1}\right]$$

$$= \frac{z - 1}{z}\left(\frac{z}{z - 1} - \frac{z}{z - e^{-T}}\right)$$

$$= \frac{1 - e^{-T}}{z - e^{-T}}$$

The feedforward pulse transfer function becomes

$$G(z) = G_D(z)\mathcal{Z}[G_h(s)G_p(s)] = \frac{Kz}{z - 1}\frac{1 - e^{-T}}{z - e^{-T}} \qquad (4\text{--}25)$$

The characteristic equation is

$$1 + G(z) = 0$$

or

$$1 + \frac{Kz(1 - e^{-T})}{(z - 1)(z - e^{-T})} = 0 \qquad (4\text{--}26)$$

1. *Sampling period T = 0.5 sec*: For this case, Equation (4–25) becomes

$$G(z) = \frac{0.3935Kz}{(z - 1)(z - 0.6065)}$$

Notice that $G(z)$ has poles at $z = 1$ and $z = 0.6065$ and a zero at $z = 0$.

To draw a root locus diagram, we first locate the poles and zero in the z plane and then find the breakaway point and break-in point. Notice that this open-loop pulse transfer function with two poles and one zero results in a circular root locus centered at the zero. The breakaway point and break-in point are determined by writing the characteristic equation in the form of Equation (4–22),

$$K = -\frac{(z - 1)(z - 0.6065)}{0.3935z} \qquad (4\text{--}27)$$

and differentiating K with respect to z and equating the result to zero:

$$\frac{dK}{dz} = -\frac{z^2 - 0.6065}{0.3935z^2} = 0$$

Hence,

$$z^2 = 0.6065$$

or

$$z = 0.7788 \qquad \text{and} \qquad z = -0.7788$$

Notice that substitution of 0.7788 for z in Equation (4–27) yields $K = 0.1244$, while letting $z = -0.7788$ yields $K = 8.041$. Since both K values are positive, $z = 0.7788$ is the actual breakaway point and $z = -0.7788$ is the actual break-in point.

Figure 4–24(a) shows the root locus diagram when $T = 0.5$ sec. The critical value of gain K for this case is obtained by use of the magnitude condition, which can be obtained from Equation (4–26) as follows:

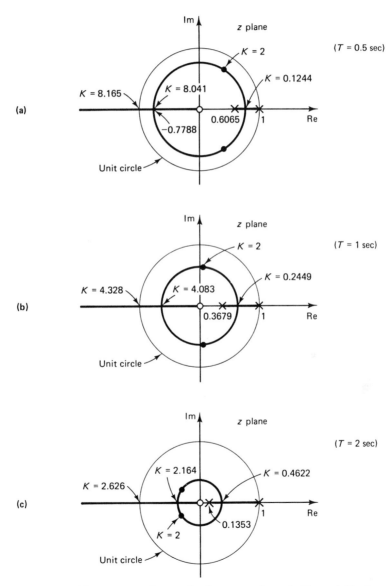

Figure 4–24 (a) Root locus diagram for the system shown in Figure 4–23 when $T = 0.5$ sec; (b) root locus diagram when $T = 1$ sec; (c) root locus diagram when $T = 2$ sec.

$$\left| \frac{z(1 - e^{-T})}{(z - 1)(z - e^{-T})} \right| = \frac{1}{K}$$

For the present case, $T = 0.5$ and this last equation becomes

$$\left| \frac{0.3935z}{(z - 1)(z - 0.6065)} \right| = \frac{1}{K} \tag{4–28}$$

Since the critical gain K_c corresponds to point $z = -1$, we substitute -1 for z in Equation (4–28):

$$\left| \frac{0.3935(-1)}{(-2)(-1.6065)} \right| = \frac{1}{K}$$

or

$$K = 8.165$$

The critical gain K_c is thus 8.165.

The closed-loop poles corresponding to $K = 2$ can be found to be

$$z_1 = 0.4098 + j0.6623 \quad \text{and} \quad z_2 = 0.4098 - j0.6623$$

These closed-loop poles are indicated by dots in the root locus diagram.

2. *Sampling period $T = 1$ sec*: For this case, Equation (4–25) becomes as follows:

$$G(z) = \frac{0.6321Kz}{(z - 1)(z - 0.3679)}$$

Hence, $G(z)$ has poles at $z = 1$ and $z = 0.3679$ and a zero at $z = 0$.

The breakaway point and break-in point are found to be $z = 0.6065$ and $z = -0.6065$, respectively. The corresponding gain values are $K = 0.2449$ and $K = 4.083$, respectively.

Figure 4–24(b) shows the root locus diagram when $T = 1$ sec. The critical value of gain K is 4.328. The closed-loop poles corresponding to $K = 2$ are found to be

$$z_1 = 0.05185 + j0.6043 \quad \text{and} \quad z_2 = 0.05185 - j0.6043$$

and are shown in the root locus diagram by dots.

3. *Sampling period $T = 2$ sec*: For this case, Equation (4–25) becomes

$$G(z) = \frac{0.8647Kz}{(z - 1)(z - 0.1353)}$$

We see that $G(z)$ has poles at $z = 1$ and $z = 0.1353$ and a zero at $z = 0$.

The breakaway point and break-in point are found to be $z = 0.3678$ and $z = -0.3678$, with corresponding gain values $K = 0.4622$ and $K = 2.164$, respectively. The critical value of gain K for this case is 2.626.

Figure 4–24(c) shows the root locus diagram when $T = 2$ sec. The closed-loop poles corresponding to $K = 2$ are found to be

$$z_1 = -0.2971 + j0.2169 \quad \text{and} \quad z_2 = -0.2971 - j0.2169$$

These closed-loop poles are shown by dots in the root locus diagram.

Effects of Sampling Period T on Transient Response Characteristics.   The transient response characteristics of the discrete-time control system depend on the sampling period T. A large sampling period has detrimental effects on the relative stability of the system. A rule of thumb is to sample eight to ten times during a cycle of the damped sinusoidal oscillations of the output of the closed-loop system, if it is underdamped. For overdamped systems, sample eight to ten times during the rise time in the step response.

As seen from the preceding analysis, for a given value of gain K, increasing the sampling period T will make the discrete-time control system less stable and eventually will make it unstable. Conversely, making the sampling period T shorter allows the critical value of gain K for stability to be larger. In fact, making the sampling period shorter and shorter tends to make the system behave more like the continuous-time system. (For the continuous-time second-order control system, the critical gain for stability is infinity, or $K = \infty$.)

For the system shown in Figure 4–23, the damping ratio ζ for the closed-loop poles for $K = 2$ for each of the preceding three cases can be found from Figure 4–25. Graphically, the damping ratios for the closed-loop poles corresponding to $T = 0.5$, $T = 1$, and $T = 2$ are determined approximately as $\zeta = 0.24$, $\zeta = 0.32$, and $\zeta = 0.37$, respectively.

The damping ratio ζ of a closed-loop pole can be analytically determined from the location of the closed-loop pole in the z plane. If the damping ratio of a closed-loop pole is ζ, then in the s plane the closed-loop pole location (in the upper half-plane) can be given by

$$s = -\zeta\omega_n + j\omega_n\sqrt{1 - \zeta^2}$$

Since $z = e^{Ts}$, the corresponding point in the z plane is

$$z = \exp[T(-\zeta\omega_n + j\omega_n\sqrt{1 - \zeta^2})]$$

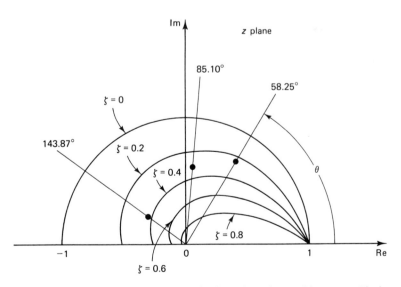

Figure 4–25   Closed-loop pole locations in the z plane shown with constant ζ loci.

from which we get

$$|z| = e^{-T\zeta\omega_n} \tag{4-29}$$

and

$$\underline{/z} = T\omega_n\sqrt{1 - \zeta^2} = T\omega_d = \theta \text{ (rad)} \tag{4-30}$$

From Equations (4-29) and (4-30), the value of ζ can be calculated. For example, in the case where the sampling period T is 0.5 sec, we have the closed-loop pole for $K = 2$ at $z = 0.4098 + j0.6623$. Hence,

$$|z| = \sqrt{0.4098^2 + 0.6623^2} = 0.7788$$

By solving

$$|z| = e^{-T\zeta\omega_n} = 0.7788$$

for the exponent, we find

$$T\zeta\omega_n = 0.25 \tag{4-31}$$

Also,

$$\underline{/z} = \tan^{-1}\frac{0.6623}{0.4098} = 58.25° = 1.0167 \text{ rad}$$

Hence,

$$\underline{/z} = T\omega_n\sqrt{1 - \zeta^2} = 1.0167 \text{ rad} \tag{4-32}$$

From Equations (4-31) and (4-32), we obtain

$$\frac{T\zeta\omega_n}{T\omega_n\sqrt{1 - \zeta^2}} = \frac{0.25}{1.0167}$$

or

$$\frac{\zeta}{\sqrt{1 - \zeta^2}} = 0.2459$$

which yields

$$\zeta = 0.2388$$

(From Figure 4-25 we graphically obtained 0.24 for ζ, which is very close to the actual ζ value of 0.2388.)

It is important to point out that in the second-order system the damping ratio ζ is indicative of the relative stability (for example, in respect to the maximum overshoot in the unit-step response) only if the sampling frequency is sufficiently high (so that there are eight to ten samplings in a cycle of oscillation). If the sampling frequency is not high enough, the maximum overshoot in the unit-step response will be much higher than would be predicted by the damping ratio ζ.

To compare the effects of different sampling periods T on the transient response, we shall compare the unit-step response sequences for the three values of T considered in the preceding analysis.

The closed-loop pulse transfer function for the system of Figure 4–23, whose feedforward pulse transfer function $G(z)$ is given by Equation (4–25), is

$$\frac{C(z)}{R(z)} = \frac{G(z)}{1 + G(z)} = \frac{Kz(1 - e^{-T})}{(z - 1)(z - e^{-T}) + Kz(1 - e^{-T})}$$

For $T = 0.5$ sec and $K = 2$, the unit-step response can be given by

$$C(z) = \frac{0.3935 \times 2z}{(z - 1)(z - 0.6065) + 0.3935 \times 2z} R(z)$$

$$= \frac{0.7870z^{-1}}{1 - 0.8195z^{-1} + 0.6065z^{-2}} \frac{1}{1 - z^{-1}}$$

from which we obtain the unit-step response sequence $c(kT)$ versus kT shown in Figure 4–26(a).

From Figure 4–25 we see that the angle θ of the line connecting the origin and the dominant closed-loop pole at $z = 0.4098 + j0.6623$ (this line is a constant ω line in the s plane) is approximately 58.25°. The angle θ of the dominant closed-loop poles determines the number of samples per cycle of sinusoidal oscillation. Note that

$$\cos \theta k = \cos \theta \left(k + \frac{360°}{\theta} \right)$$

Hence, for $\theta = 58.25°$, we have $360°/\theta = 360°/58.25° = 6.18$ samples per cycle of damped oscillation, as seen from Figure 4–26(a).

Similarly, for $T = 1$ sec and $K = 2$, the unit-step response is given by

$$C(z) = \frac{1.2642z^{-1}}{1 - 0.1037z^{-1} + 0.3679z^{-2}} \frac{1}{1 - z^{-1}}$$

The unit-step response sequence $c(kT)$ versus kT is shown in Figure 4–26(b). Since the angle of the line connecting the origin and the closed-loop pole for the present case is 85.10°, as shown in Figure 4–25, we have approximately $360°/85.10° = 4.23$ samples per cycle, which is very much less than what we normally recommend. (We recommend eight or more samples per cycle of damped sinusoidal oscillation.)

Finally, for $T = 2$ sec and $K = 2$, the unit-step response is given by

$$C(z) = \frac{1.7294z^{-1}}{1 + 0.5941z^{-1} + 0.1353z^{-2}} \frac{1}{1 - z^{-1}}$$

The unit-step response sequence $c(kT)$ versus kT is shown in Figure 4–26(c). From Figure 4–25, the angle of the line connecting the origin and the closed-loop pole for the present case is 143.87°, and consequently we have $360°/143.87° = 2.50$ samples per cycle, as seen from Figure 4–26(c). (Note that a slow sampling frequency such as 2.50 samples per cycle is unacceptable.)

Figure 4–26 has shown three different plots of the unit-step response sequence $c(kT)$ versus kT. As can be seen from these plots, if the sampling period is small, a plot of $c(kT)$ versus kT will give a fairly accurate portrait of the response $c(t)$. However, if the sampling period is not sufficiently small, then the plot of $c(kT)$ versus kT will not portray an accurate result. It is very important to choose an adequate sampling period based on the satisfaction of the sampling theorem, system

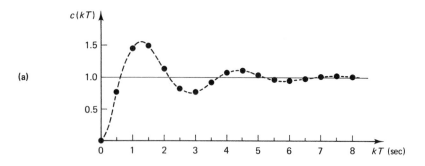

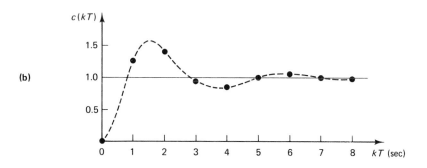

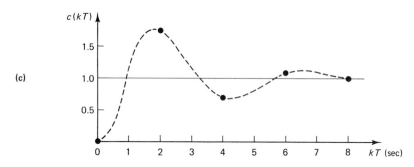

Figure 4–26　(a) Unit-step response sequence of the system shown in Figure 4–23 when $T = 0.5$ sec and $K = 2$; (b) unit-step response sequence when $T = 1$ sec and $K = 2$; (c) unit-step response sequence when $T = 2$ sec and $K = 2$.

dynamics, and actual hardware considerations. Note that barely satisfying the sampling theorem is not sufficient. An acceptable rule of thumb is to have eight to ten samples per cycle (six samples per cycle is marginal) if the system is underdamped and exhibits oscillation in the response.

Next, let us investigate the effect of the sampling period T on the steady-state accuracy. We shall consider the unit-ramp response for each of the three cases.

For the case where the sampling period T is 0.5 sec and gain K is 2, the open-loop pulse transfer function is

$$G(z) = \frac{0.7870z}{(z-1)(z-0.6065)}$$

and the static velocity error constant K_v is given by

$$K_v = \lim_{z \to 1} \frac{(1-z^{-1})G(z)}{T}$$

$$= \lim_{z \to 1} \left[\frac{z-1}{0.5z} \frac{0.7870z}{(z-1)(z-0.6065)} \right]$$

$$= 4$$

Thus, the steady-state error in response to a unit-ramp input is

$$e_{ss} = \frac{1}{K_v} = \frac{1}{4} = 0.25$$

Similarly, for the case where $T = 1$ sec and $K = 2$, the open-loop pulse transfer function is

$$G(z) = \frac{1.2642z}{(z-1)(z-0.3679)}$$

and the static velocity error constant K_v is given by

$$K_v = \lim_{z \to 1} \frac{(1-z^{-1})G(z)}{T}$$

$$= \lim_{z \to 1} \left[\frac{z-1}{z} \frac{1.2642z}{(z-1)(z-0.3679)} \right]$$

$$= 2$$

and the steady-state error in response to a unit-ramp input is

$$e_{ss} = \frac{1}{K_v} = \frac{1}{2} = 0.5$$

Finally, for the case where $T = 2$ sec and $K = 2$, the open-loop pulse transfer function is

$$G(z) = \frac{1.7294z}{(z-1)(z-0.1353)}$$

and the static velocity error constant K_v and the steady-state error in response to a unit-ramp input are obtained, respectively, as

$$K_v = 1$$

and

$$e_{ss} = \frac{1}{K_v} = 1$$

Parts (a), (b), and (c) of Figure 4–27 show, respectively, the plots of the unit-ramp response sequence $c(kT)$ versus kT for the three cases considered.

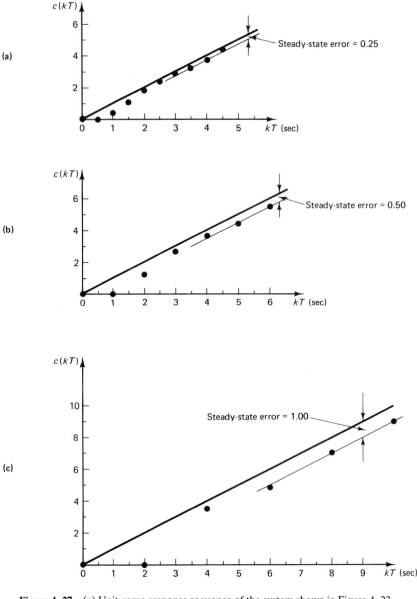

Figure 4–27   (a) Unit-ramp response sequence of the system shown in Figure 4–23 when $T = 0.5$ sec and $K = 2$; (b) unit-ramp response sequence when $T = 1$ sec and $K = 2$; (c) unit-ramp response sequence when $T = 2$ sec and $K = 2$.

The three cases we have considered demonstrate that increasing the sampling period T adversely affects the system's relative stability. (It may even cause instability in some cases.) It is important to remember that the damping ratio ζ of the closed-loop poles of the digital control system is indicative of the relative stability only if the sampling frequency is sufficiently high (that is, eight or more samples per cycle of damped sinusoidal oscillation). If the sampling frequency is low (that is, less than six samples per cycle of damped sinusoidal oscillation), then predicting the relative stability from the damping ratio value is erroneous.

Example 4-9

Consider the digital control system shown in Figure 4-28. In the z plane, design a digital controller such that the dominant closed-loop poles have a damping ratio ζ of 0.5 and a settling time of 2 sec. The sampling period is assumed to be 0.2 sec, or $T = 0.2$. Obtain the response of the designed digital control system to a unit-step input. Also, obtain the static velocity error constant K_v of the system.

For the standard second-order system having a pair of dominant closed-loop poles, the settling time of 2 sec means that

$$\text{settling time} = \frac{4}{\zeta \omega_n} = \frac{4}{0.5 \omega_n} = 2$$

which gives the undamped natural frequency ω_n of the dominant closed-loop poles as

$$\omega_n = 4$$

The damped natural frequency ω_d is determined to be

$$\omega_d = \omega_n \sqrt{1 - \zeta^2} = 4\sqrt{1 - 0.5^2} = 3.464$$

Since the sampling period T is 0.2 sec, we have

$$\omega_s = \frac{2\pi}{T} = \frac{2\pi}{0.2} = 10\pi = 31.42$$

[Notice that there are approximately nine samples per cycle of damped oscillation (31.42/3.464 = 9.07). Thus, the sampling period of 0.2 sec is satisfactory.]

We shall first locate the desired dominant closed-loop poles in the z plane. Referring to Equations (4-29) and (4-30), for a constant-damping-ratio locus we have

$$|z| = e^{-T\zeta\omega_n} = \exp\left(-\frac{2\pi\zeta}{\sqrt{1 - \zeta^2}} \frac{\omega_d}{\omega_s}\right)$$

and

$$\angle z = T\omega_d = 2\pi \frac{\omega_d}{\omega_s}$$

From the given specifications ($\zeta = 0.5$ and $\omega_d = 3.464$), the magnitude and angle of the dominant closed-loop pole in the upper half of the z plane are determined as follows:

$$|z| = \exp\left(-\frac{2\pi \times 0.5}{\sqrt{1 - 0.5^2}} \frac{3.464}{31.42}\right) = e^{-0.400} = 0.6703$$

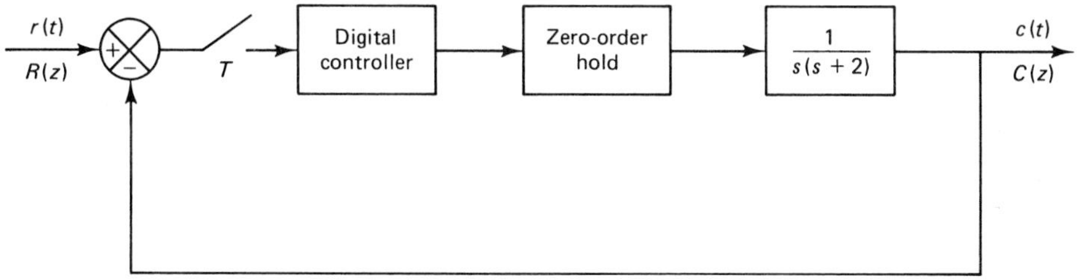

Figure 4-28   Digital control system for Example 4-9.

and

$$\underline{/z} = 2\pi\frac{3.464}{31.42} = 0.6927 \text{ rad} = 39.69°$$

We can now locate the desired dominant closed-loop pole in the upper half of the z plane, shown in Figure 4–29 as point P. Note that at point P

$$z = 0.6703\underline{/39.69°} = 0.5158 + j0.4281$$

Noting that the sampling period T is 0.2 sec, the pulse transfer function $G(z)$ of the plant preceded by the zero-order hold can be obtained as follows:

$$G(z) = \mathscr{Z}\left[\frac{1 - e^{-0.2s}}{s}\frac{1}{s(s + 2)}\right] = (1 - z^{-1})\mathscr{Z}\left[\frac{1}{s^2(s + 2)}\right]$$

This last equation can be written as

$$G(z) = \frac{0.01758(z + 0.8760)}{(z - 1)(z - 0.6703)}$$

Next, we locate the poles ($z = 1$ and $z = 0.6703$) and zero ($z = -0.8760$) of $G(z)$ on the z plane, as shown in Figure 4–29. If point P is to be the location for the desired dominant closed-loop pole in the upper half of the z plane, then the sum of the angles at point P must be equal to $\pm 180°$. However, the sum of the angle contributions at point P is

$$17.10° - 138.52° - 109.84° = -231.26°$$

Hence, the angle deficiency is

$$-231.26° + 180° = -51.26°$$

The controller pulse transfer function must provide $+51.26°$. The pulse transfer function for the controller may be assumed to be

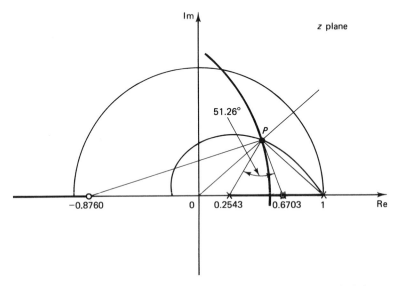

Figure 4–29 Root locus diagram of the system considered in Example 4–9.

$$G_D(z) = K\frac{z + \alpha}{z + \beta}$$

where K is the gain constant of the controller.

If we decide to cancel the pole at $z = 0.6703$ by the zero of the controller at $z = -\alpha$, then the pole of the controller can be determined (from the condition that the controller must provide $+51.26°$) as a point at $z = 0.2543$ ($\beta = -0.2543$). Thus, the pulse transfer function for the controller may be determined as

$$G_D(z) = K\frac{z - 0.6703}{z - 0.2543}$$

The open-loop pulse transfer function now becomes

$$G_D(z)G(z) = K\frac{z - 0.6703}{z - 0.2543}\frac{0.01758(z + 0.8760)}{(z - 1)(z - 0.6703)}$$

$$= K\frac{0.01758(z + 0.8760)}{(z - 0.2543)(z - 1)}$$

The gain constant K can be determined from the following magnitude condition:

$$|G_D(z)G(z)|_{z\,=\,0.5158\,+\,j0.4281} = 1$$

Hence,

$$K\left|\frac{0.01758(z + 0.8760)}{(z - 0.2543)(z - 1)}\right|_{z=0.5158+j0.4281} = 1$$

which gives

$$K = 12.67$$

The designed digital controller is

$$G_D(z) = 12.67\frac{z - 0.6703}{z - 0.2543} \tag{4–33}$$

The open-loop pulse transfer function for the present system is

$$G_D(z)G(z) = \frac{12.67 \times 0.01758(z + 0.8760)}{(z - 0.2543)(z - 1)} = \frac{0.2227(z + 0.8760)}{(z - 0.2543)(z - 1)}$$

Hence, the closed-loop pulse transfer function is

$$\frac{C(z)}{R(z)} = \frac{G_D(z)G(z)}{1 + G_D(z)G(z)} = \frac{0.2227z + 0.1951}{z^2 - 1.0316z + 0.4494}$$

The response to the unit-step input $R(z) = 1/(1 - z^{-1})$ can be obtained from

$$C(z) = \frac{0.2227z + 0.1951}{z^2 - 1.0316z + 0.4494}\frac{1}{1 - z^{-1}}$$

$$= \frac{0.2227z^{-1} + 0.1951z^{-2}}{1 - 1.0316z^{-1} + 0.4494z^{-2}}\frac{1}{1 - z^{-1}}$$

Figure 4–30 shows the unit-step response sequence $c(kT)$ versus kT. The plot shows that the maximum overshoot is approximately 16% (which means that the damping ratio is approximately 0.5) and the settling time is approximately 2 sec. The digital controller just designed satisfies the given specifications and is satisfactory.

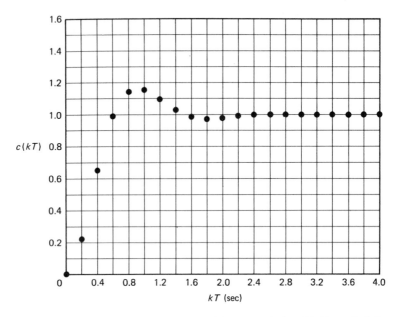

Figure 4–30   Unit-step response sequence of the system designed in Example 4–9.

The static velocity error constant K_v of the system is given by

$$K_v = \lim_{z \to 1}\left[\frac{1 - z^{-1}}{T} G_D(z)G(z)\right]$$

$$= \lim_{z \to 1}\left[\frac{z - 1}{0.2z} \frac{0.2227(z + 0.8760)}{(z - 0.2543)(z - 1)}\right]$$

$$= 2.801$$

If it is required to have a large value of K_v, then we may include a lag compensator. For example, adding a zero at $z = 0.94$ and a pole at $z = 0.98$ would raise the K_v value three times, since $(1 - 0.94)/(1 - 0.98) = 3$. (It is important that the pole and zero of the lag compensator lie on a finite number of allocable discrete points.) A lag compensator, which has a pole and a zero very close to each other, does not significantly change the root locus near the dominant closed-loop poles. The effect of a lag compensator on the transient response is to introduce a small but slowly decreasing transient component. Such a small but slow transient, however, is not desirable from the viewpoint of disturbance or noise attenuation, since the response to disturbances would not attenuate promptly.

Finally, it is noted that although the designed system is of the third order, it acts as a second-order system, since one pole of the plant has been canceled by the zero of the controller. Because of this, the present system has only two closed-loop poles. The dominant closed-loop poles are the only closed-loop poles in this case. If a pole and a zero do not cancel each other, then the system will be of the third order.

Comments.   It is important to note that the poles of the closed-loop pulse transfer function determine the natural modes of the system. The transient response and frequency response behaviors, however, are strongly influenced by the zeros of the closed-loop pulse transfer function.

Familiarity with the relationship between the z plane pole and zero locations and the time response characteristics is useful in designing discrete-time control systems. It is important to note that in the s plane adding a zero on the negative real axis near the origin increases the maximum overshoot in response to a step input. Such a zero in the s plane is mapped to a zero on the positive real axis in the z plane between 0 and 1. Therefore, in the z plane, adding a zero on the positive real axis between 0 and 1 increases the maximum overshoot. In fact, moving a zero toward point $z = 1$ will greatly increase the maximum overshoot.

Similarly, in the s plane a closed-loop pole on the negative real axis near the origin increases the settling time. In the z plane, such a closed-loop pole is mapped to a closed-loop pole on the positive real axis between 0 and 1. Thus, a closed-loop pole in the z plane between 0 and 1 (in particular, near point $z = 1$) increases the settling time. The presence of a closed-loop pole or zero on the negative real axis between 0 and -1 in the z plane, however, affects the transient response only slightly.

4-6 DESIGN BASED ON THE FREQUENCY-RESPONSE METHOD

The frequency-response concept plays the same powerful role in digital control systems as it does in continuous-time control systems. As stated earlier, it is assumed in this book that the reader is familiar with conventional frequency-response design techniques for continuous-time control systems. In fact, familiarity with Bode diagrams (logarithmic plots) is necessary in the extension of the conventional frequency-response techniques to the analysis and design of discrete-time control systems.

Frequency-response methods have very frequently been used in the compensator design. The basic reason is the simplicity of the methods. In performing frequency-response tests on a discrete-time system, it is important that the system have a low-pass filter before the sampler so that sidebands are filtered out. Then the response of the linear time-invariant system to a sinusoidal input preserves the frequency and modifies only the amplitude and phase of the input signal. Thus, the amplitude and phase are the only two quantities that must be dealt with.

In the following, we shall analyze the response of the linear time-invariant discrete-time system to a sinusoidal input; this analysis will be followed by the definition of the sinusoidal pulse transfer function. Then we discuss the design of a discrete-time control system in the w plane by use of a Bode diagram.

Response of a Linear Time-Invariant Discrete-Time System to a Sinusoidal Input.
Earlier in this book we stated that the frequency response of $G(z)$ can be obtained by substituting $z = e^{j\omega T}$ into $G(z)$. In what follows we shall show that this is indeed true.

Consider the stable linear time-invariant discrete-time system shown in Figure 4-31. The input to the system $G(z)$ before sampling is

$$u(t) = \sin \omega t$$

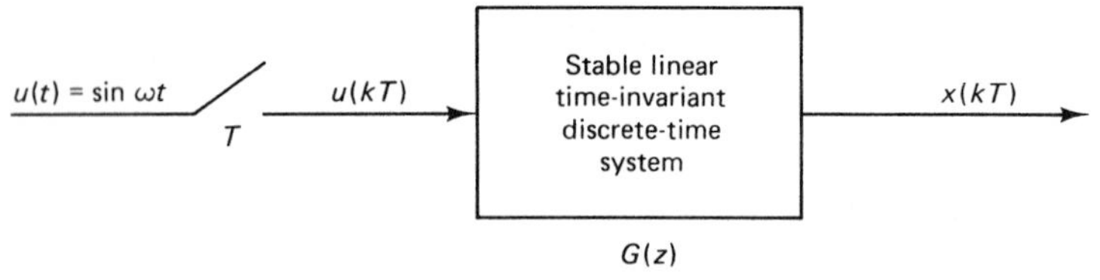

Figure 4–31   Stable linear time-invariant discrete-time system.

The sampled signal $u(kT)$ is

$$u(kT) = \sin k\omega T$$

The z transform of the sampled input is

$$U(z) = \mathcal{Z}\left[\sin k\omega T\right] = \frac{z \sin \omega T}{(z - e^{j\omega T})(z - e^{-j\omega T})}$$

The response of the system is given by

$$X(z) = G(z)U(z) = G(z)\frac{z \sin \omega T}{(z - e^{j\omega T})(z - e^{-j\omega T})}$$

$$= \frac{az}{z - e^{j\omega T}} + \frac{\bar{a}z}{z - e^{-j\omega T}} + [\text{terms due to poles of } G(z)] \qquad (4\text{–}34)$$

Multiplying both sides of Equation (4–34) by $(z - e^{j\omega T})/z$, we obtain

$$G(z)\frac{\sin \omega T}{z - e^{-j\omega T}} = a + \frac{\bar{a}(z - e^{j\omega T})}{z - e^{-j\omega T}} + \frac{z - e^{j\omega T}}{z}[\text{terms due to poles of } G(z)]$$

The second term on the right-hand side of this last equation approaches zero as z approaches $e^{j\omega T}$. Since the system considered here is stable, the third term on the right-hand side also approaches zero as z approaches $e^{j\omega T}$. Hence, by letting z approach $e^{j\omega T}$, we have

$$a = G(z)\frac{\sin \omega T}{z - e^{-j\omega T}}\bigg|_{z=e^{j\omega T}} = \frac{G(e^{j\omega T})}{2j}$$

The coefficient $\bar{a}$, the complex conjugate of a, is then obtained as follows:

$$\bar{a} = -\frac{G(e^{-j\omega T})}{2j}$$

Let us define

$$G(e^{j\omega T}) = Me^{j\theta}$$

Then

$$G(e^{-j\omega T}) = Me^{-j\theta}$$

Equation (4–34) can now be written as

$$X(z) = \frac{Me^{j\theta}}{2j}\frac{z}{z - e^{j\omega T}} - \frac{Me^{-j\theta}}{2j}\frac{z}{z - e^{-j\omega T}} + [\text{terms due to poles of } G(z)]$$

or

$$X(z) = \frac{M}{2j}\left(\frac{e^{j\theta}z}{z - e^{j\omega T}} - \frac{e^{-j\theta}z}{z - e^{-j\omega T}}\right) + [\text{terms due to poles of } G(z)]$$

The inverse z transform of this last equation is

$$x(kT) = \frac{M}{2j}(e^{jk\omega T}e^{j\theta} - e^{-jk\omega T}e^{-j\theta}) + \mathcal{Z}^{-1}[\text{terms due to poles of } G(z)] \qquad (4\text{-}35)$$

The last term on the right-hand side of Equation (4-35) represents the transient response. Since the system $G(z)$ has been assumed to be stable, all transient response terms will disappear at steady state and we will get the following steady-state response $x_{ss}(kT)$:

$$x_{ss}(kT) = \frac{M}{2j}[e^{j(k\omega T + \theta)} - e^{-j(k\omega T + \theta)}] = M\sin(k\omega T + \theta) \qquad (4\text{-}36)$$

where M, the gain of the discrete-time system when subjected to a sinusoidal input, is given by

$$M = M(\omega) = |G(e^{j\omega T})|$$

and θ, the phase angle, is given by

$$\theta = \theta(\omega) = \underline{/G(e^{j\omega T})}$$

In terms of $G(e^{j\omega T})$, Equation (4-36) can be written as follows:

$$x_{ss}(kT) = |G(e^{j\omega T})|\sin(k\omega T + \underline{/G(e^{j\omega T})})$$

We have shown that $G(e^{j\omega T})$ indeed gives the magnitude and phase of the frequency response of $G(z)$. Thus, to obtain the frequency response of $G(z)$, we need only to substitute $e^{j\omega T}$ for z in $G(z)$. The function $G(e^{j\omega T})$ is commonly called the *sinusoidal pulse transfer function*. Noting that

$$e^{j(\omega + (2\pi/T))T} = e^{j\omega T}e^{j2\pi} = e^{j\omega T}$$

we find that the sinusoidal pulse transfer function $G(e^{j\omega T})$ is periodic, with the period equal to T.

Example 4-10

Consider the system defined by

$$x(kT) = u(kT) + ax((k - 1)T), \qquad 0 < a < 1$$

where $u(kT)$ is the input and $x(kT)$ the output. Obtain the steady-state output $x(kT)$ when the input $u(kT)$ is the sampled sinusoid, or $u(kT) = A\sin k\omega T$.

The z transform of the system equation is

$$X(z) = U(z) + az^{-1}X(z)$$

By defining $G(z) = X(z)/U(z)$, we have

$$G(z) = \frac{X(z)}{U(z)} = \frac{1}{1 - az^{-1}}$$

Let us substitute $e^{j\omega T}$ for z in $G(z)$. Then the sinusoidal pulse transfer function $G(e^{j\omega T})$ can be obtained as

$$G(e^{j\omega T}) = \frac{1}{1 - ae^{-j\omega T}} = \frac{1}{1 - a \cos \omega T + ja \sin \omega T}$$

The amplitude of $G(e^{j\omega T})$ is

$$|G(e^{j\omega T})| = M = \frac{1}{\sqrt{1 + a^2 - 2a \cos \omega T}}$$

and the phase angle of $G(e^{j\omega T})$ is

$$\angle G(e^{j\omega T}) = \theta = -\tan^{-1} \frac{a \sin \omega T}{1 - a \cos \omega T}$$

Then the steady-state output $x_{ss}(kT)$ can be written as follows:

$$x_{ss}(kT) = AM \sin(k\omega T + \theta)$$

$$= \frac{A}{\sqrt{1 + a^2 - 2a \cos \omega T}} \sin\left(k\omega T - \tan^{-1} \frac{a \sin \omega T}{1 - a \cos \omega T}\right)$$

Bilinear Transformation and the w Plane. Before we can advantageously apply our well-developed frequency-response methods to the analysis and design of discrete-time control systems, certain modifications in the z plane approach are necessary. Since in the z plane the frequency appears as $z = e^{j\omega T}$, if we treat frequency response in the z plane, the simplicity of the logarithmic plots will be completely lost. Thus, the direct application of frequency-response methods is not worthy of consideration. In fact, since the z transformation maps the primary and complementary strips of the left half of the s plane into the unit circle in the z plane, conventional frequency-response methods, which deal with the entire left half plane, do not apply to the z plane.

The difficulty, however, can be overcome by transforming the pulse transfer function in the z plane into that in the w plane. The transformation, commonly called the w transformation, a bilinear transformation, is defined by

$$z = \frac{1 + (T/2)w}{1 - (T/2)w} \tag{4–37}$$

where T is the sampling period involved in the discrete-time control system under consideration. By converting a given pulse transfer function in the z plane into a rational function of w, the frequency-response methods can be extended to discrete-time control systems. By solving Equation (4–37) for w, we obtain the inverse relationship

$$w = \frac{2}{T} \frac{z - 1}{z + 1} \tag{4–38}$$

Through the z transformation and the w transformation, the primary strip of the left half of the s plane is first mapped into the inside of the unit circle in the z plane and then mapped into the entire left half of the w plane. The two mapping processes are depicted in Figure 4–32. (Note that in the s plane we consider only the primary strip.) Notice that the origin of the z plane is mapped into the point $w = -2/T$ in the w plane. Notice also that, as s varies from 0 to $j\omega_s/2$ along the $j\omega$

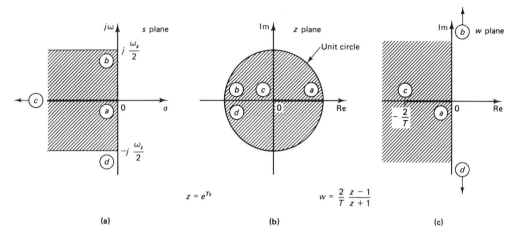

Figure 4–32 Diagrams showing mappings from the s plane to the z plane and from the z plane to the w plane. (a) Primary strip in the left half of the s plane; (b) z plane mapping of the primary strip in the s plane; (c) w plane mapping of the unit circle in the z plane.

axis in the s plane, z varies from 1 to -1 along the unit circle in the z plane and w varies from 0 to ∞ along the imaginary axis in the w plane.

Although the left half of the w plane corresponds to the left half of the s plane and the imaginary axis of the w plane corresponds to the imaginary axis of the s plane, there are differences between the two planes. The chief difference is that the behavior in the s plane over the frequency range $-\frac{1}{2}\omega_s \leq \omega \leq \frac{1}{2}\omega_s$ maps to the range $-\infty < \nu < \infty$, where ν is the fictitious frequency in the w plane. This means that, although the frequency response characteristics of the analog filter will be reproduced in the discrete or digital filter, the frequency scale on which the response occurs will be compressed from an infinite interval in the analog filter to a finite interval in the digital filter.

Once the pulse transfer function $G(z)$ is transformed into $G(w)$ by means of the w transformation, it may be treated as a conventional transfer function in w. Conventional frequency-response techniques can then be used in the w plane, and so the well-established frequency-response design techniques can be applied to the design of discrete-time control systems.

As noted earlier, ν represents the fictitious frequency. By replacing w by $j\nu$, conventional frequency-response techniques may be used to draw the Bode diagram for the transfer function in w. (In the brief review of the Bode diagrams in this section, we shall use the fictitious frequency ν as the variable.)

Although the w plane resembles the s plane geometrically, the frequency axis in the w plane is distorted. The fictitious frequency ν and the actual frequency ω are related as follows:

$$
w \bigg|_{w=j\nu} = j\nu = \frac{2}{T}\frac{z-1}{z+1}\bigg|_{z=e^{j\omega T}} = \frac{2}{T}\frac{e^{j\omega T}-1}{e^{j\omega T}+1}
$$

$$
= \frac{2}{T}\frac{e^{j(1/2)\omega T}-e^{-j(1/2)\omega T}}{e^{j(1/2)\omega T}+e^{j(1/2)\omega T}} = \frac{2}{T}j\,\tan\frac{\omega T}{2}
$$

or

$$v = \frac{2}{T} \tan \frac{\omega T}{2} \qquad (4\text{–}39)$$

Equation (4–39) gives the relationship between the actual frequency ω and the fictitious frequency v. Note that as the actual frequency ω moves from $-\frac{1}{2}\omega_s$ to 0 the fictitious frequency v moves from $-\infty$ to 0, and as ω moves from 0 to $\frac{1}{2}\omega_s$, v moves from 0 to ∞.

Referring to Equation (4–39), the actual frequency ω can be translated into the fictitious frequency v. For example, if the bandwidth is specified as ω_b, then the corresponding bandwidth in the w plane is $(2/T)\tan(\omega_b T/2)$. Similarly, $G(jv_1)$ corresponds to $G(j\omega_1)$, where $\omega_1 = (2/T)\tan^{-1}(v_1 T/2)$. Figure 4–33 shows the relationship between the fictitious frequency v times $\frac{1}{2}T$ and the actual frequency ω for the frequency range between 0 and $\frac{1}{2}\omega_s$.

Notice that in Equation (4–39) if ωT is small then

$$v \doteq \omega$$

This means that for small ωT the transfer functions $G(s)$ and $G(w)$ resemble each other. Note that this is the direct result of the inclusion of the scale factor $2/T$ in Equation (4–38). The presence of this scale factor in the transformation enables us to maintain the same error constants before and after the w transformation. (This means that the transfer function in the w plane will approach that in the s plane as T approaches zero. See Example 4–11, which follows.)

Example 4–11

Consider the transfer-function system shown in Figure 4–34. The sampling period T is assumed to be 0.1 sec. Obtain $G(w)$.

The z transform of $G(s)$ is

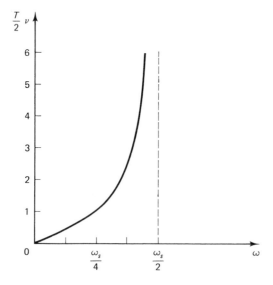

Figure 4–33 Relationship between the fictitious frequency v times $\frac{1}{2}T$ and the actual frequency ω for the frequency range between 0 and $\frac{1}{2}\omega_s$.

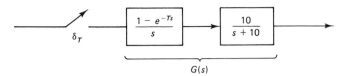

Figure 4-34   Transfer-function system of Example 4-11.

$$G(z) = \mathcal{Z}\left[\frac{1 - e^{-Ts}}{s} \frac{10}{s + 10}\right]$$

$$= (1 - z^{-1})\mathcal{Z}\left[\frac{10}{s(s + 10)}\right]$$

$$= \frac{0.6321}{z - 0.3679}$$

By use of the bilinear transformation given by Equation (4–37), or

$$z = \frac{1 + (T/2)w}{1 - (T/2)w} = \frac{1 + 0.05w}{1 - 0.05w}$$

$G(z)$ can be transformed into $G(w)$ as follows:

$$G(w) = \frac{0.6321}{\dfrac{1 + 0.05w}{1 - 0.05w} - 0.3679} = \frac{0.6321(1 - 0.05w)}{0.6321 + 0.06840w}$$

$$= 9.241\frac{1 - 0.05w}{w + 9.241}$$

Notice that the location of the pole of the plant is $s = -10$ and that of the pole in the w plane is $w = -9.241$. The gain value in the s plane is 10 and that in the w plane is 9.241. (Thus, both the pole locations and the gain values are similar in the s plane and the w plane.) However, $G(w)$ has a zero at $w = 2/T = 20$, although the plant does not have any zero. As the sampling period T becomes smaller, the w plane zero at $w = 2/T$ approaches infinity in the right half of the w plane. Note that we have

$$\lim_{w \to 0} G(w) = \lim_{s \to 0}\frac{10}{s + 10}$$

This fact is very useful in checking the numerical calculations in transforming $G(s)$ into $G(w)$.

To summarize, the w transformation, a bilinear transformation, maps the inside of the unit circle of the z plane into the left half of the w plane. The overall result due to the transformations from the s plane into the z plane and from the z plane into the w plane is that the w plane and the s plane are similar over the region of interest in the s plane. This is because some of the distortions caused by the transformation from the s plane into the z plane are partly compensated for by the transformation from the z plane into the w plane.

Note that if

$$G(z) = \frac{b_0 z^m + b_1 z^{m-1} + \cdots + b_m}{z^n + a_1 z^{n-1} + \cdots + a_n}, \qquad m \le n$$

where the a_i's and b_i's are constants, is transformed into the w plane by the transformation

$$z = \frac{1 + (T/2)w}{1 - (T/2)w}$$

then $G(w)$ takes the form

$$G(w) = \frac{\beta_0 w^n + \beta_1 w^{n-1} + \cdots + \beta_n}{\alpha_0 w^n + \alpha_1 w^{n-1} + \cdots + \alpha_n}$$

where the α_i's and the β_i's are constants (some of them may be zero). Thus, $G(w)$ is a ratio of polynomials in w, where the degrees of the numerator and denominator may or may not be equal. Since $G(jv)$ is a rational function of v, the Nyquist stability criterion can be applied to $G(jv)$. In terms of the Bode diagram, the conventional straight-line approximation to the magnitude curve as well as the concept of the phase margin and gain margin apply to $G(jv)$.

Bode Diagrams.   Design by means of Bode diagrams has been widely used in dealing with single-input–single-output continuous-time control systems. In particular, if the transfer function is in a factored form, the simplicity and ease with which the asymptotic Bode diagram can be drawn and reshaped are well known.

As stated earlier, the conventional frequency-response methods apply to the transfer functions in the w plane. Recall that the Bode diagram consists of two separate plots, the logarithmic magnitude $|G(jv)|$ versus log v and the phase angle $\underline{/G(jv)}$ versus log v. Sketching of the logarithmic magnitude is based on the factoring of $G(jv)$, so that it works on the principle of adding the individual factored terms instead of multiplying individual terms. Familiar asymptotic plotting techniques can be applied, and therefore the magnitude curve can be quickly drawn by using straight-line asymptotes. Using the Bode diagram, a digital compensator or digital controller may be designed with conventional design techniques.

It is important to note that there may be a difference in the high-frequency magnitudes for $G(j\omega)$ and $G(jv)$. The high-frequency asymptote of the logarithmic magnitude curve for $G(jv)$ may be a constant-decibel line (that is, a horizontal line). On the other hand, if $\lim_{s \to \infty} G(s) = 0$, then the magnitude of $G(j\omega)$ always approaches zero ($-\infty$ dB) as ω approaches infinity. For example, referring to Example 4–11, we obtained $G(w)$ for $G(s)$ as follows:

$$G(w) = 9.241\left(\frac{1 - 0.05w}{w + 9.241}\right)$$

The high-frequency magnitude of $G(jv)$ is

$$\lim_{v \to \infty} |G(jv)| = \lim_{v \to \infty} \left| 9.241\left(\frac{1 - 0.05jv}{jv + 9.241}\right) \right| = 0.4621$$

while the high-frequency magnitude of the plant is

$$\lim_{\omega \to \infty} \left| \frac{10}{j\omega + 10} \right| = 0$$

The difference in the Bode diagrams at the high-frequency end can be explained as follows. First, recall that we are interested only in the frequency range $0 \leq \omega \leq \frac{1}{2}\omega_s$, which corresponds to $0 \leq \nu \leq \infty$. Then, noting that $\nu = \infty$ in the w plane corresponds to $\omega = \frac{1}{2}\omega_s$ in the s plane, it can be said that $\lim_{\nu \to \infty}|G(j\nu)|$ corresponds to $\lim_{\omega \to \omega_s/2}|10/(j\omega + 10)|$, which is a constant. (It is important to note that these two values are generally not equal to each other.) From the pole–zero point of view, it can be said that when $|G(j\nu)|$ is a nonzero constant at $\nu = \infty$ it is implied that $G(w)$ contains the same number of poles and zeros.

In general, one or more zeros of $G(w)$ lie in the right half of the w plane. The presence of a zero in the right half of the w plane means that $G(w)$ is a nonminimum phase transfer function. Therefore, we must be careful in drawing the phase angle curve in the Bode diagram.

Advantages of the Bode Diagram Approach to the Design.   The Bode diagram approach to the analysis and design of control systems is particularly useful for the following reasons:

1. In the Bode diagram the low-frequency asymptote of the magnitude curve is indicative of one of the static error constants K_p, K_v, or K_a.

2. Specifications of the transient response can be translated into those of the frequency response in terms of the phase margin, gain margin, bandwidth, and so forth. These specifications can easily be handled in the Bode diagram. In particular, the phase and gain margins can be read directly from the Bode diagram.

3. The design of a digital compensator (or digital controller) to satisfy the given specifications (in terms of the phase margin and gain margin) can be carried out in the Bode diagram in a simple and straightforward manner.

Phase Lead, Phase Lag, and Phase Lag–Lead Compensation.   Before we discuss design procedures in the w plane, let us review the phase lead, phase lag, and phase lag–lead compensation techniques.

Phase lead compensation is commonly used for improving stability margins. The phase lead compensation increases the system bandwidth. Thus, the system has a faster speed to respond. However, such a system using phase lead compensation may be subjected to high-frequency noise problems due to its increased high-frequency gains.

Phase lag compensation reduces the system gain at higher frequencies without reducing the system gain at lower frequencies. The system bandwidth is reduced and thus the system has a slower speed of response. Because of the reduced high-frequency gain, the total system gain can be increased, and thereby low-frequency gain can be increased and the steady-state accuracy can be improved. Also, any high-frequency noises involved in the system can be attenuated.

In some applications, a phase lag compensator is cascaded with a phase lead compensator. The cascaded compensator is known as a *phase lag–lead* compensator. By use of the lag–lead compensator, the low-frequency gain can be increased (which means an improvement in steady-state accuracy), while at the same time the system bandwith and stability margins can be increased.

Note that the PID controller is a special case of a phase lag–lead controller. The PD control action, which affects the high-frequency region, increases the phase lead angle and improves system stability, as well as increasing the system bandwidth (and thus increasing the speed of response). That is, the PD controller behaves in much the same way as a phase lead compensator. The PI control action affects the low-frequency portion and, in fact, increases the low-frequency gain and improves steady-state accuracy. Therefore, the PI controller acts as a phase lag compensator. The PID control action is a combination of the PI and PD control actions. The design techniques for PID controllers basically follow those of phase lag–lead compensators. (In industrial control systems, however, each PID control action in the PID controller may be adjusted experimentally.)

Some Remarks on the Coefficient Quantization Problem.   From the viewpoint of microprocessor implementation of the phase lead, phase lag, and phase lag–lead compensators, phase lead compensators present no coefficient quantization problem, because the locations of poles and zeros are widely separated. However, in the case of phase lag compensators and phase lag–lead compensators, the phase lag network presents a coefficient quantization problem because the locations of poles and zeros are close to each other. (They are near the point $z = 1$.)

Since the filter coefficients must be realized by binary words that use limited numbers of bits, if the number of bits employed is insufficient, the pole and zero locations of the filter may not be realized exactly as desired and the resulting filter will not behave as expected.

Since small deviations in the pole and zero locations from the desired locations can have significant effects on the frequency-response characteristics of the compensator, the digital version of the compensator may not perform as expected. To minimize the effect of the coefficient quantization problem, it is necessary to structure the filter so that it is least subject to coefficient inaccuracies due to quantization.

Because the sensitivity of the roots of polynomials to the parameter variations becomes severe as the order of the polynomial increases, direct realization of a higher-order filter is not desirable. It is preferable to place lower-order elements in cascade or in parallel, as discussed in Section 3–6. As a matter of course, from the outset if we choose poles and zeros of the digital compensator from allowable discrete points, then the coefficient quantization problem can be avoided.

In the analog compensator, the poles and zeros of the compensator can be placed with an arbitrary accuracy. In converting an analog compensator to a digital compensator, the digital version of the lag compensator may involve considerable inaccuracies in the locations of poles and zeros. (The important thing to remember is that the poles and zeros of the filter in the z plane must lie on a finite number of allowable discrete points.)

Design Procedure in the w Plane.   Referring to the digital control system shown in Figure 4–35, the design procedure in the w plane may be stated as follows:

1. First, obtain $G(z)$, the z transform of the plant preceded by a hold. Then transform $G(z)$ into a transfer function $G(w)$ through the bilinear transformation given by Equation (4–37):

$$z = \frac{1 + (T/2)w}{1 - (T/2)w}$$

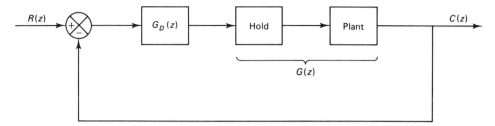

Figure 4-35   Digital control system.

That is,

$$G(w) = G(z)\big|_{z=[1+(T/2)w]/[1-(T/2)w]}$$

It is important that the sampling period T be chosen properly. A rule of thumb is to sample at the frequency 10 times that of the bandwidth of the closed-loop system. (Although digital controls and signal processing use similar approaches in sampling continuous-time signals, the sampling frequencies involved are very different. In the field of signal processing, sampling frequencies are generally very high, while in the field of digital control systems, the sampling frequencies used are generally low. Such a difference in the sampling frequencies is mainly due to the different dynamics involved and the different trade-offs in these two fields.)

2. Substitute $w = jv$ into $G(w)$ and plot the Bode diagram for $G(jv)$.

3. Read from the Bode diagram the static error constants, the phase margin, and the gain margin.

4. By assuming that the low-frequency gain of the discrete-time controller (or digital controller) transfer function $G_D(w)$ is unity, determine the system gain by satisfying the requirement for a given static error constant. Then, by using conventional design techniques for continuous-time control systems, determine the pole(s) and zero(s) of the digital controller transfer function. [$G_D(w)$ is a ratio of two polynomials in w.] Then the open-loop transfer function of the designed system is given by $G_D(w)G(w)$.

5. Transform the controller transfer function $G_D(w)$ into $G_D(z)$ through the bilinear transformation given by Equation (4–38):

$$w = \frac{2}{T}\frac{z-1}{z+1}$$

Then

$$G_D(z) = G_D(w)\big|_{w=(2/T)(z-1)/(z+1)}$$

is the pulse transfer function of the digital controller.

6. Realize the pulse transfer function $G_D(z)$ by a computational algorithm.

In following the design procedure just given, it is important to note the following:

1. The transfer function $G(w)$ is a nonminimum phase transfer function. Hence, the phase angle curve is different from that for the more typical minimum phase transfer function. It is necessary to make sure that the phase angle curve is drawn correctly by taking into consideration the nonminimum phase term.

2. The frequency axis in the w plane is distorted. The relationship between the fictitious frequency ν and the actual frequency ω is

$$\nu = \frac{2}{T} \tan \frac{\omega T}{2}$$

If, for example, a bandwidth ω_b is specified, we need to design the system for a bandwidth ν_b, where

$$\nu_b = \frac{2}{T} \tan \frac{\omega_b T}{2}$$

Example 4–12

Consider the digital control system shown in Figure 4–36. Design a digital controller in the w plane such that the phase margin is 50°, the gain margin is at least 10 dB, and the static velocity error constant K_v is 2 sec^{-1}. Assume that the sampling period is 0.2 sec, or $T = 0.2$.

First, we obtain the pulse transfer function $G(z)$ of the plant that is preceded by the zero-order hold:

$$G(z) = \mathcal{Z}\left[\frac{1 - e^{-0.2s}}{s} \frac{K}{s(s + 1)}\right]$$

$$= (1 - z^{-1})\mathcal{Z}\left[\frac{K}{s^2(s + 1)}\right]$$

$$= 0.01873\left[\frac{K(z + 0.9356)}{(z - 1)(z - 0.8187)}\right]$$

$$= \frac{K(0.01873z + 0.01752)}{z^2 - 1.8187z + 0.8187}$$

Next, we transform the pulse transfer function $G(z)$ into a transfer function $G(w)$ by means of the bilinear transformation given by Equation (4–37):

$$z = \frac{1 + (T/2)w}{1 - (T/2)w} = \frac{1 + 0.1w}{1 - 0.1w}$$

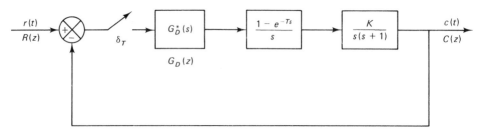

Figure 4–36 Digital control system of Example 4–12.

Thus,

$$G(w) = \frac{K\left[0.01873\left(\dfrac{1 + 0.1w}{1 - 0.1w}\right) + 0.01752\right]}{\left(\dfrac{1 + 0.1w}{1 - 0.1w}\right)^2 - 1.8187\left(\dfrac{1 + 0.1w}{1 - 0.1w}\right) + 0.8187}$$

$$= \frac{K(-0.000333w^2 - 0.09633w + 0.9966)}{w^2 + 0.9969w}$$

$$\doteq \frac{K\left(1 + \dfrac{w}{300}\right)\left(1 - \dfrac{w}{10}\right)}{w(w + 1)}$$

A simple *phase-lead compensator* will probably satisfy all requirements. There-fore, we shall try lead compensation. (If lead compensation does not satisfy all require-ments, we need to use a different type of compensation.)

Now let us assume that the transfer function of the digital controller $G_D(w)$ has unity gain for the low-frequency range and has the following form:

$$G_D(w) = \frac{1 + \tau w}{1 + \alpha \tau w}, \qquad 0 < \alpha < 1$$

(This is a phase-lead compensator.) It is one of the simplest forms of the digital controller transfer function. (Other forms may be assumed as well for this problem.) The open-loop transfer function is

$$G_D(w)G(w) = \frac{1 + \tau w}{1 + \alpha \tau w} \frac{K(-0.000333w^2 - 0.09633w + 0.9966)}{w^2 + 0.9969w}$$

The static velocity error constant K_v is specified as 2 sec^{-1}. Hence,

$$K_v = \lim_{w \to 0} wG_D(w)G(w) \doteq K = 2$$

The gain K is thus determined to be 2.

By setting $K = 2$, we plot the Bode diagram of $G(w)$:

$$G(w) = \frac{2(-0.000333w^2 - 0.09633w + 0.9966)}{w^2 + 0.9969w}$$

$$\doteq \frac{2\left(1 + \dfrac{w}{300}\right)\left(1 - \dfrac{w}{10}\right)}{w(w + 1)}$$

Figure 4–37 shows the Bode diagram for the system. For the magnitude curves we have used straight-line asymptotes. The magnitude and phase angle of $G(jv)$ are shown by dashed curves. (Note that the zero at $v = 10$, which lies in the right half of the w plane, gives phase lag.) The phase margin can be read from the Bode diagram (dashed curves) as 30° and the gain margin as 14.5 dB.

The given specifications require, in addition to $K_v = 2$, the phase margin of 50° and a gain margin of at least 10 dB. Let us design a digital controller to satisfy these specifications.

Design of lead compensator.   Since the specification calls for a phase margin of 50°, the additional phase-lead angle necessary to satisfy this requirement is 20°. To achieve a phase margin of 50° without decreasing the value of K, the lead compensator must contribute the required phase-lead angle.

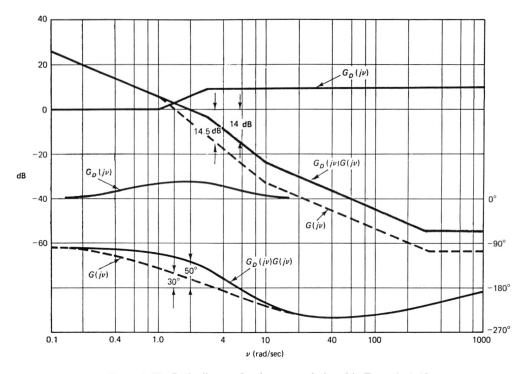

Figure 4–37 Bode diagram for the system designed in Example 4–12.

Noting that the addition of a lead compensator modifies the magnitude curve in the Bode diagram, the gain crossover frequency will be shifted to the right. Considering the shift of the gain crossover frequency, we may assume that ϕ_m, the maximum phase-lead angle required, is approximately 28°. (This means that 8° has been added to compensate for the shift in the gain crossover frequency.) Since

$$\sin \phi_m = \frac{1 - \alpha}{1 + \alpha}$$

$\phi_m = 28°$ corresponds to $\alpha = 0.361$.

Once the attenuation factor α has been determined on the basis of the required phase-lead angle, the next step is to determine the corner frequencies $\nu = 1/\tau$ and $\nu = 1/(\alpha\tau)$ of the lead compensator. To do so, we first note that the maximum phase-lead angle ϕ_m occurs at the geometric mean of the two corner frequencies, or $\nu = 1/(\sqrt{\alpha}\tau)$. The amount of the modification in the magnitude curve at $\nu = 1/(\sqrt{\alpha}\tau)$ due to the inclusion of the term $(1 + \tau j\nu)/(1 + \alpha\tau j\nu)$ is

$$\left| \frac{1 + \tau j\nu}{1 + \alpha\tau j\nu} \right|_{\nu=1/(\sqrt{\alpha}\tau)} = \frac{1}{\sqrt{\alpha}}$$

Next, we find the frequency point where the magnitude of the uncompensated system is equal to $-20 \log(1/\sqrt{\alpha})$. Note that

$$-20 \log \frac{1}{\sqrt{0.361}} = -20 \log 1.6643 = -4.425 \text{ dB}$$

To find the frequency point where the magnitude is -4.425 dB, we substitute $w = jv$ in $G(w)$ and find the magnitude of $G(jv)$:

$$|G(jv)| = \frac{2\sqrt{1 + \left(\dfrac{v}{300}\right)^2}\sqrt{1 + \left(\dfrac{v}{10}\right)^2}}{v\sqrt{1 + v^2}}$$

By trial and error, we find that at $v = 1.7$ the magnitude becomes approximately -4.4 dB. We select this frequency to be the new gain crossover frequency v_c. Noting that this frequency corresponds to $1/(\sqrt{\alpha\tau})$, or

$$v_c = \frac{1}{\sqrt{\alpha\tau}} = 1.7$$

we obtain

$$\tau = \frac{1}{1.7\sqrt{\alpha}} = 0.9790$$

and

$$\alpha\tau = 0.3534$$

The lead compensator thus determined is

$$G_D(w) = \frac{1 + \tau w}{1 + \alpha\tau w} = \frac{1 + 0.9790w}{1 + 0.3534w} \tag{4-40}$$

The magnitude and phase angle curves for $G_D(jv)$ and the magnitude and phase angle curves of the compensated open-loop transfer function $G_D(jv)G(jv)$ are shown by solid curves in Figure 4-37. From the Bode diagram we see that the phase margin is 50° and the gain margin is 14 dB.

The controller transfer function given by Equation (4-40) will now be transformed back to the z plane by the bilinear transformation given by Equation (4-38):

$$w = \frac{2}{T}\frac{z - 1}{z + 1} = \frac{2}{0.2}\frac{z - 1}{z + 1} = 10\frac{z - 1}{z + 1}$$

Thus,

$$G_D(z) = \frac{1 + 0.9790\left(10\dfrac{z - 1}{z + 1}\right)}{1 + 0.3534\left(10\dfrac{z - 1}{z + 1}\right)}$$

$$= \frac{2.3798z - 1.9387}{z - 0.5589}$$

The open-loop pulse transfer function of the compensated system is

$$G_D(z)G(z) = \frac{2.3798z - 1.9387}{z - 0.5589}\frac{0.03746(z + 0.9356)}{(z - 1)(z - 0.8187)}$$

$$= \frac{0.0891z^2 + 0.0108z - 0.0679}{z^3 - 2.3776z^2 + 1.8352z - 0.4576}$$

The closed-loop pulse transfer function of the designed system is

$$\frac{C(z)}{R(z)} = \frac{0.0891z^2 + 0.0108z - 0.0679}{z^3 - 2.2885z^2 + 1.8460z - 0.5255}$$

$$= \frac{0.0891(z + 0.9357)(z - 0.8145)}{(z - 0.8126)(z - 0.7379 - j0.3196)(z - 0.7379 + j0.3196)}$$

Notice that the closed-loop pulse transfer function involves two zeros located at $z = -0.9357$ and $z = 0.8145$. The zero at $z = 0.8145$ almost cancels with the closed-loop pole at $z = 0.8126$. The effect of another zero at $z = -0.9357$ on the transient and frequency responses is very small, since it is located on the negative real axis of the z plane between 0 and -1 and is close to point $z = -1$. The pair of complex conjugate poles acts as dominant closed-loop poles. (The system behaves as if it is a second-order system.)

To check the transient response of the designed system, we shall obtain the unit-step response of this system using MATLAB. MATLAB Program 4–2 produces the unit-step response curve as shown in Figure 4–38. The plot of the unit-step response exhibits a maximum overshoot of approximately 20% and a settling time of approximately 4 sec. From this curve we see that the number of samples per cycle of sinusoidal oscillation is approximately 15. This means that the sampling frequency ω_s is 15 times the damped natural frequency ω_d. Thus, the sampling period of 0.2 sec is satisfactory under normal operation of this system.

MATLAB Program 4–2

```
% ---------- Unit-step response of designed system ----------

num = [0  0.0891  0.0108  -0.0679];
den = [1  -2.2885  1.8460  -0.5255];
r = ones(1,41);
k = 0:40;
c = filter(num,den,r);
plot(k,c,'o')
v = [0  40  0  1.6];
axis(v);
grid
title('Unit-Step Response of Designed System')
xlabel('k    (Sampling period T = 0.2 sec)')
ylabel('Output c(k)')
```

Comments. The advantage of the w transform method is that the conventional frequency-response method using Bode diagrams can be used for the design of discrete-time control systems. In applying this method, we must carefully choose a reasonable sampling frequency. Before we conclude this section, we summarize the important facts about design in the w plane.

1. The magnitude and phase angle of $G(jv)$ are the magnitude and phase angle of $G(z)$ as z moves on the unit circle from $z = 1$ to $z = -1$. Since $z = e^{j\omega T}$,

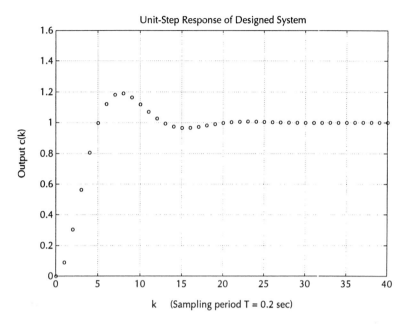

Figure 4–38   Plot of unit-step response of the designed system.

the ω value varies from 0 to $\frac{1}{2}\omega_s$. The fictitious frequency ν varies from 0 to ∞, since $\nu = (2/T)\tan(\omega T/2)$. Thus, the frequency response of the digital control system for $0 \le \omega \le \frac{1}{2}\omega_s$ is similar to the frequency response of the corresponding analog control system for $0 \le \nu \le \infty$.

2. Since $G(j\nu)$ is a rational function of ν, it is basically the same as $G(j\omega)$. In determining possible unstable zeros of the characteristic equation, the Nyquist stability criterion can be applied. Therefore, both the conventional straight-line approximation to the magnitude curve in the Bode diagram and the concept of phase margin and gain margin apply to $G(j\nu)$.

3. Compare transfer functions $G(w)$ and $G(s)$. As we mentioned earlier, because of the presence of the scale factor $2/T$ in the w transformation, the corresponding static error constants for $G(w)$ and $G(s)$ become identical. (Without the scale factor $2/T$, this will not be true.)

4. The w transformation may generate one or more right half-plane zeros in $G(w)$. If one or more right half-plane zeros exist, then $G(w)$ is a nonminimum phase transfer function. Because the zeros in the right half-plane are generated by the sample-and-hold operation, the locations of these zeros depend on the sampling period T. The effects of these zeros in the right half-plane on the response become smaller as the sampling period T becomes smaller.

In what follows, let us consider the effects on the response of the zero in the right half-plane at $w = 2/T$. The zero at $w = 2/T$ causes distortion in the frequency response as ν approaches $2/T$. Since

$$\nu = \frac{2}{T}\tan\frac{\omega T}{2}$$

then, as ν approaches $2/T$, $\tan(\omega T/2)$ approaches 1, or

$$\frac{\omega T}{2} = \frac{\pi}{4}$$

and thus

$$\omega = \frac{\pi}{2T}$$

As we stated earlier, $\omega = \frac{1}{2}\omega_s = \pi/T$ is the highest frequency that we consider in the response of the discrete-time or digital control system. Therefore, $\omega = \omega_s/4 = \pi/2T$, which is one-half the highest frequency considered, is well within the frequency range of interest. Thus, the zero at $w = 2/T$, which is in the right half of the w plane, will seriously affect the response.

5. It should be noted that the Bode diagram method in the w plane is frequently used in practice, and many successful digital control systems have been designed by this approach.

4–7 ANALYTICAL DESIGN METHOD

The main reason why the control actions of analog controllers are limited is that there are physical limitations in pneumatic, hydraulic, and electronic components. Such limitations may be completely ignored in designing digital controllers. Thus, many control schemes that have been impossible with analog controls are possible with digital controls. In fact, optimal control schemes that are not possible with analog controllers are made possible by digital control schemes.

In this section we specifically present an analytical design method for digital controllers that will force the error sequence, when subjected to a specific type of time-domain input, to become zero after a finite number of sampling periods and, in fact, to become zero and stay zero after the minimum possible number of sampling periods.

If the response of a closed-loop control system to a step input exhibits the minimum possible settling time (that is, the output reaches the final value in the minimum time and stays there), no steady-state error, and no ripples between the sampling instants, then this type of response is commonly called a *deadbeat response*. The deadbeat response will be discussed in this section. (We shall treat the deadbeat response again in Chapter 6, where we discuss the pole placement technique and the design of state observers.)

The discussions that follow are limited to the determination of the control algorithms or pulse transfer functions of digital controllers for single-input–single-output systems, given desired optimal response characteristics. For optimal control of multiple-input–multiple-output systems, see Chapter 8, where the state-space approach is used.

Design of Digital Controllers for Minimum Settling Time with Zero Steady-State Error.   Consider the digital control system shown in Figure 4–39(a). The error signal $e(t)$, which is the difference between the input $r(t)$ and the output $c(t)$, is

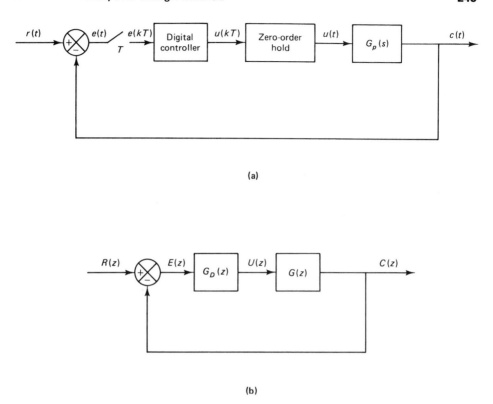

Figure 4–39   (a) A digital control system; (b) diagram showing equivalent digital control system.

sampled every time interval T. The input to the digital controller is the error signal $e(kT)$. The output of the digital controller is the control signal $u(kT)$. The control signal $u(kT)$ is fed to the zero-order hold, and the output of the hold, $u(t)$, which is a piecewise continuous-time signal, is fed to the plant. [Although the sampler at the input of the zero-order hold is not shown, the signal $u(kT)$ is first sampled and fed to the zero-order hold. As mentioned earlier, the zero-order hold shown in the diagram is a sample-and-hold device.] It is desired to design a digital controller $G_D(z)$ such that the closed-loop control system will exhibit the minimum possible settling time with zero steady-state error in response to a step, a ramp, or an acceleration input. It is required that the output not exhibit intersampling ripples after the steady state is reached. The system must satisfy any other specifications, if required, such as a specification for the static velocity error constant.

Let us define the z transform of the plant that is preceded by the zero-order hold as $G(z)$, or

$$G(z) = \mathscr{Z}\left[\frac{1 - e^{-Ts}}{s}G_p(s)\right]$$

Then the open-loop pulse transfer function becomes $G_D(z)G(z)$, as shown in Figure 4–39(b). Next, define the desired closed-loop pulse transfer function as $F(z)$:

$$\frac{C(z)}{R(z)} = \frac{G_D(z)G(z)}{1 + G_D(z)G(z)} = F(z) \tag{4-41}$$

Since it is required that the system exhibit a finite settling time with zero steady-state error, the system must exhibit a finite impulse response. Hence, the desired closed-loop pulse transfer function must be of the following form:

$$F(z) = \frac{a_0 z^N + a_1 z^{N-1} + \cdots + a_N}{z^N}$$

or

$$F(z) = a_0 + a_1 z^{-1} + \cdots + a_N z^{-N} \tag{4-42}$$

where $N \geq n$ and n is the order of the system. [Note that $F(z)$ must not contain any terms with positive powers in z, since such terms in the series expansion of $F(z)$ imply that the output precedes the input, which is not possible for a physically realizable system.] In our design approach, we solve the closed-loop pulse transfer function for the digital controller pulse transfer function $G_D(z)$. That is, we find the pulse transfer function $G_D(z)$ that will satisfy Equation (4–41). Solving Equation (4–41) for $G_D(z)$, we obtain

$$G_D(z) = \frac{F(z)}{G(z)[1 - F(z)]} \tag{4-43}$$

The designed system must be physically realizable. The conditions for physical realizability place certain constraints on the closed-loop pulse transfer function $F(z)$ and the digital controller pulse transfer function $G_D(z)$. The conditions for physical realizability may be stated as follows:

1. The order of the numerator of $G_D(z)$ must be equal to or lower than the order of the denominator. (Otherwise, the controller requires future input data to produce the current output.)

2. If the plant $G_p(s)$ involves a transportation lag e^{-Ls}, then the designed closed-loop system must involve at least the same magnitude of the transportation lag. (Otherwise, the closed-loop system would have to respond before an input was given, which is impossible for a physically realizable system.)

3. If $G(z)$ is expanded into a series in z^{-1}, the lowest-power term of the series expansion of $F(z)$ in z^{-1} must be at least as large as that of $G(z)$. For example, if an expansion of $G(z)$ into a series in z^{-1} begins with the z^{-1} term, then the first term of $F(z)$ given by Equation (4–42) must be zero, or a_0 must equal 0; that is, the expansion has to be of the form

$$F(z) = a_1 z^{-1} + a_2 z^{-2} + \cdots + a_N z^{-N}$$

where $N \geq n$ and n is the order of the system. This means that the plant cannot respond instantaneously when a control signal of finite magnitude is applied: the response comes at a delay of at least one sampling period if the series expansion of $G(z)$ begins with a term in z^{-1}.

In addition to the physical realizability conditions, we must pay attention to the stability aspects of the system. Specifically, we must avoid canceling an unstable pole of the plant by a zero of the digital controller. If such a cancellation is attempted, any error in the pole–zero cancellation will diverge as time elapses and the system will become unstable. Similarly, the digital controller pulse transfer function should not involve unstable poles to cancel plant zeros that lie outside the unit circle.

Next, let us investigate what will happen to the closed-loop pulse transfer function $F(z)$ if $G(z)$ involves an unstable (or critically stable) pole, that is, a pole $z = \alpha$ outside (or on) the unit circle. [Note that the following argument applies equally, if $G(z)$ involves two or more unstable—or critically stable—poles.] Let us define

$$G(z) = \frac{G_1(z)}{z - \alpha}$$

where $G_1(z)$ does not include a term that cancels with $z - \alpha$. Then the closed-loop pulse transfer function becomes

$$\frac{C(z)}{R(z)} = \frac{G_D(z)G(z)}{1 + G_D(z)G(z)} = \frac{G_D(z)\dfrac{G_1(z)}{z - \alpha}}{1 + G_D(z)\dfrac{G_1(z)}{z - \alpha}} = F(z) \qquad (4\text{--}44)$$

Since we require that no zero of $G_D(z)$ cancel the unstable pole of $G(z)$ at $z = \alpha$, we must have

$$1 - F(z) = \frac{1}{1 + G_D(z)\dfrac{G_1(z)}{z - \alpha}} = \frac{z - \alpha}{z - \alpha + G_D(z)G_1(z)}$$

That is, $1 - F(z)$ must have $z = \alpha$ as a zero. Also, notice that from Equation (4–44) if zeros of $G(z)$ do not cancel poles of $G_D(z)$, the zeros of $G(z)$ become zeros of $F(z)$. [$F(z)$ may involve additional zeros.]

Let us summarize what we have stated concerning stability.

1. Since the digital controller $G_D(z)$ should not cancel unstable (or critically stable) poles of $G(z)$, all unstable (or critically stable) poles of $G(z)$ must be included in $1 - F(z)$ as zeros.
2. Zeros of $G(z)$ that lie inside the unit circle may be canceled with poles of $G_D(z)$. However, zeros of $G(z)$ that lie on or outside the unit circle must not be canceled with poles of $G_D(z)$. Hence, all zeros of $G(z)$ that lie on or outside the unit circle must be included in $F(z)$ as zeros.

Now we shall proceed with the design. Since $e(kT) = r(kT) - c(kT)$, referring to Equation (4–41) we have

$$E(z) = R(z) - C(z) = R(z)[1 - F(z)] \qquad (4\text{--}45)$$

Note that for a unit-step input $r(t) = 1(t)$

$$R(z) = \frac{1}{1 - z^{-1}}$$

For a unit-ramp input $r(t) = t1(t)$,

$$R(z) = \frac{Tz^{-1}}{(1 - z^{-1})^2}$$

And for a unit-acceleration input $r(t) = \frac{1}{2}t^2 1(t)$,

$$R(z) = \frac{T^2 z^{-1}(1 + z^{-1})}{2(1 - z^{-1})^3}$$

Thus, in general, z transforms of such time-domain polynomial inputs may be written as

$$R(z) = \frac{P(z)}{(1 - z^{-1})^{q+1}} \tag{4-46}$$

where $P(z)$ is a polynomial in z^{-1}. Notice that for a unit-step input $P(z) = 1$ and $q = 0$; for a unit-ramp input, $P(z) = Tz^{-1}$ and $q = 1$; and for a unit-acceleration input, $P(z) = \frac{1}{2}T^2 z^{-1}(1 + z^{-1})$ and $q = 2$.

By substituting Equation (4-46) into Equation (4-45), we obtain

$$E(z) = \frac{P(z)[1 - F(z)]}{(1 - z^{-1})^{q+1}} \tag{4-47}$$

To ensure that the system reaches steady state in a finite number of sampling periods and maintains zero steady-state error, $E(z)$ must be a polynomial in z^{-1} with a finite number of terms. Then, by referring to Equation (4-47), we choose the function $1 - F(z)$ to be of the form

$$1 - F(z) = (1 - z^{-1})^{q+1} N(z) \tag{4-48}$$

where $N(z)$ is a polynomial in z^{-1} with a finite number of terms. Then

$$E(z) = P(z)N(z) \tag{4-49}$$

which is a polynomial in z^{-1} with a finite number of terms. This means that the error signal becomes zero in a finite number of sampling periods.

From the preceding analysis, the pulse transfer function of the digital controller can be determined as follows. By first letting $F(z)$ satisfy the physical realizability and stability conditions and then substituting Equation (4-48) into Equation (4-43), we obtain

$$G_D(z) = \frac{F(z)}{G(z)(1 - z^{-1})^{q+1} N(z)} \tag{4-50}$$

Equation (4-50) gives the pulse transfer function of the digital controller that will produce zero steady-state error after a finite number of sampling periods.

For a stable plant $G_p(s)$, the condition that the output not exhibit intersampling ripples after the settling time is reached may be written as follows:

$$c(t \geq nT) = \text{constant}, \quad \text{for step inputs}$$

$$\dot{c}(t \geq nT) = \text{constant}, \quad \text{for ramp inputs}$$

$$\ddot{c}(t \geq nT) = \text{constant}, \quad \text{for acceleration inputs}$$

The applicable condition must be satisfied when the system is designed. In designing the system, the condition on $c(t)$, $\dot{c}(t)$, or $\ddot{c}(t)$ must be interpreted in terms of $u(t)$. Note that the plant is continuous time and the input to the plant is $u(t)$, a continuous-time function; therefore, to have no ripples in the output $c(t)$, the control signal $u(t)$ at steady state must be either constant or monotonically increasing (or monotonically decreasing) for step, ramp, or acceleration inputs.

Comments

1. Since the closed-loop pulse transfer function $F(z)$ is a polynomial in z^{-1}, all the closed-loop poles are at the origin or at $z = 0$. The multiple closed-loop pole at the origin is very sensitive to system parameter variations.

2. Although a digital control system designed to exhibit minimum settling time with zero steady-state error in response to a specific type of input has excellent transient response characteristics for the input it is designed for, it may exhibit inferior or sometimes unacceptable transient response characteristics for other types of input. (This is always true in optimal control systems. An optimal control system will exhibit the best response characteristics for the type of input it is designed for, but will not exhibit optimal response characteristics for other types of input.)

3. In the case in which an analog controller is discretized, an increase in the sampling period changes the system dynamics and may lead to system instability. On the other hand, the behavior of the digital control system we are designing in this section does not depend on the choice of the sampling period. Since the inputs $r(t)$ considered here are time-domain inputs (such as step inputs, ramp inputs, and acceleration inputs), the sampling period T can be chosen arbitrarily. For a smaller sampling period, the response time (which is an integral multiple of the sampling period T) becomes smaller. However, for a very small sampling period T, the magnitude of the control signal will become excessively large, with the result that saturation phenomena will take place in the system, and the design method presented in this section will no longer apply. Hence, the sampling period T should not be too small. On the other hand, if the sampling period T is chosen too large, the system may behave unsatisfactorily or may even become unstable when it is subjected to sufficiently time varying inputs (such as frequency-domain inputs). Thus, a compromise is necessary. A rule of thumb is to choose the smallest sampling period T such that no saturation phenomena occur in the control signal.

Example 4-13

Consider the digital control system shown in Figure 4-39(a), where the plant transfer function $G_p(s)$ is given by

$$G_p(s) = \frac{1}{s(s + 1)}$$

Design a digital controller $G_D(z)$ such that the closed-loop system will exhibit a deadbeat response to a unit-step input. (In a deadbeat response the system should not exhibit intersampling ripples in the output after the settling time is reached.) The sampling period T is assumed to be 1 sec. Then, using the digital controller $G_D(z)$ so designed, investigate the response of this system to a unit-ramp input.

The first step in the design is to determine the z transform of the plant that is preceded by the zero-order hold:

$$
\begin{aligned}
G(z) &= \mathcal{Z}\left[\frac{1 - e^{-Ts}}{s} \frac{1}{s(s + 1)}\right] \\
&= (1 - z^{-1})\mathcal{Z}\left[\frac{1}{s^2(s + 1)}\right] \\
&= (1 - z^{-1})\left[\frac{z^{-1}}{(1 - z^{-1})^2} - \frac{1}{1 - z^{-1}} + \frac{1}{1 - 0.3679z^{-1}}\right] \\
&= \frac{0.3679(1 + 0.7181z^{-1})z^{-1}}{(1 - z^{-1})(1 - 0.3679z^{-1})}
\end{aligned}
\tag{4-51}
$$

Now redraw the block diagram of the system as shown in Figure 4–39(b). Define the closed-loop pulse transfer function as $F(z)$, or

$$
\frac{C(z)}{R(z)} = \frac{G_D(z)G(z)}{1 + G_D(z)G(z)} = F(z)
$$

Notice that if $G(z)$ is expanded into a series in z^{-1} then the first term will be $0.3679z^{-1}$. Hence, $F(z)$ must begin with a term in z^{-1}.

Referring to Equation (4–42) and noting that the system is of the second order ($n = 2$), we assume $F(z)$ to be of the following form:

$$
F(z) = a_1 z^{-1} + a_2 z^{-2}
\tag{4-52}
$$

Since the input is a step function, from Equation (4–48) we require that

$$
1 - F(z) = (1 - z^{-1})N(z)
\tag{4-53}
$$

Since $G(z)$ has a critically stable pole at $z = 1$, the stability requirement states that $1 - F(z)$ must have a zero at $z = 1$. However, the function $1 - F(z)$ already has a term $1 - z^{-1}$ and therefore satisfies the requirement.

Since the system should not exhibit intersampling ripples and the input is a step function, we require $c(t \geq 2T)$ to be constant. Noting that $u(t)$, the output of the zero-order hold, is a continuous-time function, a constant $c(t \geq 2T)$ requires that $u(t)$ also be constant for $t \geq 2T$. In terms of the z transform, $U(z)$ must be of the following type of series in z^{-1}:

$$
U(z) = b_0 + b_1 z^{-1} + b(z^{-2} + z^{-3} + z^{-4} + \cdots)
$$

where b is a constant. Because the plant transfer function $G_p(s)$ involves an integrator, b must be zero. (Otherwise, the output cannot stay constant.) Consequently, we have

$$
U(z) = b_0 + b_1 z^{-1}
$$

From Figure 4–39(b), $U(z)$ can be given as follows:

$$
U(z) = \frac{C(z)}{G(z)} = \frac{C(z)}{R(z)}\frac{R(z)}{G(z)} = F(z)\frac{R(z)}{G(z)}
$$

$$= F(z)\frac{1}{1-z^{-1}}\frac{(1-z^{-1})(1-0.3679z^{-1})}{0.3679(1+0.7181z^{-1})z^{-1}}$$

$$= F(z)\frac{1-0.3679z^{-1}}{0.3679(1+0.7181z^{-1})z^{-1}}$$

For $U(z)$ to be a series in z^{-1} with only two terms, $F(z)$ must be of the following form:

$$F(z) = (1+0.7181z^{-1})z^{-1}F_1 \tag{4–54}$$

where F_1 is a constant. Then $U(z)$ can be written as follows:

$$U(z) = 2.7181(1-0.3679z^{-1})F_1 \tag{4–55}$$

Equation (4–55) gives $U(z)$ in terms of F_1. Once constant F_1 is determined, $U(z)$ can be given as a series in z^{-1} with only two terms.

Now we shall determine $N(z)$, $F(z)$, and F_1. By substituting Equation (4–52) into Equation (4–53), we obtain

$$1 - a_1 z^{-1} - a_2 z^{-2} = (1-z^{-1})N(z)$$

The left-hand side of this last equation must be divisible by $1-z^{-1}$. If we divide the left-hand side by $1-z^{-1}$, the quotient is $1+(1-a_1)z^{-1}$ and the remainder is $(1-a_1-a_2)z^{-2}$. Hence, $N(z)$ is determined as

$$N(z) = 1 + (1-a_1)z^{-1} \tag{4–56}$$

and the remainder must be zero. This requires that

$$1 - a_1 - a_2 = 0 \tag{4–57}$$

Also, from Equations (4–52) and (4–54) we have

$$F(z) = a_1 z^{-1} + a_2 z^{-2} = (1+0.7181z^{-1})z^{-1}F_1$$

Hence,

$$a_1 + a_2 z^{-1} = (1+0.7181z^{-1})F_1$$

Division of the left-hand side of this last equation by $1+0.7181z^{-1}$ yields the quotient a_1 and the remainder $(a_2-0.7181a_1)z^{-1}$. By equating the quotient with F_1 and the remainder with zero, we obtain

$$F_1 = a_1$$

and

$$a_2 - 0.7181a_1 = 0 \tag{4–58}$$

Solving Equations (4–57) and (4–58) for a_1 and a_2 gives

$$a_1 = 0.5820, \qquad a_2 = 0.4180$$

Thus, $F(z)$ is determined as

$$F(z) = 0.5820z^{-1} + 0.4180z^{-2} \tag{4–59}$$

and

$$F_1 = 0.5820$$

Equation (4–56) gives

$$N(z) = 1 + 0.4180z^{-1} \tag{4–60}$$

The digital controller pulse transfer function $G_D(z)$ is then determined from Equation (4–50), as follows. By referring to Equations (4–51), (4–54), and (4–60),

$$G_D(z) = \frac{F(z)}{G(z)(1 - z^{-1})N(z)}$$

$$= \frac{(1 + 0.7181z^{-1})z^{-1}(0.5820)}{\dfrac{0.3679(1 + 0.7181z^{-1})z^{-1}}{(1 - z^{-1})(1 - 0.3679z^{-1})}(1 - z^{-1})(1 + 0.4180z^{-1})}$$

$$= \frac{1.5820 - 0.5820z^{-1}}{1 + 0.4180z^{-1}}$$

With the digital controller thus designed, the closed-loop pulse transfer function becomes as follows:

$$\frac{C(z)}{R(z)} = F(z) = 0.5820z^{-1} + 0.4180z^{-2}$$

$$= \frac{0.5820(z + 0.7181)}{z^2}$$

The system output in response to a unit-step input $r(t) = 1$ can be obtained as follows:

$$C(z) = F(z)R(z)$$

$$= (0.5820z^{-1} + 0.4180z^{-2})\frac{1}{1 - z^{-1}}$$

$$= 0.5820z^{-1} + z^{-2} + z^{-3} + z^{-4} + \cdots$$

Hence,

$$c(0) = 0$$

$$c(1) = 0.5820$$

$$c(k) = 1, \qquad k = 2, 3, 4, \ldots$$

Notice that substitution of 0.5820 for F_1 in Equation (4–55) yields

$$U(z) = 2.7181(1 - 0.3679z^{-1})(0.5820)$$

$$= 1.5820 - 0.5820z^{-1}$$

Thus, the control signal $u(k)$ becomes zero for $k \geq 2$, as required. There is no inter-sampling ripple in the output after the settling time is reached. Figure 4–40(a) shows plots of $c(k)$ versus k, $u(k)$ versus k, and $u(t)$ versus t in the unit-step response.
    Next, let us investigate the response of this system to a unit-ramp input:

$$C(z) = F(z)R(z)$$

$$= (0.5820z^{-1} + 0.4180z^{-2})\frac{z^{-1}}{(1 - z^{-1})^2}$$

$$= 0.5820z^{-2} + 1.5820z^{-3} + 2.5820z^{-4} + 3.5820z^{-5} + \cdots$$

For the unit-ramp response, the control signal $U(z)$ is obtained as follows. Referring to Equations (4–51) and (4–59),

$$U(z) = \frac{C(z)}{G(z)} = \frac{F(z)}{G(z)}R(z) = \frac{F(z)}{G(z)}\frac{z^{-1}}{(1 - z^{-1})^2}$$

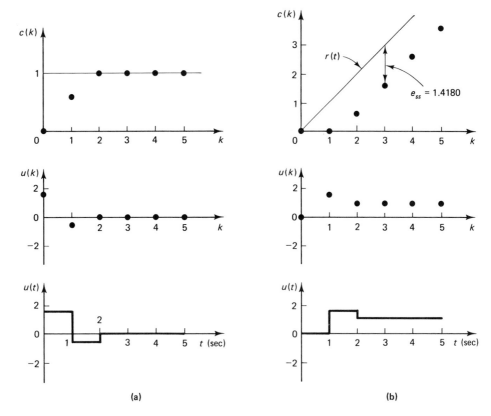

Figure 4–40   Responses of the system designed in Example 4–13.   (a) Plots of $c(k)$ versus k, $u(k)$ versus k, and $u(t)$ versus t in the unit-step response; (b) plots of $c(k)$ versus k, $u(k)$ versus k, and $u(t)$ versus t in the unit-ramp response.

$$= (1.5820 - 0.5820z^{-1})\frac{z^{-1}}{1 - z^{-1}}$$

$$= 1.5820z^{-1} + z^{-2} + z^{-3} + z^{-4} + \cdots$$

The signal $u(k)$ becomes constant ($b = 1$) for $k \geq 2$. Hence, the system output will not exhibit intersampling ripples. Figure 4–40(b) shows plots of $c(k)$ versus k, $u(k)$ versus k, and $u(t)$ versus t in the unit-ramp response.

Note that the static velocity error constant K_v for the present system is

$$K_v = \lim_{z \to 1}\left[\frac{1 - z^{-1}}{T}G_D(z)G(z)\right]$$

$$= \lim_{z \to 1}\left[(1 - z^{-1})\frac{F(z)}{(1 - z^{-1})N(z)}\right]$$

$$= \lim_{z \to 1}\frac{0.5820z^{-1} + 0.4180z^{-2}}{1 + 0.4180z^{-1}} = 0.7052$$

Thus, the steady-state error in the unit-ramp response is

$$e_{ss} = \frac{1}{K_v} = 1.4180$$

which is indicated in Figure 4–40(b).

In the present design problem, we have required that in response to a step input the system exhibit the minimum settling time with no steady-state error and no ripples in the output after the settling time is reached. If one or more additional constraints are present in the design problem (for example, if the value of the static velocity error constant K_v is arbitrarily specified), then the number of sampling periods required before reaching the steady-state must be increased. For example, the second-order system may require three or more sampling periods before the steady state is reached, depending on the additional constraints imposed. See Example 4–14, which follows.

Example 4–14

Consider a design problem the same as that of Example 4–13 except that the static velocity error constant K_v is specified. (Because of this additional constraint, the settling time will be longer than 2 sec.) The block diagram of the digital control system is shown in Figure 4–39(a). The plant transfer function $G_p(s)$ under consideration is

$$G_p(s) = \frac{1}{s(s + 1)}$$

The design specifications are (1) that the closed-loop system is to exhibit a finite settling time with zero steady-state error in the unit-step response, (2) that the output is not to exhibit intersampling ripples after the settling time is reached, (3) that the static velocity error constant K_v is to be 4 sec^{-1}, and (4) that the settling time is to be the minimum possible that will satisfy all these specifications. The sampling period T is assumed to be 1 sec. Design a digital controller $G_D(z)$ that satisfies the given specifications. After the controller is designed, investigate the response of the system to a unit-ramp input.

The z transform of the plant that is preceded by the zero-order hold was obtained in Example 4–13 as

$$G(z) = \mathcal{Z}\left[\frac{1 - e^{-Ts}}{s} \frac{1}{s(s + 1)}\right]$$

$$= \frac{0.3679(1 + 0.7181z^{-1})z^{-1}}{(1 - z^{-1})(1 - 0.3679z^{-1})}$$

Define the closed-loop pulse transfer function as $F(z)$:

$$\frac{C(z)}{R(z)} = \frac{G_D(z)G(z)}{1 + G_D(z)G(z)} = F(z)$$

Since the first term in the expansion of $G(z)$ is $0.3679z^{-1}$, $F(z)$ must begin with a term in z^{-1}:

$$F(z) = a_1 z^{-1} + a_2 z^{-2} + \cdots + a_N z^{-N}$$

where $N \geq n$ and n is the order of the system (that is, $n = 2$ in the present case). Because of the added constraint, we may assume $N > 2$. We shall try $N = 3$. Thus, we assume

$$F(z) = a_1 z^{-1} + a_2 z^{-2} + a_3 z^{-3} \tag{4–61}$$

(If a satisfactory result is not obtained, we must assume $N > 3$.) Since the input is a step function, from Equation (4–48) we require that

$$1 - F(z) = (1 - z^{-1})N(z) \qquad (4\text{–}62)$$

Note that the presence of a critically stable pole at $z = 1$ in the plant pulse transfer function $G(z)$ requires $1 - F(z)$ to have a zero at $z = 1$. However, the function $1 - F(z)$ already has a term $1 - z^{-1}$ and therefore satisfies the stability requirement.

The requirement that the static velocity error constant be 4 sec^{-1} can be written as follows:

$$K_v = \lim_{z \to 1} \left[\frac{1 - z^{-1}}{T} G_D(z)G(z) \right]$$

$$= \lim_{z \to 1} \left[(1 - z^{-1}) \frac{F(z)}{(1 - z^{-1})N(z)} \right]$$

$$= \frac{F(1)}{N(1)} = 4$$

where we used Equation (4–50) with $q = 0$. Notice that from Equation (4–62) we have $F(1) = 1$. Hence, K_v can be written as follows:

$$K_v = \frac{1}{N(1)} = 4 \qquad (4\text{–}63)$$

Since the system output should not exhibit intersampling ripples after the settling time is reached, we require $U(z)$ to be of the following form:

$$U(z) = b_0 + b_1 z^{-1} + b_2 z^{-2} + b(z^{-3} + z^{-4} + z^{-5} + \cdots)$$

Because the plant transfer function $G_p(s)$ involves an integrator, b must be zero. Consequently, we have

$$U(z) = b_0 + b_1 z^{-1} + b_2 z^{-2}$$

Also, from Figure 4–39(b), $U(z)$ can be given by

$$U(z) = \frac{C(z)}{G(z)} = \frac{C(z)}{R(z)} \frac{R(z)}{G(z)} = F(z) \frac{R(z)}{G(z)}$$

$$= F(z) \frac{1 - 0.3679z^{-1}}{0.3679(1 + 0.7181z^{-1})z^{-1}}$$

For $U(z)$ to be a series in z^{-1} with three terms, $F(z)$ must be of the following form:

$$F(z) = (1 + 0.7181z^{-1})z^{-1}F_1(z) \qquad (4\text{–}64)$$

where $F_1(z)$ is a first-degree polynomial in z^{-1}. Then $U(z)$ can be written as follows:

$$U(z) = 2.7181(1 - 0.3679z^{-1})F_1(z) \qquad (4\text{–}65)$$

From Equations (4–61) and (4–62), we have

$$1 - F(z) = 1 - a_1 z^{-1} - a_2 z^{-2} - a_3 z^{-3} = (1 - z^{-1})N(z)$$

If we divide $1 - a_1 z^{-1} - a_2 z^{-2} - a_3 z^{-3}$ by $1 - z^{-1}$, the quotient is $1 + (1 - a_1)z^{-1} + (1 - a_1 - a_2)z^{-2}$ and the remainder is $(1 - a_1 - a_2 - a_3)z^{-3}$. Hence, $N(z)$ is determined as

$$N(z) = 1 + (1 - a_1)z^{-1} + (1 - a_1 - a_2)z^{-2} \qquad (4\text{–}66)$$

and the remainder must be zero, so that

$$1 - a_1 - a_2 - a_3 = 0 \qquad (4\text{-}67)$$

Note that from Equation (4–63) we require $N(1) = \frac{1}{4}$. Therefore, by substituting $z^{-1} = 1$ into Equation (4–66), we obtain

$$2a_1 + a_2 = 2.75 \qquad (4\text{-}68)$$

Also, Equation (4–64) can be rewritten as

$$F(z) = a_1 z^{-1} + a_2 z^{-2} + a_3 z^{-3} = (1 + 0.7181z^{-1})z^{-1} F_1(z)$$

Hence,

$$a_1 + a_2 z^{-1} + a_3 z^{-2} = (1 + 0.7181z^{-1})F_1(z)$$

Division of the left-hand side of this last equation by $1 + 0.7181z^{-1}$ yields the quotient $[a_1 + (a_2 - 0.7181a_1)z^{-1}]$ and the remainder $[a_3 - 0.7181(a_2 - 0.7181a_1)]z^{-2}$. By equating the quotient with $F_1(z)$ and the remainder with zero, we obtain

$$F_1(z) = a_1 + (a_2 - 0.7181a_1)z^{-1}$$

and

$$a_3 - 0.7181(a_2 - 0.7181a_1) = 0 \qquad (4\text{-}69)$$

Solving Equations (4–67), (4–68), and (4–69) for a_1, a_2, and a_3 gives

$$a_1 = 1.26184, \qquad a_2 = 0.22633, \qquad a_3 = -0.48816$$

Thus, $F(z)$ is determined as

$$F(z) = 1.26184z^{-1} + 0.22633z^{-2} - 0.48816z^{-3}$$

and

$$F_1(z) = 1.26184 - 0.67979z^{-1}$$

Equation (4–66) gives

$$N(z) = 1 - 0.26184z^{-1} - 0.48817z^{-2}$$

The digital controller pulse transfer function $G_D(z)$ is then determined from Equation (4–50) as follows:

$$
\begin{aligned}
G_D(z) &= \frac{F(z)}{G(z)(1 - z^{-1})N(z)} \\[2mm]
&= \frac{(1 + 0.7181z^{-1})z^{-1}(1.26184 - 0.67980z^{-1})}{\dfrac{0.3679(1 + 0.7181z^{-1})z^{-1}}{(1 - z^{-1})(1 - 0.3679z^{-1})}(1 - z^{-1})(1 - 0.26184z^{-1} - 0.48817z^{-2})} \\[2mm]
&= 3.4298\frac{(1 - 0.5387z^{-1})(1 - 0.3679z^{-1})}{(1 - 0.8418z^{-1})(1 + 0.5799z^{-1})}
\end{aligned}
$$

With the digital controller thus designed, the system output in response to a unit-step input $r(t) = 1$ is obtained as follows:

$$
\begin{aligned}
C(z) &= F(z)R(z) \\[2mm]
&= (1.26184z^{-1} + 0.22633z^{-2} - 0.48816z^{-3})\frac{1}{1 - z^{-1}} \\[2mm]
&= 1.2618z^{-1} + 1.4882z^{-2} + z^{-3} + z^{-4} + \cdots
\end{aligned}
$$

Hence,

$$c(0) = 0$$

$$c(1) = 1.2618$$

$$c(2) = 1.4882$$

$$c(k) = 1, \quad k = 3, 4, 5, \ldots$$

The unit-step response sequence has a maximum overshoot of approximately 50%. The settling time is 3 sec.

Notice that from Equation (4–65) we have

$$U(z) = 2.7181(1 - 0.3679z^{-1})(1.26184 - 0.67979z^{-1})$$

$$= 3.4298 - 3.1096z^{-1} + 0.6798z^{-2}$$

Thus, the control signal $u(k)$ becomes zero for $k \geq 3$. Consequently, there are no intersampling ripples in the response. Figure 4–41 shows plots of $c(k)$ versus k, $u(k)$ versus k, and $u(t)$ versus t in the unit-step response. Notice that the assumption of $N = 3$, that is, the assumption of $F(z)$ as given by Equation (4–61), is satisfactory.

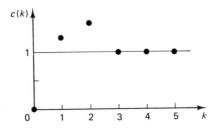

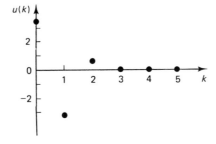

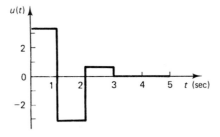

Figure 4–41 Plots of $c(k)$ versus k, $u(k)$ *versus* k, and $u(t)$ versus t in the unit-step response of the system designed in Example 4–14.

Next, let us investigate the response of this system to a unit-ramp input:

$$C(z) = F(z)R(z)$$

$$= (1.26184z^{-1} + 0.22633z^{-2} - 0.48816z^{-3})\frac{z^{-1}}{(1 - z^{-1})^2}$$

$$= 1.2618z^{-2} + 2.7500z^{-3} + 3.7500z^{-4} + \cdots$$

In the unit-ramp response, the control signal $U(z)$ is obtained as follows:

$$U(z) = \frac{C(z)}{G(z)} = \frac{F(z)}{G(z)}R(z) = \frac{F(z)}{G(z)}\frac{1}{1 - z^{-1}}\frac{z^{-1}}{1 - z^{-1}}$$

$$= (3.4298 - 3.1096z^{-1} + 0.6798z^{-2})\frac{z^{-1}}{1 - z^{-1}}$$

$$= 3.4298z^{-1} + 0.3202z^{-2} + z^{-3} + z^{-4} + z^{-5} + \cdots$$

The signal $u(k)$ becomes constant ($b = 1$) for $k \geq 3$. Hence, the system output will not exhibit intersampling ripples. Figure 4–42 shows plots of $c(k)$ versus k, $u(k)$ versus k, and $u(t)$ versus t in the unit-ramp response. Notice that the steady-state error in the unit-ramp response is $e_{ss} = 1/K_v = \frac{1}{4}$, as indicated in Figure 4–42.

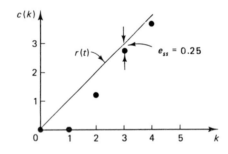

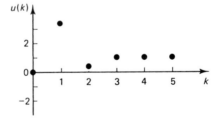

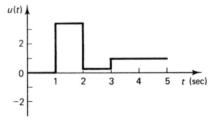

Figure 4–42   Plots of $c(k)$ versus k, $u(k)$ versus k, and $u(t)$ versus t in the unit-ramp response of the system designed in Example 4–14.

Comparing the digital control systems designed in Examples 4–13 and 4–14, we note that the latter improves the ramp response characteristics at the expense of the settling time. (The latter system requires one extra sampling period to reach the steady state.) Note also that the former has better step response characteristics, that is, a shorter settling time and no overshoot. Depending on the objectives of the system, we may choose one over the other. If good ramp response characteristics are required, then the system should be designed using the ramp input as the reference input, rather than the step input. (See Problem A–4–14.)

EXAMPLE PROBLEMS AND SOLUTIONS

Problem A–4–1

Show that geometrically the patterns of the poles near $z = 1$ in the z plane are similar to the patterns of poles in the s plane near the origin.

Solution   Note that

$$z = e^{Ts}$$

Near the origin of the s plane,

$$z = e^{Ts} = 1 + Ts + \tfrac{1}{2}T^2 s^2 + \cdots$$

or

$$z - 1 \doteq Ts$$

Thus, geometrical patterns of the poles near $z = 1$ in the z plane are similar to the patterns of poles in the s plane near the origin.

Problem A–4–2

Consider the system described by

$$y(k) - 0.6y(k - 1) - 0.81y(k - 2) + 0.67y(k - 3) - 0.12y(k - 4) = x(k)$$

where $x(k)$ is the input and $y(k)$ is the output of the system. Determine the stability of the system.

Solution   The pulse transfer function for the system is

$$\frac{Y(z)}{X(z)} = \frac{1}{1 - 0.6z^{-1} - 0.81z^{-2} + 0.67z^{-3} - 0.12z^{-4}}$$

$$= \frac{z^4}{z^4 - 0.6z^3 - 0.81z^2 + 0.67z - 0.12}$$

Define

$$P(z) = z^4 - 0.6z^3 - 0.81z^2 + 0.67z - 0.12$$

$$= a_0 z^4 + a_1 z^3 + a_2 z^2 + a_3 z + a_4, \qquad a_0 > 0$$

Then we have

$$a_0 = 1$$

$$a_1 = -0.6$$

$$a_2 = -0.81$$

$$a_3 = 0.67$$

$$a_4 = -0.12$$

The Jury stability conditions are:

1. $|a_4| < a_0$. This condition is clearly satisfied.

2. $P(1) > 0$. Since

$$P(1) = 1 - 0.6 - 0.81 + 0.67 - 0.12 = 0.14 > 0$$

the condition is satisfied.

3. $P(-1) > 0$. Since

$$P(-1) = 1 + 0.6 - 0.81 - 0.67 - 0.12 = 0$$

the condition is not satisfied. $P(-1) = 0$ implies that there is one root at $z = -1$.

4. $|b_3| > |b_0|$. Since

$$b_3 = \begin{vmatrix} a_4 & a_0 \\ a_0 & a_4 \end{vmatrix} = \begin{vmatrix} -0.12 & 1 \\ 1 & -0.12 \end{vmatrix} = -0.9856$$

$$b_0 = \begin{vmatrix} a_4 & a_3 \\ a_0 & a_1 \end{vmatrix} = \begin{vmatrix} -0.12 & 0.67 \\ 1 & -0.6 \end{vmatrix} = -0.5980$$

the condition is satisfied.

5. $|c_2| > |c_0|$. Since

$$c_2 = \begin{vmatrix} b_3 & b_0 \\ b_0 & b_3 \end{vmatrix} = \begin{vmatrix} -0.9856 & -0.5980 \\ -0.5980 & -0.9856 \end{vmatrix} = 0.6138$$

$$c_0 = \begin{vmatrix} b_3 & b_2 \\ b_0 & b_1 \end{vmatrix} = \begin{vmatrix} -0.9856 & 0.5196 \\ -0.5980 & 0.9072 \end{vmatrix} = -0.5834$$

the condition is satisfied.

From the preceding analysis, we conclude that the characteristic equation $P(z) = 0$ involves a root at $z = -1$ and the other three roots are in the unit circle centered at the origin of the z plane. The system is critically stable.

Problem A–4–3

Consider the following characteristic equation:

$$P(z) = z^3 - 1.3z^2 - 0.08z + 0.24 = 0 \tag{4-70}$$

Determine whether or not any of the roots of the characteristic equation lie outside the unit circle in the z plane. Use the bilinear transformation and the Routh stability criterion.

Solution   Let us substitute $(w + 1)/(w - 1)$ for z in the given characteristic equation, resulting in

$$\left(\frac{w + 1}{w - 1}\right)^3 - 1.3\left(\frac{w + 1}{w - 1}\right)^2 - 0.08\frac{w + 1}{w - 1} + 0.24 = 0$$

Clearing the fractions by multiplying both sides of this last equation by $(w - 1)^3$, we get

$$-0.14w^3 + 1.06w^2 + 5.10w + 1.98 = 0$$

By dividing both sides of this last equation by -0.14, we obtain

$$w^3 - 7.571w^2 - 36.43w - 14.14 = 0 \qquad (4\text{--}71)$$

The Routh array for Equation (4–71) becomes as follows:

one sign $\longrightarrow w^3$ $\qquad$ 1 $\qquad$ -36.43

change $\longrightarrow w^2$ $\qquad$ -7.571 $\qquad$ -14.14

$\qquad\qquad\quad w^1$ $\qquad$ -38.30 $\qquad$ 0

$\qquad\qquad\quad w^0$ $\qquad$ -14.14

Routh stability criterion states that the number of roots with positive real parts is equal to the number of changes in sign of the coefficients of the first column of the array. Since there is one sign change for the coefficients in the first column, there is one root in the right half of the w plane. This means that the original characteristic equation given by Equation (4–70) has one root outside the unit circle in the z plane. The system is unstable. (Compare the amount of computation needed in the present method and that needed in the Jury stability test. See in particular Example 4–6.)

Problem A–4–4

Consider the system defined by

$$\frac{Y(z)}{U(z)} = \frac{0.7870z^{-1}}{1 - 0.8195z^{-1} + 0.6065z^{-2}}$$

$$= \frac{0.7870z}{z^2 - 0.8195z + 0.6065}$$

The sampling period T is 0.5 sec. Using MATLAB, plot the unit-ramp response up to $k = 20$.

Solution   The unit-ramp input u may be written as

$$u = kT, \qquad k = 0, 1, 2, \ldots$$

In the MATLAB program, this input can be given as

$$k = 0{:}N; \qquad u = [T*k];$$

where N is the end of the process considered.

A MATLAB program for plotting the unit-ramp response of the system considered is given in MATLAB Program 4–3. The resulting plot is shown in Figure 4–43.

Problem A–4–5

Show that if the characteristic equation for a closed-loop system is written as

$$1 + \frac{KB(z)}{A(z)} = 0$$

where $A(z)$ and $B(z)$ do not contain K, then the breakaway and break-in points can be determined from the roots of

$$\frac{dK}{dz} = -\frac{A'(z)B(z) - A(z)B'(z)}{B^2(z)} = 0$$

where the primes indicate differentiation with respect to z.

MATLAB Program 4–3

```
% --------- Unit-ramp response ----------

% ***** Enter the numerator and denominator of the system *****

num = [0  0.7870  0];
den = [1  -0.8195  0.6065];

% ***** Enter k, unit-ramp input, filter command and plot
% command *****

k = 0:20;
u =  [0.5*k];
y = filter(num,den,u);
plot(k,y,'o',k,y,'-',k,0.5*k,'--')

% ***** Add grid, title, xlabel, and ylabel *****

grid
title('Unit-Ramp Response')
xlabel('k')
ylabel('y(k)')
```

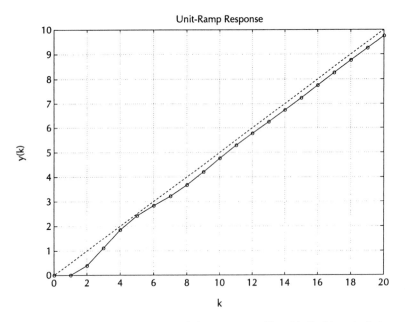

Figure 4–43 Unit-ramp response of the system considered in Problem A–4–4.

Solution Let us write the characteristic equation as

$$f(z) = A(z) + KB(z) = 0 \qquad (4\text{--}72)$$

Suppose that $f(z) = 0$ has a multiple root of order r. Then $f(z)$ may be written as

$$f(z) = (z - z_1)^r(z - z_2)\cdots(z - z_p)$$

If we differentiate this equation with respect to z and set $z = z_1$, we get

$$\left. \frac{df(z)}{dz} \right|_{z=z_1} = 0$$

This means that multiple roots of $f(z)$ will satisfy the following equation:

$$\frac{df(z)}{dz} = 0$$

or

$$\frac{df(z)}{dz} = A'(z) + KB'(z) = 0 \qquad (4\text{--}73)$$

where

$$A'(z) = \frac{dA(z)}{dz}, \qquad B'(z) = \frac{dB(z)}{dz}$$

Solving Equation (4–73) for K, we obtain

$$K = -\frac{A'(z)}{B'(z)}$$

This particular value of K will yield multiple roots of the characteristic equation. If we substitute this value of K into Equation (4–72), we obtain

$$f(z) = A(z) - \frac{A'(z)}{B'(z)} B(z) = 0$$

or

$$B'(z)A(z) - A'(z)B(z) = 0 \qquad (4\text{--}74)$$

If this last equation is solved for z, the points where multiple roots occur can be obtained. On the other hand, from Equation (4–72) we have

$$K = -\frac{A(z)}{B(z)}$$

and

$$\frac{dK}{dz} = -\frac{A'(z)B(z) - A(z)B'(z)}{B^2(z)}$$

If dK/dz is set equal to zero, we get the same equation as Equation (4–74). Therefore, the breakaway or break-in points can be simply determined from the roots of

$$\frac{dK}{dz} = 0$$

It should be noted that not all the solutions of Equation (4–74) or of $dK/dz = 0$ correspond to actual breakaway or break-in points. Such a point for which $dK/dz = 0$ is an actual breakaway or break-in point if and only if the value of K at this point is a real, positive value.

Problem A–4–6

Discuss the procedure for designing lead compensators for digital control systems by the root-locus method.

Solution   We shall consider the system shown in Figure 4–44 to discuss the procedure for designing lead compensators. Lead compensation is useful when the system is either unstable for all values of gain or is stable but has undesirable transient response characteristics. To design lead compensators, we may use the following procedure:

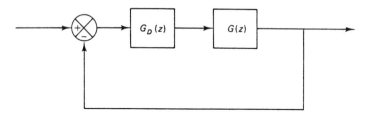

Figure 4–44   Digital control system.

1. From the performance specifications, determine the desired location for the dominant closed-loop poles.
2. By drawing the root-locus plot, ascertain whether or not the gain adjustment alone can yield the desired closed-loop poles. If not, calculate the angle deficiency ϕ. This angle must be contributed by the lead compensator if the new root locus is to pass through the desired locations for the dominant closed-loop poles.
3. Assume the lead compensator $G_D(z)$ to be

$$G_D(z) = K_D \alpha \frac{1 + \alpha z}{1 + \alpha \tau z}, \qquad 0 < \alpha < 1$$

4. If static error constants are not specified, determine the location of the pole and zero of the lead compensator so that the lead compensator will contribute the necessary angle ϕ. If no other requirements are imposed on the system, try to make the value of α as large as possible. A larger value of α generally results in a larger value of K_v, which is desirable. (If a particular static error constant is specified, it is generally simpler to use the frequency-response approach.)
5. Determine the open-loop gain of the compensated system from the magnitude condition.

Once a compensator has been designed, check to see whether or not all performance specifications have been met. If the compensated system does not meet the performance specifications, then repeat the design procedure by adjusting the compensator pole and zero until all such specifications are met. If a large static error constant is required, cascade a lag network or alter the lead compensator to a lag–lead compensator.

Problem A–4–7

Draw root locus diagrams in the z plane for the system shown in Figure 4–45 for the following three sampling periods: $T = 1$ sec, $T = 2$ sec, and $T = 4$ sec.

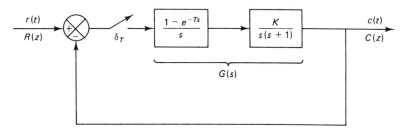

Figure 4–45   Digital control system.

Solution   We first obtain the z transform of $G(s)$. Referring to Example 3–5, we get

$$G(z) = \mathcal{Z}\left[\frac{1 - e^{-Ts}}{s}\frac{K}{s(s+1)}\right]$$

$$= (1 - z^{-1})\mathcal{Z}\left[\frac{K}{s^2(s+1)}\right]$$

$$= \frac{K[(T - 1 + e^{-T})z^{-1} + (1 - e^{-T} - Te^{-T})z^{-2}]}{(1 - z^{-1})(1 - e^{-T}z^{-1})} \qquad (4\text{-}75)$$

Next we construct root locus diagrams for the three cases considered.

1. *Sampling period T = 1:* For $T = 1$, Equation (4–75) becomes

$$G(z) = \frac{K[(1 - 1 + e^{-1})z^{-1} + (1 - e^{-1} - e^{-1})z^{-2}]}{(1 - z^{-1})(1 - e^{-1}z^{-1})}$$

$$= \frac{0.3679K(z + 0.7181)}{(z - 1)(z - 0.3679)}$$

Notice that $G(z)$ possesses a zero at $z = -0.7181$ and poles at $z = 1$ and $z = 0.3679$. The breakaway point is at $z = 0.6479$, and the break-in point is at $z = -2.0841$. The root-locus diagram for this case is shown in Figure 4–46(a). The value of gain K of any point on the root loci can be determined from the magnitude condition

$$K = \left|\frac{(z - 1)(z - 0.3679)}{0.3679(z + 0.7181)}\right|$$

If we choose a point z on the root loci, the value of K at that point can be calculated by substituting the value of z into this last equation. (This means that with this value of K that particular point becomes a closed-loop pole.) The critical gain is found to be $K = 2.3925$.

2. *Sampling period T = 2:* For the sampling period $T = 2$, we have from Equation (4–75)

$$G(z) = \frac{1.1353K(z + 0.5232)}{(z - 1)(z - 0.1353)}$$

The pulse transfer function $G(z)$ in this case possesses a zero at $z = -0.5232$ and poles at $z = 1$ and $z = 0.1353$. The breakaway point is at $z = 0.4783$, and the break-in point is at $z = -1.5247$. The root-locus diagram for this case is shown in Figure 4–46(b). The critical gain K for stability is $K = 1.4557$.

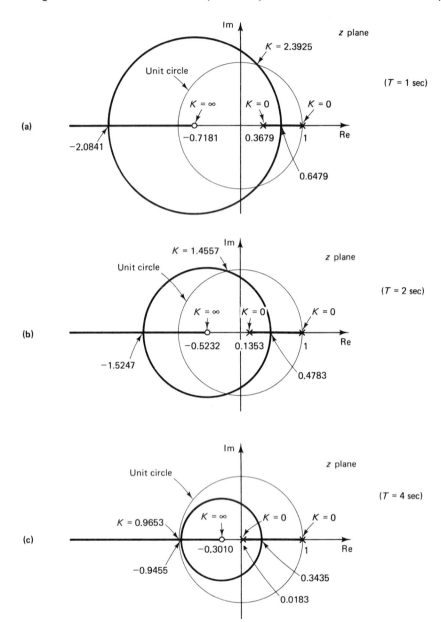

Figure 4–46 Root-locus diagrams for the system shown in Figure 4–45 when (a) $T = 1$ sec, (b) $T = 2$ sec, and (c) $T = 4$ sec.

3. *Sampling period $T = 4$:* For the case of $T = 4$, Equation (4–75) gives

$$G(z) = \frac{3.0183K(z + 0.3010)}{(z - 1)(z - 0.0183)}$$

The breakaway point is at $z = 0.3435$, and the break-in point is at $z = -0.9455$.

The root-locus diagram is shown in Figure 4–46(c). The critical gain for stability is $K = 0.9653$.

From the three cases considered, notice that the smaller the sampling period is, the larger the critical gain K for stability.

Problem A–4–8

Consider the digital control system shown in Figure 4–47, where the plant is of the first order and has a dead time of 2 sec. The sampling period is assumed to be 1 sec, or $T = 1$.

Design a digital PI controller such that the dominant closed-loop poles have a damping ratio ζ of 0.5 and the number of samples per cycle of damped sinusoidal oscillation is 10. Obtain the response of the system to a unit-step input. Also, obtain the static velocity error constant K_v and find the steady-state error in the response to a unit-ramp input.

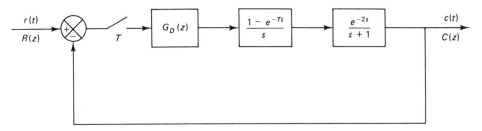

Figure 4–47   Digital control system.

Solution   The pulse transfer function of the plant that is preceded by a zero-order hold is

$$G(z) = \mathcal{Z}\left[\frac{1 - e^{-Ts}}{s}\frac{e^{-2s}}{s + 1}\right]$$

$$= (1 - z^{-1})z^{-2}\mathcal{Z}\left[\frac{1}{s(s + 1)}\right]$$

$$= (1 - z^{-1})z^{-2}\frac{(1 - e^{-1})z^{-1}}{(1 - z^{-1})(1 - e^{-1}z^{-1})}$$

$$= \frac{0.6321z^{-3}}{1 - 0.3679z^{-1}} = \frac{0.6321}{z^2(z - 0.3679)}$$

The digital PI controller has the following pulse transfer function:

$$G_D(z) = K_p + K_i\frac{1}{1 - z^{-1}}$$

$$= (K_p + K_i)\frac{z - \dfrac{K_p}{K_p + K_i}}{z - 1}$$

The open-loop pulse transfer function becomes

$$G_D(z)G(z) = \frac{(K_p + K_i)\left(z - \dfrac{K_p}{K_p + K_i}\right)}{z - 1}\frac{0.6321}{z^2(z - 0.3679)}$$

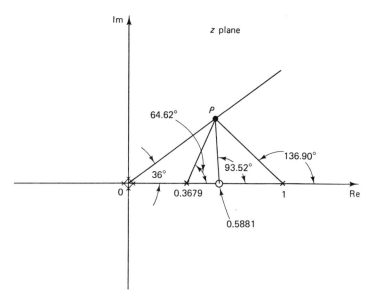

Figure 4–48 Pole and zero locations in the z plane of the system considered in Problem A–4–8.

We locate the open-loop poles in the z plane as shown in Figure 4–48. There is one open-loop zero involved in this case, but its location is unknown at this point.

Since it is required to have 10 samples per cycle of damped sinusoidal oscillation, the dominant closed-loop pole in the upper half of the z plane must lie on a line from the origin having an angle of $360°/10 = 36°$. From Equations (4–1) and (4–2), rewritten as

$$|z| = \exp\left(-\frac{2\pi\zeta}{\sqrt{1 - \zeta^2}}\frac{\omega_d}{\omega_s}\right)$$

$$\underline{/z} = 2\pi\frac{\omega_d}{\omega_s}$$

the desired closed-loop pole location can be determined as follows. Noting that $\underline{/z} = 36°$, we have

$$2\pi\frac{\omega_d}{\omega_s} = \frac{2\pi}{10}$$

or $\omega_d/\omega_s = 0.1$. Since ζ is specified as 0.5, we have

$$|z| = \exp\left(-\frac{2\pi \times 0.5}{\sqrt{1 - 0.5^2}}\frac{1}{10}\right) = e^{-0.3628} = 0.6958$$

The closed-loop pole is located at point P in Figure 4–48, where (at point P)

$$z = 0.6958 \underline{/36°}$$

$$= 0.5629 + j0.4090$$

(Note that this point is the intersection of the $\zeta = 0.5$ locus and the line from the origin having an angle of 36°.)

If point P is to be the closed-loop pole location in the upper half of the z plane, then the angle deficiency at point P is

$$-36° - 36° - 136.90° - 64.62° + 180° = -93.52°$$

The controller zero must contribute $+93.52°$. This means that the zero of the digital controller must be located at $z = 0.5881$. Therefore,

$$\frac{K_p}{K_p + K_i} = 0.5881 \tag{4-76}$$

Hence, the PI controller is determined as follows:

$$G_D(z) = K\frac{z - 0.5881}{z - 1}$$

where $K = K_p + K_i$. Gain constant K is determined from the magnitude condition:

$$K\left|\frac{z - 0.5881}{z - 1} \frac{0.6321}{z^2(z - 0.3679)}\right|_{z=0.5629+j0.4090} = 1$$

or

$$K = 0.5070$$

Thus,

$$K_p + K_i = 0.5070 \tag{4-77}$$

From Equations (4–76) and (4–77), we find that

$$K_p = 0.2982 \quad \text{and} \quad K_i = 0.2088$$

Hence, the PI controller just designed can be given by

$$G_D(z) = 0.5070 \frac{1 - 0.5881z^{-1}}{1 - z^{-1}}$$

Finally, the open-loop pulse transfer function becomes

$$G_D(z)G(z) = 0.5070\left(\frac{1 - 0.5881z^{-1}}{1 - z^{-1}} \frac{0.6321z^{-3}}{1 - 0.3679z^{-1}}\right)$$

$$= \frac{0.3205(1 - 0.5881z^{-1})z^{-3}}{(1 - z^{-1})(1 - 0.3679z^{-1})}$$

The closed-loop pulse transfer function becomes

$$\frac{C(z)}{R(z)} = \frac{0.3205z^{-3} - 0.1885z^{-4}}{1 - 1.3679z^{-1} + 0.3679z^{-2} + 0.3205z^{-3} - 0.1885z^{-4}}$$

The response $c(kT)$ to the unit-step input can be obtained easily by use of MATLAB. A MATLAB program for plotting the unit-step response is shown in MATLAB Program 4–4. The resulting plot is shown in Figure 4–49.

MATLAB Program 4–4

```
% ---------- Unit-step response ----------

num = [0  0  0  0.3205  -0.1885];
den = [1  -1.3679  0.3679  0.3205  -0.1885];
r = ones(1,51);
k = 0:50;
c = filter(num,den,r);
plot(k,c,'o')
v = [0  50  0  1.6];
axis(v);
grid
title('Unit-Step Response')
xlabel('k')
ylabel('c(k)')
```

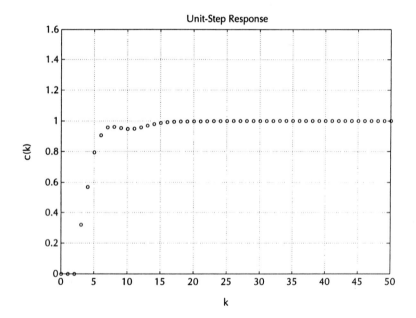

Figure 4–49 Plot of $c(kT)$ versus kT for the system designed in Problem A–4–8. (Sampling period $T = 1$ sec.)

Problem A–4–9

Consider the system shown in Figure 4–50. We wish to design a digital controller such that the dominant closed-loop poles of the system will have a damping ratio ζ of 0.5. We also want the number of samples per cycle of damped sinusoidal oscillation to be 8. Assume that the sampling period T is 0.2 sec.

Using the root-locus method in the z plane, determine the pulse transfer function of the digital controller. Obtain the response of the designed system to a unit-step input. Also obtain the static velocity error constant K_v.

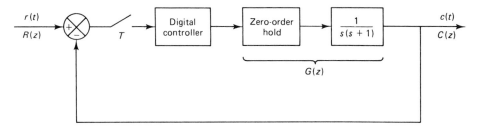

Figure 4–50 Digital control system.

Solution   We shall first locate the desired closed-loop poles in the z plane. Referring to Equations (4–1) and (4–2), for a constant-damping-ratio locus we have

$$|z| = e^{-\zeta T \omega_n} = \exp\left(-\frac{2\pi\zeta}{\sqrt{1-\zeta^2}}\frac{\omega_d}{\omega_s}\right) \tag{4–78}$$

and

$$\underline{/z} = T\omega_d = 2\pi\frac{\omega_d}{\omega_s} = \theta$$

Since we require eight samples per cycle of damped sinusoidal oscillation, the dominant closed-loop pole in the upper half of the z plane must be located on a line having an angle of 45° and passing through the origin as shown in Figure 4–51. (Note that the number of samples per cycle is 360°/θ. Hence, eight samples per cycle requires θ = 360°/8 = 45°.) Thus,

$$\underline{/z} = 45° = \frac{\pi}{4} = 2\pi\frac{\omega_d}{\omega_s}$$

which gives

$$\frac{\omega_d}{\omega_s} = \frac{1}{8} \tag{4–79}$$

Since the sampling period T is specified as 0.2 sec, we have

$$\omega_s = \frac{2\pi}{T} = \frac{2\pi}{0.2} = 10\pi$$

Therefore,

$$\omega_d = \frac{1}{8}\omega_s = \frac{10\pi}{8} = 3.9270$$

By letting $\zeta = 0.5$ and substituting Equation (4–79) into Equation (4–78), we obtain

$$|z| = e^{-0.4535} = 0.6354$$

Hence, we can locate the desired closed-loop pole in the upper half of the z plane, as shown by point P in Figure 4–51. Note that at point P

$$|z|\,\underline{/z} = 0.6354\,\underline{/45°} = 0.4493 + j0.4493$$

Next, we obtain the pulse transfer function $G(z)$ of the plant that is preceded by a zero-order hold:

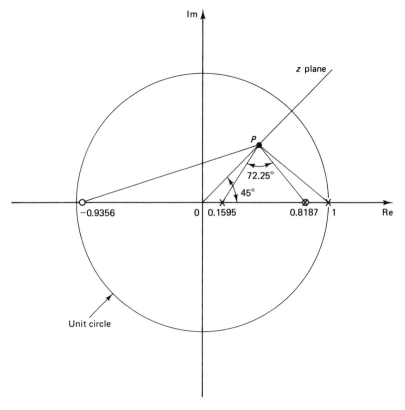

Figure 4–51   Pole and zero locations for the system considered in Problem A–4–9.

$$G(z) = \mathscr{Z}\left[\frac{1 - e^{-Ts}}{s} \frac{1}{s(s + 1)}\right]$$

$$= (1 - z^{-1})\,\mathscr{Z}\left[\frac{1}{s^2(s + 1)}\right]$$

$$= (1 - z^{-1})\left[\frac{0.2z^{-1}}{(1 - z^{-1})^2} - \frac{1}{1 - z^{-1}} + \frac{1}{1 - e^{-0.2}z^{-1}}\right]$$

$$= \frac{0.01873(1 + 0.9356z^{-1})z^{-1}}{(1 - z^{-1})(1 - 0.8187z^{-1})} = \frac{0.01873(z + 0.9356)}{(z - 1)(z - 0.8187)}$$

We can now locate the open-loop poles and a zero on the z plane as shown in Figure 4–51. Since point P is the location of the desired closed-loop pole, the angle deficiency at point P can be calculated easily as follows:

$$-140.79° - 129.43° + 17.97° + 180° = -72.25°$$

The controller pulse transfer function must contribute 72.25°.
    Let us choose the controller pulse transfer function to be

$$G_D(z) = K\frac{z + \alpha}{z + \beta}$$

and choose the zero of the controller to cancel the pole at $z = 0.8187$. Then the pole of the controller can be determined easily from the angle condition as $z = 0.1595$. Thus, we have

$$G_D(z) = K\frac{1 - 0.8187z^{-1}}{1 - 0.1595z^{-1}}$$

The open-loop pulse transfer function of the system is therefore obtained as follows:

$$G_D(z)G(z) = K\frac{1 - 0.8187z^{-1}}{1 - 0.1595z^{-1}}\frac{0.01873(1 + 0.9356z^{-1})z^{-1}}{(1 - z^{-1})(1 - 0.8187z^{-1})}$$

$$= K\frac{0.01873(1 + 0.9356z^{-1})z^{-1}}{(1 - 0.1595z^{-1})(1 - z^{-1})}$$

The gain constant K can be determined from the magnitude condition:

$$K\left|\frac{0.01873(z + 0.9356)}{(z - 0.1595)(z - 1)}\right|_{z=0.4493+j0.4493} = 1$$

or

$$K = 13.934$$

Hence, we have determined the pulse transfer function of the digital controller to be

$$G_D(z) = 13.934\left(\frac{1 - 0.8187z^{-1}}{1 - 0.1595z^{-1}}\right)$$

The open-loop pulse transfer function is

$$G_D(z)G(z) = \frac{0.2610(1 + 0.9356z^{-1})z^{-1}}{(1 - 0.1595z^{-1})(1 - z^{-1})}$$

The closed-loop pulse transfer function is

$$\frac{C(z)}{R(z)} = \frac{G_D(z)G(z)}{1 + G_D(z)G(z)}$$

$$= \frac{0.2610z^{-1} + 0.2442z^{-2}}{1 - 0.8985z^{-1} + 0.4037z^{-2}}$$

Because of the cancellation of a pole of the plant and the zero of the controller, the order of the system is reduced from third to second. The system has only a pair of conjugate complex closed-loop poles.

Figure 4–52 shows the unit-step response sequence $c(kT)$ versus kT. The plot shows the maximum overshoot to be approximately 16.5%.

Finally, the static velocity error constant K_v is determined as follows:

$$K_v = \lim_{z \to 1}\left[\frac{1 - z^{-1}}{T}G_D(z)G(z)\right]$$

$$= \lim_{z \to 1}\left[\frac{1 - z^{-1}}{0.2}\frac{0.2610(1 + 0.9356z^{-1})z^{-1}}{(1 - 0.1595z^{-1})(1 - z^{-1})}\right]$$

$$= 3.005$$

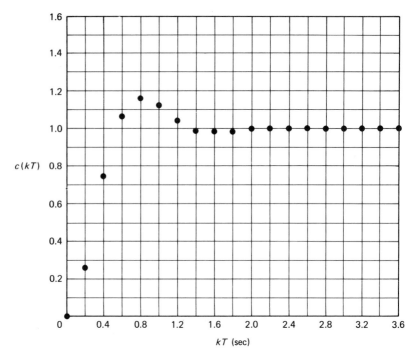

Figure 4–52   Plot of the unit-step response sequence $c(kT)$ versus kT for the system designed in Problem A–4–9.

Problem A–4–10

Consider the system shown in Figure 4–53. Assume that the performance specifications are given in terms of phase margin, gain margin, static velocity error constants, and the like. State procedures for designing lead compensators and lag compensators by the frequency-response approach.

Solution   The procedures for designing lead compensators and lag compensators may be stated as follows:

LEAD COMPENSATOR

1. Assume the following form for the lead compensator:

$$G_D(w) = K_D \frac{1 + \tau w}{1 + \alpha \tau w}, \qquad 0 < \alpha < 1$$

The open-loop transfer function of the compensated system may be written as

$$G_D(w)G(w) = K_D \frac{1 + \tau w}{1 + \alpha \tau w} G(w)$$

$$= \frac{1 + \tau w}{1 + \alpha \tau w} G_1(w)$$

where $G_1(w) = K_D\, G(w)$. Determine gain K_D to satisfy the requirement on the given static velocity error constant.

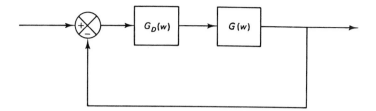

Figure 4–53 Digital control system in the w plane.

2. Using the gain K_D thus determined, draw a Bode diagram of $G_1(w)$, the gain-adjusted but uncompensated system. Evaluate the phase margin.

3. Determine the necessary phase lead angle ϕ to be added to the system.

4. Add $5° \sim 12°$ to ϕ to compensate for the shift of the gain crossover frequency. Define this added ϕ as ϕ_m. Determine the attenuation factor α from the following equation:

$$\sin \phi_m = \frac{1 - \alpha}{1 + \alpha}$$

5. Determine the frequency point where the magnitude of the uncompensated system $G_1(j\nu)$ is equal to $-20 \log (1/\sqrt{\alpha})$. Select this frequency as the new gain crossover frequency. This frequency corresponds to $\nu_m = 1/(\sqrt{\alpha}\tau)$, and the maximum phase shift ϕ_m occurs at this frequency.

6. Determine the corner frequencies of the lead compensator as follows:

$$\text{Zero of lead compensator:} \quad \nu = \frac{1}{\tau}$$

$$\text{Pole of lead compensator:} \quad \nu = \frac{1}{\alpha\tau}$$

7. Check the gain margin to be sure it is satisfactory. If not, repeat the design process by modifying the pole–zero location of the compensator until a satisfactory result is obtained.

The primary function of a lag compensator is to provide attenuation in the high-frequency range to give a system sufficient phase margin. The phase lag characteristic is of no consequence in lag compensation.

LAG COMPENSATOR

1. Assume the following form for the lag compensator:

$$G_D(w) = K_D \frac{1 + \tau w}{1 + \beta \tau w} \qquad (\beta > 1)$$

The open-loop transfer function of the compensated system may be written as

$$G_D(w)G(w) = K_D \frac{1 + \tau w}{1 + \beta \tau w} G(w)$$

$$= \frac{1 + \tau w}{1 + \beta \tau w} G_1(w)$$

where $G_1(w) = K_D G(w)$. Determine gain K_D to satisfy the requirement on the given static velocity error constant.

2. If the uncompensated system $G_1(w)$ does not satisfy the specifications on the phase and gain margins, then find the frequency point where the phase angle of the open-loop transfer function is equal to $-180°$ plus the required phase margin. The required phase margin is the specified phase margin plus 5° to 12°. (The addition of 5° to 12° compensates for the phase lag of the lag compensator.) Choose this frequency as the new gain crossover frequency.

3. To prevent detrimental effects of phase lag due to the lag compensator, the pole and zero of the lag compensator must be located substantially lower than the new gain crossover frequency. Therefore, choose the corner frequency $\nu = 1/\tau$ (corresponding to the zero of the lag compensator) one decade below the new gain crossover frequency.

4. Determine the attenuation necessary to bring the magnitude curve down to 0 dB at the new gain crossover frequency. Noting that this attenuation is $-20 \log \beta$, determine the value of β. Then the other corner frequency (corresponding to the pole of the lag compensator) is determined from $\nu = 1/(\beta\tau)$.

Caution.   Once the lag compensator is designed in the w plane, $G_D(w)$ must be transformed to the z plane lag compensator, $G_D(z)$. Note that the locations of the pole and zero of the lag compensator in the z plane are close to each other. (They are near point $z = 1$.) Since the filter coefficients must be realized by binary words that use limited number of bits, if the number of bits employed is insufficient, the pole and zero locations of the filter may not be realized exactly as desired, and the resulting compensator may not behave as expected. It is important that the pole and zero of the lag compensator lie on a finite number of allocable discrete points.

Problem A–4–11

Design a digital controller for the system shown in Figure 4–54. Use the Bode diagram approach in the w plane. The design specifications are that the phase margin be 55°, the gain margin be at least 10 dB, and the static velocity error constant be 5 sec^{-1}. The sampling period is specified as 0.1 sec, or $T = 0.1$. After the controller is designed, draw a root-locus diagram. Locate the closed-loop poles on the diagram and find the number of samples per cycle of damped sinusoidal oscillation.

Solution   The z transform of the plant that is preceded by a zero-order hold is

$$G(z) = \mathcal{Z}\left[\frac{1 - e^{-Ts}}{s} \frac{1}{s(s + 2)}\right]$$

$$= (1 - z^{-1})\mathcal{Z}\left[\frac{1}{s^2(s + 2)}\right]$$

$$= 0.004683z^{-1}\frac{1 + 0.9355z^{-1}}{(1 - z^{-1})(1 - 0.8187z^{-1})}$$

$$= (0.004683)\frac{z + 0.9355}{(z - 1)(z - 0.8187)}$$

Let us transform $G(z)$ into $G(w)$ by using the following bilinear transformation:

$$z = \frac{1 + (Tw/2)}{1 - (Tw/2)} = \frac{1 + 0.05w}{1 - 0.05w}$$

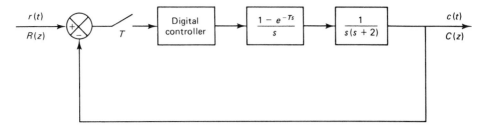

Figure 4–54   Digital control system.

By substituting this last equation into $G(z)$, we obtain

$$G(w) = \frac{0.004683\left(\dfrac{1 + 0.05w}{1 - 0.05w} + 0.9355\right)}{\left(\dfrac{1 + 0.05w}{1 - 0.05w} - 1\right)\left(\dfrac{1 + 0.05w}{1 - 0.05w} - 0.8187\right)}$$

$$= \frac{0.5(1 + 0.001666w)(1 - 0.05w)}{w(1 + 0.5016w)}$$

The Bode diagram of $G(j\nu)$ is shown in Figure 4–55.

We shall now choose the controller transfer function to be of the form

$$G_D(w) = K_D\frac{1 + \tau w}{1 + \alpha\tau w} = K_D\frac{1 + \dfrac{w}{a}}{1 + \dfrac{w}{b}}$$

where $a = 1/\tau$ and $b = 1/(\alpha\tau)$. The open-loop transfer function is

$$G_D(w)G(w) = K_D\frac{1 + (w/a)}{1 + (w/b)}\frac{0.5(1 + 0.001666w)(1 - 0.05w)}{w(1 + 0.5016w)}$$

The required static velocity error constant K_v is 5 sec^{-1}. Hence,

$$K_v = \lim_{w \to 0} [wG_D(w)G(w)] = 0.5K_D = 5$$

from which we determine that

$$K_D = 10$$

Using a conventional design technique, the digital controller transfer function is determined as

$$G_D(w) = 10\left(\frac{1 + \dfrac{w}{1.994}}{1 + \dfrac{w}{12.5}}\right)$$

Therefore, the open-loop transfer function becomes

$$G_D(w)G(w) = 10\left(\frac{1 + \dfrac{w}{1.994}}{1 + \dfrac{w}{12.5}}\right)\frac{0.5(1 + 0.001666w)(1 - 0.05w)}{w(1 + 0.5016w)}$$

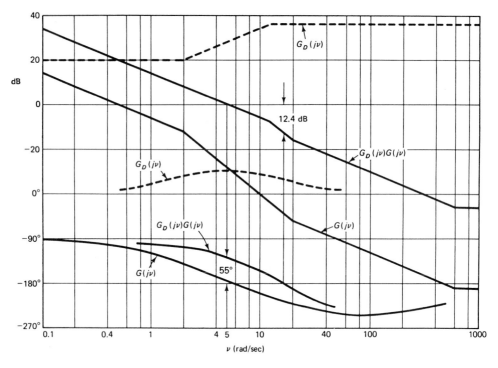

Figure 4–55 Bode diagram for the system considered in Problem A–4–11.

This open-loop transfer function gives the phase margin of approximately 55° and the gain margin of approximately 12.4 dB. The static velocity error constant K_v is 5 sec⁻¹. Hence, all requirements are satisfied and the designed controller transfer function $G_D(w)$ is satisfactory.

Next, we transform $G_D(w)$ into $G_D(z)$. The following bilinear transformation should be used:

$$w = \frac{2}{T}\frac{z-1}{z+1} = \frac{2}{0.1}\frac{z-1}{z+1} = 20\left(\frac{z-1}{z+1}\right)$$

Then

$$G_D(z) = 10\left(\frac{1 + \dfrac{20}{1.994}\dfrac{z-1}{z+1}}{1 + \dfrac{20}{12.5}\dfrac{z-1}{z+1}}\right)$$

$$= 42.423\left(\frac{z - 0.8187}{z - 0.2308}\right) = 42.423\left(\frac{1 - 0.8187z^{-1}}{1 - 0.2308z^{-1}}\right)$$

The open-loop pulse transfer function now becomes

$$G_D(z)G(z) = \frac{0.1987(z + 0.9355)}{(z - 0.2308)(z - 1)}$$

Figure 4–56 shows the root-locus diagram for the system. Using the magnitude condition, we find that the closed-loop poles are located at $z = 0.516 \pm j0.388$. On

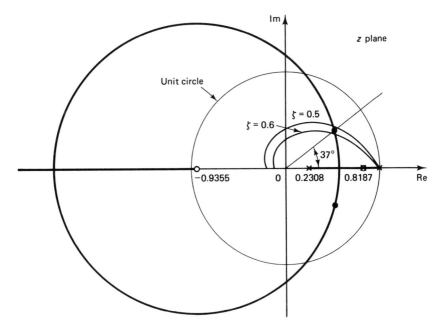

Figure 4–56   Root-locus diagram for the system designed in Problem A–4–11.

the root-locus diagram, constant ζ loci for $\zeta = 0.5$ and 0.6 are superimposed. From the diagram it can be seen that the damping ratio ζ of the closed-loop poles is approximately 0.55.

The line connecting the closed-loop pole in the upper half of the z plane and the origin has an angle of $37°$. Hence, the number of samples per cycle of damped sinusoidal oscillation is $360°/37° = 9.73$.

Problem A–4–12

Consider the digital control system shown in Figure 4–57, where the plant transfer function is $1/s^2$. Design a digital controller in the w plane such that the phase margin is $50°$ and the gain margin is at least 10 dB. The sampling period is 0.1 sec, or $T = 0.1$. After designing the controller, obtain the static velocity error constant K_v. Also, obtain the response of the designed system to a unit-step input.

Solution   We shall first obtain the z transform of the plant that is preceded by the zero-order hold:

$$G(z) = \mathcal{Z}\left[\frac{1 - e^{-Ts}}{s}\frac{1}{s^2}\right] = (1 - z^{-1})\,\mathcal{Z}\left[\frac{1}{s^3}\right]$$

$$= (1 - z^{-1})\frac{T^2(1 + z^{-1})z^{-1}}{2(1 - z^{-1})^3}$$

$$= \frac{0.005(1 + z^{-1})z^{-1}}{(1 - z^{-1})^2} = \frac{0.005(z + 1)}{(z - 1)^2}$$

Next, using the bilinear transformation given by

$$z = \frac{1 + (Tw/2)}{1 - (Tw/2)} = \frac{1 + 0.05w}{1 - 0.05w}$$

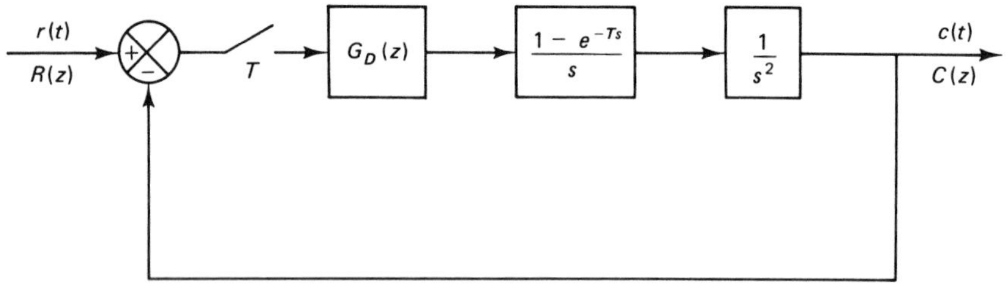

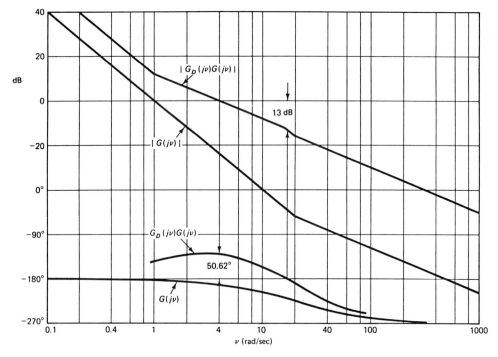

Figure 4–57 Digital control system.

we transform $G(z)$ into $G(w)$:

$$G(w) = \frac{0.005\left(\dfrac{1 + 0.05w}{1 - 0.05w} + 1\right)}{\left(\dfrac{1 + 0.05w}{1 - 0.05w} - 1\right)^2} = \frac{1 - 0.05w}{w^2}$$

Thus,

$$G(j\nu) = \frac{1 - 0.05j\nu}{(j\nu)^2}$$

Figure 4–58 shows the Bode diagram of $G(j\nu)$ thus obtained. Notice that the phase margin is $-2°$. It is necessary to add a lead network to give the required phase margin

Figure 4–58 Bode diagram for the system considered in Problem A–4–12.

and gain margin. By applying a conventional design technique, it can be seen that the following lead network will satisfy the requirements:

$$G_D(w) = 4\frac{1 + w}{1 + (w/16)} = 64\left(\frac{w + 1}{w + 16}\right)$$

The addition of this lead network modifies the Bode diagram. The gain crossover frequency is shifted to $\nu = 4$. Note that the maximum phase lead ϕ_m that this lead network can produce is 61.93°, since

$$\phi_m = \sin^{-1}\frac{1 - \frac{1}{16}}{1 + \frac{1}{16}} = \sin^{-1} 0.8824 = 61.93°$$

At the gain crossover frequency $\nu = 4$, the phase angle of $G_D(j\nu)G(j\nu)$ becomes $-191.31° + 61.93° = -129.38°$. Thus, the phase margin is 50.62°. The gain margin is found to be approximately 13 dB. Hence, the given design specifications are satisfied.

We now transform the controller transfer function $G_D(w)$ into $G_D(z)$. By using the bilinear transformation

$$w = \frac{2}{T}\frac{z - 1}{z + 1} = \frac{2}{0.1}\frac{z - 1}{z + 1} = 20\left(\frac{z - 1}{z + 1}\right)$$

we obtain

$$G_D(z) = 64\frac{20\left(\dfrac{z - 1}{z + 1}\right) + 1}{20\left(\dfrac{z - 1}{z + 1}\right) + 16} = 37.333\left(\frac{z - 0.9048}{z - 0.1111}\right)$$

Hence, the open-loop pulse transfer function becomes

$$G_D(z)G(z) = 37.333\left(\frac{z - 0.9048}{z - 0.1111}\right)\frac{0.005(z + 1)}{(z - 1)^2}$$

$$= \frac{0.1867(1 - 0.9048z^{-1})(1 + z^{-1})z^{-1}}{(1 - 0.1111z^{-1})(1 - z^{-1})^2}$$

The static velocity error constant K_v is obtained as follows:

$$K_v = \lim_{z \to 1}\left[\frac{1 - z^{-1}}{T}G_D(z)G(z)\right]$$

$$= \lim_{z \to 1}\left[\frac{1 - z^{-1}}{0.1}\frac{0.1867(1 - 0.9048z^{-1})(1 + z^{-1})z^{-1}}{(1 - 0.1111z^{-1})(1 - z^{-1})^2}\right] = \infty$$

Thus, the static velocity error constant K_v is infinity. There is no steady-state error in the ramp response.

The closed-loop pulse transfer function of the system is

$$\frac{C(z)}{R(z)} = \frac{0.1867z^{-1} + 0.0178z^{-2} - 0.1689z^{-3}}{1 - 1.9244z^{-1} + 1.2400z^{-2} - 0.2800z^{-3}}$$

Figure 4–59 shows the unit-step response. Notice that the zero of the digital controller at $z = 0.9048$ is close to the double pole at $z = 1$. A pole–zero pair near point $z = 1$ creates a long tail with small amplitude in the response.

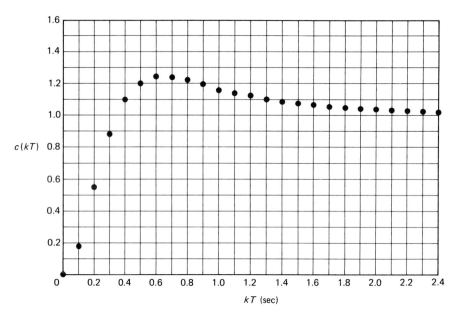

Figure 4–59   Plot of $c(kT)$ versus kT for the system designed in Problem A–4–12.

Problem A–4–13

Consider the digital control system shown in Figure 4–60. The plant transfer function involves a transportation lag e^{-5s}. The delay time is 5 sec, or $L = 5$. The desired output $c(t)$ in response to a unit-step input is as shown in Figure 4–61(a). The curve rises from zero to the final value in 10 sec (measured from $t = 5$ to $t = 15$) and there is neither overshoot nor steady-state error. The settling time is 15 sec (measured from $t = 0$ to $t = 15$). It is required that there be no intersampling ripples in the output after the settling time is reached. Design a digital controller $G_D(z)$.

Solution   Let us choose the sampling period to be 5 sec, or $T = 5$ sec. (We may, of course, choose the sampling period to be 2.5 sec, 1 sec, or another value. In this example, however, to simplify our presentation, we set the sampling period at 5 sec.)

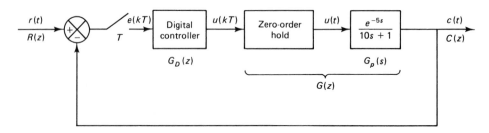

Figure 4–60   Digital control system.

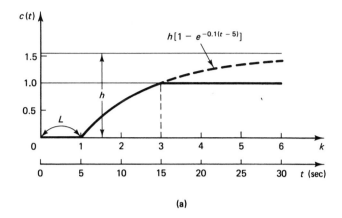

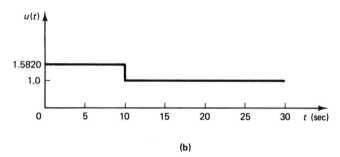

Figure 4–61   (a) Desired output $c(t)$ in response to a unit-step input; (b) plot of $u(t)$ versus t.

The z transform of the plant that is preceded by the zero-order hold is

$$G(z) = \mathcal{Z}\left[\frac{1 - e^{-Ts}}{s}\frac{e^{-5s}}{10s + 1}\right]$$

$$= (1 - z^{-1})z^{-1}\mathcal{Z}\left[\frac{1}{s(10s + 1)}\right]$$

$$= \frac{0.3935z^{-2}}{1 - 0.6065z^{-1}}$$

Notice that there is no unstable or critically stable pole involved in $G(z)$. Therefore, there is no stability problem involved in this case.

Let us define the closed-loop pulse transfer function as $F(z)$:

$$\frac{C(z)}{R(z)} = \frac{G_D(z)G(z)}{1 + G_D(z)G(z)} = F(z) \tag{4–80}$$

In the present case the output $c(t)$ in the unit-step response is specified as shown in Figure 4–61(a). Since $h[1 - e^{-0.1(15-5)}] = h(1 - e^{-1}) = 1$, we have $h = 1.5820$. from the deadbeat response curve shown in Figure 4–61(a), we obtain

$$c(0) = 0$$

$$c(1) = 0$$

$$c(2) = h(1 - e^{-0.5}) = 1.5820 \times 0.3935 = 0.6225$$
$$c(k) = 1, \quad k = 3, 4, 5, \ldots$$

from which we get

$$C(z) = 0.6225z^{-2} + z^{-3} + z^{-4} + z^{-5} + \cdots$$
$$= 0.6225z^{-2} + z^{-3}\frac{1}{1 - z^{-1}}$$
$$= \frac{0.6225z^{-2} + 0.3775z^{-3}}{1 - z^{-1}}$$

Noting that

$$C(z) = F(z)R(z) = F(z)\frac{1}{1 - z^{-1}} = \frac{0.6225z^{-2} + 0.3775z^{-3}}{1 - z^{-1}}$$

we obtain

$$F(z) = 0.6225z^{-2} + 0.3775z^{-3} = 0.6225(1 + 0.6065z^{-1})z^{-2}$$

Once $F(z)$ is determined, the pulse transfer function of the digital controller can be obtained from Equation (4–80):

$$G_D(z) = \frac{F(z)}{G(z)[1 - F(z)]}$$

Notice that from Equation (4–48) we have

$$1 - F(z) = (1 - z^{-1})N(z)$$

or

$$1 - 0.6225z^{-2} - 0.3775z^{-3} = (1 - z^{-1})N(z)$$

By dividing $(1 - 0.6225z^{-2} - 0.3775z^{-3})$ by $(1 - z^{-1})$, $N(z)$ can be determined as follows:

$$N(z) = 1 + z^{-1} + 0.3775z^{-2}$$

Consequently,

$$1 - F(z) = (1 - z^{-1})(1 + z^{-1} + 0.3775z^{-2})$$

and

$$G_D(z) = \frac{0.6225(1 + 0.6065z^{-1})z^{-2}}{\dfrac{0.3935z^{-2}}{1 - 0.6065z^{-1}}(1 - z^{-1})(1 + z^{-1} + 0.3775z^{-2})}$$
$$= \frac{1.5820(1 - 0.3678z^{-2})}{(1 - z^{-1})(1 + z^{-1} + 0.3775z^{-2})}$$

This last equation gives the pulse transfer function of the digital controller. Since $c(t)$ must be unity at steady state, $u(t)$, a continuous-time signal, must be constant after the steady state is reached.

Let us determine $U(z)$:

$$U(z) = \frac{C(z)}{G(z)} = \frac{0.6225z^{-2} + 0.3775z^{-3}}{(1 - z^{-1})\dfrac{0.3935z^{-2}}{1 - 0.6065z^{-1}}} = 1.5820\left(\frac{1 - 0.3678z^{-2}}{1 - z^{-1}}\right)$$
$$= 1.5820 + 1.5820z^{-1} + z^{-2} + z^{-3} + z^{-4} + \cdots$$

Taking the inverse z transform of $U(z)$, we find that $u(k)$ is constant for $k \geq 2$. Thus, there are no intersampling ripples in the output after the settling time is reached. The signal $u(t)$ versus t is plotted in Figure 4–61(b).

Problem A–4–14

Consider the digital control system shown in Figure 4–62. Design a digital controller $G_D(z)$ such that the closed-loop system will exhibit the minimum settling time with zero steady-state error in a unit-ramp response. The system should not exhibit intersampling ripples at steady state. The sampling period T is assumed to be 1 sec. After the controller is designed, investigate the response of the system to a Kronecker delta input and a unit-step input.

Solution   The first step in the design is to determine the z transform of the plant that is preceded by the zero-order hold:

$$G(z) = \mathcal{Z}\left[\frac{1 - e^{-Ts}}{s}\frac{1}{s^2}\right] = (1 - z^{-1})\mathcal{Z}\left[\frac{1}{s^3}\right]$$
$$= \frac{(1 + z^{-1})z^{-1}}{2(1 - z^{-1})^2}$$

Now define the closed-loop pulse transfer function as $F(z)$:

$$\frac{C(z)}{R(z)} = \frac{G_D(z)G(z)}{1 + G_D(z)G(z)} = F(z)$$

Notice that if $G(z)$ is expanded into a series in z^{-1} then the first term will be $0.5z^{-1}$. Hence, $F(z)$ must begin with a term in z^{-1}:

$$F(z) = a_1 z^{-1} + a_2 z^{-2} + \cdots + a_N z^{-N}$$

where $N \geq n$ and n is the order of the system. Since the system here is of the second order, $n = 2$.

Since the input is a unit ramp, from Equation (4–48) we require that

$$1 - F(z) = (1 - z^{-1})^2 N(z) \tag{4–81}$$

Notice that $G(z)$ has a critically stable double pole at $z = 1$. Therefore, from the stability requirement, $1 - F(z)$ must have a double zero at $z = 1$. However, the function $1 - F(z)$ already involves a term $(1 - z^{-1})^2$, and therefore it satisfies the stability requirement.

Since the system should not exhibit intersampling ripples at steady state, we require $U(z)$ to be of the following type of series in z^{-1}:

$$U(z) = b_0 + b_1 z^{-1} + b_2 z^{-2} + \cdots + b_{N-1} z^{-N+1} + b(z^{-N} + z^{-N-1} + z^{-N-2} + \cdots)$$

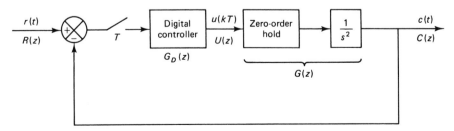

Figure 4–62   Digital control system.

Because the plant transfer function $G_p(s)$ involves a double integrator, b must be zero. (Otherwise, the output increases parabolically, instead of linearly.) Consequently, we have

$$U(z) = b_0 + b_1 z^{-1} + \cdots + b_{N-1} z^{-N+1}$$

From Figure 4–62 $U(z)$ can be given by

$$U(z) = \frac{C(z)}{G(z)} = \frac{C(z)}{R(z)} \frac{R(z)}{G(z)} = F(z) \frac{R(z)}{G(z)}$$

$$= F(z) \frac{z^{-1}}{(1 - z^{-1})^2} \frac{2(1 - z^{-1})^2}{(1 + z^{-1})z^{-1}}$$

$$= F(z) \frac{2}{1 + z^{-1}}$$

For $U(z)$ to be a series in z^{-1} with a finite number of terms, $F(z)$ must be divisible by $1 + z^{-1}$:

$$F(z) = (1 + z^{-1})F_1(z) \tag{4–82}$$

Then $U(z)$ can be written as follows:

$$U(z) = 2F_1(z) \tag{4–83}$$

where $F_1(z)$ is a polynomial in z^{-1} with a finite number of terms.

By comparing Equations (4–81) and (4–82) and by making a simple analysis, we see that $F(z)$ must involve a term with at least z^{-3}. Hence, we assume

$$F(z) = a_1 z^{-1} + a_2 z^{-2} + a_3 z^{-3}$$

This assumed form of $F(z)$ involves the minimum number of terms; the transient response will settle in three sampling periods.

We shall now determine constants a_1, a_2, and a_3. From Equation (4–81), we have

$$1 - a_1 z^{-1} - a_2 z^{-2} - a_3 z^{-3} = (1 - z^{-1})^2 N(z)$$

If we divide the left-hand side of this last equation by $(1 - z^{-1})^2$, the quotient is $1 + (2 - a_1)z^{-1}$. The remainder is $[2(2 - a_1) - (1 + a_2)]z^{-2} - [(2 - a_1) + a_3]z^{-3}$. Hence, $N(z)$ is determined as

$$N(z) = 1 + (2 - a_1)z^{-1}$$

and the remainder is set equal to zero:

$$[2(2 - a_1) - (1 + a_2)]z^{-2} - (2 - a_1 + a_3)z^{-3} = 0$$

To satisfy this last equation, we require that

$$2(2 - a_1) - (1 + a_2) = 0 \tag{4–84}$$

$$2 - a_1 + a_3 = 0 \tag{4–85}$$

From Equation (4–82), we have

$$a_1 z^{-1} + a_2 z^{-2} + a_3 z^{-3} = (1 + z^{-1})F_1(z)$$

If we divide the left-hand side of this last equation by $1 + z^{-1}$, the quotient is $a_1 z^{-1} + (a_2 - a_1)z^{-2}$. The remainder is $(a_1 - a_2 + a_3)z^{-3}$. Hence,

$$F_1(z) = a_1 z^{-1} + (a_2 - a_1)z^{-2}$$

and the remainder is set equal to zero:

$$a_1 - a_2 + a_3 = 0 \qquad (4\text{-}86)$$

By solving Equations (4–84), (4–85), and (4–86) for a_1, a_2, and a_3, we obtain

$$a_1 = 1.25, \qquad a_2 = 0.5, \qquad a_3 = -0.75$$

Hence,

$$N(z) = 1 + 0.75z^{-1}$$

and

$$F_1(z) = 1.25z^{-1} - 0.75z^{-2} = 1.25z^{-1}(1 - 0.6z^{-1})$$

and $F(z)$ is determined as follows:

$$F(z) = 1.25z^{-1} + 0.5z^{-2} - 0.75z^{-3}$$
$$= 1.25z^{-1}(1 + z^{-1})(1 - 0.6z^{-1})$$

The digital controller $G_D(z)$ is then determined from Equation (4–50):

$$G_D(z) = \frac{F(z)}{G(z)(1 - z^{-1})^2 N(z)}$$

$$= \frac{1.25z^{-1}(1 + z^{-1})(1 - 0.6z^{-1})}{\dfrac{(1 + z^{-1})z^{-1}}{2(1 - z^{-1})^2}(1 - z^{-1})^2(1 + 0.75z^{-1})}$$

$$= \frac{2.5(1 - 0.6z^{-1})}{1 + 0.75z^{-1}}$$

With the digital controller thus designed, the system output in response to a unit-ramp input is obtained as follows:

$$C(z) = F(z)R(z)$$

$$= (1.25z^{-1} + 0.5z^{-2} - 0.75z^{-3})\frac{z^{-1}}{(1 - z^{-1})^2}$$

$$= 1.25z^{-2} + 3z^{-3} + 4z^{-4} + 5z^{-5} + \cdots$$

Hence,

$$c(0) = 0$$
$$c(1) = 0$$
$$c(2) = 1.25$$
$$c(k) = k, \qquad k = 3, 4, 5, \ldots$$

Notice that from Equation (4–83) we have

$$U(z) = 2F_1(z)$$
$$= 2(1.25z^{-1})(1 - 0.6z^{-1})$$
$$= 2.5z^{-1} - 1.5z^{-2}$$

Thus, the control signal $u(k)$ becomes zero for $k \geq 3$. Consequently, there are no intersampling ripples in the response at steady state. Figure 4–63 shows plots of $c(k)$ versus k, $u(k)$ versus k, and $u(t)$ versus t in the unit-ramp response.

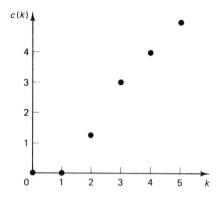

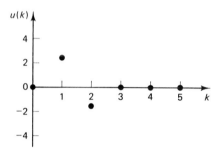

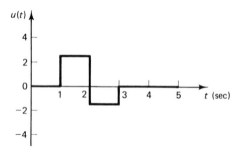

Figure 4–63 Plots of $c(k)$ versus k, $u(k)$ versus k, and $u(t)$ versus t in the unit ramp response of the system designed in Problem A–4–14.

Next, let us investigate the response of this system to a Kronecker delta input and a unit-step input. For a Kronecker delta input.

$$C(z) = F(z)R(z) = F(z) = 1.25z^{-1} + 0.5z^{-2} - 0.75z^{-3}$$

Notice that $U(z)$ in this case becomes as follows:

$$U(z) = F(z)\frac{R(z)}{G(z)} = \frac{1.25z^{-1}(1 + z^{-1})(1 - 0.6z^{-1})}{(1 + z^{-1})z^{-1}/[2(1 - z^{-1})^2]}$$

$$= 2.5(1 - 0.6z^{-1})(1 - z^{-1})^2$$

$$= 2.5 - 6.5z^{-1} + 5.5z^{-2} - 1.5z^{-3}$$

The control signal $u(k)$ becomes zero for $k \geq 4$. Hence, there are no intersampling ripples after $t \geq 4T = 4$.

For the unit-step input,

$$C(z) = F(z)R(z) = (1.25z^{-1} + 0.5z^{-2} - 0.75z^{-3})\frac{1}{1 - z^{-1}}$$

$$= 1.25z^{-1} + 1.75z^{-2} + z^{-3} + z^{-4} + z^{-5} + \cdots$$

The maximum overshoot is 75% in the unit-step response. Notice that

$$U(z) = F(z)\frac{R(z)}{G(z)} = \frac{1.25z^{-1}(1 + z^{-1})(1 - 0.6z^{-1})}{\dfrac{(1 + z^{-1})z^{-1}}{2(1 - z^{-1})^2}(1 - z^{-1})}$$

$$= 1.25(1 - 0.6z^{-1})(2)(1 - z^{-1})$$

$$= 2.5 - 4z^{-1} + 1.5z^{-2}$$

The control signal $u(k)$ becomes zero for $k \geq 3$. Consequently, there are no inter-sampling ripples after the settling time is reached. Figure 4–64(a) shows plots of $c(k)$ versus k, $u(k)$ versus k, and $u(t)$ versus t in the response to the Kronecker delta input. Figure 4–64(b) shows similar plots in the unit-step response. Notice that when the system is designed for the ramp input the response to a step input is no longer deadbeat.

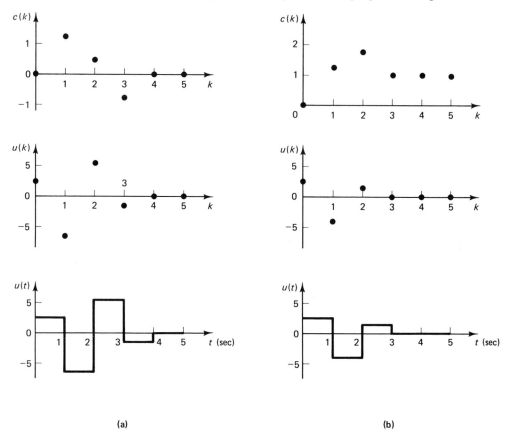

(a)   (b)

Figure 4–64   (a) Plots of $c(k)$ versus k, $u(k)$ versus k, and $u(t)$ versus t in the response to the Kronecker delta input of the system designed in Problem A–4–14; (b) plots of $c(k)$ versus k, $u(k)$ versus k, and $u(t)$ versus t in the unit-step response of the same system.

PROBLEMS

Problem B–4–1

Consider the regions in the s plane shown in Figures 4–65(a) and (b). Draw the corresponding regions in the z plane. The sampling period T is assumed to be 0.3 sec. (The sampling frequency is $\omega_s = 2\pi/T = 2\pi/0.3 = 20.9$ rad/sec.)

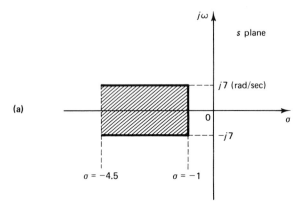

(a)

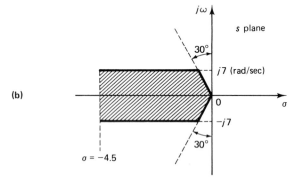

(b)

Figure 4–65   (a) Region in the s plane bounded by constant ω lines and constant σ lines; (b) region in the s plane bounded by constant ζ lines, constant ω lines, and a constant σ line.

Problem B–4–2

Consider the following characteristic equation:

$$z^3 + 2.1z^2 + 1.44z + 0.32 = 0$$

Determine whether or not any of the roots of the characteristic equation lie outside the unit circle centered at the origin of the z plane.

Problem B–4–3

Determine the stability of the following discrete-time system:

$$\frac{Y(z)}{X(z)} = \frac{z^{-3}}{1 + 0.5z^{-1} - 1.34z^{-2} + 0.24z^{-3}}$$

Problem B–4–4

Consider the discrete-time closed-loop control system shown in Figure 4–13. Determine the range of gain K for stability by use of the Jury stability criterion.

Problem B-4-5

Solve Problem B-4-4 by using the bilinear transformation coupled with the Routh stability criterion.

Problem B-4-6

Consider the system

$$\frac{Y(z)}{X(z)} = G(z) = \frac{b_0 + b_1 z^{-1} + \cdots + b_n z^{-n}}{1 + a_1 z^{-1} + \cdots + a_n z^{-n}}$$

Suppose that the input sequence $\{x(k)\}$ is bounded; that is,

$$|x(k)| \le M_1 = \text{constant}, \qquad k = 0, 1, 2, \ldots$$

Show that, if all poles of $G(z)$ lie inside the unit circle in the z plane, then the output $y(k)$ is also bounded; that is,

$$|y(k)| \le M_2 = \text{constant}, \qquad k = 0, 1, 2, \ldots$$

Problem B-4-7

State the conditions for stability, instability, and critical stability in terms of the weighting sequence $g(kT)$ of a linear time-invariant discrete-time control system.

Problem B-4-8

Consider the digital control system shown in Figure 4-66. Plot the root loci as the gain K is varied from 0 to ∞. Determine the critical value of gain K for stability. The sampling period is 0.1 sec, or $T = 0.1$. What value of gain K will yield a damping ratio ζ of the closed-loop poles equal to 0.5? With gain K set to yield $\zeta = 0.5$, determine the damped natural frequency ω_d and the number of samples per cycle of damped sinusoidal oscillation.

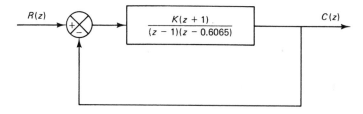

Figure 4-66   Digital control system for Problem B-4-8.

Problem B-4-9

Referring to the digital control system shown in Figure 4-67, design a digital controller $G_D(z)$ such that the damping ratio ζ of the dominant closed-loop poles is 0.5 and the number of samples per cycle of damped sinusoidal oscillation is 8. Assume that the sampling period is 0.1 sec, or $T = 0.1$. Determine the static velocity error constant. Also, determine the response of the designed system to a unit-step input.

Problem B-4-10

Consider the control system shown in Figure 4-68. Design a suitable digital controller that includes an integral control action. The design specifications are that the damping ratio ζ of the dominant closed-loop poles be 0.5 and that there be at least eight samples per cycle of damped sinusoidal oscillation. The sampling period is assumed to be 0.2

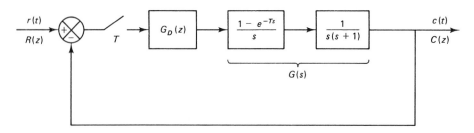

Figure 4–67 Digital control system for Problem B–4–9.

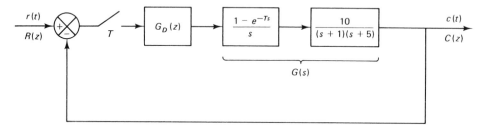

Figure 4–68 Digital control system for Problem B–4–10.

sec, or $T = 0.2$. After the digital controller is designed, determine the static velocity error constant K_v.

Problem B–4–11

Consider the digital control system shown in Figure 4–69, where the plant is of the first order and has a dead time of 5 sec. By choosing a reasonable sampling period T, design a digital PI controller such that the dominant closed-loop poles have a damping ratio ζ of 0.5 and the number of samples per cycle of damped sinusoidal oscillation is 10. After the controller is designed, determine the response of the system to a unit-step input.

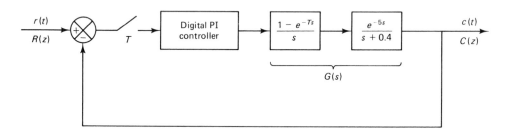

Figure 4–69 Digital control system for Problem B–4–11.

Problem B–4–12

Design a digital proportional-plus-derivative controller for the plant whose transfer function is $1/s^2$, as shown in Figure 4–70. It is desired that the damping ratio ζ of the dominant closed-loop poles be 0.5 and the undamped natural frequency be 4 rad/sec. The sampling period is 0.1 sec, or $T = 0.1$. After the controller is designed, determine the number of samples per cycle of damped sinusoidal oscillation.

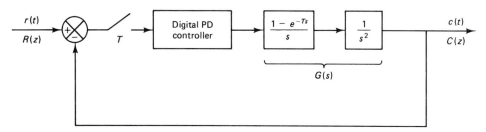

Figure 4–70   Digital control system for Problem B–4–12.

Problem B–4–13

Referring to the system considered in Problem A–4–9, redesign the digital controller so that the static velocity error constant K_v is 12 sec^{-1}, without appreciably changing the locations of the dominant closed-loop poles in the z plane. The sampling period is assumed to be 0.2 sec, or $T = 0.2$. After the controller is redesigned, obtain the unit-step response and unit-ramp response of the digital control system.

Problem B–4–14

Consider the digital control system shown in Figure 4–71. Draw a Bode diagram in the w plane. Set the gain K so that the phase margin becomes equal to 50°. With the gain K so set, determine the gain margin and the static velocity error constant K_v. The sampling period is assumed to be 0.1 sec, or $T = 0.1$.

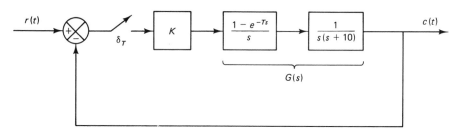

Figure 4–71   Digital control system for Problem B–4–14.

Problem B–4–15

Using the Bode diagram approach in the w plane, design a digital controller for the system shown in Figure 4–72. The design specifications are that the phase margin be 50°, the gain margin be at least 10 dB, and the static velocity error constant K_v be 20 sec^{-1}. The sampling period is assumed to be 0.1 sec, or $T = 0.1$. After the controller is designed, calculate the number of samples per cycle of damped sinusoidal oscillation.

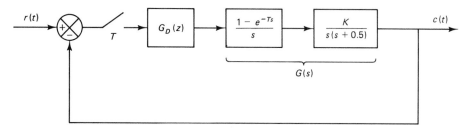

Figure 4–72   Digital control system for Problem B–4–15.

Problem B–4–16

Consider the digital control system shown in Figure 4–73. Using the Bode diagram approach in the w plane, design a digital controller such that the phase margin is 60°, the gain margin is 12 dB or more, and the static velocity error constant is 5 sec^{-1}. The sampling period is assumed to be 0.1 sec, or $T = 0.1$.

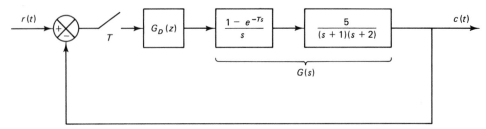

Figure 4–73   Digital control system for Problem B–4–16.

Problem B–4–17

Consider the system shown in Figure 4–74. Design a digital controller using a Bode diagram in the w plane such that the phase margin is 50° and the gain margin is at least 10 dB. It is desired that the static velocity error constant K_v be 10 sec^{-1}. The sampling period is specified as 0.1 sec, or $T = 0.1$. After the controller is designed, determine the number of samples per cycle of damped sinusoidal oscillation.

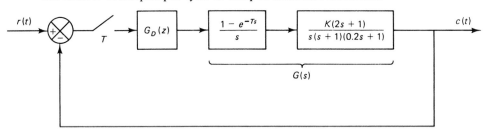

Figure 4–74   Digital control system for Problem B–4–17.

Problem B–4–18

Consider the digital control system shown in Figure 4–75. Design a digital controller $G_D(z)$ such that the system output will exhibit a deadbeat response to a unit step input (that is, the settling time will be the minimum possible and the steady-state error will be zero; also, the system output will not exhibit intersampling ripples after the settling time is reached). The sampling period T is assumed to be 1 sec, or $T = 1$.

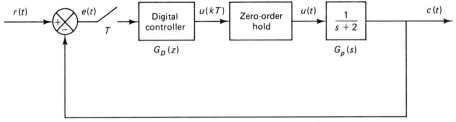

Figure 4–75   Digital control system for Problem B–4–18.

5

State-Space Analysis

5-1 INTRODUCTION

In Chapters 3 and 4 we were concerned with conventional methods for the analysis and design of control systems. Conventional methods such as the root-locus and frequency-response methods are useful for dealing with single-input–single-output systems. Conventional methods are conceptually simple and require only a reasonable number of computations, but they are applicable only to linear time-invariant systems having a single input and single output. They are based on the input–output relationship of the system, that is, the transfer function or the pulse transfer function. They do not apply to nonlinear systems except in simple cases. Also, the conventional methods do not apply to the design of optimal and adaptive control systems, which are mostly time varying and/or nonlinear.

A modern control system may have many inputs and many outputs, and these may be interrelated in a complicated manner. The state-space methods for the analysis and synthesis of control systems are best suited for dealing with multiple-input–multiple-output systems that are required to be optimal in some sense.

Concept of the State-Space Method. The state-space method is based on the description of system equations in terms of n first-order difference equations or differential equations, which may be combined into a first-order vector-matrix difference equation or differential equation. The use of the vector-matrix notation greatly simplifies the mathematical representation of the systems of equations.

System design by use of the state-space concept enables the engineer to design control systems with respect to given performance indexes. In addition, design in the state space can be carried out for a *class* of inputs, instead of a specific input function such as the impulse function, step function, or sinusoidal function. Also, state-space methods enable the engineer to include initial conditions in the design. This is a very convenient and useful feature that is not possible in the conventional design methods.

In what follows we shall first define state, state variable, state vector, and state space, and then we shall present state-space equations.

State.   The state of a dynamic system is the smallest set of variables (called *state variables*) such that the knowledge of these variables at $t = t_0$, together with the knowledge of the input for $t \geq t_0$, completely determines the behavior of the system for any time $t \geq t_0$. Note that the concept of state is by no means limited to physical systems. It is applicable to biological systems, economic systems, social systems, and others.

State Variables.   The state variables of a dynamic system are the variables making up the smallest set of variables that determine the state of the dynamic system. If at least n variables $x_1, x_2, \ldots, x_n$ are needed to completely describe the behavior of a dynamic system (so that once the input is given for $t \geq t_0$ and the initial state at $t = t_0$ is specified, the future state of the system is completely determined), then such n variables are a set of state variables.

Note that state variables need not be physically measurable or observable quantities. Variables that do not represent physical quantities and those that are neither measurable nor observable can be chosen as state variables. Such freedom in choosing state variables is an advantage of the state-space methods. Practically speaking, however, it is convenient to choose easily measurable quantities for the state variables, if this is possible at all, because optimal control laws will require the feedback of all state variables with suitable weighting.

State Vector.   If n state variables are needed to completely describe the behavior of a given system, then these n state variables can be considered the n components of a vector **x**. Such a vector is called a *state vector*. A state vector is thus a vector that determines uniquely the system state $\mathbf{x}(t)$ for any time $t \geq t_0$, once the state at $t = t_0$ is given and the input $\mathbf{u}(t)$ for $t \geq t_0$ is specified.

State Space.   The n-dimensional space whose coordinate axes consist of the x_1 axis, x_2 axis, $\ldots$, x_n axis is called a *state space*. Any state can be represented by a point in the state space.

State-Space Equations.   In state-space analysis we are concerned with three types of variables that are involved in the modeling of dynamic systems: input variables, output variables, and state variables. As we shall see in Section 5–2, the state-space representation for a given system is not unique, except that the number of state variables is the same for any of the different state-space representations of the same system.

For time-varying (linear or nonlinear) discrete-time systems, the state equation may be written as

$$\mathbf{x}(k + 1) = \mathbf{f}[\mathbf{x}(k), \mathbf{u}(k), k]$$

and the output equation as

$$\mathbf{y}(k) = \mathbf{g}[\mathbf{x}(k), \mathbf{u}(k), k]$$

For linear time-varying discrete-time systems, the state equation and output equation may be simplified to

$$\mathbf{x}(k + 1) = \mathbf{G}(k)\mathbf{x}(k) + \mathbf{H}(k)\mathbf{u}(k)$$

$$\mathbf{y}(k) = \mathbf{C}(k)\mathbf{x}(k) + \mathbf{D}(k)\mathbf{u}(k)$$

where

$\mathbf{x}(k) = n$-vector	(state vector)
$\mathbf{y}(k) = m$-vector	(output vector)
$\mathbf{u}(k) = r$-vector	(input vector)
$\mathbf{G}(k) = n \times n$ matrix	(state matrix)
$\mathbf{H}(k) = n \times r$ matrix	(input matrix)
$\mathbf{C}(k) = m \times n$ matrix	(output matrix)
$\mathbf{D}(k) = m \times r$ matrix	(direct transmission matrix)

The appearance of the variable k in the arguments of matrices $\mathbf{G}(k)$, $\mathbf{H}(k)$, $\mathbf{C}(k)$, and $\mathbf{D}(k)$ implies that these matrices are time varying. If the variable k does not appear explicitly in the matrices, they are assumed to be time invariant, or constant. That is, if the system is time invariant, then the last two equations can be simplified to

$$\mathbf{x}(k + 1) = \mathbf{Gx}(k) + \mathbf{Hu}(k) \qquad (5\text{--}1)$$

$$\mathbf{y}(k) = \mathbf{Cx}(k) + \mathbf{Du}(k) \qquad (5\text{--}2)$$

As in the discrete-time case, continuous-time (linear or nonlinear) systems may be represented by the following state equation and output equation:

$$\dot{\mathbf{x}}(t) = \mathbf{f}[\mathbf{x}(t), \mathbf{u}(t), t]$$

$$\mathbf{y}(t) = \mathbf{g}[\mathbf{x}(t), \mathbf{u}(t), t]$$

For linear time-varying continuous-time systems, the state equation and output equation are given by

$$\dot{\mathbf{x}}(t) = \mathbf{A}(t)\mathbf{x}(t) + \mathbf{B}(t)\mathbf{u}(t)$$

$$\mathbf{y}(t) = \mathbf{C}(t)\mathbf{x}(t) + \mathbf{D}(t)\mathbf{u}(t)$$

If the system is time invariant, then the last two equations are simplified to

$$\dot{\mathbf{x}}(t) = \mathbf{Ax}(t) + \mathbf{Bu}(t) \qquad (5\text{--}3)$$

$$\mathbf{y}(t) = \mathbf{Cx}(t) + \mathbf{Du}(t) \qquad (5\text{--}4)$$

Figure 5-1(a) shows the block diagram representation of the discrete-time control system defined by Equations (5-1) and (5-2), and Figure 5-1(b) shows the continuous-time control system defined by Equations (5-3) and (5-4). Notice that the basic configurations of the discrete-time and continuous-time systems are the same.

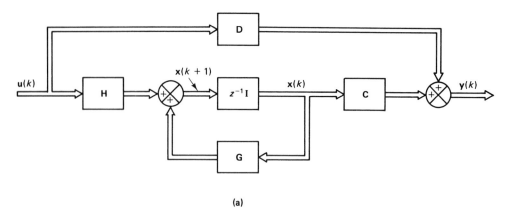

(a)

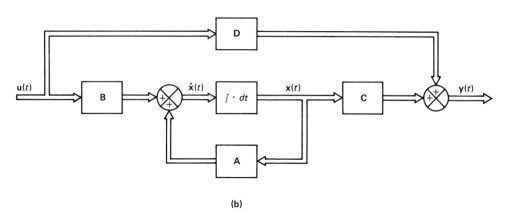

(b)

Figure 5–1   (a) Block diagram of the linear time-invariant discrete-time control system represented in state space; (b) block diagram of the linear time-invariant continuous-time control system represented in state space.

Note that in this book $\mathbf{u}(k)$ [or $\mathbf{u}(t)$] denotes both the input vector to a system and the control vector (control input to a plant). Hence, we need to interpret $\mathbf{u}(k)$ [or $\mathbf{u}(t)$] as either the input vector or the control vector depending on circumstances.

Outline of the Chapter.   Section 5–1 has introduced the state-space method and defined some basic terms. Section 5–2 presents various state-space representations of linear time-invariant discrete-time systems. Section 5–3 first treats the solution of the linear time-invariant discrete-time state equation by the recursion procedure and by the z transform approach. Then it presents a method for computing $(z\mathbf{I} - \mathbf{G})^{-1}$. Section 5–3 concludes with discussions of the solution of the linear time-varying discrete-time state equation. Section 5–4 deals with the pulse transfer function matrix. Section 5–5 first treats the discretization of linear continuous-time state-space equations. Then it discusses time response between two consecutive sampling instants. The final section, Section 5–6, presents Liapunov stability analysis. It begins with discussions of the Liapunov function and definitions of stability of dynamic systems. Then it presents Liapunov's main stability theorem, followed by its applications to stability analysis of linear continuous-time systems and discrete-time systems.

5–2 STATE-SPACE REPRESENTATIONS OF DISCRETE-TIME SYSTEMS

Canonical Forms for Discrete-Time State-Space Equations.   Many techniques are available for obtaining state-space representations of discrete-time systems. Consider the discrete-time system described by

$$y(k) + a_1 y(k-1) + a_2 y(k-2) + \cdots + a_n y(k-n)$$
$$= b_0 u(k) + b_1 u(k-1) + \cdots + b_n u(k-n) \qquad (5\text{--}5)$$

where $u(k)$ is the input and $y(k)$ is the output of the system at the kth sampling instant. Note that some of the coefficients a_i $(i = 1, 2, \ldots, n)$ and b_j $(j = 0, 1, 2, \ldots, n)$ may be zero. Equation (5–5) can be written in the form of the pulse transfer function as

$$\frac{Y(z)}{U(z)} = \frac{b_0 + b_1 z^{-1} + \cdots + b_n z^{-n}}{1 + a_1 z^{-1} + \cdots + a_n z^{-n}} \qquad (5\text{--}6)$$

or

$$\frac{Y(z)}{U(z)} = \frac{b_0 z^n + b_1 z^{n-1} + \cdots + b_n}{z^n + a_1 z^{n-1} + \cdots + a_n} \qquad (5\text{--}7)$$

There are many ways to realize state-space representations for the discrete-time system described by Equation (5–5), (5–6), or (5–7). Here we shall present the following representations:

1. Controllable canonical form
2. Observable canonical form
3. Diagonal canonical form
4. Jordan canonical form

(For the meaning of the terms *controllable* and *observable*, see Sections 6–2 and 6–3.) The controllable canonical form can be derived by the direct programming method. (See Problem A–5–1.) The observable canonical form can be obtained by the nested programming method. (See Problem A–5–2.) The diagonal canonical form and Jordan canonical form may be obtained by use of the partial-fraction-expansion method. (See Problems A–5–3 and A–5–4.)

Controllable Canonical Form.   The state-space representation of the discrete-time system given by Equation (5–5), (5–6), or (5–7) may be put in the form given by the following equations:

$$
\begin{bmatrix} x_1(k+1) \\ x_2(k+1) \\ \vdots \\ x_{n-1}(k+1) \\ x_n(k+1) \end{bmatrix} = \begin{bmatrix} 0 & 1 & 0 & \cdots & 0 \\ 0 & 0 & 1 & \cdots & 0 \\ \vdots & \vdots & \vdots & & \vdots \\ 0 & 0 & 0 & \cdots & 1 \\ -a_n & -a_{n-1} & -a_{n-2} & \cdots & -a_1 \end{bmatrix} \begin{bmatrix} x_1(k) \\ x_2(k) \\ \vdots \\ x_{n-1}(k) \\ x_n(k) \end{bmatrix} + \begin{bmatrix} 0 \\ 0 \\ \vdots \\ 0 \\ 1 \end{bmatrix} u(k) \qquad (5\text{–}8)
$$

$$
y(k) = [b_n - a_n b_0 \vdots b_{n-1} - a_{n-1} b_0 \vdots \cdots \vdots b_1 - a_1 b_0] \begin{bmatrix} x_1(k) \\ x_2(k) \\ \vdots \\ x_n(k) \end{bmatrix} + b_0 u(k) \qquad (5\text{–}9)
$$

Equations (5–8) and (5–9) are the state equation and output equation, respectively. The state-space representation given by Equations (5–8) and (5–9) is commonly called a *controllable canonical form*. [For the derivation of Equations (5–8) and (5–9), see Problem A–5–1.]

Note that if we reverse the order of the state variables, that is, we define new state variables according to the fashion

$$
\begin{bmatrix} \hat{x}_1(k) \\ \hat{x}_2(k) \\ \vdots \\ \hat{x}_n(k) \end{bmatrix} = \begin{bmatrix} 0 & 0 & \cdots & 0 & 1 \\ 0 & 0 & \cdots & 1 & 0 \\ \vdots & \vdots & & \vdots & \vdots \\ 1 & 0 & \cdots & 0 & 0 \end{bmatrix} \begin{bmatrix} x_1(k) \\ x_2(k) \\ \vdots \\ x_n(k) \end{bmatrix}
$$

then the state equation can be written as follows:

$$
\begin{bmatrix} \hat{x}_1(k+1) \\ \hat{x}_2(k+1) \\ \hat{x}_3(k+1) \\ \vdots \\ \hat{x}_n(k+1) \end{bmatrix} = \begin{bmatrix} -a_1 & -a_2 & \cdots & -a_{n-1} & -a_n \\ 1 & 0 & \cdots & 0 & 0 \\ 0 & 1 & \cdots & 0 & 0 \\ \vdots & \vdots & & \vdots & \vdots \\ 0 & 0 & \cdots & 1 & 0 \end{bmatrix} \begin{bmatrix} \hat{x}_1(k) \\ \hat{x}_2(k) \\ \hat{x}_3(k) \\ \vdots \\ \hat{x}_n(k) \end{bmatrix} + \begin{bmatrix} 1 \\ 0 \\ 0 \\ \vdots \\ 0 \end{bmatrix} u(k) \qquad (5\text{–}10)
$$

The output equation can be given by

$$
y(k) = [b_1 - a_1 b_0 \vdots b_2 - a_2 b_0 \vdots \cdots \vdots b_n - a_n b_0] \begin{bmatrix} \hat{x}_1(k) \\ \hat{x}_2(k) \\ \vdots \\ \hat{x}_n(k) \end{bmatrix} + b_0 u(k) \qquad (5\text{–}11)
$$

Equations (5–10) and (5–11) are also in the controllable canonical form.

Observable Canonical Form.   The state-space representation of the discrete-time system given by Equation (5–5), (5–6), or (5–7) may be put in the following form:

$$
\begin{bmatrix} x_1(k+1) \\ x_2(k+1) \\ \vdots \\ x_{n-1}(k+1) \\ x_n(k+1) \end{bmatrix} = \begin{bmatrix} 0 & 0 & \cdots & 0 & 0 & -a_n \\ 1 & 0 & \cdots & 0 & 0 & -a_{n-1} \\ \vdots & \vdots & & \vdots & \vdots & \vdots \\ 0 & 0 & \cdots & 1 & 0 & -a_2 \\ 0 & 0 & \cdots & 0 & 1 & -a_1 \end{bmatrix} \begin{bmatrix} x_1(k) \\ x_2(k) \\ \vdots \\ x_{n-1}(k) \\ x_n(k) \end{bmatrix} + \begin{bmatrix} b_n - a_n b_0 \\ b_{n-1} - a_{n-1} b_0 \\ \vdots \\ b_2 - a_2 b_0 \\ b_1 - a_1 b_0 \end{bmatrix} u(k) \qquad (5\text{–}12)
$$

$$y(k) = [0 \quad 0 \quad \cdots \quad 0 \quad 1] \begin{bmatrix} x_1(k) \\ x_2(k) \\ \vdots \\ x_{n-1}(k) \\ x_n(k) \end{bmatrix} + b_0 u(k) \qquad (5\text{-}13)$$

The state-space representation given by Equations (5–12) and (5–13) is called an *observable canonical form*. [For the derivation of Equations (5–12) and (5–13), see Problem A–5–2.] Notice that the $n \times n$ state matrix of the state equation given by Equation (5–12) is the transpose of that of the state equation defined by Equation (5–8).

Note that if we reverse the order of the state variables, that is, if we define

$$\begin{bmatrix} \hat{x}_1(k) \\ \hat{x}_2(k) \\ \vdots \\ \hat{x}_n(k) \end{bmatrix} = \begin{bmatrix} 0 & 0 & \cdots & 0 & 1 \\ 0 & 0 & \cdots & 1 & 0 \\ \vdots & \vdots & & \vdots & \vdots \\ 1 & 0 & \cdots & 0 & 0 \end{bmatrix} \begin{bmatrix} x_1(k) \\ x_2(k) \\ \vdots \\ x_n(k) \end{bmatrix}$$

then the state equation and the output equation become as follows:

$$\begin{bmatrix} \hat{x}_1(k + 1) \\ \hat{x}_2(k + 1) \\ \vdots \\ \hat{x}_{n-1}(k + 1) \\ \hat{x}_n(k + 1) \end{bmatrix}$$

$$= \begin{bmatrix} -a_1 & 1 & 0 & \cdots & 0 & 0 \\ -a_2 & 0 & 1 & \cdots & 0 & 0 \\ \vdots & \vdots & \vdots & & \vdots & \vdots \\ -a_{n-1} & 0 & 0 & \cdots & 0 & 1 \\ -a_n & 0 & 0 & \cdots & 0 & 0 \end{bmatrix} \begin{bmatrix} \hat{x}_1(k) \\ \hat{x}_2(k) \\ \vdots \\ \hat{x}_{n-1}(k) \\ \hat{x}_n(k) \end{bmatrix} + \begin{bmatrix} b_1 - a_1 b_0 \\ b_2 - a_2 b_0 \\ \vdots \\ b_{n-1} - a_{n-1} b_0 \\ b_n - a_n b_0 \end{bmatrix} u(k) \qquad (5\text{-}14)$$

$$y(k) = [1 \quad 0 \quad \cdots \quad 0 \quad 0] \begin{bmatrix} \hat{x}_1(k) \\ \hat{x}_2(k) \\ \vdots \\ \hat{x}_{n-1}(k) \\ \hat{x}_n(k) \end{bmatrix} + b_0 u(k) \qquad (5\text{-}15)$$

Equations (5–14) and (5–15) are also in the observable canonical form.

Diagonal Canonical Form.   If the poles of the pulse transfer function given by Equation (5–5), (5–6), or (5–7) are all distinct, then the state-space representation may be put in the diagonal canonical form as follows:

$$\begin{bmatrix} x_1(k + 1) \\ x_2(k + 1) \\ \vdots \\ x_n(k + 1) \end{bmatrix} = \begin{bmatrix} p_1 & 0 & \cdots & 0 \\ 0 & p_2 & \cdots & 0 \\ \vdots & \vdots & & \vdots \\ 0 & 0 & \cdots & p_n \end{bmatrix} \begin{bmatrix} x_1(k) \\ x_2(k) \\ \vdots \\ x_n(k) \end{bmatrix} + \begin{bmatrix} 1 \\ 1 \\ \vdots \\ 1 \end{bmatrix} u(k) \qquad (5\text{-}16)$$

$$y(k) = [c_1 \quad c_2 \quad \cdots \quad c_n]\begin{bmatrix} x_1(k) \\ x_2(k) \\ \vdots \\ x_n(k) \end{bmatrix} + b_0 u(k) \tag{5-17}$$

[For the derivation of Equations (5-16) and (5-17), see Problem A-5-3.]

Jordan Canonical Form.   If the pulse transfer function given by Equation (5-5), (5-6), or (5-7) involves a multiple pole of order m at $z = p_1$ and all other poles are distinct, then the state equation and output equation may be given as follows:

$$\begin{bmatrix} x_1(k+1) \\ x_2(k+1) \\ \vdots \\ x_m(k+1) \\ \hline x_{m+1}(k+1) \\ \vdots \\ x_n(k+1) \end{bmatrix} = \left[\begin{array}{cccc|ccc} p_1 & 1 & 0 & \cdots & 0 & 0 & \cdots & 0 \\ 0 & p_1 & 1 & \cdots & 0 & 0 & \cdots & 0 \\ \vdots & \vdots & \vdots & & \vdots & \vdots & & \vdots \\ 0 & 0 & 0 & \cdots & p_1 & 0 & \cdots & 0 \\ \hline 0 & 0 & 0 & \cdots & 0 & p_{m+1} & \cdots & 0 \\ \vdots & \vdots & \vdots & & \vdots & \vdots & & \vdots \\ 0 & 0 & 0 & \cdots & 0 & 0 & \cdots & p_n \end{array}\right]\begin{bmatrix} x_1(k) \\ x_2(k) \\ \vdots \\ x_m(k) \\ \hline x_{m+1}(k) \\ \vdots \\ x_n(k) \end{bmatrix} + \begin{bmatrix} 0 \\ 0 \\ \vdots \\ 1 \\ \hline 1 \\ \vdots \\ 1 \end{bmatrix} u(k) \tag{5-18}$$

$$y(k) = [c_1 \quad c_2 \quad \cdots \quad c_n]\begin{bmatrix} x_1(k) \\ x_2(k) \\ \vdots \\ x_n(k) \end{bmatrix} + b_0 u(k) \tag{5-19}$$

[For the derivation of Equations (5-18) and (5-19), see Problem A-5-4.] The $n \times n$ state matrix is in a Jordan canonical form. (For details of Jordan canonical forms, see Appendix A.)

Example 5-1

Consider the following system:

$$\frac{Y(z)}{U(z)} = \frac{z + 1}{z^2 + 1.3z + 0.4}$$

The state-space representations in the controllable canonical form, observable canonical form, and diagonal canonical form become as follows:

CONTROLLABLE CANONICAL FORM:

$$\begin{bmatrix} x_1(k+1) \\ x_2(k+1) \end{bmatrix} = \begin{bmatrix} 0 & 1 \\ -0.4 & -1.3 \end{bmatrix}\begin{bmatrix} x_1(k) \\ x_2(k) \end{bmatrix} + \begin{bmatrix} 0 \\ 1 \end{bmatrix} u(k)$$

$$y(k) = [1 \quad 1]\begin{bmatrix} x_1(k) \\ x_2(k) \end{bmatrix}$$

OBSERVABLE CANONICAL FORM:

$$\begin{bmatrix} x_1(k+1) \\ x_2(k+1) \end{bmatrix} = \begin{bmatrix} 0 & -0.4 \\ 1 & -1.3 \end{bmatrix}\begin{bmatrix} x_1(k) \\ x_2(k) \end{bmatrix} + \begin{bmatrix} 1 \\ 1 \end{bmatrix} u(k)$$

$$y(k) = [0 \quad 1]\begin{bmatrix} x_1(k) \\ x_2(k) \end{bmatrix}$$

DIAGONAL CANONICAL FORM:

The given pulse transfer function $Y(z)/U(z)$ can be expanded as follows:

$$\frac{Y(z)}{U(z)} = \frac{5/3}{z + 0.5} + \frac{-2/3}{z + 0.8}$$

Hence,

$$\begin{bmatrix} x_1(k + 1) \\ x_2(k + 1) \end{bmatrix} = \begin{bmatrix} -0.5 & 0 \\ 0 & -0.8 \end{bmatrix} \begin{bmatrix} x_1(k) \\ x_2(k) \end{bmatrix} + \begin{bmatrix} 1 \\ 1 \end{bmatrix} u(k)$$

$$y(k) = \begin{bmatrix} \dfrac{5}{3} & -\dfrac{2}{3} \end{bmatrix} \begin{bmatrix} x_1(k) \\ x_2(k) \end{bmatrix}$$

Nonuniqueness of State-Space Representations.   For a given pulse-transfer-function system the state-space representation is not unique. We have demonstrated that different state-space representations for a given pulse-transfer-function system are possible. The state equations, however, are related to each other by the similarity transformation.

Consider the system defined by

$$\mathbf{x}(k + 1) = \mathbf{G}\mathbf{x}(k) + \mathbf{H}u(k) \qquad (5\text{-}20)$$

$$\mathbf{y}(k) = \mathbf{C}\mathbf{x}(k) + \mathbf{D}u(k) \qquad (5\text{-}21)$$

Let us define a new state vector $\hat{\mathbf{x}}(k)$ by

$$\mathbf{x}(k) = \mathbf{P}\hat{\mathbf{x}}(k) \qquad (5\text{-}22)$$

where **P** is a nonsingular matrix. [Note that, since both $\mathbf{x}(k)$ and $\hat{\mathbf{x}}(k)$ are n-dimensional vectors, they are related to each other by a nonsingular matrix.]

Then, by substituting Equation (5-22) into Equation (5-20), we obtain

$$\mathbf{P}\hat{\mathbf{x}}(k + 1) = \mathbf{G}\mathbf{P}\hat{\mathbf{x}}(k) + \mathbf{H}u(k) \qquad (5\text{-}23)$$

Premultiplying both sides of Equation (5-23) by $\mathbf{P}^{-1}$ yields

$$\hat{\mathbf{x}}(k + 1) = \mathbf{P}^{-1}\mathbf{G}\mathbf{P}\hat{\mathbf{x}}(k) + \mathbf{P}^{-1}\mathbf{H}u(k) \qquad (5\text{-}24)$$

Let us define

$$\mathbf{P}^{-1}\mathbf{G}\mathbf{P} = \hat{\mathbf{G}}, \qquad \mathbf{P}^{-1}\mathbf{H} = \hat{\mathbf{H}}$$

Then Equation (5-24) can be written as follows:

$$\hat{\mathbf{x}}(k + 1) = \hat{\mathbf{G}}\hat{\mathbf{x}}(k) + \hat{\mathbf{H}}u(k) \qquad (5\text{-}25)$$

Similarly, by substituting Equation (5-22) into Equation (5-21), we obtain

$$\mathbf{y}(k) = \mathbf{C}\mathbf{P}\hat{\mathbf{x}}(k) + \mathbf{D}u(k) \qquad (5\text{-}26)$$

By defining

$$\mathbf{C}\mathbf{P} = \hat{\mathbf{C}}, \qquad \mathbf{D} = \hat{\mathbf{D}}$$

we can write Equation (5-26) as

$$\mathbf{y}(k) = \hat{\mathbf{C}}\hat{\mathbf{x}}(k) + \hat{\mathbf{D}}u(k) \qquad (5\text{-}27)$$

We have thus shown that the state-space representation given by Equations (5-20) and (5-21),

$$\mathbf{x}(k + 1) = \mathbf{Gx}(k) + \mathbf{Hu}(k)$$
$$\mathbf{y}(k) = \mathbf{Cx}(k) + \mathbf{Du}(k)$$

is equivalent to the state-space representation given by Equations (5–25) and (5–27),

$$\hat{\mathbf{x}}(k + 1) = \hat{\mathbf{G}}\hat{\mathbf{x}}(k) + \hat{\mathbf{H}}\mathbf{u}(k)$$
$$\mathbf{y}(k) = \hat{\mathbf{C}}\hat{\mathbf{x}}(k) + \hat{\mathbf{D}}\mathbf{u}(k)$$

The state vectors $\mathbf{x}(k)$ and $\hat{\mathbf{x}}(k)$ are related to each other by Equation (5–22). Since matrix $\mathbf{P}$ can be any nonsingular $n \times n$ matrix, there are infinitely many state-space representations for a given system.

In some applications, we may desire to diagonalize the state matrix $\mathbf{G}$. This may be done by properly choosing a matrix $\mathbf{P}$ such that

$$\mathbf{P}^{-1}\mathbf{GP} = \text{diagonal matrix}$$

In the case where diagonalization is not possible, $\mathbf{P}^{-1}\mathbf{GP}$ may be transformed into a Jordan canonical form:

$$\mathbf{P}^{-1}\mathbf{GP} = \mathbf{J} = \text{Jordan canonical form}$$

For methods for transforming matrix $\mathbf{G}$ into a diagonal matrix or into a matrix in a Jordan canonical form, refer to Appendix A. [Note that if the partial-fraction-expansion programming method is used the state matrix becomes diagonal if all poles involved are distinct, and it becomes a Jordan canonical form if multiple poles are involved in $Y(z)/U(z)$.]

5–3 SOLVING DISCRETE-TIME STATE-SPACE EQUATIONS

In this section, we first present the solution of the linear time-invariant discrete-time state equation

$$\mathbf{x}(k + 1) = \mathbf{Gx}(k) + \mathbf{Hu}(k)$$

by a recursion procedure and then by the z transform method. Then we discuss methods for computing $(z\mathbf{I} - \mathbf{G})^{-1}$. Finally, we treat the solution of the linear time-varying discrete-time state equation

$$\mathbf{x}(k + 1) = \mathbf{G}(k)\mathbf{x}(k) + \mathbf{H}(k)\mathbf{u}(k)$$

Solution of the Linear Time-Invariant Discrete-Time State Equation. In general, discrete-time equations are easier to solve than differential equations because the former can be solved easily by means of a recursion procedure. The recursion procedure is quite simple and convenient for digital computations.

Consider the following state equation and output equation:

$$\mathbf{x}(k + 1) = \mathbf{Gx}(k) + \mathbf{Hu}(k) \tag{5–28}$$
$$\mathbf{y}(k) = \mathbf{Cx}(k) + \mathbf{Du}(k) \tag{5–29}$$

The solution of Equation (5–28) for any positive integer k may be obtained directly by recursion, as follows:

$$\mathbf{x}(1) = \mathbf{Gx}(0) + \mathbf{Hu}(0)$$

$$\mathbf{x}(2) = \mathbf{Gx}(1) + \mathbf{Hu}(1) = \mathbf{G}^2\mathbf{x}(0) + \mathbf{GHu}(0) + \mathbf{Hu}(1)$$

$$\mathbf{x}(3) = \mathbf{Gx}(2) + \mathbf{Hu}(2) = \mathbf{G}^3\mathbf{x}(0) + \mathbf{G}^2\mathbf{Hu}(0) + \mathbf{GHu}(1) + \mathbf{Hu}(2)$$

$$\vdots$$

By repeating this procedure, we obtain

$$\mathbf{x}(k) = \mathbf{G}^k\mathbf{x}(0) + \sum_{j=0}^{k-1} \mathbf{G}^{k-j-1}\mathbf{Hu}(j), \qquad k = 1, 2, 3, \ldots \tag{5–30}$$

Clearly, $\mathbf{x}(k)$ consists of two parts, one representing the contribution of the initial state $\mathbf{x}(0)$ and the other the contribution of the input $\mathbf{u}(j)$, where $j = 0, 1, 2, \ldots, k - 1$. The output $\mathbf{y}(k)$ is given by

$$\mathbf{y}(k) = \mathbf{CG}^k\mathbf{x}(0) + \mathbf{C}\sum_{j=0}^{k-1} \mathbf{G}^{k-j-1}\mathbf{Hu}(j) + \mathbf{Du}(k) \tag{5–31}$$

State Transition Matrix.   Notice that it is possible to write the solution of the homogeneous state equation

$$\mathbf{x}(k + 1) = \mathbf{Gx}(k) \tag{5–32}$$

as

$$\mathbf{x}(k) = \mathbf{\Psi}(k)\mathbf{x}(0) \tag{5–33}$$

where $\mathbf{\Psi}(k)$ is a unique $n \times n$ matrix satisfying the condition

$$\mathbf{\Psi}(k + 1) = \mathbf{G\Psi}(k), \qquad \mathbf{\Psi}(0) = \mathbf{I} \tag{5–34}$$

Clearly, $\mathbf{\Psi}(k)$ can be given by

$$\mathbf{\Psi}(k) = \mathbf{G}^k \tag{5–35}$$

From Equation (5–33), we see that the solution of Equation (5–32) is simply a transformation of the initial state. Therefore, the unique matrix $\mathbf{\Psi}(k)$ is called the *state transition matrix*. It is also called the *fundamental matrix*. The state transition matrix contains all the information about the free motions of the system defined by Equation (5–32).

In terms of the state transition matrix $\mathbf{\Psi}(k)$, Equation (5–30) can be written in the form

$$\mathbf{x}(k) = \mathbf{\Psi}(k)\mathbf{x}(0) + \sum_{j=0}^{k-1} \mathbf{\Psi}(k - j - 1)\mathbf{Hu}(j) \tag{5–36}$$

$$= \mathbf{\Psi}(k)\mathbf{x}(0) + \sum_{j=0}^{k-1} \mathbf{\Psi}(j)\mathbf{Hu}(k - j - 1) \tag{5–37}$$

Substituting Equation (5–36) or Equation (5–37) into Equation (5–31), the following output equation can be obtained:

$$y(k) = C\Psi(k)x(0) + C\sum_{j=0}^{k-1}\Psi(k-j-1)Hu(j) + Du(k) \qquad (5\text{--}38)$$

$$= C\Psi(k)x(0) + C\sum_{j=0}^{k-1}\Psi(j)Hu(k-j-1) + Du(k) \qquad (5\text{--}39)$$

z Transform Approach to the Solution of Discrete-Time State Equations. We next present the solution of a discrete-time state equation by the z transform method. Consider the discrete-time system described by Equation (5–28):

$$x(k+1) = Gx(k) + Hu(k) \qquad (5\text{--}40)$$

Taking the z transform of both sides of Equation (5–40), we get

$$zX(z) - zx(0) = GX(z) + HU(z)$$

where $X(z) = \mathcal{Z}[x(k)]$ and $U(z) = \mathcal{Z}[u(k)]$. Then

$$(zI - G)X(z) = zx(0) + HU(z)$$

Premultiplying both sides of this last equation by $(zI - G)^{-1}$, we obtain

$$X(z) = (zI - G)^{-1}zx(0) + (zI - G)^{-1}HU(z) \qquad (5\text{--}41)$$

Taking the inverse z transform of both sides of Equation (5–41) gives

$$x(k) = \mathcal{Z}^{-1}[(zI - G)^{-1}z]x(0) + \mathcal{Z}^{-1}[(zI - G)^{-1}HU(z)] \qquad (5\text{--}42)$$

Comparing Equation (5–30) with Equation (5–42), we obtain

$$G^k = \mathcal{Z}^{-1}[(zI - G)^{-1}z] \qquad (5\text{--}43)$$

and

$$\sum_{j=0}^{k-1}G^{k-j-1}Hu(j) = \mathcal{Z}^{-1}[(zI - G)^{-1}HU(z)] \qquad (5\text{--}44)$$

where $k = 1, 2, 3, \ldots$.

Notice that the solution by the z transform method involves the process of inverting the matrix $(zI - G)$, which may be accomplished by analytical means or by use of a computer routine. The solution also requires the inverse z transforms of $(zI - G)^{-1}z$ and $(zI - G)^{-1}HU(z)$.

Example 5–2

Obtain the state transition matrix of the following discrete-time system:

$$x(k+1) = Gx(k) + Hu(k)$$

$$y(k) = Cx(k)$$

where

$$G = \begin{bmatrix} 0 & 1 \\ -0.16 & -1 \end{bmatrix}, \qquad H = \begin{bmatrix} 1 \\ 1 \end{bmatrix}, \qquad C = [1 \quad 0]$$

Then obtain the state $x(k)$ and the output $y(k)$ when the input $u(k) = 1$ for $k = 0, 1, 2, \ldots$. Assume that the initial state is given by

$$x(0) = \begin{bmatrix} x_1(0) \\ x_2(0) \end{bmatrix} = \begin{bmatrix} 1 \\ -1 \end{bmatrix}$$

From Equations (5–35) and (5–43) the state transition matrix $\Psi(k)$ is

$$\Psi(k) = \mathbf{G}^k = \mathscr{Z}^{-1}[(z\mathbf{I} - \mathbf{G})^{-1}z]$$

Therefore, we first obtain $(z\mathbf{I} - \mathbf{G})^{-1}$:

$$(z\mathbf{I} - \mathbf{G})^{-1} = \begin{bmatrix} z & -1 \\ 0.16 & z + 1 \end{bmatrix}^{-1}$$

$$= \begin{bmatrix} \dfrac{z + 1}{(z + 0.2)(z + 0.8)} & \dfrac{1}{(z + 0.2)(z + 0.8)} \\[3mm] \dfrac{-0.16}{(z + 0.2)(z + 0.8)} & \dfrac{z}{(z + 0.2)(z + 0.8)} \end{bmatrix}$$

$$= \begin{bmatrix} \dfrac{\frac{4}{3}}{z + 0.2} + \dfrac{-\frac{1}{3}}{z + 0.8} & \dfrac{\frac{5}{3}}{z + 0.2} + \dfrac{-\frac{5}{3}}{z + 0.8} \\[3mm] \dfrac{-\frac{0.8}{3}}{z + 0.2} + \dfrac{\frac{0.8}{3}}{z + 0.8} & \dfrac{-\frac{1}{3}}{z + 0.2} + \dfrac{\frac{4}{3}}{z + 0.8} \end{bmatrix}$$

The state transition matrix $\Psi(k)$ is now obtained as follows:

$$\Psi(k) = \mathbf{G}^k = \mathscr{Z}^{-1}[(z\mathbf{I} - \mathbf{G})^{-1}z]$$

$$= \mathscr{Z}^{-1} \begin{bmatrix} \dfrac{4}{3}\dfrac{z}{z + 0.2} - \dfrac{1}{3}\dfrac{z}{z + 0.8} & \dfrac{5}{3}\dfrac{z}{z + 0.2} - \dfrac{5}{3}\dfrac{z}{z + 0.8} \\[3mm] -\dfrac{0.8}{3}\dfrac{z}{z + 0.2} + \dfrac{0.8}{3}\dfrac{z}{z + 0.8} & -\dfrac{1}{3}\dfrac{z}{z + 0.2} + \dfrac{4}{3}\dfrac{z}{z + 0.8} \end{bmatrix}$$

$$= \begin{bmatrix} \frac{4}{3}(-0.2)^k - \frac{1}{3}(-0.8)^k & \frac{5}{3}(-0.2)^k - \frac{5}{3}(-0.8)^k \\[2mm] -\frac{0.8}{3}(-0.2)^k + \frac{0.8}{3}(-0.8)^k & -\frac{1}{3}(-0.2)^k + \frac{4}{3}(-0.8)^k \end{bmatrix} \tag{5–45}$$

Equation (5–45) gives the state transition matrix.

Next, compute $\mathbf{x}(k)$. The z transform of $\mathbf{x}(k)$ is given by

$$\mathscr{Z}[\mathbf{x}(k)] = \mathbf{X}(z) = (z\mathbf{I} - \mathbf{G})^{-1}z\mathbf{x}(0) + (z\mathbf{I} - \mathbf{G})^{-1}\mathbf{H}U(z)$$

$$= (z\mathbf{I} - \mathbf{G})^{-1}[z\mathbf{x}(0) + \mathbf{H}U(z)]$$

Since

$$U(z) = \frac{1}{1 - z^{-1}} = \frac{z}{z - 1}$$

we obtain

$$z\mathbf{x}(0) + \mathbf{H}U(z) = \begin{bmatrix} z \\ -z \end{bmatrix} + \begin{bmatrix} \dfrac{z}{z - 1} \\[3mm] \dfrac{z}{z - 1} \end{bmatrix} = \begin{bmatrix} \dfrac{z^2}{z - 1} \\[3mm] \dfrac{-z^2 + 2z}{z - 1} \end{bmatrix}$$

Hence

$$\mathbf{X}(z) = (z\mathbf{I} - \mathbf{G})^{-1}[z\mathbf{x}(0) + \mathbf{H}U(z)]$$

$$= \begin{bmatrix} \dfrac{(z^2 + 2)z}{(z + 0.2)(z + 0.8)(z - 1)} \\[4mm] \dfrac{(-z^2 + 1.84z)z}{(z + 0.2)(z + 0.8)(z - 1)} \end{bmatrix}$$

$$
= \begin{bmatrix} \dfrac{-\frac{17}{6}z}{z + 0.2} + \dfrac{\frac{22}{9}z}{z + 0.8} + \dfrac{\frac{25}{18}z}{z - 1} \\[4mm] \dfrac{\frac{3.4}{6}z}{z + 0.2} + \dfrac{-\frac{17.6}{9}z}{z + 0.8} + \dfrac{\frac{7}{18}z}{z - 1} \end{bmatrix}
$$

Thus, the state vector $\mathbf{x}(k)$ is given by

$$
\mathbf{x}(k) = \mathcal{Z}^{-1}[\mathbf{X}(z)] = \begin{bmatrix} -\frac{17}{6}(-0.2)^k + \frac{22}{9}(-0.8)^k + \frac{25}{18} \\[3mm] \frac{3.4}{6}(-0.2)^k - \frac{17.6}{9}(-0.8)^k + \frac{7}{18} \end{bmatrix}
$$

Finally, the output $y(k)$ is obtained as follows:

$$
y(k) = \mathbf{C}\mathbf{x}(k) = \begin{bmatrix} 1 & 0 \end{bmatrix} \begin{bmatrix} -\frac{17}{6}(-0.2)^k + \frac{22}{9}(-0.8)^k + \frac{25}{18} \\[3mm] \frac{3.4}{6}(-0.2)^k - \frac{17.6}{9}(-0.8)^k + \frac{7}{18} \end{bmatrix}
$$

$$
= -\tfrac{17}{6}(-0.2)^k + \tfrac{22}{9}(-0.8)^k + \tfrac{25}{18}
$$

Computation of $(z\mathbf{I} - \mathbf{G})^{-1}$. The solution of the state equation given by Equation (5–28) by the z transform method requires the computation of $(z\mathbf{I} - \mathbf{G})^{-1}$. Computing $(z\mathbf{I} - \mathbf{G})^{-1}$ is, except in simple cases, generally a time-consuming task. There are both analytical and computational methods available for computing $(z\mathbf{I} - \mathbf{G})^{-1}$. We shall present one method here.

Method for Computing $(z\mathbf{I} - \mathbf{G})^{-1}$. The method presented here is based on the expansion of the adjoint of $(z\mathbf{I} - \mathbf{G})$. The inverse of $(z\mathbf{I} - \mathbf{G})$ can be written in terms of the adjoint of $(z\mathbf{I} - \mathbf{G})$, as follows:

$$
(z\mathbf{I} - \mathbf{G})^{-1} = \frac{\operatorname{adj}(z\mathbf{I} - \mathbf{G})}{|z\mathbf{I} - \mathbf{G}|} \tag{5–46}
$$

Note that the determinant $|z\mathbf{I} - \mathbf{G}|$ may be written as follows:

$$
|z\mathbf{I} - \mathbf{G}| = z^n + a_1 z^{n-1} + a_2 z^{n-2} + \cdots + a_n \tag{5–47}
$$

It can be shown (see Problem A–5–13) that $\operatorname{adj}(z\mathbf{I} - \mathbf{G})$ may be given by

$$
\operatorname{adj}(z\mathbf{I} - \mathbf{G}) = \mathbf{I}z^{n-1} + \mathbf{H}_1 z^{n-2} + \mathbf{H}_2 z^{n-3} + \cdots + \mathbf{H}_{n-1} \tag{5–48}
$$

where

$$
\mathbf{H}_1 = \mathbf{G} + a_1\mathbf{I}
$$

$$
\mathbf{H}_2 = \mathbf{G}\mathbf{H}_1 + a_2\mathbf{I}
$$

$$
\vdots \tag{5–49}
$$

$$
\mathbf{H}_{n-1} = \mathbf{G}\mathbf{H}_{n-2} + a_{n-1}\mathbf{I}
$$

$$
\mathbf{H}_n = \mathbf{G}\mathbf{H}_{n-1} + a_n\mathbf{I} = \mathbf{0}
$$

Note that $a_1, a_2, \ldots, a_n$ are the coefficients appearing in the determinant given by Equation (5–47). The a_i's can also be given (see Problem A–5–13) by use of the trace, as follows:

$$a_1 = -\operatorname{tr} \mathbf{G}$$

$$a_2 = -\tfrac{1}{2} \operatorname{tr} \mathbf{GH}_1$$

$$a_3 = -\tfrac{1}{3} \operatorname{tr} \mathbf{GH}_2 \tag{5-50}$$

$$\vdots$$

$$a_n = -\frac{1}{n} \operatorname{tr} \mathbf{GH}_{n-1}$$

(The trace of an $n \times n$ matrix is the sum of its diagonal elements.)

For a higher-order determinant ($n > 3$), expanding the determinant $|z\mathbf{I} - \mathbf{G}|$ into the form given by Equation (5–47) may be time consuming; in this case, using Equation (5–50) to compute the a_i's proves to be useful, since $a_1, \mathbf{H}_1, a_2, \mathbf{H}_2, \dots, a_{n-1}, \mathbf{H}_{n-1}$ can easily be computed sequentially.

By substituting Equation (5–49) into Equation (5–48) and substituting the resulting equation into Equation (5–46), we obtain the inverse of $(z\mathbf{I} - \mathbf{G})$. The present method is convenient for computer solution; a standard program is available.

Example 5–3

Determine the inverse of the matrix $(z\mathbf{I} - \mathbf{G})$, where

$$\mathbf{G} = \begin{bmatrix} 0.1 & 0.1 & 0 \\ 0.3 & -0.1 & -0.2 \\ 0 & 0 & -0.3 \end{bmatrix}$$

Also, obtain $\mathbf{G}^k$.

From Equation (5–46), we have

$$(z\mathbf{I} - \mathbf{G})^{-1} = \frac{\operatorname{adj}(z\mathbf{I} - \mathbf{G})}{|z\mathbf{I} - \mathbf{G}|}$$

Although the determinant $|z\mathbf{I} - \mathbf{G}|$ can be expanded easily, here for demonstration purposes let us use Equation (5–50) to compute a_1, a_2, and a_3. First, notice that

$$a_1 = -\operatorname{tr} \mathbf{G} = -\operatorname{tr} \begin{bmatrix} 0.1 & 0.1 & 0 \\ 0.3 & -0.1 & -0.2 \\ 0 & 0 & -0.3 \end{bmatrix} = 0.3$$

Then, from Equation (5–49), we obtain

$$\mathbf{H}_1 = \mathbf{G} + a_1\mathbf{I} = \begin{bmatrix} 0.1 & 0.1 & 0 \\ 0.3 & -0.1 & -0.2 \\ 0 & 0 & -0.3 \end{bmatrix} + \begin{bmatrix} 0.3 & 0 & 0 \\ 0 & 0.3 & 0 \\ 0 & 0 & 0.3 \end{bmatrix}$$

$$= \begin{bmatrix} 0.4 & 0.1 & 0 \\ 0.3 & 0.2 & -0.2 \\ 0 & 0 & 0 \end{bmatrix}$$

Hence

$$a_2 = -\tfrac{1}{2} \operatorname{tr} \mathbf{GH}_1 = -\tfrac{1}{2} \operatorname{tr} \left\{ \begin{bmatrix} 0.1 & 0.1 & 0 \\ 0.3 & -0.1 & -0.2 \\ 0 & 0 & -0.3 \end{bmatrix} \begin{bmatrix} 0.4 & 0.1 & 0 \\ 0.3 & 0.2 & -0.2 \\ 0 & 0 & 0 \end{bmatrix} \right\}$$

$$= -\tfrac{1}{2} \operatorname{tr} \begin{bmatrix} 0.07 & 0.03 & -0.02 \\ 0.09 & 0.01 & 0.02 \\ 0 & 0 & 0 \end{bmatrix} = -0.04$$

By substituting the matrix $\mathbf{H}_1$ and the value of a_2 just obtained into Equation (5–49), we get

$$\mathbf{H}_2 = \mathbf{G}\mathbf{H}_1 + a_2\mathbf{I} = \begin{bmatrix} 0.03 & 0.03 & -0.02 \\ 0.09 & -0.03 & 0.02 \\ 0 & 0 & -0.04 \end{bmatrix}$$

and

$$a_3 = -\tfrac{1}{3}\operatorname{tr}\mathbf{G}\mathbf{H}_2 = -\tfrac{1}{3}\operatorname{tr}\begin{bmatrix} 0.012 & 0 & 0 \\ 0 & 0.012 & 0 \\ 0 & 0 & 0.012 \end{bmatrix} = -0.012$$

Notice that

$$\mathbf{H}_3 = \mathbf{G}\mathbf{H}_2 - 0.012\mathbf{I} = \mathbf{0}$$

The adjoint of $(z\mathbf{I} - \mathbf{G})$ can now be given by Equation (5–48), or

$$\operatorname{adj}(z\mathbf{I} - \mathbf{G}) = \mathbf{I}z^2 + \mathbf{H}_1 z + \mathbf{H}_2$$

$$= \begin{bmatrix} 1 & 0 & 0 \\ 0 & 1 & 0 \\ 0 & 0 & 1 \end{bmatrix}z^2 + \begin{bmatrix} 0.4 & 0.1 & 0 \\ 0.3 & 0.2 & -0.2 \\ 0 & 0 & 0 \end{bmatrix}z + \begin{bmatrix} 0.03 & 0.03 & -0.02 \\ 0.09 & -0.03 & 0.02 \\ 0 & 0 & -0.04 \end{bmatrix}$$

$$= \begin{bmatrix} z^2 + 0.4z + 0.03 & 0.1z + 0.03 & -0.02 \\ 0.3z + 0.09 & z^2 + 0.2z - 0.03 & -0.2z + 0.02 \\ 0 & 0 & z^2 - 0.04 \end{bmatrix}$$

Also,

$$|z\mathbf{I} - \mathbf{G}| = z^3 + a_1 z^2 + a_2 z + a_3 = z^3 + 0.3z^2 - 0.04z - 0.012$$

$$= (z + 0.2)(z - 0.2)(z + 0.3)$$

Hence,

$$(z\mathbf{I} - \mathbf{G})^{-1} = \frac{\operatorname{adj}(z\mathbf{I} - \mathbf{G})}{|z\mathbf{I} - \mathbf{G}|}$$

$$= \begin{bmatrix} \dfrac{z + 0.1}{(z + 0.2)(z - 0.2)} & \dfrac{0.1}{(z + 0.2)(z - 0.2)} & \dfrac{-0.02}{(z + 0.2)(z - 0.2)(z + 0.3)} \\[2ex] \dfrac{0.3}{(z + 0.2)(z - 0.2)} & \dfrac{z - 0.1}{(z + 0.2)(z - 0.2)} & \dfrac{-0.2(z - 0.1)}{(z + 0.2)(z - 0.2)(z + 0.3)} \\[2ex] 0 & 0 & \dfrac{1}{z + 0.3} \end{bmatrix}$$

This last equation gives the inverse of $(z\mathbf{I} - \mathbf{G})$.

Next, we shall obtain $\mathbf{G}^k$. From Equation (5–43), we have

$$\mathbf{G}^k = \mathscr{Z}^{-1}[(z\mathbf{I} - \mathbf{G})^{-1}z]$$

$$= \mathscr{Z}^{-1}\begin{bmatrix} \dfrac{0.25z}{z + 0.2} + \dfrac{0.75z}{z - 0.2} & -\dfrac{0.25z}{z + 0.2} + \dfrac{0.25z}{z - 0.2} & \dfrac{0.5z}{z + 0.2} - \dfrac{0.1z}{z - 0.2} - \dfrac{0.4z}{z + 0.3} \\[2ex] -\dfrac{0.75z}{z + 0.2} + \dfrac{0.75z}{z - 0.2} & \dfrac{0.75z}{z + 0.2} + \dfrac{0.25z}{z - 0.2} & -\dfrac{1.5z}{z + 0.2} - \dfrac{0.1z}{z - 0.2} + \dfrac{1.6z}{z + 0.3} \\[2ex] 0 & 0 & \dfrac{z}{z + 0.3} \end{bmatrix}$$

$$= \begin{bmatrix} 0.25(-0.2)^k + 0.75(0.2)^k & -0.25(-0.2)^k + 0.25(0.2)^k \\ -0.75(-0.2)^k + 0.75(0.2)^k & 0.75(-0.2)^k + 0.25(0.2)^k \\ 0 & 0 \end{bmatrix}$$

$$\begin{bmatrix} 0.5(-0.2)^k - 0.1(0.2)^k - 0.4(-0.3)^k \\ -1.5(-0.2)^k - 0.1(0.2)^k + 1.6(-0.3)^k \\ (-0.3)^k \end{bmatrix} \quad (5\text{--}51)$$

Solution of Linear Time-Varying Discrete-Time State Equations. Consider the following linear time-varying discrete-time state equation and output equation:

$$\mathbf{x}(k + 1) = \mathbf{G}(k)\mathbf{x}(k) + \mathbf{H}(k)\mathbf{u}(k) \qquad (5\text{--}52)$$

$$\mathbf{y}(k) = \mathbf{C}(k)\mathbf{x}(k) + \mathbf{D}(k)\mathbf{u}(k) \qquad (5\text{--}53)$$

The solution of Equation (5–52) may be found easily by recursion, as follows:

$$\mathbf{x}(h + 1) = \mathbf{G}(h)\mathbf{x}(h) + \mathbf{H}(h)\mathbf{u}(h)$$

$$\mathbf{x}(h + 2) = \mathbf{G}(h + 1)\mathbf{x}(h + 1) + \mathbf{H}(h + 1)\mathbf{u}(h + 1)$$

$$= \mathbf{G}(h + 1)\mathbf{G}(h)\mathbf{x}(h) + \mathbf{G}(h + 1)\mathbf{H}(h)\mathbf{u}(h) + \mathbf{H}(h + 1)\mathbf{u}(h + 1)$$

$$\vdots$$

Let us define the state transition matrix (fundamental matrix) for the system defined by Equation (5–52) as $\mathbf{\Psi}(k, h)$. It is a unique matrix satisfying the conditions

$$\mathbf{\Psi}(k + 1, h) = \mathbf{G}(k)\mathbf{\Psi}(k, h), \qquad \mathbf{\Psi}(h, h) = \mathbf{I}$$

where $k = h, h + 1, h + 2, \ldots$. It can be seen that the state transition matrix $\mathbf{\Psi}(k, h)$ is given by the equation

$$\mathbf{\Psi}(k, h) = \mathbf{G}(k - 1)\mathbf{G}(k - 2)\cdots\mathbf{G}(h), \qquad k > h \qquad (5\text{--}54)$$

Using $\mathbf{\Psi}(k, h)$, the solution of Equation (5–52) becomes

$$\mathbf{x}(k) = \mathbf{\Psi}(k, h)\mathbf{x}(h) + \sum_{j=h}^{k-1} \mathbf{\Psi}(k, j + 1)\mathbf{H}(j)\mathbf{u}(j), \qquad k > h \qquad (5\text{--}55)$$

Notice that the first term on the right-hand side of Equation (5–55) is the contribution of the initial state $\mathbf{x}(h)$ to the current state $\mathbf{x}(k)$ and that the second term is the contribution of the input $\mathbf{u}(h), \mathbf{u}(h + 1), \ldots, \mathbf{u}(k - 1)$.

Equation (5–55) can be verified easily. Referring to Equation (5–54), we have

$$\mathbf{\Psi}(k + 1, h) = \mathbf{G}(k)\mathbf{G}(k - 1)\cdots\mathbf{G}(h) = \mathbf{G}(k)\mathbf{\Psi}(k, h) \qquad (5\text{--}56)$$

If we substitute Equation (5–56) into

$$\mathbf{x}(k + 1) = \mathbf{\Psi}(k + 1, h)\mathbf{x}(h) + \sum_{j=h}^{k} \mathbf{\Psi}(k + 1, j + 1)\mathbf{H}(j)\mathbf{u}(j)$$

we obtain

$$\mathbf{x}(k + 1) = \mathbf{G}(k)\mathbf{\Psi}(k, h)\mathbf{x}(h) + \sum_{j=h}^{k-1} \mathbf{\Psi}(k + 1, j + 1)\mathbf{H}(j)\mathbf{u}(j)$$

$$+ \mathbf{\Psi}(k + 1, k + 1)\mathbf{H}(k)\mathbf{u}(k)$$

$$= \mathbf{G}(k)\left[\mathbf{\Psi}(k,h)\mathbf{x}(h) + \sum_{j=h}^{k-1}\mathbf{\Psi}(k,j+1)\mathbf{H}(j)\mathbf{u}(j)\right] + \mathbf{H}(k)\mathbf{u}(k)$$

$$= \mathbf{G}(k)\mathbf{x}(k) + \mathbf{H}(k)\mathbf{u}(k)$$

Thus, we have shown that Equation (5–55) is the solution of Equation (5–52).

Once we get the solution $\mathbf{x}(k)$, the output equation, Equation (5–53), becomes as follows:

$$\mathbf{y}(k) = \mathbf{C}(k)\mathbf{\Psi}(k,h)\mathbf{x}(h) + \sum_{j=h}^{k-1}\mathbf{C}(k)\mathbf{\Psi}(k,j+1)\mathbf{H}(j)\mathbf{u}(j) + \mathbf{D}(k)\mathbf{u}(k), \qquad k > h$$

If $\mathbf{G}(k)$ is nonsingular for all k values considered, so that the inverse of $\mathbf{\Psi}(k,h)$ exists, then the inverse of $\mathbf{\Psi}(k,h)$, denoted by $\mathbf{\Psi}(h,k)$, is given as follows:

$$\mathbf{\Psi}^{-1}(k,h) = \mathbf{\Psi}(h,k)$$

$$= [\mathbf{G}(k-1)\mathbf{G}(k-2)\cdots\mathbf{G}(h)]^{-1}$$

$$= \mathbf{G}^{-1}(h)\mathbf{G}^{-1}(h+1)\cdots\mathbf{G}^{-1}(k-1) \qquad (5\text{–}57)$$

Summary on $\mathbf{\Psi}(k, h)$.   A summary on the state transition matrix $\mathbf{\Psi}(k,h)$ gives the following:

1. $\mathbf{\Psi}(k,k) = \mathbf{I}$
2. $\mathbf{\Psi}(k,h) = \mathbf{G}(k-1)\mathbf{G}(k-2)\cdots\mathbf{G}(h), \qquad k > h$
3. If the inverse of $\mathbf{\Psi}(k,h)$ exists, then

$$\mathbf{\Psi}^{-1}(k,h) = \mathbf{\Psi}(h,k)$$

4. If $\mathbf{G}(k)$ is nonsingular for all k values considered, then

$$\mathbf{\Psi}(k,i) = \mathbf{\Psi}(k,j)\mathbf{\Psi}(j,i), \qquad \text{for any } i, j, k$$

If $\mathbf{G}(k)$ is singular for any value of k, then

$$\mathbf{\Psi}(k,i) = \mathbf{\Psi}(k,j)\mathbf{\Psi}(j,i), \qquad \text{for } k > j > i$$

5–4 PULSE-TRANSFER-FUNCTION MATRIX

A single-input–single-output discrete-time system may be modeled by a pulse trans-fer function. Extension of the pulse-transfer-function concept to a multiple-input–multiple-output discrete-time system gives us the pulse-transfer-function matrix. In this section we shall investigate the relationship between state-space representation and representation by the pulse-transfer-function matrix.

Pulse-Transfer-Function Matrix.   The state-space representation of an nth-order linear time-invariant discrete-time system with r inputs and m outputs can be given by

$$\mathbf{x}(k+1) = \mathbf{G}\mathbf{x}(k) + \mathbf{H}\mathbf{u}(k) \qquad (5\text{–}58)$$

$$\mathbf{y}(k) = \mathbf{C}\mathbf{x}(k) + \mathbf{D}\mathbf{u}(k) \qquad (5\text{–}59)$$

where $\mathbf{x}(k)$ is an n-vector, $\mathbf{u}(k)$ is an r-vector, $\mathbf{y}(k)$ is an m-vector, $\mathbf{G}$ is an $n \times n$ matrix, $\mathbf{H}$ is an $n \times r$ matrix, $\mathbf{C}$ is an $m \times n$ matrix, and $\mathbf{D}$ is an $m \times r$ matrix. Taking the z transforms of Equations (5–58) and (5–59), we obtain

$$z\mathbf{X}(z) - z\mathbf{x}(0) = \mathbf{G}\mathbf{X}(z) + \mathbf{H}\mathbf{U}(z)$$

$$\mathbf{Y}(z) = \mathbf{C}\mathbf{X}(z) + \mathbf{D}\mathbf{U}(z)$$

Noting that the definition of the pulse transfer function calls for the assumption of zero initial conditions, here we also assume that the initial state $\mathbf{x}(0)$ is zero. Then we obtain

$$\mathbf{X}(z) = (z\mathbf{I} - \mathbf{G})^{-1}\mathbf{H}\mathbf{U}(z)$$

and

$$\mathbf{Y}(z) = [\mathbf{C}(z\mathbf{I} - \mathbf{G})^{-1}\mathbf{H} + \mathbf{D}]\mathbf{U}(z) = \mathbf{F}(z)\mathbf{U}(z)$$

where

$$\mathbf{F}(z) = \mathbf{C}(z\mathbf{I} - \mathbf{G})^{-1}\mathbf{H} + \mathbf{D} \qquad (5–60)$$

$\mathbf{F}(z)$ is called the *pulse-transfer-function matrix*. It is an $m \times r$ matrix. The pulse transfer function matrix $\mathbf{F}(z)$ characterizes the input–output dynamics of the given discrete-time system.

Since the inverse of matrix $(z\mathbf{I} - \mathbf{G})$ can be written as

$$(z\mathbf{I} - \mathbf{G})^{-1} = \frac{\text{adj}\,(z\mathbf{I} - \mathbf{G})}{|z\mathbf{I} - \mathbf{G}|}$$

the pulse-transfer-function matrix $\mathbf{F}(z)$ can be given by the equation

$$\mathbf{F}(z) = \frac{\mathbf{C}\,\text{adj}\,(z\mathbf{I} - \mathbf{G})\mathbf{H}}{|z\mathbf{I} - \mathbf{G}|} + \mathbf{D}$$

Clearly, the poles of $\mathbf{F}(z)$ are the zeros of $|z\mathbf{I} - \mathbf{G}| = 0$. This means that the characteristic equation of the discrete-time system is given by

$$|z\mathbf{I} - \mathbf{G}| = 0$$

or

$$z^n + a_1 z^{n-1} + a_2 z^{n-2} + \cdots + a_{n-1}z + a_n = 0$$

where the coefficients a_i depend on the elements of $\mathbf{G}$.

Similarity Transformation.   We have shown that for the system defined by

$$\mathbf{x}(k + 1) = \mathbf{G}\mathbf{x}(k) + \mathbf{H}\mathbf{u}(k)$$

$$\mathbf{y}(k) = \mathbf{C}\mathbf{x}(k) + \mathbf{D}\mathbf{u}(k)$$

the pulse-transfer-function matrix is

$$\mathbf{F}(z) = \mathbf{C}(z\mathbf{I} - \mathbf{G})^{-1}\mathbf{H} + \mathbf{D}$$

In Section 5–2 we showed that various different state-space representations for a given system are related by the similarity transformation. By defining a new state vector $\hat{\mathbf{x}}(k)$ by using a similarity transformation matrix $\mathbf{P}$, or

$$\mathbf{x}(k) = \mathbf{P}\hat{\mathbf{x}}(k)$$

where $\mathbf{P}$ is a nonsingular $n \times n$ matrix, we have

$$\hat{\mathbf{x}}(k + 1) = \hat{\mathbf{G}}\hat{\mathbf{x}}(k) + \hat{\mathbf{H}}u(k) \qquad (5\text{--}61)$$

$$\mathbf{y}(k) = \hat{\mathbf{C}}\hat{\mathbf{x}}(k) + \hat{\mathbf{D}}u(k) \qquad (5\text{--}62)$$

where $\mathbf{G}$, $\mathbf{H}$, $\mathbf{C}$, $\mathbf{D}$ and $\hat{\mathbf{G}}$, $\hat{\mathbf{H}}$, $\hat{\mathbf{C}}$, $\hat{\mathbf{D}}$ are related, respectively, by

$$\mathbf{P}^{-1}\mathbf{G}\mathbf{P} = \hat{\mathbf{G}}$$

$$\mathbf{P}^{-1}\mathbf{H} = \hat{\mathbf{H}}$$

$$\mathbf{C}\mathbf{P} = \hat{\mathbf{C}}$$

$$\mathbf{D} = \hat{\mathbf{D}}$$

The pulse-transfer-function matrix $\hat{\mathbf{F}}(z)$ for the system defined by Equations (5–61) and (5–62) is

$$\hat{\mathbf{F}}(z) = \hat{\mathbf{C}}(z\mathbf{I} - \hat{\mathbf{G}})^{-1}\hat{\mathbf{H}} + \hat{\mathbf{D}}$$

Notice that the pulse-transfer-function matrices $\mathbf{F}(z)$ and $\hat{\mathbf{F}}(z)$ are the same, since

$$\hat{\mathbf{F}}(z) = \hat{\mathbf{C}}(z\mathbf{I} - \hat{\mathbf{G}})^{-1}\hat{\mathbf{H}} + \hat{\mathbf{D}} = \mathbf{C}\mathbf{P}(z\mathbf{I} - \mathbf{P}^{-1}\mathbf{G}\mathbf{P})^{-1}\mathbf{P}^{-1}\mathbf{H} + \mathbf{D}$$

$$= \mathbf{C}\mathbf{P}(z\mathbf{P} - \mathbf{G}\mathbf{P})^{-1}\mathbf{H} + \mathbf{D} = \mathbf{C}(z\mathbf{P}\mathbf{P}^{-1} - \mathbf{G}\mathbf{P}\mathbf{P}^{-1})^{-1}\mathbf{H} + \mathbf{D}$$

$$= \mathbf{C}(z\mathbf{I} - \mathbf{G})^{-1}\mathbf{H} + \mathbf{D} = \mathbf{F}(z)$$

Thus, the pulse-transfer-function matrix is invariant under similarity transformation. That is, it does not depend on the particular state vector $\mathbf{x}(k)$ chosen for the system representation.

The characteristic equation $|z\mathbf{I} - \mathbf{G}| = 0$ is also invariant under similarity transformation, since

$$|z\mathbf{I} - \mathbf{G}| = |\mathbf{P}^{-1}||z\mathbf{I} - \mathbf{G}||\mathbf{P}| = |z\mathbf{I} - \mathbf{P}^{-1}\mathbf{G}\mathbf{P}| = |z\mathbf{I} - \hat{\mathbf{G}}|$$

Thus, the eigenvalues of $\mathbf{G}$ are invariant under similarity transformation.

5–5 DISCRETIZATION OF CONTINUOUS-TIME STATE-SPACE EQUATIONS

In digital control of continuous-time plants, we need to convert continuous-time state-space equations into discrete-time state-space equations. Such conversion can be done by introducing fictitious samplers and fictitious holding devices into continuous-time systems. The error introduced by discretization may be made negligible by using a sufficiently small sampling period compared with the significant time constant of the system.

Review of Solution of Continuous-Time State Equations. We shall first review the matrix exponential, $e^{\mathbf{A}t}$. The matrix exponential is defined by

$$e^{\mathbf{A}t} = \mathbf{I} + \mathbf{A}t + \frac{1}{2!}\mathbf{A}^2 t^2 + \cdots + \frac{1}{k!}\mathbf{A}^k t^k + \cdots = \sum_{k=0}^{\infty} \frac{\mathbf{A}^k t^k}{k!}$$

Because of the convergence of the infinite series $\sum_{k=0}^{\infty} \mathbf{A}^k t^k/k!$, the series can be differentiated term by term to give

$$\frac{d}{dt}e^{\mathbf{A}t} = \mathbf{A} + \mathbf{A}^2 t + \frac{\mathbf{A}^3 t^2}{2!} + \cdots + \frac{\mathbf{A}^k t^{k-1}}{(k-1)!} + \cdots$$

$$= \mathbf{A}\left[\mathbf{I} + \mathbf{A}t + \frac{\mathbf{A}^2 t^2}{2!} + \cdots + \frac{\mathbf{A}^{k-1} t^{k-1}}{(k-1)!} + \cdots\right] = \mathbf{A}e^{\mathbf{A}t}$$

$$= \left[\mathbf{I} + \mathbf{A}t + \frac{\mathbf{A}^2 t^2}{2!} + \cdots + \frac{\mathbf{A}^{k-1} t^{k-1}}{(k-1)!} + \cdots\right]\mathbf{A} = e^{\mathbf{A}t}\mathbf{A}$$

The matrix exponential has the property that

$$e^{\mathbf{A}(t+s)} = e^{\mathbf{A}t} e^{\mathbf{A}s}$$

This can be proved as follows:

$$e^{\mathbf{A}t} e^{\mathbf{A}s} = \left(\sum_{k=0}^{\infty} \frac{\mathbf{A}^k t^k}{k!}\right)\left(\sum_{k=0}^{\infty} \frac{\mathbf{A}^k s^k}{k!}\right) = \sum_{k=0}^{\infty} \mathbf{A}^k\left[\sum_{i=0}^{k} \frac{t^i s^{k-i}}{i!(k-i)!}\right]$$

$$= \sum_{k=0}^{\infty} \mathbf{A}^k \frac{(t+s)^k}{k!} = e^{\mathbf{A}(t+s)}$$

In particular, if $s = -t$, then

$$e^{\mathbf{A}t} e^{-\mathbf{A}t} = e^{-\mathbf{A}t} e^{\mathbf{A}t} = e^{\mathbf{A}(t-t)} = \mathbf{I}$$

Thus, the inverse of $e^{\mathbf{A}t}$ is $e^{-\mathbf{A}t}$. Since the inverse of $e^{\mathbf{A}t}$ always exists, $e^{\mathbf{A}t}$ is nonsingular.

It is important to point out that

$$e^{(\mathbf{A}+\mathbf{B})t} = e^{\mathbf{A}t} e^{\mathbf{B}t}, \qquad \text{if } \mathbf{A}\mathbf{B} = \mathbf{B}\mathbf{A}$$

$$e^{(\mathbf{A}+\mathbf{B})t} \neq e^{\mathbf{A}t} e^{\mathbf{B}t}, \qquad \text{if } \mathbf{A}\mathbf{B} \neq \mathbf{B}\mathbf{A}$$

We shall next obtain the solution of the continuous-time state equation

$$\dot{\mathbf{x}} = \mathbf{A}\mathbf{x} + \mathbf{B}\mathbf{u} \tag{5–63}$$

where $\mathbf{x}$ is the state vector (n-vector), $\mathbf{u}$ the input vector (r-vector), $\mathbf{A}$ an $n \times n$ constant matrix, and $\mathbf{B}$ an $n \times r$ constant matrix.

By writing Equation (5–63) as

$$\dot{\mathbf{x}}(t) - \mathbf{A}\mathbf{x}(t) = \mathbf{B}\mathbf{u}(t)$$

and premultiplying both sides of this last equation by $e^{-\mathbf{A}t}$, we obtain

$$e^{-\mathbf{A}t}[\dot{\mathbf{x}}(t) - \mathbf{A}\mathbf{x}(t)] = \frac{d}{dt}[e^{-\mathbf{A}t}\mathbf{x}(t)] = e^{-\mathbf{A}t}\mathbf{B}\mathbf{u}(t)$$

Integrating the preceding equation between 0 and t gives

$$e^{-\mathbf{A}t}\mathbf{x}(t) = \mathbf{x}(0) + \int_0^t e^{-\mathbf{A}\tau}\mathbf{B}\mathbf{u}(\tau)\,d\tau$$

or

$$\mathbf{x}(t) = e^{\mathbf{A}t}\mathbf{x}(0) + \int_0^t e^{\mathbf{A}(t-\tau)}\mathbf{B}\mathbf{u}(\tau)\,d\tau \tag{5–64}$$

Equation (5–64) is the solution of Equation (5–63). Note that the solution of the state equation starting with the initial state $\mathbf{x}(t_0)$ is

$$\mathbf{x}(t) = e^{\mathbf{A}(t-t_0)}\mathbf{x}(t_0) + \int_{t_0}^{t} e^{\mathbf{A}(t-\tau)}\mathbf{B}\mathbf{u}(\tau)\,d\tau \tag{5–65}$$

Discretization of Continuous-Time State-Space Equations.   In what follows we shall present a procedure for discretizing continuous-time state-space equations. We assume that the input vector $\mathbf{u}(t)$ changes only at equally spaced sampling instants. Note that the sampling operation here is fictitious. We shall derive the discrete-time state equation and output equation that yield the exact values at $t = kT$, where $k = 0, 1, 2, \ldots$.

Consider the continuous-time state equation and output equation

$$\dot{\mathbf{x}} = \mathbf{A}\mathbf{x} + \mathbf{B}\mathbf{u} \tag{5–66}$$

$$\mathbf{y} = \mathbf{C}\mathbf{x} + \mathbf{D}\mathbf{u} \tag{5–67}$$

In the following analysis, to clarify the presentation, we use the notation kT and $(k + 1)T$ instead of k and $k + 1$. The discrete-time representation of Equation (5–66) will take the form

$$\mathbf{x}((k + 1)T) = \mathbf{G}(T)\mathbf{x}(kT) + \mathbf{H}(T)\mathbf{u}(kT) \tag{5–68}$$

Note that the matrices $\mathbf{G}$ and $\mathbf{H}$ depend on the sampling period T. Once the sampling period T is fixed, $\mathbf{G}$ and $\mathbf{H}$ are constant matrices.

To determine $\mathbf{G}(T)$ and $\mathbf{H}(T)$, we use Equation (5–64), the solution of Equation (5–66). We assume that the input $\mathbf{u}(t)$ is sampled and fed to a zero-order hold so that all the components of $\mathbf{u}(t)$ are constant over the interval between any two consecutive sampling instants, or

$$\mathbf{u}(t) = \mathbf{u}(kT), \qquad \text{for } kT \leq t < kT + T \tag{5–69}$$

Since

$$\mathbf{x}((k + 1)T) = e^{\mathbf{A}(k+1)T}\mathbf{x}(0) + e^{\mathbf{A}(k+1)T}\int_{0}^{(k+1)T} e^{-\mathbf{A}\tau}\mathbf{B}\mathbf{u}(\tau)\,d\tau \tag{5–70}$$

and

$$\mathbf{x}(kT) = e^{\mathbf{A}kT}\mathbf{x}(0) + e^{\mathbf{A}kT}\int_{0}^{kT} e^{-\mathbf{A}\tau}\mathbf{B}\mathbf{u}(\tau)\,d\tau \tag{5–71}$$

multiplying Equation (5–71) by $e^{\mathbf{A}T}$ and subtracting it from Equation (5–70) gives us

$$\mathbf{x}((k + 1)T) = e^{\mathbf{A}T}\mathbf{x}(kT) + e^{\mathbf{A}(k+1)T}\int_{kT}^{(k+1)T} e^{-\mathbf{A}\tau}\mathbf{B}\mathbf{u}(\tau)\,d\tau$$

Since from Equation (5–69) $\mathbf{u}(t) = \mathbf{u}(kT)$ for $kT \leq t < kT + T$, we may substitute $\mathbf{u}(\tau) = \mathbf{u}(kT) = \text{constant}$ in this last equation. [Note that $\mathbf{u}(t)$ may jump at $t = kT + T$ and thus $\mathbf{u}(kT + T)$ may be different from $\mathbf{u}(kT)$. Such a jump in $\mathbf{u}(\tau)$ at $\tau = kT + T$, the upper limit of integration, does not affect the value of the integral in this last equation, because the integrand does not involve impulse functions.] Hence, we may write

$$\mathbf{x}((k + 1)T) = e^{\mathbf{A}T}\mathbf{x}(kT) + e^{\mathbf{A}T}\int_0^T e^{-\mathbf{A}t}\mathbf{B}\mathbf{u}(kT)\, dt$$

$$= e^{\mathbf{A}T}\mathbf{x}(kT) + \int_0^T e^{\mathbf{A}\lambda}\mathbf{B}\mathbf{u}(kT)\, d\lambda \tag{5-72}$$

where $\lambda = T - t$. If we define

$$\mathbf{G}(T) = e^{\mathbf{A}T} \tag{5-73}$$

$$\mathbf{H}(T) = \left(\int_0^T e^{\mathbf{A}\lambda}\, d\lambda\right)\mathbf{B} \tag{5-74}$$

then Equation (5-72) becomes

$$\mathbf{x}((k + 1)T) = \mathbf{G}(T)\mathbf{x}(kT) + \mathbf{H}(T)\mathbf{u}(kT) \tag{5-75}$$

which is Equation (5-68). Thus, Equations (5-73) and (5-74) give the desired matrices $\mathbf{G}(T)$ and $\mathbf{H}(T)$. Note that $\mathbf{G}(T)$ and $\mathbf{H}(T)$ depend on the sampling period T. Referring to Equation (5-67), the output equation becomes

$$\mathbf{y}(kT) = \mathbf{C}\mathbf{x}(kT) + \mathbf{D}\mathbf{u}(kT) \tag{5-76}$$

where matrices $\mathbf{C}$ and $\mathbf{D}$ are constant matrices and do not depend on the sampling period T.

If matrix $\mathbf{A}$ is nonsingular, then $\mathbf{H}(T)$ given by Equation (5-74) can be simplified to

$$\mathbf{H}(T) = \left(\int_0^T e^{\mathbf{A}\lambda}\, d\lambda\right)\mathbf{B} = \mathbf{A}^{-1}(e^{\mathbf{A}T} - \mathbf{I})\mathbf{B} = (e^{\mathbf{A}T} - \mathbf{I})\mathbf{A}^{-1}\mathbf{B}$$

Comments

1. In the state-space approach, notice that by assuming the input vector $\mathbf{u}(t)$ to be constant between any two consecutive sampling instants, the discrete-time model can be obtained simply by integrating the continuous-time state equation over one sampling period. The discrete-time state equation given by Equation (5-68) is called the *zero-order hold equivalent* of the continuous-time state equation given by Equation (5-66).

2. In general, in converting the continuous-time system equation into a discrete-time system equation, some sort of approximation is necessary. It is important to point out that Equation (5-75) involves no approximation, provided the input vector $\mathbf{u}(t)$ is constant between any two consecutive sampling instants, as assumed in the derivation.

3. Notice that for $T \ll 1$, $\mathbf{G}(T) \doteq \mathbf{G}(0) = e^{\mathbf{A}0} = \mathbf{I}$. Thus, as the sampling period T becomes very small, $\mathbf{G}(T)$ approaches the identity matrix.

Example 5-4

Consider the continuous-time system given by

$$G(s) = \frac{Y(s)}{U(s)} = \frac{1}{s + a}$$

Obtain the continuous-time state-space representation of the system. Then discretize
the state equation and output equation and obtain the discrete-time state-space representation of the system. Also, obtain the pulse transfer function for the system by using
Equation (5–60).

The continuous-time state-space representation of the system is simply

$$\dot{x} = -ax + u$$

$$y = x$$

Now we discretize the state equation and the output equation. Referring to Equations
(5–73) and (5–74), we have

$$G(T) = e^{-aT}$$

$$H(T) = \int_0^T e^{-a\lambda}\, d\lambda = \frac{1 - e^{-aT}}{a}$$

Hence, the discretized version of the system equations is

$$x(k + 1) = e^{-aT}x(k) + \frac{1 - e^{-aT}}{a}u(k)$$

$$y(k) = x(k)$$

Referring to Equation (5–60), the pulse transfer function for this system is

$$F(z) = C(zI - G)^{-1}H$$

$$= (z - e^{-aT})^{-1}\frac{1 - e^{-aT}}{a} = \frac{(1 - e^{-aT})z^{-1}}{a(1 - e^{-aT}z^{-1})}$$

This result, of course, agrees with the z transform of $G(s)$ where it is preceded
by a sampler and zero-order hold [that is, where the signal $u(t)$ is sampled and fed to
a zero-order hold before being applied to $G(s)$]:

$$G(z) = \mathscr{Z}\left[\frac{1 - e^{-Ts}}{s}\frac{1}{s + a}\right] = (1 - z^{-1})\mathscr{Z}\left[\frac{1}{s(s + a)}\right]$$

$$= \frac{(1 - e^{-aT})z^{-1}}{a(1 - e^{-aT}z^{-1})}$$

Example 5–5

Obtain the discrete-time state and output equations and the pulse transfer function
(when the sampling period $T = 1$) of the following continuous-time system:

$$G(s) = \frac{Y(s)}{U(s)} = \frac{1}{s(s + 2)}$$

which may be represented in state space by the equations

$$\begin{bmatrix} \dot{x}_1 \\ \dot{x}_2 \end{bmatrix} = \begin{bmatrix} 0 & 1 \\ 0 & -2 \end{bmatrix}\begin{bmatrix} x_1 \\ x_2 \end{bmatrix} + \begin{bmatrix} 0 \\ 1 \end{bmatrix}u$$

$$y = \begin{bmatrix} 1 & 0 \end{bmatrix}\begin{bmatrix} x_1 \\ x_2 \end{bmatrix}$$

The desired discrete-time state equation will have the form

$$\mathbf{x}((k + 1)T) = \mathbf{G}(T)\mathbf{x}(kT) + \mathbf{H}(T)u(kT)$$

where matrices $\mathbf{G}(T)$ and $\mathbf{H}(T)$ are obtained from Equations (5–73) and (5–74) as
follows:

$$\mathbf{G}(T) = e^{\mathbf{A}T} = \begin{bmatrix} 1 & \frac{1}{2}(1 - e^{-2T}) \\ 0 & e^{-2T} \end{bmatrix}$$

$$\mathbf{H}(T) = \left(\int_0^T e^{\mathbf{A}t} dt \right) \mathbf{B} = \left\{ \int_0^T \begin{bmatrix} 1 & \frac{1}{2}(1 - e^{-2t}) \\ 0 & e^{-2t} \end{bmatrix} dt \right\} \begin{bmatrix} 0 \\ 1 \end{bmatrix} = \begin{bmatrix} \frac{1}{2}\left(T + \dfrac{e^{-2T} - 1}{2}\right) \\ \frac{1}{2}(1 - e^{-2T}) \end{bmatrix}$$

Thus,

$$\begin{bmatrix} x_1((k+1)T) \\ x_2((k+1)T) \end{bmatrix} = \begin{bmatrix} 1 & \frac{1}{2}(1 - e^{-2T}) \\ 0 & e^{-2T} \end{bmatrix} \begin{bmatrix} x_1(kT) \\ x_2(kT) \end{bmatrix} + \begin{bmatrix} \frac{1}{2}\left(T + \dfrac{e^{-2T} - 1}{2}\right) \\ \frac{1}{2}(1 - e^{-2T}) \end{bmatrix} u(kT)$$

The output equation becomes

$$y(kT) = \begin{bmatrix} 1 & 0 \end{bmatrix} \begin{bmatrix} x_1(kT) \\ x_2(kT) \end{bmatrix}$$

When the sampling period is 1 sec, or $T = 1$, the discrete-time state equation and the output equation become, respectively,

$$\begin{bmatrix} x_1(k+1) \\ x_2(k+1) \end{bmatrix} = \begin{bmatrix} 1 & 0.4323 \\ 0 & 0.1353 \end{bmatrix} \begin{bmatrix} x_1(k) \\ x_2(k) \end{bmatrix} + \begin{bmatrix} 0.2838 \\ 0.4323 \end{bmatrix} u(k)$$

and

$$y(k) = \begin{bmatrix} 1 & 0 \end{bmatrix} \begin{bmatrix} x_1(k) \\ x_2(k) \end{bmatrix}$$

The pulse-transfer-function representation of this system can be obtained from Equation (5–60), as follows:

$$F(z) = \mathbf{C}(z\mathbf{I} - \mathbf{G})^{-1}\mathbf{H} + D$$

$$= \begin{bmatrix} 1 & 0 \end{bmatrix} \begin{bmatrix} z - 1 & -0.4323 \\ 0 & z - 0.1353 \end{bmatrix}^{-1} \begin{bmatrix} 0.2838 \\ 0.4323 \end{bmatrix} + 0$$

$$= \begin{bmatrix} 1 & 0 \end{bmatrix} \begin{bmatrix} \dfrac{1}{z - 1} & \dfrac{0.4323}{(z - 1)(z - 0.1353)} \\ 0 & \dfrac{1}{z - 0.1353} \end{bmatrix} \begin{bmatrix} 0.2838 \\ 0.4323 \end{bmatrix}$$

$$= \frac{0.2838z + 0.1485}{(z - 1)(z - 0.1353)}$$

$$= \frac{0.2838z^{-1} + 0.1485z^{-2}}{(1 - z^{-1})(1 - 0.1353z^{-1})}$$

Note that the same pulse transfer function can be obtained by taking the z transform of $G(s)$ when it is preceded by a sampler and zero-order hold. Assuming $T = 1$, we obtain

$$G(z) = \mathscr{Z}\left[\frac{1 - e^{Ts}}{s} \frac{1}{s(s + 2)} \right] = (1 - z^{-1})\mathscr{Z}\left[\frac{1}{s^2(s + 2)} \right]$$

$$= (1 - z^{-1})\mathscr{Z}\left[\frac{0.5}{s^2} - \frac{0.25}{s} + \frac{0.25}{s + 2} \right]$$

$$= (1 - z^{-1})\left[\frac{0.5z^{-1}}{(1 - z^{-1})^2} - \frac{0.25}{1 - z^{-1}} + \frac{0.25}{1 - 0.1353z^{-1}} \right]$$

$$= \frac{0.2838z^{-1} + 0.1485z^{-2}}{(1 - z^{-1})(1 - 0.1353z^{-1})}$$

MATLAB Approach to the Discretization of Continuous-Time State Equations.
MATLAB has a convenient command to discretize the continuous-time state
equation

$$\dot{\mathbf{x}} = \mathbf{Ax} + \mathbf{Bu}$$

into

$$\mathbf{x}(k + 1) = \mathbf{Gx}(k) + \mathbf{Hu}(k)$$

The MATLAB command for discretization is

$$[G,H] = c2d(A,B,T)$$

where T is the sampling period of the discrete-time system. T should be specified in
seconds.

If good accuracy is needed in obtaining **G** and **H**, use *format long*. If only four
decimal places are needed, use *format short*. If no format statement is included in
the program, MATLAB will produce **G** and **H** in *format short*.

Consider the following example: If the continuous-time system is given by

$$\begin{bmatrix} \dot{x}_1 \\ \dot{x}_2 \end{bmatrix} = \begin{bmatrix} 0 & 1 \\ -25 & -4 \end{bmatrix} \begin{bmatrix} x_1 \\ x_2 \end{bmatrix} + \begin{bmatrix} 0 \\ 1 \end{bmatrix} u \tag{5-77}$$

then, assuming the sampling period to be 0.05 sec, we obtain **G** and **H** as follows:

```
A = [0   1;-25   -4];
B = [0;1];
[G,H] = c2d(A,B,0.05)

G =

         0.9709      0.0448
        -1.1212      0.7915

H =

         0.0012
         0.0448
```

Note that the state matrix **G** and input matrix **H** of the discrete-time state-
space equation

$$\mathbf{x}(k + 1) = \mathbf{Gx}(k) + \mathbf{Hu}(k)$$

depend on the sampling period T. For example, consider discretization of the
continuous-time system given by Equation (5–77) with two more different sampling
periods: $T = 0.2$ sec and $T = 1$ sec. As seen in the previous and following MATLAB
outputs, a set of matrices **G** and **H** differs for a different sampling period T.

```
A = [0   1;−25   −4];
B = [0;1];
[G,H] = c2d(A,B,0.2)

G =

          0.6401        0.1161
         −2.9017        0.1758

H =

          0.0144
          0.1161
```

```
A = [0   1;−25   −4];
B = [0;1];
[G,H] = c2d(A,B,1)

G =

         −0.0761       −0.0293
          0.7321        0.0410

H =

          0.0430
         −0.0293
```

As another example, consider the following system:

$$\dot{\mathbf{x}} = \mathbf{A}\mathbf{x} + \mathbf{B}u$$

where

$$\mathbf{A} = \begin{bmatrix} 0 & 1 & 0 & 0 \\ 20.601 & 0 & 0 & 0 \\ 0 & 0 & 0 & 1 \\ -0.4905 & 0 & 0 & 0 \end{bmatrix}, \qquad \mathbf{B} = \begin{bmatrix} 0 \\ -1 \\ 0 \\ 0.5 \end{bmatrix}$$

Assuming that the sampling period T is 0.05 sec and without specifying the format, we get the following discrete-time state equation:

$$\mathbf{x}(k + 1) = \mathbf{G}\mathbf{x}(k) + \mathbf{H}u(k)$$

where matrices **G** and **H** can be found in the following computer output:

```
A = [0           1  0  0
     20.601      0  0  0
     0           0  0  1
     −0.4905  0  0  0];
B = [0;−1;0;0.5];
[G,H] = c2d(A,B,0.05)

G =

          1.0259     0.0504        0          0
          1.0389     1.0259        0          0
         −0.0006    −0.0000     1.0000     0.0500
         −0.0247    −0.0006        0       1.0000

H =

         −0.0013
         −0.0504
          0.0006
          0.0250
```

Time Response Between Two Consecutive Sampling Instants. In a sampled continuous-time system the output is continuous in time. As seen in Chapters 3 and 4, the z transform solution of the discrete-time system equation gives the output response only at the sampling instants. In practice, we may wish to determine the output between two consecutive sampling instants. There are a few methods available for finding the response (output) between two consecutive sampling instants, such as the Laplace transform method and the modified z transform method (see Appendix B). Here we shall show that the state-space method can be easily modified to obtain the output between any two consecutive sampling instants. In what follows, we shall demonstrate this.

Consider the time-invariant continuous-time system defined by

$$\dot{\mathbf{x}} = \mathbf{Ax} + \mathbf{Bu}$$

$$\mathbf{y} = \mathbf{Cx} + \mathbf{Du}$$

Let us assume that the input $\mathbf{u}$ is sampled and fed to a zero-order hold. Then $\mathbf{u}(\tau) = \mathbf{u}(kT)$ for $kT \le \tau < kT + T$. Referring to Equation (5–65), the solution of the state equation starting with the initial state $\mathbf{x}(t_0)$ is

$$\mathbf{x}(t) = e^{\mathbf{A}(t-t_0)} \mathbf{x}(t_0) + \int_{t_0}^{t} e^{\mathbf{A}(t-\tau)} \mathbf{Bu}(\tau) \, d\tau$$

To obtain the response of the sampled system at $t = kT + \Delta T$, where $0 < \Delta T < T$, we put $t = kT + \Delta T$, $t_0 = kT$, and $\mathbf{u}(\tau) = \mathbf{u}(kT)$ in the solution $\mathbf{x}(t)$. Then

$$\mathbf{x}(kT + \Delta T) = e^{\mathbf{A}\Delta T} \mathbf{x}(kT) + \int_{kT}^{kT+\Delta T} e^{\mathbf{A}(kT+\Delta T - \tau)} \mathbf{Bu}(kT) \, d\tau$$

$$= e^{\mathbf{A}\Delta T} \mathbf{x}(kT) + \int_{0}^{\Delta T} e^{\mathbf{A}\lambda} \mathbf{Bu}(kT) \, d\lambda$$

where $\lambda = kT + \Delta T - \tau$. Let us define

$$\mathbf{G}(\Delta T) = e^{\mathbf{A}\Delta T} \tag{5–78}$$

$$\mathbf{H}(\Delta T) = \left(\int_{0}^{\Delta T} e^{\mathbf{A}\lambda} \, d\lambda \right) \mathbf{B} \tag{5–79}$$

Then we obtain

$$\mathbf{x}(kT + \Delta T) = \mathbf{G}(\Delta T)\mathbf{x}(kT) + \mathbf{H}(\Delta T)\mathbf{u}(kT) \tag{5–80}$$

The output $\mathbf{y}(kT + \Delta T)$ can be given by

$$\mathbf{y}(kT + \Delta T) = \mathbf{Cx}(kT + \Delta T) + \mathbf{Du}(kT)$$

$$= \mathbf{CG}(\Delta T)\mathbf{x}(kT) + [\mathbf{CH}(\Delta T) + \mathbf{D}]\mathbf{u}(kT) \tag{5–81}$$

Thus, the values of $\mathbf{x}(kT + \Delta T)$ and $\mathbf{y}(kT + \Delta T)$ between any two consecutive sampling instants can be obtained by computing $\mathbf{G}(\Delta T)$ and $\mathbf{H}(\Delta T)$ for various values of ΔT, where $0 < \Delta T < T$, and substituting the computed values into Equations (5–80) and (5–81). (Such computations can easily be programmed for digital computer calculation.)

Example 5-6

Consider the system discussed in Example 5–5. Obtain the discrete-time state equation and the output equation at $t = kT + \Delta T$. Also, obtain the specific expressions for the state equation and output equation when $T = 1$ sec and $\Delta T = 0.5$ sec.

In Example 5–5, the matrices $\mathbf{G}(T)$ and $\mathbf{H}(T)$ were obtained as follows:

$$\mathbf{G}(T) = \begin{bmatrix} 1 & \frac{1}{2}(1 - e^{-2T}) \\ 0 & e^{-2T} \end{bmatrix}$$

$$\mathbf{H}(T) = \begin{bmatrix} \frac{1}{2}\left(T + \dfrac{e^{-2T} - 1}{2}\right) \\ \frac{1}{2}(1 - e^{-2T}) \end{bmatrix}$$

To obtain the state equation and the output equation at $t = kT + \Delta T$, where $0 < \Delta T < T$, we first convert $\mathbf{G}(T)$ to $\mathbf{G}(\Delta T)$ and $\mathbf{H}(T)$ to $\mathbf{H}(\Delta T)$ and then substitute $\mathbf{G}(\Delta T)$ and $\mathbf{H}(\Delta T)$ into Equations (5–80) and (5–81), as follows:

$$\begin{bmatrix} x_1(kT + \Delta T) \\ x_2(kT + \Delta T) \end{bmatrix} = \begin{bmatrix} 1 & \frac{1}{2}(1 - e^{-2\Delta T}) \\ 0 & e^{-2\Delta T} \end{bmatrix} \begin{bmatrix} x_1(kT) \\ x_2(kT) \end{bmatrix} + \begin{bmatrix} \frac{1}{2}\left(\Delta T + \dfrac{e^{-2\Delta T} - 1}{2}\right) \\ \frac{1}{2}(1 - e^{-2\Delta T}) \end{bmatrix} u(kT)$$

$$y(kT + \Delta T) = \begin{bmatrix} 1 & 0 \end{bmatrix} \begin{bmatrix} 1 & \frac{1}{2}(1 - e^{-2\Delta T}) \\ 0 & e^{-2\Delta T} \end{bmatrix} \begin{bmatrix} x_1(kT) \\ x_2(kT) \end{bmatrix}$$

$$+ \begin{bmatrix} 1 & 0 \end{bmatrix} \begin{bmatrix} \frac{1}{2}\left(\Delta T + \dfrac{e^{-2\Delta T} - 1}{2}\right) \\ \frac{1}{2}(1 - e^{-2\Delta T}) \end{bmatrix} u(kT)$$

For $T = 1$ and $\Delta T = 0.5$ we obtain the state equation and output equation as follows:

$$\begin{bmatrix} x_1(k + 0.5) \\ x_2(k + 0.5) \end{bmatrix} = \begin{bmatrix} 1 & 0.3161 \\ 0 & 0.3679 \end{bmatrix} \begin{bmatrix} x_1(k) \\ x_2(k) \end{bmatrix} + \begin{bmatrix} 0.0920 \\ 0.3161 \end{bmatrix} u(k)$$

$$y(k + 0.5) = \begin{bmatrix} 1 & 0.3161 \end{bmatrix} \begin{bmatrix} x_1(k) \\ x_2(k) \end{bmatrix} + (0.0920)u(k)$$

5-6 LIAPUNOV STABILITY ANALYSIS

Liapunov stability analysis plays an important role in the stability analysis of control systems described by state-space equations. There are two methods of stability analysis due to Liapunov, called the *first method* and the *second method*; both apply to the determination of the stability of dynamic systems described by ordinary differential or difference equations. The first method consists entirely of procedures in which the explicit forms of the solutions of the differential equations or difference equations are used for the analysis. The second method, on the other hand, does not require the solutions of the differential or difference equations. This is the reason the second method is so useful in practice.

Although there are many powerful stability criteria available for control systems, such as the Jury stability criterion and the Routh–Hurwitz stability criteria, they are limited to linear time-invariant systems. The second method of Liapunov, on the other hand, is not limited to linear time-invariant systems: it is applicable to both linear and nonlinear systems, time invariant or time varying. In particular, we find that the second method of Liapunov is indispensable for the stability analysis of nonlinear systems for which exact solutions may be unobtainable. (It is cautioned, however, that although the second method of Liapunov is applicable to any non-linear system, obtaining successful results may not be an easy task. Experience and imagination may be necessary to carry out the stability analysis of most nonlinear systems.)

The second method of Liapunov is also called the *direct method* of Liapunov.

Second Method of Liapunov.   From the classical theory of mechanics, we know that a vibratory system is stable if its total energy is continually decreasing until an equilibrium state is reached.

The second method of Liapunov is based on a generalization of this fact: If the system has an asymptotically stable equilibrium state, then the stored energy of the system displaced within a domain of attraction decays with increasing time until it finally assumes its minimum value at the equilibrium state. For purely mathematical systems, however, there is no simple way of defining an "energy function." To circumvent this difficulty, Liapunov introduced the Liapunov function, a fictitious energy function. This idea is, however, more general than that of energy and is more widely applicable. In fact, any scalar function satisfying the hypotheses of Liapunov's stability theorems (see Theorems 5–1 through 5–6) can serve as a Liapunov function.

Before we discuss the Liapunov function further, it is necessary to define the positive definiteness of scalar functions.

Positive Definiteness of Scalar Functions.   A scalar function $V(\mathbf{x})$ is said to be *positive definite* in a region Ω (which includes the origin of the state space) if $V(\mathbf{x}) > 0$ for all nonzero states $\mathbf{x}$ in the region Ω and if $V(\mathbf{0}) = 0$.

A time-varying function $V(\mathbf{x}, t)$ is said to be positive definite in a region Ω (which includes the origin of the state space) if it is bounded from below by a time-invariant positive definite function, that is, if there exists a positive definite function $V(\mathbf{x})$ such that

$$V(\mathbf{x}, t) > V(\mathbf{x}), \qquad \text{for all } t \geq t_0$$

$$V(\mathbf{0}, t) = 0, \qquad \text{for all } t \geq t_0$$

Negative Definiteness of Scalar Functions.   A scalar function $V(\mathbf{x})$ is said to be *negative definite* if $-V(\mathbf{x})$ is positive definite.

Positive Semidefiniteness of Scalar Functions.   A scalar function $V(\mathbf{x})$ is said to be *positive semidefinite* if it is positive at all states in the region Ω except at the origin and at certain other states, where it is zero.

Negative Semidefiniteness of Scalar Functions.   A scalar function $V(\mathbf{x})$ is said to be *negative semidefinite* if $-V(\mathbf{x})$ is positive semidefinite.

Indefiniteness of Scalar Functions.   A scalar function $V(\mathbf{x})$ is said to be *indefinite* if in the region Ω it assumes both positive and negative values, no matter how small the region Ω is.

Example 5–7

In this example, we give several scalar functions and their classifications according to the foregoing definitions. Here we assume **x** to be a two-dimensional vector.

1. $V(\mathbf{x}) = x_1^2 + x_2^2$ positive definite
2. $V(\mathbf{x}) = x_1^2 + \dfrac{x_2^2}{1 + x_2^2}$ positive definite
3. $V(\mathbf{x}) = (x_1 + x_2)^2$ positive semidefinite
4. $V(\mathbf{x}) = -x_1^2 - (x_1 + x_2)^2$ negative definite
5. $V(\mathbf{x}) = x_1 x_2 + x_2^2$ indefinite

Liapunov Functions.   The Liapunov function, a scalar function, is a positive definite function, and it is continuous together with its first partial derivatives (with respect to its arguments) in the region Ω about the origin and has a time derivative that, when taken along the trajectory, is negative definite (or negative semidefinite). Liapunov functions involve $x_1, x_2, \ldots, x_n$, and possibly t. We denote them by $V(x_1, x_2, \ldots, x_n, t)$, or simply by $V(\mathbf{x}, t)$. If Liapunov functions do not include t explicitly, then we denote them by $V(x_1, x_2, \ldots, x_n)$, or $V(\mathbf{x})$.

Notice that $\dot{V}(\mathbf{x}, t)$ is actually the total derivative of $V(\mathbf{x}, t)$ with respect to t along a solution of the system. Hence, $\dot{V}(\mathbf{x}, t) < 0$ implies that $V(\mathbf{x}, t)$ is a decreasing function of t. A Liapunov function is not unique for a given system. (For this reason, the second method of Liapunov is a more powerful tool than conventional energy considerations. Note that a system whose energy E decreases on the average but not necessarily at each instant is stable, but that E is not a Liapunov function.)

Later in this section we shall show that in the second method of Liapunov the sign behavior of $V(\mathbf{x}, t)$ and that of its time derivative $\dot{V}(\mathbf{x}, t) = dV(\mathbf{x}, t)/dt$ give information about the stability of an equilibrium state without having the solution.

Note that the simplest positive definite function is of a quadratic form:

$$V(\mathbf{x}) = \sum_{i=1}^{n} \sum_{j=1}^{n} q_{ij} x_i x_j, \qquad i, j = 1, 2, \ldots, n$$

In general, Liapunov functions may not be of a simple quadratic form. For any Liapunov function, however, the lowest-degree terms in V must be even. This can be seen as follows. If we define

$$\frac{x_1}{x_n} = \hat{x}_1, \qquad \frac{x_2}{x_n} = \hat{x}_2, \qquad \ldots, \qquad \frac{x_{n-1}}{x_n} = \hat{x}_{n-1}$$

then in the neighborhood of the origin the lowest-degree terms alone will become dominant and we can write $V(\mathbf{x})$ as

$$V(\mathbf{x}) = x_n^p V(\hat{x}_1, \hat{x}_2, \ldots, \hat{x}_{n-1}, 1)$$

If we keep the $\hat{x}_i$'s fixed, $V(\hat{x}_1, \hat{x}_2, \ldots, \hat{x}_{n-1}, 1)$ is a fixed quantity. For p odd, x_n^p can assume both positive and negative values near the origin, which means that $V(\mathbf{x})$ is not positive definite. Hence, p must be even.

In what follows, we give definitions of a system, an equilibrium state, stability, asymptotic stability, and instability.

System.   The system we consider here is defined by

$$\dot{\mathbf{x}} = \mathbf{f}(\mathbf{x}, t) \tag{5-82}$$

where $\mathbf{x}$ is a state vector (an n-vector) and $\mathbf{f}(\mathbf{x}, t)$ is an n-vector whose elements are functions of $x_1, x_2, \ldots, x_n$, and t. (Note that we use the continuous-time system as a model to present basic materials on stability analysis by the second method of Liapunov. Then we extend the results obtained to the discrete-time system.) We assume that the system of Equation (5–82) has a unique solution starting at the given initial condition. We shall denote the solution of Equation (5–82) as $\boldsymbol{\phi}(t; \mathbf{x}_0, t_0)$, where $\mathbf{x} = \mathbf{x}_0$ at $t = t_0$ and t is the observed time. Thus,

$$\boldsymbol{\phi}(t_0; \mathbf{x}_0, t_0) = \mathbf{x}_0$$

Equilibrium State.   In the system of Equation (5–82), a state $\mathbf{x}_e$, where

$$\mathbf{f}(\mathbf{x}_e, t) = \mathbf{0}, \qquad \text{for all } t \tag{5-83}$$

is called an *equilibrium state* of the system. If the system is linear and time invariant, that is, if $\mathbf{f}(\mathbf{x}, t) = \mathbf{A}\mathbf{x}$, then there exists only one equilibrium state if $\mathbf{A}$ is nonsingular, and there exist infinitely many equilibrium states if $\mathbf{A}$ is singular. For nonlinear systems, there may be one or more equilibrium states. These states correspond to the constant solutions of the system ($\mathbf{x} = \mathbf{x}_e$ for all t). Determination of the equilibrium states does not involve the solution of the differential equation of the system, Equation (5–82), but only the solution of Equation (5–83).

Any isolated equilibrium state (that is, where isolated from each other) can be shifted to the origin of the coordinates, or $\mathbf{f}(\mathbf{0}, t) = \mathbf{0}$, by a translation of coordinates. In this section, we shall treat the stability analysis only of such states.

Stability in the Sense of Liapunov.   In the following, we shall denote a spherical region of radius r about an equilibrium state $\mathbf{x}_e$ as

$$\|\mathbf{x} - \mathbf{x}_e\| \leq r$$

where $\|\mathbf{x} - \mathbf{x}_e\|$ is called the *Euclidean norm* and is defined as follows:

$$\|\mathbf{x} - \mathbf{x}_e\| = [(x_1 - x_{1e})^2 + (x_2 - x_{2e})^2 + \cdots + (x_n - x_{ne})^2]^{1/2}$$

Let $S(\delta)$ consist of all points such that

$$\|\mathbf{x}_0 - \mathbf{x}_e\| \leq \delta$$

and let $S(\epsilon)$ consist of all points such that

$$\|\boldsymbol{\phi}(t; \mathbf{x}_0, t_0) - \mathbf{x}_e\| \leq \epsilon, \qquad \text{for all } t \geq t_0$$

An equilibrium state $\mathbf{x}_e$ of the system of Equation (5–82) is said to be *stable in the sense of Liapunov* if, corresponding to each $S(\epsilon)$, there is an $S(\delta)$ such that trajectories starting in $S(\delta)$ do not leave $S(\epsilon)$ as t increases indefinitely. The real number δ depends on ϵ and, in general, also depends on t_0. If δ does not depend on t_0, the equilibrium state is said to be *uniformly stable*.

What we have stated here is that we first choose the region $S(\epsilon)$, and for each $S(\epsilon)$, there must be a region $S(\delta)$ such that trajectories starting within $S(\delta)$ do not leave $S(\epsilon)$ as t increases indefinitely.

Asymptotic Stability.   An equilibrium state $\mathbf{x}_e$ of the system of Equation (5–82) is said to be *asymptotically stable* if it is stable in the sense of Liapunov and if every solution starting within $S(\delta)$ converges, without leaving $S(\epsilon)$, to $\mathbf{x}_e$ as t increases indefinitely.

In practice, asymptotic stability is more important than mere stability. Also, since asymptotic stability is a local concept, simply to establish asymptotic stability does not necessarily mean that the system will operate properly. Some knowledge of the size of the largest region of asymptotic stability is usually necessary. This region is called the *domain of attraction*. It is that part of the state space in which asymptotically stable trajectories originate. In other words, every trajectory originating in the domain of attraction is asymptotically stable.

Asymptotic Stability in the Large.   If asymptotic stability holds for all states (all points in the state space) from which trajectories originate, the equilibrium state is said to be *asymptotically stable in the large*. That is, the equilibrium state $\mathbf{x}_e$ of the system given by Equation (5–82) is said to be asymptotically stable in the large if it is stable and if every solution converges to $\mathbf{x}_e$ as t increases indefinitely. Obviously, a necessary condition for asymptotic stability in the large is that there be only one equilibrium state in the whole state space.

In control engineering problems, asymptotic stability in the large is a desirable feature. If the equilibrium state is not asymptotically stable in the large, then the problem becomes one of determining the largest region of asymptotic stability. This is usually very difficult. For all practical purposes, however, it is sufficient to determine a region of asymptotic stability large enough that no disturbance will exceed it.

Instability.   An equilibrium state $\mathbf{x}_e$ is said to be unstable if for some real number $\epsilon > 0$ and any real number $\delta > 0$, no matter how small, there is always a state $\mathbf{x}_0$ in $S(\delta)$ such that the trajectory starting at this state leaves $S(\epsilon)$.

Graphical Representation of Stability, Asymptotic Stability, and Instability.   A graphical representation of the foregoing definitions will clarify their meanings.

Let us consider the two-dimensional case. Figures 5–2(a), (b), and (c) show equilibrium states and typical trajectories corresponding to stability, asymptotic

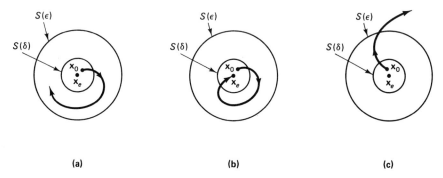

(a)   (b)   (c)

Figure 5–2   (a) Stable equilibrium state and a representative trajectory; (b) asymptotically stable equilibrium state and a representative trajectory; (c) unstable equilibrium state and a representative trajectory.

stability, and instability, respectively. In Figure 5–2(a), (b), or (c), the region $S(\delta)$
bounds the initial state $\mathbf{x}_0$, and the region $S(\epsilon)$ corresponds to the boundary for the
trajectory starting from any initial state $\mathbf{x}_0$ in the region $S(\delta)$.

Note that the foregoing definitions do not specify the exact region of allowable
initial conditions. Thus, the definitions apply to the neighborhood of the equilibrium
state, unless $S(\epsilon)$ corresponds to the entire state plane.

Note that in Figure 5–2(c) the trajectory leaves $S(\epsilon)$ and therefore the equilibrium state is unstable. We cannot, however, say that the trajectory will go to
infinity, since it may approach a limit cycle outside the region $S(\epsilon)$. (If a linear
time-invariant system is unstable, trajectories starting even near the unstable equilibrium state go to infinity. But in the case of nonlinear systems, this is not necessarily
true.)

It is important to point out that the definitions presented here are not the only
ones defining the concepts of the stability of an equilibrium state. In fact, various
other ways to define stability are available in the literature. For example, in conventional control theory, only systems that are asymptotically stable are called stable
systems, and those systems that are stable in the sense of Liapunov but are not
asymptotically stable are called unstable. Another example is BIBO stability. A
linear time-invariant system is called bounded-input–bounded-output stable (BIBO
stable) if the output starting from an arbitrary initial state is bounded when the input
is bounded. It is noted, however, that in this book when the word "stability" is
casually used, it normally means asymptotic stability in the sense of Liapunov.

Liapunov Theorem on Asymptotic Stability.   It can be shown that if a scalar
function $V(\mathbf{x})$, where $\mathbf{x}$ is an n-vector, is positive definite then the states $\mathbf{x}$ that satisfy

$$V(\mathbf{x}) = C$$

where C is a positive constant, lie on a closed hypersurface in the n-dimensional state
space, at least in the neighborhood of the origin. If $V(\mathbf{x}) \to \infty$ as $\|\mathbf{x}\| \to \infty$, then such
closed surfaces extend over the entire state space. The hypersurface $V(\mathbf{x}) = C_1$ lies
entirely inside the hypersurface $V(\mathbf{x}) = C_2$ if $C_1 < C_2$.

For a given system, if a positive definite scalar function $V(\mathbf{x})$ can be found such
that its time derivative taken along a trajectory is always negative, then as time
increases, $V(\mathbf{x})$ takes smaller and smaller values of C. As time increases, $V(\mathbf{x})$ finally
shrinks to zero, and therefore $\mathbf{x}$ also shrinks to zero. This implies the asymptotic
stabiity of the origin of the state space. Liapunov's main stability theorem, which
is a generalization of the foregoing, provides a sufficient condition for asymptotic
stability. This theorem may be stated as follows.

Theorem 5–1.   Suppose a system is described by

$$\dot{\mathbf{x}} = \mathbf{f}(\mathbf{x}, t)$$

where

$$\mathbf{f}(\mathbf{0}, t) = \mathbf{0}, \qquad \text{for all } t$$

If there exists a scalar function $V(\mathbf{x}, t)$ having continuous first partial derivatives and
satisfying the conditions

1. $V(\mathbf{x}, t)$ is positive definite.
2. $\dot{V}(\mathbf{x}, t)$ is negative definite.

then the equilibrium state at the origin is uniformly asymptotically stable.

If, in addition, $V(\mathbf{x}, t) \rightarrow \infty$ as $\|\mathbf{x}\| \rightarrow \infty$, then the equilibrium state at the origin is uniformly asymptotically stable in the large. (For the proof of this theorem, see Problem A–5–18.)

The conditions of this theorem may be modified as follows:

1′. $V(\mathbf{x}, t)$ is positive definite.
2′. $\dot{V}(\mathbf{x}, t)$ is negative semidefinite.
3′. $\dot{V}(\boldsymbol{\phi}(t; \mathbf{x}_0, t_0), t)$ does not vanish identically in $t \geq t_0$ for any t_0 and any $\mathbf{x}_0 \neq \mathbf{0}$, where $\boldsymbol{\phi}(t; \mathbf{x}_0, t_0)$ denotes the solution starting from $\mathbf{x}_0$ at $t = t_0$.

Then the origin of the system is uniformly asymptotically stable in the large.

The equivalence of condition 2 in the theorem and the modified conditions 2′ and 3′ may be seen as follows. If $\dot{V}(\mathbf{x}, t)$ is not negative definite but only negative semidefinite, then the trajectory of the representative point can become tangent to some particular surface $V(\mathbf{x}, t) = C$. Since $\dot{V}(\boldsymbol{\phi}(t; \mathbf{x}_0, t_0), t)$ does not vanish identically in $t \geq t_0$ for any t_0 and any $\mathbf{x}_0 \neq \mathbf{0}$, the representative point cannot remain at the tangent point [the point that corresponds to $\dot{V}(\mathbf{x}, t) = 0$] and therefore must move toward the origin.

Liapunov Theorem on Stability.   To prove stability (but not asymptotic stability) of the origin of the system defined by Equation (5–82), the following theorem may be applied.

Theorem 5–2.   Suppose a system is described by

$$\dot{\mathbf{x}} = \mathbf{f}(\mathbf{x}, t)$$

where $\mathbf{f}(\mathbf{0}, t) = \mathbf{0}$ for all t. If there exists a scalar function $V(\mathbf{x}, t)$ having continuous first partial derivatives and satisfying the conditions

1. $V(\mathbf{x}, t)$ is positive definite.
2. $\dot{V}(\mathbf{x}, t)$ is negative semidefinite.

then the equilibrium state at the origin is uniformly stable.

It should be noted that the negative semidefiniteness of $\dot{V}(\mathbf{x}, t)$ [$\dot{V}(\mathbf{x}, t) \leq 0$ along the trajectories] means that the origin is uniformly stable but not necessarily uniformly asymptotically stable. Hence, in this case the system may exhibit a limit cycle operation.

Instability.   If an equilibrium state $\mathbf{x} = \mathbf{0}$ of a system is unstable, then there exists a scalar function $W(\mathbf{x}, t)$ that determines the instability of the equilibrium state. We shall present a theorem on instability in the following.

Theorem 5–3.   Suppose a system is described by

$$\dot{\mathbf{x}} = \mathbf{f}(\mathbf{x}, t)$$

where

$$\mathbf{f}(\mathbf{0}, t) = \mathbf{0}, \qquad \text{for all } t \geq t_0$$

If there exists a scalar function $W(\mathbf{x}, t)$ having continuous first partial derivatives and satisfying the conditions

1. $W(\mathbf{x}, t)$ is positive definite in some region about the origin.
2. $\dot{W}(\mathbf{x}, t)$ is positive definite in the same region.

then the equilibrium state at the origin is unstable.

Remarks.   A few comments are in order when the Liapunov stability analysis is applied to nonlinear systems.

1. In applying Liapunov stability theorems to a nonlinear system, the stability conditions obtained from a particular Liapunov function are sufficient conditions but are not necessary conditions.
2. A Liapunov function for a particular system is not unique. Therefore, it is important to note that failure in finding a suitable Liapunov function to show stability or asymptotic stability or instability of the equilibrium state under consideration can give no information on stability.
3. Although a particular Liapunov function may prove that the equilibrium state under consideration is stable or asymptotically stable in the region Ω that includes this equilibrium state, it does not necessarily mean that the motions are unstable outside the region Ω.
4. For a stable or asymptotically stable equilibrium state, a Liapunov function with the required properties always exists.

Stability Analysis of Linear Time-Invariant Systems.   There are many approaches to the investigation of the asymptotic stability of linear time-invariant systems. For example, for a continuous-time system described by the equation

$$\dot{\mathbf{x}} = \mathbf{A}\mathbf{x}$$

it can be stated that a necessary and sufficient condition for the asymptotic stability of the origin of the system is that all eigenvalues of $\mathbf{A}$ have negative real parts, or that the zeros of the characteristic polynomial

$$|s\mathbf{I} - \mathbf{A}| = s^n + a_1 s^{n-1} + \cdots + a_{n-1}s + a_n$$

have negative real parts.

Similarly, for a discrete-time system represented by the equation

$$\mathbf{x}(k + 1) = \mathbf{G}\mathbf{x}(k)$$

a necessary and sufficient condition that can be stated for the asymptotic stability of the origin is that all eigenvalues of $\mathbf{G}$ be less than unity in their magnitudes, or that the zeros of the characteristic polynomial

$$|z\mathbf{I} - \mathbf{G}| = z^n + a_1 z^{n-1} + \cdots + a_{n-1} z + a_n$$

lie within the unit circle centered at the origin of the z plane.

Finding the eigenvalues, however, may become difficult in the case of higher-order systems or in the case where some of the coefficients of the characteristic polynomial are nonnumerical. In such a case, the Jury stability criterion or the Routh–Hurwitz stability criteria may be applied. The Liapunov approach, which provides an alternative approach to the stability analysis of linear time-invariant systems, is algebraic and does not require factoring of the characteristic polynomial, as will be seen later. It is important to note that for linear time-invariant systems the second method of Liapunov gives not just sufficient conditions, but the necessary and sufficient conditions for stability or asymptotic stability.

In the following stability analysis of linear time-invariant systems, it is assumed that if an eigenvalue λ_i of matrix $\mathbf{A}$ is a complex quantity then $\mathbf{A}$ must have $\bar{\lambda}_i$, the complex conjugate of λ_i, as its eigenvalue. Thus, any complex eigenvalues of $\mathbf{A}$ will appear as conjugate complex pairs. Also, in the following discussions on stability, we shall use the conjugate transpose expression, rather than the transpose expression, of matrix $\mathbf{A}$, since the elements of matrix $\mathbf{A}$ may include complex conjugates. The conjugate transpose of $\mathbf{A}$ is denoted by $\mathbf{A}^*$. It is a conjugate of the transpose:

$$\mathbf{A}^* = \overline{\mathbf{A}^T}$$

Liapunov Stability Analysis of Linear Time-Invariant Continuous-Time Systems.
Consider the following linear time-invariant system:

$$\dot{\mathbf{x}} = \mathbf{A}\mathbf{x} \qquad (5-84)$$

where $\mathbf{x}$ is a state vector (an n-vector) and $\mathbf{A}$ is an $n \times n$ constant matrix. We assume that $\mathbf{A}$ is nonsingular. Then the only equilibrium state is the origin, $\mathbf{x} = \mathbf{0}$. The stability of the equilibrium state of the linear time-invariant system can be investigated easily with the second method of Liapunov.

For the system defined by Equation (5–84), let us choose as a possible Liapunov function

$$V(\mathbf{x}) = \mathbf{x}^*\mathbf{P}\mathbf{x}$$

where $\mathbf{P}$ is a positive definite Hermitian matrix. (If $\mathbf{x}$ is a real vector, then $\mathbf{P}$ can be chosen to be a positive definite real symmetric matrix.) The time derivative of $V(\mathbf{x})$ along any trajectory is

$$\dot{V}(\mathbf{x}) = \dot{\mathbf{x}}^*\mathbf{P}\mathbf{x} + \mathbf{x}^*\mathbf{P}\dot{\mathbf{x}}$$
$$= (\mathbf{A}\mathbf{x})^*\mathbf{P}\mathbf{x} + \mathbf{x}^*\mathbf{P}\mathbf{A}\mathbf{x}$$
$$= \mathbf{x}^*\mathbf{A}^*\mathbf{P}\mathbf{x} + \mathbf{x}^*\mathbf{P}\mathbf{A}\mathbf{x}$$
$$= \mathbf{x}^*(\mathbf{A}^*\mathbf{P} + \mathbf{P}\mathbf{A})\mathbf{x}$$

Since $V(\mathbf{x})$ was chosen to be positive definite, we require, for asymptotic stability, that $\dot{V}(\mathbf{x})$ be negative definite. Therefore, we require that

$$\dot{V}(\mathbf{x}) = -\mathbf{x}^*\mathbf{Q}\mathbf{x}$$

where

$$\mathbf{Q} = -(\mathbf{A}^*\mathbf{P} + \mathbf{PA}) = \text{positive definite}$$

Hence, for the asymptotic stability of the system of Equation (5–84), it is sufficient that $\mathbf{Q}$ be positive definite.

For a test of positive definiteness of an $n \times n$ matrix, we apply Sylvester's criterion, which states that a necessary and sufficient condition for the matrix to be positive definite is that the determinants of all the successive principal minors of the matrix be positive. Consider, for example, the following $n \times n$ Hermitian matrix $\mathbf{P}$ (if the elements of $\mathbf{P}$ are all real, then the Hermitian matrix becomes a real symmetric matrix):

$$\mathbf{P} = \begin{bmatrix} p_{11} & p_{12} & \cdots & p_{1n} \\ \bar{p}_{12} & p_{22} & \cdots & p_{2n} \\ \vdots & \vdots & & \vdots \\ \bar{p}_{1n} & \bar{p}_{2n} & \cdots & p_{nn} \end{bmatrix}$$

where $\bar{p}_{ij}$ denotes the complex conjugate of p_{ij}. The matrix $\mathbf{P}$ is positive definite if all the successive principal minors are positive, that is, if

$$p_{11} > 0, \qquad \begin{vmatrix} p_{11} & p_{12} \\ \bar{p}_{12} & p_{22} \end{vmatrix} > 0, \qquad \cdots, \qquad \begin{vmatrix} p_{11} & p_{12} & \cdots & p_{1n} \\ \bar{p}_{12} & p_{22} & \cdots & p_{2n} \\ \vdots & \vdots & & \vdots \\ \bar{p}_{1n} & \bar{p}_{2n} & \cdots & p_{nn} \end{vmatrix} > 0$$

Instead of first specifying a positive definite matrix $\mathbf{P}$ and examining whether or not $\mathbf{Q}$ is positive definite, it is convenient to specify a positive definite matrix $\mathbf{Q}$ first and then examine whether or not $\mathbf{P}$ determined from

$$\mathbf{A}^*\mathbf{P} + \mathbf{PA} = -\mathbf{Q}$$

is positive definite. Note that positive definite $\mathbf{P}$ is a necessary and sufficient condition. We shall summarize what we have just stated in the form of a theorem.

Theorem 5–4.   Consider the system described by

$$\dot{\mathbf{x}} = \mathbf{Ax}$$

where $\mathbf{x}$ is a state vector (an n-vector) and $\mathbf{A}$ is an $n \times n$ constant nonsingular matrix. A necessary and sufficient condition for the equilibrium state $\mathbf{x} = \mathbf{0}$ to be asymptotically stable in the large is that, given any positive definite Hermitian (or any positive definite real symmetric) matrix $\mathbf{Q}$, there exists a positive definite Hermitian (or a positive definite real symmetric) matrix $\mathbf{P}$ such that

$$\mathbf{A}^*\mathbf{P} + \mathbf{PA} = -\mathbf{Q} \tag{5–85}$$

The scalar function $\mathbf{x}^*\mathbf{Px}$ is a Liapunov function for this system. [Note that in the linear system considered, if the equilibrium state (the origin) is asymptotically stable, then it is asymptotically stable in the large.]

Remarks.   In applying Theorem 5–4 to the stability analysis of linear time-invariant continuous-time systems, several important remarks may be made.

1. If $\dot{V}(x) = -x*Qx$ does not vanish identically along any trajectory, then Q may be chosen to be positive semidefinite.

2. If an arbitrary positive definite matrix is chosen for Q [or an arbitrary positive-semidefinite matrix if $\dot{V}(x)$ does not vanish identically along any trajectory] and the matrix equation

$$A*P + PA = -Q$$

is solved to determine P, then the positive definiteness of P is a necessary and sufficient condition for the asymptotic stability of the equilibrium state x = 0.

3. The final result does not depend on the particular Q matrix chosen so long as Q is positive definite (or positive semidefinite, as the case may be).

4. To determine the elements of the P matrix, we equate the matrices A*P + PA and −Q element by element. This results in $n(n + 1)/2$ linear equations for the determination of the elements $p_{ij} = \bar{p}_{ji}$ of P. If we denote the eigenvalues of A by $\lambda_1, \lambda_2, \ldots, \lambda_n$, each repeated a number of times equal to its multiplicity as a root of the characteristic equation, and if for every sum of two roots

$$\lambda_j + \lambda_k \neq 0$$

then the elements of P are uniquely determined. (Note that for a stable matrix A the sum $\lambda_j + \lambda_k$ is always nonzero.)

5. In determining whether or not there exists a positive definite Hermitian or positive definite real symmetric matrix P, it is convenient to choose Q = I, where I is the identity matrix. Then the elements of P are determined from

$$A*P + PA = -I$$

and the matrix P is tested for positive definiteness.

Example 5–8

Determine the stability of the equilibrium state of the following system:

$$\dot{x}_1 = -x_1 - 2x_2$$
$$\dot{x}_2 = x_1 - 4x_2$$

The system has only one equilibrium state at the origin. By choosing Q = I and substituting I into Equation (5–85), we have

$$A*P + PA = -I$$

Noting that A is a real matrix, P must be a real symmetric matrix. This last equation may then be written as follows:

$$\begin{bmatrix} -1 & 1 \\ -2 & -4 \end{bmatrix}\begin{bmatrix} p_{11} & p_{12} \\ p_{12} & p_{22} \end{bmatrix} + \begin{bmatrix} p_{11} & p_{12} \\ p_{12} & p_{22} \end{bmatrix}\begin{bmatrix} -1 & -2 \\ 1 & -4 \end{bmatrix} = -\begin{bmatrix} 1 & 0 \\ 0 & 1 \end{bmatrix} \tag{5-86}$$

where we have noted that $p_{21} = p_{12}$ and made the appropriate substitution. If the matrix P turns out to be positive definite, then x*Px is a Liapunov function and the origin is asymptotically stable.

Equation (5–86) yields the following three equations:

$$-2p_{11} + 2p_{12} = -1$$

$$-2p_{11} - 5p_{12} + p_{22} = 0$$

$$-4p_{12} - 8p_{22} = -1$$

Solving for the p's, we obtain

$$p_{11} = \tfrac{23}{60}, \qquad p_{12} = -\tfrac{7}{60}, \qquad p_{22} = \tfrac{11}{60}$$

Hence,

$$\mathbf{P} = \begin{bmatrix} \frac{23}{60} & -\frac{7}{60} \\ -\frac{7}{60} & \frac{11}{60} \end{bmatrix}$$

By Sylvester's criterion, this matrix is positive definite. Hence, we conclude that the origin of the system is asymptotically stable in the large.

It is noted that a Liapunov function for this system is

$$V(\mathbf{x}) = \mathbf{x}^*\mathbf{P}\mathbf{x} = \begin{bmatrix} x_1 & x_2 \end{bmatrix} \begin{bmatrix} \frac{23}{60} & -\frac{7}{60} \\ -\frac{7}{60} & \frac{11}{60} \end{bmatrix} \begin{bmatrix} x_1 \\ x_2 \end{bmatrix}$$

$$= \tfrac{1}{60}(23x_1^2 - 14x_1 x_2 + 11x_2^2)$$

and $\dot{V}(\mathbf{x})$ is given by

$$\dot{V}(\mathbf{x}) = -x_1^2 - x_2^2$$

Liapunov Stability Analysis of Discrete-Time Systems. In what follows, we extend the Liapunov stability analysis presented thus far in this section to discrete-time systems. As in the case of continuous-time systems, asymptotic stability is the most important concept in the stability of equilibrium states of discrete-time systems.

We shall now present a stability theorem for linear or nonlinear time-invariant discrete-time systems based on the second method of Liapunov. It is noted that for discrete-time systems, instead of $\dot{V}(\mathbf{x})$, we use the forward difference $V(\mathbf{x}(k + 1)T) - V(\mathbf{x}(kT))$, or

$$\Delta V(\mathbf{x}(kT)) = V(\mathbf{x}(k + 1)T) - V(\mathbf{x}(kT)) \tag{5–87}$$

Theorem 5–5. Consider the discrete-time system

$$\mathbf{x}((k + 1)T) = \mathbf{f}(\mathbf{x}(kT)) \tag{5–88}$$

where

$$\mathbf{x} = n\text{-vector}$$

$$\mathbf{f}(\mathbf{x}) = n\text{-vector with property that } \mathbf{f}(\mathbf{0}) = \mathbf{0}$$

$$T = \text{sampling period}$$

Suppose there exists a scalar function $V(\mathbf{x})$ continuous in $\mathbf{x}$ such that

1. $V(\mathbf{x}) > 0$ for $\mathbf{x} \neq \mathbf{0}$.
2. $\Delta V(\mathbf{x}) < 0$ for $\mathbf{x} \neq \mathbf{0}$, where

$$\Delta V(\mathbf{x}(kT)) = V(\mathbf{x}(k + 1)T) - V(\mathbf{x}(kT)) = V(\mathbf{f}(\mathbf{x}(kT))) - V(\mathbf{x}(kT))$$

3. $V(0) = 0$.

4. $V(\mathbf{x}) \to \infty$ as $\|\mathbf{x}\| \to \infty$.

Then the equilibrium state $\mathbf{x} = \mathbf{0}$ is asymptotically stable in the large and $V(\mathbf{x})$ is a Liapunov function.

Note that in this theorem condition 2 may be replaced by

2′. $\Delta V(\mathbf{x}) \leq 0$ for all $\mathbf{x}$, and $\Delta V(\mathbf{x})$ does not vanish identically for any solution sequence $\{\mathbf{x}(kT)\}$ satisfying Equation (5–88).

This means that $\Delta V(\mathbf{x})$ need not be negative definite if it does not vanish identically on any solution sequence of the difference equation.

Liapunov Stability Analysis of Linear Time-Invariant Discrete-Time Systems. Consider the discrete-time system described by

$$\mathbf{x}(k + 1) = \mathbf{G}\mathbf{x}(k) \tag{5–89}$$

where $\mathbf{x}$ is a state vector (an n-vector) and $\mathbf{G}$ is an $n \times n$ constant nonsingular matrix. The origin $\mathbf{x} = \mathbf{0}$ is the equilibrium state. We shall investigate the stability of this state by use of the second method of Liapunov.

Let us choose as a possible Liapunov function

$$V(\mathbf{x}(k)) = \mathbf{x}^*(k)\mathbf{P}\mathbf{x}(k)$$

where $\mathbf{P}$ is a positive definite Hermitian (or a positive definite real symmetric) matrix. Then

$$\Delta V(\mathbf{x}(k)) = V(\mathbf{x}(k + 1)) - V(\mathbf{x}(k))$$
$$= \mathbf{x}^*(k + 1)\mathbf{P}\mathbf{x}(k + 1) - \mathbf{x}^*(k)\mathbf{P}\mathbf{x}(k)$$
$$= [\mathbf{G}\mathbf{x}(k)]^*\mathbf{P}[\mathbf{G}\mathbf{x}(k)] - \mathbf{x}^*(k)\mathbf{P}\mathbf{x}(k)$$
$$= \mathbf{x}^*(k)\mathbf{G}^*\mathbf{P}\mathbf{G}\mathbf{x}(k) - \mathbf{x}^*(k)\mathbf{P}\mathbf{x}(k)$$
$$= \mathbf{x}^*(k)(\mathbf{G}^*\mathbf{P}\mathbf{G} - \mathbf{P})\mathbf{x}(k)$$

Since $V(\mathbf{x}(k))$ is chosen to be positive definite, we require, for asymptotic stability, that $\Delta V(\mathbf{x}(k))$ be negative definite. Therefore,

$$\Delta V(\mathbf{x}(k)) = -\mathbf{x}^*(k)\mathbf{Q}\mathbf{x}(k)$$

where

$$\mathbf{Q} = -(\mathbf{G}^*\mathbf{P}\mathbf{G} - \mathbf{P}) = \text{positive definite}$$

Hence, for the asymptotic stability of the discrete-time system of Equation (5–89), it is sufficient that $\mathbf{Q}$ be positive definite.

As in the case of linear continuous-time systems, it is convenient to specify first a positive definite Hermitian (or a positive definite real symmetric) matrix $\mathbf{Q}$ and then to see whether or not the $\mathbf{P}$ matrix determined from

$$\mathbf{G}^*\mathbf{P}\mathbf{G} - \mathbf{P} = -\mathbf{Q}$$

is positive definite. Note that a positive definite $\mathbf{P}$ is a necessary and sufficient condition. We shall summarize in a theorem what we have stated here.

Theorem 5–6.   Consider the discrete-time system

$$\mathbf{x}(k + 1) = \mathbf{Gx}(k)$$

where $\mathbf{x}$ is a state vector (an n-vector) and $\mathbf{G}$ is an $n \times n$ constant nonsingular matrix. A necessary and sufficient condition for the equilibrium state $\mathbf{x} = \mathbf{0}$ to be asymptotically stable in the large is that, given any positive-definite Hermitian (or any positive definite real symmetric) matrix $\mathbf{Q}$, there exists a positive definite Hermitian (or a positive definite real symmetric) matrix $\mathbf{P}$ such that

$$\mathbf{G*PG} - \mathbf{P} = -\mathbf{Q} \qquad\qquad (5\text{–}90)$$

The scalar function $\mathbf{x*Px}$ is a Liapunov function for this system.

If $\Delta V(\mathbf{x}(k)) = -\mathbf{x*}(k)\mathbf{Qx}(k)$ does not vanish identically along any solution series, then $\mathbf{Q}$ may be chosen to be positive semidefinite.

Stability of a Discrete-Time System Obtained by Discretizing a Continuous-Time System.   If the system is described in terms of state-space equations, the asymptotic stability of an equilibrium state of a discrete-time system obtained by discretizing a continuous-time system is equivalent to that of the original continuous-time system.

Consider a continuous-time system

$$\dot{\mathbf{x}} = \mathbf{Ax}$$

and the corresponding discrete-time system

$$\mathbf{x}((k + 1)T) = \mathbf{Gx}(kT)$$

where

$$\mathbf{G} = e^{\mathbf{A}T}$$

If the continuous-time system is asymptotically stable, that is, if all the eigenvalues of the matrix $\mathbf{A}$ have negative real parts, then

$$\|\mathbf{G}^n\| \to 0, \qquad \text{as } n \to \infty$$

and the discretized system is also asymptotically stable. This is because if the λ_i's are the eigenvalues of $\mathbf{A}$ then the $e^{\lambda_i T}$'s are the eigenvalues of $\mathbf{G}$. (Note that $|e^{\lambda_i T}| < 1$ if $\lambda_i T$ is negative.)

It should be noted here that, if a continuous-time system having complex poles is discretized, then in certain exceptional cases hidden instability may occur, depending on the choice of the sampling period T. That is, in some cases where a continuous-time system is not asymptotically stable, the equivalent discretized system may seem to be asymptotically stable if we look at the values of the output only at the sampling instants. This phenomenon occurs only at certain values of the sampling period T. If the value of T is varied, then such hidden instability shows up as explicit instability. See Problem A–5–15.

Contraction.   A norm of $\mathbf{x}$ denoted by $\|\mathbf{x}\|$ may be thought of as a measure of the length of the vector. There are several different definitions of a norm. Any norm, however, has the following properties:

$$\|\mathbf{x}\| = 0, \qquad\qquad \text{for } \mathbf{x} = \mathbf{0}$$

$$\|\mathbf{x}\| > 0, \qquad\qquad \text{for } \mathbf{x} \neq \mathbf{0}$$

$$\|\mathbf{x} + \mathbf{y}\| \leq \|\mathbf{x}\| + \|\mathbf{y}\|, \qquad \text{for all } \mathbf{x} \text{ and } \mathbf{y}$$

$$\|k\mathbf{x}\| = |k|\,\|\mathbf{x}\|, \qquad\qquad \text{for all } \mathbf{x} \text{ and real constant } k$$

A function $\mathbf{f}(\mathbf{x})$ is said to be a contraction if $\mathbf{f}(\mathbf{0}) = \mathbf{0}$ and

$$\|\mathbf{f}(\mathbf{x})\| < \|\mathbf{x}\|$$

for some set of values of $\mathbf{x} \neq \mathbf{0}$ and some norm.

For discrete-time systems a norm $\|\mathbf{x}\|$ may be used as a Liapunov function. Consider the following discrete-time system:

$$\mathbf{x}(k + 1) = \mathbf{f}(\mathbf{x}(k)), \qquad \mathbf{f}(\mathbf{0}) = \mathbf{0} \tag{5-91}$$

where $\mathbf{x}$ is an n-vector and $\mathbf{f}(\mathbf{x})$ is also an n-vector. Assume that $\mathbf{f}(\mathbf{x})$ is a contraction for all $\mathbf{x}$ and some norm. Then the origin of the system of Equation (5-91) is asymptotically stable in the large, and one of its Liapunov functions is

$$V(\mathbf{x}) = \|\mathbf{x}\|$$

This can be seen as follows. Since $V(\mathbf{x}) = \|\mathbf{x}\|$ is positive definite and

$$\Delta V(\mathbf{x}(k)) = V(\mathbf{f}(\mathbf{x}(k))) - V(\mathbf{x}(k)) = \|\mathbf{f}(\mathbf{x})\| - \|\mathbf{x}\|$$

is negative definite because $\mathbf{f}(\mathbf{x})$ is a contraction for all $\mathbf{x}$, we find $V(\mathbf{x}) = \|\mathbf{x}\|$ is a Liapunov function, and by Theorem 5-5 the origin of the system is asymptotically stable in the large. (See Problem A-5-20.)

Example 5-9

Consider the following system:

$$\begin{bmatrix} x_1(k + 1) \\ x_2(k + 1) \end{bmatrix} = \begin{bmatrix} 0 & 1 \\ -0.5 & -1 \end{bmatrix} \begin{bmatrix} x_1(k) \\ x_2(k) \end{bmatrix}$$

Determine the stability of the origin of the system.

Let us choose $\mathbf{Q}$ to be $\mathbf{I}$. Then, referring to Equation (5-90), the Liapunov stability equation becomes

$$\begin{bmatrix} 0 & -0.5 \\ 1 & -1 \end{bmatrix} \begin{bmatrix} p_{11} & p_{12} \\ p_{12} & p_{22} \end{bmatrix} \begin{bmatrix} 0 & 1 \\ -0.5 & -1 \end{bmatrix} - \begin{bmatrix} p_{11} & p_{12} \\ p_{12} & p_{22} \end{bmatrix} = -\begin{bmatrix} 1 & 0 \\ 0 & 1 \end{bmatrix} \tag{5-92}$$

If matrix $\mathbf{P}$ is found to be positive definite, then the origin $\mathbf{x} = \mathbf{0}$ is asymptotically stable in the large.

From Equation (5-92) we obtain the following three equations:

$$0.25 p_{22} - p_{11} = -1$$

$$0.5(-p_{12} + p_{22}) - p_{12} = 0$$

$$p_{11} - 2p_{12} = -1$$

from which we get

$$p_{11} = \tfrac{11}{5}, \qquad p_{12} = \tfrac{8}{5}, \qquad p_{22} = \tfrac{24}{5}$$

Consequently,

$$\mathbf{P} = \begin{bmatrix} \frac{11}{5} & \frac{8}{5} \\ \frac{8}{5} & \frac{24}{5} \end{bmatrix}$$

By applying Sylvester's criterion for the positive definiteness of matrix $\mathbf{P}$, we find $\mathbf{P}$ is positive definite. Hence, the equilibrium state, the origin $\mathbf{x} = \mathbf{0}$, is asymptotically stable in the large.

Note that instead of choosing $\mathbf{Q}$ to be $\mathbf{I}$ we could choose $\mathbf{Q}$ to be a positive-semidefinite matrix, such as

$$\mathbf{Q} = \begin{bmatrix} 0 & 0 \\ 0 & 1 \end{bmatrix}$$

as long as $\Delta V(\mathbf{x}) = -\mathbf{x}^*(k)\mathbf{Q}\mathbf{x}(k)$ does not vanish identically along any solution series. For the positive semidefinite matrix $\mathbf{Q}$ just given, we have

$$\Delta V(\mathbf{x}) = -x_2^2(k)$$

For the present system, $x_2(k)$ identically zero implies that $x_1(k)$ is identically zero. Hence, $\Delta V(\mathbf{x})$ does not vanish identically along any solution series, except at the origin. Therefore, we can choose this positive semidefinite matrix $\mathbf{Q}$ for the determination of matrix $\mathbf{P}$ of the Liapunov stability equation. The Liapunov stability equation in this case becomes

$$\begin{bmatrix} 0 & -0.5 \\ 1 & -1 \end{bmatrix}\begin{bmatrix} p_{11} & p_{12} \\ p_{12} & p_{22} \end{bmatrix}\begin{bmatrix} 0 & 1 \\ -0.5 & -1 \end{bmatrix} - \begin{bmatrix} p_{11} & p_{12} \\ p_{12} & p_{22} \end{bmatrix} = -\begin{bmatrix} 0 & 0 \\ 0 & 1 \end{bmatrix}$$

By solving this last equation, we obtain

$$\mathbf{P} = \begin{bmatrix} \frac{3}{5} & \frac{4}{5} \\ \frac{4}{5} & \frac{12}{5} \end{bmatrix}$$

By applying the Sylvester criterion, we find $\mathbf{P}$ to be positive definite. Hence, we get the same conclusion as before: The origin is asymptotically stable in the large.

EXAMPLE PROBLEMS AND SOLUTIONS

Problem A–5–1

(Direct programming method)   Consider the discrete-time system defined by

$$\frac{Y(z)}{U(z)} = \frac{b_0 z^n + b_1 z^{n-1} + \cdots + b_n}{z^n + a_1 z^{n-1} + \cdots + a_n}$$

Show that a state-space representation of this system may be given by

$$\begin{bmatrix} x_1(k+1) \\ x_2(k+1) \\ \vdots \\ x_{n-1}(k+1) \\ x_n(k+1) \end{bmatrix} = \begin{bmatrix} 0 & 1 & 0 & \cdots & 0 \\ 0 & 0 & 1 & \cdots & 0 \\ \vdots & \vdots & \vdots & & \vdots \\ 0 & 0 & 0 & \cdots & 1 \\ -a_n & -a_{n-1} & -a_{n-2} & \cdots & -a_1 \end{bmatrix}\begin{bmatrix} x_1(k) \\ x_2(k) \\ \vdots \\ x_{n-1}(k) \\ x_n(k) \end{bmatrix} + \begin{bmatrix} 0 \\ 0 \\ \vdots \\ 0 \\ 1 \end{bmatrix}u(k) \qquad (5\text{–}93)$$

$$y(k) = [b_n - a_n b_0 \,\vdots\, b_{n-1} - a_{n-1} b_0 \,\vdots\, \cdots \,\vdots\, b_1 - a_1 b_0]\begin{bmatrix} x_1(k) \\ x_2(k) \\ \vdots \\ x_n(k) \end{bmatrix} + b_0 u(k) \qquad (5\text{–}94)$$

Solution   The given system can be modified to

$$\frac{Y(z)}{U(z)} = \frac{b_0 + b_1 z^{-1} + b_2 z^{-2} + \cdots + b_n z^{-n}}{1 + a_1 z^{-1} + a_2 z^{-2} + \cdots + a_n z^{-n}}$$

$$= b_0 + \frac{(b_1 - a_1 b_0)z^{-1} + (b_2 - a_2 b_0)z^{-2} + \cdots + (b_n - a_n b_0)z^{-n}}{1 + a_1 z^{-1} + a_2 z^{-2} + \cdots + a_n z^{-n}}$$

This last equation can be written as follows:

$$Y(z) = b_0 U(z)$$

$$+ \frac{(b_1 - a_1 b_0)z^{-1} + (b_2 - a_2 b_0)z^{-2} + \cdots + (b_n - a_n b_0)z^{-n}}{1 + a_1 z^{-1} + a_2 z^{-2} + \cdots + a_n z^{-n}} U(z) \tag{5-95}$$

Let us define

$$\tilde{Y}(z) = \frac{(b_1 - a_1 b_0)z^{-1} + (b_2 - a_2 b_0)z^{-2} + \cdots + (b_n - a_n b_0)z^{-n}}{1 + a_1 z^{-1} + a_2 z^{-2} + \cdots + a_n z^{-n}} U(z) \tag{5-96}$$

Then Equation (5–95) becomes

$$Y(z) = b_0 U(z) + \tilde{Y}(z) \tag{5-97}$$

Let us rewrite Equation (5–96) in the following form:

$$\frac{\tilde{Y}(z)}{(b_1 - a_1 b_0)z^{-1} + (b_2 - a_2 b_0)z^{-2} + \cdots + (b_n - a_n b_0)z^{-n}}$$

$$= \frac{U(z)}{1 + a_1 z^{-1} + a_2 z^{-2} + \cdots + a_n z^{-n}} = Q(z)$$

From this last equation the following two equations may be obtained:

$$Q(z) = -a_1 z^{-1} Q(z) - a_2 z^{-2} Q(z) - \cdots - a_n z^{-n} Q(z) + U(z) \tag{5-98}$$

and

$$\tilde{Y}(z) = (b_1 - a_1 b_0)z^{-1} Q(z) + (b_2 - a_2 b_0)z^{-2} Q(z) + \cdots$$

$$+ (b_n - a_n b_0)z^{-n} Q(z) \tag{5-99}$$

Now we define the state variables as follows:

$$X_1(z) = z^{-n} Q(z)$$

$$X_2(z) = z^{-n+1} Q(z)$$

$$\vdots \tag{5-100}$$

$$X_{n-1}(z) = z^{-2} Q(z)$$

$$X_n(z) = z^{-1} Q(z)$$

Then clearly we have

$$z X_1(z) = X_2(z)$$

$$z X_2(z) = X_3(z)$$

$$\vdots$$

$$z X_{n-1}(z) = X_n(z)$$

In terms of difference equations, the preceding $n - 1$ equations become

$$x_1(k + 1) = x_2(k)$$
$$x_2(k + 1) = x_3(k)$$
$$\vdots$$
$$x_{n-1}(k + 1) = x_n(k)$$

(5–101)

By substituting Equation (5–100) into Equation (5–98), we obtain

$$zX_n(z) = -a_1 X_n(z) - a_2 X_{n-1}(z) - \cdots - a_n X_1(z) + U(z)$$

which may be transformed into a difference equation:

$$x_n(k + 1) = -a_n x_1(k) - a_{n-1} x_2(k) - \cdots - a_1 x_n(k) + u(k)$$

(5–102)

Also, Equation (5–99) can be rewritten as follows:

$$\tilde{Y}(z) = (b_1 - a_1 b_0)X_n(z) + (b_2 - a_2 b_0)X_{n-1}(z) + \cdots + (b_n - a_n b_0)X_1(z)$$

By use of this last equation, Equation (5–97) can be written in the form

$$y(k) = (b_n - a_n b_0)x_1(k) + (b_{n-1} - a_{n-1} b_0)x_2(k)$$
$$+ \cdots + (b_1 - a_1 b_0)x_n(k) + b_0 u(k)$$

(5–103)

Combining Equations (5–101) and (5–102) results in the state equation given by Equation (5–93). The output equation, Equation (5–103), can be rewritten in the form given by Equation (5–94).

Problem A–5–2

(Nested programming method)   Consider the pulse transfer function system defined by

$$G(z) = \frac{Y(z)}{U(z)} = \frac{b_0 + b_1 z^{-1} + \cdots + b_n z^{-n}}{1 + a_1 z^{-1} + \cdots + a_n z^{-n}}$$

Show that a state-space representation of this system may be given as follows:

$$
\begin{bmatrix}
x_1(k + 1) \\
x_2(k + 1) \\
\vdots \\
x_{n-1}(k + 1) \\
x_n(k + 1)
\end{bmatrix}
=
\begin{bmatrix}
0 & 0 & \cdots & 0 & 0 & -a_n \\
1 & 0 & \cdots & 0 & 0 & -a_{n-1} \\
\vdots & \vdots & & \vdots & \vdots & \vdots \\
0 & 0 & \cdots & 1 & 0 & -a_2 \\
0 & 0 & \cdots & 0 & 1 & -a_1
\end{bmatrix}
\begin{bmatrix}
x_1(k) \\
x_2(k) \\
\vdots \\
x_{n-1}(k) \\
x_n(k)
\end{bmatrix}
$$

$$
+
\begin{bmatrix}
b_n - a_n b_0 \\
b_{n-1} - a_{n-1} b_0 \\
\vdots \\
b_2 - a_2 b_0 \\
b_1 - a_1 b_0
\end{bmatrix}
u(k)
$$

(5–104)

$$
y(k) = \begin{bmatrix} 0 & 0 & \cdots & 0 & 1 \end{bmatrix}
\begin{bmatrix}
x_1(k) \\
x_2(k) \\
\vdots \\
x_{n-1}(k) \\
x_n(k)
\end{bmatrix}
+ b_0 u(k)
$$

(5–105)

Solution   Rewrite the pulse transfer function as follows:

$$Y(z) - b_0 U(z) + z^{-1}[a_1 Y(z) - b_1 U(z)]$$
$$+ z^{-2}[a_2 Y(z) - b_2 U(z)] + \cdots + z^{-n}[a_n Y(z) - b_n U(z)] = 0$$

or

$$Y(z) = b_0 U(z) + z^{-1}\big(b_1 U(z) - a_1 Y(z) + z^{-1}\{b_2 U(z) - a_2 Y(z)$$
$$+ z^{-1}[b_3 U(z) - a_3 Y(z) + \cdots]\}\big) \qquad (5\text{--}106)$$

Now define the state variables as follows:

$$X_n(z) = z^{-1}[b_1 U(z) - a_1 Y(z) + X_{n-1}(z)]$$
$$X_{n-1}(z) = z^{-1}[b_2 U(z) - a_2 Y(z) + X_{n-2}(z)]$$
$$\vdots \qquad\qquad\qquad (5\text{--}107)$$
$$X_2(z) = z^{-1}[b_{n-1} U(z) - a_{n-1} Y(z) + X_1(z)]$$
$$X_1(z) = z^{-1}[b_n U(z) - a_n Y(z)]$$

Then Equation (5–106) can be written in the form

$$Y(z) = b_0 U(z) + X_n(z) \qquad (5\text{--}108)$$

By substituting Equation (5–108) into Equation (5–107) and multiplying both sides of the equations by z, we obtain

$$zX_n(z) = X_{n-1}(z) - a_1 X_n(z) + (b_1 - a_1 b_0)U(z)$$
$$zX_{n-1}(z) = X_{n-2}(z) - a_2 X_n(z) + (b_2 - a_2 b_0)U(z)$$
$$\vdots$$
$$zX_2(z) = X_1(z) - a_{n-1} X_n(z) + (b_{n-1} - a_{n-1} b_0)U(z)$$
$$zX_1(z) = -a_n X_n(z) + (b_n - a_n b_0)U(z)$$

Taking the inverse z transforms of the preceding n equations and writing the resulting equations in the reverse order, we obtain

$$x_1(k + 1) = -a_n x_n(k) + (b_n - a_n b_0)u(k)$$
$$x_2(k + 1) = x_1(k) - a_{n-1} x_n(k) + (b_{n-1} - a_{n-1} b_0)u(k)$$
$$\vdots$$
$$x_{n-1}(k + 1) = x_{n-2}(k) - a_2 x_n(k) + (b_2 - a_2 b_0)u(k)$$
$$x_n(k + 1) = x_{n-1}(k) - a_1 x_n(k) + (b_1 - a_1 b_0)u(k)$$

Also, the inverse z transform of Equation (5–108) yields

$$y(k) = x_n(k) + b_0 u(k)$$

Rewriting the state equation and the output equation in the standard vector-matrix form gives Equations (5–104) and (5–105), respectively.

Problem A–5–3

(Partial-fraction-expansion programming method)   Consider the pulse transfer function system given by

$$\frac{Y(z)}{U(z)} = \frac{b_0 z^n + b_1 z^{n-1} + \cdots + b_n}{z^n + a_1 z^{n-1} + \cdots + a_n}$$

Show that the state equation and output equation can be given in the following diagonal canonical form if all poles are distinct.

$$\begin{bmatrix} x_1(k+1) \\ x_2(k+1) \\ \vdots \\ x_n(k+1) \end{bmatrix} = \begin{bmatrix} p_1 & 0 & \cdots & 0 \\ 0 & p_2 & \cdots & 0 \\ \vdots & \vdots & & \vdots \\ 0 & 0 & \cdots & p_n \end{bmatrix} \begin{bmatrix} x_1(k) \\ x_2(k) \\ \vdots \\ x_n(k) \end{bmatrix} + \begin{bmatrix} 1 \\ 1 \\ \vdots \\ 1 \end{bmatrix} u(k) \qquad (5\text{--}109)$$

and

$$y(k) = \begin{bmatrix} c_1 & c_2 & \cdots & c_n \end{bmatrix} \begin{bmatrix} x_1(k) \\ x_2(k) \\ \vdots \\ x_n(k) \end{bmatrix} + b_0 u(k) \qquad (5\text{--}110)$$

Solution   The system pulse transfer function can be modified as follows:

$$\frac{Y(z)}{U(z)} = \frac{b_0 z^n + b_1 z^{n-1} + \cdots + b_n}{z^n + a_1 z^{n-1} + \cdots + a_n}$$

$$= b_0 + \frac{(b_1 - a_1 b_0)z^{n-1} + (b_2 - a_2 b_0)z^{n-2} + \cdots + (b_n - a_n b_0)}{(z - p_1)(z - p_2)\cdots(z - p_n)} \qquad (5\text{--}111)$$

Since all poles of the pulse transfer function $Y(z)/U(z)$ are distinct, $Y(z)/U(z)$ can be expanded into the following form:

$$\frac{Y(z)}{U(z)} = b_0 + \frac{c_1}{z - p_1} + \frac{c_2}{z - p_2} + \cdots + \frac{c_n}{z - p_n} \qquad (5\text{--}112)$$

where

$$c_i = \lim_{z \to p_i} \left[\frac{Y(z)}{U(z)}(z - p_i) \right]$$

Equation (5–112) can be written in the form

$$Y(z) = b_0 U(z) + \frac{c_1}{z - p_1} U(z) + \frac{c_2}{z - p_2} U(z) + \cdots + \frac{c_n}{z - p_n} U(z) \qquad (5\text{--}113)$$

Let us define the state variables as follows:

$$X_1(z) = \frac{1}{z - p_1} U(z)$$

$$X_2(z) = \frac{1}{z - p_2} U(z)$$

$$\vdots$$

$$X_n(z) = \frac{1}{z - p_n} U(z)$$

$$(5\text{--}114)$$

Then Equation (5–114) can be rewritten as

$$zX_1(z) = p_1 X_1(z) + U(z)$$

$$zX_2(z) = p_2 X_2(z) + U(z)$$

$$\vdots$$

$$zX_n(z) = p_n X_n(z) + U(z)$$

$$(5\text{--}115)$$

Also, Equation (5–113) can be written as

$$Y(z) = b_0 U(z) + c_1 X_1(z) + c_2 X_2(z) + \cdots + c_n X_n(z) \qquad (5\text{–}116)$$

The inverse z transforms of Equations (5–115) and (5–116) become

$$x_1(k + 1) = p_1 x_1(k) + u(k)$$
$$x_2(k + 1) = p_2 x_2(k) + u(k)$$
$$\vdots \qquad\qquad\qquad (5\text{–}117)$$
$$x_n(k + 1) = p_n x_n(k) + u(k)$$

and

$$y(k) = c_1 x_1(k) + c_2 x_2(k) + \cdots + c_n x_n(k) + b_0 u(k) \qquad (5\text{–}118)$$

Rewriting the state equation and the output equation in the form of vector-matrix equations, we obtain Equations (5–109) and (5–110).

Problem A–5–4

(Partial-fraction-expansion programming method)   Consider the pulse transfer function system defined by

$$\frac{Y(z)}{U(z)} = \frac{b_0 z^n + b_1 z^{n-1} + \cdots + b_n}{z^n + a_1 z^{n-1} + \cdots + a_n}$$

Assume that the system involves a multiple pole of order m at $z = p_i$ and that all other poles are distinct.

Show that this system may be represented by the following state equation and output equation:

$$
\begin{bmatrix} x_1(k+1) \\ x_2(k+1) \\ \vdots \\ x_m(k+1) \\ \hline x_{m+1}(k+1) \\ \vdots \\ x_n(k+1) \end{bmatrix}
=
\left[\begin{array}{ccccc|ccc}
p_1 & 1 & 0 & \cdots & 0 & 0 & \cdots & 0 \\
0 & p_1 & 1 & \cdots & 0 & 0 & \cdots & 0 \\
\vdots & \vdots & \vdots & & \vdots & \vdots & & \vdots \\
0 & 0 & 0 & \cdots & p_1 & 0 & \cdots & 0 \\
\hline
0 & 0 & 0 & \cdots & 0 & p_{m+1} & \cdots & 0 \\
\vdots & \vdots & \vdots & & \vdots & \vdots & & \vdots \\
0 & 0 & 0 & \cdots & 0 & 0 & \cdots & p_n
\end{array}\right]
\begin{bmatrix} x_1(k) \\ x_2(k) \\ \vdots \\ x_m(k) \\ \hline x_{m+1}(k) \\ \vdots \\ x_n(k) \end{bmatrix}
+
\begin{bmatrix} 0 \\ 0 \\ \vdots \\ 1 \\ \hline 1 \\ \vdots \\ 1 \end{bmatrix} u(k) \qquad (5\text{–}119)
$$

$$y(k) = \begin{bmatrix} c_1 & c_2 & \cdots & c_n \end{bmatrix} \begin{bmatrix} x_1(k) \\ x_2(k) \\ \vdots \\ x_n(k) \end{bmatrix} + b_0 u(k) \qquad (5\text{–}120)$$

Solution   Since the system pulse transfer function can be written in the form

$$\frac{Y(z)}{U(z)} = \frac{b_0 z^n + b_1 z^{n-1} + \cdots + b_{n-1} z + b_n}{(z - p_1)^m (z - p_{m+1})(z - p_{m+2}) \cdots (z - p_n)}$$

$$= b_0 + \frac{(b_1 - a_1 b_0)z^{n-1} + (b_2 - a_2 b_0)z^{n-2} + \cdots + (b_n - a_n b_0)}{(z - p_1)^m (z - p_{m+1}) \cdots (z - p_n)}$$

$$= b_0 + \frac{c_1}{(z - p_1)^m} + \frac{c_2}{(z - p_1)^{m-1}} + \cdots + \frac{c_m}{z - p_1}$$

$$+ \frac{c_{m+1}}{z - p_{m+1}} + \frac{c_{m+2}}{z - p_{m+2}} + \cdots + \frac{c_n}{z - p_n} \qquad (5\text{–}121)$$

we obtain

$$Y(z) = b_0 U(z) + \frac{c_1}{(z - p_1)^m} U(z) + \frac{c_2}{(z - p_1)^{m-1}} U(z) + \cdots + \frac{c_m}{z - p_1} U(z)$$

$$+ \frac{c_{m+1}}{z - p_{m+1}} U(z) + \frac{c_{m+2}}{z - p_{m+2}} U(z) + \cdots + \frac{c_n}{z - p_n} U(z) \tag{5-122}$$

Let us define the first m state variables $X_1(z), X_2(z), \ldots, X_m(z)$ by the equations

$$X_1(z) = \frac{1}{(z - p_1)^m} U(z)$$

$$X_2(z) = \frac{1}{(z - p_1)^{m-1}} U(z) \tag{5-123}$$

$$\vdots$$

$$X_m(z) = \frac{1}{z - p_1} U(z)$$

and the remaining $n - m$ state variables $X_{m+1}(z), X_{m+2}(z), \ldots, X_n(z)$ by the equations

$$X_{m+1}(z) = \frac{1}{z - p_{m+1}} U(z)$$

$$X_{m+2}(z) = \frac{1}{z - p_{m+2}} U(z) \tag{5-124}$$

$$\vdots$$

$$X_n(z) = \frac{1}{z - p_n} U(z)$$

Notice that the m state variables defined by Equation (5–123) are related each to the next by the following equations:

$$\frac{X_1(z)}{X_2(z)} = \frac{1}{z - p_1}$$

$$\frac{X_2(z)}{X_3(z)} = \frac{1}{z - p_1} \tag{5-125}$$

$$\vdots$$

$$\frac{X_{m-1}(z)}{X_m(z)} = \frac{1}{z - p_1}$$

By taking the inverse z transforms of all of Equation (5–125), the last equation in Equation (5–123), and all of Equation (5–124), we obtain

$$x_1(k + 1) = p_1 x_1(k) + x_2(k)$$

$$x_2(k + 1) = p_1 x_2(k) + x_3(k)$$

$$\vdots$$

$$x_{m-1}(k + 1) = p_1 x_{m-1}(k) + x_m(k) \tag{5-126}$$

$$x_m(k + 1) = p_1 x_m(k) + u(k)$$

$$x_{m+1}(k + 1) = p_{m+1} x_{m+1}(k) + u(k)$$

$$\vdots$$

$$x_n(k + 1) = p_n x_n(k) + u(k)$$

The output equation given by Equation (5–122) can be rewritten as follows:

$$Y(z) = c_1 X_1(z) + c_2 X_2(z) + \cdots + c_m X_m(z) + c_{m+1} X_{m+1}(z)$$
$$+ c_{m+2} X_{m+2}(z) + \cdots + c_n X_n(z) + b_0 U(z)$$

By taking the inverse z transform of this last equation, we get

$$y(k) = c_1 x_1(k) + c_2 x_2(k) + \cdots + c_m x_m(k) + c_{m+1} x_{m+1}(k)$$
$$+ c_{m+2} x_{m+2}(k) + \cdots + c_n x_n(k) + b_0 u(k) \qquad (5\text{–}127)$$

Rewriting Equations (5–126) and (5–127) in the standard vector matrix form, we obtain Equations (5–119) and (5–120), respectively.

Problem A–5–5

Using the nested programming method (refer to Problem A–5–2), obtain the state equation and output equation for the system defined by

$$\frac{Y(z)}{U(z)} = \frac{z^{-1} + 5z^{-2}}{1 + 4z^{-1} + 3z^{-2}}$$

Then draw a block diagram for the system showing all state variables.

Solution   The given pulse transfer function can be written as

$$Y(z) = z^{-1}\{U(z) - 4Y(z) + z^{-1}[5U(z) - 3Y(z)]\}$$

Define

$$X_1(z) = z^{-1}[U(z) - 4Y(z) + X_2(z)]$$
$$X_2(z) = z^{-1}[5U(z) - 3Y(z)]$$
$$Y(z) = X_1(z)$$

Then we obtain

$$zX_1(z) = -4X_1(z) + X_2(z) + U(z)$$
$$zX_2(z) = -3X_1(z) + 5U(z)$$

The state equation can therefore be given by

$$\begin{bmatrix} x_1(k + 1) \\ x_2(k + 1) \end{bmatrix} = \begin{bmatrix} -4 & 1 \\ -3 & 0 \end{bmatrix} \begin{bmatrix} x_1(k) \\ x_2(k) \end{bmatrix} + \begin{bmatrix} 1 \\ 5 \end{bmatrix} u(k)$$

and the output equation becomes

$$y(k) = \begin{bmatrix} 1 & 0 \end{bmatrix} \begin{bmatrix} x_1(k) \\ x_2(k) \end{bmatrix}$$

Figure 5–3 shows the block diagram for the system defined by the state-space equations. The output of each delay element constitutes a state variable.

Problem A–5–6

Obtain a state-space representation of the system shown in Figure 5–4. The sampling period T is 1 sec.

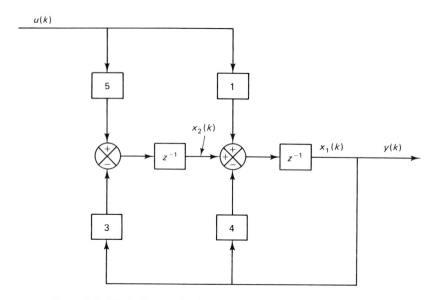

Figure 5–3   Block diagram for the system considered in Problem A–5–5.

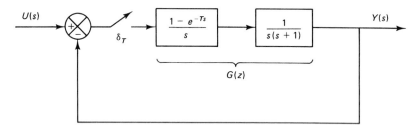

Figure 5–4   Block diagram of the control system of Problem A–5–6.

Solution   We shall first obtain the z transform of the feedforward transfer function:

$$G(z) = \mathcal{Z}\left[\frac{1 - e^{-s}}{s} \frac{1}{s(s+1)}\right] = (1 - z^{-1})\mathcal{Z}\left[\frac{1}{s^2(s+1)}\right]$$

$$= \frac{0.3679(z + 0.7181)}{(z - 1)(z - 0.3679)}$$

Then the closed-loop pulse transfer function can be obtained easily. There are many ways to obtain the state-space representation for such a system, as discussed in Section 5–2. In this problem, we shall demonstrate another approach, based on block diagram modification.

Let us expand $G(z)$ into partial fractions:

$$G(z) = \frac{1}{z - 1} - \frac{0.6321}{z - 0.3679} = \frac{z^{-1}}{1 - z^{-1}} - \frac{0.6321 z^{-1}}{1 - 0.3679 z^{-1}}$$

Figure 5–5 shows the block diagram for the system. Let us choose the output of each unit delay element as a state variable, as shown in Figure 5–5. Then we obtain

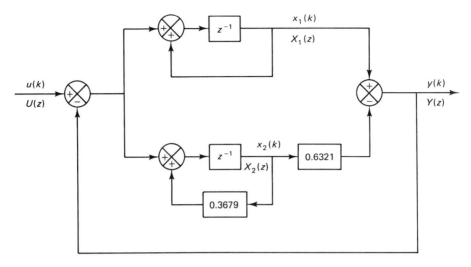

Figure 5–5   Modified block diagram for the system shown in Figure 5–4.

$$zX_1(z) = X_1(z) - [X_1(z) - 0.6321X_2(z)] + U(z)$$
$$zX_2(z) = 0.3679X_2(z) - [X_1(z) - 0.6321X_2(z)] + U(z)$$
$$Y(z) = X_1(z) - 0.6321X_2(z)$$

from which we get

$$x_1(k + 1) = 0.6321x_2(k) + u(k)$$
$$x_2(k + 1) = -x_1(k) + x_2(k) + u(k)$$
$$y(k) = x_1(k) - 0.6321x_2(k)$$

or

$$\begin{bmatrix} x_1(k + 1) \\ x_2(k + 1) \end{bmatrix} = \begin{bmatrix} 0 & 0.6321 \\ -1 & 1 \end{bmatrix} \begin{bmatrix} x_1(k) \\ x_2(k) \end{bmatrix} + \begin{bmatrix} 1 \\ 1 \end{bmatrix} u(k)$$
$$y(k) = \begin{bmatrix} 1 & -0.6321 \end{bmatrix} \begin{bmatrix} x_1(k) \\ x_2(k) \end{bmatrix}$$

Problem A–5–7

Obtain a state-space representation of the following pulse-transfer-function system:

$$\frac{Y(z)}{U(z)} = \frac{5}{(z + 1)^2(z + 2)}$$

Use the partial-fraction-expansion programming method. Also, obtain the initial values of the state variables in terms of $y(0)$, $y(1)$, and $y(2)$. Then draw a block diagram for the system.

Solution   Because we need the initial values of the state variables in terms of $y(0)$, $y(1)$, and $y(2)$, we slightly modify the partial-fraction-expansion programming method presented in Section 5–2. Let us expand $Y(z)/U(z)$, $zY(z)/U(z)$, and $z^2 Y(z)/U(z)$ into partial fractions as follows:

$$\frac{Y(z)}{U(z)} = \frac{5}{(z+1)^2} - \frac{5}{z+1} + \frac{5}{z+2}$$

$$\frac{zY(z)}{U(z)} = -\frac{5}{(z+1)^2} + \frac{10}{z+1} - \frac{10}{z+2}$$

$$\frac{z^2 Y(z)}{U(z)} = \frac{5}{(z+1)^2} - \frac{15}{z+1} + \frac{20}{z+2}$$

Then we have

$$\begin{bmatrix} \dfrac{Y(z)}{U(z)} \\[2ex] \dfrac{zY(z)}{U(z)} \\[2ex] \dfrac{z^2 Y(z)}{U(z)} \end{bmatrix} = \begin{bmatrix} 5 & -5 & 5 \\ -5 & 10 & -10 \\ 5 & -15 & 20 \end{bmatrix} \begin{bmatrix} \dfrac{1}{(z+1)^2} \\[2ex] \dfrac{1}{z+1} \\[2ex] \dfrac{1}{z+2} \end{bmatrix}$$

Now let us define the state variables by the following equation:

$$\begin{bmatrix} \dfrac{X_1(z)}{U(z)} \\[2ex] \dfrac{X_2(z)}{U(z)} \\[2ex] \dfrac{X_3(z)}{U(z)} \end{bmatrix} = \begin{bmatrix} \dfrac{1}{(z+1)^2} \\[2ex] \dfrac{1}{z+1} \\[2ex] \dfrac{1}{z+2} \end{bmatrix} \qquad (5\text{--}128)$$

Then the state variables $X_1(z)$, $X_2(z)$, and $X_3(z)$ are related to $Y(z)$, $zY(z)$, and $z^2 Y(z)$ as follows:

$$\begin{bmatrix} Y(z) \\ zY(z) \\ z^2 Y(z) \end{bmatrix} = \begin{bmatrix} 5 & -5 & 5 \\ -5 & 10 & -10 \\ 5 & -15 & 20 \end{bmatrix} \begin{bmatrix} X_1(z) \\ X_2(z) \\ X_3(z) \end{bmatrix} \qquad (5\text{--}129)$$

From Equation (5–128), we obtain

$$(z+1)^2 X_1(z) = U(z)$$

$$(z+1)X_2(z) = U(z)$$

$$(z+2)X_3(z) = U(z)$$

Noting that

$$(z+1)X_1(z) = X_2(z)$$

we get

$$zX_1(z) = -X_1(z) + X_2(z)$$

$$zX_2(z) = -X_2(z) + U(z)$$

$$zX_3(z) = -2X_3(z) + U(z)$$

The output $Y(z)$ is given by the equation

$$Y(z) = 5X_1(z) - 5X_2(z) + 5X_3(z)$$

Consequently, we have the state-space equations as follows:

$$x_1(k + 1) = -x_1(k) + x_2(k)$$
$$x_2(k + 1) = -x_2(k) + u(k)$$
$$x_3(k + 1) = -2x_3(k) + u(k)$$
$$y(k) = 5x_1(k) - 5x_2(k) + 5x_3(k)$$

or

$$\begin{bmatrix} x_1(k+1) \\ x_2(k+1) \\ x_3(k+1) \end{bmatrix} = \begin{bmatrix} -1 & 1 & 0 \\ 0 & -1 & 0 \\ 0 & 0 & -2 \end{bmatrix} \begin{bmatrix} x_1(k) \\ x_2(k) \\ x_3(k) \end{bmatrix} + \begin{bmatrix} 0 \\ 1 \\ 1 \end{bmatrix} u(k)$$

$$y(k) = \begin{bmatrix} 5 & -5 & 5 \end{bmatrix} \begin{bmatrix} x_1(k) \\ x_2(k) \\ x_3(k) \end{bmatrix}$$

The initial data are obtained by use of Equation (5–129), as follows:

$$\begin{bmatrix} x_1(0) \\ x_2(0) \\ x_3(0) \end{bmatrix} = \begin{bmatrix} 5 & -5 & 5 \\ -5 & 10 & -10 \\ 5 & -15 & 20 \end{bmatrix}^{-1} \begin{bmatrix} y(0) \\ y(1) \\ y(2) \end{bmatrix}$$

$$= \begin{bmatrix} \frac{2}{5} & \frac{1}{5} & 0 \\ \frac{2}{5} & \frac{3}{5} & \frac{1}{5} \\ \frac{1}{5} & \frac{2}{5} & \frac{1}{5} \end{bmatrix} \begin{bmatrix} y(0) \\ y(1) \\ y(2) \end{bmatrix}$$

The block diagram for this system is shown in Figure 5–6.

Problem A–5–8

Obtain a state-space representation of the following pulse-transfer-function system such that the state matrix is diagonal:

$$\frac{Y(z)}{U(z)} = \frac{z^3 + 8z^2 + 17z + 8}{(z + 1)(z + 2)(z + 3)}$$

Then obtain the initial state $\mathbf{x}(0)$ in terms of $y(0), y(1), y(2)$ and $u(0), u(1), u(2)$.

Solution   Let us first divide the numerators of the right-hand sides of $Y(z)/U(z)$, $zY(z)/U(z)$, and $z^2 Y(z)/U(z)$ by the respective denominators and expand the remaining terms into partial fractions, as follows:

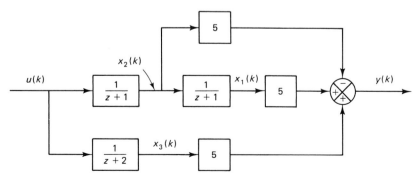

Figure 5–6   Block diagram for the system considered in Problem A–5–7.

$$\frac{Y(z)}{U(z)} = 1 - \frac{1}{z+1} + \frac{2}{z+2} + \frac{1}{z+3}$$

$$\frac{zY(z)}{U(z)} = z + 2 + \frac{1}{z+1} - \frac{4}{z+2} - \frac{3}{z+3}$$

$$\frac{z^2 Y(z)}{U(z)} = z^2 + 2z - 6 - \frac{1}{z+1} + \frac{8}{z+2} + \frac{9}{z+3}$$

Rewriting, we have

$$\frac{Y(z) - U(z)}{U(z)} = -\frac{1}{z+1} + \frac{2}{z+2} + \frac{1}{z+3}$$

$$\frac{zY(z) - zU(z) - 2U(z)}{U(z)} = \frac{1}{z+1} - \frac{4}{z+2} - \frac{3}{z+3}$$

$$\frac{z^2 Y(z) - z^2 U(z) - 2zU(z) + 6U(z)}{U(z)} = -\frac{1}{z+1} + \frac{8}{z+2} + \frac{9}{z+3}$$

or

$$\begin{bmatrix} \dfrac{Y(z) - U(z)}{U(z)} \\[2mm] \dfrac{zY(z) - zU(z) - 2U(z)}{U(z)} \\[2mm] \dfrac{z^2 Y(z) - z^2 U(z) - 2zU(z) + 6U(z)}{U(z)} \end{bmatrix} = \begin{bmatrix} -1 & 2 & 1 \\ 1 & -4 & -3 \\ -1 & 8 & 9 \end{bmatrix} \begin{bmatrix} \dfrac{1}{z+1} \\[2mm] \dfrac{1}{z+2} \\[2mm] \dfrac{1}{z+3} \end{bmatrix}$$

Let us define the state variables $X_1(z)$, $X_2(z)$, and $X_3(z)$ as follows:

$$\begin{bmatrix} \dfrac{X_1(z)}{U(z)} \\[2mm] \dfrac{X_2(z)}{u(z)} \\[2mm] \dfrac{X_3(z)}{U(z)} \end{bmatrix} = \begin{bmatrix} \dfrac{1}{z+1} \\[2mm] \dfrac{1}{z+2} \\[2mm] \dfrac{1}{z+3} \end{bmatrix} \qquad (5\text{--}130)$$

Then we have

$$\begin{bmatrix} Y(z) - U(z) \\ zY(z) - zU(z) - 2U(z) \\ z^2 Y(z) - z^2 U(z) - 2zU(z) + 6U(z) \end{bmatrix} = \begin{bmatrix} -1 & 2 & 1 \\ 1 & -4 & -3 \\ -1 & 8 & 9 \end{bmatrix} \begin{bmatrix} X_1(z) \\ X_2(z) \\ X_3(z) \end{bmatrix} \qquad (5\text{--}131)$$

Notice that Equation (5–130) can be written as

$$zX_1(z) = -X_1(z) + U(z)$$

$$zX_2(z) = -2X_2(z) + U(z)$$

$$zX_3(z) = -3X_3(z) + U(z)$$

from which we obtain

$$x_1(k+1) = -x_1(k) + u(k)$$

$$x_2(k+1) = -2x_2(k) + u(k)$$

$$x_3(k+1) = -3x_3(k) + u(k)$$

The output $Y(z)$ is given by

$$Y(z) = -X_1(z) + 2X_2(z) + X_3(z) + U(z)$$

or

$$y(k) = -x_1(k) + 2x_2(k) + x_3(k) + u(k)$$

In vector-matrix notation, the state space equations become

$$\begin{bmatrix} x_1(k+1) \\ x_2(k+1) \\ x_3(k+1) \end{bmatrix} = \begin{bmatrix} -1 & 0 & 0 \\ 0 & -2 & 0 \\ 0 & 0 & -3 \end{bmatrix} \begin{bmatrix} x_1(k) \\ x_2(k) \\ x_3(k) \end{bmatrix} + \begin{bmatrix} 1 \\ 1 \\ 1 \end{bmatrix} u(k)$$

$$y(k) = \begin{bmatrix} -1 & 2 & 1 \end{bmatrix} \begin{bmatrix} x_1(k) \\ x_2(k) \\ x_3(k) \end{bmatrix} + u(k)$$

The initial data are obtained from Equation (5–131) as follows:

$$\begin{bmatrix} x_1(0) \\ x_2(0) \\ x_3(0) \end{bmatrix} = \begin{bmatrix} -1 & 2 & 1 \\ 1 & -4 & -3 \\ -1 & 8 & 9 \end{bmatrix}^{-1} \begin{bmatrix} y(0) - u(0) \\ y(1) - u(1) - 2u(0) \\ y(2) - u(2) - 2u(1) + 6u(0) \end{bmatrix}$$

$$= \begin{bmatrix} -3 & -\frac{5}{2} & -\frac{1}{2} \\ -\frac{3}{2} & -2 & -\frac{1}{2} \\ 1 & \frac{3}{2} & \frac{1}{2} \end{bmatrix} \begin{bmatrix} y(0) - u(0) \\ y(1) - u(1) - 2u(0) \\ y(2) - u(2) - 2u(1) + 6u(0) \end{bmatrix}$$

Figure 5–7 shows the block diagram for the present system.

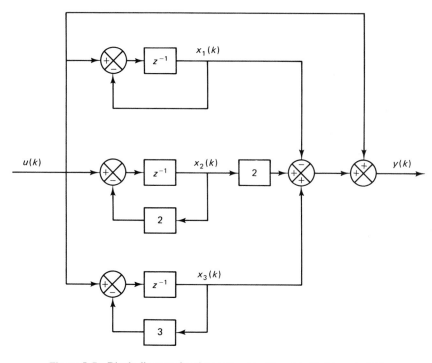

Figure 5–7 Block diagram for the system considered in Problem A–5–8.

Problem A–5–9

Let $\mathbf{A}$ be an $n \times n$ matrix, and let its characteristic equation be

$$|\lambda\mathbf{I} - \mathbf{A}| = \lambda^n + a_1\lambda^{n-1} + \cdots + a_{n-1}\lambda + a_n = 0$$

Show that matrix $\mathbf{A}$ satisfies its characteristic equation, or that

$$\mathbf{A}^n + a_1\mathbf{A}^{n-1} + \cdots + a_{n-1}\mathbf{A} + a_n\mathbf{I} = \mathbf{0}$$

(This is the Cayley–Hamilton theorem.)

Solution   We first note that $\operatorname{adj}(\lambda\mathbf{I} - \mathbf{A})$ is a polynomial in λ of degree $n - 1$. That is,

$$\operatorname{adj}(\lambda\mathbf{I} - \mathbf{A}) = \mathbf{B}_1\lambda^{n-1} + \mathbf{B}_2\lambda^{n-2} + \cdots + \mathbf{B}_{n-1}\lambda + \mathbf{B}_n$$

where

$$\mathbf{B}_1 = \mathbf{I}$$

Note also that

$$(\lambda\mathbf{I} - \mathbf{A})\operatorname{adj}(\lambda\mathbf{I} - \mathbf{A}) = [\operatorname{adj}(\lambda\mathbf{I} - \mathbf{A})](\lambda\mathbf{I} - \mathbf{A}) = |\lambda\mathbf{I} - \mathbf{A}|\mathbf{I}$$

Hence, we obtain

$$|\lambda\mathbf{I} - \mathbf{A}|\mathbf{I} = \mathbf{I}\lambda^n + a_1\mathbf{I}\lambda^{n-1} + \cdots + a_{n-1}\mathbf{I}\lambda + a_n\mathbf{I}$$

$$= (\lambda\mathbf{I} - \mathbf{A})(\mathbf{B}_1\lambda^{n-1} + \mathbf{B}_2\lambda^{n-2} + \cdots + \mathbf{B}_{n-1}\lambda + \mathbf{B}_n)$$

$$= (\mathbf{B}_1\lambda^{n-1} + \mathbf{B}_2\lambda^{n-2} + \cdots + \mathbf{B}_{n-1}\lambda + \mathbf{B}_n)(\lambda\mathbf{I} - \mathbf{A})$$

From this equation we see that $\mathbf{A}$ and $\mathbf{B}_i$ $(i = 1, 2, \ldots, n)$ commute. Hence, the product of $(\lambda\mathbf{I} - \mathbf{A})$ and $\operatorname{adj}(\lambda\mathbf{I} - \mathbf{A})$ becomes zero if either of these is zero. If $\mathbf{A}$ is substituted for λ in this last equation, then clearly $\lambda\mathbf{I} - \mathbf{A}$ becomes zero. Hence,

$$\mathbf{A}^n + a_1\mathbf{A}^{n-1} + \cdots + a_{n-1}\mathbf{A} + a_n\mathbf{I} = \mathbf{0}$$

This proves the Cayley–Hamilton theorem.

Problem A–5–10

Referring to Problem A–5–9, it has been shown that every $n \times n$ matrix $\mathbf{A}$ satisfies its own characteristic equation. The characteristic equation is not, however, necessarily the scalar equation of least degree that $\mathbf{A}$ satisfies. The least-degree polynomial having $\mathbf{A}$ as a root is called the *minimal polynomial*. That is, the minimal polynomial of an $n \times n$ matrix $\mathbf{A}$ is defined as the polynomial $\phi(\lambda)$ of least degree:

$$\phi(\lambda) = \lambda^m + a_1\lambda^{m-1} + \cdots + a_{m-1}\lambda + a_m, \qquad m \le n$$

such that $\phi(\mathbf{A}) = \mathbf{0}$, or

$$\phi(\mathbf{A}) = \mathbf{A}^m + a_1\mathbf{A}^{m-1} + \cdots + a_{m-1}\mathbf{A} + a_m\mathbf{I} = \mathbf{0}$$

The minimal polynomial plays an important role in the computation of polynomials in an $n \times n$ matrix.

Let us suppose that $d(\lambda)$, a polynomial in λ, is the greatest common divisor of all the elements of $\operatorname{adj}(\lambda\mathbf{I} - \mathbf{A})$. Show that if the coefficient of the highest-degree term in λ of $d(\lambda)$ is chosen as 1 then the minimal polynomial $\phi(\lambda)$ is given by

$$\phi(\lambda) = \frac{|\lambda\mathbf{I} - \mathbf{A}|}{d(\lambda)}$$

Solution   By assumption, the greatest common divisor of the matrix adj $(\lambda\mathbf{I} - \mathbf{A})$ is $d(\lambda)$. Therefore,

$$\text{adj}\,(\lambda\mathbf{I} - \mathbf{A}) = d(\lambda)\mathbf{B}(\lambda)$$

where the greatest common divisor of the n^2 elements (which are functions of λ) of $\mathbf{B}(\lambda)$ is unity. Since

$$(\lambda\mathbf{I} - \mathbf{A})\,\text{adj}\,(\lambda\mathbf{I} - \mathbf{A}) = |\lambda\mathbf{I} - \mathbf{A}|\mathbf{I}$$

we obtain

$$d(\lambda)(\lambda\mathbf{I} - \mathbf{A})\mathbf{B}(\lambda) = |\lambda\mathbf{I} - \mathbf{A}|\mathbf{I} \tag{5–132}$$

from which we find that $|\lambda\mathbf{I} - \mathbf{A}|$ is divisible by $d(\lambda)$. Let us put

$$|\lambda\mathbf{I} - \mathbf{A}| = d(\lambda)\psi(\lambda) \tag{5–133}$$

Then the coefficient of the highest-degree term in λ of $\psi(\lambda)$ is unity. From Equations (5–132) and (5–133), we have

$$(\lambda\mathbf{I} - \mathbf{A})\mathbf{B}(\lambda) = \psi(\lambda)\mathbf{I}$$

Hence,

$$\psi(\mathbf{A}) = \mathbf{0}$$

Note that $\psi(\lambda)$ can be written as follows:

$$\psi(\lambda) = g(\lambda)\phi(\lambda) + \alpha(\lambda)$$

where $\alpha(\lambda)$ is of lower degree than $\phi(\lambda)$. Since $\psi(\mathbf{A}) = \mathbf{0}$ and $\phi(\mathbf{A}) = \mathbf{0}$, we must have $\alpha(\mathbf{A}) = \mathbf{0}$. Since $\phi(\lambda)$ is the minimal polynomial, $\alpha(\lambda)$ must be identically zero, or

$$\psi(\lambda) = g(\lambda)\phi(\lambda)$$

Note that because $\phi(\mathbf{A}) = \mathbf{0}$ we can write

$$\phi(\lambda)\mathbf{I} = (\lambda\mathbf{I} - \mathbf{A})\mathbf{C}(\lambda)$$

Hence,

$$\psi(\lambda)\mathbf{I} = g(\lambda)\phi(\lambda)\mathbf{I} = g(\lambda)(\lambda\mathbf{I} - \mathbf{A})\mathbf{C}(\lambda)$$

and we obtain

$$\mathbf{B}(\lambda) = g(\lambda)\mathbf{C}(\lambda)$$

Note that the greatest common divisor of the n^2 elements of $\mathbf{B}(\lambda)$ is unity. Hence,

$$g(\lambda) = 1$$

Therefore,

$$\psi(\lambda) = \phi(\lambda)$$

Then, from this last equation and Equation (5–133), we obtain

$$\phi(\lambda) = \frac{|\lambda\mathbf{I} - \mathbf{A}|}{d(\lambda)}$$

It is noted that the minimal polynomial $\phi(\lambda)$ of an $n \times n$ matrix $\mathbf{A}$ can be determined by the following procedure:

1. Form adj $(\lambda\mathbf{I} - \mathbf{A})$ and write the elements of adj $(\lambda\mathbf{I} - \mathbf{A})$ as factored polynomials in λ.

2. Determine $d(\lambda)$ as the greatest common divisor of all the elements of adj $(\lambda I - A)$. Choose the coefficient of the highest-degree term in λ of $d(\lambda)$ to be 1. If there is no common divisor, $d(\lambda) = 1$.

3. The minimal polynomial $\phi(\lambda)$ is then given as $|\lambda I - A|$ divided by $d(\lambda)$.

Problem A–5–11

If an $n \times n$ matrix A has n distinct eigenvalues, then the minimal polynomial of A is identical with the characteristic polynomial. Also, if the multiple eigenvalues of A are linked in a Jordan chain, the minimal polynomial and the characteristic polynomial are identical. If, however, the multiple eigenvalues of A are not linked in a Jordan chain, the minimal polynomial is of lower degree than the characteristic polynomial.

Using the following matrices A and B as examples, verify the foregoing statements about the minimal polynomial when multiple eigenvalues are involved.

$$A = \begin{bmatrix} 2 & 1 & 4 \\ 0 & 2 & 0 \\ 0 & 3 & 1 \end{bmatrix}, \quad B = \begin{bmatrix} 2 & 0 & 0 \\ 0 & 2 & 0 \\ 0 & 3 & 1 \end{bmatrix}$$

Solution   First, consider the matrix A. The characteristic polynomial is given by

$$|\lambda I - A| = \begin{vmatrix} \lambda - 2 & -1 & -4 \\ 0 & \lambda - 2 & 0 \\ 0 & -3 & \lambda - 1 \end{vmatrix} = (\lambda - 2)^2(\lambda - 1)$$

Thus, the eigenvalues of A are 2, 2, and 1. It can be shown that the Jordan canonical form of A is

$$\begin{bmatrix} 2 & 1 & 0 \\ 0 & 2 & 0 \\ 0 & 0 & 1 \end{bmatrix}$$

and the multiple eigenvalues are linked in the Jordan chain as shown. (For the procedure for deriving the Jordan canonical form of A, refer to Appendix A.)

To determine the minimal polynomial, let us first obtain adj $(\lambda I - A)$. It is given by

$$\text{adj} (\lambda I - A) = \begin{bmatrix} (\lambda - 2)(\lambda - 1) & (\lambda + 11) & 4(\lambda - 2) \\ 0 & (\lambda - 2)(\lambda - 1) & 0 \\ 0 & 3(\lambda - 2) & (\lambda - 2)^2 \end{bmatrix}$$

Notice that there is no common divisor of all the elements of adj $(\lambda I - A)$. Hence, $d(\lambda) = 1$. Thus, the minimal polynomial $\phi(\lambda)$ is identical with the characteristic polynomial, or

$$\phi(\lambda) = |\lambda I - A| = (\lambda - 2)^2(\lambda - 1)$$
$$= \lambda^3 - 5\lambda^2 + 8\lambda - 4$$

A simple calculation proves that

$$A^3 - 5A^2 + 8A - 4I = 0$$

but

$$A^2 - 3A + 2I \neq 0$$

Thus, we have shown that the minimal polynomial and the characteristic polynomial of this matrix A are the same.

Next, consider the matrix **B**. The characteristic polynomial is given by

$$|\lambda\mathbf{I} - \mathbf{B}| = \begin{vmatrix} \lambda - 2 & 0 & 0 \\ 0 & \lambda - 2 & 0 \\ 0 & -3 & \lambda - 1 \end{vmatrix} = (\lambda - 2)^2(\lambda - 1)$$

A simple computation reveals that matrix **B** has three eigenvectors, and the Jordan canonical form of **B** is given by

$$\begin{bmatrix} 2 & 0 & 0 \\ 0 & 2 & 0 \\ 0 & 0 & 1 \end{bmatrix}$$

Thus, the multiple eigenvalues are not linked. To obtain the minimal polynomial, we first compute adj $(\lambda\mathbf{I} - \mathbf{B})$:

$$\text{adj}\,(\lambda\mathbf{I} - \mathbf{B}) = \begin{bmatrix} (\lambda - 2)(\lambda - 1) & 0 & 0 \\ 0 & (\lambda - 2)(\lambda - 1) & 0 \\ 0 & 3(\lambda - 2) & (\lambda - 2)^2 \end{bmatrix}$$

from which it is evident that

$$d(\lambda) = \lambda - 2$$

Hence,

$$\phi(\lambda) = \frac{|\lambda\mathbf{I} - \mathbf{B}|}{d(\lambda)} = \frac{(\lambda - 2)^2(\lambda - 1)}{\lambda - 2} = \lambda^2 - 3\lambda + 2$$

As a check, let us compute $\phi(\mathbf{B})$:

$$\phi(\mathbf{B}) = \mathbf{B}^2 - 3\mathbf{B} + 2\mathbf{I} = \begin{bmatrix} 4 & 0 & 0 \\ 0 & 4 & 0 \\ 0 & 9 & 1 \end{bmatrix} - 3\begin{bmatrix} 2 & 0 & 0 \\ 0 & 2 & 0 \\ 0 & 3 & 1 \end{bmatrix} + 2\begin{bmatrix} 1 & 0 & 0 \\ 0 & 1 & 0 \\ 0 & 0 & 1 \end{bmatrix} = \begin{bmatrix} 0 & 0 & 0 \\ 0 & 0 & 0 \\ 0 & 0 & 0 \end{bmatrix}$$

For the given matrix **B**, the degree of the minimal polynomial is lower by 1 than that of the characteristic polynomial. As shown here, if the multiple eigenvalues of an $n \times n$ matrix are not linked in a Jordan chain, the minimal polynomial is of lower degree than the characteristic polynomial.

Problem A–5–12

Show that by use of the minimal polynomial the inverse of a nonsingular matrix **A** can be expressed as a polynomial in **A** with scalar coefficients as follows:

$$\mathbf{A}^{-1} = -\frac{1}{a_m}(\mathbf{A}^{m-1} + a_1\mathbf{A}^{m-2} + \cdots + a_{m-2}\mathbf{A} + a_{m-1}\mathbf{I}) \qquad (5\text{–}134)$$

where $a_1, a_2, \ldots, a_m$ are coefficients of the minimal polynomial

$$\phi(\lambda) = \lambda^m + a_1\lambda^{m-1} + \cdots + a_{m-1}\lambda + a_m$$

Then, obtain the inverse of the following matrix **A**:

$$\mathbf{A} = \begin{bmatrix} 1 & 2 & 0 \\ 3 & -1 & -2 \\ 1 & 0 & -3 \end{bmatrix}$$

Solution For a nonsingular matrix **A**, its minimal polynomial $\phi(\mathbf{A})$ can be written as follows:

$$\phi(\mathbf{A}) = \mathbf{A}^m + a_1 \mathbf{A}^{m-1} + \cdots + a_{m-1}\mathbf{A} + a_m \mathbf{I} = \mathbf{0}$$

where $a_m \neq 0$. Hence,

$$\mathbf{I} = -\frac{1}{a_m}(\mathbf{A}^m + a_1 \mathbf{A}^{m-1} + \cdots + a_{m-2}\mathbf{A}^2 + a_{m-1}\mathbf{A})$$

Premultiplying by $\mathbf{A}^{-1}$, we obtain

$$\mathbf{A}^{-1} = -\frac{1}{a_m}(\mathbf{A}^{m-1} + a_1 \mathbf{A}^{m-2} + \cdots + a_{m-2}\mathbf{A} + a_{m-1}\mathbf{I})$$

which is Equation (5–134).

For the given matrix $\mathbf{A}$, $\text{adj}(\lambda\mathbf{I} - \mathbf{A})$ can be given as follows:

$$\text{adj}(\lambda\mathbf{I} - \mathbf{A}) = \begin{bmatrix} \lambda^2 + 4\lambda + 3 & 2\lambda + 6 & -4 \\ 3\lambda + 7 & \lambda^2 + 2\lambda - 3 & -2\lambda + 2 \\ \lambda + 1 & 2 & \lambda^2 - 7 \end{bmatrix}$$

Clearly, there is no common divisor $d(\lambda)$ of all elements of $\text{adj}(\lambda\mathbf{I} - \mathbf{A})$. Hence, $d(\lambda) = 1$. Consequently, the minimal polynomial $\phi(\lambda)$ is given by the equation

$$\phi(\lambda) = \frac{|\lambda\mathbf{I} - \mathbf{A}|}{d(\lambda)} = |\lambda\mathbf{I} - \mathbf{A}|$$

Thus, the minimal polynomial $\phi(\lambda)$ is the same as the characteristic polynomial.

Since the characteristic equation is

$$|\lambda\mathbf{I} - \mathbf{A}| = \lambda^3 + 3\lambda^2 - 7\lambda - 17 = 0$$

we obtain

$$\phi(\lambda) = \lambda^3 + 3\lambda^2 - 7\lambda - 17$$

By identifying the coefficients a_i of the minimal polynomial (which is the same as the characteristic polynomial in this case), we have

$$a_1 = 3, \qquad a_2 = -7, \qquad a_3 = -17$$

The inverse of $\mathbf{A}$ can then be obtained from Equation (5–134) as follows:

$$\mathbf{A}^{-1} = -\frac{1}{a_3}(\mathbf{A}^2 + a_1 \mathbf{A} + a_2 \mathbf{I}) = \frac{1}{17}(\mathbf{A}^2 + 3\mathbf{A} - 7\mathbf{I})$$

$$= \frac{1}{17}\left\{ \begin{bmatrix} 7 & 0 & -4 \\ -2 & 7 & 8 \\ -2 & 2 & 9 \end{bmatrix} + 3\begin{bmatrix} 1 & 2 & 0 \\ 3 & -1 & -2 \\ 1 & 0 & -3 \end{bmatrix} - 7\begin{bmatrix} 1 & 0 & 0 \\ 0 & 1 & 0 \\ 0 & 0 & 1 \end{bmatrix} \right\}$$

$$= \frac{1}{17}\begin{bmatrix} 3 & 6 & -4 \\ 7 & -3 & 2 \\ 1 & 2 & -7 \end{bmatrix}$$

$$= \begin{bmatrix} \frac{3}{17} & \frac{6}{17} & -\frac{4}{17} \\ \frac{7}{17} & -\frac{3}{17} & \frac{2}{17} \\ \frac{1}{17} & \frac{2}{17} & -\frac{7}{17} \end{bmatrix}$$

Problem A–5–13

Show that the inverse of $z\mathbf{I} - \mathbf{G}$ can be given by the equation

$$(z\mathbf{I} - \mathbf{G})^{-1} = \frac{\text{adj}(z\mathbf{I} - \mathbf{G})}{|z\mathbf{I} - \mathbf{G}|}$$

$$= \frac{\mathbf{I}z^{n-1} + \mathbf{H}_1 z^{n-2} + \mathbf{H}_2 z^{n-3} + \cdots + \mathbf{H}_{n-1}}{|z\mathbf{I} - \mathbf{G}|} \qquad (5\text{--}135)$$

where

$$\mathbf{H}_1 = \mathbf{G} + a_1 \mathbf{I}$$

$$\mathbf{H}_2 = \mathbf{GH}_1 + a_2 \mathbf{I}$$

$$\vdots$$

$$\mathbf{H}_{n-1} = \mathbf{GH}_{n-2} + a_{n-1} \mathbf{I}$$

$$\mathbf{H}_n = \mathbf{GH}_{n-1} + a_n \mathbf{I} = \mathbf{0}$$

and $a_1, a_2, \ldots, a_n$ are the coefficients appearing in the characteristic polynomial given by

$$|z\mathbf{I} - \mathbf{G}| = z^n + a_1 z^{n-1} + a_2 z^{n-2} + \cdots + a_n$$

Show also that

$$a_1 = -\operatorname{tr} \mathbf{G}$$

$$a_2 = -\tfrac{1}{2} \operatorname{tr} \mathbf{GH}_1$$

$$\vdots$$

$$a_n = -\frac{1}{n} \operatorname{tr} \mathbf{GH}_{n-1}$$

To simplify the derivation, assume that $n = 3$. (The derivation can be easily extended to the case of any positive integer n.)

Solution   Note that

$$(z\mathbf{I} - \mathbf{G})(\mathbf{I}z^2 + \mathbf{H}_1 z + \mathbf{H}_2) = z^3\mathbf{I} - z^2\mathbf{G} + z^2\mathbf{H}_1 - z\mathbf{GH}_1 + z\mathbf{H}_2 - \mathbf{GH}_2$$

$$= z^3\mathbf{I} - z^2\mathbf{G} + z^2(\mathbf{G} + a_1\mathbf{I}) - z\mathbf{G}(\mathbf{G} + a_1\mathbf{I})$$

$$+ z[\mathbf{G}(\mathbf{G} + a_1\mathbf{I}) + a_2\mathbf{I}] - \mathbf{G}[\mathbf{G}(\mathbf{G} + a_1\mathbf{I}) + a_2\mathbf{I}]$$

$$= z^3\mathbf{I} + a_1 z^2\mathbf{I} + a_2 z\mathbf{I} + a_3\mathbf{I} - \mathbf{G}^3$$

$$- a_1\mathbf{G}^2 - a_2\mathbf{G} - a_3\mathbf{I} \qquad (5\text{-}136)$$

The Cayley–Hamilton theorem (see Problem A–5–9) states that an $n \times n$ matrix $\mathbf{G}$ satisfies its own characteristic equation. Since $n = 3$ in the present case, the characteristic equation is

$$|z\mathbf{I} - \mathbf{G}| = z^3 + a_1 z^2 + a_2 z + a_3 = 0$$

and $\mathbf{G}$ satisfies the following equation:

$$\mathbf{G}^3 + a_1\mathbf{G}^2 + a_2\mathbf{G} + a_3\mathbf{I} = \mathbf{0}$$

Hence Equation (5–136) simplifies to

$$(z\mathbf{I} - \mathbf{G})(\mathbf{I}z^2 + \mathbf{H}_1 z + \mathbf{H}_2) = (z^3 + a_1 z^2 + a_2 z + a_3)\mathbf{I} = |z\mathbf{I} - \mathbf{G}|\mathbf{I}$$

Consequently,

$$\mathbf{I} = \frac{(z\mathbf{I} - \mathbf{G})(\mathbf{I}z^2 + \mathbf{H}_1 z + \mathbf{H}_2)}{|z\mathbf{I} - \mathbf{G}|}$$

or

$$(z\mathbf{I} - \mathbf{G})^{-1} = \frac{\mathbf{I}z^2 + \mathbf{H}_1 z + \mathbf{H}_2}{|z\mathbf{I} - \mathbf{G}|}$$

which is Equation (5–135) when $n = 3$.

Next, we shall show that

$$a_1 = -\operatorname{tr}\mathbf{G}$$

$$a_2 = -\tfrac{1}{2}\operatorname{tr}\mathbf{GH}_1$$

$$a_3 = -\tfrac{1}{3}\operatorname{tr}\mathbf{GH}_2$$

We shall transform $\mathbf{G}$ into a diagonal matrix if $\mathbf{G}$ involves n linearly independent eigenvectors (where $n = 3$ in the present case) or into a matrix in a Jordan canonical form if $\mathbf{G}$ involves fewer than n linearly independent eigenvectors. That is,

$$\mathbf{P}^{-1}\mathbf{GP} = \mathbf{D} = \text{matrix in diagonal form}$$

or

$$\mathbf{S}^{-1}\mathbf{GS} = \mathbf{J} = \text{matrix in a Jordan canonical form}$$

where matrices $\mathbf{P}$ and $\mathbf{S}$ are nonsingular transformation matrices.

Since the following derivation applies regardless of whether matrix $\mathbf{G}$ can be transformed into a diagonal matrix or into a matrix in a Jordan canonical form, we shall use the notation

$$\mathbf{T}^{-1}\mathbf{GT} = \hat{\mathbf{D}}$$

where $\hat{\mathbf{D}}$ represents either a diagonal matrix or a matrix in a Jordan canonical form, as the case may be.

In the following we shall first show that

$$\operatorname{tr}\mathbf{G} = \operatorname{tr}\hat{\mathbf{D}}$$

$$\operatorname{tr}\mathbf{GH}_1 = \operatorname{tr}\hat{\mathbf{D}}\hat{\mathbf{H}}_1$$

$$\operatorname{tr}\mathbf{GH}_2 = \operatorname{tr}\hat{\mathbf{D}}\hat{\mathbf{H}}_2$$

where

$$\hat{\mathbf{H}}_1 = \hat{\mathbf{D}} + a_1\mathbf{I}$$

$$\hat{\mathbf{H}}_2 = \hat{\mathbf{D}}\hat{\mathbf{H}}_1 + a_2\mathbf{I}$$

Then we shall show that

$$a_1 = -\operatorname{tr}\hat{\mathbf{D}}$$

$$a_2 = -\tfrac{1}{2}\operatorname{tr}\hat{\mathbf{D}}\hat{\mathbf{H}}_1$$

$$a_3 = -\tfrac{1}{3}\operatorname{tr}\hat{\mathbf{D}}\hat{\mathbf{H}}_2$$

Notice that since

$$\operatorname{tr}\mathbf{AB} = \operatorname{tr}\mathbf{BA}$$

we have

$$\operatorname{tr}\mathbf{T}\hat{\mathbf{D}}\mathbf{T}^{-1} = \operatorname{tr}(\mathbf{T}\hat{\mathbf{D}})(\mathbf{T}^{-1}) = \operatorname{tr}(\mathbf{T}^{-1})(\mathbf{T}\hat{\mathbf{D}}) = \operatorname{tr}\hat{\mathbf{D}}$$

Notice also that

$$\operatorname{tr}(\mathbf{A} + \mathbf{B}) = \operatorname{tr}\mathbf{A} + \operatorname{tr}\mathbf{B}$$

Now we have

$$\operatorname{tr}\mathbf{G} = \operatorname{tr}\mathbf{T}\hat{\mathbf{D}}\mathbf{T}^{-1} = \operatorname{tr}\hat{\mathbf{D}}$$

$$\operatorname{tr}\mathbf{GH}_1 = \operatorname{tr}\mathbf{G}(\mathbf{G} + a_1\mathbf{I}) = \operatorname{tr}\mathbf{G}^2 + \operatorname{tr}a_1\mathbf{G}$$

$$= \operatorname{tr} \mathbf{T} \hat{\mathbf{D}}^2 \mathbf{T}^{-1} + \operatorname{tr} a_1 \mathbf{T} \hat{\mathbf{D}} \mathbf{T}^{-1}$$

$$= \operatorname{tr} \hat{\mathbf{D}}^2 + \operatorname{tr} a_1 \hat{\mathbf{D}} = \operatorname{tr} (\hat{\mathbf{D}}^2 + a_1 \hat{\mathbf{D}}) = \operatorname{tr} \hat{\mathbf{D}} \hat{\mathbf{H}}_1$$

$$\operatorname{tr} \mathbf{G} \mathbf{H}_2 = \operatorname{tr} \mathbf{G} (\mathbf{G} \mathbf{H}_1 + a_2 \mathbf{I}) = \operatorname{tr} \mathbf{G} [\mathbf{G} (\mathbf{G} + a_1 \mathbf{I}) + a_2 \mathbf{I}]$$

$$= \operatorname{tr} (\mathbf{G}^3 + a_1 \mathbf{G}^2 + a_2 \mathbf{G})$$

$$= \operatorname{tr} \mathbf{T} \hat{\mathbf{D}}^3 \mathbf{T}^{-1} + \operatorname{tr} a_1 \mathbf{T} \hat{\mathbf{D}}^2 \mathbf{T}^{-1} + \operatorname{tr} a_2 \mathbf{T} \hat{\mathbf{D}} \mathbf{T}^{-1}$$

$$= \operatorname{tr} \hat{\mathbf{D}}^3 + \operatorname{tr} a_1 \hat{\mathbf{D}}^2 + \operatorname{tr} a_2 \hat{\mathbf{D}}$$

$$= \operatorname{tr} (\hat{\mathbf{D}}^3 + a_1 \hat{\mathbf{D}}^2 + a_2 \hat{\mathbf{D}}) = \operatorname{tr} \hat{\mathbf{D}} \hat{\mathbf{H}}_2$$

Let us write

$$\mathbf{T}^{-1} \mathbf{G} \mathbf{T} = \hat{\mathbf{D}} = \begin{bmatrix} p_1 & * & 0 \\ 0 & p_2 & * \\ 0 & 0 & p_3 \end{bmatrix}$$

where an asterisk denotes "either 0 or 1." Then

$$|z\mathbf{I} - \hat{\mathbf{D}}| = z^3 - (p_1 + p_2 + p_3)z^2 + (p_1 p_2 + p_2 p_3 + p_3 p_1)z - p_1 p_2 p_3$$

$$= z^3 + a_1 z^2 + a_2 z + a_3$$

where

$$a_1 = -(p_1 + p_2 + p_3)$$

$$a_2 = p_1 p_2 + p_2 p_3 + p_3 p_1$$

$$a_3 = -p_1 p_2 p_3$$

Notice that

$$\operatorname{tr} \hat{\mathbf{D}} = p_1 + p_2 + p_3 = -a_1$$

$$\operatorname{tr} \hat{\mathbf{D}} \hat{\mathbf{H}}_1 = \operatorname{tr} \hat{\mathbf{D}} (\hat{\mathbf{D}} + a_1 \mathbf{I}) = \operatorname{tr} \hat{\mathbf{D}}^2 + \operatorname{tr} a_1 \hat{\mathbf{D}}$$

$$= p_1^2 + p_2^2 + p_3^2 - (p_1 + p_2 + p_3)(p_1 + p_2 + p_3)$$

$$= -2(p_1 p_2 + p_2 p_3 + p_3 p_1) = -2a_2$$

$$\operatorname{tr} \hat{\mathbf{D}} \hat{\mathbf{H}}_2 = \operatorname{tr} \hat{\mathbf{D}} (\hat{\mathbf{D}} \hat{\mathbf{H}}_1 + a_2 \mathbf{I}) = \operatorname{tr} (\hat{\mathbf{D}}^3 + a_1 \hat{\mathbf{D}}^2 + a_2 \hat{\mathbf{D}})$$

$$= \operatorname{tr} \hat{\mathbf{D}}^3 + \operatorname{tr} a_1 \hat{\mathbf{D}}^2 + \operatorname{tr} a_2 \hat{\mathbf{D}}$$

$$= (p_1^3 + p_2^3 + p_3^3) - (p_1 + p_2 + p_3)(p_1^2 + p_2^2 + p_3^2)$$

$$+ (p_1 p_2 + p_2 p_3 + p_3 p_1)(p_1 + p_2 + p_3)$$

$$= 3 p_1 p_2 p_3 = -3 a_3$$

Thus, we have shown that

$$a_1 = -\operatorname{tr} \hat{\mathbf{D}} = -\operatorname{tr} \mathbf{G}$$

$$a_2 = -\tfrac{1}{2} \operatorname{tr} \hat{\mathbf{D}} \hat{\mathbf{H}}_1 = -\tfrac{1}{2} \operatorname{tr} \mathbf{G} \mathbf{H}_1$$

$$a_3 = -\tfrac{1}{3} \operatorname{tr} \hat{\mathbf{D}} \hat{\mathbf{H}}_2 = -\tfrac{1}{3} \operatorname{tr} \mathbf{G} \mathbf{H}_2$$

Problem A–5–14

Consider the following oscillator system:

$$\frac{Y(s)}{U(s)} = \frac{\omega^2}{s^2 + \omega^2}$$

Obtain the continuous-time state-space representation of the system. Then discretize the system and obtain the discrete-time state-space representation. Also obtain the pulse transfer function of the discretized system.

Solution   From the given transfer function, we have

$$\ddot{y} + \omega^2 y = \omega^2 u$$

Define

$$x_1 = y$$

$$x_2 = \frac{1}{\omega}\dot{y}$$

Then we obtain the following continuous-time state-space representation of the system:

$$\begin{bmatrix} \dot{x}_1 \\ \dot{x}_2 \end{bmatrix} = \begin{bmatrix} 0 & \omega \\ -\omega & 0 \end{bmatrix}\begin{bmatrix} x_1 \\ x_2 \end{bmatrix} + \begin{bmatrix} 0 \\ \omega \end{bmatrix}u$$

$$y = \begin{bmatrix} 1 & 0 \end{bmatrix}\begin{bmatrix} x_1 \\ x_2 \end{bmatrix}$$

The discrete-time state-space representation of the system is obtained as follows. Noting that

$$\mathbf{A} = \begin{bmatrix} 0 & \omega \\ -\omega & 0 \end{bmatrix}, \qquad \mathbf{B} = \begin{bmatrix} 0 \\ \omega \end{bmatrix}$$

we have

$$\mathbf{G} = e^{\mathbf{A}T} = \mathcal{L}^{-1}[(s\mathbf{I} - \mathbf{A})^{-1}] = \mathcal{L}^{-1}\left\{\begin{bmatrix} s & -\omega \\ \omega & s \end{bmatrix}^{-1}\right\}$$

$$= \mathcal{L}^{-1}\begin{bmatrix} \dfrac{s}{s^2 + \omega^2} & \dfrac{\omega}{s^2 + \omega^2} \\ \dfrac{-\omega}{s^2 + \omega^2} & \dfrac{s}{s^2 + \omega^2} \end{bmatrix} = \begin{bmatrix} \cos\omega T & \sin\omega T \\ -\sin\omega T & \cos\omega T \end{bmatrix}$$

and

$$\mathbf{H} = \left(\int_0^T e^{\mathbf{A}\lambda}\,d\lambda\right)\mathbf{B} = \left(\int_0^T \begin{bmatrix} \cos\omega\lambda & \sin\omega\lambda \\ -\sin\omega\lambda & \cos\omega\lambda \end{bmatrix} d\lambda\right)\begin{bmatrix} 0 \\ \omega \end{bmatrix}$$

$$= \begin{bmatrix} 1 - \cos\omega T \\ \sin\omega T \end{bmatrix}$$

Hence, the discrete-time state-space representation of the oscillator system becomes as follows:

$$\begin{bmatrix} x_1((k+1)T) \\ x_2((k+1)T) \end{bmatrix} = \begin{bmatrix} \cos\omega T & \sin\omega T \\ -\sin\omega T & \cos\omega T \end{bmatrix}\begin{bmatrix} x_1(kT) \\ x_2(kT) \end{bmatrix} + \begin{bmatrix} 1 - \cos\omega T \\ \sin\omega T \end{bmatrix}u(kT)$$

$$y(kT) = \begin{bmatrix} 1 & 0 \end{bmatrix}\begin{bmatrix} x_1(kT) \\ x_2(kT) \end{bmatrix}$$

The pulse transfer function of the discretized system can be obtained from Equation (5–60):

$$F(z) = \mathbf{C}(z\mathbf{I} - \mathbf{G})^{-1}\mathbf{H} + D$$

Noting that D is zero, we have

$$F(z) = [1 \ 0] \begin{bmatrix} z - \cos \omega T & -\sin \omega T \\ \sin \omega T & z - \cos \omega T \end{bmatrix}^{-1} \begin{bmatrix} 1 - \cos \omega T \\ \sin \omega T \end{bmatrix}$$

$$= \frac{1}{z^2 - 2z \cos \omega T + 1} [1 \ 0] \begin{bmatrix} z - \cos \omega T & \sin \omega T \\ -\sin \omega T & z - \cos \omega T \end{bmatrix} \begin{bmatrix} 1 - \cos \omega T \\ \sin \omega T \end{bmatrix}$$

$$= \frac{(1 - \cos \omega T)(z + 1)}{z^2 - 2z \cos \omega T + 1}$$

Hence,

$$\frac{Y(z)}{U(z)} = F(z) = \frac{(1 - \cos \omega T)(1 + z^{-1})z^{-1}}{1 - 2z^{-1} \cos \omega T + z^{-2}}$$

Note that the pulse transfer function obtained in this way is the same as that obtained by taking the z transform of the system that is preceded by a zero-order hold. That is,

$$\frac{Y(z)}{U(z)} = \mathcal{Z} \left[\frac{1 - e^{-Ts}}{s} \frac{\omega^2}{s^2 + \omega^2} \right] = (1 - z^{-1}) \mathcal{Z} \left[\frac{1}{s} - \frac{s}{s^2 + \omega^2} \right]$$

$$= (1 - z^{-1}) \left(\frac{1}{1 - z^{-1}} - \frac{1 - z^{-1} \cos \omega T}{1 - 2z^{-1} \cos \omega T + z^{-2}} \right)$$

$$= \frac{(1 - \cos \omega T)(1 + z^{-1})z^{-1}}{1 - 2z^{-1} \cos \omega T + z^{-2}}$$

Thus, we get the same expression for the pulse transfer function. The reason for this is that discretization in the state space yields the zero-order hold equivalent of the continuous-time system.

Problem A–5–15

Consider the system shown in Figure 5–8(a). This system involves complex poles. It is stable but not asymptotically stable in the sense of Liapunov. Figure 5–8(b) shows a

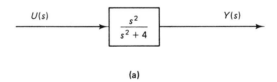

(a)

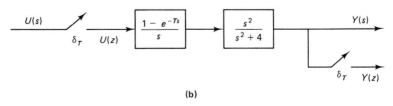

(b)

Figure 5–8   (a) Continuous-time system of Problem A–5–15; (b) discretized version of the system.

discretized version of the continuous-time system. The discretized system is also stable but not asymptotically stable.

Assuming a unit-step input, show that the discretized system may exhibit hidden oscillations when the sampling period T assumes a certain value.

Solution   The unit-step response of the continuous-time system shown in Figure 5–8(a) is

$$Y(s) = \frac{s^2}{s^2 + 4} \frac{1}{s} = \frac{s}{s^2 + 4}$$

Hence,

$$y(t) = \cos 2t$$

[Notice that the average value of the output $y(t)$ is zero, not unity.] The response $y(t)$ versus t is shown in Figure 5–9(a).

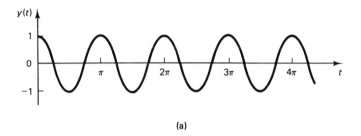

(a)

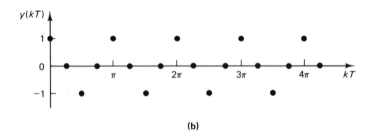

(b)

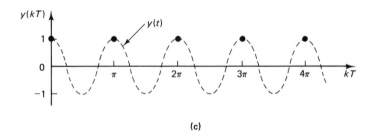

(c)

Figure 5–9   (a) Unit-step response $y(t)$ of the continuous-time system shown in Figure 5–8(a); (b) plot of $y(kT)$ versus kT of the discretized system shown in Figure 5–8(b) when $T = \frac{1}{4}\pi$ sec; (c) plot of $y(kT)$ versus kT of the discretized system when $T = \pi$ sec. (Hidden oscillations are shown in the diagram.)

The pulse transfer function of the discretized system shown in Figure 5–8(b) is

$$\frac{Y(z)}{U(z)} = \mathcal{Z}\left[\frac{1 - e^{-Ts}}{s}\frac{s^2}{s^2 + 4}\right] = (1 - z^{-1})\mathcal{Z}\left[\frac{s}{s^2 + 4}\right]$$

$$= (1 - z^{-1})\frac{1 - z^{-1}\cos 2T}{1 - 2z^{-1}\cos 2T + z^{-2}}$$

Hence, the unit-step response is obtained as follows:

$$Y(z) = \frac{(1 - z^{-1})(1 - z^{-1}\cos 2T)}{1 - 2z^{-1}\cos 2T + z^{-2}}\frac{1}{1 - z^{-1}}$$

$$= \frac{1 - z^{-1}\cos 2T}{1 - 2z^{-1}\cos 2T + z^{-2}}$$

The response $y(kT)$ becomes oscillatory if $T \ne n\pi \sec (n = 1, 2, 3, \dots)$. For example, the response of the discretized system when $T = \frac{1}{4}\pi$ sec becomes as follows:

$$Y(z) = \frac{1}{1 + z^{-2}} = 1 - z^{-2} + z^{-4} - z^{-6} + \cdots$$

Hence,

$$y(0) = 1$$

$$y(T) = 0$$

$$y(2T) = -1$$

$$y(3T) = 0$$

$$y(4T) = 1$$

$$\vdots$$

A plot of $y(kT)$ versus kT when $T = \frac{1}{4}\pi$ sec is shown in Figure 5–9(b). Clearly, the response is oscillatory. If, however, the sampling period T is π sec, or $T = \pi$, then

$$Y(z) = \frac{(1 - z^{-1})(1 - z^{-1})}{1 - 2z^{-1} + z^{-2}}\frac{1}{1 - z^{-1}} = \frac{1}{1 - z^{-1}}$$

$$= 1 + z^{-1} + z^{-2} + z^{-3} + \cdots$$

The response $y(kT)$ for $k = 0, 1, 2, \dots$ is constant at unity. A plot of $y(kT)$ versus kT when $T = \pi$ is shown in Figure 5–9(c).

Notice that if $T = \pi$ sec (in fact, if $T = n\pi$ sec, where $n = 1, 2, 3, \dots$) the unit-step response sequence stays at unity. Such a response may give us an impression that $y(t)$ is constant. The actual response is not unity but oscillates between 1 and -1. Thus, the output of the discretized system when $T = \pi$ sec (or when $T = n\pi$ sec, where $n = 1, 2, 3, \dots$) exhibits hidden oscillations.

Note that such hidden oscillations (hidden instability) occur only at certain particular values of the sampling period T. If the value of T is varied, such hidden oscillations (hidden instability) show up in the output as explicit oscillations.

Problem A–5–16

Even though the double-integrator system is dynamically simple, it represents an important class of systems. An example of double-integrator systems is a satellite attitude control system, which can be described by

$$J\ddot{\theta} = u + v$$

where J is the moment of inertia, θ is the attitude angle, u is the control torque, and v is the disturbance torque.

Consider the double-integrator system in the absence of disturbance input. Define $J\theta = y$. Then the system equation becomes

$$\ddot{y} = u$$

Obtain a continuous-time state-space representation of the system. Then obtain a discrete-time equivalent. Also obtain the pulse transfer function for the discrete-time system.

Solution   Define

$$x_1 = y$$

$$x_2 = \dot{y}$$

Then the continuous-time state equation and output equation become

$$\begin{bmatrix} \dot{x}_1 \\ \dot{x}_2 \end{bmatrix} = \begin{bmatrix} 0 & 1 \\ 0 & 0 \end{bmatrix} \begin{bmatrix} x_1 \\ x_2 \end{bmatrix} + \begin{bmatrix} 0 \\ 1 \end{bmatrix} u$$

$$y = \begin{bmatrix} 1 & 0 \end{bmatrix} \begin{bmatrix} x_1 \\ x_2 \end{bmatrix}$$

The discrete-time equivalent of this system can be given by

$$\mathbf{x}((k + 1)T) = \mathbf{G}\mathbf{x}(kT) + \mathbf{H}u(kT)$$

$$y(kT) = \mathbf{C}\mathbf{x}(kT)$$

Matrices $\mathbf{G}$ and $\mathbf{H}$ are obtained from Equations (5–73) and (5–74). Noting that

$$\mathbf{A} = \begin{bmatrix} 0 & 1 \\ 0 & 0 \end{bmatrix}, \qquad \mathbf{B} = \begin{bmatrix} 0 \\ 1 \end{bmatrix}$$

we have

$$\mathbf{G} = e^{\mathbf{A}T} = \begin{bmatrix} 1 & T \\ 0 & 1 \end{bmatrix}$$

and

$$\mathbf{H} = \left(\int_0^T e^{\mathbf{A}\lambda} \, d\lambda \right) \mathbf{B} = \left(\int_0^T \begin{bmatrix} 1 & \lambda \\ 0 & 1 \end{bmatrix} d\lambda \right) \begin{bmatrix} 0 \\ 1 \end{bmatrix} = \begin{bmatrix} \dfrac{T^2}{2} \\ T \end{bmatrix}$$

Hence, the discrete-time state equation and output equation become

$$\begin{bmatrix} x_1((k + 1)T) \\ x_2((k + 1)T) \end{bmatrix} = \begin{bmatrix} 1 & T \\ 0 & 1 \end{bmatrix} \begin{bmatrix} x_1(kT) \\ x_2(kT) \end{bmatrix} + \begin{bmatrix} \dfrac{T^2}{2} \\ T \end{bmatrix} u(kT)$$

$$y(kT) = \begin{bmatrix} 1 & 0 \end{bmatrix} \begin{bmatrix} x_1(kT) \\ x_2(kT) \end{bmatrix}$$

The pulse transfer function of the discrete-time system is obtained from Equation (5–60) as follows:

$$\frac{Y(z)}{U(z)} = F(z) = \mathbf{C}(z\mathbf{I} - \mathbf{G})^{-1}\mathbf{H} + D$$

$$= [1 \quad 0] \begin{bmatrix} z - 1 & -T \\ 0 & z - 1 \end{bmatrix}^{-1} \begin{bmatrix} \dfrac{T^2}{2} \\ T \end{bmatrix} + 0$$

$$= \frac{T^2(z + 1)}{2(z - 1)^2} = \frac{T^2 z^{-1}(1 + z^{-1})}{2(1 - z^{-1})^2}$$

Problem A–5–17

Show that the following quadratic form is positive definite:

$$V(\mathbf{x}) = 10x_1^2 + 4x_2^2 + x_3^2 + 2x_1 x_2 - 2x_2 x_3 - 4x_1 x_3$$

Solution   The quadratic form $V(\mathbf{x})$ can be written as follows:

$$V(\mathbf{x}) = \mathbf{x}^T \mathbf{P} \mathbf{x} = [x_1 \quad x_2 \quad x_3] \begin{bmatrix} 10 & 1 & -2 \\ 1 & 4 & -1 \\ -2 & -1 & 1 \end{bmatrix} \begin{bmatrix} x_1 \\ x_2 \\ x_3 \end{bmatrix}$$

Applying Sylvester's criterion, we obtain

$$10 > 0, \qquad \begin{vmatrix} 10 & 1 \\ 1 & 4 \end{vmatrix} > 0, \qquad \begin{vmatrix} 10 & 1 & -2 \\ 1 & 4 & -1 \\ -2 & -1 & 1 \end{vmatrix} > 0$$

Since all the successive principal minors of the matrix $\mathbf{P}$ are positive, $V(\mathbf{x})$ is positive definite.

Problem A–5–18

Consider the system defined by

$$\dot{\mathbf{x}} = \mathbf{f}(\mathbf{x}, t)$$

Suppose that

$$\mathbf{f}(\mathbf{0}, t) = \mathbf{0}, \qquad \text{for all } t$$

Suppose also that there exists a scalar function $V(\mathbf{x}, t)$ that has continuous first partial derivatives. If $V(\mathbf{x}, t)$ satisfies the conditions

1. $V(\mathbf{x}, t)$ is positive definite. That is, $V(\mathbf{0}, t) = 0$ and $V(\mathbf{x}, t) \geq \alpha(\|\mathbf{x}\|) > 0$ for all $\mathbf{x} \neq \mathbf{0}$ and all t, where α is a continuous nondecreasing scalar function such that $\alpha(0) = 0$.
2. The total derivative $\dot{V}(\mathbf{x}, t)$ is negative for all $\mathbf{x} \neq \mathbf{0}$ and all t, or $\dot{V}(\mathbf{x}, t) \leq -\gamma(\|\mathbf{x}\|) < 0$ for all $\mathbf{x} \neq \mathbf{0}$ and all t, where γ is a continuous nondecreasing scalar function such that $\gamma(0) = 0$.
3. There exists a continuous nondecreasing scalar function β such that $\beta(0) = 0$ and, for all t, $V(\mathbf{x}, t) \leq \beta(\|\mathbf{x}\|)$.
4. $\alpha(\|\mathbf{x}\|)$ approaches infinity as $\|\mathbf{x}\|$ increases indefinitely, or

$$\alpha(\|\mathbf{x}\|) \to \infty, \qquad \text{as } \|\mathbf{x}\| \to \infty$$

then the origin of the system, $\mathbf{x} = \mathbf{0}$, is uniformly asymptotically stable in the large. (This is Liapunov's main stability theorem.)

Prove this theorem.

Solution To prove uniform asymptotic stability in the large, we need to prove the following:

1. The origin is uniformly stable.
2. Every solution is uniformly bounded.
3. Every solution converges to the origin when $t \to \infty$ uniformly in t_0 and $\|\mathbf{x}_0\| \le \delta$, where δ is fixed but arbitrarily large. That is, given two real numbers $\delta > 0$ and $\mu > 0$, there is a real number $T(\mu, \delta)$ such that

$$\|\mathbf{x}_0\| \le \delta$$

implies

$$\|\boldsymbol{\phi}(t; \mathbf{x}_0, t_0)\| \le \mu, \qquad \text{for all } t \ge t_0 + T(\mu, \delta)$$

where $\boldsymbol{\phi}(t; \mathbf{x}_0, t_0)$ is the solution to the given differential equation.

Since β is continuous and $\beta(0) = 0$, we can take a $\delta(\epsilon) > 0$ such that $\beta(\delta) < \alpha(\epsilon)$ for any $\epsilon > 0$. Figure 5–10 shows the curves $\alpha(\|\mathbf{x}\|)$, $\beta(\|\mathbf{x}\|)$, and $V(\mathbf{x}, t)$. Noting that

$$V(\boldsymbol{\phi}(t; \mathbf{x}_0, t_0), t) - V(\mathbf{x}_0, t_0) = \int_{t_0}^{t} \dot{V}(\boldsymbol{\phi}(\tau; \mathbf{x}_0, t_0), \tau) \, d\tau < 0, \qquad t > t_0$$

if $\|\mathbf{x}_0\| \le \delta$, t_0 being arbitrary, we have

$$\alpha(\epsilon) > \beta(\delta) \ge V(\mathbf{x}_0, t_0) \ge V(\boldsymbol{\phi}(t; \mathbf{x}_0, t_0), t) \ge \alpha(\|\boldsymbol{\phi}(t; \mathbf{x}_0, t_0)\|)$$

for all $t \ge t_0$. Since α is nondecreasing and positive, this implies that

$$\|\boldsymbol{\phi}(t; \mathbf{x}_0, t_0)\| < \epsilon, \qquad \text{for } t \ge t_0, \|\mathbf{x}_0\| \le \delta$$

Hence, we have shown that for each real number $\epsilon > 0$ there is a real number $\delta > 0$ such that $\|\mathbf{x}_0\| \le \delta$ implies $\|\boldsymbol{\phi}(t; \mathbf{x}_0, t_0)\| \le \epsilon$ for all $t \ge t_0$. Thus, we have proved uniform stability.

Next, we shall prove that $\|\boldsymbol{\phi}(t; \mathbf{x}_0, t_0)\| \to 0$ when $t \to \infty$ uniformly in t_0 and $\|\mathbf{x}_0\| \le \delta$. Let us take any $0 < \mu < \|\mathbf{x}_0\|$ and find a $\nu(\mu) > 0$ such that $\beta(\nu) < \alpha(\mu)$. Let

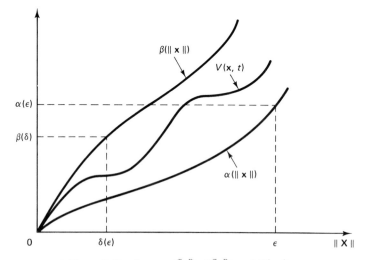

Figure 5–10 Curves $\alpha(\|\mathbf{x}\|)$, $\beta(\|\mathbf{x}\|)$, and $V(\mathbf{x}, t)$.

us denote by $\epsilon'(\mu, \delta) > 0$ the minimum of the continuous nondecreasing function $\gamma(\|\mathbf{x}\|)$ on the compact set $\nu(\mu) \le \|\mathbf{x}\| \le \epsilon(\delta)$. Let us define

$$T(\mu, \delta) = \frac{\beta(\delta)}{\epsilon'(\mu, \delta)} > 0$$

Suppose that $\|\boldsymbol{\phi}(t; \mathbf{x}_0, t_0)\| > \nu$ over the time interval $t_0 \le t \le t_1 = t_0 + T$. Then we have

$$0 < \alpha(\nu) \le V(\boldsymbol{\phi}(t_1; \mathbf{x}_0, t_0), t_1) \le V(\mathbf{x}_0, t_0) - (t_1 - t_0)\epsilon' \le \beta(\delta) - T\epsilon' = 0$$

which is a contradiction. Hence, for some t in the interval $t_0 \le t \le t_1$, such as an arbitrary t_2, we have

$$\|\mathbf{x}_2\| = \|\boldsymbol{\phi}(t_2; \mathbf{x}_0, t_0)\| = \nu$$

Therefore,

$$\alpha(\|\boldsymbol{\phi}(t; \mathbf{x}_2, t_2)\|) < V(\boldsymbol{\phi}(t; \mathbf{x}_2, t_2), t) \le V(\mathbf{x}_2, t_2) \le \beta(\nu) < \alpha(\mu)$$

for all $t \ge t_2$. Hence,

$$\|\boldsymbol{\phi}(t; \mathbf{x}_0, t_0)\| < \mu$$

for all $t \ge t_0 + T(\mu, \delta) \ge t_2$, which proves uniform asymptotic stability. Since $\alpha(\|\mathbf{x}\|) \to \infty$ as $\|\mathbf{x}\| \to \infty$, there exists for arbitrarily large δ a constant $\epsilon(\delta)$ such that $\beta(\delta) < \alpha(\epsilon)$. Moreover, since $\epsilon(\delta)$ does not depend on t_0, the solution $\boldsymbol{\phi}(t; \mathbf{x}_0, t_0)$ is uniformly bounded. We thus have proved uniform asymptotic stability in the large.

Problem A–5–19

In z plane analysis, an $n \times n$ matrix $\mathbf{G}$ whose n eigenvalues are less than unity in magnitude is called a stable matrix. Consider an $n \times n$ Hermitian (or real symmetric) matrix $\mathbf{P}$ that satisfies the following matrix equation:

$$\mathbf{G*PG} - \mathbf{P} = -\mathbf{Q} \qquad (5\text{–}137)$$

where $\mathbf{Q}$ is a positive definite $n \times n$ Hermitian (or real symmetric) matrix. Prove that if matrix $\mathbf{G}$ is a stable matrix then a matrix $\mathbf{P}$ that satisfies Equation (5–137) is unique and is positive definite. Prove that matrix $\mathbf{P}$ can be given by

$$\mathbf{P} = \sum_{k=0}^{\infty} (\mathbf{G*})^k \mathbf{QG}^k$$

Prove also that although the right-hand side of this last equation is an infinite series the matrix is finite. Finally, prove that if Equation (5–137) is satisfied by positive definite matrices $\mathbf{P}$ and $\mathbf{Q}$, then matrix $\mathbf{G}$ is a stable matrix. Assume that all eigenvalues of $\mathbf{G}$ are distinct and all eigenvectors of $\mathbf{G}$ are linearly independent.

Solution   Let us assume that there exist two matrices $\mathbf{P}_1$ and $\mathbf{P}_2$ that satisfy Equation (5–137). Then

$$\mathbf{G*P}_1\mathbf{G} - \mathbf{P}_1 = -\mathbf{Q} \qquad (5\text{–}138)$$

and

$$\mathbf{G*P}_2\mathbf{G} - \mathbf{P}_2 = -\mathbf{Q} \qquad (5\text{–}139)$$

By subtracting Equation (5–139) from Equation (5–138), we obtain

$$\mathbf{G*\hat{P}G} - \hat{\mathbf{P}} = 0 \qquad (5\text{–}140)$$

where

$$\hat{P} = P_1 - P_2$$

Notice that if $\hat{P} \neq 0$, then there exists an eigenvector x_i of matrix G such that

$$\hat{P}x_i \neq 0$$

Let us define the eigenvalue that is associated with the eigenvector x_i to be λ_i. Then

$$Gx_i = \lambda_i x_i$$

Hence, from Equation (5–140), we obtain

$$G^*\hat{P}Gx_i - \hat{P}x_i = G^*\hat{P}\lambda_i x_i - \hat{P}x_i = (\lambda_i G^* - I)\hat{P}x_i = 0 \qquad (5\text{–}141)$$

Equation (5–141) implies that λ_i^{-1} is an eigenvalue of G^*. Since $|\lambda_i| < 1$, we have $|\lambda_i^{-1}| > 1$. This contradicts the assumption that G is a stable matrix. Hence, $\hat{P}$ must be a zero matrix, or it is necessary that

$$P_1 = P_2$$

Thus, we have proved the uniqueness of the matrix P, the solution to Equation (5–137).
   To prove that a matrix P that satisfies Equation (5–137) can be given by

$$P = \sum_{k=0}^{\infty} (G^*)^k QG^k \qquad (5\text{–}142)$$

we may rewrite Equation (5–142) as follows:

$$P = (G^*)^0 QG^0 + \sum_{k=1}^{\infty} (G^*)^k QG^k = Q + G^* \left[\sum_{k=0}^{\infty} (G^*)^k QG^k \right] G$$

$$= Q + G^*PG$$

Thus, Equation (5–137) is satisfied. Since matrix Q is a positive definite matrix, from Equation (5–142) matrix P is also positive definite.
   We shall now prove that, although matrix P given by Equation (5–142) is the sum of an infinite series, it is a finite matrix. Because of the assumptions made in the problem statement, the eigenvalues λ_i are distinct and the eigenvectors of G are linearly independent. For the eigenvalue λ_i that is associated with the eigenvector x_i, we have

$$Gx_i = \lambda_i x_i$$

By using this relationship, we may simplify $x_i^* \left[\sum_{k=0}^{\infty} (G^*)^k QG^k \right] x_i$. First note that

$$x_i^* (G^*)^2 QG^2 x_i = (x_i^* G^*)(G^*QG)(Gx_i) = \bar{\lambda}_i x_i^* G^*QG\lambda_i x_i$$

$$= |\lambda_i|^2 (x_i^* G^*)Q(Gx_i) = |\lambda_i|^2 (\bar{\lambda}_i x_i^*)Q(\lambda_i x_i)$$

$$= |\lambda_i|^2 |\lambda_i|^2 x_i^* Qx_i$$

Then, by using this type of simplification, we have

$$x_i^* \left[\sum_{k=0}^{\infty} (G^*)^k QG^k \right] x_i = x_i^* Qx_i + x_i^* G^*QGx_i + x_i^* (G^*)^2 QG^2 x_i + x_i^* (G^*)^3 QG^3 x_i + \cdots$$

$$= x_i^* Qx_i + \bar{\lambda}_i \lambda_i x_i^* Qx_i + |\lambda_i|^2 |\lambda_i|^2 x_i^* Qx_i + |\lambda_i|^4 |\lambda_i|^2 x_i^* Qx_i$$
$$+ \cdots$$

$$= x_i^* Qx_i(1 + |\lambda_i|^2 + |\lambda_i|^4 + |\lambda_i|^6 + \cdots)$$

$$= x_i^* Qx_i \frac{1}{1 - |\lambda_i|^2}$$

This proves that

$$\sum_{k=0}^{\infty} (\mathbf{G}^*)^k \mathbf{Q} \mathbf{G}^k$$

is a finite matrix.

Finally, we shall prove that if Equation (5–137) is satisfied by positive definite matrices $\mathbf{P}$ and $\mathbf{Q}$ then matrix $\mathbf{G}$ is a stable matrix. Let us define the eigenvector associated with an eigenvalue λ_i of $\mathbf{G}$ as $\mathbf{x}_i$. Then

$$\mathbf{G}\mathbf{x}_i = \lambda_i \mathbf{x}_i$$

By premultiplying both sides of Equation (5–137) by $\mathbf{x}_i^*$ and postmultiplying both sides by $\mathbf{x}_i$, we obtain

$$\mathbf{x}_i^* \, \mathbf{G}^*\mathbf{P}\mathbf{G}\mathbf{x}_i - \mathbf{x}_i^* \, \mathbf{P}\mathbf{x}_i = -\mathbf{x}_i^* \, \mathbf{Q}\mathbf{x}_i$$

Hence,

$$\overline{\lambda}_i \mathbf{x}_i^* \, \mathbf{P}\lambda_i \mathbf{x}_i - \mathbf{x}_i^* \, \mathbf{P}\mathbf{x}_i = -\mathbf{x}_i^* \, \mathbf{Q}\mathbf{x}_i$$

or

$$(|\lambda_i|^2 - 1)\mathbf{x}_i^* \, \mathbf{P}\mathbf{x}_i = -\mathbf{x}_i^* \, \mathbf{Q}\mathbf{x}_i$$

Since both $\mathbf{x}_i^* \, \mathbf{P}\mathbf{x}_i$ and $\mathbf{x}_i^* \, \mathbf{Q}\mathbf{x}_i$ are positive-definite, we have

$$|\lambda_i|^2 - 1 < 0$$

or

$$|\lambda_i| < 1$$

Hence, we have proved that matrix $\mathbf{G}$ is a stable matrix.

It is noted that the proofs and derivations presented here can be extended to the case where matrix $\mathbf{G}$ involves multiple eigenvalues and multiple eigenvectors.

Problem A–5–20

Consider the system

$$\mathbf{x}(k + 1) = \mathbf{H}(\mathbf{x}(k))\mathbf{x}(k)$$

Assume that there exist positive constants $c_1, c_2, \ldots, c_n$ such that either

$$(1) \qquad \max_i \left\{ \sum_{j=1}^{n} \frac{c_i}{c_j} |h_{ij}(\mathbf{x})| \right\} < 1, \qquad \text{for all } \mathbf{x}$$

or

$$(2) \qquad \max_j \left\{ \sum_{i=1}^{n} \frac{c_i}{c_j} |h_{ij}(\mathbf{x})| \right\} < 1, \qquad \text{for all } \mathbf{x}$$

Show that in either case $\mathbf{H}(\mathbf{x})\mathbf{x}$ is a contraction for all $\mathbf{x}$ and therefore the equilibrium state of the system is asymptotically stable in the large.

Solution   In case 1, define the norm by

$$\|\mathbf{x}\| = \max_i \{c_i |x_i|\}$$

Then

$$\|\mathbf{H}(\mathbf{x})\mathbf{x}\| = \max_i \left\{ c_i \left| \sum_{j=1}^{n} h_{ij}(\mathbf{x})x_j \right| \right\} \le \max_i \left[\sum_{j=1}^{n} \frac{c_i}{c_j} |h_{ij}(\mathbf{x})| \cdot c_j |x_j| \right]$$

$$\leq \max_i \left\{ \sum_{j=1}^n \frac{c_i}{c_j} |h_{ij}(\mathbf{x})| \right\} \max_j \{c_j |x_j|\} < \max_j \{c_j |x_j|\}$$

$$= \|\mathbf{x}\|$$

which verifies that $\mathbf{H}(\mathbf{x})\mathbf{x}$ is a contraction.

In case 2, define the norm by

$$\|\mathbf{x}\| = \sum_{i=1}^n c_i |x_i|$$

Then

$$\|\mathbf{H}(\mathbf{x})\mathbf{x}\| = \sum_{i=1}^n c_i \left| \sum_{j=1}^n h_{ij}(\mathbf{x})x_j \right|$$

$$\leq \sum_{i=1}^n \frac{c_i}{c_j} \sum_{j=1}^n |h_{ij}(\mathbf{x})| \cdot c_j |x_j|$$

$$\leq \max_j \left\{ \sum_{i=1}^n \frac{c_i}{c_j} |h_{ij}(\mathbf{x})| \right\} \cdot \sum_{j=1}^n c_j |x_j| < \sum_{j=1}^n c_j |x_j|$$

$$= \|\mathbf{x}\|$$

which shows that $\mathbf{H}(\mathbf{x})\mathbf{x}$ is a contraction.

Now consider a scalar function $V(\mathbf{x}) = \|\mathbf{x}\|$. Clearly, $V(\mathbf{x}) = \|\mathbf{x}\|$ is positive definite and

$$\Delta V(\mathbf{x}(k)) = V(\mathbf{x}(k + 1)) - V(\mathbf{x}(k))$$

$$= \|\mathbf{H}(\mathbf{x}(k))\mathbf{x}(k)\| - \|\mathbf{x}(k)\| < 0$$

and

$$\Delta V(\mathbf{0}) = 0$$

Thus, $\Delta V(\mathbf{x})$ is negative definite. Hence, $V(\mathbf{x}) = \|\mathbf{x}\|$ is a Liapunov function for the system considered, and by Theorem 5–5 the origin of the system is asymptotically stable in the large.

Problem A–5–21

Prove that if all solutions of

$$\mathbf{x}(k + 1) = \mathbf{G}\mathbf{x}(k) \tag{5-143}$$

where $\mathbf{x}$ is an n vector and $\mathbf{G}$ is an $n \times n$ constant matrix, tend to zero as k approaches infinity, then all solutions of the system

$$\mathbf{x}(k + 1) = \mathbf{G}\mathbf{x}(k) + \mathbf{H}\mathbf{u}(k) \tag{5-144}$$

where $\mathbf{H}$ is an $n \times r$ constant matrix, are bounded, provided that the input vector $\mathbf{u}(k)$, an r vector, is bounded.

Solution   Since $\mathbf{u}(k)$ is bounded, there exists a positive constant c such that

$$\|\mathbf{u}(k)\| < c, \qquad k = 0, 1, 2, \ldots$$

The solution of Equation (5–144) is given by

$$\mathbf{x}(k) = \mathbf{G}^k \mathbf{x}(0) + \sum_{j=1}^k \mathbf{G}^{k-j} \mathbf{H}\mathbf{u}(j - 1)$$

Hence,

$$\|\mathbf{x}(k)\| < \|\mathbf{G}\|^k \|\mathbf{x}(0)\| + c \sum_{j=1}^{k} \|\mathbf{G}\|^{k-j} \|\mathbf{H}\| \le \|\mathbf{G}\|^k \|\mathbf{x}(0)\| + \lim_{k \to \infty} c \sum_{j=1}^{k} \|\mathbf{G}\|^{k-j} \|\mathbf{H}\|$$

Since the origin of the homogeneous system given by Equation (5–143) is asymptotically stable, there exist positive constants a and b $(0 < b < 1)$ such that

$$0 < \|\mathbf{G}\|^k < ab^k$$

Then

$$\lim_{k \to \infty} \sum_{j=1}^{k} \|\mathbf{G}\|^{k-j} < \lim_{k \to \infty} \sum_{j=1}^{k} ab^{k-j} = a \frac{1}{1-b}$$

Therefore,

$$\|\mathbf{x}(k)\| < a\|\mathbf{x}(0)\| + c\|\mathbf{H}\|a \frac{1}{1-b}$$

We have thus proved that $\|\mathbf{x}(k)\|$ is bounded.

Problem A–5–22

Consider the system defined by the equations

$$x_1(k + 1) = 2x_1(k) + 0.5x_2(k) - 5$$
$$x_2(k + 1) = 0.8x_2(k) + 2$$

Determine the stability of the equilibrium state.

Solution   Define the equilibrium state as

$$x_1(k) = x_{1e}, \qquad x_2(k) = x_{2e}$$

Then such an equilibrium state can be determined from the following two simultaneous equations:

$$x_{1e} = 2x_{1e} + 0.5x_{2e} - 5$$
$$x_{2e} = 0.8x_{2e} + 2$$

or

$$x_{1e} = 0, \qquad x_{2e} = 10$$

The equilibrium state is thus $(0, 10)$.

Now let us consider a new coordinate system with the origin at the equilibrium state. Define

$$\hat{x}_1(k) = x_1(k)$$
$$\hat{x}_2(k) = x_2(k) - 10$$

Then the system equations become

$$\hat{x}_1(k + 1) = 2\hat{x}_1(k) + 0.5[\hat{x}_2(k) + 10] - 5$$
$$\hat{x}_2(k + 1) + 10 = 0.8[\hat{x}_2(k) + 10] + 2$$

or

$$\begin{bmatrix} \hat{x}_1(k + 1) \\ \hat{x}_2(k + 1) \end{bmatrix} = \begin{bmatrix} 2 & 0.5 \\ 0 & 0.8 \end{bmatrix} \begin{bmatrix} \hat{x}_1(k) \\ \hat{x}_2(k) \end{bmatrix}$$

To determine the stability of the origin of the system in the new coordinate system, let us apply the Liapunov stability equation given by Equation (5–90):

$$\begin{bmatrix} 2 & 0 \\ 0.5 & 0.8 \end{bmatrix} \begin{bmatrix} p_{11} & p_{12} \\ p_{12} & p_{22} \end{bmatrix} \begin{bmatrix} 2 & 0.5 \\ 0 & 0.8 \end{bmatrix} - \begin{bmatrix} p_{11} & p_{12} \\ p_{12} & p_{22} \end{bmatrix} = - \begin{bmatrix} 9 & 0 \\ 0 & 0.35 \end{bmatrix}$$

where we choose $\mathbf{Q}$ to be a positive definite matrix having elements that simplify the computation involved. Solving this last equation for matrix $\mathbf{P}$, we obtain

$$\mathbf{P} = \begin{bmatrix} p_{11} & p_{12} \\ p_{12} & p_{22} \end{bmatrix} = \begin{bmatrix} -3 & 5 \\ 5 & 10 \end{bmatrix}$$

By applying the Sylvester criterion for positive definiteness, we find that matrix $\mathbf{P}$ is not positive definite. Therefore, the origin (equilibrium state) is not stable.

The instability of the equilibrium state can, of course, be determined by the z transform approach. Let us first eliminate $\hat{x}_2$ from the state equation. Then we have

$$\hat{x}_1(k + 2) - 2.8\hat{x}_1(k + 1) + 1.6\hat{x}_1(k) = 0$$

The characteristic equation for the system in the z plane is

$$z^2 - 2.8z + 1.6 = 0$$

or

$$(z - 2)(z - 0.8) = 0$$

Hence,

$$z = 2, \qquad z = 0.8$$

Since pole $z = 2$ is located outside the unit circle in the z plane, the origin (equilibrium state) is unstable.

PROBLEMS

Problem B–5–1

Obtain a state-space representation of the following pulse-transfer-function system in the controllable canonical form.

$$\frac{Y(z)}{U(z)} = \frac{z^{-1} + 2z^{-2}}{1 + 4z^{-1} + 3z^{-2}}$$

Problem B–5–2

Obtain a state-space representation of the following pulse-transfer-function system in the observable canonical form.

$$\frac{Y(z)}{U(z)} = \frac{z^{-2} + 4z^{-3}}{1 + 6z^{-1} + 11z^{-2} + 6z^{-3}}$$

Problem B–5–3

Obtain a state-space representation of the following pulse-transfer-function system in the diagonal canonical form.

$$\frac{Y(z)}{U(z)} = \frac{1 + 6z^{-1} + 8z^{-2}}{1 + 4z^{-1} + 3z^{-2}}$$

Problem B–5–4

Obtain a state-space representation of the system described by the equation

$$y(k + 2) + y(k + 1) + 0.16y(k) = u(k + 1) + 2u(k)$$

Problem B–5–5

Obtain the state equation and output equation for the system shown in Figure 5–11.

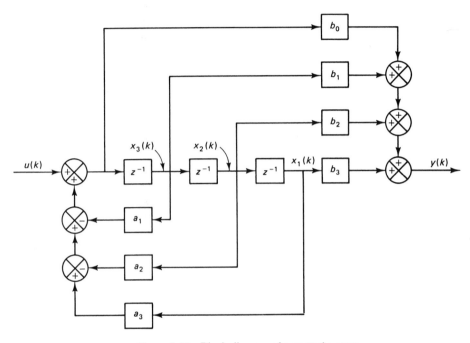

Figure 5–11 Block diagram of a control system.

Problem B–5–6

Obtain the state equation and output equation for the system shown in Figure 5–12.

Problem B–5–7

Obtain the state-space representation of the system shown in Figure 5–13.

Problem B–5–8

Figure 5–14 shows a block diagram of a discrete-time multiple-input–multiple-output system. Obtain state-space equations for the system by considering $x_1(k)$, $x_2(k)$, and $x_3(k)$ as shown in the diagram to be state variables. Then define new state variables such that the state matrix becomes a diagonal matrix.

Problem B–5–9

Obtain the state equation and output equation for the system shown in Figure 5–15.

Problem B–5–10

Obtain a state-space representation of the discrete-time control system shown in Figure 5–16.

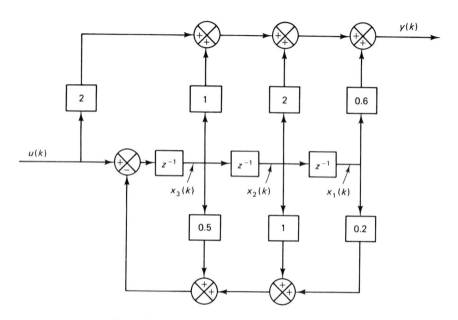

Figure 5–12 Block diagram of a control system.

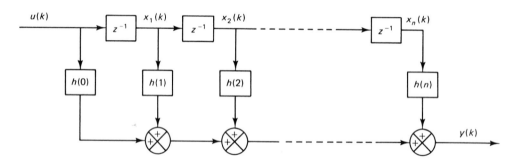

Figure 5–13 Block diagram of the control system of Problem B–5–7.

Problem B–5–11

Obtain a state-space representation of the following system in the diagonal canonical form.

$$\frac{Y(z)}{U(z)} = \frac{z^{-1} + 2z^{-2}}{1 + 0.7z^{-1} + 0.12z^{-2}}$$

Problem B–5–12

Obtain a state-space representation of the following pulse-transfer-function system such that the state matrix is a diagonal matrix:

$$\frac{Y(z)}{U(z)} = \frac{1}{(z + 1)(z + 2)(z + 3)}$$

Then obtain the initial state variables $x_1(0)$, $x_2(0)$, and $x_3(0)$ in terms of $y(0)$, $y(1)$, and $y(2)$.

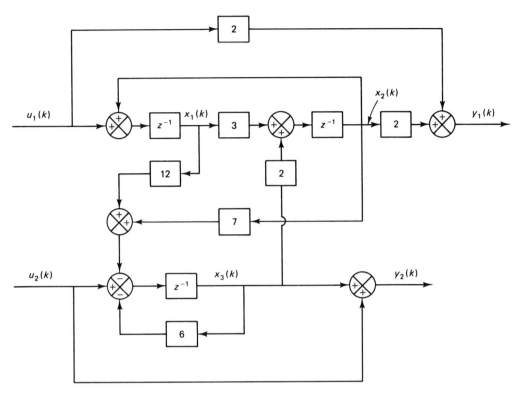

Figure 5–14 Block diagram of the discrete-time multiple-input–multiple-output system of Problem B–5–8.

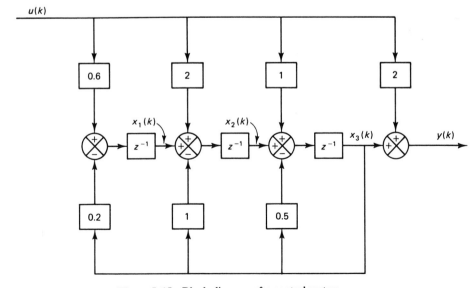

Figure 5–15 Block diagram of a control system.

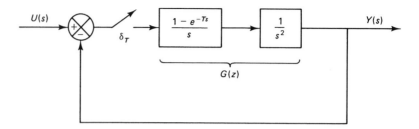

Figure 5–16 Discrete-time control system.

Problem B–5–13

A state-space representation of the scalar difference equation system

$$y(k + n) + a_1(k)y(k + n - 1) + \cdots + a_n(k)y(k)$$
$$= b_0(k)u(k + n) + b_1(k)u(k + n - 1) + \cdots + b_n(k)u(k)$$

where $k = 0, 1, 2, \ldots$, may be given by

$$\begin{bmatrix} x_1(k + 1) \\ x_2(k + 1) \\ \vdots \\ x_{n-1}(k + 1) \\ x_n(k + 1) \end{bmatrix} = \begin{bmatrix} 0 & 1 & \cdots & 0 & 0 \\ 0 & 0 & \cdots & 0 & 0 \\ \vdots & \vdots & & \vdots & \vdots \\ 0 & 0 & \cdots & 0 & 1 \\ -a_n(k) & -a_{n-1}(k) & \cdots & -a_2(k) & -a_1(k) \end{bmatrix} \begin{bmatrix} x_1(k) \\ x_2(k) \\ \vdots \\ x_{n-1}(k) \\ x_n(k) \end{bmatrix} + \begin{bmatrix} h_1(k) \\ h_2(k) \\ \vdots \\ h_{n-1}(k) \\ h_n(k) \end{bmatrix} u(k)$$

$$y(k) = x_1(k) + b_0(k - n)u(k)$$

Determine $h_1(k), h_2(k), \ldots, h_n(k)$ in terms of $a_i(k)$ and $b_j(k)$, where $i = 1, 2, \ldots, n$ and $j = 0, 1, \ldots, n$. Determine also the initial values of the state variables $x_1(0)$, $x_2(0), \ldots, x_n(0)$ in terms of the input sequence $u(0), u(1), \ldots, u(n - 1)$ and the output sequence $y(0), y(1), \ldots, y(n - 1)$.

Problem B–5–14

If the minimal polynomial of an $n \times n$ matrix **G** involves only distinct roots, then the inverse of $z\mathbf{I} - \mathbf{G}$ can be given by the following expression:

$$(z\mathbf{I} - \mathbf{G})^{-1} = \sum_{k=1}^{m} \frac{\mathbf{X}_k}{z - z_k} \qquad (5\text{–}145)$$

where m is the degree of the minimal polynomial of **G** and the $\mathbf{X}_k$'s are $n \times n$ matrices determined from

$$g_j(\mathbf{G}) = g_j(z_1)\mathbf{X}_1 + g_j(z_2)\mathbf{X}_2 + \cdots + g_j(z_m)\mathbf{X}_m$$

where

$$g_j(\mathbf{G}) = (\mathbf{G} - z_k\,\mathbf{I})^{j-1}, \qquad g_j(z) = (z - z_k)^{j-1}$$

where $j = 1, 2, \ldots, m$ and z_k is any one of the roots of the minimal polynomial of **G**.
Using Equation (5–145), obtain $(z\mathbf{I} - \mathbf{G})^{-1}$ for the following 2×2 matrix **G**:

$$\mathbf{G} = \begin{bmatrix} 0 & 1 \\ 0 & -2 \end{bmatrix}$$

Problem B–5–15

Obtain the pulse transfer function of the system defined by the equations

$$\mathbf{x}(k + 1) = \mathbf{G}\mathbf{x}(k) + \mathbf{H}u(k)$$
$$y(k) = \mathbf{C}\mathbf{x}(k) + Du(k)$$

where

$$\mathbf{G} = \begin{bmatrix} -a_1 & -a_2 & -a_3 \\ 1 & 0 & 0 \\ 0 & 1 & 0 \end{bmatrix}, \quad \mathbf{H} = \begin{bmatrix} 1 \\ 0 \\ 0 \end{bmatrix}$$

$$\mathbf{C} = [b_1 - a_1 b_0 \,\vdots\, b_2 - a_2 b_0 \,\vdots\, b_3 - a_3 b_0], \qquad D = b_0$$

Problem B–5–16

Find the pulse transfer function of the system defined by

$$\mathbf{x}(k + 1) = \mathbf{G}\mathbf{x}(k) + \mathbf{H}u(k)$$
$$y(k) = \mathbf{C}\mathbf{x}(k) + Du(k)$$

where

$$\mathbf{G} = \begin{bmatrix} -a_1 & 1 & 0 \\ -a_2 & 0 & 1 \\ -a_3 & 0 & 0 \end{bmatrix}, \quad \mathbf{H} = \begin{bmatrix} h_1 \\ h_2 \\ h_3 \end{bmatrix}$$

$$\mathbf{C} = [1 \quad 0 \quad 0], \qquad D = b_0$$

Problem B–5–17

Obtain a state-space representation for the system defined by the following pulse-transfer-function matrix:

$$\begin{bmatrix} Y_1(z) \\ Y_2(z) \end{bmatrix} = \begin{bmatrix} \dfrac{1}{1 - z^{-1}} & \dfrac{1 + z^{-1}}{1 - z^{-1}} \\ \dfrac{1}{1 + 0.6z^{-1}} & \dfrac{1 + z^{-1}}{1 + 0.6z^{-1}} \end{bmatrix} \begin{bmatrix} U_1(z) \\ U_2(z) \end{bmatrix}$$

Problem B–5–18

Consider the discrete-time state equation

$$\begin{bmatrix} x_1(k + 1) \\ x_2(k + 1) \end{bmatrix} = \begin{bmatrix} 0 & 1 \\ -0.24 & -1 \end{bmatrix} \begin{bmatrix} x_1(k) \\ x_2(k) \end{bmatrix}$$

Obtain the state transition matrix $\mathbf{\Psi}(k)$.

Problem B–5–19

Consider the system defined by

$$\mathbf{x}(k + 1) = \mathbf{G}\mathbf{x}(k) + \mathbf{H}u(k)$$
$$y(k) = \mathbf{C}\mathbf{x}(k) + Du(k)$$

where matrix $\mathbf{G}$ is a stable matrix.
      Obtain the steady-state values of $\mathbf{x}(k)$ and $\mathbf{y}(k)$ when $\mathbf{u}(k)$ is a constant vector.

Problem B–5–20

Consider the system defined by

$$\mathbf{x}(k + 1) = \mathbf{Gx}(k)$$

where **G** is a stable matrix.

Show that for a positive definite (or positive semidefinite) matrix **Q**

$$J = \sum_{k=0}^{\infty} \mathbf{x}^*(k)\mathbf{Qx}(k)$$

can be given by

$$J = \mathbf{x}^*(0)\mathbf{Px}(0)$$

where $\mathbf{P} = \mathbf{Q} + \mathbf{G}^*\mathbf{PG}$.

Problem B–5–21

Determine a Liapunov function $V(\mathbf{x})$ for the following system:

$$\begin{bmatrix} x_1(k + 1) \\ x_2(k + 1) \end{bmatrix} = \begin{bmatrix} 1 & -1.2 \\ 0.5 & 0 \end{bmatrix} \begin{bmatrix} x_1(k) \\ x_2(k) \end{bmatrix}$$

Problem B–5–22

Determine the stability of the origin of the following discrete-time system:

$$\begin{bmatrix} x_1(k + 1) \\ x_2(k + 1) \\ x_3(k + 1) \end{bmatrix} = \begin{bmatrix} 1 & 3 & 0 \\ -3 & -2 & -3 \\ 1 & 0 & 0 \end{bmatrix} \begin{bmatrix} x_1(k) \\ x_2(k) \\ x_3(k) \end{bmatrix}$$

Problem B–5–23

Determine the stability of the origin of the following discrete-time system:

$$\begin{bmatrix} x_1((k + 1)T) \\ x_2((k + 1)T) \end{bmatrix} = \begin{bmatrix} \cos T & \sin T \\ -\sin T & \cos T \end{bmatrix} \begin{bmatrix} x_1(kT) \\ x_2(kT) \end{bmatrix}$$

Problem B–5–24

Consider the system defined by the equations

$$x_1(k + 1) = x_1(k) + 0.2x_2(k) + 0.4$$

$$x_2(k + 1) = 0.5x_1(k) - 0.5$$

Determine the stability of the equilibrium state.

6

Pole Placement and Observer Design

6–1 INTRODUCTION

In the first part of this chapter we present two fundamental concepts of control systems: controllability and observability. Controllability is concerned with the problem of whether it is possible to steer a system from a given initial state to an arbitrary state: a system is said to be controllable if it is possible by means of an unbounded control vector to transfer the system from any initial state to any other state in a finite number of sampling periods. (Thus, the concept of controllability is concerned with the existence of a control vector that can cause the system's state to reach some arbitrary state.)

Observability is concerned with the problem of determining the state of a dynamic system from observations of the output and control vectors in a finite number of sampling periods. A system is said to be observable if, with the system in state $\mathbf{x}(0)$, it is possible to determine this state from the observation of the output and control vectors over a finite number of sampling periods.

The concepts of controllability and observability were introduced by R. E. Kalman. They play an important role in the optimal control of multivariable systems. In fact, the conditions of controllability and observability may govern the existence of a complete solution to an optimal control problem.

In the second part of this chapter we discuss the pole placement design method and state observers. Note that the concept of controllability is the basis for the solutions of the pole placement problem and the concept of observability plays an important role for the design of state observers. The design method based on pole placement coupled with state observers is one of the fundamental design methods available to control engineers. If the system is completely state controllable, then the desired closed-loop poles in the z plane (or the roots of the characteristic equation) can be selected and the system that will give such closed-loop poles can be designed. The design approach of placing the closed-loop poles in the desired locations in the z plane is called the *pole placement design technique*; that is, in the pole placement design technique we feed back all state variables so that all poles of the closed-loop system are placed at desired locations. In practical control systems, however, measurement of all state variables may not be possible; in that case, not all state variables will be available for feedback. To implement a design based on state feedback, it becomes necessary to estimate the unmeasurable state variables. Such estimation can be done by use of state observers, which will be discussed in detail in this chapter.

The pole placement design process of control systems may be separated into two phases. In the first phase, we design the system assuming that all state variables are available for feedback. In the second phase, we design the state observer that estimates all state variables (or only those that are not directly measurable) that are required for feedback to complete the design.

Note that in the preceding design approach the design parameters are the locations of desired closed-loop poles and the sampling period T. (The sampling period T effectively determines the settling time for response.)

In the analysis in this chapter we assume that the disturbances are impulses that take place randomly. The effect of such impulses is to change the system state. Therefore, a disturbance may be represented as an initial state. It is assumed further that the spacing between adjacent disturbances is sufficiently wide that any response to such a disturbance settles down before the next disturbance takes place, so the system is always ready for the next round.

Although our concern in this chapter is primarily with the regulator problem, we discuss control problems also. Our problem is to reduce the error vector to zero with sufficient speed. In both the regulator problem and control problem the pole placement formulation of the design boils down to the determination of the desired state feedback gain matrix. The procedure for determining the state feedback gain matrix is first to select suitable locations for all closed-loop poles and then to determine the state feedback gain matrix that yields the specified closed-loop poles so that the errors due to disturbances or command inputs can be reduced to zero with sufficient speed. In the final state of the design process the state feedback is accomplished by use of the estimated state variables rather than the actual state variables, which are probably not available for direct measurement. If some of the state variables are measurable, then we may use those available state variables and use estimated state variables for those not actually measurable.

In the last part of this chapter we treat a servo design problem that uses integral control coupled with the pole placement technique and the state observer. Note that in the regulator problem we desire to transfer the nonzero error vector (due to disturbance) to the origin. In the servo problem, we require the output to follow the command input. Note that the servo system must follow the command input and at the same time must solve any regulator problem. Consequently, in the design of a servo system we may begin with the design of a regulator system and then modify the regulator system to a servo system.

Outline of the Chapter.   Section 6-1 has presented an introduction to the material to be presented in this chapter. Section 6-2 discusses the controllability of linear time-invariant control systems. Section 6-3 treats the observability of such systems. Section 6-4 reviews useful transformations in state-space analysis and design that we shall use in the remaining sections of this chapter. The basic state-space design method is presented in Sections 6-5 and 6-6. Section 6-5 presents the pole placement method, the first phase of the design. In the pole placement method we assume that all state variables can be measured and are available for feedback. Section 6-6 discusses the second phase of the design, the design of state observers that estimate the state variables that are not actually measurable. Estimation is based on the measurements of the output and control signals. The estimated state variables can be used for state feedback based on the pole placement design. The final section, Section 6-7, treats servo systems and discusses the design of such systems; the section concludes with a design example.

6-2 CONTROLLABILITY

A control system is said to be completely state controllable if it is possible to transfer the system from any arbitrary initial state to any desired state (also an arbitrary state) in a finite time period. That is, a control system is controllable if every state variable can be controlled in a finite time period by some unconstrained control signal. If any state variable is independent of the control signal, then it is impossible to control this state variable and therefore the system is uncontrollable.

The solution to an optimal control problem may not exist if the system considered is not controllable. Although most physical systems are controllable, the corresponding mathematical models may not possess the property of controllability. Therefore, it is necessary to know the condition under which a system is controllable. We shall see later in Section 6-5 that the concept of controllability plays an important role in arbitrary pole placement of control systems. Now we shall derive this condition in the following.

Complete State Controllability for a Linear Time-Invariant Discrete-Time Control System.   Consider the discrete-time control system defined by

$$\mathbf{x}((k + 1)T) = \mathbf{G}\mathbf{x}(kT) + \mathbf{H}u(kT) \qquad (6-1)$$

where

$\mathbf{x}(kT)$ = state vector (n-vector) at kth sampling instant

$u(kT)$ = control signal at kth sampling instant

$\mathbf{G} = n \times n$ matrix

$\mathbf{H} = n \times 1$ matrix

T = sampling period

We assume that $u(kT)$ is constant for $kT \leq t < (k + 1)T$.

The discrete-time control system given by Equation (6-1) is said to be completely state controllable or simply state controllable if there exists a piecewise-constant control signal $u(kT)$ defined over a finite number of sampling periods such that, starting from any initial state, the state $\mathbf{x}(kT)$ can be transferred to the desired state $\mathbf{x}_f$ in at most n sampling periods. (In discussing controllability, the desired state $\mathbf{x}_f$ may be specified as the origin, or $\mathbf{x}_f = \mathbf{0}$. See Problem A–6–1. Here, however, we assume that $\mathbf{x}_f$ is an arbitrary state in the n-dimensional space, including the origin.)

Using the definition just given, we shall now derive the condition for complete state controllability. Since the solution of Equation (6–1) is

$$\mathbf{x}(nT) = \mathbf{G}^n \mathbf{x}(0) + \sum_{j=0}^{n-1} \mathbf{G}^{n-j-1} \mathbf{H} u(jT)$$

$$= \mathbf{G}^n \mathbf{x}(0) + \mathbf{G}^{n-1} \mathbf{H} u(0) + \mathbf{G}^{n-2} \mathbf{H} u(T) + \cdots + \mathbf{H} u((n-1)T)$$

we obtain

$$\mathbf{x}(nT) - \mathbf{G}^n \mathbf{x}(0) = [\mathbf{H} \vdots \mathbf{GH} \vdots \cdots \vdots \mathbf{G}^{n-1} \mathbf{H}] \begin{bmatrix} u((n-1))T) \\ u((n-2)T) \\ \vdots \\ u(0) \end{bmatrix} \tag{6–2}$$

Since $\mathbf{H}$ is an $n \times 1$ matrix, we find that each of the matrices $\mathbf{H}, \mathbf{GH}, \ldots, \mathbf{G}^{n-1}\mathbf{H}$ is an $n \times 1$ matrix or column vector. If the rank of the following matrix is n, or

$$\operatorname{rank}[\mathbf{H} \vdots \mathbf{GH} \vdots \cdots \vdots \mathbf{G}^{n-1}\mathbf{H}] = n \tag{6–3}$$

then n vectors $\mathbf{H}, \mathbf{GH}, \ldots, \mathbf{G}^{n-1}\mathbf{H}$ can span the n-dimensional space. The matrix

$$[\mathbf{H} \vdots \mathbf{GH} \vdots \cdots \vdots \mathbf{G}^{n-1}\mathbf{H}]$$

is commonly called the *controllability matrix*. (Note that all states that can be reached from the origin are spanned by the columns of the controllability matrix.) Thus, if the rank of the controllability matrix is n, then for an arbitrary state $\mathbf{x}(nT) = \mathbf{x}_f$, there exists a sequence of unbounded control signals $u(0), u(T), \ldots, u((n-1)T)$ that satisfies Equation (6–2). Hence, the condition that the rank of the controllability matrix be n gives a sufficient condition for complete state controllability.

To prove that Equation (6–3) is also a necessary condition for complete state controllability, let us assume that

$$\operatorname{rank}[\mathbf{H} \vdots \mathbf{GH} \vdots \cdots \vdots \mathbf{G}^{n-1}\mathbf{H}] < n$$

Then, by use of the Cayley–Hamilton theorem, it can be shown that, for an arbitrary i, $\mathbf{G}^i \mathbf{H}$ can be expressed as a linear combination of $\mathbf{H}, \mathbf{GH}, \ldots, \mathbf{G}^{n-1}\mathbf{H}$. Consequently, we have for any i

$$\operatorname{rank}[\mathbf{H} \vdots \mathbf{GH} \vdots \cdots \vdots \mathbf{G}^{i-1}\mathbf{H}] < n$$

and so the vectors $\mathbf{H}, \mathbf{GH}, \ldots, \mathbf{G}^{i-1}\mathbf{H}$ cannot span the n-dimensional space; and therefore, for some $\mathbf{x}_f$, it is not possible to have $\mathbf{x}(iT) = \mathbf{x}_f$ for all i. Thus, the condition given by Equation (6–3) is necessary. Consequently, we find the rank condition given by Equation (6–3) to be a necessary and sufficient condition for complete state controllability.

If the system defined by Equation (6–1) is completely state controllable, then it is possible to transfer any initial state to any arbitrary state in at most n sampling periods. Note, however, that this is true if and only if the magnitude of $u(kT)$ is unbounded. If the magnitude of $u(kT)$ is bounded, it may take more than n sampling periods. (See Problem A–6–2.)

Complete State Controllability in the Case Where* u(kT) *Is a Vector. If the system is defined by

$$\mathbf{x}((k+1)T) = \mathbf{G}\mathbf{x}(kT) + \mathbf{H}\mathbf{u}(kT)$$

where $\mathbf{x}(kT)$ is an n-vector, $\mathbf{u}(kT)$ is an r-vector, $\mathbf{G}$ is an $n \times n$ matrix, and $\mathbf{H}$ is an $n \times r$ matrix, then it can be proved that the condition for complete state controllability is that the $n \times nr$ matrix

$$[\mathbf{H} \vdots \mathbf{GH} \vdots \cdots \vdots \mathbf{G}^{n-1}\mathbf{H}]$$

be of rank n, or that

$$\text{rank}\,[\mathbf{H} \vdots \mathbf{GH} \vdots \cdots \vdots \mathbf{G}^{n-1}\mathbf{H}] = n$$

Determination of Control Sequence to Bring the Initial State to a Desired State. If the matrix

$$[\mathbf{H} \vdots \mathbf{GH} \vdots \cdots \vdots \mathbf{G}^{n-1}\mathbf{H}]$$

is of rank n and $u(kT)$ is a scalar, then it is possible to find n linearly independent scalar equations from which a sequence of unbounded control signals $u(kT)$ $(k = 0, 1, 2, \ldots, n - 1)$ can be uniquely determined such that any initial state $\mathbf{x}(0)$ is transferred to the desired state in n sampling periods. See Equation (6–2).

Note also that, if the control signal is not a scalar but a vector, then the sequence of $\mathbf{u}(kT)$ is not unique. Then there exists more than one sequence of control vector $\mathbf{u}(kT)$ to bring the initial state $\mathbf{x}(0)$ to a desired state in not more than n sampling periods.

Alternative Form of the Condition for Complete State Controllability. Consider the system defined by

$$\mathbf{x}((k+1)T) = \mathbf{G}\mathbf{x}(kT) + \mathbf{H}\mathbf{u}(kT) \qquad (6\text{–}4)$$

where

$$\mathbf{x}(kT) = \text{state vector (n-vector) at kth sampling instant}$$

$$\mathbf{u}(kT) = \text{control vector (r-vector) at kth sampling instant}$$

$$\mathbf{G} = n \times n \text{ matrix}$$

$$\mathbf{H} = n \times r \text{ matrix}$$

$$T = \text{sampling period}$$

If the eigenvectors of $\mathbf{G}$ are distinct, then it is possible to find a transformation matrix $\mathbf{P}$ such that

$$\mathbf{P}^{-1}\mathbf{GP} = \begin{bmatrix} \lambda_1 & & & & \\ & \lambda_2 & & & 0 \\ & & \cdot & & \\ & & & \cdot & \\ & & & & \cdot \\ 0 & & & & \lambda_n \end{bmatrix}$$

Note that if the eigenvalues of $\mathbf{G}$ are distinct then the eigenvectors of $\mathbf{G}$ are distinct. However, the converse is not true. (For example, an $n \times n$ real symmetric matrix having multiple eigenvalues has n distinct eigenvectors.) Note also that the ith column of the $\mathbf{P}$ matrix is an eigenvector of $\mathbf{G}$ associated with ith eigenvalue λ_i $(i = 1, 2, \dots, n)$. Let us define

$$\mathbf{x}(kT) = \mathbf{P}\hat{\mathbf{x}}(kT) \tag{6-5}$$

Substituting Equation (6–5) into Equation (6–4), we obtain

$$\hat{\mathbf{x}}((k + 1)T) = \mathbf{P}^{-1}\mathbf{GP}\hat{\mathbf{x}}(kT)) + \mathbf{P}^{-1}\mathbf{H}\mathbf{u}(kT) \tag{6-6}$$

Let us define

$$\mathbf{P}^{-1}\mathbf{H} = \mathbf{F} = \begin{bmatrix} f_{11} & f_{12} & \cdots & f_{1r} \\ f_{21} & f_{22} & \cdots & f_{2r} \\ \vdots & \vdots & & \vdots \\ f_{n1} & f_{n2} & \cdots & f_{nr} \end{bmatrix}$$

Then, Equation (6–6) may be written as follows:

$$\hat{x}_1((k + 1)T) = \lambda_1 \hat{x}_1(kT) + f_{11} u_1(kT) + f_{12} u_2(kT) + \cdots + f_{1r} u_r(kT)$$
$$\hat{x}_2((k + 1)T) = \lambda_2 \hat{x}_2(kT) + f_{21} u_1(kT) + f_{22} u_2(kT) + \cdots + f_{2r} u_r(kT)$$
$$\vdots$$
$$\hat{x}_n((k + 1)T) = \lambda_n \hat{x}_n(kT) + f_{n1} u_1(kT) + f_{n2} u_2(kT) + \cdots + f_{nr} u_r(kT)$$

If the elements of any one row of the $n \times r$ matrix $\mathbf{F}$ are all zero, then the corresponding state variable cannot be controlled by any of the $u_i(kT)$. Hence, the condition for complete state controllability is that, if the eigenvectors of $\mathbf{G}$ are distinct, then the system is completely state controllable if and only if no row of $\mathbf{P}^{-1}\mathbf{H}$ has all zero elements. It is important to note that to apply this condition for complete state controllability we must put the matrix $\mathbf{P}^{-1}\mathbf{GP}$ in Equation (6–6) into diagonal form.

   If the $\mathbf{G}$ matrix in Equation (6–4) does not possess distinct eigenvectors, then diagonalization is impossible. In such a case, we may transform $\mathbf{G}$ into a Jordan canonical form. If, for example, $\mathbf{G}$ has eigenvalues $\lambda_1, \lambda_1, \lambda_1, \lambda_4, \lambda_4, \lambda_6, \dots, \lambda_n$ and has $n - 3$ distinct eigenvectors, then the Jordan canonical form of $\mathbf{G}$ is

$$
J = \begin{bmatrix}
\lambda_1 & 1 & 0 & & & & & & 0 \\
0 & \lambda_1 & 1 & & & & & & \\
0 & 0 & \lambda_1 & & & & & & \\
& & & \lambda_4 & 1 & & & & \\
& & & 0 & \lambda_4 & & & & \\
& & & & & \lambda_6 & & & 0 \\
& & & & & & \cdot & & \\
& & & & & & & \cdot & \\
0 & & & & 0 & & & & \lambda_n
\end{bmatrix}
$$

The 3×3 and 2×2 submatrices on the main diagonal are called *Jordan blocks*. Suppose it is possible to find a transformation matrix $\mathbf{S}$ such that

$$\mathbf{S}^{-1}\mathbf{GS} = \mathbf{J}$$

If we define a new state vector $\hat{\mathbf{x}}$ by

$$\mathbf{x}(kT) = \mathbf{S}\hat{\mathbf{x}}(kT) \tag{6-7}$$

then substituting Equation (6–7) into Equation (6–4) gives

$$\hat{\mathbf{x}}((k+1)T) = \mathbf{S}^{-1}\mathbf{GS}\hat{\mathbf{x}}(kT) + \mathbf{S}^{-1}\mathbf{H}\mathbf{u}(kT)$$

$$= \mathbf{J}\hat{\mathbf{x}}(kT) + \mathbf{S}^{-1}\mathbf{H}\mathbf{u}(kT) \tag{6-8}$$

The conditions for the complete state controllability of the system of Equation (6–8) may then be stated as follows: The system is completely state controllable if and only if (1) no two Jordan blocks in $\mathbf{J}$ of Equation (6–8) are associated with the same eigenvalues, (2) the elements of any row of $\mathbf{S}^{-1}\mathbf{H}$ that corresponds to the last row of each Jordan block are not all zero, and (3) the elements of each row of $\mathbf{S}^{-1}\mathbf{H}$ that correspond to distinct eigenvalues are not all zero.

Comments.   In the preceding conditions for state controllability, it is stated that no two Jordan blocks in $\mathbf{J}$ of Equation (6–8) should be associated with the same eigenvalues. This point is elaborated as follows.

Consider the following system where the two Jordan blocks are associated with the same eigenvalues λ_1:

$$
\begin{bmatrix} x_1(k+1) \\ x_2(k+1) \\ x_3(k+1) \end{bmatrix} = \begin{bmatrix} \lambda_1 & 0 & 0 \\ 0 & \lambda_1 & 1 \\ 0 & 0 & \lambda_1 \end{bmatrix} \begin{bmatrix} x_1(k) \\ x_2(k) \\ x_3(k) \end{bmatrix} + \begin{bmatrix} 1 \\ 1 \\ 1 \end{bmatrix} u(k)
$$

Although every state variable is affected by $u(k)$, this system is uncontrollable, since the rank of the controllability matrix

$$
[\mathbf{H} \vdots \mathbf{GH} \vdots \mathbf{G}^2\mathbf{H}] = \begin{bmatrix} 1 & \lambda_1 & \lambda_1^2 \\ 1 & \lambda_1 + 1 & \lambda_1^2 + 2\lambda_1 \\ 1 & \lambda_1 & \lambda_1^2 \end{bmatrix}
$$

is 2. Hence, in applying the preceding criterion for state controllability, no two Jordan blocks in $\mathbf{J}$ of Equation (6–8) should be associated with the same eigenvalues.

Example 6–1

The following systems are completely state controllable:

1. $\begin{bmatrix} x_1(k+1) \\ x_2(k+1) \end{bmatrix} = \begin{bmatrix} -1 & 0 \\ 0 & -2 \end{bmatrix} \begin{bmatrix} x_1(k) \\ x_2(k) \end{bmatrix} + \begin{bmatrix} 2 \\ 3 \end{bmatrix} [u(k)]$

2. $\begin{bmatrix} x_1(k+1) \\ x_2(k+1) \\ x_3(k+1) \\ x_4(k+1) \\ x_5(k+1) \end{bmatrix} = \begin{bmatrix} -2 & 1 & 0 & \vdots & 0 \\ 0 & -2 & 1 & \vdots & \\ 0 & 0 & -2 & \vdots & \\ & & \vdots & -5 & 1 \\ 0 & & \vdots & 0 & -5 \end{bmatrix} \begin{bmatrix} x_1(k) \\ x_2(k) \\ x_3(k) \\ x_4(k) \\ x_5(k) \end{bmatrix} + \begin{bmatrix} 0 & 1 \\ 0 & 0 \\ 3 & 0 \\ 0 & 0 \\ 2 & 1 \end{bmatrix} \begin{bmatrix} u_1(k) \\ u_2(k) \end{bmatrix}$

The following systems are not completely state controllable:

1. $\begin{bmatrix} x_1(k+1) \\ x_2(k+1) \end{bmatrix} = \begin{bmatrix} -1 & 0 \\ 0 & -2 \end{bmatrix} \begin{bmatrix} x_1(k) \\ x_2(k) \end{bmatrix} + \begin{bmatrix} 2 \\ 0 \end{bmatrix} [u(k)]$

2. $\begin{bmatrix} x_1(k+1) \\ x_2(k+1) \\ x_3(k+1) \\ x_4(k+1) \\ x_5(k+1) \end{bmatrix} = \begin{bmatrix} -2 & 1 & 0 & \vdots & 0 \\ 0 & -2 & 1 & \vdots & \\ 0 & 0 & -2 & \vdots & \\ & & \vdots & -5 & 1 \\ 0 & & \vdots & 0 & -5 \end{bmatrix} \begin{bmatrix} x_1(k) \\ x_2(k) \\ x_3(k) \\ x_4(k) \\ x_5(k) \end{bmatrix} + \begin{bmatrix} 0 & 1 \\ 3 & 0 \\ 0 & 0 \\ 2 & 1 \\ 0 & 0 \end{bmatrix} \begin{bmatrix} u_1(k) \\ u_2(k) \end{bmatrix}$

Condition for Complete State Controllability in the z Plane.   The condition for complete state controllability can be stated in terms of pulse transfer functions.

A necessary and sufficient condition for complete state controllability is that no cancellation occur in the pulse transfer function. If cancellation occurs, the system cannot be controlled in the direction of the canceled mode. (See Problem A–6–4.)

Example 6–2

Consider the following pulse transfer function:

$$\frac{Y(z)}{U(z)} = \frac{z + 0.2}{(z + 0.8)(z + 0.2)}$$

Clearly, cancellation of factors $(z + 0.2)$ in the numerator and denominator occurs. Thus, one degree of freedom is lost. Because of this cancellation, this system is not completely state controllable.

The same conclusion can be obtained, of course, by writing this pulse transfer function in the form of state equations. A possible state-space representation for this system is

$$\begin{bmatrix} x_1(k+1) \\ x_2(k+1) \end{bmatrix} = \begin{bmatrix} 0 & 1 \\ -0.16 & -1 \end{bmatrix} \begin{bmatrix} x_1(k) \\ x_2(k) \end{bmatrix} + \begin{bmatrix} 1 \\ -0.8 \end{bmatrix} u(k)$$

$$y(k) = \begin{bmatrix} 1 & 0 \end{bmatrix} \begin{bmatrix} x_1(k) \\ x_2(k) \end{bmatrix}$$

Since

$$[\mathbf{H} \vdots \mathbf{GH}] = \begin{bmatrix} 1 & -0.8 \\ -0.8 & 0.64 \end{bmatrix}$$

the rank of $[\mathbf{H} \vdots \mathbf{GH}]$ is 1. Therefore, we arrive at the same conclusion: that the system is not completely state controllable.

Complete Output Controllability.   In the practical design of a control system, we may want to control the output rather than the state of the system. Complete state controllability is neither necessary nor sufficient for controlling the output of the system. For this reason, it is necessary to define separately complete output controllability.

Consider the system defined by the equations

$$\mathbf{x}((k + 1)T) = \mathbf{Gx}(kT) + \mathbf{H}u(kT) \tag{6–9}$$

$$\mathbf{y}(kT) = \mathbf{Cx}(kT) \tag{6–10}$$

where

$\mathbf{x}(kT)$ = state vector (n-vector) at kth sampling instant

$u(kT)$ = control signal (scalar) at kth sampling instant

$\mathbf{y}(kT)$ = output vector (m-vector) at kth sampling instant

$\mathbf{G} = n \times n$ matrix

$\mathbf{H} = n \times 1$ matrix

$\mathbf{C} = m \times n$ matrix

The system defined by Equations (6–9) and (6–10) is said to be completely output controllable, or simply output controllable, if it is possible to construct an unconstrained control signal $u(kT)$ defined over a finite number of sampling periods $0 \le kT < nT$ such that, starting from any initial output $\mathbf{y}(0)$, the output $\mathbf{y}(kT)$ can be transferred to the desired point (an arbitrary point) $\mathbf{y}_f$ in the output space in at most n sampling periods.

In what follows we derive the condition for complete output controllability. Note that, if a system is completely output controllable, then a piecewise constant control signal exists that will transfer any initial output to any desired point $\mathbf{y}_f$ in the output space in at most n sampling periods. Since the solution of Equation (6–9) is

$$\mathbf{x}(nT) = \mathbf{G}^n \mathbf{x}(0) + \sum_{j=0}^{n-1} \mathbf{G}^{n-j-1} \mathbf{H}u(jT)$$

we have

$$\mathbf{y}(nT) = \mathbf{Cx}(nT)$$

$$= \mathbf{CG}^n \mathbf{x}(0) + \sum_{j=0}^{n-1} \mathbf{CG}^{n-j-1} \mathbf{H}u(jT)$$

or

$$\mathbf{y}(nT) - \mathbf{CG}^n \mathbf{x}(0) = \sum_{j=0}^{n-1} \mathbf{CG}^{n-j-1} \mathbf{H}u(jT)$$

$$= \mathbf{CG}^{n-1} \mathbf{H}u(0) + \mathbf{CG}^{n-2} \mathbf{H}u(T) + \cdots + \mathbf{CH}u((n - 1)T)$$

$$= [\mathbf{CH} \vdots \mathbf{CGH} \vdots \cdots \vdots \mathbf{CG}^{n-1}\mathbf{H}] \begin{bmatrix} u((n - 1)T) \\ u((n - 2)T) \\ \vdots \\ u(0) \end{bmatrix}$$

Note that $\mathbf{y}(nT) - \mathbf{CG}^n\mathbf{x}(0) = \mathbf{y}_f - \mathbf{CG}^n\mathbf{x}(0)$ represents an arbitrary point in the m-dimensional output space. Thus, as in the case of complete state controllability, a necessary and sufficient condition for the system to be completely output controllable is that vectors $\mathbf{CH}, \mathbf{CGH}, \ldots, \mathbf{CG}^{n-1}\mathbf{H}$ span the m-dimensional output space, or that

$$\text{rank}\,[\mathbf{CH} \vdots \mathbf{CGH} \vdots \cdots \vdots \mathbf{CG}^{n-1}\mathbf{H}] = m \qquad (6\text{–}11)$$

From this analysis it can be seen that, in the system defined by Equations (6–9) and (6–10), complete state controllability implies complete output controllability if and only if the m rows of $\mathbf{C}$ are linearly independent.

Next, consider the system defined by the equations

$$\mathbf{x}((k+1)T) = \mathbf{Gx}(kT) + \mathbf{Hu}(kT) \qquad (6\text{–}12)$$

$$\mathbf{y}(kT) = \mathbf{Cx}(kT) + \mathbf{Du}(kT) \qquad (6\text{–}13)$$

where

$\mathbf{x}(kT)$ = state vector (n-vector) at kth sampling instant

$\mathbf{u}(kT)$ = control vector (r-vector) at kth sampling instant

$\mathbf{y}(kT)$ = output vector (m-vector) at kth sampling instant

$\mathbf{G}$ = $n \times n$ matrix

$\mathbf{H}$ = $n \times r$ matrix

$\mathbf{C}$ = $m \times n$ matrix

$\mathbf{D}$ = $m \times r$ matrix

The condition for complete output controllability for this system can be derived as follows. Since the output $\mathbf{y}(nT)$ can be given by the equation

$$\mathbf{y}(nT) = \mathbf{Cx}(nT) + \mathbf{Du}(nT)$$

$$= \mathbf{CG}^n\mathbf{x}(0) + \sum_{j=0}^{n-1} \mathbf{CG}^{n-j-1}\mathbf{Hu}(jT) + \mathbf{Du}(nT)$$

we obtain

$$\mathbf{y}(nT) - \mathbf{CG}^n\mathbf{x}(0) = \sum_{j=0}^{n-1} \mathbf{CG}^{n-j-1}\mathbf{Hu}(jT) + \mathbf{Du}(nT)$$

$$= \mathbf{CG}^{n-1}\mathbf{Hu}(0) + \mathbf{CG}^{n-2}\mathbf{Hu}(T) + \cdots + \mathbf{CHu}((n-1)T)$$

$$+ \mathbf{Du}(nT)$$

$$= [\mathbf{D} \vdots \mathbf{CH} \vdots \mathbf{CGH} \vdots \cdots \vdots \mathbf{CG}^{n-1}\mathbf{H}] \begin{bmatrix} \mathbf{u}(nT) \\ \mathbf{u}((n-1)T) \\ \vdots \\ \mathbf{u}(0) \end{bmatrix}$$

A necessary and sufficient condition for the system defined by Equations (6–12) and (6–13) to be completely output controllable is that the $m \times (n+1)r$ matrix

$$[\mathbf{D} \vdots \mathbf{CH} \vdots \mathbf{CGH} \vdots \cdots \vdots \mathbf{CG}^{n-1}\mathbf{H}]$$

be of rank m:

$$\text{rank} [\mathbf{D} \vdots \mathbf{CH} \vdots \mathbf{CGH} \vdots \cdots \vdots \mathbf{CG}^{n-1}\mathbf{H}] = m \qquad (6\text{–}14)$$

It is noted that the presence of matrix $\mathbf{D}$ in the system output equation always helps to establish complete output controllability.

Controllability of a Linear Time-Invariant Continuous-Time Control System.   In what follows, we shall briefly state the conditions for complete state controllability and output controllability of linear time-invariant continuous-time control systems. Consider the system defined by

$$\dot{\mathbf{x}} = \mathbf{Ax} + \mathbf{Bu}$$

$$\mathbf{y} = \mathbf{Cx} + \mathbf{Du}$$

where

$$\mathbf{x} = \text{state vector } (n\text{-vector})$$

$$\mathbf{u} = \text{control vector } (r\text{-vector})$$

$$\mathbf{y} = \text{output vector } (m\text{-vector})$$

$$\mathbf{A} = n \times n \text{ matrix}$$

$$\mathbf{B} = n \times r \text{ matrix}$$

$$\mathbf{C} = m \times n \text{ matrix}$$

$$\mathbf{D} = m \times r \text{ matrix}$$

Complete State Controllability.   A necessary and sufficient condition for complete state controllability for this system can be derived in a way similar to what was used in the case of the discrete-time system. Here, we shall present only the result. The condition for complete state controllability is that the $n \times nr$ matrix

$$[\mathbf{B} \vdots \mathbf{AB} \vdots \cdots \vdots \mathbf{A}^{n-1}\mathbf{B}]$$

be of rank n or that it contains n linearly independent column vectors. (This matrix is commonly called the controllability matrix for the continuous-time system.)

The condition for complete state controllability can also be stated in terms of transfer functions or transfer matrices. A necessary and sufficient condition for complete state controllability is that no cancellation occur in the transfer function or transfer matrix. If cancellation occurs, the system cannot be controlled in the direction of the canceled mode.

Output Controllability.   As in the case of the discrete-time control system, complete state controllability is neither necessary nor sufficient for controlling the output of a linear time-invariant continuous-time control system. It can be proved that the condition for complete output controllability is that the rank of the $m \times (n + 1)r$ matrix

$$[\mathbf{D} \vdots \mathbf{CB} \vdots \mathbf{CAB} \vdots \mathbf{CA}^2\mathbf{B} \vdots \cdots \vdots \mathbf{CA}^{n-1}\mathbf{B}]$$

be m.

6–3 OBSERVABILITY

In this section, we shall discuss the observability of linear time-invariant discrete-time control systems. Consider the unforced discrete-time control system defined by

$$\mathbf{x}((k + 1)T) = \mathbf{Gx}(kT) \tag{6–15}$$

$$\mathbf{y}(kT) = \mathbf{Cx}(kT) \tag{6–16}$$

where

$\mathbf{x}(kT)$ = state vector (n-vector) at kth sampling instant

$\mathbf{y}(kT)$ = output vector (m-vector) at kth sampling instant

$\mathbf{G}$ = $n \times n$ matrix

$\mathbf{C}$ = $m \times n$ matrix

The system is said to be completely observable if every initial state $\mathbf{x}(0)$ can be determined from the observation of $\mathbf{y}(kT)$ over a finite number of sampling periods. The system, therefore, is completely observable if every transition of the state eventually affects every element of the output vector.

The concept of observability is useful in solving the problem of reconstructing unmeasurable state variables. It will be seen later that state feedback control systems designed by the pole placement method will require feedback of weighted state variables. In practice, however, the difficulty encountered with state feedback control systems is that some of the state variables are not accessible for direct measurement. Then it becomes necessary to estimate the unmeasurable state variables in order to construct the feedback control signals. In Section 6–6 we shall see that the concept of observability plays a dominant role in the design of state observers.

The reason we are considering the unforced system is as follows. If the system is described by the equations

$$\mathbf{x}((k + 1)T) = \mathbf{Gx}(kT) + \mathbf{Hu}(kT)$$

$$\mathbf{y}(kT) = \mathbf{Cx}(kT) + \mathbf{Du}(kT)$$

then

$$\mathbf{x}(kT) = \mathbf{G}^k \mathbf{x}(0) + \sum_{j=0}^{k-1} \mathbf{G}^{k-j-1} \mathbf{Hu}(jT)$$

and $\mathbf{y}(kT)$ is

$$\mathbf{y}(kT) = \mathbf{CG}^k \mathbf{x}(0) + \sum_{j=0}^{k-1} \mathbf{CG}^{k-j-1} \mathbf{Hu}(jT) + \mathbf{Du}(kT)$$

Since the matrices $\mathbf{G}$, $\mathbf{H}$, $\mathbf{C}$, and $\mathbf{D}$ are known and $\mathbf{u}(kT)$ is also known, the second and third terms on the right-hand side of this last equation are known quantities. Therefore, they may be subtracted from the observed value of $\mathbf{y}(kT)$. Hence, for investigating a necessary and sufficient condition for complete observability, it suffices to consider the system described by Equations (6–15) and (6–16).

Once $\mathbf{x}(0)$ can be determined from the observation of the output, $\mathbf{x}(kT)$ can also be determined, since $u(0), u(T), \ldots, u((k-1)T)$ are known.

Complete Observability of Discrete-Time Systems.   Consider the system defined by Equations (6-15) and (6-16). The system is completely observable if, given the output $\mathbf{y}(kT)$ over a finite number of sampling periods, it is possible to determine the initial state vector $\mathbf{x}(0)$.

In what follows we shall derive the condition for the complete observability of the discrete-time system described by Equations (6-15) and (6-16). Since the solution $\mathbf{x}(kT)$ of Equation (6-15) is

$$\mathbf{x}(kT) = \mathbf{G}^k \mathbf{x}(0)$$

we obtain

$$\mathbf{y}(kT) = \mathbf{C}\mathbf{G}^k \mathbf{x}(0)$$

Complete observability means that, given $\mathbf{y}(0), \mathbf{y}(T), \mathbf{y}(2T), \ldots$, it is possible to determine $x_1(0), x_2(0), \ldots, x_n(0)$. To determine n unknowns, we need only n values of $\mathbf{y}(kT)$. Hence, we may use the first n values of $\mathbf{y}(kT)$, or $\mathbf{y}(0), \mathbf{y}(T), \ldots, \mathbf{y}((n-1)T)$ for the determination of $x_1(0), x_2(0), \ldots, x_n(0)$.

For a completely observable system, given

$$\mathbf{y}(0) = \mathbf{C}\mathbf{x}(0)$$

$$\mathbf{y}(T) = \mathbf{C}\mathbf{G}\mathbf{x}(0)$$

$$\vdots$$

$$\mathbf{y}((n-1)T) = \mathbf{C}\mathbf{G}^{n-1}\mathbf{x}(0)$$

we must be able to determine $x_1(0), x_2(0), \ldots, x_n(0)$. Noting that $\mathbf{y}(kT)$ is an m vector, the preceding n simultaneous equations yield nm equations, all involving $x_1(0), x_2(0), \ldots, x_n(0)$. To obtain a unique set of solutions $x_1(0), x_2(0), \ldots, x_n(0)$ from these nm equations, we must be able to write exactly n linearly independent equations among them. This requires that the $nm \times n$ matrix

$$\begin{bmatrix} \mathbf{C} \\ \hline \mathbf{C}\mathbf{G} \\ \hline \vdots \\ \hline \mathbf{C}\mathbf{G}^{n-1} \end{bmatrix}$$

be of rank n.

Noting that the rank of a matrix and that of the conjugate transpose of the matrix are the same, it is possible to state the condition for complete observability as follows. A necessary and sufficient condition for the system defined by Equations (6-15) and (6-16) to be completely observable is that the rank of the $n \times nm$ matrix

$$[\mathbf{C}^* \vdots \mathbf{G}^*\mathbf{C}^* \vdots \cdots \vdots (\mathbf{G}^*)^{n-1} \mathbf{C}^*] \tag{6-17}$$

be n. The matrix given by (6-17) is commonly called the *observability matrix*. [Note that in (6-17), asterisks indicate conjugate transposes. If matrices $\mathbf{C}$ and $\mathbf{G}$ are real, then the conjugate transpose notation such as $\mathbf{G}^*\mathbf{C}^*$ may be changed to the transpose notation such as $\mathbf{G}^T\mathbf{C}^T$.]

Alternative Form of the Condition for Complete Observability.   Consider the system defined by Equations (6–15) and (6–16), repeated here:

$$\mathbf{x}((k + 1)T) = \mathbf{Gx}(kT) \qquad\qquad (6\text{–}18)$$

$$\mathbf{y}(kT) = \mathbf{Cx}(kT) \qquad\qquad (6\text{–}19)$$

Suppose the eigenvalues of $\mathbf{G}$ are distinct, and a transformation matrix $\mathbf{P}$ transforms $\mathbf{G}$ into a diagonal matrix, so that $\mathbf{P}^{-1}\mathbf{GP}$ is a diagonal matrix. Let us define

$$\mathbf{x}(kT) = \mathbf{P}\hat{\mathbf{x}}(kT)$$

Then Equations (6–18) and (6–19) can be written as follows:

$$\hat{\mathbf{x}}((k + 1)T) = \mathbf{P}^{-1}\mathbf{GP}\hat{\mathbf{x}}(kT)$$

$$\mathbf{y}(kT) = \mathbf{CP}\hat{\mathbf{x}}(kT)$$

Hence,

$$\mathbf{y}(nT) = \mathbf{CP}(\mathbf{P}^{-1}\mathbf{GP})^n\,\hat{\mathbf{x}}(0)$$

or

$$\mathbf{y}(nT) = \mathbf{CP}\begin{bmatrix} \lambda_1^n & & & \\ & \lambda_2^n & & \\ & & \ddots & \\ & & & \lambda_n^n \end{bmatrix}\hat{\mathbf{x}}(0) = \mathbf{CP}\begin{bmatrix} \lambda_1^n \hat{x}_1(0) \\ \lambda_2^n \hat{x}_2(0) \\ \vdots \\ \lambda_n^n \hat{x}_n(0) \end{bmatrix}$$

where $\lambda_1, \lambda_2, \ldots, \lambda_n$ are n distinct eigenvalues of $\mathbf{G}$. The system is completely observable if and only if none of the columns of the $m \times n$ matrix $\mathbf{CP}$ consists of all zero elements. This is because, if the ith column of $\mathbf{CP}$ consists of all zero elements, then the state variable $\hat{x}_i(0)$ will not appear in the output equation and therefore cannot be determined from observation of $\mathbf{y}(kT)$. Thus, $\mathbf{x}(0)$, which is related to $\hat{\mathbf{x}}(0)$ by the nonsingular matrix $\mathbf{P}$, cannot be determined.

If the matrix $\mathbf{G}$ involves multiple eigenvalues and cannot be transformed into a diagonal matrix, then by using a suitable transformation matrix $\mathbf{S}$ we may transform $\mathbf{G}$ into the Jordan canonical form:

$$\mathbf{S}^{-1}\mathbf{GS} = \mathbf{J}$$

where $\mathbf{J}$ is in the Jordan canonical form. Let us define

$$\mathbf{x}(kT) = \mathbf{S}\hat{\mathbf{x}}(kT)$$

Then Equations (6–18) and (6–19) can be written as follows:

$$\hat{\mathbf{x}}((k + 1)T) = \mathbf{S}^{-1}\mathbf{GS}\hat{\mathbf{x}}(kT) = \mathbf{J}\hat{\mathbf{x}}(kT)$$

$$\mathbf{y}(kT) = \mathbf{CS}\hat{\mathbf{x}}(kT)$$

Hence,

$$\mathbf{y}(nT) = \mathbf{CS}(\mathbf{S}^{-1}\mathbf{GS})^n\,\hat{\mathbf{x}}(0)$$

The system is completely observable if and only if (1) no two Jordan blocks in **J** are associated with the same eigenvalue, (2) none of the columns of **CS** that corresponds to the first row of each Jordan block consists of all zero elements, and (3) no columns of **CS** that correspond to distinct eigenvalues consist of all zero elements.

To clarify condition 2, in Example 6–3, which follows, we have enclosed in dashed lines the columns of **CS** that correspond to the first row of each Jordan block.

Example 6–3

The following systems are completely observable.

1. $\begin{bmatrix} x_1((k+1)T) \\ x_2((k+1)T) \end{bmatrix} = \begin{bmatrix} -1 & 0 \\ 0 & -2 \end{bmatrix} \begin{bmatrix} x_1(kT) \\ x_2(kT) \end{bmatrix},$ $y(kT) = \begin{bmatrix} 1 & 5 \end{bmatrix} \begin{bmatrix} x_1(kT) \\ x_2(kT) \end{bmatrix}$

2. $\begin{bmatrix} x_1((k+1)T) \\ x_2((k+1)T) \\ x_3((k+1)T) \\ x_4((k+1)T) \\ x_5((k+1)T) \end{bmatrix} = \begin{bmatrix} 2 & 1 & 0 & & 0 \\ 0 & 2 & 1 & & \\ 0 & 0 & 2 & & \\ \hline & & & -3 & 1 \\ 0 & & & 0 & -3 \end{bmatrix} \begin{bmatrix} x_1(kT) \\ x_2(kT) \\ x_3(kT) \\ x_4(kT) \\ x_5(kT) \end{bmatrix}$

$\begin{bmatrix} y_1(kT) \\ y_2(kT) \end{bmatrix} = \begin{bmatrix} 1 & 1 & 1 & 0 & 1 \\ 0 & 1 & 1 & 1 & 0 \end{bmatrix} \begin{bmatrix} x_1(kT) \\ x_2(kT) \\ x_3(kT) \\ x_4(kT) \\ x_5(kT) \end{bmatrix}$

The following systems are not completely observable:

1. $\begin{bmatrix} x_1((k+1)T) \\ x_2((k+1)T) \end{bmatrix} = \begin{bmatrix} -1 & 0 \\ 0 & -2 \end{bmatrix} \begin{bmatrix} x_1(kT) \\ x_2(kT) \end{bmatrix},$ $y(kT) = \begin{bmatrix} 0 & 1 \end{bmatrix} \begin{bmatrix} x_1(kT) \\ x_2(kT) \end{bmatrix}$

2. $\begin{bmatrix} x_1((k+1)T) \\ x_2((k+1)T) \\ x_3((k+1)T) \\ x_4((k+1)T) \\ x_5((k+1)T) \end{bmatrix} = \begin{bmatrix} 2 & 1 & 0 & & 0 \\ 0 & 2 & 1 & & \\ 0 & 0 & 2 & & \\ \hline & & & -3 & 1 \\ 0 & & & 0 & -3 \end{bmatrix} \begin{bmatrix} x_1(kT) \\ x_2(kT) \\ x_3(kT) \\ x_4(kT) \\ x_5(kT) \end{bmatrix}$

$\begin{bmatrix} y_1(kT) \\ y_2(kT) \end{bmatrix} = \begin{bmatrix} 1 & 1 & 1 & 0 & 1 \\ 0 & 1 & 1 & 0 & 0 \end{bmatrix} \begin{bmatrix} x_1(kT) \\ x_2(kT) \\ x_3(kT) \\ x_4(kT) \\ x_5(kT) \end{bmatrix}$

Condition for Complete Observability in the z Plane.   The condition for complete observability can also be stated in terms of pulse transfer functions. A necessary and sufficient condition for complete observability is that no pole–zero cancellation occur in the pulse transfer function. If cancellation occurs, the canceled mode cannot be observed in the output.

Example 6–4

Show that the following system is not completely observable:

$$\mathbf{x}((k + 1)T) = \mathbf{G}\mathbf{x}(kT) + \mathbf{H}u(kT)$$

$$y(kT) = \mathbf{C}\mathbf{x}(kT)$$

where

$$\mathbf{G} = \begin{bmatrix} 0 & 1 & 0 \\ 0 & 0 & 1 \\ -6 & -11 & -6 \end{bmatrix}, \quad \mathbf{H} = \begin{bmatrix} 0 \\ 0 \\ 1 \end{bmatrix}, \quad \mathbf{C} = [4 \ \ 5 \ \ 1]$$

Note that the control signal $u(kT)$ does not affect the complete observability of the system. To examine complete observability, we may simply set $u(kT) = 0$. For this system, we have

$$[\mathbf{C}^* \vdots \mathbf{G}^*\mathbf{C}^* \vdots (\mathbf{G}^*)^2\mathbf{C}^*] = \begin{bmatrix} 4 & -6 & 6 \\ 5 & -7 & 5 \\ 1 & -1 & -1 \end{bmatrix}$$

Notice that

$$\begin{vmatrix} 4 & -6 & 6 \\ 5 & -7 & 5 \\ 1 & -1 & -1 \end{vmatrix} = 0$$

Hence, the rank of the matrix $[\mathbf{C}^* \vdots \mathbf{G}^*\mathbf{C}^* \vdots (\mathbf{G}^*)^2\mathbf{C}^*]$ is less than 3. Therefore, the system is not completely observable.

In fact, in this system a pole–zero cancellation occurs in the pulse transfer function of the system. The pulse transfer function between $X_1(z)$ and $U(z)$ is

$$\frac{X_1(z)}{U(z)} = \frac{1}{(z + 1)(z + 2)(z + 3)}$$

and the pulse transfer function between $Y(z)$ and $X_1(z)$ is

$$\frac{Y(z)}{X_1(z)} = (z + 1)(z + 4)$$

Therefore, the pulse transfer function between output $Y(z)$ and input $U(z)$ is

$$\frac{Y(z)}{U(z)} = \frac{(z + 1)(z + 4)}{(z + 1)(z + 2)(z + 3)}$$

Clearly, the $(z + 1)$ factors in the numerator and denominator cancel each other. This means that there are nonzero initial states $\mathbf{x}(0)$ that cannot be determined from the measurement of $y(kT)$.

Comments. The pulse transfer function has no cancellation if and only if the system is completely state controllable and completely observable. (See Problem A–6–4.) This means that a canceled transfer function does not carry along all the information characterizing the dynamic system.

Principle of Duality. In what follows, we shall examine the relationship between controllability and observability. Consider the system S_1 defined by the equations

$$\mathbf{x}((k + 1)T) = \mathbf{G}\mathbf{x}(kT) + \mathbf{H}u(kT) \tag{6–20}$$

$$\mathbf{y}(kT) = \mathbf{C}\mathbf{x}(kT) \tag{6–21}$$

where

$\mathbf{x}(kT)$ = state vector (n-vector) at kth sampling instant

$\mathbf{u}(kT)$ = control vector (r-vector) at kth sampling instant

$\mathbf{y}(kT)$ = output vector (m-vector) at kth sampling instant

$\mathbf{G}$ = $n \times n$ matrix

$\mathbf{H}$ = $n \times r$ matrix

$\mathbf{C}$ = $m \times n$ matrix

and its dual counterpart, which we call system S_2, defined by the equations

$$\hat{\mathbf{x}}((k + 1)T) = \mathbf{G}^*\hat{\mathbf{x}}(kT) + \mathbf{C}^*\hat{\mathbf{u}}(kT) \tag{6–22}$$

$$\hat{\mathbf{y}}(kT) = \mathbf{H}^*\hat{\mathbf{x}}(kT) \tag{6–23}$$

where

$\hat{\mathbf{x}}(kT)$ = state vector (n-vector) at kth sampling instant

$\hat{\mathbf{u}}(kT)$ = control vector (m-vector) at kth sampling instant

$\hat{\mathbf{y}}(kT)$ = output vector (r-vector) at kth sampling instant

$\mathbf{G}^*$ = conjugate transpose of $\mathbf{G}$

$\mathbf{H}^*$ = conjugate transpose of $\mathbf{H}$

$\mathbf{C}^*$ = conjugate transpose of $\mathbf{C}$

We shall now examine an analogy between controllability and observability. This analogy is referred to as the *principle of duality*, due to Kalman.

The principle of duality states that system S_1 defined by Equations (6–20) and (6–21) is completely state controllable (observable) if and only if system S_2 defined by Equations (6–22) and (6–23) is completely observable (state controllable). To verify this principle, let us write down the necessary and sufficient conditions for complete state controllability and complete observability for systems S_1 and S_2, respectively.

FOR SYSTEM S_1:

1. A necessary and sufficient condition for complete state controllability is that

$$\text{rank}\,[\mathbf{H} \vdots \mathbf{GH} \vdots \cdots \vdots \mathbf{G}^{n-1}\mathbf{H}] = n$$

2. A necessary and sufficient condition for complete observability is that

$$\text{rank}[\mathbf{C}^* \vdots \mathbf{G}^*\mathbf{C}^* \vdots \cdots \vdots (\mathbf{G}^*)^{n-1}\mathbf{C}^*] = n$$

FOR SYSTEM S_2:

1. A necessary and sufficient condition for complete state controllability is that

$$\text{rank}\,[\mathbf{C}^* \vdots \mathbf{G}^*\mathbf{C}^* \vdots \cdots \vdots (\mathbf{G}^*)^{n-1}\mathbf{C}^*] = n$$

2. A necessary and sufficient condition for complete observability is that

$$\text{rank}\,[\mathbf{H} \vdots \mathbf{GH} \vdots \cdots \vdots \mathbf{G}^{n-1}\mathbf{H}] = n$$

By comparing these conditions, the truth of the principle of duality is apparent.

We see that system S_1 being completely state controllable is equivalent to system S_2 being completely observable. And system S_1 being completely observable is equivalent to system S_2 being completely state controllable. By use of this principle, the observability of a given system can be checked by testing the state controllability of its dual.

Complete Observability of Linear Time-Invariant Continuous-Time Control Systems. Finally, we shall briefly state the complete observability condition for the linear time-invariant continuous-time control system. The system is said to be completely observable if every initial state $\mathbf{x}(0)$ can be determined from the observation of $\mathbf{y}(t)$ over a finite time interval. Similar to the case of the discrete-time control system, we need to consider only an unforced system. Consider the system defined by the equations:

$$\dot{\mathbf{x}} = \mathbf{A}\mathbf{x}$$

$$\mathbf{y} = \mathbf{C}\mathbf{x}$$

where

$$\mathbf{x} = \text{state vector } (n\text{-vector})$$

$$\mathbf{y} = \text{output vector } (m\text{-vector})$$

$$\mathbf{A} = n \times n \text{ matrix}$$

$$\mathbf{C} = m \times n \text{ matrix}$$

As in the case of the discrete-time control system, it can be stated that the condition for complete observability is that the rank of the $n \times nm$ matrix

$$[\mathbf{C}^* \vdots \mathbf{A}^*\mathbf{C}^* \vdots \cdots \vdots (\mathbf{A}^*)^{n-1}\mathbf{C}^*]$$

be n. (This $n \times nm$ matrix is commonly called the observability matrix for the continuous-time system.)

Effects of the Discretization of a Continuous-Time Control System on Controllability and Observability. When a continuous-time control system with complex poles is discretized, the introduction of sampling may impair the controllability and observability of the resulting discretized system. That is, pole–zero cancellation may take place in passing from the continuous-time case to the discrete-time case. Thus, the discretized system may lose controllability and observability.

It can be shown that a system that is completely state controllable and completely observable in the absence of sampling remains completely state controllable and completely observable after the introduction of sampling if and only if, for every eigenvalue of the characteristic equation for the continuous-time control system, the relationship

$$\text{Re}\,\lambda_i = \text{Re}\,\lambda_j \qquad\qquad (6\text{--}24)$$

implies

$$\text{Im}\,(\lambda_i - \lambda_j) \neq \frac{2n\pi}{T} \tag{6–25}$$

where T is the sampling period and $n = \pm 1, \pm 2, \ldots$. It is noted that, unless the system contains complex poles, pole–zero cancellation will not occur in passing from the continuous-time to the discrete-time case.

Example 6–5

Consider the following continuous-time control system:

$$\begin{bmatrix} \dot{x}_1 \\ \dot{x}_2 \end{bmatrix} = \begin{bmatrix} 0 & 1 \\ -1 & 0 \end{bmatrix}\begin{bmatrix} x_1 \\ x_2 \end{bmatrix} + \begin{bmatrix} 0 \\ 1 \end{bmatrix} u \tag{6–26}$$

$$y = \begin{bmatrix} 1 & 0 \end{bmatrix}\begin{bmatrix} x_1 \\ x_2 \end{bmatrix} \tag{6–27}$$

This system is completely state controllable and completely observable, since the rank of the controllability matrix

$$[\mathbf{B} \,\vdots\, \mathbf{AB}] = \begin{bmatrix} 0 & 1 \\ 1 & 0 \end{bmatrix}$$

is 2 and the rank of the observability matrix

$$[\mathbf{C}^* \,\vdots\, \mathbf{A}^*\mathbf{C}^*] = \begin{bmatrix} 1 & 0 \\ 0 & 1 \end{bmatrix}$$

is also 2. Notice that the eigenvalues of the state matrix are

$$\lambda_1 = j, \qquad \lambda_2 = -j$$

The discrete-time control system obtained by discretizing the continuous-time control system defined by Equations (6–26) and (6–27) may be given as follows:

$$\begin{bmatrix} x_1((k+1)T) \\ x_2((k+1)T) \end{bmatrix} = \begin{bmatrix} \cos T & \sin T \\ -\sin T & \cos T \end{bmatrix}\begin{bmatrix} x_1(kT) \\ x_2(kT) \end{bmatrix} + \begin{bmatrix} 1 - \cos T \\ \sin T \end{bmatrix} u(kT) \tag{6–28}$$

$$y(kT) = \begin{bmatrix} 1 & 0 \end{bmatrix}\begin{bmatrix} x_1(kT) \\ x_2(kT) \end{bmatrix} \tag{6–29}$$

where T is the sampling period.

Let us show that the discretized system given by Equations (6–28) and (6–29) is completely state controllable and completely observable if and only if

$$\text{Im}\,(\lambda_1 - \lambda_2) = 1 + 1 \neq \frac{2n\pi}{T}$$

or

$$T \neq n\pi, \qquad n = 1, 2, 3, \ldots$$

For the discrete-time control system obtained by discretizing the continuous-time control system, we have the following controllability matrix:

$$[\mathbf{H} \,\vdots\, \mathbf{GH}] = \begin{bmatrix} 1 - \cos T & \cos T + 1 - 2\cos^2 T \\ \sin T & -\sin T + 2\cos T \sin T \end{bmatrix}$$

Notice that the rank of $[\mathbf{H} \,\vdots\, \mathbf{GH}]$ is 2 if and only if $T \neq n\pi$ (where $n = 1, 2, 3, \ldots$). Also, the rank of the observability matrix

$$[\mathbf{C}^* \,\vdots\, \mathbf{G}^*\mathbf{C}^*] = \begin{bmatrix} 1 & \cos T \\ 0 & \sin T \end{bmatrix}$$

is 2 if and only if $T \neq n\pi$ (where $n = 1, 2, 3, \dots$).

From the foregoing analysis, we conclude that the discretized system is completely state controllable and completely observable if and only if $T \neq n\pi$, where $n = 1, 2, 3, \dots$.

Note that it is always possible to avoid the loss of controllability and observability by choosing a sampling period T not equal to $n\pi$.

6–4 USEFUL TRANSFORMATIONS IN STATE-SPACE ANALYSIS AND DESIGN

In this section we shall first review techniques for transforming state-space equations into canonical forms. Then we shall review the invariance property of the rank conditions for the controllability matrix and observability matrix.

Transforming State-Space Equations Into Canonical forms.   Consider the discrete-time state equation and output equation

$$\mathbf{x}(k + 1) = \mathbf{G}\mathbf{x}(k) + \mathbf{H}u(k) \tag{6–30}$$

$$y(k) = \mathbf{C}\mathbf{x}(k) + Du(k) \tag{6–31}$$

We shall review techniques for transforming the state-space equations defined by Equations (6–30) and (6–31) into the following three canonical forms:

1. Controllable canonical form
2. Observable canonical form
3. Diagonal or Jordan canonical form

(Note that the diagonal canonical form is a special case of the Jordan canonical form.) It is assumed that the system defined by Equations (6–30) and (6–31) is completely state controllable and completely observable.

Controllable Canonical Form.   The system defined by Equations (6–30) and (6–31) can be transformed into a controllable canonical form by means of the transformation matrix

$$\mathbf{T} = \mathbf{MW} \tag{6–32}$$

where

$$\mathbf{M} = [\mathbf{H} \,\vdots\, \mathbf{GH} \,\vdots\, \cdots \,\vdots\, \mathbf{G}^{n-1}\mathbf{H}] \tag{6–33}$$

and

$$\mathbf{W} = \begin{bmatrix} a_{n-1} & a_{n-2} & \cdots & a_1 & 1 \\ a_{n-2} & a_{n-3} & \cdots & 1 & 0 \\ \vdots & \vdots & & \vdots & \vdots \\ a_1 & 1 & \cdots & 0 & 0 \\ 1 & 0 & \cdots & 0 & 0 \end{bmatrix} \tag{6–34}$$

The elements a_i shown in matrix $\mathbf{W}$ are coefficients of the characteristic equation

$$|z\mathbf{I} - \mathbf{G}| = z^n + a_1 z^{n-1} + \cdots + a_{n-1} z + a_n = 0$$

It can be shown that

$$\mathbf{T}^{-1}\mathbf{GT} = (\mathbf{MW})^{-1}\mathbf{G}(\mathbf{MW}) = \mathbf{W}^{-1}\mathbf{M}^{-1}\mathbf{GMW}$$

$$= \begin{bmatrix} 0 & 1 & 0 & \cdots & 0 \\ 0 & 0 & 1 & \cdots & 0 \\ \vdots & \vdots & \vdots & & \vdots \\ 0 & 0 & 0 & \cdots & 1 \\ -a_n & -a_{n-1} & -a_{n-2} & \cdots & -a_1 \end{bmatrix} \tag{6-35}$$

and

$$\mathbf{T}^{-1}\mathbf{H} = \begin{bmatrix} 0 \\ 0 \\ \vdots \\ 0 \\ 1 \end{bmatrix} \tag{6-36}$$

(For details of the derivations of the preceding two equations, see Problems A–6–5 and A–6–6.)

Now let us define

$$\mathbf{x}(k) = \mathbf{T}\hat{\mathbf{x}}(k)$$

where the transformation matrix $\mathbf{T}$ is given by Equation (6–32). Then Equations (6–30) and (6–31) become

$$\hat{\mathbf{x}}(k + 1) = \mathbf{T}^{-1}\mathbf{GT}\hat{\mathbf{x}}(k) + \mathbf{T}^{-1}\mathbf{H}u(k) = \hat{\mathbf{G}}\hat{\mathbf{x}}(k) + \hat{\mathbf{H}}u(k)$$

$$y(k) = \mathbf{CT}\hat{\mathbf{x}}(k) + Du(k) = \hat{\mathbf{C}}\hat{\mathbf{x}}(k) + \hat{D}u(k)$$

where $\hat{\mathbf{G}} = \mathbf{T}^{-1}\mathbf{GT}$, $\hat{\mathbf{H}} = \mathbf{T}^{-1}\mathbf{H}$, $\hat{\mathbf{C}} = \mathbf{CT}$, and $\hat{D} = D$, or

$$\begin{bmatrix} \hat{x}_1(k+1) \\ \hat{x}_2(k+1) \\ \vdots \\ \hat{x}_{n-1}(k+1) \\ \hat{x}_n(k+1) \end{bmatrix} = \begin{bmatrix} 0 & 1 & 0 & \cdots & 0 \\ 0 & 0 & 1 & \cdots & 0 \\ \vdots & \vdots & \vdots & & \vdots \\ 0 & 0 & 0 & \cdots & 1 \\ -a_n & -a_{n-1} & -a_{n-2} & \cdots & -a_1 \end{bmatrix} \begin{bmatrix} \hat{x}_1(k) \\ \hat{x}_2(k) \\ \vdots \\ \hat{x}_{n-1}(k) \\ \hat{x}_n(k) \end{bmatrix} + \begin{bmatrix} 0 \\ 0 \\ \vdots \\ 0 \\ 1 \end{bmatrix} u(k)$$

$$\tag{6-37}$$

$$y(k) = [b_n - a_n b_0 \,\vdots\, b_{n-1} - a_{n-1} b_0 \,\vdots\, \cdots \,\vdots\, b_1 - a_1 b_0] \begin{bmatrix} \hat{x}_1(k) \\ \hat{x}_2(k) \\ \vdots \\ \hat{x}_n(k) \end{bmatrix} + \hat{D}u(k)$$

$$\tag{6-38}$$

where the b_k's are those coefficients appearing in the numerator of the following pulse transfer function:

$$\mathbf{C}(z\mathbf{I} - \mathbf{G})^{-1}\mathbf{H} + \mathbf{D} = \hat{\mathbf{C}}(z\mathbf{I} - \hat{\mathbf{G}})^{-1}\hat{\mathbf{H}} + \hat{\mathbf{D}}$$

$$= \frac{b_0 z^n + b_1 z^{n-1} + \cdots + b_{n-1} z + b_n}{z^n + a_1 z^{n-1} + \cdots + a_{n-1} z + a_n} \qquad (6\text{–}39)$$

Note that $D = \hat{D} = b_0$. The system given by Equations (6–37) and (6–38) is in a controllable canonical form.

Observable Canonical Form.   The system defined by Equations (6–30) and (6–31) can be transformed into an observable canonical form by means of the transformation matrix

$$\mathbf{Q} = (\mathbf{WN^*})^{-1}$$

where

$$\mathbf{N} = [\mathbf{C^*} \,\vdots\, \mathbf{G^*C^*} \,\vdots\, \cdots \,\vdots\, (\mathbf{G^*})^{n-1}\mathbf{C^*}] \qquad (6\text{–}40)$$

and $\mathbf{W}$ is given by Equation (6–34). It can be shown that

$$\mathbf{Q}^{-1}\mathbf{GQ} = \hat{\mathbf{G}} = \begin{bmatrix} 0 & 0 & \cdots & 0 & -a_n \\ 1 & 0 & \cdots & 0 & -a_{n-1} \\ 0 & 1 & \cdots & 0 & -a_{n-2} \\ \vdots & \vdots & & \vdots & \vdots \\ 0 & 0 & \cdots & 1 & -a_1 \end{bmatrix}$$

$$\mathbf{Q}^{-1}\mathbf{H} = \hat{\mathbf{H}} = \begin{bmatrix} b_n - a_n b_0 \\ b_{n-1} - a_{n-1} b_0 \\ \vdots \\ b_1 - a_1 b_0 \end{bmatrix}$$

and

$$\mathbf{CQ} = \hat{\mathbf{C}} = [0 \quad 0 \quad \cdots \quad 0 \quad 1]$$

where the b_k's are those coefficients appearing in the numerator of the pulse transfer function given by Equation (6–39). (For details of the derivations of the preceding equations, see Problems A–6–8 and A–6–9.) Hence, by defining

$$\mathbf{x}(k) = \mathbf{Q}\hat{\mathbf{x}}(k)$$

Equations (6–30) and (6–31) become as follows:

$$\hat{\mathbf{x}}(k + 1) = \hat{\mathbf{G}}\hat{\mathbf{x}}(k) + \hat{\mathbf{H}}u(k)$$

$$y(k) = \hat{\mathbf{C}}\hat{\mathbf{x}}(k) + \hat{D}u(k)$$

or

$$\begin{bmatrix} \hat{x}_1(k+1) \\ \hat{x}_2(k+1) \\ \vdots \\ \hat{x}_{n-1}(k+1) \\ \hat{x}_n(k+1) \end{bmatrix} = \begin{bmatrix} 0 & 0 & \cdots & 0 & -a_n \\ 1 & 0 & \cdots & 0 & -a_{n-1} \\ \vdots & \vdots & & \vdots & \vdots \\ 0 & 0 & \cdots & 0 & -a_2 \\ 0 & 0 & \cdots & 1 & -a_1 \end{bmatrix} \begin{bmatrix} \hat{x}_1(k) \\ \hat{x}_2(k) \\ \vdots \\ \hat{x}_{n-1}(k) \\ \hat{x}_n(k) \end{bmatrix} + \begin{bmatrix} b_n - a_n b_0 \\ b_{n-1} - a_{n-1} b_0 \\ \vdots \\ b_2 - a_2 b_0 \\ b_1 - a_1 b_0 \end{bmatrix} u(k) \qquad (6\text{–}41)$$

$$y(k) = [0 \quad 0 \quad \cdots \quad 0 \quad 1] \begin{bmatrix} \hat{x}_1(k) \\ \hat{x}_2(k) \\ \vdots \\ \hat{x}_{n-1}(k) \\ \hat{x}_n(k) \end{bmatrix} + Du(k) \tag{6-42}$$

The system defined by Equations (6–41) and (6–42) is in an observable canonical form.

Diagonal or Jordan Canonical Form. If the eigenvalues p_i of matrix **G** are distinct, then the corresponding eigenvectors $\xi_1, \xi_2, \ldots, \xi_n$ are distinct. Define the transformation matrix **P** as follows:

$$\mathbf{P} = [\xi_1 \vdots \xi_2 \vdots \cdots \vdots \xi_n]$$

Then

$$\mathbf{P}^{-1}\mathbf{GP} = \begin{bmatrix} p_1 & 0 & \cdots & 0 \\ 0 & p_2 & \cdots & 0 \\ \vdots & \vdots & & \vdots \\ 0 & 0 & \cdots & p_n \end{bmatrix}$$

Thus, if we define

$$\mathbf{x}(k) = \mathbf{P}\hat{\mathbf{x}}(k)$$

then Equations (6–30) and (6–31) can be given by the equations

$$\hat{\mathbf{x}}(k + 1) = \hat{\mathbf{G}}\hat{\mathbf{x}}(k) + \hat{\mathbf{H}}u(k) \tag{6-43}$$

$$y(k) = \hat{\mathbf{C}}\hat{\mathbf{x}}(k) + \hat{D}u(k) \tag{6-44}$$

where $\hat{\mathbf{G}} = \mathbf{P}^{-1}\mathbf{GP}$, $\hat{\mathbf{H}} = \mathbf{P}^{-1}\mathbf{H}$, $\hat{\mathbf{C}} = \mathbf{CP}$, and $\hat{D} = D$. Thus, Equations (6–43) and (6–44) can be written in the form

$$\begin{bmatrix} \hat{x}_1(k + 1) \\ \hat{x}_2(k + 1) \\ \vdots \\ \hat{x}_n(k + 1) \end{bmatrix} = \begin{bmatrix} p_1 & 0 & \cdots & 0 \\ 0 & p_2 & \cdots & 0 \\ \vdots & \vdots & & \vdots \\ 0 & 0 & \cdots & p_n \end{bmatrix} \begin{bmatrix} \hat{x}_1(k) \\ \hat{x}_2(k) \\ \vdots \\ \hat{x}_n(k) \end{bmatrix} + \begin{bmatrix} \alpha_1 \\ \alpha_2 \\ \vdots \\ \alpha_n \end{bmatrix} u(k) \tag{6-45}$$

$$y(k) = [\beta_1 \quad \beta_2 \quad \cdots \quad \beta_n] \begin{bmatrix} \hat{x}_1(k) \\ \hat{x}_2(k) \\ \vdots \\ \hat{x}_n(k) \end{bmatrix} + Du(k) \tag{6-46}$$

where the α_i's and the β_i's are constants such that $\alpha_i \beta_i$ is the residue at the pole $z = p_i$, that is, such that $\alpha_i \beta_i$ will appear in the numerator of the term $1/(z - p_i)$ when the pulse transfer function is expanded into partial fractions as follows:

$$\mathbf{C}(z\mathbf{I} - \mathbf{G})^{-1}\mathbf{H} + D = \hat{\mathbf{C}}(z\mathbf{I} - \hat{\mathbf{G}})^{-1}\hat{\mathbf{H}} + \hat{D}$$

$$= \frac{\alpha_1 \beta_1}{z - p_1} + \frac{\alpha_2 \beta_2}{z - p_2} + \cdots + \frac{\alpha_n \beta_n}{z - p_n} + D \tag{6-47}$$

In many cases we choose $\alpha_1 = \alpha_2 = \cdots = \alpha_n = 1$. [Note that the necessary and sufficient condition for the system to be completely state controllable is that $\alpha_i \neq 0$ $(i = 1, 2, \ldots, n)$ and that the condition for it to be completely observable is that $\beta_i \neq 0$ $(i = 1, 2, \ldots, n)$.]

If there are multiple eigenvalues p_i of matrix $\mathbf{G}$, then we choose the transformation matrix $\mathbf{S}$ defined as follows:

$$\mathbf{S} = [\boldsymbol{\eta}_1 \vdots \boldsymbol{\eta}_2 \vdots \cdots \vdots \boldsymbol{\eta}_n]$$

where the $\boldsymbol{\eta}_i$'s are eigenvectors (which correspond to distinct eigenvalues) or generalized eigenvectors (which correspond to multiple eigenvalues). (For details of generalized eigenvectors, see Appendix A.) Then

$$\mathbf{S}^{-1}\mathbf{G}\mathbf{S} = \text{matrix in Jordan canonical form}$$

Now if we define

$$\mathbf{x}(k) = \mathbf{S}\hat{\mathbf{x}}(k)$$

then Equations (6–30) and (6–31) can be given as follows:

$$\hat{\mathbf{x}}(k + 1) = \hat{\mathbf{G}}\hat{\mathbf{x}}(k) + \hat{\mathbf{H}}u(k)$$

$$y(k) = \hat{\mathbf{C}}\hat{\mathbf{x}}(k) + \hat{D}\hat{u}(k)$$

where $\hat{\mathbf{G}} = \mathbf{S}^{-1}\mathbf{G}\mathbf{S}$, $\hat{\mathbf{H}} = \mathbf{S}^{-1}\mathbf{H}$, $\hat{\mathbf{C}} = \mathbf{C}\mathbf{S}$, and $\hat{D} = D$. If, for example, matrix $\mathbf{G}$ involves an m-multiple eigenvalue p_1 and other eigenvalues $p_{m+1}, p_{m+2}, \ldots, p_n$ that are all distinct and different from p_1 and, in addition, if the rank of $p_1\mathbf{I} - \mathbf{G}$ is $n - 1$ (which implies that the minimal polynomial is identical to the characteristic polynomial), then the state-space equations in the Jordan canonical form are given as follows:

$$
\begin{bmatrix}
\hat{x}_1(k+1) \\
\hat{x}_2(k+1) \\
\vdots \\
\hat{x}_m(k+1) \\
\hline
\hat{x}_{m+1}(k+1) \\
\vdots \\
\hat{x}_n(k+1)
\end{bmatrix}
=
\left[
\begin{array}{ccccc|ccc}
p_1 & 1 & & & 0 & & & 0 \\
 & p_1 & 1 & & & & & \\
 & & \ddots & \ddots & 1 & & & \\
0 & & & & p_1 & & & \\
\hline
 & & & & & p_{m+1} & & 0 \\
 & & & & & & \ddots & \\
0 & & & & & 0 & & p_n
\end{array}
\right]
\begin{bmatrix}
\hat{x}_1(k) \\
\hat{x}_2(k) \\
\vdots \\
\hat{x}_m(k) \\
\hline
\hat{x}_{m+1}(k) \\
\vdots \\
\hat{x}_n(k)
\end{bmatrix}
+
\begin{bmatrix}
0 \\
0 \\
\vdots \\
a_m \\
\hline
a_{m+1} \\
\vdots \\
a_n
\end{bmatrix}
u(k)
$$

$$(6\text{–}48)$$

$$y(k) = [\beta_1 \quad \beta_2 \quad \cdots \quad \beta_n]
\begin{bmatrix}
\hat{x}_1(k) \\
\hat{x}_2(k) \\
\vdots \\
\hat{x}_n(k)
\end{bmatrix}
+ Du(k) \qquad\qquad (6\text{–}49)$$

where the α_i's and the β_i's are constants appearing in the pulse transfer function for this system:

$$C(z\mathbf{I} - \mathbf{G})^{-1}\mathbf{H} + D = \hat{\mathbf{C}}(z\mathbf{I} - \hat{\mathbf{G}})^{-1}\hat{\mathbf{H}} + \hat{D}$$

$$= \frac{\alpha_m \beta_1}{(z - p_1)^m} + \frac{\alpha_m \beta_2}{(z - p_1)^{m-1}} + \cdots + \frac{\alpha_m \beta_m}{z - p_1}$$

$$+ \frac{\alpha_{m+1}\beta_{m+1}}{z - p_{m+1}} + \cdots + \frac{\alpha_n \beta_n}{z - p_n} + D$$

In many cases we choose $\alpha_m = \alpha_{m+1} = \cdots = \alpha_n = 1$. [Note that the necessary and sufficient condition for the system to be completely state controllable is that $\alpha_i \neq 0$ ($i = m, m + 1, \ldots, n$) and that to be completely observable is that $\beta_i \neq 0$ ($i = 1$, $m + 1, m + 2, \ldots, n$).]

Note that if the rank of $p_1\mathbf{I} - \mathbf{G}$ is $n - s$ (where $2 \leq s \leq n$), that is, if the minimal polynomial is $s - 1$ degree lower than the characteristic polynomial, then $\mathbf{S}^{-1}\mathbf{GS}$ will have a different Jordan canonical form. (For details, see Appendix A.)

Invariance Property of the Rank Conditions for the Controllability Matrix and Observability Matrix.   Consider systems related by similarity transformations. Let us define the controllability matrix as $\mathbf{M}$:

$$\mathbf{M} = [\mathbf{H} \vdots \mathbf{GH} \vdots \cdots \vdots \mathbf{G}^{n-1}\mathbf{H}]$$

Let $\mathbf{P}$ (an arbitrary $n \times n$ nonsingular matrix) be a similarity transformation matrix and write

$$\mathbf{P}^{-1}\mathbf{GP} = \tilde{\mathbf{G}}, \qquad \mathbf{P}^{-1}\mathbf{H} = \tilde{\mathbf{H}}$$

Then

$$\mathbf{P}^{-1}\mathbf{G}^2\mathbf{P} = \mathbf{P}^{-1}\mathbf{GPP}^{-1}\mathbf{GP} = \tilde{\mathbf{G}}\tilde{\mathbf{G}} = \tilde{\mathbf{G}}^2$$

$$\mathbf{P}^{-1}\mathbf{G}^3\mathbf{P} = \mathbf{P}^{-1}\mathbf{GPP}^{-1}\mathbf{GPP}^{-1}\mathbf{GP} = \tilde{\mathbf{G}}^3$$

$$\vdots$$

$$\mathbf{P}^{-1}\mathbf{G}^{n-1}\mathbf{P} = \tilde{\mathbf{G}}^{n-1}$$

Hence,

$$\mathbf{P}^{-1}\mathbf{M} = \mathbf{P}^{-1}[\mathbf{H} \vdots \mathbf{GH} \vdots \cdots \vdots \mathbf{G}^{n-1}\mathbf{H}]$$

$$= [\mathbf{P}^{-1}\mathbf{H} \vdots \mathbf{P}^{-1}\mathbf{GH} \vdots \cdots \vdots \mathbf{P}^{-1}\mathbf{G}^{n-1}\mathbf{H}]$$

$$= [\mathbf{P}^{-1}\mathbf{H} \vdots \mathbf{P}^{-1}\mathbf{GPP}^{-1}\mathbf{H} \vdots \cdots \vdots \mathbf{P}^{-1}\mathbf{G}^{n-1}\mathbf{PP}^{-1}\mathbf{H}]$$

$$= [\tilde{\mathbf{H}} \vdots \tilde{\mathbf{G}}\tilde{\mathbf{H}} \vdots \cdots \vdots \tilde{\mathbf{G}}^{n-1}\tilde{\mathbf{H}}] = \tilde{\mathbf{M}}$$

Since matrix $\mathbf{P}$ is nonsingular, we have

$$\text{rank } \mathbf{M} = \text{rank } \tilde{\mathbf{M}}$$

Similarly, for the observability matrix define

$$\mathbf{N} = [\mathbf{C}^* \vdots \mathbf{G}^*\mathbf{C}^* \vdots \cdots \vdots (\mathbf{G}^*)^{n-1}\mathbf{C}^*]$$

Let $\mathbf{P}$ be an arbitrary $n \times n$ nonsingular matrix and write

$$\mathbf{P}^{-1}\mathbf{GP} = \tilde{\mathbf{G}}, \qquad \mathbf{CP} = \tilde{\mathbf{C}}$$

Then

$$\mathbf{P}*\mathbf{N} = \mathbf{P}*[\mathbf{C}* \vdots \mathbf{G}*\mathbf{C}* \vdots \cdots \vdots (\mathbf{G}*)^{n-1}\mathbf{C}*]$$

$$= [\mathbf{P}*\mathbf{C}* \vdots \mathbf{P}*\mathbf{G}*\mathbf{C}* \vdots \cdots \vdots \mathbf{P}*(\mathbf{G}*)^{n-1}\mathbf{C}*]$$

$$= [\tilde{\mathbf{C}}* \vdots (\mathbf{P}^{-1}\mathbf{G}\mathbf{P})*\tilde{\mathbf{C}}* \vdots \cdots \vdots (\mathbf{P}^{-1}\mathbf{G}^{n-1}\mathbf{P})*\tilde{\mathbf{C}}*]$$

$$= [\tilde{\mathbf{C}}* \vdots \tilde{\mathbf{G}}*\tilde{\mathbf{C}}* \vdots \cdots \vdots (\tilde{\mathbf{G}}*)^{n-1}\tilde{\mathbf{C}}*] = \tilde{\mathbf{N}}$$

Hence,

$$\text{rank } \mathbf{N} = \text{rank } \tilde{\mathbf{N}}$$

6–5 DESIGN VIA POLE PLACEMENT

In this section we shall present a design method commonly called the *pole placement* or *pole assignment technique*. We assume that all state variables are measurable and are available for feedback. It will be shown that, if the system considered is completely state controllable, then poles of the closed-loop system may be placed at any desired locations by means of state feedback through an appropriate state feedback gain matrix.

The present design technique begins with a determination of the desired closed-loop poles based on transient-response and/or frequency-response requirements such as speed, damping ratio, or bandwidth. Given such considerations, let us assume that we decide that the desired closed-loop poles are to be at $z = \mu_1$, $z = \mu_2, \ldots, z = \mu_n$. (In choosing the sampling period, care must be exercised so that the desired system will not require unusually large control signals. Otherwise, saturation phenomena will occur in the system. If saturation takes place in the system, the system will become nonlinear, and the design method presented here will no longer apply, since the method is applicable only to linear time-invariant systems.) Then, by choosing an appropriate gain matrix for state feedback, it is possible to force the system to have closed-loop poles at the desired locations, provided that the original system is completely state controllable.

In what follows, we shall treat the case where the control signal is a scalar and prove that a necessary and sufficient condition that the closed-loop poles can be placed at any arbitrary locations in the z plane is that the system be completely state controllable. Then we shall discuss a few methods for determining the required state feedback gain matrix.

Necessary and Sufficient Condition for Arbitrary Pole Placement.   Consider the open-loop control system shown in Figure 6–1(a). The state equation is

$$\mathbf{x}(k + 1) = \mathbf{G}\mathbf{x}(k) + \mathbf{H}u(k) \tag{6–50}$$

where

$$\mathbf{x}(k) = \text{state vector } (n\text{-vector}) \text{ at } k\text{th sampling instant}$$

$$u(k) = \text{control signal (scalar) at } k\text{th sampling instant}$$

$$\mathbf{G} = n \times n \text{ matrix}$$

$$\mathbf{H} = n \times 1 \text{ matrix}$$

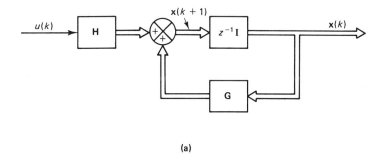

(a)

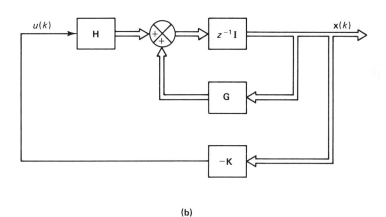

(b)

Figure 6–1   (a) Open-loop control system; (b) closed-loop control system with $u(k) = -\mathbf{K}\mathbf{x}(k)$.

We assume that the magnitude of the control signal $u(k)$ is unbounded. If the control signal $u(k)$ is chosen as

$$u(k) = -\mathbf{K}\mathbf{x}(k)$$

where $\mathbf{K}$ is the state feedback gain matrix (a $1 \times n$ matrix), then the system becomes a closed-loop control system as shown in Figure 6–1(b), and its state equation becomes

$$\mathbf{x}(k + 1) = (\mathbf{G} - \mathbf{H}\mathbf{K})\mathbf{x}(k) \tag{6–51}$$

Note that we choose matrix $\mathbf{K}$ such that the eigenvalues of $\mathbf{G} - \mathbf{H}\mathbf{K}$ are the desired closed-loop poles, $\mu_1, \mu_2, \ldots, \mu_n$.

We shall now prove that a necessary and sufficient condition for arbitrary pole placement is that the system be completely state controllable. We shall first derive the necessary condition. We begin by proving that, if the system is not completely state controllable, then there are eigenvalues of $\mathbf{G} - \mathbf{H}\mathbf{K}$ that cannot be controlled by state feedback.

Suppose the system of Equation (6–50) is not completely state controllable. Then the rank of the controllability matrix is less than n, or

$$\text{rank}[\mathbf{H} \vdots \mathbf{GH} \vdots \cdots \vdots \mathbf{G}^{n-1}\mathbf{H}] = q < n \qquad (6\text{–}52)$$

This means that there are q linearly independent column vectors in the controllability matrix. Let us define such q linearly independent column vectors as $\mathbf{f}_1, \mathbf{f}_2, \ldots, \mathbf{f}_q$. Also, let us choose $n - q$ additional n-vectors $\mathbf{v}_{q+1}, \mathbf{v}_{q+2}, \ldots, \mathbf{v}_n$ such that

$$\mathbf{P} = [\mathbf{f}_1 \vdots \mathbf{f}_2 \vdots \cdots \vdots \mathbf{f}_q \vdots \mathbf{v}_{q+1} \vdots \mathbf{v}_{q+2} \vdots \cdots \vdots \mathbf{v}_n]$$

is of rank n. By using matrix $\mathbf{P}$ as the transformation matrix, let us define

$$\mathbf{P}^{-1}\mathbf{GP} = \hat{\mathbf{G}}, \qquad \mathbf{P}^{-1}\mathbf{H} = \hat{\mathbf{H}}$$

Then we have

$$\mathbf{GP} = \mathbf{P}\hat{\mathbf{G}}$$

or

$$[\mathbf{Gf}_1 \vdots \cdots \vdots \mathbf{Gf}_q \vdots \mathbf{Gv}_{q+1} \vdots \cdots \vdots \mathbf{Gv}_n] = [\mathbf{f}_1 \vdots \mathbf{f}_2 \vdots \cdots \vdots \mathbf{f}_q \vdots \mathbf{v}_{q+1} \vdots \cdots \vdots \mathbf{v}_n]\hat{\mathbf{G}} \qquad (6\text{–}53)$$

Also,

$$\mathbf{H} = \mathbf{P}\hat{\mathbf{H}} = [\mathbf{f}_1 \vdots \mathbf{f}_2 \vdots \cdots \vdots \mathbf{f}_q \vdots \mathbf{v}_{q+1} \vdots \cdots \vdots \mathbf{v}_n]\hat{\mathbf{H}} \qquad (6\text{–}54)$$

Since we have here q linearly independent column vectors $\mathbf{f}_1, \mathbf{f}_2, \ldots, \mathbf{f}_q$, we can use the Cayley–Hamilton theorem to express matrices $\mathbf{Gf}_1, \mathbf{Gf}_2, \ldots, \mathbf{Gf}_q$ in terms of these q vectors. That is,

$$\mathbf{Gf}_1 = g_{11}\mathbf{f}_1 + g_{21}\mathbf{f}_2 + \cdots + g_{q1}\mathbf{f}_q$$

$$\mathbf{Gf}_2 = g_{12}\mathbf{f}_1 + g_{22}\mathbf{f}_2 + \cdots + g_{q2}\mathbf{f}_q$$

$$\vdots$$

$$\mathbf{Gf}_q = g_{1q}\mathbf{f}_1 + g_{2q}\mathbf{f}_2 + \cdots + g_{qq}\mathbf{f}_q$$

Hence, Equation (6–53) may be written as follows:

$$[\mathbf{Gf}_1 \vdots \cdots \vdots \mathbf{Gf}_q \vdots \mathbf{Gv}_{q+1} \vdots \cdots \vdots \mathbf{Gv}_n]$$

$$= [\mathbf{f}_1 \vdots \cdots \vdots \mathbf{f}_q \vdots \mathbf{v}_{q+1} \vdots \cdots \vdots \mathbf{v}_n] \begin{bmatrix} g_{11} & g_{12} & \cdots & g_{1q} & g_{1\,q+1} & g_{1\,q+2} & \cdots & g_{1n} \\ g_{21} & g_{22} & \cdots & g_{2q} & g_{2\,q+1} & g_{2\,q+2} & \cdots & g_{2n} \\ \vdots & \vdots & & \vdots & \vdots & \vdots & & \vdots \\ g_{q1} & g_{q2} & \cdots & g_{qq} & g_{q\,q+1} & g_{q\,q+2} & \cdots & g_{qn} \\ \hline 0 & 0 & \cdots & 0 & g_{q+1\,q+1} & q_{q+1\,q+2} & \cdots & g_{q+1\,n} \\ \vdots & \vdots & & \vdots & \vdots & \vdots & & \vdots \\ 0 & 0 & \cdots & 0 & g_{n\,q+1} & q_{n\,q+2} & \cdots & g_{nn} \end{bmatrix}$$

To simplify the notation, let us define

$$\begin{bmatrix} g_{11} & g_{12} & \cdots & g_{1q} \\ g_{21} & g_{22} & \cdots & g_{2q} \\ \vdots & \vdots & & \vdots \\ g_{q1} & g_{q2} & \cdots & g_{qq} \end{bmatrix} = \mathbf{G}_{11}$$

$$\begin{bmatrix} g_{1\,q+1} & g_{1\,q+2} & \cdots & g_{1n} \\ g_{2\,q+1} & g_{2\,q+2} & \cdots & g_{2n} \\ \vdots & \vdots & & \vdots \\ g_{q\,q+1} & g_{q\,q+2} & \cdots & g_{qn} \end{bmatrix} = \mathbf{G}_{12}$$

$$\begin{bmatrix} 0 & 0 & \cdots & 0 \\ \vdots & \vdots & & \vdots \\ 0 & 0 & \cdots & 0 \end{bmatrix} = \mathbf{G}_{21} = (n - q) \times q \text{ zero matrix}$$

$$\begin{bmatrix} g_{q+1\,q+1} & g_{q+1\,q+2} & \cdots & g_{q+1\,n} \\ g_{q+2\,q+1} & g_{q+2\,q+2} & \cdots & g_{q+2\,n} \\ \vdots & \vdots & & \vdots \\ g_{n\,q+1} & g_{n\,q+2} & \cdots & g_{nn} \end{bmatrix} = \mathbf{G}_{22}$$

Then Equation (6–53) can be written as follows:

$$[\mathbf{Gf}_1 \vdots \cdots \vdots \mathbf{Gf}_q \vdots \mathbf{Gv}_{q+1} \vdots \cdots \vdots \mathbf{Gv}_n] = [\mathbf{f}_1 \vdots \cdots \vdots \mathbf{f}_q \vdots \mathbf{v}_{q+1} \vdots \cdots \vdots \mathbf{v}_n]\begin{bmatrix} \mathbf{G}_{11} & \mathbf{G}_{12} \\ \mathbf{0} & \mathbf{G}_{22} \end{bmatrix}$$

Thus,

$$\hat{\mathbf{G}} = \begin{bmatrix} \mathbf{G}_{11} & \mathbf{G}_{12} \\ \mathbf{0} & \mathbf{G}_{22} \end{bmatrix} \tag{6–55}$$

Next, referring to Equation (6–54), we have

$$\mathbf{H} = [\mathbf{f}_1 \vdots \mathbf{f}_2 \vdots \cdots \vdots \mathbf{f}_q \vdots \mathbf{v}_{q+1} \vdots \cdots \vdots \mathbf{v}_n]\hat{\mathbf{H}} \tag{6–56}$$

Referring to Equation (6–52), notice that vector $\mathbf{H}$ can be written in terms of q linearly independent column vectors $\mathbf{f}_1, \mathbf{f}_2, \ldots, \mathbf{f}_q$. Thus, we have

$$\mathbf{H} = h_{11}\mathbf{f}_1 + h_{21}\mathbf{f}_2 + \cdots + h_{q1}\mathbf{f}_q$$

Consequently, Equation (6–56) may be written as follows:

$$h_{11}\mathbf{f}_1 + h_{21}\mathbf{f}_2 + \cdots + h_{q1}\mathbf{f}_q = [\mathbf{f}_1 \vdots \mathbf{f}_2 \vdots \cdots \vdots \mathbf{f}_q \vdots \mathbf{v}_{q+1} \vdots \cdots \vdots \mathbf{v}_n]\begin{bmatrix} h_{11} \\ h_{21} \\ \vdots \\ h_{q1} \\ \hline 0 \\ \vdots \\ 0 \end{bmatrix}$$

Thus,

$$\hat{\mathbf{H}} = \begin{bmatrix} \mathbf{H}_{11} \\ \mathbf{0} \end{bmatrix} \tag{6–57}$$

where

$$\mathbf{H}_{11} = \begin{bmatrix} h_{11} \\ h_{21} \\ \vdots \\ h_{q1} \end{bmatrix}$$

Now consider the closed-loop system equation given by Equation (6–51). The characteristic equation is

$$|z\mathbf{I} - \mathbf{G} + \mathbf{HK}| = 0$$

Let us define

$$\tilde{\mathbf{K}} = \mathbf{KP}$$

and partition the matrix $\tilde{\mathbf{K}}$ to give

$$\tilde{\mathbf{K}} = [\mathbf{K}_{11} \vdots \mathbf{K}_{12}] \tag{6–58}$$

where $\mathbf{K}_{11}$ is a $1 \times q$ matrix and $\mathbf{K}_{12}$ is a $1 \times (n - q)$ matrix. Now, $1 \times n$ matrix $\mathbf{K}$ can be written as follows:

$$\mathbf{K} = \tilde{\mathbf{K}}\mathbf{P}^{-1} = [\mathbf{K}_{11} \vdots \mathbf{K}_{12}]\mathbf{P}^{-1}$$

Then the characteristic equation for the closed-loop system can be written as follows:

$$|z\mathbf{I} - \mathbf{G} + \mathbf{HK}| = |\mathbf{P}^{-1}||z\mathbf{I} - \mathbf{G} + \mathbf{HK}||\mathbf{P}|$$

$$= |z\mathbf{I} - \mathbf{P}^{-1}\mathbf{GP} + \mathbf{P}^{-1}\mathbf{HKP}|$$

$$= |z\mathbf{I} - \hat{\mathbf{G}} + \hat{\mathbf{H}}\tilde{\mathbf{K}}|$$

Substituting Equations (6–55), (6–57), and (6–58) into this last equation, we obtain

$$|z\mathbf{I} - \hat{\mathbf{G}} + \hat{\mathbf{H}}\tilde{\mathbf{K}}| = \left| z\left[\begin{array}{c|c} \mathbf{I}_q & \mathbf{0} \\ \hline \mathbf{0} & \mathbf{I}_{n-q} \end{array}\right] - \left[\begin{array}{c|c} \mathbf{G}_{11} & \mathbf{G}_{12} \\ \hline \mathbf{0} & \mathbf{G}_{22} \end{array}\right] + \left[\begin{array}{c} \mathbf{H}_{11} \\ \hline \mathbf{0} \end{array}\right][\mathbf{K}_{11} \vdots \mathbf{K}_{12}] \right|$$

$$= \left| \begin{array}{c|c} z\mathbf{I}_q - \mathbf{G}_{11} + \mathbf{H}_{11}\mathbf{K}_{11} & -\mathbf{G}_{12} + \mathbf{H}_{11}\mathbf{K}_{12} \\ \hline \mathbf{0} & z\mathbf{I}_{n-q} - \mathbf{G}_{22} \end{array} \right|$$

$$= |z\mathbf{I}_q - \mathbf{G}_{11} + \mathbf{H}_{11}\mathbf{K}_{11}||z\mathbf{I}_{n-q} - \mathbf{G}_{22}| \tag{6–59}$$

Equation (6–59) shows that matrix $\mathbf{K} = \tilde{\mathbf{K}}\mathbf{P}^{-1}$ has control over the q eigenvalues of $\mathbf{G}_{11} - \mathbf{H}_{11}\mathbf{K}_{11}$, but not over the $n - q$ eigenvalues of $\mathbf{G}_{22}$. That is, there are $n - q$ eigenvalues of $\mathbf{G} - \mathbf{HK}$ that do not depend on matrix $\mathbf{K}$. Hence, we have proved that complete state controllability is a necessary condition for controlling all eigenvalues (closed-loop pole locations) of matrix $\mathbf{G} - \mathbf{HK}$.

We shall next derive a sufficient condition. We shall prove that if the system is completely state controllable then there exists a matrix $\mathbf{K}$ that will make the eigenvalues of $\mathbf{G} - \mathbf{HK}$ as desired, or place the closed-loop poles at the desired locations.

The desired eigenvalues of $\mathbf{G} - \mathbf{HK}$ are $\mu_1, \mu_2, \ldots, \mu_n$; any complex eigenvalues are to occur as conjugate pairs. Noting that the characteristic equation of the original system given by Equation (6–50) is

$$|z\mathbf{I} - \mathbf{G}| = z^n + a_1 z^{n-1} + a_2 z^{n-2} + \cdots + a_{n-1}z + a_n = 0$$

we define a transformation matrix $\mathbf{T}$ as follows:

$$\mathbf{T} = \mathbf{MW}$$

where

$$\mathbf{M} = [\mathbf{H} \vdots \mathbf{GH} \vdots \cdots \vdots \mathbf{G}^{n-1}\mathbf{H}] \tag{6–60}$$

which is of rank n, and where

$$\mathbf{W} = \begin{bmatrix} a_{n-1} & a_{n-2} & \cdots & a_1 & 1 \\ a_{n-2} & a_{n-3} & \cdots & 1 & 0 \\ \vdots & \vdots & & \vdots & \vdots \\ a_1 & 1 & \cdots & 0 & 0 \\ 1 & 0 & \cdots & 0 & 0 \end{bmatrix} \tag{6-61}$$

Then, referring to Equations (6–35) and (6–36), we have

$$\mathbf{T}^{-1}\mathbf{GT} = \hat{\mathbf{G}} = \begin{bmatrix} 0 & 1 & 0 & \cdots & 0 \\ 0 & 0 & 1 & \cdots & 0 \\ \vdots & \vdots & \vdots & & \vdots \\ 0 & 0 & 0 & \cdots & 1 \\ -a_n & -a_{n-1} & -a_{n-2} & \cdots & -a_1 \end{bmatrix}$$

and

$$\mathbf{T}^{-1}\mathbf{H} = \hat{\mathbf{H}} = \begin{bmatrix} 0 \\ 0 \\ \vdots \\ 0 \\ 1 \end{bmatrix}$$

Next we define

$$\hat{\mathbf{K}} = \mathbf{KT} = [\delta_n \quad \delta_{n-1} \quad \cdots \quad \delta_1] \tag{6-62}$$

Then

$$\hat{\mathbf{H}}\hat{\mathbf{K}} = \begin{bmatrix} 0 \\ 0 \\ \vdots \\ 0 \\ 1 \end{bmatrix} [\delta_n \quad \delta_{n-1} \quad \cdots \quad \delta_1] = \begin{bmatrix} 0 & 0 & \cdots & 0 \\ 0 & 0 & \cdots & 0 \\ \vdots & \vdots & & \vdots \\ 0 & 0 & \cdots & 0 \\ \delta_n & \delta_{n-1} & \cdots & \delta_1 \end{bmatrix}$$

The characteristic equation $|z\mathbf{I} - \mathbf{G} + \mathbf{HK}|$ becomes as follows:

$$|z\mathbf{I} - \mathbf{G} + \mathbf{HK}| = |z\mathbf{I} - \hat{\mathbf{G}} + \hat{\mathbf{H}}\hat{\mathbf{K}}|$$

$$= \left| z\begin{bmatrix} 1 & 0 & \cdots & 0 \\ 0 & 1 & \cdots & 0 \\ \vdots & \vdots & & \vdots \\ 0 & 0 & \cdots & 0 \\ 0 & 0 & \cdots & 1 \end{bmatrix} - \begin{bmatrix} 0 & 1 & \cdots & 0 \\ 0 & 0 & \cdots & 0 \\ \vdots & \vdots & & \vdots \\ 0 & 0 & \cdots & 1 \\ -a_n & -a_{n-1} & \cdots & -a_1 \end{bmatrix} + \begin{bmatrix} 0 & 0 & \cdots & 0 \\ 0 & 0 & \cdots & 0 \\ \vdots & \vdots & & \vdots \\ 0 & 0 & \cdots & 0 \\ \delta_n & \delta_{n-1} & \cdots & \delta_1 \end{bmatrix} \right|$$

$$= \begin{vmatrix} z & -1 & \cdots & 0 \\ 0 & z & \cdots & 0 \\ \vdots & \vdots & & \vdots \\ 0 & 0 & \cdots & -1 \\ a_n + \delta_n & a_{n-1} + \delta_{n-1} & \cdots & z + a_1 + \delta_1 \end{vmatrix}$$

$$= z^n + (a_1 + \delta_1)z^{n-1} + \cdots + (a_{n-1} + \delta_{n-1})z + a_n + \delta_n = 0 \tag{6-63}$$

The characteristic equation with the desired eigenvalues is given by

$$(z - \mu_1)(z - \mu_2) \cdots (z - \mu_n)$$

$$= z^n + \alpha_1 z^{n-1} + \alpha_2 z^{n-2} + \cdots + \alpha_{n-1} z + \alpha_n = 0 \qquad (6\text{-}64)$$

Equating the coefficients of equal powers of z of Equations (6–63) and (6–64) we obtain

$$\alpha_1 = a_1 + \delta_1$$

$$\alpha_2 = a_2 + \delta_2$$

$$\vdots$$

$$\alpha_n = a_n + \delta_n$$

Hence, from Equation (6–62) we have

$$\mathbf{K} = \hat{\mathbf{K}}\mathbf{T}^{-1}$$

$$= [\delta_n \quad \delta_{n-1} \quad \cdots \quad \delta_1]\mathbf{T}^{-1}$$

$$= [\alpha_n - a_n \mathbin{\vdots} \alpha_{n-1} - a_{n-1} \mathbin{\vdots} \cdots \mathbin{\vdots} \alpha_1 - a_1]\mathbf{T}^{-1} \qquad (6\text{-}65)$$

where the a_i's and the α_i's are known coefficients and $\mathbf{T}$ is a known matrix. Hence, we have determined the required feedback gain matrix $\mathbf{K}$ in terms of known coefficients and a known matrix of the system. This proves the sufficient condition; that is, if the system defined by Equation (6–50) is completely state controllable, then it is always possible to determine the required state feedback gain matrix $\mathbf{K}$ for arbitrary pole placement. Hence, we have proved that a necessary and sufficient condition for arbitrary pole placement is that the system be completely state controllable.

  Ackermann's Formula. The expression given by Equation (6–65) is not the only one used for the determination of the state feedback gain matrix $\mathbf{K}$. There are other expressions available. In the following, we shall present one such expression, commonly called *Ackermann's formula*.

  Consider the system defined by Equation (6–50). It is assumed that the system is completely state controllable. By using the state feedback $u(k) = -\mathbf{K}\mathbf{x}(k)$, we wish to place closed-loop poles at $z = \mu_1, z = \mu_2, \ldots, z = \mu_n$. That is, we desire the characteristic equation to be

$$|z\mathbf{I} - \mathbf{G} + \mathbf{H}\mathbf{K}| = (z - \mu_1)(z - \mu_2) \cdots (z - \mu_n)$$

$$= z^n + \alpha_1 z^{n-1} + \alpha_2 z^{n-2} + \cdots + \alpha_{n-1} z + \alpha_n = 0$$

Let us define

$$\tilde{\mathbf{G}} = \mathbf{G} - \mathbf{H}\mathbf{K}$$

Since the Cayley–Hamilton theorem states that $\tilde{\mathbf{G}}$ satisfies its own characteristic equation, we have

$$\tilde{\mathbf{G}}^n + \alpha_1 \tilde{\mathbf{G}}^{n-1} + \alpha_2 \tilde{\mathbf{G}}^{n-2} + \cdots + \alpha_{n-1} \tilde{\mathbf{G}} + \alpha_n \mathbf{I} = \phi(\tilde{\mathbf{G}}) = \mathbf{0}$$

We shall utilize this last equation to derive Ackermann's formula.

Consider now the following identities:

$$\mathbf{I} = \mathbf{I}$$

$$\tilde{\mathbf{G}} = \mathbf{G} - \mathbf{HK}$$

$$\tilde{\mathbf{G}}^2 = (\mathbf{G} - \mathbf{HK})^2 = \mathbf{G}^2 - \mathbf{GHK} - \mathbf{HK}\tilde{\mathbf{G}}$$

$$\tilde{\mathbf{G}}^3 = (\mathbf{G} - \mathbf{HK})^3 = \mathbf{G}^3 - \mathbf{G}^2\mathbf{HK} - \mathbf{GHK}\tilde{\mathbf{G}} - \mathbf{HK}\tilde{\mathbf{G}}^2$$

$$\vdots$$

$$\tilde{\mathbf{G}}^n = (\mathbf{G} - \mathbf{HK})^n = \mathbf{G}^n - \mathbf{G}^{n-1}\mathbf{HK} - \cdots - \mathbf{HK}\tilde{\mathbf{G}}^{n-1}$$

Multiplying the preceding equations in order by $\alpha_n, \alpha_{n-1}, \ldots, \alpha_0$ (where $\alpha_0 = 1$), respectively, and adding the results, we obtain

$$\alpha_n \mathbf{I} + \alpha_{n-1}\tilde{\mathbf{G}} + \alpha_{n-2}\tilde{\mathbf{G}}^2 + \cdots + \tilde{\mathbf{G}}^n = \alpha_n \mathbf{I} + \alpha_{n-1}\mathbf{G} + \alpha_{n-2}\mathbf{G}^2$$

$$+ \cdots + \mathbf{G}^n - \alpha_{n-1}\mathbf{HK} - \alpha_{n-2}\mathbf{GHK} - \alpha_{n-2}\mathbf{HK}\tilde{\mathbf{G}} - \cdots - \mathbf{G}^{n-1}\mathbf{HK} - \cdots$$

$$- \mathbf{HK}\tilde{\mathbf{G}}^{n-1}$$

which can be written as follows:

$$\phi(\tilde{\mathbf{G}}) = \phi(\mathbf{G}) - \alpha_{n-1}\mathbf{HK} - \alpha_{n-2}\mathbf{GHK} - \alpha_{n-2}\mathbf{HK}\tilde{\mathbf{G}} - \cdots - \mathbf{HK}\tilde{\mathbf{G}}^{n-1} - \mathbf{G}^{n-1}\mathbf{HK}$$

$$= \phi(\mathbf{G}) - [\mathbf{H} \vdots \mathbf{GH} \vdots \cdots \vdots \mathbf{G}^{n-1}\mathbf{H}] \begin{bmatrix} \alpha_{n-1}\mathbf{K} + \alpha_{n-2}\mathbf{K}\tilde{\mathbf{G}} + \cdots + \mathbf{K}\tilde{\mathbf{G}}^{n-1} \\ \alpha_{n-2}\mathbf{K} + \alpha_{n-3}\mathbf{K}\tilde{\mathbf{G}} + \cdots + \mathbf{K}\tilde{\mathbf{G}}^{n-2} \\ \vdots \\ \mathbf{K} \end{bmatrix} \quad (6\text{--}66)$$

Notice that

$$\phi(\tilde{\mathbf{G}}) = \mathbf{0}$$

Hence, Equation (6–66) may be modified to read

$$\phi(\mathbf{G}) = [\mathbf{H} \vdots \mathbf{GH} \vdots \cdots \vdots \mathbf{G}^{n-1}\mathbf{H}] \begin{bmatrix} \alpha_{n-1}\mathbf{K} + \alpha_{n-2}\mathbf{K}\tilde{\mathbf{G}} + \cdots + \mathbf{K}\tilde{\mathbf{G}}^{n-1} \\ \alpha_{n-2}\mathbf{K} + \alpha_{n-3}\mathbf{K}\tilde{\mathbf{G}} + \cdots + \mathbf{K}\tilde{\mathbf{G}}^{n-2} \\ \vdots \\ \mathbf{K} \end{bmatrix} \quad (6\text{--}67)$$

Since the system is completely state controllable, the controllability matrix

$$[\mathbf{H} \vdots \mathbf{GH} \vdots \cdots \vdots \mathbf{G}^{n-1}\mathbf{H}]$$

is of rank n and its inverse exists. Then Equation (6–67) can be modified to the form

$$\begin{bmatrix} \alpha_{n-1}\mathbf{K} + \alpha_{n-2}\mathbf{K}\tilde{\mathbf{G}} + \cdots + \mathbf{K}\tilde{\mathbf{G}}^{n-1} \\ \alpha_{n-2}\mathbf{K} + \alpha_{n-3}\mathbf{K}\tilde{\mathbf{G}} + \cdots + \mathbf{K}\tilde{\mathbf{G}}^{n-2} \\ \vdots \\ \mathbf{K} \end{bmatrix} = [\mathbf{H} \vdots \mathbf{GH} \vdots \cdots \vdots \mathbf{G}^{n-1}\mathbf{H}]^{-1}\phi(\mathbf{G})$$

Premultiplying both sides of this last equation by $[0 \quad 0 \quad \cdots \quad 0 \quad 1]$, we obtain

$$[0 \quad 0 \quad \cdots \quad 0 \quad 1] \begin{bmatrix} \alpha_{n-1}\mathbf{K} + \alpha_{n-2}\mathbf{K\tilde{G}} + \cdots + \mathbf{K\tilde{G}}^{n-1} \\ \alpha_{n-2}\mathbf{K} + \alpha_{n-3}\mathbf{K\tilde{G}} + \cdots + \mathbf{K\tilde{G}}^{n-2} \\ \vdots \\ \mathbf{K} \end{bmatrix}$$

$$= [0 \quad 0 \quad \cdots \quad 0 \quad 1][\mathbf{H} \vdots \mathbf{GH} \vdots \cdots \vdots \mathbf{G}^{n-1}\mathbf{H}]^{-1} \phi(\mathbf{G})$$

which can be simplified to

$$\mathbf{K} = [0 \quad 0 \quad \cdots \quad 0 \quad 1][\mathbf{H} \vdots \mathbf{GH} \vdots \cdots \vdots \mathbf{G}^{n-1}\mathbf{H}]^{-1} \phi(\mathbf{G}) \qquad (6\text{-}68)$$

Equation (6–68) gives the required state feedback gain matrix $\mathbf{K}$. It is this particular expression for matrix $\mathbf{K}$ that is commonly called Ackermann's formula. [See Problem A–6–12 for the derivation of Equation (6–68) when $n = 3$.]

Comments. The state feedback gain matrix $\mathbf{K}$ is determined in such a way that the error (caused by disturbances) will reduce to zero with sufficient speed. Note that the matrix $\mathbf{K}$ is not unique for a given system, but depends on the desired closed-loop pole locations (which determine the speed of response) selected. The selection of the desired closed-loop poles or the desired characteristic equation is a compromise between the rapidity of the response of the error vector and the sensitivity to disturbances and measurement noises. That is, if we increase the speed of error response, then the adverse effects of disturbances and measurement noises generally increase. In determining the state feedback gain matrix $\mathbf{K}$ for a given system, it is desirable to examine several matrices $\mathbf{K}$ based on several different desired characteristic equations and to choose the one that gives the best overall system performance.

Once the desired characteristic equation is selected, there are several different ways to determine the corresponding state feedback gain matrix $\mathbf{K}$ for the system defined by Equation (6–50) which is assumed to be completely state controllable. Four of them are listed in the following.

1. As shown in the preceding discussion, matrix $\mathbf{K}$ can be given by Equation (6–65):

$$\mathbf{K} = [\alpha_n - a_n \vdots \alpha_{n-1} - a_{n-1} \vdots \cdots \vdots \alpha_1 - a_1]\mathbf{T}^{-1}$$

$$= [\alpha_n - a_n \vdots \alpha_{n-1} - a_{n-1} \vdots \cdots \vdots \alpha_1 - a_1](\mathbf{MW})^{-1} \qquad (6\text{-}69)$$

where the a_i's are the coefficients of the original system characteristic equation

$$|z\mathbf{I} - \mathbf{G}| = z^n + a_1 z^{n-1} + \cdots + a_{n-1}z + a_n = 0$$

and the α_i's are the coefficients of the desired characteristic equation for the state feedback control system; that is,

$$|z\mathbf{I} - \mathbf{G} + \mathbf{HK}| = z^n + \alpha_1 z^{n-1} + \cdots + \alpha_{n-1}z + \alpha_n = 0$$

Matrix $\mathbf{T}$ is given by

$$\mathbf{T} = \mathbf{MW} \qquad (6\text{-}70)$$

where $\mathbf{M}$ and $\mathbf{W}$ are given by Equations (6–60) and (6–61), respectively.

If the system state equation is already in the controllable canonical form, the determination of the state feedback gain matrix $\mathbf{K}$ can be made simple, because the transformation matrix $\mathbf{T}$ becomes the identity matrix. In this case the desired matrix $\mathbf{K}$ is obtained by substituting $\mathbf{T} = \mathbf{MW} = \mathbf{I}$ into Equation (6-69).

2. The desired state feedback gain matrix $\mathbf{K}$ can be given by Ackermann's formula:

$$\mathbf{K} = [0 \quad 0 \quad \cdots \quad 0 \quad 1][\mathbf{H} \vdots \mathbf{GH} \vdots \cdots \vdots \mathbf{G}^{n-1}\mathbf{H}]^{-1} \phi(\mathbf{G}) \qquad (6\text{-}71)$$

where

$$\phi(\mathbf{G}) = \mathbf{G}^n + \alpha_1 \mathbf{G}^{n-1} + \cdots + \alpha_{n-1}\mathbf{G} + \alpha_n \mathbf{I}$$

3. If the desired eigenvalues $\mu_1, \mu_2, \ldots, \mu_n$ are distinct, then the desired state feedback gain matrix $\mathbf{K}$ can be given as follows:

$$\mathbf{K} = [1 \quad 1 \quad \cdots \quad 1][\boldsymbol{\xi}_1 \vdots \boldsymbol{\xi}_2 \vdots \cdots \vdots \boldsymbol{\xi}_n]^{-1} \qquad (6\text{-}72)$$

where vectors $\boldsymbol{\xi}_1, \boldsymbol{\xi}_2, \ldots, \boldsymbol{\xi}_n$ satisfy the equation

$$\boldsymbol{\xi}_i = (\mathbf{G} - \mu_i \mathbf{I})^{-1}\mathbf{H}, \qquad i = 1, 2, \ldots, n$$

Note that the $\boldsymbol{\xi}_i$'s are eigenvectors of matrix $\mathbf{G} - \mathbf{HK}$; that is, $\boldsymbol{\xi}_i$ satisfies the equation

$$(\mathbf{G} - \mathbf{HK})\boldsymbol{\xi}_i = \mu_i \boldsymbol{\xi}_i, \qquad i = 1, 2, \ldots, n$$

For the deadbeat response, $\mu_1 = \mu_2 = \cdots = \mu_n = 0$. Equation (6-72) for this case can be simplified as follows:

$$\mathbf{K} = [1 \quad 0 \quad \cdots \quad 0][\boldsymbol{\xi}_1 \vdots \boldsymbol{\xi}_2 \vdots \cdots \vdots \boldsymbol{\xi}_n]^{-1} \qquad (6\text{-}73)$$

where

$$\boldsymbol{\xi}_1 = \mathbf{G}^{-1}\mathbf{H}, \qquad \boldsymbol{\xi}_2 = \mathbf{G}^{-2}\mathbf{H}, \qquad \ldots, \qquad \boldsymbol{\xi}_n = \mathbf{G}^{-n}\mathbf{H}$$

[For detailed derivations of Equations (6-72) and (6-73), see Problems A-6-12 and A-6-13.]

4. If the order n of the system is low, substitute $\mathbf{K} = [k_1 \vdots k_2 \vdots \cdots \vdots k_n]$ into the characteristic equation

$$|z\mathbf{I} - \mathbf{G} + \mathbf{HK}| = 0$$

and then match the coefficients of powers in z of this characteristic equation with equal powers in z of the desired characteristic equation

$$z^n + \alpha_1 z^{n-1} + \cdots + \alpha_{n-1} z + \alpha_n = 0$$

Such a direct calculation of matrix $\mathbf{K}$ may be simpler for low-order systems.

Example 6-6

Consider the system

$$\mathbf{x}(k + 1) = \mathbf{Gx}(k) + \mathbf{H}u(k)$$

where

$$\mathbf{G} = \begin{bmatrix} 0 & 1 \\ -0.16 & -1 \end{bmatrix}, \qquad \mathbf{H} = \begin{bmatrix} 0 \\ 1 \end{bmatrix}$$

Note that

$$|z\mathbf{I} - \mathbf{G}| = \begin{vmatrix} z & -1 \\ 0.16 & z+1 \end{vmatrix} = z^2 + z + 0.16$$

Hence,

$$a_1 = 1, \qquad a_2 = 0.16$$

Determine a suitable state feedback gain matrix $\mathbf{K}$ such that the system will have the closed-loop poles at

$$z = 0.5 + j0.5, \qquad z = 0.5 - j0.5$$

Let us first examine the rank of the controllability matrix. The rank of

$$[\mathbf{H} \vdots \mathbf{GH}] = \begin{bmatrix} 0 & 1 \\ 1 & -1 \end{bmatrix}$$

is 2. Thus, the system is completely state controllable, and therefore arbitrary pole placement is possible. The characteristic equation for the desired system is

$$|z\mathbf{I} - \mathbf{G} + \mathbf{HK}| = (z - 0.5 - j0.5)(z - 0.5 + j0.5) = z^2 - z + 0.5 = 0$$

Hence,

$$\alpha_1 = -1, \qquad \alpha_2 = 0.5$$

We shall demonstrate four different ways to determine matrix $\mathbf{K}$.

Method 1.   From Equation (6–69), the state feedback gain matrix $\mathbf{K}$ is given as follows:

$$\mathbf{K} = [\alpha_2 - a_2 \vdots \alpha_1 - a_1]\mathbf{T}^{-1}$$

Notice that the original system is already in a controllable canonical form, and therefore the transformation matrix $\mathbf{T}$ becomes $\mathbf{I}$:

$$\mathbf{T} = \mathbf{MW} = [\mathbf{H} \vdots \mathbf{GH}]\begin{bmatrix} a_1 & 1 \\ 1 & 0 \end{bmatrix} = \begin{bmatrix} 0 & 1 \\ 1 & -1 \end{bmatrix}\begin{bmatrix} 1 & 1 \\ 1 & 0 \end{bmatrix} = \begin{bmatrix} 1 & 0 \\ 0 & 1 \end{bmatrix}$$

Hence,

$$\mathbf{K} = [\alpha_2 - a_2 \vdots \alpha_1 - a_1] = [0.5 - 0.16 \vdots -1 - 1]$$

$$= [0.34 \quad -2]$$

Method 2.   Referring to Ackermann's formula given by Equation (6–71), we have

$$\mathbf{K} = [0 \quad 1][\mathbf{H} \vdots \mathbf{GH}]^{-1} \phi(\mathbf{G})$$

where

$$\phi(\mathbf{G}) = \mathbf{G}^2 - \mathbf{G} + 0.5\mathbf{I} = \begin{bmatrix} -0.16 & -1 \\ 0.16 & 0.84 \end{bmatrix} - \begin{bmatrix} 0 & 1 \\ -0.16 & -1 \end{bmatrix} + \begin{bmatrix} 0.5 & 0 \\ 0 & 0.5 \end{bmatrix}$$

$$= \begin{bmatrix} 0.34 & -2 \\ 0.32 & 2.34 \end{bmatrix}$$

Thus,

$$\mathbf{K} = [0 \quad 1]\begin{bmatrix} 0 & 1 \\ 1 & -1 \end{bmatrix}^{-1}\begin{bmatrix} 0.34 & -2 \\ 0.32 & 2.34 \end{bmatrix}$$

$$= [0.34 \quad -2]$$

Method 3.   From Equation (6–72), the desired state feedback gain matrix **K** is determined as follows:

$$\mathbf{K} = [1 \quad 1][\xi_1 \quad \xi_2]^{-1}$$

where

$$\xi_i = (\mathbf{G} - \mu_i \mathbf{I})^{-1}\mathbf{H}, \qquad i = 1, 2$$

Since $\mu_1 = 0.5 + j0.5$, we have

$$\xi_1 = [\mathbf{G} - (0.5 + j0.5)\mathbf{I}]^{-1}\mathbf{H}$$

$$= \begin{bmatrix} -0.5 - j0.5 & 1 \\ -0.16 & -1.5 - j0.5 \end{bmatrix}^{-1}\begin{bmatrix} 0 \\ 1 \end{bmatrix} = \begin{bmatrix} \dfrac{-1}{0.66 + j} \\ \dfrac{-0.5 - j0.5}{0.66 + j} \end{bmatrix}$$

Similarly, for $\mu_2 = 0.5 - j0.5$, we have

$$\xi_2 = [\mathbf{G} - (0.5 - j0.5)\mathbf{I}]^{-1}\mathbf{H}$$

$$= \begin{bmatrix} -0.5 + j0.5 & 1 \\ -0.16 & -1.5 + j0.5 \end{bmatrix}^{-1}\begin{bmatrix} 0 \\ 1 \end{bmatrix} = \begin{bmatrix} \dfrac{-1}{0.66 - j} \\ \dfrac{-0.5 + j0.5}{0.66 - j} \end{bmatrix}$$

Consequently, we have

$$[\xi_1 \quad \xi_2]^{-1} = \begin{bmatrix} \dfrac{-1}{0.66 + j} & \dfrac{-1}{0.66 - j} \\ \dfrac{-0.5 - j0.5}{0.66 + j} & \dfrac{-0.5 + j0.5}{0.66 - j} \end{bmatrix}^{-1}$$

$$= \begin{bmatrix} \dfrac{0.7178(1 - j)}{1 + j0.66} & \dfrac{-1.4356}{1 + j0.66} \\ \dfrac{-0.7178(1 + j)}{-1 + j0.66} & \dfrac{1.4356}{-1 + j0.66} \end{bmatrix}$$

Hence, the desired state feedback gain matrix **K** is determined to be

$$\mathbf{K} = [1 \quad 1][\xi_1 \quad \xi_2]^{-1}$$

$$= [1 \quad 1]\begin{bmatrix} \dfrac{0.7178(1 - j)}{1 + j0.66} & \dfrac{-1.4356}{1 + j0.66} \\ \dfrac{-0.7178(1 + j)}{-1 + j0.66} & \dfrac{1.4356}{-1 + j0.66} \end{bmatrix}$$

$$= [0.34 \quad -2]$$

Method 4.   It is noted that for lower-order systems such as this one it may be simpler to substitute

$$\mathbf{K} = [k_1 \quad k_2]$$

into the characteristic equation and to write the equation in terms of undetermined k's. Then this characteristic equation is equated with the desired characteristic equation. The procedure is as follows:

$$|z\mathbf{I} - \mathbf{G} + \mathbf{HK}| = \left| \begin{bmatrix} z & 0 \\ 0 & z \end{bmatrix} - \begin{bmatrix} 0 & 1 \\ -0.16 & -1 \end{bmatrix} + \begin{bmatrix} 0 \\ 1 \end{bmatrix} [k_1 \quad k_2] \right|$$

$$= \left| \begin{matrix} z & -1 \\ 0.16 + k_1 & z + 1 + k_2 \end{matrix} \right|$$

$$= z^2 + (1 + k_2)z + 0.16 + k_1 = 0$$

Now we equate this characteristic equation with the desired characteristic equation

$$(z - 0.5 - j0.5)(z - 0.5 + j0.5) = 0$$

so that

$$z^2 + (1 + k_2)z + 0.16 + k_1 = z^2 - z + 0.5$$

By comparing the coefficients of equal powers of z, we obtain

$$1 + k_2 = -1, \qquad 0.16 + k_1 = 0.5$$

from which we get

$$k_1 = 0.34, \qquad k_2 = -2$$

Thus, the desired state feedback gain matrix $\mathbf{K}$ is given by

$$\mathbf{K} = [k_1 \quad k_2] = [0.34 \quad -2]$$

It is noted that for higher-order systems the calculations involved by this approach may become laborious. For such a case, other methods may be preferred.

Deadbeat Response.   Consider the system defined by

$$\mathbf{x}(k + 1) = \mathbf{Gx}(k) + \mathbf{H}u(k)$$

With state feedback $u(k) = -\mathbf{Kx}(k)$, the state equation becomes

$$\mathbf{x}(k + 1) = (\mathbf{G} - \mathbf{HK})\mathbf{x}(k)$$

Note that the solution of this last equation is given by

$$\mathbf{x}(k) = (\mathbf{G} - \mathbf{HK})^k \mathbf{x}(0) \tag{6–74}$$

If the eigenvalues μ_i of matrix $\mathbf{G} - \mathbf{HK}$ lie inside the unit circle, then the system is asymptotically stable.

In what follows, we shall show that, by choosing all eigenvalues of $\mathbf{G} - \mathbf{HK}$ to be zero, it is possible to get the deadbeat response, or

$$\mathbf{x}(k) = \mathbf{0}, \qquad \text{for } k \geq q, q \leq n$$

In discussing deadbeat response, the nilpotent matrix

$$N = \begin{bmatrix} 0 & 1 & 0 & \cdots & 0 \\ 0 & 0 & 1 & \cdots & 0 \\ \vdots & \vdots & \vdots & & \vdots \\ 0 & 0 & 0 & \cdots & 1 \\ 0 & 0 & 0 & \cdots & 0 \end{bmatrix}$$

plays an important role. Consider, for example, a 4×4 nilpotent matrix:

$$N = \begin{bmatrix} 0 & 1 & 0 & 0 \\ 0 & 0 & 1 & 0 \\ 0 & 0 & 0 & 1 \\ 0 & 0 & 0 & 0 \end{bmatrix}$$

Notice that

$$N^2 = \begin{bmatrix} 0 & 0 & 1 & 0 \\ 0 & 0 & 0 & 1 \\ 0 & 0 & 0 & 0 \\ 0 & 0 & 0 & 0 \end{bmatrix}, \quad N^3 = \begin{bmatrix} 0 & 0 & 0 & 1 \\ 0 & 0 & 0 & 0 \\ 0 & 0 & 0 & 0 \\ 0 & 0 & 0 & 0 \end{bmatrix}, \quad N^4 = \begin{bmatrix} 0 & 0 & 0 & 0 \\ 0 & 0 & 0 & 0 \\ 0 & 0 & 0 & 0 \\ 0 & 0 & 0 & 0 \end{bmatrix}$$

Similarly, for an $n \times n$ nilpotent matrix N, we have

$$N^n = 0$$

Now consider the completely state controllable system given by

$$x(k + 1) = Gx(k) + Hu(k) \tag{6-75}$$

Let us choose the desired pole locations to be at the origin, or choose the desired eigenvalues to be zero: $\mu_1 = \mu_2 = \cdots = \mu_n = 0$. Then we shall show that the response to any initial state $x(0)$ is deadbeat. Since the characteristic equation with the desired eigenvalues can be given by

$$(z - \mu_1)(z - \mu_2) \cdots (z - \mu_n) = z^n + \alpha_1 z^{n-1} + \cdots + \alpha_{n-1} z + \alpha_n = z^n$$

we obtain

$$\alpha_1 = \alpha_2 = \cdots = \alpha_n = 0$$

and matrix K given by Equation (6-65) can be simplified to the following:

$$K = [\alpha_n - a_n \vdots \alpha_{n-1} - a_{n-1} \vdots \cdots \vdots \alpha_1 - a_1]T^{-1}$$
$$= [-a_n \quad -a_{n-1} \quad \cdots \quad -a_1]T^{-1} \tag{6-76}$$

By using the transformation matrix T given by Equation (6-32), define

$$x(k) = T\hat{x}(k)$$

Define also

$$T^{-1}GT = \hat{G}, \qquad T^{-1}H = \hat{H}$$

Then Equation (6-75) can be written as

$$\hat{x}(k + 1) = T^{-1}GT\hat{x}(k) + T^{-1}Hu(k) = \hat{G}\hat{x}(k) + \hat{H}u(k)$$

If we use the state feedback $u(k) = -\mathbf{K}\mathbf{x}(k) = -\mathbf{K}\mathbf{T}\hat{\mathbf{x}}(k)$, then this last equation becomes

$$\hat{\mathbf{x}}(k + 1) = (\hat{\mathbf{G}} - \hat{\mathbf{H}}\mathbf{K}\mathbf{T})\hat{\mathbf{x}}(k)$$

Referring to Equation (6–76), we have

$$\hat{\mathbf{G}} - \hat{\mathbf{H}}\mathbf{K}\mathbf{T} = \hat{\mathbf{G}} - \hat{\mathbf{H}}[-a_n \quad -a_{n-1} \quad \cdots \quad -a_1]$$

$$= \begin{bmatrix} 0 & 1 & 0 & \cdots & 0 \\ 0 & 0 & 1 & \cdots & 0 \\ \vdots & \vdots & \vdots & & \vdots \\ 0 & 0 & 0 & \cdots & 1 \\ -a_n & -a_{n-1} & -a_{n-2} & \cdots & -a_1 \end{bmatrix} - \begin{bmatrix} 0 \\ 0 \\ \vdots \\ 0 \\ 1 \end{bmatrix}[-a_n \quad -a_{n-1} \quad \cdots \quad -a_1]$$

$$= \begin{bmatrix} 0 & 1 & 0 & \cdots & 0 \\ 0 & 0 & 1 & \cdots & 0 \\ \vdots & \vdots & \vdots & & \vdots \\ 0 & 0 & 0 & \cdots & 1 \\ -a_n & -a_{n-1} & -a_{n-2} & \cdots & -a_1 \end{bmatrix} - \begin{bmatrix} 0 & 0 & 0 & \cdots & 0 \\ 0 & 0 & 0 & \cdots & 0 \\ \vdots & \vdots & \vdots & & \vdots \\ 0 & 0 & 0 & \cdots & 0 \\ -a_n & -a_{n-1} & -a_{n-2} & \cdots & -a_1 \end{bmatrix}$$

$$= \begin{bmatrix} 0 & 1 & 0 & \cdots & 0 \\ 0 & 0 & 1 & \cdots & 0 \\ \vdots & \vdots & \vdots & & \vdots \\ 0 & 0 & 0 & \cdots & 1 \\ 0 & 0 & 0 & \cdots & 0 \end{bmatrix}$$

Thus, $\hat{\mathbf{G}} - \hat{\mathbf{H}}\mathbf{K}\mathbf{T}$ is a nilpotent matrix. Therefore, we have

$$(\hat{\mathbf{G}} - \hat{\mathbf{H}}\mathbf{K}\mathbf{T})^n = \mathbf{0}$$

In terms of the original state $\mathbf{x}(k)$, we have

$$\mathbf{x}(n) = (\mathbf{G} - \mathbf{H}\mathbf{K})^n \mathbf{x}(0) = (\mathbf{T}\hat{\mathbf{G}}\mathbf{T}^{-1} - \mathbf{T}\hat{\mathbf{H}}\mathbf{K})^n \mathbf{x}(0) = [\mathbf{T}(\hat{\mathbf{G}} - \hat{\mathbf{H}}\mathbf{K}\mathbf{T})\mathbf{T}^{-1}]^n \mathbf{x}(0)$$

$$= \mathbf{T}(\hat{\mathbf{G}} - \hat{\mathbf{H}}\mathbf{K}\mathbf{T})^n \mathbf{T}^{-1}\mathbf{x}(0) = \mathbf{0}$$

Thus, we have shown that if the desired eigenvalues are all zeros then any initial state $\mathbf{x}(0)$ can be brought to the origin in at most n sampling periods and the response is deadbeat, provided the control signal $u(k)$ is unbounded.

Comments on Deadbeat Control.   The concept of deadbeat response is unique to discrete-time control systems. There is no such thing as deadbeat response in continuous-time control systems. In deadbeat control, any nonzero error vector will be driven to zero in at most n sampling periods if the magnitude of the scalar control $u(k)$ is unbounded. The settling time depends on the sampling period, since the response settles down in at most n sampling periods. If the sampling period T is chosen very small, the settling time will also be very small, which implies that the control signal must have an extremely large magnitude. Otherwise, it will not be possible to bring the error response to zero in a short time period.

In deadbeat control, the sampling period is the only design parameter. Thus, if the deadbeat response is desired, the designer must choose the sampling period carefully so that an extremely large control magnitude is not required in normal operation of the system. Note that it is not physically possible to increase the magnitude of the control signal without bound. If the magnitude is increased sufficiently, the saturation phenomenon always takes place. If saturation occurs in the magnitude of the control signal, then the response can no longer be deadbeat. The settling time will be more than n sampling periods. In the actual design of deadbeat control systems, the designer must be aware of the trade-off that must be made between the magnitude of the control signal and the response speed.

Example 6-7

Consider the system given by

$$\begin{bmatrix} x_1(k+1) \\ x_2(k+1) \end{bmatrix} = \begin{bmatrix} 0 & 1 \\ -0.16 & -1 \end{bmatrix} \begin{bmatrix} x_1(k) \\ x_2(k) \end{bmatrix} + \begin{bmatrix} 0 \\ 1 \end{bmatrix} u(k) \tag{6-77}$$

Determine the state feedback gain matrix $\mathbf{K}$ such that when the control signal is given by

$$u(k) = -\mathbf{K}\mathbf{x}(k)$$

the closed-loop system (regulator system) exhibits the deadbeat response to an initial state $\mathbf{x}(0)$. Assume that the control signal $u(k)$ is unbounded.

Referring to Equation (6-76), for the deadbeat response we have

$$\mathbf{K} = [-a_2 \quad -a_1]\mathbf{T}^{-1} \tag{6-78}$$

The system given by Equation (6-77) is already in the controllable canonical form. Therefore, in Equation (6-78), $\mathbf{T} = \mathbf{I}$. The characteristic equation for the system given by Equation (6-77) is

$$|z\mathbf{I} - \mathbf{G}| = \begin{vmatrix} z & -1 \\ 0.16 & z+1 \end{vmatrix} = z^2 + z + 0.16 = z^2 + a_1 z + a_2$$

Thus,

$$a_1 = 1, \qquad a_2 = 0.16$$

Consequently, Equation (6-78) becomes

$$\mathbf{K} = [-a_2 \quad -a_1] = [-0.16 \quad -1]$$

This gives the desired state feedback gain matrix.

Let us verify that the response of this system to an arbitrary initial state $\mathbf{x}(0)$ is indeed the deadbeat response. Since the closed-loop state equation becomes

$$\begin{bmatrix} x_1(k+1) \\ x_2(k+1) \end{bmatrix} = \begin{bmatrix} 0 & 1 \\ -0.16 & -1 \end{bmatrix} \begin{bmatrix} x_1(k) \\ x_2(k) \end{bmatrix} + \begin{bmatrix} 0 \\ 1 \end{bmatrix}[0.16 \quad 1]\begin{bmatrix} x_1(k) \\ x_2(k) \end{bmatrix}$$

$$= \begin{bmatrix} 0 & 1 \\ 0 & 0 \end{bmatrix} \begin{bmatrix} x_1(k) \\ x_2(k) \end{bmatrix}$$

if the initial state is given by

$$\begin{bmatrix} x_1(0) \\ x_2(0) \end{bmatrix} = \begin{bmatrix} a \\ b \end{bmatrix}$$

where a and b are arbitrary constants, then we have

$$\begin{bmatrix} x_1(1) \\ x_2(1) \end{bmatrix} = \begin{bmatrix} 0 & 1 \\ 0 & 0 \end{bmatrix} \begin{bmatrix} x_1(0) \\ x_2(0) \end{bmatrix} = \begin{bmatrix} 0 & 1 \\ 0 & 0 \end{bmatrix} \begin{bmatrix} a \\ b \end{bmatrix} = \begin{bmatrix} b \\ 0 \end{bmatrix}$$

$$\begin{bmatrix} x_1(2) \\ x_2(2) \end{bmatrix} = \begin{bmatrix} 0 & 1 \\ 0 & 0 \end{bmatrix} \begin{bmatrix} x_1(1) \\ x_2(1) \end{bmatrix} = \begin{bmatrix} 0 & 1 \\ 0 & 0 \end{bmatrix} \begin{bmatrix} b \\ 0 \end{bmatrix} = \begin{bmatrix} 0 \\ 0 \end{bmatrix}$$

Thus, the state $\mathbf{x}(k)$ for $k = 2, 3, 4, \ldots$ becomes zero and the response is indeed deadbeat.

Pole Placement When the Control Signal Is a Vector.   Thus far, we have considered the pole placement design problem when the control signal is a scalar. If the control signal is a vector quantity (r-vector), the response can be speeded up, because we have more freedom to choose control signals $u_1(k), u_2(k), \ldots, u_r(k)$ to speed up the response. For example, in the case of the nth-order system with a scalar control, the deadbeat response can be achieved in at most n sampling periods. In the case of the vector control $\mathbf{u}(k)$, the deadbeat response can be achieved in less than n sampling periods.

In the case of the vector control, however, the determination of the state feedback gain matrix $\mathbf{K}$ becomes more complex. We shall present such a case in Appendix C.

Control System With Reference Input.   Thus far, we have considered regulator systems. In the regulator system, the reference input is fixed for a long period, and external disturbances create nonzero states. The characteristic equation for the system determines the speed by which the nonzero states approach the origin. In what follows, we shall consider the case where the system has a reference input.

Consider the system shown in Figure 6–2. The plant is described by the following state and output equations:

$$\mathbf{x}(k + 1) = \mathbf{Gx}(k) + \mathbf{H}u(k)$$

$$y(k) = \mathbf{Cx}(k)$$

The control signal $u(k)$ is given by

$$u(k) = K_0 r(k) - \mathbf{Kx}(k)$$

By eliminating $u(k)$ from the state equation, we have

$$\mathbf{x}(k + 1) = (\mathbf{G} - \mathbf{HK})\mathbf{x}(k) + \mathbf{H}K_0 r(k)$$

The characteristic equation for the system is

$$|z\mathbf{I} - \mathbf{G} + \mathbf{HK}| = 0$$

As stated earlier, if the system is completely state controllable, then the feedback gain matrix $\mathbf{K}$ can be determined to yield the desired closed-loop poles.

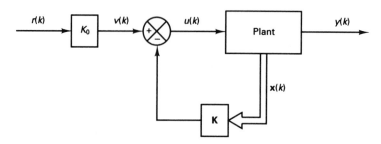

Figure 6–2   State feedback control system.

It is important to point out that state feedback can change the characteristic equation for the system, but in doing so the steady-state gain of the entire system is changed. Therefore, it is necessary to have an adjustable gain K_0 in the system. This gain K_0 should be adjusted such that the unit-step response of the system at steady state is unity, or $y(\infty) = 1$. To clarify the details, consider the following example problem.

Example 6-8

Consider the system defined by

$$\mathbf{x}(k + 1) = \mathbf{Gx}(k) + \mathbf{H}u(k)$$

$$y(k) = \mathbf{Cx}(k)$$

$$u(k) = K_0 r(k) - \mathbf{Kx}(k)$$

where

$$\mathbf{G} = \begin{bmatrix} 0 & 1 \\ -0.16 & -1 \end{bmatrix}, \quad \mathbf{H} = \begin{bmatrix} 0 \\ 1 \end{bmatrix}, \quad \mathbf{C} = \begin{bmatrix} 1 & 0 \end{bmatrix}$$

Design a control system such that the desired closed-loop poles of the characteristic equation are at

$$z_1 = 0.5 + j0.5, \quad z_2 = 0.5 - j0.5$$

Thus, the desired characteristic polynomial is given by

$$|z\mathbf{I} - \mathbf{G} + \mathbf{HK}| = (z - 0.5 - j0.5)(z - 0.5 + j0.5)$$

$$= z^2 - z + 0.5$$

The state feedback gain matrix $\mathbf{K}$ can be determined as

$$\mathbf{K} = \begin{bmatrix} 0.34 & -2 \end{bmatrix}$$

(See Example 6-6 for the computation for determining matrix $\mathbf{K}$.)

Using this $\mathbf{K}$ matrix, the state equation becomes

$$\mathbf{x}(k + 1) = (\mathbf{G} - \mathbf{HK})\mathbf{x}(k) + \mathbf{H}K_0 r(k)$$

$$= \hat{\mathbf{G}}\mathbf{x}(k) + \hat{\mathbf{H}}r(k)$$

where

$$\hat{\mathbf{G}} = \mathbf{G} - \mathbf{HK}, \quad \hat{\mathbf{H}} = \mathbf{H}K_0$$

The gain constant K_0 can be determined in state space or can be determined in the z plane using the pulse transfer function. In this example, we shall use the latter approach.

The pulse transfer function $Y(z)/R(z)$ for this system is given by

$$G(z) = \mathbf{C}(z\mathbf{I} - \hat{\mathbf{G}})^{-1}\hat{\mathbf{H}}$$

where

$$\hat{\mathbf{G}} = \mathbf{G} - \mathbf{HK} = \begin{bmatrix} 0 & 1 \\ -0.16 & -1 \end{bmatrix} - \begin{bmatrix} 0 \\ 1 \end{bmatrix}\begin{bmatrix} 0.34 & -2 \end{bmatrix} = \begin{bmatrix} 0 & 1 \\ -0.5 & 1 \end{bmatrix}$$

$$\hat{\mathbf{H}} = \begin{bmatrix} 0 \\ 1 \end{bmatrix}K_0 = \begin{bmatrix} 0 \\ K_0 \end{bmatrix}$$

Hence,

$$G(z) = \begin{bmatrix} 1 & 0 \end{bmatrix} \begin{bmatrix} z & -1 \\ 0.5 & z-1 \end{bmatrix}^{-1} \begin{bmatrix} 0 \\ K_0 \end{bmatrix}$$

$$= \frac{K_0}{z^2 - z + 0.5}$$

Thus,

$$\frac{Y(z)}{R(z)} = G(z) = \frac{K_0}{z^2 - z + 0.5}$$

To determine gain constant K_0, we use the condition that the steady-state output $y(\infty)$ for the unit-step input is unity, or

$$\lim_{k \to \infty} y(k) = \lim_{z \to 1} (1 - z^{-1})Y(z)$$

$$= \lim_{z \to 1} \frac{z-1}{z} \frac{K_0}{z^2 - z + 0.5} \frac{z}{z-1}$$

$$= 2K_0 = 1$$

Hence, we have determined K_0 as

$$K_0 = 0.5$$

Figure 6–3 shows block diagrams of the designed system. The unit-step response of this system can be obtained easily by use of MATLAB. MATLAB Program 6–1 is a sample program for obtaining the unit-step response. Figure 6–4 shows the resulting response curve.

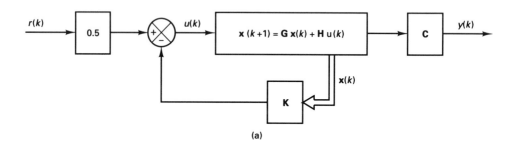

(a)

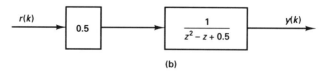

(b)

Figure 6–3 (a) Block diagram of the control system designed in Example 6–8; (b) simplified block diagram.

```
MATLAB Program 6–1

num = [0   0    0.5];
den = [1   -1    0.5];
r = ones(1,41);
k = 0: 40;
y = filter(num,den,r);
plot(k,y,'o')
v = [0   40   0   1.6];
axis(v);
grid
title('Unit-Step Response')
xlabel('k')
ylabel('y(k)')
```

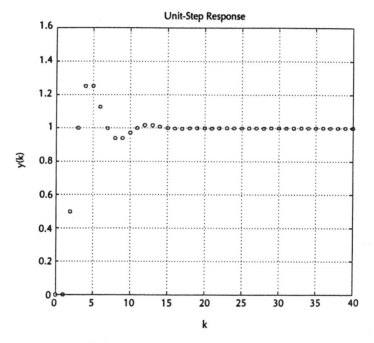

Figure 6–4   Unit-step response of the system shown in Figure 6–3(b).

6–6 STATE OBSERVERS

In Section 6–5 we discussed a pole placement design method that utilizes the feedback of all state variables to form the desired control vector. In practice, however, not all state variables are available for direct measurement. In many practical cases, only a few state variables of a given system are measurable, and the rest are not measurable. For instance, it may be that only the output variables are

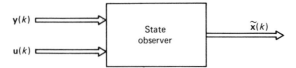

Figure 6–5 Schematic diagram of the state observer.

measurable. Hence, it is necessary to estimate the state variables that are not directly measurable. Such estimation is commonly called *observation*. In a practical system it is necessary to observe or estimate the unmeasurable state variables from the output and control variables.

A *state observer*, also called a *state estimator*, is a subsystem in the control system that performs an estimation of the state variables based on the measurements of the output and control variables. Here, the concept of observability discussed in Section 6–3 plays an important role. As we shall see later, state observers can be designed if and only if the observability condition is satisfied.

In the following discussions of state observers, we shall use the notation $\tilde{\mathbf{x}}(k)$ to designate the observed state vector. In many cases the observed state vector $\tilde{\mathbf{x}}(k)$ is used in the state feedback to generate the optimal control vector. Figure 6–5 shows a schematic diagram of a state observer. The state observer will have $\mathbf{y}(k)$ and $\mathbf{u}(k)$ as inputs and $\tilde{\mathbf{x}}(k)$ as output.

In what follows, we shall first discuss the necessary and sufficient condition for state observation and then treat the *full-order state observer*. Full-order state observation means that we observe (estimate) all n state variables regardless of whether some state variables are available for direct measurement. There are times when this will be unnecessary, when we will need observation of only the unmeasurable state variables but not of those that are directly measurable. Observation of only the unmeasurable state variables is referred to as *minimum-order state observation*, and we shall discuss it later in this section. Observation of all unmeasurable state variables plus some (but not all) of the measurable state variables is referred to as *reduced-order state observation*.

Necessary and Sufficient Condition for State Observation.   Figure 6–6 shows a regulator system with a state observer. We shall discuss a necessary and sufficient condition under which the state vector can be observed (estimated). From Figure 6–6 we obtain the state and output equations as follows:

$$\mathbf{x}(k + 1) = \mathbf{Gx}(k) + \mathbf{Hu}(k) \tag{6–79}$$

$$\mathbf{y}(k) = \mathbf{Cx}(k) \tag{6–80}$$

where

$$\mathbf{x}(k) = \text{state vector (}n\text{-vector)}$$

$$\mathbf{u}(k) = \text{control vector (}r\text{-vector)}$$

$$\mathbf{y}(k) = \text{output vector (}m\text{-vector)}$$

$$\mathbf{G} = n \times n \text{ nonsingular matrix}$$

$$\mathbf{H} = n \times r \text{ matrix}$$

$$\mathbf{C} = m \times n \text{ matrix}$$

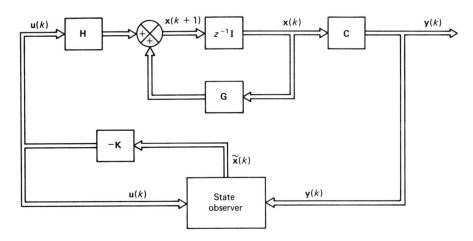

Figure 6-6 Regulator system with a state observer.

To be able to observe (estimate) state variables, we must be able to obtain $\mathbf{x}(k + 1)$ in terms of $\mathbf{y}(k), \mathbf{y}(k - 1), \ldots, \mathbf{y}(k - n + 1)$ and $\mathbf{u}(k), \mathbf{u}(k - 1), \ldots, \mathbf{u}(k - n + 1)$. From Equation (6-79), we have

$$\mathbf{G}^{-1}\mathbf{x}(k + 1) = \mathbf{x}(k) + \mathbf{G}^{-1}\mathbf{H}\mathbf{u}(k)$$

or

$$\mathbf{x}(k) = \mathbf{G}^{-1}\mathbf{x}(k + 1) - \mathbf{G}^{-1}\mathbf{H}\mathbf{u}(k) \tag{6-81}$$

By shifting k by 1, we get

$$\mathbf{x}(k - 1) = \mathbf{G}^{-1}\mathbf{x}(k) - \mathbf{G}^{-1}\mathbf{H}\mathbf{u}(k - 1) \tag{6-82}$$

By substituting Equation (6-81) into Equation (6-82), we obtain

$$\mathbf{x}(k - 1) = \mathbf{G}^{-1}[\mathbf{G}^{-1}\mathbf{x}(k + 1) - \mathbf{G}^{-1}\mathbf{H}\mathbf{u}(k)] - \mathbf{G}^{-1}\mathbf{H}\mathbf{u}(k - 1)$$
$$= \mathbf{G}^{-2}\mathbf{x}(k + 1) - \mathbf{G}^{-2}\mathbf{H}\mathbf{u}(k) - \mathbf{G}^{-1}\mathbf{H}\mathbf{u}(k - 1)$$

Similarly,

$$\mathbf{x}(k - 2) = \mathbf{G}^{-2}\mathbf{x}(k) - \mathbf{G}^{-2}\mathbf{H}\mathbf{u}(k - 1) - \mathbf{G}^{-1}\mathbf{H}\mathbf{u}(k - 2)$$
$$= \mathbf{G}^{-3}\mathbf{x}(k + 1) - \mathbf{G}^{-3}\mathbf{H}\mathbf{u}(k) - \mathbf{G}^{-2}\mathbf{H}\mathbf{u}(k - 1) - \mathbf{G}^{-1}\mathbf{H}\mathbf{u}(k - 2)$$
$$\vdots$$
$$\mathbf{x}(k - n + 1) = \mathbf{G}^{-n}\mathbf{x}(k + 1) - \mathbf{G}^{-n}\mathbf{H}\mathbf{u}(k) - \mathbf{G}^{-n+1}\mathbf{H}\mathbf{u}(k - 1)$$
$$- \cdots - \mathbf{G}^{-1}\mathbf{H}\mathbf{u}(k - n + 1)$$

By substituting Equation (6-81) into Equation (6-80), we obtain

$$\mathbf{y}(k) = \mathbf{C}\mathbf{x}(k) = \mathbf{C}\mathbf{G}^{-1}\mathbf{x}(k + 1) - \mathbf{C}\mathbf{G}^{-1}\mathbf{H}\mathbf{u}(k)$$

Similarly,

$$y(k - 1) = \mathbf{C}\mathbf{x}(k - 1) = \mathbf{C}\mathbf{G}^{-2}\mathbf{x}(k + 1) - \mathbf{C}\mathbf{G}^{-2}\mathbf{H}\mathbf{u}(k) - \mathbf{C}\mathbf{G}^{-1}\mathbf{H}\mathbf{u}(k - 1)$$

$$y(k - 2) = \mathbf{C}\mathbf{x}(k - 2) = \mathbf{C}\mathbf{G}^{-3}\mathbf{x}(k + 1) - \mathbf{C}\mathbf{G}^{-3}\mathbf{H}\mathbf{u}(k) - \mathbf{C}\mathbf{G}^{-2}\mathbf{H}\mathbf{u}(k - 1)$$

$$- \mathbf{C}\mathbf{G}^{-1}\mathbf{H}\mathbf{u}(k - 2)$$

$$\vdots$$

$$y(k - n + 1) = \mathbf{C}\mathbf{x}(k - n + 1) = \mathbf{C}\mathbf{G}^{-n}\mathbf{x}(k + 1) - \mathbf{C}\mathbf{G}^{-n}\mathbf{H}\mathbf{u}(k)$$

$$- \mathbf{C}\mathbf{G}^{-n+1}\mathbf{H}\mathbf{u}(k - 1) - \cdots - \mathbf{C}\mathbf{G}^{-1}\mathbf{H}\mathbf{u}(k - n + 1)$$

By combining the preceding n equations into one matrix equation, we get

$$\begin{bmatrix} y(k) \\ y(k - 1) \\ \vdots \\ y(k - n + 1) \end{bmatrix} = \begin{bmatrix} \mathbf{C}\mathbf{G}^{-1} \\ \mathbf{C}\mathbf{G}^{-2} \\ \vdots \\ \mathbf{C}\mathbf{G}^{-n} \end{bmatrix} \mathbf{x}(k + 1)$$

$$- \begin{bmatrix} \mathbf{C}\mathbf{G}^{-1}\mathbf{H} & \mathbf{0} & \cdots & \mathbf{0} \\ \mathbf{C}\mathbf{G}^{-2}\mathbf{H} & \mathbf{C}\mathbf{G}^{-1}\mathbf{H} & \cdots & \mathbf{0} \\ \vdots & \vdots & & \vdots \\ \mathbf{C}\mathbf{G}^{-n}\mathbf{H} & \mathbf{C}\mathbf{G}^{-n+1}\mathbf{H} & \cdots & \mathbf{C}\mathbf{G}^{-1}\mathbf{H} \end{bmatrix} \begin{bmatrix} \mathbf{u}(k) \\ \mathbf{u}(k - 1) \\ \vdots \\ \mathbf{u}(k - n + 1) \end{bmatrix}$$

or

$$\begin{bmatrix} \mathbf{C}\mathbf{G}^{-1} \\ \mathbf{C}\mathbf{G}^{-2} \\ \vdots \\ \mathbf{C}\mathbf{G}^{-n} \end{bmatrix} \mathbf{x}(k + 1) = \begin{bmatrix} y(k) \\ y(k - 1) \\ \vdots \\ y(k - n + 1) \end{bmatrix}$$

$$+ \begin{bmatrix} \mathbf{C}\mathbf{G}^{-1}\mathbf{H} & \mathbf{0} & \cdots & \mathbf{0} \\ \mathbf{C}\mathbf{G}^{-2}\mathbf{H} & \mathbf{C}\mathbf{G}^{-1}\mathbf{H} & \cdots & \mathbf{0} \\ \vdots & \vdots & & \vdots \\ \mathbf{C}\mathbf{G}^{-n}\mathbf{H} & \mathbf{C}\mathbf{G}^{-n+1}\mathbf{H} & \cdots & \mathbf{C}\mathbf{G}^{-1}\mathbf{H} \end{bmatrix} \begin{bmatrix} \mathbf{u}(k) \\ \mathbf{u}(k - 1) \\ \vdots \\ \mathbf{u}(k - n + 1) \end{bmatrix}$$

$$(6–83)$$

Notice that the right-hand side of Equation (6–83) is entirely known. Hence, $\mathbf{x}(k + 1)$ can be determined if and only if

$$\text{rank} \begin{bmatrix} \mathbf{C}\mathbf{G}^{-1} \\ \mathbf{C}\mathbf{G}^{-2} \\ \vdots \\ \mathbf{C}\mathbf{G}^{-n} \end{bmatrix} = n \qquad (6–84)$$

Since matrix $\mathbf{G}$ is nonsingular, multiplication of each row of the left-hand side of Equation (6–84) by $\mathbf{G}^n$ does not change the rank condition. Hence, Equation (6–84) is equivalent to

$$\text{rank} \begin{bmatrix} \mathbf{C}\mathbf{G}^{n-1} \\ \mathbf{C}\mathbf{G}^{n-2} \\ \vdots \\ \mathbf{C} \end{bmatrix} = n$$

which is also equivalent to

$$\text{rank}\,[\mathbf{C}^* \vdots \mathbf{G}^*\mathbf{C}^* \vdots \cdots \vdots (\mathbf{G}^*)^{n-1}\mathbf{C}^*] = n \tag{6–85}$$

Clearly, this is the complete observability condition of the system defined by Equations (6–79) and (6–80). [Refer to Equation (6–17).] This means that if Equation (6–85) is satisfied (that is, if the system is completely observable) then $\mathbf{x}(k + 1)$ can be determined from $\mathbf{y}(k), \mathbf{y}(k - 1), \ldots, \mathbf{y}(k - n + 1)$ and $\mathbf{u}(k), \mathbf{u}(k - 1), \ldots,$ $\mathbf{u}(k - n + 1)$. Thus, we have shown that the necessary and sufficient condition for state observation is that the system be completely observable.

As a special case, if $y(k)$ is a scalar and matrix $\mathbf{C}$ is a $1 \times n$ matrix, then $\mathbf{x}(k + 1)$ can be obtained by premultiplying both sides of Equation (6–83) by the inverse of the matrix given in Equation (6–84), as follows:

$$\mathbf{x}(k + 1) = \begin{bmatrix} \mathbf{CG}^{-1} \\ \mathbf{CG}^{-2} \\ \vdots \\ \mathbf{CG}^{-n} \end{bmatrix}^{-1} \begin{bmatrix} y(k) \\ y(k - 1) \\ \vdots \\ y(k - n + 1) \end{bmatrix}$$

$$+ \begin{bmatrix} \mathbf{CG}^{-1} \\ \mathbf{CG}^{-2} \\ \vdots \\ \mathbf{CG}^{-n} \end{bmatrix}^{-1} \begin{bmatrix} \mathbf{CG}^{-1}\mathbf{H} & \mathbf{0} & \cdots & \mathbf{0} \\ \mathbf{CG}^{-2}\mathbf{H} & \mathbf{CG}^{-1}\mathbf{H} & \cdots & \mathbf{0} \\ \vdots & \vdots & & \vdots \\ \mathbf{CG}^{-n}\mathbf{H} & \mathbf{CG}^{-n+1}\mathbf{H} & \cdots & \mathbf{CG}^{-1}\mathbf{H} \end{bmatrix} \begin{bmatrix} \mathbf{u}(k) \\ \mathbf{u}(k - 1) \\ \vdots \\ \mathbf{u}(k - n + 1) \end{bmatrix}$$

$$\tag{6–86}$$

Equation (6–86) gives $\mathbf{x}(k + 1)$ when $y(k)$ is a scalar.

As shown in the foregoing analysis, the state $\mathbf{x}(k + 1)$ can be determined from Equation (6–83) provided the system is completely observable. Thus, for a completely observable system, the state vector can be determined in at most n sampling periods. In the presence of external disturbances and measurement noises, however, this approach may not give an accurate determination of the state vector. Hence, to determine the state vector in the presence of disturbances and measurement noises, a different approach is necessary. Also, if matrix $\mathbf{C}$ is not a $1 \times n$ matrix but is an $m \times n$ matrix ($m > 1$), then the inverse of the matrix of Equation (6–84) cannot be defined and Equation (6–86) does not apply. To cope with such cases, one very powerful approach for estimating the state vector is to use a dynamic model of the original system, as follows.

Consider the control system defined by Equations (6–79) and (6–80). Let us assume that the state $\mathbf{x}(k)$ is to be approximated by the state $\tilde{\mathbf{x}}(k)$ of the dynamic model:

$$\tilde{\mathbf{x}}(k + 1) = \mathbf{G}\tilde{\mathbf{x}}(k) + \mathbf{H}\mathbf{u}(k) \tag{6–87}$$

$$\tilde{\mathbf{y}}(k) = \mathbf{C}\tilde{\mathbf{x}}(k) \tag{6–88}$$

where matrices $\mathbf{G}$, $\mathbf{H}$, and $\mathbf{C}$ are the same as those of the original system. Also, let us assume that the dynamic model is subjected to the same control signal $\mathbf{u}(k)$ as the original system. If the initial conditions for the actual system defined by Equations (6–79) and (6–80) and the dynamic model defined by Equations (6–87) and

(6–88) are the same, then the state $\tilde{\mathbf{x}}(k)$ and the state $\mathbf{x}(k)$ will be the same. If the initial conditions are different, then the state $\tilde{\mathbf{x}}(k)$ and the state $\mathbf{x}(k)$ will be different.

If the matrix $\mathbf{G}$ is a stable one, however, $\tilde{\mathbf{x}}(k)$ will approach $\mathbf{x}(k)$ even for different initial conditions, as we shall see. If we denote the difference between $\mathbf{x}(k)$ and $\tilde{\mathbf{x}}(k)$ as $\mathbf{e}(k)$, or define

$$\mathbf{e}(k) = \mathbf{x}(k) - \tilde{\mathbf{x}}(k)$$

then by subtracting Equation (6–87) from Equation (6–79), we obtain

$$\mathbf{x}(k + 1) - \tilde{\mathbf{x}}(k + 1) = \mathbf{G}[\mathbf{x}(k) - \tilde{\mathbf{x}}(k)]$$

or

$$\mathbf{e}(k + 1) = \mathbf{G}\mathbf{e}(k)$$

If matrix $\mathbf{G}$ is a stable matrix, then $\mathbf{e}(k)$ will approach zero and $\tilde{\mathbf{x}}(k)$ will approach $\mathbf{x}(k)$. However, the behavior of the error vector, which depends solely on matrix $\mathbf{G}$, may not be acceptable. Also, if matrix $\mathbf{G}$ is not a stable matrix, then the error $\mathbf{e}(k)$ will not approach zero. It is therefore desirable to modify the dynamic model defined by Equations (6–87) and (6–88).

It is noted that although the state $\mathbf{x}(k)$ may not be measurable the output $\mathbf{y}(k)$ is measurable. The dynamic model defined by Equations (6–87) and (6–88) does not make use of the measured output $\mathbf{y}(k)$. The performance of the dynamic model can be improved if the difference between the measured output $\mathbf{y}(k)$ and the estimated output $\mathbf{C}\tilde{\mathbf{x}}(k)$ is used to monitor the state $\tilde{\mathbf{x}}(k)$, that is, if the dynamic model of Equation (6–87) is modified into the following form:

$$\tilde{\mathbf{x}}(k + 1) = \mathbf{G}\tilde{\mathbf{x}}(k) + \mathbf{H}\mathbf{u}(k) + \mathbf{K}_e[\mathbf{y}(k) - \mathbf{C}\tilde{\mathbf{x}}(k)]$$

where matrix $\mathbf{K}_e$ serves as a weighting matrix. (This means that the dynamics of the state observer shown in Figure 6–6 must be given by this last equation.) In the presence of discrepancies between the $\mathbf{G}$ and $\mathbf{H}$ matrices used in the model and those of the actual system, the addition of the difference between the measured output and the estimated output will help reduce the differences between the dynamic model and the actual system.

In what follows we shall discuss details of the observer whose dynamics are characterized by $\mathbf{G}$ and $\mathbf{H}$ matrices and by the additional correction term, which consists of the difference between the measured output and the estimated output.

Full-Order State Observer.   The order of the state observer that will be discussed here is the same as that of the system. As stated earlier, such a state observer is called a full-order state observer.

In the following analysis we assume that the actual state $\mathbf{x}(k)$ cannot be measured directly. If the state $\mathbf{x}(k)$ is to be estimated, it is desirable that the observed state or estimated state $\tilde{\mathbf{x}}(k)$ be as close to the actual state $\mathbf{x}(k)$ as possible. Although it is not necessary, it is convenient if the state observer has the same $\mathbf{G}$ and $\mathbf{H}$ matrices as the original system.

It is important to note that, in the present analysis, state $\mathbf{x}(k)$ is not available for direct measurement, and consequently the observed state $\tilde{\mathbf{x}}(k)$ cannot be com-

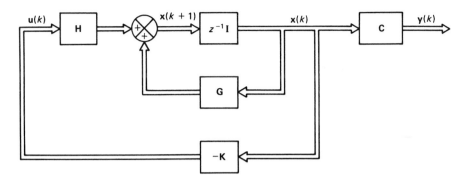

Figure 6-7   State feedback control system.

pared with the actual state $\mathbf{x}(k)$. Since the output $\mathbf{y}(k) = \mathbf{Cx}(k)$ can be measured, however, it is possible to compare $\tilde{\mathbf{y}}(k) = \mathbf{C\tilde{x}}(k)$ with $\mathbf{y}(k)$.

Consider the state feedback control system shown in Figure 6-7. The system equations are

$$\mathbf{x}(k + 1) = \mathbf{Gx}(k) + \mathbf{Hu}(k) \tag{6-89}$$

$$\mathbf{y}(k) = \mathbf{Cx}(k) \tag{6-90}$$

$$\mathbf{u}(k) = -\mathbf{Kx}(k)$$

where

$\mathbf{x}(k)$ = state vector (n-vector)

$\mathbf{u}(k)$ = control vector (r-vector)

$\mathbf{y}(k)$ = output vector (m-vector)

$\mathbf{G}$ = $n \times n$ nonsingular matrix

$\mathbf{H}$ = $n \times r$ matrix

$\mathbf{C}$ = $m \times n$ matrix

$\mathbf{K}$ = state feedback gain matrix ($n \times r$ matrix)

We assume that the system is completely state controllable and completely observable, but $\mathbf{x}(k)$ is not available for direct measurement. Figure 6-8 shows a state observer incorporated into the system of Figure 6-7. The observed state $\tilde{\mathbf{x}}(k)$ is used to form the control vector $\mathbf{u}(k)$, or

$$\mathbf{u}(k) = -\mathbf{K\tilde{x}}(k) \tag{6-91}$$

From Figure 6-8, we have

$$\tilde{\mathbf{x}}(k + 1) = \mathbf{G\tilde{x}}(k) + \mathbf{Hu}(k) + \mathbf{K}_e[\mathbf{y}(k) - \tilde{\mathbf{y}}(k)] \tag{6-92}$$

where $\mathbf{K}_e$ is the observer feedback gain matrix (an $n \times m$ matrix). This last equation can be modified to read

$$\tilde{\mathbf{x}}(k + 1) = (\mathbf{G} - \mathbf{K}_e \mathbf{C})\tilde{\mathbf{x}}(k) + \mathbf{Hu}(k) + \mathbf{K}_e \mathbf{y}(k) \tag{6-93}$$

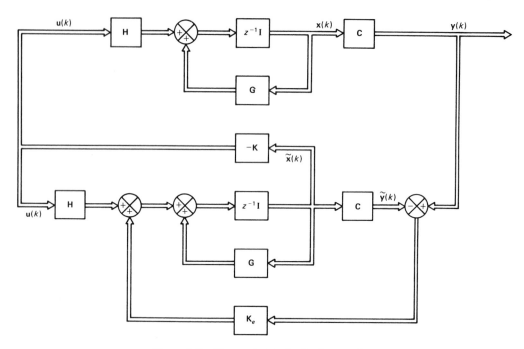

Figure 6–8   Observed-state feedback control system.

The state observer given by Equation (6–93) is called a *prediction observer*, since the estimate $\tilde{\mathbf{x}}(k + 1)$ is one sampling period ahead of the measurement $\mathbf{y}(k)$. The eigenvalues of $\mathbf{G} - \mathbf{K}_e \mathbf{C}$ are commonly called the *observer poles*.

Error Dynamics of the Full-Order State Observer.   Notice that if $\tilde{\mathbf{x}}(k) = \mathbf{x}(k)$, then Equation (6–93) becomes

$$\tilde{\mathbf{x}}(k + 1) = \mathbf{G}\tilde{\mathbf{x}}(k) + \mathbf{H}\mathbf{u}(k)$$

which is identical to the state equation of the system. Thus, if $\tilde{\mathbf{x}}(k) = \mathbf{x}(k)$, then the response of the state observer system is identical to the response of the original system.

To obtain the observer error equation, let us subtract Equation (6–93) from Equation (6–89):

$$\mathbf{x}(k + 1) - \tilde{\mathbf{x}}(k + 1) = (\mathbf{G} - \mathbf{K}_e \mathbf{C})[\mathbf{x}(k) - \tilde{\mathbf{x}}(k)] \qquad (6\text{–}94)$$

Now let us define the difference between $\mathbf{x}(k)$ and $\tilde{\mathbf{x}}(k)$ as the error $\mathbf{e}(k)$:

$$\mathbf{e}(k) = \mathbf{x}(k) - \tilde{\mathbf{x}}(k)$$

Then Equation (6–94) becomes

$$\mathbf{e}(k + 1) = (\mathbf{G} - \mathbf{K}_e \mathbf{C})\mathbf{e}(k) \qquad (6\text{–}95)$$

From Equation (6–95) we see that the dynamic behavior of the error signal is determined by the eigenvalues of $\mathbf{G} - \mathbf{K}_e \mathbf{C}$. If matrix $\mathbf{G} - \mathbf{K}_e \mathbf{C}$ is a stable matrix, the error vector will converge to zero for any initial error $\mathbf{e}(0)$. That is, $\tilde{\mathbf{x}}(k)$ will converge to $x(k)$ regardless of the values of $x(0)$ and $\tilde{x}(0)$. If the eigenvalues of $G - K_e C$ are located in such a way that the dynamic behavior of the error vector is adequately fast, then any error will tend to zero with adequate speed. One way to obtain fast response is to use deadbeat response. This can be achieved if all eigenvalues of $G - K_e C$ are chosen to be zero.

Comments.   Since the system defined by Equations (6-89) and (6-90) is assumed to be completely observable, an arbitrary placement of the eigenvalues of $G - K_e C$ is possible. To explain this further, notice that the eigenvalues of $G - K_e C$ and those of $G^* - C^*K_e^*$ are the same. By use of the principle of duality presented in Section 6-3, the condition for complete observability for the system defined by Equations (6-89) and (6-90) is the same as the complete state controllability condition for the system

$$x(k + 1) = G^*x(k) + C^*u(k) \qquad (6-96)$$

In Section 6-5 we saw that arbitrary pole placement is possible for the system of Equation (6-96) provided it is completely state controllable or provided the rank of the matrix

$$[C^* \vdots G^*C^* \vdots \cdots \vdots (G^*)^{n-1} C^*]$$

is n. [This is the condition for complete observability of the system defined by Equations (6-89) and (6-90).] For the system defined by Equation (6-96), by selecting a set of n desired eigenvalues of $G^* - C^*K$, the state feedback gain matrix K may be determined. The desired matrix; K_e, such that the eigenvalues of $G - K_e C$ are the same as those of $G^* - C^*K$, is related to matrix K by the equation $K_e = K^*$.

Example 6-9

Consider the system

$$x(k + 1) = Gx(k) + Hu(k)$$
$$y(k) = Cx(k)$$

where

$$G = \begin{bmatrix} 0 & -0.16 \\ 1 & -1 \end{bmatrix}, \qquad H = \begin{bmatrix} 0 \\ 1 \end{bmatrix}, \qquad C = [0 \quad 1]$$

Design a full-order state observer, assuming that the system configuration is identical to that shown in Figure 6-8. The desired eigenvalues of the observer matrix are

$$z = 0.5 + j0.5, \qquad z = 0.5 - j0.5$$

and so the desired characteristic equation is

$$(z - 0.5 - j0.5)(z - 0.5 + j0.5) = z^2 - z + 0.5 = 0$$

Since the configuration of the state observer is specified as shown in Figure 6-8, the design of the state observer reduces to the determination of an appropriate observer feedback gain matrix K_e. Before we proceed further, let us examine the observability matrix. The rank of

$$[C^* \vdots G^*C^*] = \begin{bmatrix} 0 & 1 \\ 1 & -1 \end{bmatrix}$$

is 2. Hence, the system is completely observable and determination of the desired observer feedback gain matrix is possible.

Referring to Equation (6–95)

$$e(k + 1) = (G - K_e C)e(k)$$

where

$$e(k) = x(k) - \tilde{x}(k)$$

the characteristic equation of the observer becomes

$$|zI - G + K_e C| = 0$$

Let us denote the observer feedback gain matrix K_e as follows:

$$K_e = \begin{bmatrix} k_1 \\ k_2 \end{bmatrix}$$

Then the characteristic equation becomes

$$\left| z\begin{bmatrix} 1 & 0 \\ 0 & 1 \end{bmatrix} - \begin{bmatrix} 0 & -0.16 \\ 1 & -1 \end{bmatrix} + \begin{bmatrix} k_1 \\ k_2 \end{bmatrix}[0 \;\; 1] \right| = \begin{vmatrix} z & 0.16 + k_1 \\ -1 & z + 1 + k_2 \end{vmatrix} = 0$$

which reduces to

$$z^2 + (1 + k_2)z + k_1 + 0.16 = 0 \tag{6–97}$$

Since the desired characteristic equation is

$$z^2 - z + 0.5 = 0$$

by comparing Equation (6–97) with this last equation, we obtain

$$k_1 = 0.34, \qquad k_2 = -2$$

or

$$K_e = \begin{bmatrix} 0.34 \\ -2 \end{bmatrix}$$

Note that the dual relationship exists between the system state equation considered in Example 6–6 and that of the present system. The state feedback gain matrix K obtained in Example 6–6 was $K = [0.34 \;\; -2]$. The observer feedback gain matrix K_e obtained here is related to matrix K by the relationship $K_e = K^*$.

Design of Prediction Observers.   We have thus far discussed full-order prediction observers. They are prediction observers because the estimate $\tilde{x}(k + 1)$ is one sampling period ahead of the measurement $y(k)$. We solved a simple example problem by assuming the matrix K_e to exist and determined the characteristic equation $|zI - G + K_e C| = 0$ to have prescribed eigenvalues. In what follows we shall discuss a more general approach to determine the observer feedback gain matrix K_e.

Consider the system defined by

$$x(k + 1) = Gx(k) + Hu(k) \tag{6–98}$$

$$y(k) = Cx(k) \tag{6–99}$$

where

$$\mathbf{x}(k) = \text{state vector } (n\text{-vector})$$

$$\mathbf{u}(k) = \text{control vector } (r\text{-vector})$$

$$y(k) = \text{output signal (scalar)}$$

$$\mathbf{G} = n \times n \text{ nonsingular matrix}$$

$$\mathbf{H} = n \times r \text{ matrix}$$

$$\mathbf{C} = 1 \times n \text{ matrix}$$

The system is assumed to be completely state controllable and completely observable. Thus, the inverse of

$$[\mathbf{C^*} \vdots \mathbf{G^*C^*} \vdots \cdots \vdots (\mathbf{G^*})^{n-1} \mathbf{C^*}]$$

exists. We also assume that the control law to be used is

$$\mathbf{u}(k) = -\mathbf{K}\tilde{\mathbf{x}}(k)$$

where $\tilde{\mathbf{x}}(k)$ is the observed state and $\mathbf{K}$ is an $r \times n$ matrix. Assume further that the system configuration is the same as that shown in Figure 6–8.

The state observer dynamics are given by the equation

$$\tilde{\mathbf{x}}(k + 1) = \mathbf{G}\tilde{\mathbf{x}}(k) + \mathbf{H}\mathbf{u}(k) + \mathbf{K}_e[y(k) - \tilde{y}(k)]$$

$$= (\mathbf{G} - \mathbf{K}_e\,\mathbf{C})\tilde{\mathbf{x}}(k) + \mathbf{H}\mathbf{u}(k) + \mathbf{K}_e\,\mathbf{C}\mathbf{x}(k) \qquad (6\text{-}100)$$

First, define

$$\mathbf{Q} = (\mathbf{WN^*})^{-1} \qquad (6\text{-}101)$$

where

$$\mathbf{N} = [\mathbf{C^*} \vdots \mathbf{G^*C^*} \vdots \cdots \vdots (\mathbf{G^*})^{n-1} \mathbf{C^*}] \qquad (6\text{-}102)$$

and

$$\mathbf{W} = \begin{bmatrix} a_{n-1} & a_{n-2} & \cdots & a_1 & 1 \\ a_{n-2} & a_{n-3} & \cdots & 1 & 0 \\ \vdots & \vdots & & \vdots & \vdots \\ a_1 & 1 & \cdots & 0 & 0 \\ 1 & 0 & \cdots & 0 & 0 \end{bmatrix} \qquad (6\text{-}103)$$

where $a_1, a_2, \ldots, a_{n-1}$ are coefficients in the characteristic equation of the original state equation given by Equation (6–98),

$$|z\mathbf{I} - \mathbf{G}| = z^n + a_1 z^{n-1} + \cdots + a_{n-1}z + a_n = 0$$

Next, define

$$\mathbf{x}(k) = \mathbf{Q}\boldsymbol{\xi}(k) \qquad (6\text{-}104)$$

where $\boldsymbol{\xi}(k)$ is an n-vector. By use of Equation (6–104), Equations (6–98) and (6–99) can be modified to read

$$\boldsymbol{\xi}(k + 1) = \mathbf{Q}^{-1}\mathbf{G}\mathbf{Q}\boldsymbol{\xi}(k) + \mathbf{Q}^{-1}\mathbf{H}u(k) \tag{6–105}$$

$$y(k) = \mathbf{C}\mathbf{Q}\boldsymbol{\xi}(k) \tag{6–106}$$

where

$$\mathbf{Q}^{-1}\mathbf{G}\mathbf{Q} = \begin{bmatrix} 0 & 0 & \cdots & 0 & -a_n \\ 1 & 0 & \cdots & 0 & -a_{n-1} \\ \vdots & \vdots & & \vdots & \vdots \\ 0 & 0 & \cdots & 1 & -a_1 \end{bmatrix} \tag{6–107}$$

$$\mathbf{C}\mathbf{Q} = \begin{bmatrix} 0 & 0 & \cdots & 0 & 1 \end{bmatrix} \tag{6–108}$$

[Refer to Problem A–6–9 for the derivations of Equations (6–107) and (6–108).]
    Now define

$$\tilde{\mathbf{x}}(k) = \mathbf{Q}\tilde{\boldsymbol{\xi}}(k) \tag{6–109}$$

By substituting Equation (6–109) into Equation (6–100), we have

$$\tilde{\boldsymbol{\xi}}(k + 1) = \mathbf{Q}^{-1}(\mathbf{G} - \mathbf{K}_e\mathbf{C})\mathbf{Q}\tilde{\boldsymbol{\xi}}(k) + \mathbf{Q}^{-1}\mathbf{H}u(k) + \mathbf{Q}^{-1}\mathbf{K}_e\mathbf{C}\mathbf{Q}\boldsymbol{\xi}(k) \tag{6–110}$$

Subtracting Equation (6–110) from Equation (6–105), we obtain

$$\boldsymbol{\xi}(k + 1) - \tilde{\boldsymbol{\xi}}(k + 1) = (\mathbf{Q}^{-1}\mathbf{G}\mathbf{Q} - \mathbf{Q}^{-1}\mathbf{K}_e\mathbf{C}\mathbf{Q})[\boldsymbol{\xi}(k) - \tilde{\boldsymbol{\xi}}(k)] \tag{6–111}$$

Define

$$\mathbf{e}(k) = \boldsymbol{\xi}(k) - \tilde{\boldsymbol{\xi}}(k)$$

Then Equation (6–111) becomes

$$\mathbf{e}(k + 1) = \mathbf{Q}^{-1}(\mathbf{G} - \mathbf{K}_e\mathbf{C})\mathbf{Q}\mathbf{e}(k) \tag{6–112}$$

We require the error dynamics to be stable and $\mathbf{e}(k)$ to reach zero with sufficient speed. The procedure for determining matrix $\mathbf{K}_e$ is first to select the desired observer poles (the eigenvalues of $\mathbf{G} - \mathbf{K}_e\mathbf{C}$) and then to determine matrix $\mathbf{K}_e$ so that it will give the desired poles. If we require $\mathbf{e}(k)$ to reach zero as fast as possible, then we require the error response to be deadbeat, so we must select all eigenvalues of $\mathbf{G} - \mathbf{K}_e\mathbf{C}$ to be zero.
    Notice that

$$\mathbf{Q}^{-1}\mathbf{K}_e = \begin{bmatrix} a_{n-1} & a_{n-2} & \cdots & a_1 & 1 \\ a_{n-2} & a_{n-3} & \cdots & 1 & 0 \\ \vdots & \vdots & & \vdots & \vdots \\ a_1 & 1 & \cdots & 0 & 0 \\ 1 & 0 & \cdots & 0 & 0 \end{bmatrix} \begin{bmatrix} \mathbf{C} \\ \mathbf{C}\mathbf{G} \\ \vdots \\ \mathbf{C}\mathbf{G}^{n-2} \\ \mathbf{C}\mathbf{G}^{n-1} \end{bmatrix} \begin{bmatrix} k_1 \\ k_2 \\ \vdots \\ k_{n-1} \\ k_n \end{bmatrix}$$

where

$$\mathbf{K}_e = \begin{bmatrix} k_1 \\ k_2 \\ \vdots \\ k_n \end{bmatrix}$$

Since $\mathbf{Q}^{-1}\mathbf{K}_e$ is an n-vector, let us write

$$\mathbf{Q}^{-1}\mathbf{K}_e = \begin{bmatrix} \delta_n \\ \delta_{n-1} \\ \vdots \\ \delta_1 \end{bmatrix} \tag{6-113}$$

Then, referring to Equation (6-108), we have

$$\mathbf{Q}^{-1}\mathbf{K}_e\,\mathbf{CQ} = \begin{bmatrix} \delta_n \\ \delta_{n-1} \\ \vdots \\ \delta_1 \end{bmatrix}\begin{bmatrix} 0 & 0 & \cdots & 1\end{bmatrix} = \begin{bmatrix} 0 & 0 & \cdots & 0 & \delta_n \\ 0 & 0 & \cdots & 0 & \delta_{n-1} \\ \vdots & \vdots & & \vdots & \vdots \\ 0 & 0 & \cdots & 0 & \delta_1 \end{bmatrix}$$

and

$$\mathbf{Q}^{-1}(\mathbf{G} - \mathbf{K}_e\,\mathbf{C})\mathbf{Q} = \mathbf{Q}^{-1}\mathbf{GQ} - \mathbf{Q}^{-1}\mathbf{K}_e\,\mathbf{CQ} = \begin{bmatrix} 0 & 0 & \cdots & 0 & -a_n - \delta_n \\ 1 & 0 & \cdots & 0 & -a_{n-1} - \delta_{n-1} \\ 0 & 1 & \cdots & 0 & -a_{n-2} - \delta_{n-2} \\ \vdots & \vdots & & \vdots & \vdots \\ 0 & 0 & \cdots & 1 & -a_1 - \delta_1 \end{bmatrix}$$

The characteristic equation

$$|z\mathbf{I} - \mathbf{Q}^{-1}(\mathbf{G} - \mathbf{K}_e\,\mathbf{C})\mathbf{Q}| = 0$$

becomes

$$\begin{vmatrix} z & 0 & 0 & \cdots & 0 & a_n + \delta_n \\ -1 & z & 0 & \cdots & 0 & a_{n-1} + \delta_{n-1} \\ 0 & -1 & z & \cdots & 0 & a_{n-2} + \delta_{n-2} \\ \vdots & \vdots & \vdots & & \vdots & \vdots \\ 0 & 0 & 0 & \cdots & -1 & z + a_1 + \delta_1 \end{vmatrix} = 0$$

or

$$z^n + (a_1 + \delta_1)z^{n-1} + (a_2 + \delta_2)z^{n-2} + \cdots + (a_n + \delta_n) = 0 \tag{6-114}$$

It can be seen that each of $\delta_n, \delta_{n-1}, \ldots, \delta_1$ is associated with only one of the coefficients of the characteristic equation.

Suppose the desired characteristic equation for the error dynamics is

$$(z - \mu_1)(z - \mu_2)\cdots(z - \mu_n)$$
$$= z^n + \alpha_1 z^{n-1} + \alpha_2 z^{n-2} + \cdots + \alpha_{n-1}z + \alpha_n = 0 \tag{6-115}$$

Note that the desired eigenvalues μ_i (or the locations of the desired closed-loop poles) determine how fast the observed state converges to the actual state of the plant. Comparing the coefficients of equal powers of z in Equations (6-114) and (6-115), we obtain

$$a_1 + \delta_1 = \alpha_1$$
$$a_2 + \delta_2 = \alpha_2$$
$$\vdots$$
$$a_n + \delta_n = \alpha_n$$

from which we get

$$\delta_1 = \alpha_1 - a_1$$

$$\delta_2 = \alpha_2 - a_2$$

$$\vdots$$

$$\delta_n = \alpha_n - a_n$$

Then, from Equation (6–113), we have

$$\mathbf{Q}^{-1}\mathbf{K}_e = \begin{bmatrix} \delta_n \\ \delta_{n-1} \\ \vdots \\ \delta_1 \end{bmatrix} = \begin{bmatrix} \alpha_n - a_n \\ \alpha_{n-1} - a_{n-1} \\ \vdots \\ \alpha_1 - a_1 \end{bmatrix} \tag{6–116}$$

Hence,

$$\mathbf{K}_e = \mathbf{Q}\begin{bmatrix} \alpha_n - a_n \\ \alpha_{n-1} - a_{n-1} \\ \vdots \\ \alpha_1 - a_1 \end{bmatrix} = (\mathbf{WN}^*)^{-1}\begin{bmatrix} \alpha_n - a_n \\ \alpha_{n-1} - a_{n-1} \\ \vdots \\ \alpha_1 - a_1 \end{bmatrix} \tag{6–117}$$

Equation (6–117) specifies the necessary observer feedback gain matrix $\mathbf{K}_e$. Figure 6–9 shows an alternative representation of the observed-state feedback control system.

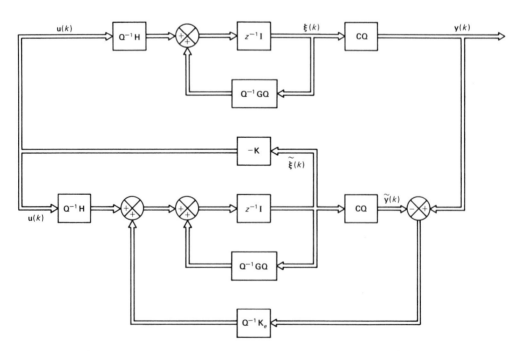

Figure 6–9   Alternative representation of the observed-state feedback control system.

Once we select the desired eigenvalues (or desired characteristic equation), the observer can be designed in a way similar to the method used in the case of the pole placement problem. The desired characteristic equation may be chosen so that the observer responds at least four or five times faster than the closed-loop system; or in some applications, deadbeat response may be desired.

If we wish to have deadbeat response, the desired characteristic equation becomes

$$z^n = 0 \tag{6-118}$$

Comparing Equation (6-114) with Equation (6-118), we require that

$$a_1 + \delta_1 = 0$$

$$a_2 + \delta_2 = 0$$

$$\vdots$$

$$a_n + \delta_n = 0$$

Hence, for the deadbeat response,

$$\mathbf{K}_e = \begin{bmatrix} k_1 \\ k_2 \\ \vdots \\ k_n \end{bmatrix} = \mathbf{Q} \begin{bmatrix} \delta_n \\ \delta_{n-1} \\ \vdots \\ \delta_1 \end{bmatrix} = (\mathbf{WN}^*)^{-1} \begin{bmatrix} -a_n \\ -a_{n-1} \\ \vdots \\ -a_1 \end{bmatrix} \tag{6-119}$$

Ackermann's Formula.   The expression given by Equation (6-117) is not the only one commonly available for the determination of the observer feedback gain matrix $\mathbf{K}_e$. In what follows, we shall derive Ackermann's formula for the determination of $\mathbf{K}_e$.

Consider the completely observable system defined by Equations (6-98) and (6-99). Note that in this system the output $y(k)$ is a scalar. Referring to Equation (6-112), the characteristic equation for the error dynamics is

$$|z\mathbf{I} - \mathbf{Q}^{-1}\mathbf{GQ} + \mathbf{Q}^{-1}\mathbf{K}_e\,\mathbf{CQ}| = 0 \tag{6-120}$$

where $\mathbf{K}_e$ is an $n \times 1$ matrix. Define

$$\mathbf{Q}^{-1}\mathbf{GQ} = \hat{\mathbf{G}}, \qquad \mathbf{Q}^{-1}\mathbf{K}_e = \hat{\mathbf{K}}_e, \qquad \mathbf{CQ} = \hat{\mathbf{C}}$$

Then Equation (6-120) becomes

$$|z\mathbf{I} - \hat{\mathbf{G}} + \hat{\mathbf{K}}_e\,\hat{\mathbf{C}}| = 0$$

In the observer design we determine matrix $\hat{\mathbf{K}}_e$ so that this last characteristic equation is identical to the desired characteristic equation for the error vector, which is

$$z^n + \alpha_1 z^{n-1} + \cdots + \alpha_{n-1} z + \alpha_n = 0 \tag{6-121}$$

That is,

$$|z\mathbf{I} - \hat{\mathbf{G}} + \hat{\mathbf{K}}_e\,\hat{\mathbf{C}}| = (z - \mu_1)(z - \mu_2) \cdots (z - \mu_n)$$

$$= z^n + \alpha_1 z^{n-1} + \cdots + \alpha_{n-1} z + \alpha_n = 0$$

where $\mu_1, \mu_2, \ldots, \mu_n$ are the eigenvalues of matrix $(\hat{\mathbf{G}} - \hat{\mathbf{K}}_e \hat{\mathbf{C}})$. For physical systems, complex eigenvalues always occur as conjugate complex pairs. In the present analysis we assume that all complex eigenvalues occur as conjugate complex pairs so that the coefficients $\alpha_1, \alpha_2, \ldots, \alpha_n$ of the characteristic equation are real. Then the characteristic equation for the matrix $(\hat{\mathbf{G}}^* - \hat{\mathbf{C}}^*\hat{\mathbf{K}}_e^*)$ can be given by

$$|z\mathbf{I} - \hat{\mathbf{G}}^* + \hat{\mathbf{C}}^*\hat{\mathbf{K}}_e^*| = (z - \overline{\mu}_1)(z - \overline{\mu}_2)\cdots(z - \overline{\mu}_n)$$

$$= z^n + \alpha_1 z^{n-1} + \cdots + \alpha_{n-1}z + \alpha_n = 0$$

where $\overline{\mu}_i$ is the complex conjugate of μ_i.

In Section 6–5 we derived Ackermann's formula for the determination of the state feedback gain matrix $\mathbf{K}$ for the pole placement design problem. There we determined matrix $\mathbf{K}$ so that the characteristic equation

$$|z\mathbf{I} - \mathbf{G} + \mathbf{H}\mathbf{K}| = 0$$

would be the same as the desired characteristic equation, Equation (6–121). Here, in the observer design problem, we wish to determine matrix $\hat{\mathbf{K}}_e^*$ so that the characteristic equation

$$|z\mathbf{I} - \hat{\mathbf{G}}^* + \hat{\mathbf{C}}^*\hat{\mathbf{K}}_e^*| = 0$$

will be the same as the desired characteristic equation given by Equation (6–121). Clearly, we can see that these two problems are a dual problem. (That is, mathematically, the determination of matrix $\hat{\mathbf{K}}_e^*$ is the same as the determination of the feedback gain matrix $\mathbf{K}$ in the pole placement problem.) Therefore, it is possible to utilize the results obtained in Section 6–5 toward the determination of matrix $\mathbf{K}_e$ for the present problem, as will be shown.

In the pole placement design problem discussed in Section 6–5, for the system equation

$$\mathbf{x}(k + 1) = \mathbf{G}\mathbf{x}(k) + \mathbf{H}u(k)$$

with state feedback

$$u(k) = -\mathbf{K}\mathbf{x}(k)$$

the desired matrix $\mathbf{K}$ was obtained as given by Equation (6–68), repeated here:

$$\mathbf{K} = [0 \quad 0 \quad \cdots \quad 0 \quad 1][\mathbf{H} \vdots \mathbf{G}\mathbf{H} \vdots \cdots \vdots \mathbf{G}^{n-1}\mathbf{H}]^{-1}\phi(\mathbf{G}) \qquad (6\text{--}122)$$

Here, in the observer design problem, for the state equation

$$\mathbf{x}(k + 1) = \hat{\mathbf{G}}^*\mathbf{x}(k) + \hat{\mathbf{C}}^*u(k)$$

with state feedback

$$u(k) = -\hat{\mathbf{K}}_e^* \, \mathbf{x}(k)$$

the desired matrix $\hat{\mathbf{K}}_e^*$ can therefore be obtained in a form similar to Equation (6–122), as follows:

$$\hat{\mathbf{K}}_e^* = [0 \quad 0 \quad \cdots \quad 0 \quad 1][\hat{\mathbf{C}}^* \vdots \hat{\mathbf{G}}^*\hat{\mathbf{C}}^* \vdots \cdots \vdots (\hat{\mathbf{G}}^*)^{n-1}\hat{\mathbf{C}}^*]^{-1}\phi(\hat{\mathbf{G}}^*) \qquad (6\text{--}123)$$

which may be modified to

$$\hat{\mathbf{K}}_e^* = [0 \quad 0 \quad \cdots \quad 0 \quad 1][\mathbf{C}^* \vdots \mathbf{G}^*\mathbf{C}^* \vdots \cdots \vdots (\mathbf{G}^*)^{n-1}\mathbf{C}^*]^{-1}(\mathbf{Q}^*)^{-1}\phi(\hat{\mathbf{G}}^*)$$

By taking the conjugate transpose of both sides of this last equation, we have

$$\hat{\mathbf{K}}_e = [\phi(\hat{\mathbf{G}}^*)]^*\mathbf{Q}^{-1}\begin{bmatrix} \mathbf{C} \\ \mathbf{CG} \\ \vdots \\ \mathbf{CG}^{n-1} \end{bmatrix}^{-1}\begin{bmatrix} 0 \\ 0 \\ \vdots \\ 1 \end{bmatrix}$$

Noting that

$$\mathbf{Q}^{-1}\mathbf{K}_e = \hat{\mathbf{K}}_e$$

we obtain

$$\mathbf{K}_e = \mathbf{Q}\phi(\hat{\mathbf{G}})\mathbf{Q}^{-1}\begin{bmatrix} \mathbf{C} \\ \mathbf{CG} \\ \vdots \\ \mathbf{CG}^{n-1} \end{bmatrix}^{-1}\begin{bmatrix} 0 \\ 0 \\ \vdots \\ 1 \end{bmatrix} \tag{6-124}$$

Notice that, since $\hat{\mathbf{G}} = \mathbf{Q}^{-1}\mathbf{GQ}$, we have

$$\mathbf{Q}\hat{\mathbf{G}}^k\mathbf{Q}^{-1} = \mathbf{G}^k, \quad k = 0, 1, 2, \ldots, n$$

Consequently,

$$\begin{aligned} \mathbf{Q}\phi(\hat{\mathbf{G}})\mathbf{Q}^{-1} &= \mathbf{Q}[\hat{\mathbf{G}}^n + \alpha_1\hat{\mathbf{G}}^{n-1} + \cdots + \alpha_{n-1}\hat{\mathbf{G}} + \alpha_n\mathbf{I}]\mathbf{Q}^{-1} \\ &= \mathbf{Q}\hat{\mathbf{G}}^n\mathbf{Q}^{-1} + \alpha_1\mathbf{Q}\hat{\mathbf{G}}^{n-1}\mathbf{Q}^{-1} + \cdots + \alpha_{n-1}\mathbf{Q}\hat{\mathbf{G}}\mathbf{Q}^{-1} + \alpha_n\mathbf{I} \\ &= \mathbf{G}^n + \alpha_1\mathbf{G}^{n-1} + \cdots + \alpha_{n-1}\mathbf{G} + \alpha_n\mathbf{I} = \phi(\mathbf{G}) \end{aligned} \tag{6-125}$$

By use of Equation (6-125), the desired observer feedback gain matrix $\mathbf{K}_e$, given by Equation (6-124), can be rewritten as follows:

$$\mathbf{K}_e = \phi(\mathbf{G})\begin{bmatrix} \mathbf{C} \\ \mathbf{CG} \\ \vdots \\ \mathbf{CG}^{n-1} \end{bmatrix}^{-1}\begin{bmatrix} 0 \\ 0 \\ \vdots \\ 1 \end{bmatrix} \tag{6-126}$$

where $\phi(\mathbf{G})$ is the desired characteristic polynomial of the error dynamics. The expression for $\mathbf{K}_e$ given by Equation (6-126) is commonly called Ackermann's formula for the determination of the observer feedback gain matrix $\mathbf{K}_e$.

Summary.   The full-order prediction observer is given by Equation (6-92):

$$\tilde{\mathbf{x}}(k+1) = (\mathbf{G} - \mathbf{K}_e\mathbf{C})\tilde{\mathbf{x}}(k) + \mathbf{H}\mathbf{u}(k) + \mathbf{K}_e\mathbf{y}(k)$$

The observed-state feedback is given by

$$\mathbf{u}(k) = -\mathbf{K}\tilde{\mathbf{x}}(k)$$

If this last equation is substituted into the observer equation, we obtain

$$\tilde{\mathbf{x}}(k+1) = (\mathbf{G} - \mathbf{K}_e\mathbf{C} - \mathbf{H}\mathbf{K})\tilde{\mathbf{x}}(k) + \mathbf{K}_e\mathbf{y}(k)$$

This equation defines the full-order prediction observer when the observed-state feedback control is incorporated.

As in the case of the pole placement design four methods, that are commonly available for the determination of the observer feedback gain matrix $\mathbf{K}_e$ for the completely observable system, are summarized as follows:

1. Referring to Equation (6–117), the observer feedback gain matrix $\mathbf{K}_e$ can be given by

$$\mathbf{K}_e = \mathbf{Q} \begin{bmatrix} \alpha_n - a_n \\ \alpha_{n-1} - a_{n-1} \\ \vdots \\ \alpha_1 - a_1 \end{bmatrix} = (\mathbf{WN^*})^{-1} \begin{bmatrix} \alpha_n - a_n \\ \alpha_{n-1} - a_{n-1} \\ \vdots \\ \alpha_1 - a_1 \end{bmatrix} \qquad (6\text{–}127)$$

where matrices $\mathbf{N}$ and $\mathbf{W}$ are defined by Equations (6–102) and (6–103), respectively. The α_i's are the coefficients of the desired characteristic equation

$$z^n + \alpha_1 z^{n-1} + \cdots + \alpha_{n-1} z + \alpha_n = 0$$

and the a_i's are coefficients of the characteristic equation of the original state equation

$$|z\mathbf{I} - \mathbf{G}| = z^n + a_1 z^{n-1} + \cdots + a_{n-1} z + a_n = 0$$

Note that if the system is already in an observable canonical form then the matrix $\mathbf{K}_e$ can be determined easily, because matrix $\mathbf{WN^*}$ becomes an identity matrix, and thus $(\mathbf{WN^*})^{-1} = \mathbf{I}$.

2. The observer feedback gain matrix $\mathbf{K}_e$ may be given by Ackermann's formula, given by Equation (6–126):

$$\mathbf{K}_e = \phi(\mathbf{G}) \begin{bmatrix} \mathbf{C} \\ \mathbf{CG} \\ \vdots \\ \mathbf{CG}^{n-1} \end{bmatrix}^{-1} \begin{bmatrix} 0 \\ 0 \\ \vdots \\ 1 \end{bmatrix} \qquad (6\text{–}128)$$

where

$$\phi(\mathbf{G}) = \mathbf{G}^n + \alpha_1 \mathbf{G}^{n-1} + \cdots + \alpha_{n-1} \mathbf{G} + \alpha_n \mathbf{I}$$

3. If the desired eigenvalues $\mu_1, \mu_2, \ldots, \mu_n$ of matrix $\mathbf{G} - \mathbf{K}_e \mathbf{C}$ are distinct, then the observer feedback gain matrix $\mathbf{K}_e$ may be given by the equation

$$\mathbf{K}_e = \begin{bmatrix} \boldsymbol{\eta}_1 \\ \boldsymbol{\eta}_2 \\ \vdots \\ \boldsymbol{\eta}_n \end{bmatrix}^{-1} \begin{bmatrix} 1 \\ 1 \\ \vdots \\ 1 \end{bmatrix} \qquad (6\text{–}129)$$

where the $\boldsymbol{\eta}_i$'s are defined as follows:

$$\boldsymbol{\eta}_i = \mathbf{C}(\mathbf{G} - \mu_i \mathbf{I})^{-1}$$

Note that the $\boldsymbol{\eta}_i^*$ are the eigenvectors of matrix $(\mathbf{G} - \mathbf{K}_e \mathbf{C})^*$.

In the special case where we desire the error vector to exhibit deadbeat response, so that $\mu_1 = \mu_2 = \cdots = \mu_n = 0$, Equation (6–129) can be simplified. The following equation will give the matrix $\mathbf{K}_e$ for the deadbeat response:

$$\mathbf{K}_e = \begin{bmatrix} \boldsymbol{\eta}_1 \\ \boldsymbol{\eta}_2 \\ \vdots \\ \boldsymbol{\eta}_n \end{bmatrix}^{-1} \begin{bmatrix} 1 \\ 0 \\ \vdots \\ 0 \end{bmatrix} \tag{6–130}$$

where the $\boldsymbol{\eta}_i$'s are given by the equation

$$\boldsymbol{\eta}_i = \mathbf{C}\mathbf{G}^{-i}, \qquad i = 1, 2, 3, \ldots, n$$

[For details of the derivations of Equations (6–129) and (6–130), see Problems A–6–12 and A–6–13.]

4. If the order of the system is low, assume an observer feedback gain matrix $\mathbf{K}_e$ with unknown elements. Then the elements of matrix $\mathbf{K}_e$ may be determined by equating the coefficients of like powers of z of

$$|z\mathbf{I} - \mathbf{G} + \mathbf{K}_e\mathbf{C}|$$

and of the desired characteristic polynomial, which is given by

$$(z - \mu_1)(z - \mu_2) \cdots (z - \mu_n) = z^n + \alpha_1 z^{n-1} + \cdots + \alpha_{n-1} z + \alpha_n$$

where the μ_i's are the desired eigenvalues of $\mathbf{G} - \mathbf{K}_e\mathbf{C}$.

Example 6–10

Consider the double integrator system given by the equations

$$\mathbf{x}(k + 1) = \mathbf{G}\mathbf{x}(k) + \mathbf{H}u(k)$$

$$y(k) = \mathbf{C}\mathbf{x}(k)$$

where

$$\mathbf{G} = \begin{bmatrix} 1 & T \\ 0 & 1 \end{bmatrix}, \qquad \mathbf{H} = \begin{bmatrix} T^2/2 \\ T \end{bmatrix}, \qquad \mathbf{C} = [1 \quad 0]$$

and T is the sampling period. (See Problem A–5–16 for the derivation of the discrete-time state-space equations for the double integrator system.) Assuming that the observer configuration is the same as that shown in Figure 6–8, design a state observer for this system. It is desired that the error vector exhibit deadbeat response. Use the four different methods listed in the foregoing discussion.

First we check the observability condition. Notice that the rank of

$$[\mathbf{C}^* \vdots \mathbf{G}^*\mathbf{C}^*] = \begin{bmatrix} 1 & 1 \\ 0 & T \end{bmatrix}$$

is 2. Hence, the system is completely observable. Next we examine the characteristic equation for the system:

$$|z\mathbf{I} - \mathbf{G}| = \left| \begin{bmatrix} z & 0 \\ 0 & z \end{bmatrix} - \begin{bmatrix} 1 & T \\ 0 & 1 \end{bmatrix} \right| = \begin{vmatrix} z - 1 & -T \\ 0 & z - 1 \end{vmatrix} = z^2 - 2z + 1 = 0$$

Comparing this characteristic equation with

$$z^2 + a_1 z + a_2 = 0$$

we obtain

$$a_1 = -2, \qquad a_2 = 1$$

Since the deadbeat response is desired, the desired characteristic equation for the error dynamics is

$$z^2 + \alpha_1 z + \alpha_2 = z^2 = 0$$

Thus,

$$\alpha_1 = 0, \qquad \alpha_2 = 0$$

Method 1. Referring to Equation (6–127), we have

$$\mathbf{K}_e = (\mathbf{W}\mathbf{N}^*)^{-1} \begin{bmatrix} \alpha_2 - a_2 \\ \alpha_1 - a_1 \end{bmatrix} = (\mathbf{W}\mathbf{N}^*)^{-1} \begin{bmatrix} -1 \\ 2 \end{bmatrix}$$

where $\mathbf{N}$ and $\mathbf{W}$, defined by Equations (6–102) and (6–103), respectively, are

$$\mathbf{N} = [\mathbf{C}^* \vdots \mathbf{G}^*\mathbf{C}^*] = \begin{bmatrix} 1 & 1 \\ 0 & T \end{bmatrix}$$

and

$$\mathbf{W} = \begin{bmatrix} a_1 & 1 \\ 1 & 0 \end{bmatrix} = \begin{bmatrix} -2 & 1 \\ 1 & 0 \end{bmatrix}$$

Hence, the observer feedback gain matrix $\mathbf{K}_e$ is obtained as follows:

$$\mathbf{K}_e = \left\{ \begin{bmatrix} -2 & 1 \\ 1 & 0 \end{bmatrix} \begin{bmatrix} 1 & 0 \\ 1 & T \end{bmatrix} \right\}^{-1} \begin{bmatrix} -1 \\ 2 \end{bmatrix} = \begin{bmatrix} 0 & 1 \\ \dfrac{1}{T} & \dfrac{1}{T} \end{bmatrix} \begin{bmatrix} -1 \\ 2 \end{bmatrix} = \begin{bmatrix} 2 \\ \dfrac{1}{T} \end{bmatrix}$$

Method 2. From Equation (6–128), Ackermann's formula, $\mathbf{K}_e$ is given by

$$\mathbf{K}_e = \phi(\mathbf{G}) \begin{bmatrix} \mathbf{C} \\ \overline{\mathbf{C}\mathbf{G}} \end{bmatrix}^{-1} \begin{bmatrix} 0 \\ 1 \end{bmatrix}$$

where

$$\phi(\mathbf{G}) = \mathbf{G}^2 + \alpha_1 \mathbf{G} + \alpha_2 \mathbf{I} = \mathbf{G}^2$$

Hence, the observer feedback gain matrix $\mathbf{K}_e$ is obtained as follows:

$$\mathbf{K}_e = \begin{bmatrix} 1 & T \\ 0 & 1 \end{bmatrix}^2 \begin{bmatrix} 1 & 0 \\ 1 & T \end{bmatrix}^{-1} \begin{bmatrix} 0 \\ 1 \end{bmatrix}$$

$$= \begin{bmatrix} 1 & 2T \\ 0 & 1 \end{bmatrix} \begin{bmatrix} 1 & 0 \\ -\dfrac{1}{T} & \dfrac{1}{T} \end{bmatrix} \begin{bmatrix} 0 \\ 1 \end{bmatrix} = \begin{bmatrix} 2 \\ \dfrac{1}{T} \end{bmatrix}$$

Method 3. Since deadbeat response is desired, from Equation (6–130) we have

$$\mathbf{K}_e = \begin{bmatrix} \boldsymbol{\eta}_1 \\ \boldsymbol{\eta}_2 \end{bmatrix}^{-1} \begin{bmatrix} 1 \\ 0 \end{bmatrix}$$

where

$$\boldsymbol{\eta}_1 = \mathbf{C}\mathbf{G}^{-1}, \qquad \boldsymbol{\eta}_2 = \mathbf{C}\mathbf{G}^{-2}$$

Notice that

$$\mathbf{G}^{-1} = \begin{bmatrix} 1 & -T \\ 0 & 1 \end{bmatrix}$$

The vectors $\boldsymbol{\eta}_1$ and $\boldsymbol{\eta}_2$ are obtained as follows:

$$\boldsymbol{\eta}_1 = \begin{bmatrix} 1 & 0 \end{bmatrix} \begin{bmatrix} 1 & -T \\ 0 & 1 \end{bmatrix} = \begin{bmatrix} 1 & -T \end{bmatrix}$$

$$\boldsymbol{\eta}_2 = \begin{bmatrix} 1 & 0 \end{bmatrix} \begin{bmatrix} 1 & -T \\ 0 & 1 \end{bmatrix}^2 = \begin{bmatrix} 1 & 0 \end{bmatrix} \begin{bmatrix} 1 & -2T \\ 0 & 1 \end{bmatrix} = \begin{bmatrix} 1 & -2T \end{bmatrix}$$

Thus, the observer feedback gain matrix $\mathbf{K}_e$ is obtained as follows:

$$\mathbf{K}_e = \begin{bmatrix} 1 & -T \\ 1 & -2T \end{bmatrix}^{-1} \begin{bmatrix} 1 \\ 0 \end{bmatrix} = \begin{bmatrix} 2 & -1 \\ \dfrac{1}{T} & -\dfrac{1}{T} \end{bmatrix} \begin{bmatrix} 1 \\ 0 \end{bmatrix} = \begin{bmatrix} 2 \\ \dfrac{1}{T} \end{bmatrix}$$

Method 4.   We assume

$$\mathbf{K}_e = \begin{bmatrix} k_1 \\ k_2 \end{bmatrix}$$

and expand the characteristic equation as follows:

$$|z\mathbf{I} - \mathbf{G} + \mathbf{K}_e \mathbf{C}| = \left| z \begin{bmatrix} 1 & 0 \\ 0 & 1 \end{bmatrix} - \begin{bmatrix} 1 & T \\ 0 & 1 \end{bmatrix} + \begin{bmatrix} k_1 \\ k_2 \end{bmatrix} \begin{bmatrix} 1 & 0 \end{bmatrix} \right|$$

$$= \begin{vmatrix} z - 1 + k_1 & -T \\ k_2 & z - 1 \end{vmatrix} = z^2 + (k_1 - 2)z + 1 - k_1 + k_2 T = 0$$

Since we desire the deadbeat response, this characteristic equation must be equal to

$$z^2 = 0$$

Thus,

$$k_1 = 2, \qquad k_2 = \frac{1}{T}$$

or

$$\mathbf{K}_e = \begin{bmatrix} k_1 \\ k_2 \end{bmatrix} = \begin{bmatrix} 2 \\ 1/T \end{bmatrix}$$

Let us verify that the error vector reduces to zero in at most two sampling periods. Note that the coefficient matrix for the error equation becomes

$$\mathbf{G} - \mathbf{K}_e \mathbf{C} = \begin{bmatrix} 1 & T \\ 0 & 1 \end{bmatrix} - \begin{bmatrix} 2 \\ \dfrac{1}{T} \end{bmatrix} \begin{bmatrix} 1 & 0 \end{bmatrix} = \begin{bmatrix} -1 & T \\ -\dfrac{1}{T} & 1 \end{bmatrix}$$

If the initial state $\mathbf{x}(0)$ is given as

$$\mathbf{x}(0) = \begin{bmatrix} a_1 \\ b_1 \end{bmatrix}$$

where a_1 and b_1 are arbitrary and $\tilde{\mathbf{x}}(0)$ is assumed as

$$\tilde{\mathbf{x}}(0) = \begin{bmatrix} a_2 \\ b_2 \end{bmatrix}$$

where a_2 and b_2 are arbitrary, then

$$\mathbf{e}(0) = \mathbf{x}(0) - \tilde{\mathbf{x}}(0) = \begin{bmatrix} a_1 - a_2 \\ b_1 - b_2 \end{bmatrix} = \begin{bmatrix} a \\ b \end{bmatrix}$$

where a and b are arbitrary constants. Now Equation (6–95) becomes

$$\begin{bmatrix} e_1(k+1) \\ e_2(k+1) \end{bmatrix} = \begin{bmatrix} -1 & T \\ -\dfrac{1}{T} & 1 \end{bmatrix} \begin{bmatrix} e_1(k) \\ e_2(k) \end{bmatrix}, \qquad \begin{bmatrix} e_1(0) \\ e_2(0) \end{bmatrix} = \begin{bmatrix} a \\ b \end{bmatrix}$$

The vectors $\mathbf{e}(1)$ and $\mathbf{e}(2)$ are found as follows:

$$\begin{bmatrix} e_1(1) \\ e_2(1) \end{bmatrix} = \begin{bmatrix} -1 & T \\ -\dfrac{1}{T} & 1 \end{bmatrix} \begin{bmatrix} a \\ b \end{bmatrix} = \begin{bmatrix} -a + bT \\ -\dfrac{1}{T}a + b \end{bmatrix}$$

and

$$\begin{bmatrix} e_1(2) \\ e_2(2) \end{bmatrix} = \begin{bmatrix} -1 & T \\ -\dfrac{1}{T} & 1 \end{bmatrix} \begin{bmatrix} -a + bT \\ -\dfrac{1}{T}a + b \end{bmatrix} = \begin{bmatrix} a - bT - a + bT \\ \dfrac{1}{T}a - b - \dfrac{1}{T}a + b \end{bmatrix} = \begin{bmatrix} 0 \\ 0 \end{bmatrix}$$

Clearly, the error vector $\mathbf{e}(k)$ becomes zero in at most two-sampling periods. Thus, the response is deadbeat. Note that for any initial state $\mathbf{x}(0)$ the observed-state vector becomes identical to the actual state vector in at most two sampling periods.

Finally, the observer equation is

$$\begin{bmatrix} \tilde{x}_1(k+1) \\ \tilde{x}_2(k+1) \end{bmatrix} = \begin{bmatrix} -1 & T \\ -\dfrac{1}{T} & 1 \end{bmatrix} \begin{bmatrix} \tilde{x}_1(k) \\ \tilde{x}_2(k) \end{bmatrix} + \begin{bmatrix} \dfrac{T^2}{2} \\ T \end{bmatrix} u(kT) + \begin{bmatrix} 2 \\ \dfrac{1}{T} \end{bmatrix} y(k)$$

[Note that this is the equation given by Equation (6–93).]

Comments on Selecting the Best $\mathbf{K}_e$.   Referring to Figure 6–8, notice that the feedback signal through the observer feedback gain matrix $\mathbf{K}_e$ serves as a correction signal to the plant model to account for the unknowns in the plant. If significant unknowns are involved, the feedback signal through the matrix $\mathbf{K}_e$ should be relatively large. However, if the output signal is contaminated significantly by disturbances and measurement noises, then the output $\mathbf{y}(k)$ is not reliable and the feedback signal through the matrix $\mathbf{K}_e$ should be relatively small. In determining the matrix $\mathbf{K}_e$ (which depends on the desired eigenvalues $\mu_1, \mu_2, \ldots, \mu_n$), we should carefully examine the effects of disturbances and noises involved in the output $\mathbf{y}(k)$.

Remember that the observer feedback gain matrix $\mathbf{K}_e$ depends on the desired characteristic equation

$$\phi(z) = (z - \mu_1)(z - \mu_2) \cdots (z - \mu_n) = 0$$

The choice of a set of $\mu_1, \mu_2, \ldots, \mu_n$ is not unique. Hence, many different characteristic equations might be chosen as desired characteristic equations. For each desired characteristic equation, we have a different matrix $\mathbf{K}_e$.

In the design of the observer, it is desirable to determine several observer feedback gain matrices $\mathbf{K}_e$ based on several different desired characteristic equations. For each of the several different matrices $\mathbf{K}_e$, simulation tests must be run to evaluate the resulting system performance. Then we select the best $\mathbf{K}_e$ from the viewpoint of overall system performance. In many practical cases the selection of the best matrix $\mathbf{K}_e$ boils down to a compromise between speedy response and sensitivity to disturbances and noises.

Effects of the Addition of the Observer on a Closed-Loop System.   In the pole placement design process, we assumed that the true state $\mathbf{x}(k)$ was available for feedback. But in practice the true state $\mathbf{x}(k)$ may not be measurable, so we will need to use the observed state $\tilde{\mathbf{x}}(k)$. Let us now investigate the effects of the use of the observed state $\tilde{\mathbf{x}}(k)$ rather than the true state $\mathbf{x}(k)$ upon the characteristic equation of a closed-loop control system.

Consider the completely state controllable and completely observable system defined by the equations

$$\mathbf{x}(k + 1) = \mathbf{Gx}(k) + \mathbf{Hu}(k)$$

$$\mathbf{y}(k) = \mathbf{Cx}(k)$$

For the state feedback control based on the observed state $\tilde{\mathbf{x}}(k)$, we have

$$\mathbf{u}(k) = -\mathbf{K}\tilde{\mathbf{x}}(k)$$

With this control the state equation becomes

$$\mathbf{x}(k + 1) = \mathbf{Gx}(k) - \mathbf{HK}\tilde{\mathbf{x}}(k) = (\mathbf{G} - \mathbf{HK})\mathbf{x}(k) + \mathbf{HK}[\mathbf{x}(k) - \tilde{\mathbf{x}}(k)] \quad (6\text{–}131)$$

The difference between the actual state $\mathbf{x}(k)$ and the observed state $\tilde{\mathbf{x}}(k)$ has been defined as the error $\mathbf{e}(k)$:

$$\mathbf{e}(k) = \mathbf{x}(k) - \tilde{\mathbf{x}}(k)$$

By substitution of the error vector $\mathbf{e}(k)$, Equation (6–131) becomes

$$\mathbf{x}(k + 1) = (\mathbf{G} - \mathbf{HK})\mathbf{x}(k) + \mathbf{HKe}(k) \quad (6\text{–}132)$$

Note that the observer error equation was given by Equation (6–95), repeated here:

$$\mathbf{e}(k + 1) = (\mathbf{G} - \mathbf{K}_e\mathbf{C})\mathbf{e}(k) \quad (6\text{–}133)$$

Combining Equations (6–132) and (6–133), we obtain

$$\begin{bmatrix} \mathbf{x}(k + 1) \\ \mathbf{e}(k + 1) \end{bmatrix} = \begin{bmatrix} \mathbf{G} - \mathbf{HK} & \mathbf{HK} \\ \mathbf{0} & \mathbf{G} - \mathbf{K}_e\mathbf{C} \end{bmatrix} \begin{bmatrix} \mathbf{x}(k) \\ \mathbf{e}(k) \end{bmatrix}$$

This equation describes the dynamics of the observed-state feedback control system.

The characteristic equation for the system is

$$\begin{vmatrix} z\mathbf{I} - \mathbf{G} + \mathbf{HK} & -\mathbf{HK} \\ \mathbf{0} & z\mathbf{I} - \mathbf{G} + \mathbf{K}_e\mathbf{C} \end{vmatrix} = 0$$

or

$$|z\mathbf{I} - \mathbf{G} + \mathbf{HK}|\,|z\mathbf{I} - \mathbf{G} + \mathbf{K}_e\mathbf{C}| = 0 \quad (6\text{–}134)$$

Notice that the closed-loop poles of the observed-state feedback control system consist of the poles due to the pole placement design alone plus the poles due to the observer design alone. This means that the pole placement design and the observer design are independent of each other. They can be designed separately and combined to form the observed-state feedback control system.

The desired closed-loop poles to be generated by state feedback (pole placement) are chosen in such a way that the system satisfies the performance requirements. The poles of the observer are usually chosen so that the observer response is much faster than the system response. A rule of thumb is to choose an observer response at least four to five times faster than the system response or in some cases to choose all observer poles at the origin (for deadbeat response). Since the observer is, in general, not a hardware structure but is programmed on the computer, it is possible to increase the response speed or achieve deadbeat response so that the observed state quickly converges to the true state. The maximum response speed of the observer is generally limited only by noise and sensitivity problems involved in the control system.

Current Observer.   In the prediction observer the observed state $\tilde{x}(k)$ is obtained from measurements of the output vector up to $y(k-1)$ and of the control vector up to $u(k-1)$. Hence, the control vector $u(k) = -K\tilde{x}(k)$ does not utilize the information on the current output $y(k)$. A different formulation of the state observer is to use $y(k)$ for the estimation of $\tilde{x}(k)$. This can be done by separating the observation process into two steps. In the first step we determine $z(k+1)$, an approximation of $x(k+1)$ based on $\tilde{x}(k)$ and $u(k)$. In the second step, we use $y(k+1)$ to improve $z(k+1)$. The improved $z(k+1)$ is $\tilde{x}(k+1)$. The state observer based on this formulation is called the *current observer*.

Consider the completely state controllable and completely observable system defined by the equations

$$x(k+1) = Gx(k) + Hu(k)$$

$$y(k) = Cx(k)$$

where

$$x(k) = \text{state vector } (n\text{-vector})$$

$$u(k) = \text{control vector } (r\text{-vector})$$

$$y(k) = \text{output vector } (m\text{-vector})$$

$$G = n \times n \text{ matrix}$$

$$H = n \times r \text{ matrix}$$

$$C = m \times n \text{ matrix}$$

The current observer equations are given by

$$\tilde{x}(k+1) = z(k+1) + K_e[y(k+1) - Cz(k+1)] \qquad (6\text{-}135)$$

$$z(k+1) = G\tilde{x}(k) + Hu(k) \qquad (6\text{-}136)$$

Equation (6–136) gives the prediction $z(k+1)$ based on $\tilde{x}(k)$ and $u(k)$ at stage k. Equation (6–135) states that by measuring $y(k+1)$ we can improve $z(k+1)$ to obtain $\tilde{x}(k+1)$.

Define the observer error $\mathbf{e}(k)$ as follows:

$$\mathbf{e}(k) = \mathbf{x}(k) - \tilde{\mathbf{x}}(k)$$

Then

$$
\begin{aligned}
\mathbf{e}(k + 1) &= \mathbf{x}(k + 1) - \tilde{\mathbf{x}}(k + 1) \\
&= \mathbf{Gx}(k) + \mathbf{Hu}(k) - \bigl(\mathbf{G\tilde{x}}(k) + \mathbf{Hu}(k) + \mathbf{K}_e\{\mathbf{C}[\mathbf{Gx}(k) + \mathbf{Hu}(k)] \\
&\quad - \mathbf{C}[\mathbf{G\tilde{x}}(k) + \mathbf{Hu}(k)]\}\bigr) \\
&= (\mathbf{G} - \mathbf{K}_e\mathbf{CG})[\mathbf{x}(k) - \tilde{\mathbf{x}}(k)] \\
&= (\mathbf{G} - \mathbf{K}_e\mathbf{CG})\mathbf{e}(k)
\end{aligned}
$$

Thus, the observer error equation for the current observer is similar to that for the prediction observer given by Equation (6–95). However, a difference appears in the error dynamics. The matrix $\mathbf{K}_e$ can be obtained exactly as in the case of the prediction observer except that matrix $\mathbf{C}$ is replaced by matrix $\mathbf{CG}$. To make it possible for the eigenvalues of $(\mathbf{G} - \mathbf{K}_e\mathbf{CG})$ to be arbitrarily placed, the rank of the matrix

$$
\begin{bmatrix} \mathbf{CG} \\ \mathbf{CG}^2 \\ \vdots \\ \mathbf{CG}^n \end{bmatrix} = \begin{bmatrix} \mathbf{C} \\ \mathbf{CG} \\ \vdots \\ \mathbf{CG}^{n-1} \end{bmatrix} \mathbf{G}
$$

must be n. Notice that if matrix $\mathbf{G}$ is nonsingular then this condition is equivalent to the observability condition, or

$$\operatorname{rank}[\mathbf{C}^* \vdots \mathbf{G}^*\mathbf{C}^* \vdots \cdots \vdots (\mathbf{G}^*)^{n-1}\mathbf{C}^*] = n$$

If the rank of the observability matrix is n, then the eigenvalues of $\mathbf{G} - \mathbf{K}_e\mathbf{CG}$ can be arbitrarily located by a proper choice of $\mathbf{K}_e$, and matrix $\mathbf{K}_e$ can be determined in a way similar to what was done in the case of the prediction observer. In determining matrix $\mathbf{K}_e$, we replace matrix $\mathbf{C}$ by $\mathbf{CG}$ in the computations involved. For example, if the output $\mathbf{y}(k)$ is a scalar, then Ackermann's formula as given by Equation (6–126) is altered to the corresponding form:

$$
\mathbf{K}_e = \phi(\mathbf{G}) \begin{bmatrix} \mathbf{CG} \\ \mathbf{CG}^2 \\ \vdots \\ \mathbf{CG}^{n-1} \\ \mathbf{CG}^n \end{bmatrix}^{-1} \begin{bmatrix} 0 \\ 0 \\ \vdots \\ 0 \\ 1 \end{bmatrix} \tag{6–137}
$$

However, if matrix $\mathbf{G}$ is singular, then the rank of

$$[\mathbf{G}^*\mathbf{C}^* \vdots (\mathbf{G}^*)^2\mathbf{C}^* \vdots \cdots \vdots (\mathbf{G}^*)^n\mathbf{C}^*]$$

is q, which is less than n. In this case, let us write

$$(\mathbf{G} - \mathbf{K}_e\mathbf{CG})^* = \mathbf{G}^* - \mathbf{G}^*\mathbf{C}^*\mathbf{K}_e^* = \mathbf{G}^* - \mathbf{BK}_e^*$$

where $\mathbf{B} = \mathbf{G}^*\mathbf{C}^*$. Note that matrix $(\mathbf{G}^* - \mathbf{BK}_e^*)$ is of the same form as matrix $(\mathbf{G} - \mathbf{HK})$, which played an important role in the pole placement design.

With an analysis similar to that given in Section 6–5 [refer to Equations (6–55) and (6–57), it is possible by use of a suitable transformation matrix $\mathbf{T}$ to transform matrices $\mathbf{G}$ and $\mathbf{B}$ into $\hat{\mathbf{G}}^*$ and $\hat{\mathbf{B}}$, where

$$\hat{\mathbf{G}}^* = \mathbf{T}^{-1}\mathbf{G}^*\mathbf{T} = \begin{bmatrix} \mathbf{G}_{11}^* & \mathbf{G}_{12}^* \\ \hline \mathbf{0} & \mathbf{G}_{22}^* \end{bmatrix}, \qquad \hat{\mathbf{B}} = \mathbf{T}^{-1}\mathbf{B} = \begin{bmatrix} \mathbf{B}_{11} \\ \mathbf{0} \end{bmatrix}$$

and where all eigenvalues of the uncontrollable $(n - q) \times (n - q)$ matrix $\mathbf{G}_{22}^*$ can be made zero. (Hence, the system can be stabilized.) Next, define

$$\mathbf{K}_e^* \, \mathbf{T} = \hat{\mathbf{K}}_e^* = [\hat{\mathbf{K}}_{e11}^* \vdots \hat{\mathbf{K}}_{e12}^*]$$

Then

$$\begin{aligned} \hat{\mathbf{G}}^* - \hat{\mathbf{B}}\hat{\mathbf{K}}_e^* &= \begin{bmatrix} \mathbf{G}_{11}^* & \mathbf{G}_{12}^* \\ \hline \mathbf{0} & \mathbf{G}_{22}^* \end{bmatrix} - \begin{bmatrix} \mathbf{B}_{11} \\ \mathbf{0} \end{bmatrix}[\hat{\mathbf{K}}_{e11}^* \vdots \hat{\mathbf{K}}_{e12}^*] \\ &= \begin{bmatrix} \mathbf{G}_{11}^* - \mathbf{B}_{11}\hat{\mathbf{K}}_{e11}^* & \mathbf{G}_{12}^* - \mathbf{B}_{11}\hat{\mathbf{K}}_{e12}^* \\ \hline \mathbf{0} & \mathbf{G}_{22}^* \end{bmatrix} \end{aligned}$$

Hence, if matrix $\mathbf{G}$ is singular and the rank of the observability matrix is q, then we need to specify only q eigenvalues of the $q \times q$ matrix $\mathbf{G}_{11}^* - \mathbf{B}_{11}\hat{\mathbf{K}}_{e11}^*$.

Minimum-Order Observer.   The observers discussed thus far are designed to reconstruct all the state variables. In practice, some of the state variables may be accurately measured. Such accurately measurable state variables need not be estimated. An observer that estimates fewer than n state variables, where n is the dimension of the state vector, is called a *reduced-order observer*. If the order of the reduced-order observer is the minimum possible, the observer is called a *minimum-order observer*.

Suppose the state vector $\mathbf{x}(k)$ is an n-vector and the output vector $\mathbf{y}(k)$ is an m-vector that can be measured. Since m output variables are linear combinations of the state variables, m state variables need not be estimated. We need to estimate only $n - m$ state variables. Then the reduced-order observer becomes an $(n - m)$th-order observer. Such an $(n - m)$th-order observer is the minimum-order observer. Figure 6–10 shows the block diagram of a system with a minimum-order observer.

It is important to note, however, that if the measurement of output variables involves significant noises and is relatively inaccurate then the use of the full-order observer may result in a better system performance.

The minimum-order observer can be designed by first partitioning the state vector $\mathbf{x}(k)$ into two parts, as follows:

$$\mathbf{x}(k) = \begin{bmatrix} \mathbf{x}_a(k) \\ \hline \mathbf{x}_b(k) \end{bmatrix}$$

where $\mathbf{x}_a(k)$ is that portion of the state vector that can be directly measured [thus, $\mathbf{x}_a(k)$ is an m-vector] and $\mathbf{x}_b(k)$ is the unmeasurable portion of the state vector [thus, $\mathbf{x}_b(k)$ is an $(n - m)$-vector]. Then the partitioned state equations become as follows:

$$\begin{bmatrix} \mathbf{x}_a(k + 1) \\ \hline \mathbf{x}_b(k + 1) \end{bmatrix} = \begin{bmatrix} \mathbf{G}_{aa} & \mathbf{G}_{ab} \\ \hline \mathbf{G}_{ba} & \mathbf{G}_{bb} \end{bmatrix}\begin{bmatrix} \mathbf{x}_a(k) \\ \hline \mathbf{x}_b(k) \end{bmatrix} + \begin{bmatrix} \mathbf{H}_a \\ \hline \mathbf{H}_b \end{bmatrix}\mathbf{u}(k) \qquad (6\text{–}138)$$

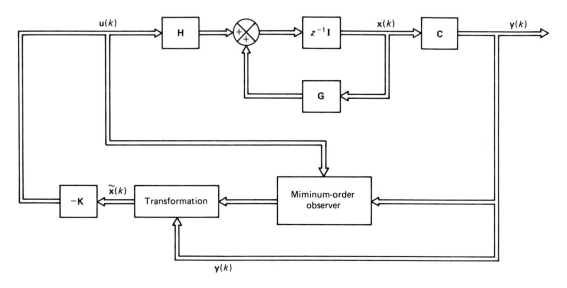

Figure 6–10 Observed-state feedback control system with a minimum-order observer.

$$y(k) = [\mathbf{I} \vdots \mathbf{0}]\left[\frac{\mathbf{x}_a(k)}{\mathbf{x}_b(k)}\right] \tag{6–139}$$

where

$$\mathbf{G}_{aa} = m \times m \text{ matrix}$$

$$\mathbf{G}_{ab} = m \times (n - m) \text{ matrix}$$

$$\mathbf{G}_{ba} = (n - m) \times m \text{ matrix}$$

$$\mathbf{G}_{bb} = (n - m) \times (n - m) \text{ matrix}$$

$$\mathbf{H}_a = m \times r \text{ matrix}$$

$$\mathbf{H}_b = (n - m) \times r \text{ matrix}$$

By rewriting Equation (6–138), the equation for the measured portion of the state becomes

$$\mathbf{x}_a(k + 1) = \mathbf{G}_{aa}\,\mathbf{x}_a(k) + \mathbf{G}_{ab}\,\mathbf{x}_b(k) + \mathbf{H}_a\,\mathbf{u}(k)$$

or

$$\mathbf{x}_a(k + 1) - \mathbf{G}_{aa}\,\mathbf{x}_a(k) - \mathbf{H}_a\,\mathbf{u}(k) = \mathbf{G}_{ab}\,\mathbf{x}_b(k) \tag{6–140}$$

where the terms on the left-hand side of the equation can be measured. Equation (6–140) acts as the output equation. In designing the minimum-order observer, we consider the left-hand side of Equation (6–140) to be known quantities. In fact, Equation (6–140) relates the measurable quantities and the unmeasurable quantities of the state.

From Equation (6–138), the equation for the unmeasured portion of the state becomes

$$\mathbf{x}_b(k + 1) = \mathbf{G}_{ba}\,\mathbf{x}_a(k) + \mathbf{G}_{bb}\,\mathbf{x}_b(k) + \mathbf{H}_b\,\mathbf{u}(k) \qquad (6\text{–}141)$$

Equation (6–141) describes the dynamics of the unmeasured portion of the state. Notice that the terms $\mathbf{G}_{ba}\,\mathbf{x}_a(k)$ and $\mathbf{H}_b\,\mathbf{u}(k)$ are known quantities.

The design of the minimum-order observer can be facilitated if we utilize the design technique developed for the full-order observer. Let us now compare the state equation for the full-order observer with that for the minimum-order observer. The state equation for the full-order observer is

$$\mathbf{x}(k + 1) = \mathbf{G}\mathbf{x}(k) + \mathbf{H}\mathbf{u}(k)$$

and the "state equation" for the minimum-order observer is

$$\mathbf{x}_b(k + 1) = \mathbf{G}_{bb}\,\mathbf{x}_b(k) + [\mathbf{G}_{ba}\,\mathbf{x}_a(k) + \mathbf{H}_b\,\mathbf{u}(k)]$$

The output equation for the full-order observer is

$$\mathbf{y}(k) = \mathbf{C}\mathbf{x}(k)$$

and the "output equation" for the minimum-order observer is

$$\mathbf{x}_a(k + 1) - \mathbf{G}_{aa}\,\mathbf{x}_a(k) - \mathbf{H}_a\,\mathbf{u}(k) = \mathbf{G}_{ab}\,\mathbf{x}_b(k)$$

The design of the minimum-order observer can be carried out by making the substitutions given in Table 6–1 into the observer equation for the full-order observer given by Equation (6–93), which we repeat here:

$$\tilde{\mathbf{x}}(k + 1) = (\mathbf{G} - \mathbf{K}_e\,\mathbf{C})\tilde{\mathbf{x}}(k) + \mathbf{H}\mathbf{u}(k) + \mathbf{K}_e\,\mathbf{y}(k) \qquad (6\text{–}142)$$

Making the substitution of Table 6–1 into Equation (6–142), we obtain

$$\tilde{\mathbf{x}}_b(k + 1) = (\mathbf{G}_{bb} - \mathbf{K}_e\,\mathbf{G}_{ab})\tilde{\mathbf{x}}_b(k) + \mathbf{G}_{ba}\,\mathbf{x}_a(k) + \mathbf{H}_b\,\mathbf{u}(k)$$
$$+ \mathbf{K}_e[\mathbf{x}_a(k + 1) - \mathbf{G}_{aa}\,\mathbf{x}_a(k) - \mathbf{H}_a\,\mathbf{u}(k)] \qquad (6\text{–}143)$$

TABLE 6–1 LIST OF NECESSARY SUBSTITUTIONS FOR WRITING THE OBSERVER EQUATION FOR THE MINIMUM-ORDER STATE OBSERVER

Full-order state observer	Minimum-order state observer
$\tilde{\mathbf{x}}(k)$	$\tilde{\mathbf{x}}_b(k)$
$\mathbf{G}$	$\mathbf{G}_{bb}$
$\mathbf{H}\mathbf{u}(k)$	$\mathbf{G}_{ba}\,\mathbf{x}_a(k) + \mathbf{H}_b\,\mathbf{u}(k)$
$\mathbf{y}(k)$	$\mathbf{x}_a(k + 1) - \mathbf{G}_{aa}\,\mathbf{x}_a(k) - \mathbf{H}_a\,\mathbf{u}(k)$
$\mathbf{C}$	$\mathbf{G}_{ab}$
$\mathbf{K}_e,\quad n \times m$ matrix	$\mathbf{K}_e,\quad (n - m) \times m$ matrix

where the observer feedback gain matrix $\mathbf{K}_e$ is an $(n - m) \times m$ matrix. Equation (6–143) defines the minimum-order observer.

Referring to Equation (6–139), we have

$$y(k) = \mathbf{x}_a(k) \tag{6–144}$$

Substituting Equation (6–144) into Equation (6–143), we obtain

$$\tilde{\mathbf{x}}_b(k + 1) = (\mathbf{G}_{bb} - \mathbf{K}_e \mathbf{G}_{ab})\tilde{\mathbf{x}}_b(k) + \mathbf{K}_e y(k + 1)$$
$$+ (\mathbf{G}_{ba} - \mathbf{K}_e \mathbf{G}_{aa})y(k) + (\mathbf{H}_b - \mathbf{K}_e \mathbf{H}_a)\mathbf{u}(k) \tag{6–145}$$

Notice that to estimate $\tilde{\mathbf{x}}_b(k + 1)$ we need the measured value of $y(k + 1)$. This is inconvenient, and so we may desire some modifications. [In the case of the full-order observer, $\tilde{\mathbf{x}}(k + 1)$ can be estimated by use of measurement $y(k)$ and does not require measurement of $y(k + 1)$. See Equation (6–93).] Let us rewrite Equation (6–145) as follows:

$$\tilde{\mathbf{x}}_b(k + 1) - \mathbf{K}_e y(k + 1) = (\mathbf{G}_{bb} - \mathbf{K}_e \mathbf{G}_{ab})\tilde{\mathbf{x}}_b(k) + (\mathbf{G}_{ba} - \mathbf{K}_e \mathbf{G}_{aa})y(k)$$
$$+ (\mathbf{H}_b - \mathbf{K}_e \mathbf{H}_a)\mathbf{u}(k)$$
$$= (\mathbf{G}_{bb} - \mathbf{K}_e \mathbf{G}_{ab})[\tilde{\mathbf{x}}_b(k) - \mathbf{K}_e y(k)] + (\mathbf{G}_{bb} - \mathbf{K}_e \mathbf{G}_{ab})\mathbf{K}_e y(k)$$
$$+ (\mathbf{G}_{ba} - \mathbf{K}_e \mathbf{G}_{aa})y(k) + (\mathbf{H}_b - \mathbf{K}_e \mathbf{H}_a)\mathbf{u}(k)$$
$$= (\mathbf{G}_{bb} - \mathbf{K}_e \mathbf{G}_{ab})[\tilde{\mathbf{x}}_b(k) - \mathbf{K}_e y(k)] + [(\mathbf{G}_{bb} - \mathbf{K}_e \mathbf{G}_{ab})\mathbf{K}_e$$
$$+ \mathbf{G}_{ba} - \mathbf{K}_e \mathbf{G}_{aa}]y(k) + (\mathbf{H}_b - \mathbf{K}_e \mathbf{H}_a)\mathbf{u}(k) \tag{6–146}$$

Define

$$\mathbf{x}_b(k) - \mathbf{K}_e y(k) = \mathbf{x}_b(k) - \mathbf{K}_e \mathbf{x}_a(k) = \boldsymbol{\eta}(k) \tag{6–147}$$

and

$$\tilde{\mathbf{x}}_b(k) - \mathbf{K}_e y(k) = \tilde{\mathbf{x}}_b(k) - \mathbf{K}_e \mathbf{x}_a(k) = \tilde{\boldsymbol{\eta}}(k) \tag{6–148}$$

Then Equation (6–146) can be written as follows:

$$\tilde{\boldsymbol{\eta}}(k + 1) = (\mathbf{G}_{bb} - \mathbf{K}_e \mathbf{G}_{ab})\tilde{\boldsymbol{\eta}}(k) + [(\mathbf{G}_{bb} - \mathbf{K}_e \mathbf{G}_{ab})\mathbf{K}_e + \mathbf{G}_{ba}$$
$$- \mathbf{K}_e \mathbf{G}_{aa}]y(k) + (\mathbf{H}_b - \mathbf{K}_e \mathbf{H}_a)\mathbf{u}(k) \tag{6–149}$$

Equations (6–148) and (6–149) define the dynamics of the minimum-order observer. Notice that to obtain $\tilde{\boldsymbol{\eta}}(k + 1)$ we do not need the measured value of $y(k + 1)$.

Let us next obtain the observer error equation. Define

$$\mathbf{e}(k) = \boldsymbol{\eta}(k) - \tilde{\boldsymbol{\eta}}(k) = \mathbf{x}_b(k) - \tilde{\mathbf{x}}_b(k) \tag{6–150}$$

Subtracting Equation (6–143) from Equation (6–141), we obtain

$$\mathbf{x}_b(k + 1) - \tilde{\mathbf{x}}_b(k + 1) = \mathbf{G}_{bb}[\mathbf{x}_b(k) - \tilde{\mathbf{x}}_b(k)] + \mathbf{K}_e \mathbf{G}_{ab} \tilde{\mathbf{x}}_b(k)$$
$$- \mathbf{K}_e[\mathbf{x}_a(k + 1) - \mathbf{G}_{aa} \mathbf{x}_a(k) - \mathbf{H}_a \mathbf{u}(k)]$$

By substituting Equation (6–140) into this last equation, we obtain

$$\mathbf{x}_b(k + 1) - \tilde{\mathbf{x}}_b(k + 1) = \mathbf{G}_{bb}[\mathbf{x}_b(k) - \tilde{\mathbf{x}}_b(k)] + \mathbf{K}_e \mathbf{G}_{ab} \tilde{\mathbf{x}}_b(k) - \mathbf{K}_e \mathbf{G}_{ab} \mathbf{x}_b(k)$$
$$= (\mathbf{G}_{bb} - \mathbf{K}_e \mathbf{G}_{ab})[\mathbf{x}_b(k) - \tilde{\mathbf{x}}_b(k)]$$

This last equation can be written in the form

$$\mathbf{e}(k + 1) = (\mathbf{G}_{bb} - \mathbf{K}_e \mathbf{G}_{ab})\mathbf{e}(k) \tag{6–151}$$

This is the observer error equation. Note that $\mathbf{e}(k)$ is an $(n - m)$-vector. The error dynamics can be determined as desired by following the technique developed for the full-order observer, provided that the rank of matrix

$$\begin{bmatrix} \mathbf{G}_{ab} \\ \mathbf{G}_{ab}\mathbf{G}_{bb} \\ \vdots \\ \mathbf{G}_{ab}\mathbf{G}_{bb}^{n-m-1} \end{bmatrix}$$

is $n - m$. (This is the complete observability condition applicable to the minimum-order observer.)

The characteristic equation for the minimum-order observer is obtained from Equation (6–151) as follows:

$$|z\mathbf{I} - \mathbf{G}_{bb} + \mathbf{K}_e \mathbf{G}_{ab}| = 0 \tag{6–152}$$

The observer feedback gain matrix $\mathbf{K}_e$ can be determined from Equation (6–152) by first choosing the desired closed-loop pole locations for the minimum-order observer [that is, by placing the roots of the characteristic equation, Equation (6–152), at the desired locations] and then using the procedure developed for the full-order prediction observer.

If, for example, the output $y(k)$ is a scalar, then $x_a(k)$ is a scalar, $\mathbf{G}_{ab}$ is a $1 \times (n - 1)$ matrix, and $\mathbf{G}_{bb}$ is an $(n - 1) \times (n - 1)$ matrix. For this case, Ackermann's formula as given by Equation (6–126) may be modified to read

$$\mathbf{K}_e = \phi(\mathbf{G}_{bb}) \begin{bmatrix} \mathbf{G}_{ab} \\ \mathbf{G}_{ab}\mathbf{G}_{bb} \\ \vdots \\ \mathbf{G}_{ab}\mathbf{G}_{bb}^{n-3} \\ \mathbf{G}_{ab}\mathbf{G}_{bb}^{n-2} \end{bmatrix}^{-1} \begin{bmatrix} 0 \\ 0 \\ \vdots \\ 0 \\ 1 \end{bmatrix} \tag{6–153}$$

where

$$\phi(\mathbf{G}_{bb}) = \mathbf{G}_{bb}^{n-1} + \alpha_1 \mathbf{G}_{bb}^{n-2} + \cdots + \alpha_{n-2}\mathbf{G}_{bb} + \alpha_{n-1}\mathbf{I} \tag{6–154}$$

Summary.   Once the observer feedback gain matrix $\mathbf{K}_e$, which is an $(n - m) \times m$ matrix, is determined, then the minimum-order observer can be defined by Equations (6–148) and (6–149):

$$\tilde{\mathbf{x}}_b(k) = \tilde{\boldsymbol{\eta}}(k) + \mathbf{K}_e \mathbf{x}_a(k)$$

$$\tilde{\boldsymbol{\eta}}(k + 1) = (\mathbf{G}_{bb} - \mathbf{K}_e \mathbf{G}_{ab})\tilde{\boldsymbol{\eta}}(k) + [(\mathbf{G}_{bb} - \mathbf{K}_e \mathbf{G}_{ab})\mathbf{K}_e + \mathbf{G}_{ba} - \mathbf{K}_e \mathbf{G}_{aa}]\mathbf{y}(k)$$
$$+ (\mathbf{H}_b - \mathbf{K}_e \mathbf{H}_a)\mathbf{u}(k)$$

Equivalently, in terms of $\mathbf{e}(k)$ rather than $\tilde{\boldsymbol{\eta}}(k)$, the minimum-order observer can be defined by Equations (6–150) and (6–151):

$$\tilde{\mathbf{x}}_b(k) = \mathbf{x}_b(k) - \mathbf{e}(k) \tag{6–155}$$

$$\mathbf{e}(k + 1) = (\mathbf{G}_{bb} - \mathbf{K}_e \mathbf{G}_{ab})\mathbf{e}(k) \tag{6–156}$$

Observed-State Feedback Control System With Minimum-Order Observer.   Consider the completely state controllable and completely observable system given by

$$\mathbf{x}(k + 1) = \mathbf{G}\mathbf{x}(k) + \mathbf{H}\mathbf{u}(k) \qquad (6\text{--}157)$$

$$\mathbf{y}(k) = \mathbf{C}\mathbf{x}(k) \qquad (6\text{--}158)$$

where $\mathbf{x}(k)$ is an n-vector, $\mathbf{u}(k)$ is an r-vector, and $\mathbf{y}(k)$ is an m-vector. Matrices $\mathbf{G}$, $\mathbf{H}$, and $\mathbf{C}$ are given by

$$\mathbf{G} = \begin{bmatrix} \mathbf{G}_{aa} & \mathbf{G}_{ab} \\ \hline \mathbf{G}_{ba} & \mathbf{G}_{bb} \end{bmatrix}, \qquad \mathbf{H} = \begin{bmatrix} \mathbf{H}_a \\ \hline \mathbf{H}_b \end{bmatrix}, \qquad \mathbf{C} = [\mathbf{I}_m \vdots \mathbf{0}]$$

Consider the state feedback control scheme where the fed back state consists of the measured portion of the state and the observed (estimated) portion of the state obtained by use of the minimum-order observer. Figure 6–11 shows the block diagram for the system. In this system the control vector $\mathbf{u}(k)$ is given by

$$\mathbf{u}(k) = -\mathbf{K}\tilde{\mathbf{x}}(k) \qquad (6\text{--}159)$$

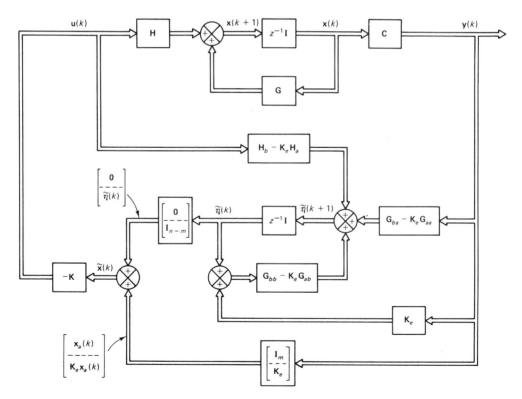

Figure 6–11   State feedback control scheme where the fed back state consists of the measured portion of the state and the observed portion of the state obtained by use of the minimum-order observer.

where $\bar{\mathbf{x}}(k)$ consists of the measurable state $\mathbf{x}_a(k)$ and unmeasurable (observed) state $\tilde{\mathbf{x}}_b(k)$:

$$\bar{\mathbf{x}}(k) = \left[\frac{\mathbf{x}_a(k)}{\tilde{\mathbf{x}}_b(k)}\right] = \left[\frac{\mathbf{x}_a(k)}{\boldsymbol{\eta}(k) + \mathbf{K}_e\mathbf{x}_a(k)}\right] \qquad (6\text{--}160)$$

By substituting Equation (6–159) into Equation (6–157), we obtain

$$\mathbf{x}(k + 1) = \mathbf{G}\mathbf{x}(k) - \mathbf{HK}\bar{\mathbf{x}}(k) = (\mathbf{G} - \mathbf{HK})\mathbf{x}(k) + \mathbf{HK}[\mathbf{x}(k) - \bar{\mathbf{x}}(k)] \qquad (6\text{--}161)$$

Notice that

$$\mathbf{x}(k) - \bar{\mathbf{x}}(k) = \left[\frac{\mathbf{x}_a(k)}{\mathbf{x}_b(k)}\right] - \left[\frac{\mathbf{x}_a(k)}{\tilde{\mathbf{x}}_b(k)}\right] = \left[\frac{\mathbf{0}}{\mathbf{x}_b(k) - \tilde{\mathbf{x}}_b(k)}\right] = \left[\frac{\mathbf{0}}{\mathbf{e}(k)}\right]$$

where $\mathbf{e}(k) = \mathbf{x}_b(k) - \tilde{\mathbf{x}}_b(k)$. Define

$$\boldsymbol{\Gamma} = \left[\frac{\mathbf{0}}{\mathbf{I}_{n-m}}\right]$$

Then, by use of this matrix $\boldsymbol{\Gamma}$, Equation (6–161) can be rewritten as follows:

$$\mathbf{x}(k + 1) = (\mathbf{G} - \mathbf{HK})\mathbf{x}(k) + \mathbf{HK}\boldsymbol{\Gamma}\mathbf{e}(k) \qquad (6\text{--}162)$$

Equations (6–162) and (6–156) characterize the state feedback control system where the fed back state consists of the measured portion of the state, $\mathbf{x}_a(k)$, and the observed portion of the state, $\tilde{\mathbf{x}}_b(k)$, obtained by use of the minimum-order observer. Combining Equations (6–162) and (6–156), we have

$$\left[\frac{\mathbf{x}(k + 1)}{\mathbf{e}(k + 1)}\right] = \left[\begin{array}{c:c} \mathbf{G} - \mathbf{HK} & \mathbf{HK}\boldsymbol{\Gamma} \\ \hdashline \mathbf{0} & \mathbf{G}_{bb} - \mathbf{K}_e\mathbf{G}_{ab} \end{array}\right]\left[\frac{\mathbf{x}(k)}{\mathbf{e}(k)}\right] \qquad (6\text{--}163)$$

Equation (6–163) characterizes the dynamics of the system with observed-state feedback using a minimum-order observer. The characteristic equation for the system is

$$\left|\begin{array}{c:c} z\mathbf{I} - \mathbf{G} + \mathbf{HK} & -\mathbf{HK}\boldsymbol{\Gamma} \\ \hdashline \mathbf{0} & z\mathbf{I} - \mathbf{G}_{bb} + \mathbf{K}_e\mathbf{G}_{ab} \end{array}\right|$$

$$= |z\mathbf{I} - \mathbf{G} + \mathbf{HK}||z\mathbf{I} - \mathbf{G}_{bb} + \mathbf{K}_e\mathbf{G}_{ab}| = 0 \qquad (6\text{--}164)$$

Equation (6–164) implies that the closed-loop poles of the system comprise the closed-loop poles due to pole placement [the eigenvalues of matrix $(\mathbf{G} - \mathbf{HK})$] and the closed-loop poles due to the minimum-order observer [the eigenvalues of matrix $(\mathbf{G}_{bb} - \mathbf{K}_e\mathbf{G}_{ab})$].

Example 6–11

Consider the discrete-time double integrator system defined by the equations

$$\mathbf{x}(k + 1) = \mathbf{G}\mathbf{x}(k) + \mathbf{H}u(k) \qquad (6\text{--}165)$$

$$y(k) = \mathbf{C}\mathbf{x}(k) \qquad (6\text{--}166)$$

where the sampling period T is assumed to be 0.2 sec, or $T = 0.2$, and

$$\mathbf{G} = \begin{bmatrix} 1 & T \\ 0 & 1 \end{bmatrix} = \begin{bmatrix} 1 & 0.2 \\ 0 & 1 \end{bmatrix}, \qquad \mathbf{H} = \begin{bmatrix} \dfrac{T^2}{2} \\ T \end{bmatrix} = \begin{bmatrix} 0.02 \\ 0.2 \end{bmatrix}, \qquad \mathbf{C} = \begin{bmatrix} 1 & 0 \end{bmatrix}$$

By use of the pole placement design technique, determine the state feedback gain matrix **K** to be such that the closed-loop poles of the system are located at

$$z_1 = 0.6 + j0.4, \qquad z_2 = 0.6 - j0.4$$

Assuming that the output $y(k) = x_1(k)$ is the only state variable that can be measured, design a minimum-order observer such that the error signal will exhibit a deadbeat response to an arbitrary initial error. Determine the pulse transfer function for the controller (which consists of the state feedback control and the minimum-order observer).

We shall first examine the controllability and observability of the system. Since the rank of the matrices

$$[\mathbf{H} \vdots \mathbf{GH}] = \begin{bmatrix} 0.02 & 0.06 \\ 0.2 & 0.2 \end{bmatrix}, \qquad [\mathbf{C}^* \vdots \mathbf{G}^*\mathbf{C}^*] = \begin{bmatrix} 1 & 1 \\ 0 & 0.2 \end{bmatrix}$$

is 2 in both cases, the system is completely state controllable and observable.

We shall now solve the pole placement portion of the problem. Since

$$|z\mathbf{I} - \mathbf{G}| = \begin{vmatrix} z - 1 & -0.2 \\ 0 & z - 1 \end{vmatrix} = z^2 - 2z + 1 = z^2 + a_1 z + a_2 = 0$$

we have

$$a_1 = -2, \qquad a_2 = 1$$

The desired characteristic equation is given by

$$|z\mathbf{I} - \mathbf{G} + \mathbf{HK}| = (z - 0.6 - j0.4)(z - 0.6 + j0.4) = z^2 - 1.2z + 0.52$$

$$= z^2 + \alpha_1 z + \alpha_2 = 0$$

Hence,

$$\alpha_1 = -1.2, \qquad \alpha_2 = 0.52$$

From Equation (6-65), the state feedback gain matrix **K** is obtained as follows:

$$\mathbf{K} = [\alpha_2 - a_2 \vdots \alpha_1 - a_1]\mathbf{T}^{-1} = [-0.48 \quad 0.8]\mathbf{T}^{-1} \qquad (6\text{-}167)$$

where

$$\mathbf{T} = [\mathbf{H} \vdots \mathbf{GH}]\begin{bmatrix} a_1 & 1 \\ 1 & 0 \end{bmatrix} = \begin{bmatrix} 0.02 & 0.06 \\ 0.2 & 0.2 \end{bmatrix}\begin{bmatrix} -2 & 1 \\ 1 & 0 \end{bmatrix}$$

$$= \begin{bmatrix} 0.02 & 0.02 \\ -0.2 & 0.2 \end{bmatrix}$$

and

$$\mathbf{T}^{-1} = \begin{bmatrix} 25 & -2.5 \\ 25 & 2.5 \end{bmatrix}$$

Thus, the state feedback gain matrix **K** given by Equation (6-167) becomes

$$\mathbf{K} = [-0.48 \quad 0.8]\begin{bmatrix} 25 & -2.5 \\ 25 & 2.5 \end{bmatrix} = [8 \quad 3.2]$$

The feedback control signal can then be given by

$$u(k) = -\mathbf{K}\tilde{\mathbf{x}}(k)$$

$$= -[8 \quad 3.2]\begin{bmatrix} x_1(k) \\ \tilde{x}_2(k) \end{bmatrix} = -[8 \quad 3.2]\begin{bmatrix} y(k) \\ \tilde{x}_2(k) \end{bmatrix} \qquad (6\text{-}168)$$

Next, we shall solve the observer portion of the problem. Since the state $\mathbf{x}(k)$ is a 2-vector and the output $y(k)$ is a scalar, the minimum-order observer is of the first order. Notice that

$$\left[\begin{array}{c|c} G_{aa} & G_{ab} \\ \hline G_{ba} & G_{bb} \end{array}\right] = \left[\begin{array}{c|c} 1 & 0.2 \\ \hline 0 & 1 \end{array}\right], \qquad \left[\frac{H_a}{H_b}\right] = \left[\frac{0.02}{0.2}\right]$$

Since we desire deadbeat response, the desired characteristic equation for the observer is

$$\phi(z) = z = 0$$

Referring to Ackermann's formula as given by Equation (6–153), we obtain

$$K_e = \phi(G_{bb})[G_{ab}]^{-1}[1] = (1)(0.2)^{-1}(1) = 5$$

Referring to the minimum-order observer equation given by Equation (6–149), we have

$$\tilde{\eta}(k + 1) = (G_{bb} - K_e G_{ab})\tilde{\eta}(k) + [(G_{bb} - K_e G_{ab})K_e + G_{ba} - K_e G_{aa}]y(k)$$
$$+ (H_b - K_e H_a)u(k)$$
$$= (1 - 5 \times 0.2)\tilde{\eta}(k) + [(1 - 5 \times 0.2) \times 5 + 0 - 5 \times 1]y(k)$$
$$+ (0.2 - 5 \times 0.02)u(k)$$

which can be simplified to read

$$\tilde{\eta}(k + 1) = -5y(k) + 0.1u(k) \tag{6–169}$$

Equation (6–169) defines the minimum-order observer.

The observed-state feedback control $u(k)$ is now given by

$$u(k) = -\mathbf{K}\tilde{\mathbf{x}}(k) = -8x_1(k) - 3.2\tilde{x}_2(k) = -8y(k) - 3.2\tilde{x}_2(k) \tag{6–170}$$

where, referring to Equation (6–148),

$$\tilde{x}_2(k) = K_e y(k) + \tilde{\eta}(k) = 5y(k) + \tilde{\eta}(k) \tag{6–171}$$

The block diagram for the system is shown in Figure 6–12. From Equations (6–169), (6–170), and (6–171), we obtain

$$u(k + 1) = -8y(k + 1) - 3.2[5y(k + 1) + \tilde{\eta}(k + 1)]$$
$$= -24y(k + 1) + 16y(k) - 0.32u(k)$$

or

$$u(k + 1) + 0.32u(k) = -24y(k + 1) + 16y(k)$$

By taking the z transform of this last equation, assuming zero initial conditions, we obtain

$$zU(z) + 0.32U(z) = -24zY(z) + 16Y(z)$$

The pulse transfer function of the regulator is

$$G_D(z) = -\frac{U(z)}{Y(z)} = 24\left(\frac{z - 0.6667}{z + 0.32}\right) = 24\left(\frac{1 - 0.6667z^{-1}}{1 + 0.32z^{-1}}\right) \tag{6–172}$$

By referring to Equation (5–60), the pulse transfer function of the system defined by Equations (6–165) and (6–166) can be obtained as follows:

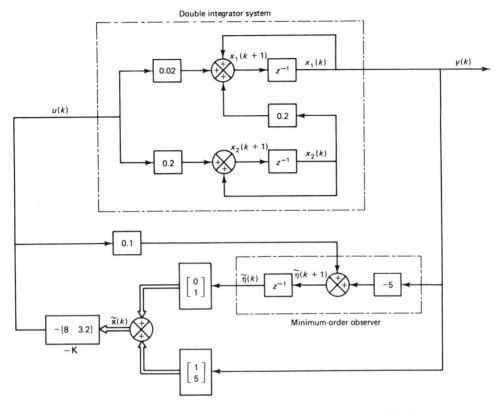

Figure 6–12   Block diagram of the system designed in Example 6–11.

$$\frac{Y(z)}{U(z)} = G_p(z) = \mathbf{C}(z\mathbf{I} - \mathbf{G})^{-1}\mathbf{H}$$

$$= [1 \quad 0]\begin{bmatrix} z - 1 & -0.2 \\ 0 & z - 1 \end{bmatrix}^{-1}\begin{bmatrix} 0.02 \\ 0.2 \end{bmatrix}$$

$$= \frac{0.02(z + 1)}{(z - 1)^2} = \frac{0.02(1 + z^{-1})z^{-1}}{(1 - z^{-1})^2} \qquad (6\text{–}173)$$

By using the pulse transfer functions of Equations (6–172) and (6–173), the block diagram of Figure 6–12 may be modified to the form shown in Figure 6–13.

Using the form given by Equation (6–164),

$$|z\mathbf{I} - \mathbf{G} + \mathbf{HK}||z - G_{bb} + K_e G_{ab}| = 0$$

we have obtained the following characteristic equation for the system:

$$(z^2 - 1.2z + 0.52)(z - 1 + 5 \times 0.2) = (z^2 - 1.2z + 0.52)z = 0 \qquad (6\text{–}174)$$

The characteristic equation for the closed-loop system shown in Figure 6–13 is

$$1 + G_p(z)G_D(z) = 0$$

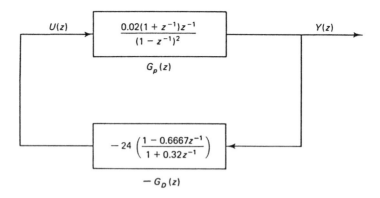

Figure 6–13 Modified form of the block diagram of the system designed in Example 6–11.

or

$$1 + \left[\frac{0.02(1 + z^{-1})z^{-1}}{(1 - z^{-1})^2}\right]\left[24\left(\frac{1 - 0.6667z^{-1}}{1 + 0.32z^{-1}}\right)\right] = 0$$

which can be written as follows:

$$1 + \left[\frac{0.02(z + 1)}{(z - 1)^2}\right]\left[24\left(\frac{z - 0.6667}{z + 0.32}\right)\right] = 0$$

And, as a matter of course, this characteristic equation can be simplified to

$$(z^2 - 1.2z + 0.52)z = 0$$

which is the same as Equation (6–174), obtained by use of Equation (6–164).

Control System With Reference Input. We shall apply the observed-state feedback method to design control systems that must follow changing reference inputs.

It is important to point out that the pole placement with observed state approach has no control over the numerator dynamics of the closed-loop system. (To control the numerator dynamics, refer to the polynomial equations approach presented in Chapter 7.) However, it is possible to modify an observed-state feedback regulator system to a control system, as shown in Figure 6–14.

As stated earlier, the pole placement part determines the desired nth-degree characteristic equation for the nth-order system. The state observer part determines the observer error characteristic equation of degree n or less. As given by Equation (6–134) or (6–164), the product of the nth-degree characteristic equation and the state observer error characteristic equation gives the characteristic equation for the entire system.

In modifying the regulator system to the control system, it is necessary to provide an adjustable gain K_0 in the input path so that the gain of the entire control system can be determined such that the steady-state output to a unit-step input is unity. This is because the pole placement with state observer modifies the gain of the entire system. Therefore, unless K_0 is properly adjusted, the system will not behave properly.

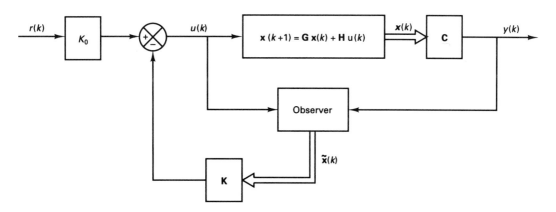

Figure 6–14 Block diagram of a control system with observed-state feedback.

Example 6–12

Modify the regulator system considered in Example 6–11 to a control system such that the output will follow the reference input. Then obtain the unit-step response and unit-ramp response of the control system. (Assume that the sampling period T is 0.2 sec, or $T = 0.2$.)

Figure 6–15 shows a possible block diagram of the control system. In this control system, it is necessary to set gain K_0 so that there will be no offset in the output to step input. (Note that if $K_0 = 1$ then the steady-state output to a unit-step input will not be equal to unity, except in special cases.)

From the block diagram the closed-loop pulse transfer function $Y(z)/R(z)$ is

$$\frac{Y(z)}{R(z)} = \frac{K_0(0.02)(z + 0.32)(z + 1)}{(z - 1)^2(z + 0.32) + 0.48(z + 1)(z - 0.6667)}$$

$$= \frac{K_0(0.02)(z + 0.32)(z + 1)}{z^3 - 1.2z^2 + 0.52z}$$

The system is of third order.

Before we examine the system behavior, it is necessary to determine the gain constant K_0. Let us assume that $R(z)$ is the z transform of the unit-step sequence. Then the steady-state output is given by

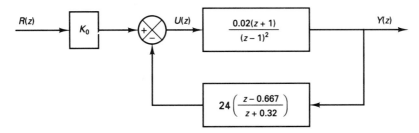

Figure 6–15 Control system obtained by modifying the block diagram shown in Figure 6–13.

$$\lim_{k \to \infty} y(k) = \lim_{z \to 1} [(1 - z^{-1})Y(z)]$$

$$= \lim_{z \to 1} \frac{z - 1}{z} \frac{K_0(0.02)(z + 0.32)(z + 1)}{z^3 - 1.2z^2 + 0.52z} \frac{z}{z - 1}$$

$$= 0.165K_0$$

We set the gain K_0 such that

$$y(\infty) = 0.165K_0 = 1$$

or

$$K_0 = 6.0606$$

By substituting $K_0 = 6.0606$ into the closed-loop pulse transfer function, we obtain

$$\frac{Y(z)}{R(z)} = \frac{0.1212z^2 + 0.16z + 0.03879}{z^3 - 1.2z^2 + 0.52z}$$

The unit-step response of this system can be obtained easily with MATLAB. A sample MATLAB program for obtaining the unit-step response is shown in MATLAB Program 6–2. The resulting unit-step response is shown in Figure 6–16.

MATLAB Program 6–2

```
num = [0  0.1212  0.1600  0.03879];
den = [1  -1.2  0.52  0];
r = ones(1,41);
k = 0: 40;
y = filter(num,den,r);
plot(k,y,'o')
v = [0  40  0  1.6];
axis(v);
grid
title('Unit-Step Response')
xlabel('k')
ylabel('y(k)')
```

Also, the unit-ramp response can be obtained by entering MATLAB Program 6–3 into the computer. The resulting unit-ramp response is shown in Figure 6–17.

The error following the unit-ramp input is obtained as follows: Noting that the sampling period T is 0.2 sec, the unit-ramp input is given by

$$R(z) = \frac{0.2z^{-1}}{(1 - z^{-1})^2}$$

Then we obtain

$$E(z) = R(z) - Y(z) = \left[1 - \frac{Y(z)}{R(z)} \right] R(z)$$

$$= \frac{z^3 - 1.3212z^2 + 0.36z - 0.03879}{z^3 - 1.2z^2 + 0.52z} \frac{0.2z}{(z - 1)^2}$$

$$= \frac{(z - 1)(z^2 - 0.3212z + 0.03879)}{z^3 - 1.2z^2 + 0.52z} \frac{0.2z}{(z - 1)^2}$$

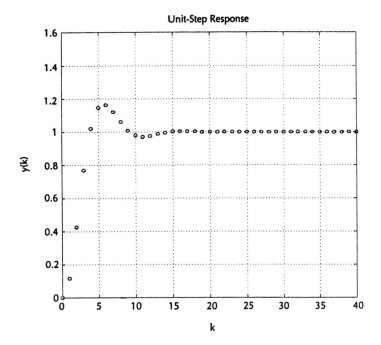

Figure 6–16   Unit-step response of the control system shown in Figure 6–15 with $K_0 = 6.0606$.

```
MATLAB Program 6–3

num = [0  0.1212  0.1600  0.03879];
den = [1  -1.2  0.52  0];
k = 0:20;
r = [0.2*k];
y = filter(num,den,r );
plot(k,y,'o',k,y,'-',k,0.2*k,'--')
v = [0  20  0  4];
axis(v);
grid
title('Unit-Ramp Response')
xlabel('k')
ylabel('y(k)')
```

Thus,

$$\lim_{k \to \infty} e(k) = \lim_{z \to 1} \frac{z-1}{z} \frac{(z-1)(z^2 - 0.3212z + 0.03879)}{z^3 - 1.2z^2 + 0.52z} \frac{0.2z}{(z-1)^2}$$

$$= 0.4485$$

The steady-state error in following the unit-ramp input is 0.4485. (See Figure 6–17.)

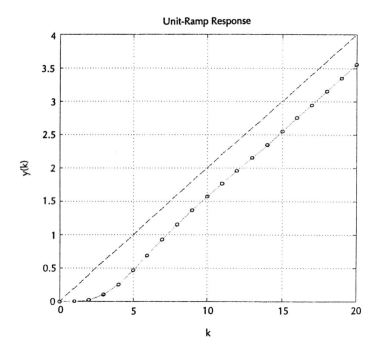

Unit-Ramp Response

Figure 6–17   Unit-ramp response of the control system shown in Figure 6–15 with $K_0 = 6.0606$.

6–7 SERVO SYSTEMS

In the servo system it is generally required that the system have one or more integrators within the closed loop. (Unless the plant to be controlled has an integrating property, it is necessary to add one or more integrators within the loop to eliminate steady-state error to step inputs.)

One way to introduce an integrator in the mathematical model of a closed-loop system is to introduce a new state vector that integrates the difference between the command vector **r** and the output vector **y**. Figure 6–18 shows a possible block diagram configuration for a servo system with state feedback and integral control. The integral controller consists of m integrating elements, one for each command input component. (The command input is an m-vector and has m components.) The integrator can be included as part of the pole placement formulation that was presented in Section 6–5.

Servo System With Integrator.   Consider the servo system shown in Figure 6–18. The plant is assumed to be completely state controllable and completely observable. Assume that the plant does not have an integrator. The plant state equation and output equation are

$$\mathbf{x}(k+1) = \mathbf{G}\mathbf{x}(k) + \mathbf{H}\mathbf{u}(k) \qquad (6\text{–}175)$$

$$\mathbf{y}(k) = \mathbf{C}\mathbf{x}(k) \qquad (6\text{–}176)$$

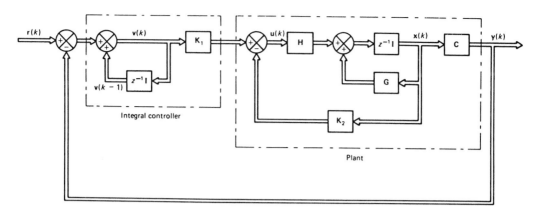

Figure 6–18 Servo system with state feedback and integral control.

where

$$\mathbf{x}(k) = \text{plant state vector } (n\text{-vector})$$

$$\mathbf{u}(k) = \text{control vector } (m\text{-vector})$$

$$\mathbf{y}(k) = \text{output vector } (m\text{-vector})$$

$$\mathbf{G} = n \times n \text{ matrix}$$

$$\mathbf{H} = n \times m \text{ matrix}$$

$$\mathbf{C} = m \times n \text{ matrix}$$

(Note that in the present analysis we assume that the dimensions of the output vector and the control vector are the same; they are both m-vectors.) The integrator state equation is

$$\mathbf{v}(k) = \mathbf{v}(k-1) + \mathbf{r}(k) - \mathbf{y}(k) \tag{6–177}$$

where

$$\mathbf{v}(k) = \text{actuating error vector } (m\text{-vector})$$

$$\mathbf{r}(k) = \text{command input vector } (m\text{-vector})$$

Equation (6–177) can be rewritten as follows:

$$\mathbf{v}(k+1) = \mathbf{v}(k) + \mathbf{r}(k+1) - \mathbf{y}(k+1)$$

$$= \mathbf{v}(k) + \mathbf{r}(k+1) - \mathbf{C}[\mathbf{G}\mathbf{x}(k) + \mathbf{H}\mathbf{u}(k)]$$

$$= -\mathbf{C}\mathbf{G}\mathbf{x}(k) + \mathbf{v}(k) - \mathbf{C}\mathbf{H}\mathbf{u}(k) + \mathbf{r}(k+1) \tag{6–178}$$

The control vector $\mathbf{u}(k)$ is given by

$$\mathbf{u}(k) = -\mathbf{K}_2\mathbf{x}(k) + \mathbf{K}_1\mathbf{v}(k) \tag{6–179}$$

In our servo system the system configuration is specified in Figure 6–18. Our design parameters are matrices $\mathbf{K}_1$ and $\mathbf{K}_2$.

In what follows we shall discuss the procedure for determining matrices $\mathbf{K}_1$ and $\mathbf{K}_2$ such that the system has the desired closed-loop poles. From Equations (6–175), (6–178), and (6–179), we obtain

$$\begin{aligned}
\mathbf{u}(k+1) &= -\mathbf{K}_2\mathbf{x}(k+1) + \mathbf{K}_1\mathbf{v}(k+1)\\
&= (\mathbf{K}_2 - \mathbf{K}_2\mathbf{G} - \mathbf{K}_1\mathbf{CG})\mathbf{x}(k)\\
&\quad + (\mathbf{I}_m - \mathbf{K}_2\mathbf{H} - \mathbf{K}_1\mathbf{CH})\mathbf{u}(k) + \mathbf{K}_1\mathbf{r}(k+1)
\end{aligned} \qquad (6\text{–}180)$$

Noting that $\mathbf{u}(k)$ is a linear combination of state vectors $\mathbf{x}(k)$ and $\mathbf{v}(k)$, define a new state vector consisting of $\mathbf{x}(k)$ and $\mathbf{u}(k)$ [rather than $\mathbf{x}(k)$ and $\mathbf{v}(k)$]. Then we obtain from Equations (6–175) and (6–180) the following state equation:

$$\begin{bmatrix} \mathbf{x}(k+1) \\ \mathbf{u}(k+1) \end{bmatrix}
= \begin{bmatrix} \mathbf{G} & \mathbf{H} \\ \mathbf{K}_2 - \mathbf{K}_2\mathbf{G} - \mathbf{K}_1\mathbf{CG} & \mathbf{I}_m - \mathbf{K}_2\mathbf{H} - \mathbf{K}_1\mathbf{CH} \end{bmatrix} \begin{bmatrix} \mathbf{x}(k) \\ \mathbf{u}(k) \end{bmatrix} + \begin{bmatrix} \mathbf{0} \\ \mathbf{K}_1 \end{bmatrix}\mathbf{r}(k+1) \qquad (6\text{–}181)$$

The output equation, Equation (6–176), can be written as follows:

$$\mathbf{y}(k) = [\mathbf{C} \quad \mathbf{0}]\begin{bmatrix} \mathbf{x}(k) \\ \mathbf{u}(k) \end{bmatrix} \qquad (6\text{–}182)$$

Note that the closed-loop poles of the system are determined by the system itself and do not depend on the command input $\mathbf{r}(k)$. The eigenvalues of the state matrix in Equation (6–181) determine the closed-loop poles of the system.

To apply the pole placement technique of Section 6–5 directly to the design of the present servo system, consider the case where the command vector $\mathbf{r}(k)$ is a constant vector (step input) so that

$$\mathbf{r}(k) = \mathbf{r}$$

Then Equation (6–181) becomes

$$\begin{bmatrix} \mathbf{x}(k+1) \\ \mathbf{u}(k+1) \end{bmatrix}
= \begin{bmatrix} \mathbf{G} & \mathbf{H} \\ \mathbf{K}_2 - \mathbf{K}_2\mathbf{G} - \mathbf{K}_1\mathbf{CG} & \mathbf{I}_m - \mathbf{K}_2\mathbf{H} - \mathbf{K}_1\mathbf{CH} \end{bmatrix} \begin{bmatrix} \mathbf{x}(k) \\ \mathbf{u}(k) \end{bmatrix} + \begin{bmatrix} \mathbf{0} \\ \mathbf{K}_1\mathbf{r} \end{bmatrix} \qquad (6\text{–}183)$$

Notice that, for the step input, $\mathbf{x}(k)$, $\mathbf{u}(k)$, and $\mathbf{v}(k)$ approach the constant vector values $\mathbf{x}(\infty)$, $\mathbf{u}(\infty)$, and $\mathbf{v}(\infty)$, respectively. Thus, from Equation (6–177), we obtain the following equation at steady state:

$$\mathbf{v}(\infty) = \mathbf{v}(\infty) + \mathbf{r} - \mathbf{y}(\infty)$$

or

$$\mathbf{y}(\infty) = \mathbf{r}$$

There is no steady-state error in the output when the command input is a step vector. Also, at steady state Equation (6–183) becomes

$$\begin{bmatrix} \mathbf{x}(\infty) \\ \mathbf{u}(\infty) \end{bmatrix}
= \begin{bmatrix} \mathbf{G} & \mathbf{H} \\ \mathbf{K}_2 - \mathbf{K}_2\mathbf{G} - \mathbf{K}_1\mathbf{CG} & \mathbf{I}_m - \mathbf{K}_2\mathbf{H} - \mathbf{K}_1\mathbf{CH} \end{bmatrix} \begin{bmatrix} \mathbf{x}(\infty) \\ \mathbf{u}(\infty) \end{bmatrix} + \begin{bmatrix} \mathbf{0} \\ \mathbf{K}_1\mathbf{r} \end{bmatrix} \qquad (6\text{–}184)$$

Let us define the error vectors by

$$\mathbf{x}_e(k) = \mathbf{x}(k) - \mathbf{x}(\infty)$$

$$\mathbf{u}_e(k) = \mathbf{u}(k) - \mathbf{u}(\infty)$$

Then, subtracting Equation (6–184) from Equation (6–183), we obtain

$$\begin{bmatrix} \mathbf{x}_e(k+1) \\ \mathbf{u}_e(k+1) \end{bmatrix} = \begin{bmatrix} \mathbf{G} & \mathbf{H} \\ \mathbf{K}_2 - \mathbf{K}_2\mathbf{G} - \mathbf{K}_1\mathbf{CG} & \mathbf{I}_m - \mathbf{K}_2\mathbf{H} - \mathbf{K}_1\mathbf{CH} \end{bmatrix} \begin{bmatrix} \mathbf{x}_e(k) \\ \mathbf{u}_e(k) \end{bmatrix} \qquad (6\text{–}185)$$

The dynamics of the system are determined by the eigenvalues of the state matrix appearing in Equation (6–185). Equation (6–185) can be modified to read

$$\begin{bmatrix} \mathbf{x}_e(k+1) \\ \mathbf{u}_e(k+1) \end{bmatrix} = \begin{bmatrix} \mathbf{G} & \mathbf{H} \\ \mathbf{0} & \mathbf{0} \end{bmatrix} \begin{bmatrix} \mathbf{x}_e(k) \\ \mathbf{u}_e(k) \end{bmatrix} + \begin{bmatrix} \mathbf{0} \\ \mathbf{I}_m \end{bmatrix} \mathbf{w}(k) \qquad (6\text{–}186)$$

where

$$\mathbf{w}(k) = [\mathbf{K}_2 - \mathbf{K}_2\mathbf{G} - \mathbf{K}_1\mathbf{CG} \vdots \mathbf{I}_m - \mathbf{K}_2\mathbf{H} - \mathbf{K}_1\mathbf{CH}] \begin{bmatrix} \mathbf{x}_e(k) \\ \mathbf{u}_e(k) \end{bmatrix} \qquad (6\text{–}187)$$

If we define

$$\boldsymbol{\xi}(k) = \begin{bmatrix} \mathbf{x}_e(k) \\ \mathbf{u}_e(k) \end{bmatrix} = (n+m)\text{-vector}$$

$$\hat{\mathbf{G}} = \begin{bmatrix} \mathbf{G} & \mathbf{H} \\ \mathbf{0} & \mathbf{0} \end{bmatrix} = (n+m) \times (n+m) \text{ matrix}$$

$$\hat{\mathbf{H}} = \begin{bmatrix} \mathbf{0} \\ \mathbf{I}_m \end{bmatrix} = (n+m) \times m \text{ matrix}$$

$$\hat{\mathbf{K}} = -[\mathbf{K}_2 - \mathbf{K}_2\mathbf{G} - \mathbf{K}_1\mathbf{CG} \vdots \mathbf{I}_m - \mathbf{K}_2\mathbf{H} - \mathbf{K}_1\mathbf{CH}]$$

$$= m \times (n+m) \text{ matrix} \qquad (6\text{–}188)$$

then Equations (6–186) and (6–187) become, respectively,

$$\boldsymbol{\xi}(k+1) = \hat{\mathbf{G}}\boldsymbol{\xi}(k) + \hat{\mathbf{H}}\mathbf{w}(k) \qquad (6\text{–}189)$$

and

$$\mathbf{w}(k) = -\hat{\mathbf{K}}\boldsymbol{\xi}(k) \qquad (6\text{–}190)$$

Notice that the controllability matrix for the system defined by Equation (6–189) is

$$[\hat{\mathbf{H}} \vdots \hat{\mathbf{G}}\hat{\mathbf{H}} \vdots \cdots \vdots \hat{\mathbf{G}}^{n+m-1}\hat{\mathbf{H}}] = (n+m) \times m(n+m) \text{ matrix}$$

In terms of **G** and **H**, this controllability matrix can be written as follows:

$$[\hat{\mathbf{H}} \vdots \hat{\mathbf{G}}\hat{\mathbf{H}} \vdots \cdots \vdots \hat{\mathbf{G}}^{n+m-1}\hat{\mathbf{H}}] = \begin{bmatrix} \mathbf{0} & \mathbf{H} & \mathbf{GH} & \cdots & \mathbf{G}^{n-1}\mathbf{H} & \cdots & \mathbf{G}^{n+m-2}\mathbf{H} \\ \mathbf{I}_m & \mathbf{0} & \mathbf{0} & \cdots & \mathbf{0} & \cdots & \mathbf{0} \end{bmatrix} \qquad (6\text{–}191)$$

Since the plant state equation given by Equation (6–175) is assumed to be completely state controllable, the rank of the matrix

$$[\mathbf{H} \vdots \mathbf{GH} \vdots \cdots \vdots \mathbf{G}^{n-1}\mathbf{H}]$$

is n. Hence, the rank of the matrix given by Equation (6–191) is $n + m$. Consequently, if the plant is completely state controllable, then the system defined by Equation (6–189) is completely state controllable and therefore the pole placement technique discussed in Section 6–5 applies to this case.

Once the desired closed-loop poles are specified, matrix $\hat{\mathbf{K}}$ can be determined by the pole placement technique. Using matrix $\hat{\mathbf{K}}$ thus determined, we can obtain matrices $\mathbf{K}_1$ and $\mathbf{K}_2$ as follows. First, note that

$$[\mathbf{K}_2 \vdots \mathbf{K}_1]\left[\begin{array}{c|c} \mathbf{G} - \mathbf{I}_n & \mathbf{H} \\ \hline \mathbf{CG} & \mathbf{CH} \end{array}\right] = [\mathbf{K}_2\mathbf{G} - \mathbf{K}_2 + \mathbf{K}_1\mathbf{CG} \vdots \mathbf{K}_2\mathbf{H} + \mathbf{K}_1\mathbf{CH}] \qquad (6\text{–}192)$$

Then, from Equations (6–188) and (6–192), we have

$$\hat{\mathbf{K}} = [\mathbf{K}_2\mathbf{G} - \mathbf{K}_2 + \mathbf{K}_1\mathbf{CG} \vdots -\mathbf{I}_m + \mathbf{K}_2\mathbf{H} + \mathbf{K}_1\mathbf{CH}]$$

$$= [\mathbf{K}_2 \vdots \mathbf{K}_1]\left[\begin{array}{c|c} \mathbf{G} - \mathbf{I}_n & \mathbf{H} \\ \hline \mathbf{CG} & \mathbf{CH} \end{array}\right] + [\mathbf{0} \vdots -\mathbf{I}_m]$$

Hence, we obtain

$$[\mathbf{K}_2 \vdots \mathbf{K}_1]\left[\begin{array}{c|c} \mathbf{G} - \mathbf{I}_n & \mathbf{H} \\ \hline \mathbf{CG} & \mathbf{CH} \end{array}\right] = \hat{\mathbf{K}} + [\mathbf{0} \vdots \mathbf{I}_m] \qquad (6\text{–}193)$$

The desired matrices $\mathbf{K}_1$ and $\mathbf{K}_2$ may be determined from Equation (6–193).

It is noted that, when $\mathbf{u}(k)$ is an m-vector and $m > 1$, matrix $\hat{\mathbf{K}}$ is not unique. Consequently, more than one set of matrices $\mathbf{K}_1$ and $\mathbf{K}_2$ can be determined. (Each possible $\hat{\mathbf{K}}$ yields a set of matrices $\mathbf{K}_1$ and $\mathbf{K}_2$.) In general, the set of $\mathbf{K}_1$ and $\mathbf{K}_2$ that gives the best overall system performance must be chosen.

Finally, it is noted that, if not all state variables are measurable, then we need to substitute the observed-state variables for the unmeasurable state variables for state feedback purposes. (Also, if the measured state variables are contaminated by noises and therefore are not accurate, then we prefer to use the observed-state variables, rather than the actual state variables, for state feedback purposes.) Figure 6–19 shows a block diagram for the servo system with state feedback where the observed state is used in place of the actual state.

Example 6–13

Consider the digital control of a plant by use of state feedback and integral control. Assume that the system configuration is the same as that shown in Figure 6–18. Assume also that the pulse transfer function of the plant is

$$\frac{Y(z)}{U(z)} = \frac{z^{-2} + 0.5z^{-3}}{1 - z^{-1} + 0.01z^{-2} + 0.12z^{-3}} \qquad (6\text{–}194)$$

where $Y(z)$ and $U(z)$ are the z transforms of the plant output $y(k)$ and plant input (control signal) $u(k)$, respectively.

Determine an integral gain constant K_1 and a state feedback gain matrix $\mathbf{K}_2$ such that the response to a unit-step command input is deadbeat. Assuming that not all state variables are available for direct measurement and using the system configuration shown in Figure 6–19 as an example for a block diagram of a system with a state observer, design a state observer such that the observed state approaches the true state as fast as possible.

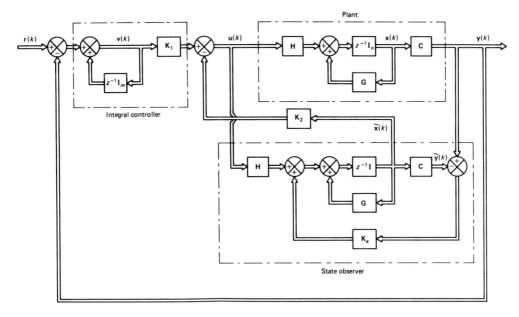

Figure 6-19   Servo system with observed-state feedback.

We shall first obtain a state-space representation for the plant pulse transfer function. By comparing the given pulse transfer function with the standard form

$$\frac{Y(z)}{U(z)} = \frac{b_0 + b_1 z^{-1} + b_2 z^{-2} + b_3 z^{-3}}{1 + a_1 z^{-1} + a_2 z^{-2} + a_3 z^{-3}}$$

we find

$$b_0 = 0, \qquad b_1 = 0, \qquad b_2 = 1, \qquad b_3 = 0.5$$

$$a_1 = -1, \qquad a_2 = 0.01, \qquad a_3 = 0.12$$

Then, by referring to Equations (5-8) and (5-9), we can obtain the following state-space equations for the plant:

$$\mathbf{x}(k + 1) = \mathbf{G}\mathbf{x}(k) + \mathbf{H}u(k) \tag{6-195}$$

$$y(k) = \mathbf{C}\mathbf{x}(k) \tag{6-196}$$

where

$$\mathbf{G} = \begin{bmatrix} 0 & 1 & 0 \\ 0 & 0 & 1 \\ -0.12 & -0.01 & 1 \end{bmatrix}, \qquad \mathbf{H} = \begin{bmatrix} 0 \\ 0 \\ 1 \end{bmatrix}, \qquad \mathbf{C} = [0.5 \quad 1 \quad 0]$$

Note that this plant is completely state controllable and completely observable.

Determination of Integral Gain Constant K_1 and State Feedback Gain Matrix $\mathbf{K}_2$ for Deadbeat Response.   We shall now determine the integral gain constant K_1 and the state feedback gain matrix $\mathbf{K}_2$. In the present system we require the response to the step command input to be deadbeat. (Thus, we must place the closed-loop poles of the system at the origin.)

Referring to Equations (6–189) and (6–190), we have

$$\xi(k + 1) = \hat{\mathbf{G}}\xi(k) + \hat{\mathbf{H}}w(k)$$

$$w(k) = -\hat{\mathbf{K}}\xi(k)$$

where

$$\hat{\mathbf{G}} = \left[\begin{array}{c|c} \mathbf{G} & \mathbf{H} \\ \hline \mathbf{0} & 0 \end{array}\right] = \left[\begin{array}{ccc|c} 0 & 1 & 0 & 0 \\ 0 & 0 & 1 & 0 \\ -0.12 & -0.01 & 1 & 1 \\ \hline 0 & 0 & 0 & 0 \end{array}\right]$$

$$\hat{\mathbf{H}} = \left[\begin{array}{c} 0 \\ 0 \\ 0 \\ \hline 1 \end{array}\right]$$

Our problem here is to determine matrix $\hat{\mathbf{K}}$ so that the closed-loop poles of the system are at the origin, or the desired characteristic equation is

$$z^4 = 0$$

By using the pole placement technique discussed in Section 6–5, matrix $\hat{\mathbf{K}}$ can be determined easily. Referring to Ackermann's formula as given by Equation (6–71), we obtain

$$\hat{\mathbf{K}} = [0 \quad 0 \quad 0 \quad 1][\hat{\mathbf{H}} \;\vdots\; \hat{\mathbf{G}}\hat{\mathbf{H}} \;\vdots\; \hat{\mathbf{G}}^2\hat{\mathbf{H}} \;\vdots\; \hat{\mathbf{G}}^3\hat{\mathbf{H}}]^{-1} \phi(\hat{\mathbf{G}})$$

where

$$\phi(\hat{\mathbf{G}}) = \hat{\mathbf{G}}^4$$

Thus,

$$\hat{\mathbf{K}} = [0 \quad 0 \quad 0 \quad 1] \left[\begin{array}{cccc} 0 & 0 & 0 & 1 \\ 0 & 0 & 1 & 1 \\ 0 & 1 & 1 & 0.99 \\ 1 & 0 & 0 & 0 \end{array}\right]^{-1} \left[\begin{array}{cccc} 0 & 1 & 0 & 0 \\ 0 & 0 & 1 & 0 \\ -0.12 & -0.01 & 1 & 1 \\ 0 & 0 & 0 & 0 \end{array}\right]^4$$

$$= [0 \quad 0 \quad 0 \quad 1] \left[\begin{array}{cccc} 0 & 0 & 0 & 1 \\ 0.01 & -1 & 1 & 0 \\ -1 & 1 & 0 & 0 \\ 1 & 0 & 0 & 0 \end{array}\right] \left[\begin{array}{cccc} -0.12 & -0.13 & 0.99 & 1 \\ -0.1188 & -0.1299 & 0.86 & 0.99 \\ -0.1032 & -0.1274 & 0.7301 & 0.86 \\ 0 & 0 & 0 & 0 \end{array}\right]$$

$$= [-0.12 \quad -0.13 \quad 0.99 \quad 1] \tag{6–197}$$

Equation (6–197) gives the matrix $\hat{\mathbf{K}}$.

The desired integral gain constant K_1 and the state feedback gain matrix $\mathbf{K}_2$ are obtained from Equation (6–193). Noting that

$$\left[\begin{array}{c|c} \mathbf{G} - \mathbf{I}_n & \mathbf{H} \\ \hline \mathbf{CG} & \mathbf{CH} \end{array}\right] = \left[\begin{array}{ccc|c} -1 & 1 & 0 & 0 \\ 0 & -1 & 1 & 0 \\ -0.12 & -0.01 & 0 & 1 \\ \hline 0 & 0.5 & 1 & 0 \end{array}\right]$$

is nonsingular (to see this fact, use row operations and column operations and make the matrix a triangular matrix), we obtain

$$[\mathbf{K}_2 \vdots K_1] = [\hat{\mathbf{K}} + [0 \vdots 1]]\left[\frac{\mathbf{G} - \mathbf{I}_3 \mid \mathbf{H}}{\mathbf{CG} \mid \mathbf{CH}}\right]^{-1}$$

$$= [-0.12 \quad -0.13 \quad 0.99 \vdots 2]\begin{bmatrix} -1 & 1 & 0 & 0 \\ 0 & -1 & 1 & 0 \\ -0.12 & -0.01 & 0 & 1 \\ \hline 0 & 0.5 & 1 & 0 \end{bmatrix}^{-1}$$

$$= [-0.12 \quad -0.13 \quad 0.99 \vdots 2]\begin{bmatrix} -1 & -\frac{2}{3} & 0 & \frac{1}{1.5} \\ 0 & -\frac{2}{3} & 0 & \frac{1}{1.5} \\ 0 & \frac{1}{3} & 0 & \frac{1}{1.5} \\ \hline -0.12 & -\frac{0.26}{3} & 1 & \frac{0.13}{1.5} \end{bmatrix}$$

$$= [-0.12 \quad 0.3233 \quad 2 \vdots 0.6667] \tag{6–198}$$

From Equation (6–198) we obtain the integral gain constant K_1:

$$K_1 = 0.6667 = \tfrac{2}{3} \tag{6–199}$$

The state feedback gain matrix $\mathbf{K}_2$ is given by

$$\mathbf{K}_2 = [-0.12 \quad 0.3233 \quad 2] \tag{6–200}$$

Determining Output $y(k)$.   Next, let us determine the output $y(k)$. From Equation (6–196) we have

$$y(k) = \mathbf{Cx}(k) = [0.5 \quad 1 \quad 0]\begin{bmatrix} x_1(k) \\ x_2(k) \\ x_3(k) \end{bmatrix}$$

To obtain output $y(k)$, we shall first determine the state vector $\mathbf{x}(k)$ and signal $v(k)$. From Figure 6–18, we have

$$\mathbf{x}(k + 1) = \mathbf{Gx}(k) + \mathbf{H}u(k) \tag{6–201}$$

$$y(k) = \mathbf{Cx}(k) \tag{6–202}$$

$$v(k) = v(k - 1) + r(k) - y(k) \tag{6–203}$$

$$u(k) = -\mathbf{K}_2\mathbf{x}(k) + K_1 v(k) \tag{6–204}$$

Hence, from Equations (6–201) and (6–204) we obtain

$$\mathbf{x}(k + 1) = \mathbf{Gx}(k) + \mathbf{H}u(k)$$

$$= (\mathbf{G} - \mathbf{HK}_2)\mathbf{x}(k) + \mathbf{H}K_1 v(k) \tag{6–205}$$

Also, from Equations (6–202), (6–203), and (6–205) we get

$$v(k + 1) = v(k) + r(k + 1) - y(k + 1)$$

$$= v(k) + r(k + 1) - \mathbf{Cx}(k + 1)$$

$$= v(k) + r(k + 1) - \mathbf{C}[(\mathbf{G} - \mathbf{HK}_2)\mathbf{x}(k) + \mathbf{H}K_1 v(k)]$$

$$= -(\mathbf{CG} - \mathbf{CHK}_2)\mathbf{x}(k) + (1 - \mathbf{CH}K_1)v(k) + r(k + 1) \tag{6–206}$$

Combining Equation (6–205) and (6–206), we get

$$\begin{bmatrix} \mathbf{x}(k + 1) \\ v(k + 1) \end{bmatrix} = \left[\frac{\mathbf{G} - \mathbf{HK}_2 \mid \mathbf{H}K_1}{-\mathbf{CG} + \mathbf{CHK}_2 \mid 1 - \mathbf{CH}K_1}\right]\begin{bmatrix} \mathbf{x}(k) \\ v(k) \end{bmatrix} + \begin{bmatrix} 0 \\ 1 \end{bmatrix}r(k + 1) \tag{6–207}$$

which can be rewritten as

$$
\begin{bmatrix} x_1(k+1) \\ x_2(k+1) \\ x_3(k+1) \\ v(k+1) \end{bmatrix} = \begin{bmatrix} 0 & 1 & 0 & 0 \\ 0 & 0 & 1 & 0 \\ 0 & -\frac{1}{3} & -1 & \frac{2}{3} \\ 0 & -\frac{1}{2} & -1 & 1 \end{bmatrix} \begin{bmatrix} x_1(k) \\ x_2(k) \\ x_3(k) \\ v(k) \end{bmatrix} + \begin{bmatrix} 0 \\ 0 \\ 0 \\ 1 \end{bmatrix} r(k+1) \qquad (6\text{--}208)
$$

Since the command input $r(k)$ is a unit-step input, we have

$$
r(k) = 1, \qquad k = 0, 1, 2, \ldots
$$

Let us assume that the initial state is

$$
\begin{bmatrix} x_1(0) \\ x_2(0) \\ x_3(0) \\ v(0) \end{bmatrix} = \begin{bmatrix} a \\ b \\ c \\ d \end{bmatrix}
$$

where a, b, c, and d are arbitrary. Then, from Equation (6–208), we have

$$
\begin{bmatrix} x_1(1) \\ x_2(1) \\ x_3(1) \\ v(1) \end{bmatrix} = \begin{bmatrix} 0 & 1 & 0 & 0 \\ 0 & 0 & 1 & 0 \\ 0 & -\frac{1}{3} & -1 & \frac{2}{3} \\ 0 & -\frac{1}{2} & -1 & 1 \end{bmatrix} \begin{bmatrix} a \\ b \\ c \\ d \end{bmatrix} + \begin{bmatrix} 0 \\ 0 \\ 0 \\ 1 \end{bmatrix} [1] = \begin{bmatrix} b \\ c \\ -\frac{1}{3}b - c + \frac{2}{3}d \\ -\frac{1}{2}b - c + d + 1 \end{bmatrix}
$$

Similarly,

$$
\begin{bmatrix} x_1(2) \\ x_2(2) \\ x_3(2) \\ v(2) \end{bmatrix} = \begin{bmatrix} c \\ -\frac{1}{3}b - c + \frac{2}{3}d \\ \frac{2}{3} \\ -\frac{1}{6}b - \frac{1}{2}c + \frac{1}{3}d + 2 \end{bmatrix}
$$

$$
\begin{bmatrix} x_1(3) \\ x_2(3) \\ x_3(3) \\ v(3) \end{bmatrix} = \begin{bmatrix} -\frac{1}{3}b - c + \frac{2}{3}d \\ \frac{2}{3} \\ \frac{2}{3} \\ \frac{7}{3} \end{bmatrix}
$$

and

$$
\begin{bmatrix} x_1(k) \\ x_2(k) \\ x_3(k) \\ v(k) \end{bmatrix} = \begin{bmatrix} \frac{2}{3} \\ \frac{2}{3} \\ \frac{2}{3} \\ \frac{7}{3} \end{bmatrix}, \qquad k = 4, 5, 6, \ldots
$$

The output $y(k)$ is obtained as follows:

$$
y(0) = [0.5 \quad 1 \quad 0] \begin{bmatrix} x_1(0) \\ x_2(0) \\ x_3(0) \end{bmatrix} = [0.5 \quad 1 \quad 0] \begin{bmatrix} a \\ b \\ c \end{bmatrix} = \tfrac{1}{2}a + b
$$

Similarly,

$$
y(1) = \tfrac{1}{2}b + c
$$

$$
y(2) = -\tfrac{1}{3}b - \tfrac{1}{2}c + \tfrac{2}{3}d
$$

$$
y(3) = -\tfrac{1}{6}b - \tfrac{1}{2}c + \tfrac{1}{3}d + \tfrac{2}{3}
$$

$$
y(k) = 1, \qquad k = 4, 5, 6, \ldots
$$

Notice that

$$u(k) = -\mathbf{K}_2\,\mathbf{x}(k) + K_1\,v(k)$$

$$= -[-0.12 \quad 0.3233 \quad 2]\begin{bmatrix} \frac{2}{3} \\ \frac{2}{3} \\ \frac{2}{3} \\ \frac{2}{3} \end{bmatrix} + (\tfrac{2}{3})(\tfrac{7}{3}) = 0.08670$$

where $k = 4, 5, 6, \dots$. Since $u(t)$ for $t \geq 4T$ (where T is the sampling period) is constant, there is no intersampling oscillation in the output. Thus, the response of the system is deadbeat.

Note that the output $y(k)$ reaches unity in at most four sampling periods and will stay there in the absence of disturbances or new command inputs. [See, for example, the sample unit-step response sequence shown in Figure 6–20(a).] Under special initial conditions, for example, $a = b = c = 0$ and $d = 1$, the output reaches unity in three sampling periods and stays there, or $y(k) = 1$ for $k = 3, 4, 5, \dots$ [see Figure 6–20(b)].

Design of the State Observer.   Next, we shall design a state observer for the system. Since the plant output $y(k)$ is measurable, let us design a minimum-order

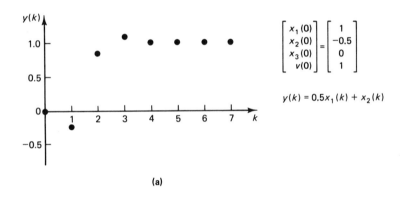

(a)

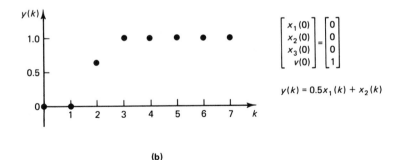

(b)

Figure 6–20   Sample unit-step response sequences for the servo system with actual-(measured-) state feedback and integral control designed in Example 6–13.

observer. Let us assume that we desire the deadbeat response. In the present system, output matrix $\mathbf{C}$ is given by

$$\mathbf{C} = [0.5 \quad 1 \quad 0]$$

To change the output matrix $\mathbf{C}$ from $[0.5 \quad 1 \quad 0]$ to $[1 \quad 0 \quad 0]$, let us make the following transformation:

$$\mathbf{x}(k) = \mathbf{T}\boldsymbol{\xi}(k) \tag{6-209}$$

where

$$\mathbf{T} = \begin{bmatrix} 0 & 0 & 1 \\ 1 & 0 & -0.5 \\ 0 & 1 & 0 \end{bmatrix} \tag{6-210}$$

Note that

$$\mathbf{T}^{-1} = \begin{bmatrix} 0.5 & 1 & 0 \\ 0 & 0 & 1 \\ 1 & 0 & 0 \end{bmatrix} \tag{6-211}$$

Then the plant state equations become

$$\boldsymbol{\xi}(k + 1) = \mathbf{T}^{-1}\mathbf{G}\mathbf{T}\boldsymbol{\xi}(k) + \mathbf{T}^{-1}\mathbf{H}u(k) \tag{6-212}$$

$$y(k) = \mathbf{C}\mathbf{T}\boldsymbol{\xi}(k) \tag{6-213}$$

where

$$\mathbf{T}^{-1}\mathbf{G}\mathbf{T} = \begin{bmatrix} 0.5 & 1 & 0 \\ 0 & 0 & 1 \\ 1 & 0 & 0 \end{bmatrix} \begin{bmatrix} 0 & 1 & 0 \\ 0 & 0 & 1 \\ -0.12 & -0.01 & 1 \end{bmatrix} \begin{bmatrix} 0 & 0 & 1 \\ 1 & 0 & -0.5 \\ 0 & 1 & 0 \end{bmatrix}$$

$$= \begin{bmatrix} 0.5 & 1 & -0.25 \\ -0.01 & 1 & -0.115 \\ 1 & 0 & -0.5 \end{bmatrix} = \hat{\mathbf{G}}$$

$$\mathbf{T}^{-1}\mathbf{H} = \begin{bmatrix} 0.5 & 1 & 0 \\ 0 & 0 & 1 \\ 1 & 0 & 0 \end{bmatrix} \begin{bmatrix} 0 \\ 0 \\ 1 \end{bmatrix} = \begin{bmatrix} 0 \\ 1 \\ 0 \end{bmatrix} = \hat{\mathbf{H}}$$

$$\mathbf{C}\mathbf{T} = [0.5 \quad 1 \quad 0] \begin{bmatrix} 0 & 0 & 1 \\ 1 & 0 & -0.5 \\ 0 & 1 & 0 \end{bmatrix} = [1 \quad 0 \quad 0] = \hat{\mathbf{C}}$$

The transformed system equations are as follows:

$$\begin{bmatrix} \xi_1(k + 1) \\ \xi_2(k + 1) \\ \xi_3(k + 1) \end{bmatrix} = \begin{bmatrix} 0.5 & 1 & -0.25 \\ -0.01 & 1 & -0.115 \\ 1 & 0 & -0.5 \end{bmatrix} \begin{bmatrix} \xi_1(k) \\ \xi_2(k) \\ \xi_3(k) \end{bmatrix} + \begin{bmatrix} 0 \\ 1 \\ 0 \end{bmatrix} u(k) \tag{6-214}$$

$$y(k) = [1 \vdots 0 \quad 0] \begin{bmatrix} \xi_1(k) \\ \xi_2(k) \\ \xi_3(k) \end{bmatrix} \tag{6-215}$$

Since only one state variable can be measured, we need to observe two state variables. Hence, the order of the minimum-order observer is 2. From Equations (6-138) and (6-214), we have

$$\hat{\mathbf{G}}_{bb} - \mathbf{K}_e \hat{\mathbf{G}}_{ab} = \begin{bmatrix} 1 & -0.115 \\ 0 & -0.5 \end{bmatrix} - \begin{bmatrix} k_{e_1} \\ k_{e_2} \end{bmatrix} \begin{bmatrix} 1 & -0.25 \end{bmatrix}$$

$$= \begin{bmatrix} 1 - k_{e_1} & -0.115 + 0.25k_{e_1} \\ -k_{e_2} & -0.5 + 0.25k_{e_2} \end{bmatrix}$$

The observer characteristic equation is

$$|z\mathbf{I} - \hat{\mathbf{G}}_{bb} + \mathbf{K}_e \hat{\mathbf{G}}_{ab}| = \begin{vmatrix} z - 1 + k_{e_1} & 0.115 - 0.25k_{e_1} \\ k_{e_2} & z + 0.5 - 0.25k_{e_2} \end{vmatrix}$$

$$= (z - 1 + k_{e_1})(z + 0.5 - 0.25k_{e_2}) - k_{e_2}(0.115 - 0.25k_{e_1})$$

$$= z^2 + (k_{e_1} - 0.25k_{e_2} - 0.5)z + (0.5k_{e_1} + 0.135k_{e_2} - 0.5)$$

$$= 0 \qquad\qquad\qquad\qquad (6\text{–}216)$$

Since we desire the deadbeat response, the desired characteristic equation is

$$z^2 = 0$$

Hence, we require

$$k_{e_1} - 0.25k_{e_2} - 0.5 = 0$$

$$0.5k_{e_1} + 0.135k_{e_2} - 0.5 = 0$$

Solving these two simultaneous equations for k_{e_1} and k_{e_2}, we obtain

$$\mathbf{K}_e = \begin{bmatrix} k_{e_1} \\ k_{e_2} \end{bmatrix} = \begin{bmatrix} 0.7404 \\ 0.9615 \end{bmatrix} \qquad\qquad (6\text{–}217)$$

Integral Control with State Observer.   We have thus considered a design problem in which the observed state variables are fed back in a minor loop and an integral controller is used in the main loop.

In the pole placement part of the design, we used the actual state rather than the observed state. In what follows we shall obtain the system equations for the case where the integral controller and the state observer are used.

The use of the observed state $\tilde{\mathbf{x}}(k)$, where $\tilde{\mathbf{x}}(k) = \mathbf{T}\tilde{\boldsymbol{\xi}}(k)$, in the state feedback control modifies the control signal $u(k)$ as follows. From Equation (6–204) we have

$$u(k) = -\mathbf{K}_2\tilde{\mathbf{x}}(k) + K_1 v(k) = -\mathbf{K}_2\mathbf{T}\tilde{\boldsymbol{\xi}}(k) + K_1 v(k) \qquad (6\text{–}218)$$

Define

$$\boldsymbol{\xi}(k) - \tilde{\boldsymbol{\xi}}(k) = \boldsymbol{\epsilon}(k)$$

Then Equation (6–218) can be written as follows:

$$u(k) = -\mathbf{K}_2\mathbf{T}\boldsymbol{\xi}(k) + K_1 v(k) + \mathbf{K}_2\mathbf{T}\boldsymbol{\epsilon}(k) \qquad (6\text{–}219)$$

which can be rewritten as

$$u(k) = \begin{bmatrix} -0.3233 & -2 & 0.2817 \end{bmatrix} \begin{bmatrix} \xi_1(k) \\ \xi_2(k) \\ \xi_3(k) \end{bmatrix}$$

$$+ \tfrac{2}{3}v(k) + \begin{bmatrix} 0.3233 & 2 & -0.2817 \end{bmatrix} \begin{bmatrix} \epsilon_1(k) \\ \epsilon_2(k) \\ \epsilon_3(k) \end{bmatrix} \qquad (6\text{–}220)$$

By substituting Equation (6–220) into Equation (6–214), we obtain

$$
\begin{bmatrix} \xi_1(k+1) \\ \xi_2(k+1) \\ \xi_3(k+1) \end{bmatrix} = \begin{bmatrix} 0.5 & 1 & -0.25 \\ -\frac{1}{3} & -1 & \frac{1}{6} \\ 1 & 0 & -0.5 \end{bmatrix} \begin{bmatrix} \xi_1(k) \\ \xi_2(k) \\ \xi_3(k) \end{bmatrix}
$$
$$
+ \begin{bmatrix} 0 \\ \frac{2}{3} \\ 0 \end{bmatrix} v(k) + \begin{bmatrix} 0 & 0 & 0 \\ 0.3233 & 2 & -0.2817 \\ 0 & 0 & 0 \end{bmatrix} \begin{bmatrix} \epsilon_1(k) \\ \epsilon_2(k) \\ \epsilon_3(k) \end{bmatrix} \qquad (6\text{–}221)
$$

Also, Equation (6–206) can be modified to read

$$
v(k+1) = -(\mathbf{CGT} - \mathbf{CHK_2\,T})\boldsymbol{\xi}(k) + (1 - \mathbf{CHK_1})v(k) + r(k+1)
$$

or

$$
v(k+1) = -[0.5 \quad 1 \quad -0.25] \begin{bmatrix} \xi_1(k) \\ \xi_2(k) \\ \xi_3(k) \end{bmatrix} + v(k) + r(k+1) \qquad (6\text{–}222)
$$

Referring to Equation (6–156) and noting that $\epsilon_1(k) = 0$, we can give the observer error dynamics by

$$
\begin{bmatrix} \epsilon_2(k+1) \\ \epsilon_3(k+1) \end{bmatrix} = [\hat{\mathbf{G}}_{bb} - \mathbf{K}_e\,\hat{\mathbf{G}}_{ab}] \begin{bmatrix} \epsilon_2(k) \\ \epsilon_3(k) \end{bmatrix}
$$

Therefore,

$$
\begin{bmatrix} \epsilon_1(k+1) \\ \epsilon_2(k+1) \\ \epsilon_3(k+1) \end{bmatrix} = \left[\begin{array}{c:cc} 1 & 0 & 0 \\ \hdashline 0 & 0.2596 & 0.0701 \\ 0 & -0.9615 & -0.2596 \end{array} \right] \begin{bmatrix} \epsilon_1(k) \\ \epsilon_2(k) \\ \epsilon_3(k) \end{bmatrix} \qquad (6\text{–}223)
$$

Combining Equations (6–221), (6–222), and (6–223) into one state equation, we obtain

$$
\begin{bmatrix} \xi_1(k+1) \\ \xi_2(k+1) \\ \xi_3(k+1) \\ v(k+1) \\ \epsilon_1(k+1) \\ \epsilon_2(k+1) \\ \epsilon_3(k+1) \end{bmatrix} = \begin{bmatrix} 0.5 & 1 & -0.25 & 0 & 0 & 0 & 0 \\ -\frac{1}{3} & -1 & \frac{1}{6} & \frac{2}{3} & 0.3233 & 2 & -0.2817 \\ 1 & 0 & -0.5 & 0 & 0 & 0 & 0 \\ -0.5 & -1 & 0.25 & 1 & 0 & 0 & 0 \\ 0 & 0 & 0 & 0 & 1 & 0 & 0 \\ 0 & 0 & 0 & 0 & 0 & 0.2596 & 0.0701 \\ 0 & 0 & 0 & 0 & 0 & -0.9615 & -0.2596 \end{bmatrix} \begin{bmatrix} \xi_1(k) \\ \xi_2(k) \\ \xi_3(k) \\ v(k) \\ \epsilon_1(k) \\ \epsilon_2(k) \\ \epsilon_3(k) \end{bmatrix}
$$
$$
+ \begin{bmatrix} 0 \\ 0 \\ 0 \\ 1 \\ 0 \\ 0 \\ 0 \end{bmatrix} r(k+1) \qquad (6\text{–}224)
$$

The plant output $y(k)$ can be given by

$$
y(k) = [1 \quad 0 \quad 0] \begin{bmatrix} \xi_1(k) \\ \xi_2(k) \\ \xi_3(k) \end{bmatrix} = \xi_1(k) \qquad (6\text{–}225)
$$

It can be shown that for the unit-step command input, the response of the system under an arbitrary initial condition,

$$\begin{bmatrix} \xi_1(0) \\ \xi_2(0) \\ \xi_3(0) \\ v(0) \\ \epsilon_1(0) \\ \epsilon_2(0) \\ \epsilon_3(0) \end{bmatrix} = \begin{bmatrix} a \\ b \\ c \\ d \\ 0 \\ \alpha \\ \beta \end{bmatrix}$$

requires at most six sampling periods to complete. (That is, the output reaches unity in at most six sampling periods and thereafter it will stay there in the absence of disturbances and new command inputs.)

Notice that, as we have seen earlier, if the actual state can be fed back, the system requires at most four sampling periods to complete the unit-step response. [The system requires only three sampling periods to complete the unit-step response if $\xi_1(0) = 0$, $\xi_2(0) = 0$, $\xi_3(0) = 0$, and $v(0) = 1$.] However, if the state observer is used, the response time increases. If the minimum-order observer (second-order observer, in the present case) is used, then the system requires at most six sampling periods to complete the unit-step response. [See, for example, the sample unit-step response sequence shown in Figure 6–21(a).] This means that in special cases the response time is much shorter.

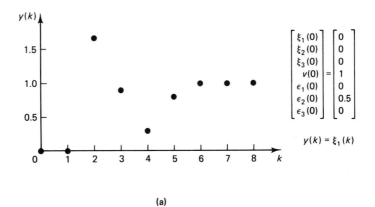

(a)

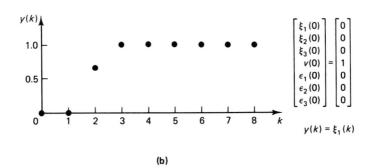

(b)

Figure 6–21 Sample unit-step response sequences for the servo system with observed-state feedback and integral control designed in Example 6–13.

For example, if the initial conditions are $\xi_1(0) = 0$, $\xi_2(0) = 0$, $\xi_3(0) = 0$, $v(0) = 1$, $\epsilon_2(0) = 0$, and $\epsilon_3(0) = 0$, then the system requires only three sampling periods to complete the unit-step response. See Figure 6–21(b).

EXAMPLE PROBLEMS AND SOLUTIONS

Problem A–6–1

The definition of controllability given in Section 6–2 is not the only one used in the literature. Sometimes the following definition is used: A control system is defined to be state controllable if, given an arbitrary initial state $\mathbf{x}(0)$, it is possible to bring the state to the origin of the state space in a finite time interval, provided the control vector is unconstrained (unbounded).

The concept of reachability, similar to the concept of controllability, is available in the literature and is used in the following way: A control system is defined to be reachable if, starting from the origin of the state space, the state can be brought to an arbitrary point in the state space in a finite time period, provided the control vector is unconstrained.

Show that the system

$$\begin{bmatrix} x_1(k+1) \\ x_2(k+1) \end{bmatrix} = \begin{bmatrix} 0 & 0 \\ -1 & 1 \end{bmatrix} \begin{bmatrix} x_1(k) \\ x_2(k) \end{bmatrix} + \begin{bmatrix} 0 \\ 1 \end{bmatrix} u(k)$$

is controllable (in the sense defined in this problem), but is not reachable.

Solution   Rewriting the system state equation, we obtain

$$x_1(k+1) = 0$$

$$x_2(k+1) = -x_1(k) + x_2(k) + u(k)$$

Starting from an arbitrary initial state, we have

$$x_1(1) = 0$$

$$x_2(1) = -x_1(0) + x_2(0) + u(0)$$

Hence, by choosing

$$u(0) = x_1(0) - x_2(0)$$

the state can be brought to the origin in one step. Thus, the system is controllable in the sense defined in this problem.

If the state starts from the origin, we have

$$x_1(1) = 0$$

$$x_2(1) = -x_1(0) + x_2(0) + u(0) = -0 + 0 + u(0) = u(0)$$

Although $x_2(1)$ can be brought to an arbitrary point in one step, $x_1(1)$ cannot be controlled. Consequently, the system is not reachable.

Notice that the present system is not controllable in the sense defined in Section 6–2 (where the final state is an arbitrary point in the state space including the origin), because the required rank condition is not satisfied, as the following shows:

$$\text{rank}\,[\mathbf{H} \,\vdots\, \mathbf{GH}] = \text{rank} \begin{bmatrix} 0 & 0 \\ 1 & 1 \end{bmatrix} = 1 < 2$$

As seen in this problem, controllability and reachability (both defined in this problem) are different. However, if the state matrix **G** is nonsingular, then complete state controllability in the sense defined in this problem and complete state reachability mean the same thing. That is, for the system with a nonsingular matrix **G**, complete state controllability means complete reachability, and vice versa.

Problem A–6–2

Consider the completely state controllable system defined by

$$\mathbf{x}(k + 1) = \mathbf{G}\mathbf{x}(k) + \mathbf{H}u(k)$$

where

$$\mathbf{G} = \begin{bmatrix} 1 & 0.6321 \\ 0 & 0.3679 \end{bmatrix}, \qquad \mathbf{H} = \begin{bmatrix} 0.3679 \\ 0.6321 \end{bmatrix}$$

The sampling period is 1 sec. If the control signal $u(k)$ is unbounded, or

$$-\infty \le u(k) \le \infty$$

then an arbitrary initial state $\mathbf{x}(0)$ can be brought to the origin in at most two sampling periods by using a piecewise-constant control signal.

Derive the control law to transfer an arbitrary initial state $\mathbf{x}(0)$ to the origin. Determine the region in the state space in which the initial state can be brought to the origin in one sampling period.

If the magnitude of $u(k)$ is bounded, then some initial state cannot be transferred to the origin in two sampling periods. (Three, four, or more sampling periods may be required.) Suppose that

$$|u(k)| \le 1$$

Determine the region of the initial states in the $x_1 x_2$ plane that can be transferred to the origin in one sampling period and two sampling periods, respectively, by using the bounded control signal $|u(k)| \le 1$.

Solution

For the case where $u(k)$ is unbounded.   Since the system is of the second order, we need at most two sampling periods to transfer any initial state $\mathbf{x}(0)$ to the origin. Noting that

$$\mathbf{x}(1) = \mathbf{G}\mathbf{x}(0) + \mathbf{H}u(0) \tag{6–226}$$

$$\mathbf{x}(2) = \mathbf{0} = \mathbf{G}\mathbf{x}(1) + \mathbf{H}u(1) = \mathbf{G}^2\mathbf{x}(0) + \mathbf{G}\mathbf{H}u(0) + \mathbf{H}u(1)$$

and **G** is nonsingular, we obtain

$$\mathbf{x}(0) = -\mathbf{G}^{-1}\mathbf{H}u(0) - \mathbf{G}^{-2}\mathbf{H}u(1) \tag{6–227}$$

Substituting Equation (6–227) into Equation (6–226), we get

$$\mathbf{x}(1) = -\mathbf{G}^{-1}\mathbf{H}u(1) \tag{6–228}$$

Noting that

$$\mathbf{G}^{-1}\mathbf{H} = \begin{bmatrix} 1 & -1.7181 \\ 0 & 2.7181 \end{bmatrix} \begin{bmatrix} 0.3679 \\ 0.6321 \end{bmatrix} = \begin{bmatrix} -0.7181 \\ 1.7181 \end{bmatrix}$$

$$\mathbf{G}^{-2}\mathbf{H} = \begin{bmatrix} 1 & -6.3881 \\ 0 & 7.3881 \end{bmatrix} \begin{bmatrix} 0.3679 \\ 0.6321 \end{bmatrix} = \begin{bmatrix} -3.6700 \\ 4.6700 \end{bmatrix}$$

we obtain from Equations (6–227) and (6–228) the following two equations:

$$\begin{bmatrix} x_1(0) \\ x_2(0) \end{bmatrix} = -\begin{bmatrix} -0.7181 \\ 1.7181 \end{bmatrix} u(0) - \begin{bmatrix} -3.6700 \\ 4.6700 \end{bmatrix} u(1) \qquad (6\text{–}229)$$

$$\begin{bmatrix} x_1(1) \\ x_2(1) \end{bmatrix} = -\begin{bmatrix} -0.7181 \\ 1.7181 \end{bmatrix} u(1) \qquad (6\text{–}230)$$

Combining Equations (6–229) and (6–230), we have

$$\begin{bmatrix} x_1(0) & x_1(1) \\ x_2(0) & x_2(1) \end{bmatrix} = \begin{bmatrix} 0.7181 & 3.6700 \\ -1.7181 & -4.6700 \end{bmatrix} \begin{bmatrix} u(0) & u(1) \\ u(1) & 0 \end{bmatrix}$$

which can be modified to

$$\begin{bmatrix} u(0) & u(1) \\ u(1) & 0 \end{bmatrix} = \begin{bmatrix} -1.5820 & -1.2433 \\ 0.5820 & 0.2433 \end{bmatrix} \begin{bmatrix} x_1(0) & x_1(1) \\ x_2(0) & x_2(1) \end{bmatrix} \qquad (6\text{–}231)$$

from which we obtain

$$u(k) = -1.5820x_1(k) - 1.2433x_2(k), \qquad k = 0, 1$$

This equation gives the required control law. With this control law, any initial state $\mathbf{x}(0)$ can be transferred to the origin in at most two sampling periods.

Let us next find the initial states from which the system state can be transferred to the origin in one sampling period. By equating $\mathbf{x}(1)$ with $\mathbf{0}$ in Equation (6–226), we obtain

$$\mathbf{x}(1) = \mathbf{0} = \mathbf{G}\mathbf{x}(0) + \mathbf{H}u(0)$$

from which we get

$$\mathbf{x}(0) = -\mathbf{G}^{-1}\mathbf{H}u(0)$$

or

$$\begin{bmatrix} x_1(0) \\ x_2(0) \end{bmatrix} = \begin{bmatrix} 0.7181 \\ -1.7181 \end{bmatrix} u(0) \qquad (6\text{–}232)$$

From Equation (6–232), we find that if the initial state lies on the line

$$1.7181x_1(0) + 0.7181x_2(0) = 0$$

then it can be transferred to the origin in one sampling period. (Otherwise, we require two sampling periods to bring the initial state to the origin.)

For the case where $u(k)$ is bounded, or $|u(k)| \le 1$. If we require $\mathbf{x}(1) = \mathbf{0}$, then from Equation (6–232) we have

$$x_1(0) = 0.7181u(0), \qquad x_2(0) = -1.7181u(0)$$

Since $|u(0)| \le 1$, we obtain

$$|x_1(0)| \le 0.7181, \qquad |x_2(0)| \le 1.7181$$

Hence, if the initial state lies on the line segment

$$1.7181x_1(0) + 0.7181x_2(0) = 0, \qquad -0.7181 \le x_1(0) \le 0.7181$$

it can be brought to the origin in one sampling period. This line segment is shown in Figure 6–22 as line AOB.

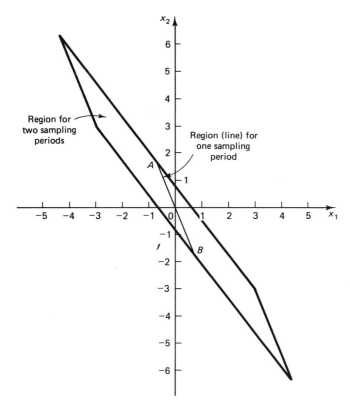

Figure 6–22   Regions from which initial states can be brought to the origin in one or two sampling periods when $u(k)$ is bounded so that $|u(k)| \le 1$.

If we require $\mathbf{x}(2) = \mathbf{0}$, then from Equation (6–231) we obtain

$$u(0) = -1.5820x_1(0) - 1.2433x_2(0)$$

$$u(1) = 0.5820x_1(0) + 0.2433x_2(0)$$

Since $|u(0)| \le 1$ and $|u(1)| \le 1$, we obtain the following four relationships:

$$1.5820x_1(0) + 1.2433x_2(0) \le 1$$

$$1.5820x_1(0) + 1.2433x_2(0) \ge -1$$

$$0.5820x_1(0) + 0.2433x_2(0) \le 1$$

$$0.5820x_1(0) + 0.2433x_2(0) \ge -1$$

The region bounded by these four inequalities is shown in Figure 6–22. If the initial state lies in this region, except on line AOB, then it can be transferred to the origin in two sampling periods. If the initial state lies outside this region, then it will take more than two sampling periods to bring the state to the origin.

Problem A–6–3

Consider the following pulse-transfer-function system:

$$\frac{Y(z)}{U(z)} = \frac{z^{-1}(1 + 0.8z^{-1})}{1 + 1.3z^{-1} + 0.4z^{-2}}$$

A state-space representation for this system can be given by

$$\begin{bmatrix} x_1(k + 1) \\ x_2(k + 1) \end{bmatrix} = \begin{bmatrix} 0 & 1 \\ -0.4 & -1.3 \end{bmatrix}\begin{bmatrix} x_1(k) \\ x_2(k) \end{bmatrix} + \begin{bmatrix} 0 \\ 1 \end{bmatrix}u(k) \qquad (6\text{–}233)$$

$$y(k) = [0.8 \quad 1]\begin{bmatrix} x_1(k) \\ x_2(k) \end{bmatrix} \qquad (6\text{–}234)$$

A different state-space representation for the same system can be given by

$$\begin{bmatrix} x_1(k + 1) \\ x_2(k + 1) \end{bmatrix} = \begin{bmatrix} 0 & -0.4 \\ 1 & -1.3 \end{bmatrix}\begin{bmatrix} x_1(k) \\ x_2(k) \end{bmatrix} + \begin{bmatrix} 0.8 \\ 1 \end{bmatrix}u(k) \qquad (6\text{–}235)$$

$$y(k) = [0 \quad 1]\begin{bmatrix} x_1(k) \\ x_2(k) \end{bmatrix} \qquad (6\text{–}236)$$

Show that the state-space representation defined by Equations (6–233) and (6–234) gives a system which is state controllable but not observable. Show, on the other hand, that the state-space representation defined by Equations (6–235) and (6–236) gives a system that is not completely state controllable but is observable. Explain what causes the apparent difference in the controllability and observability of the same system.

Solution   Consider the discrete-time control system defined by Equations (6–233) and (6–234). The rank of the controllability matrix

$$[\mathbf{H} \vdots \mathbf{GH}] = \begin{bmatrix} 0 & 1 \\ 1 & -1.3 \end{bmatrix}$$

is 2. Hence, the system is completely state controllable. The rank of the observability matrix

$$[\mathbf{C}^* \vdots \mathbf{G}^*\mathbf{C}^*] = \begin{bmatrix} 0.8 & -0.4 \\ 1 & -0.5 \end{bmatrix}$$

is 1. Hence, the system is not observable.

Next, consider the system defined by Equations (6–235) and (6–236). The rank of the controllability matrix

$$[\mathbf{H} \vdots \mathbf{GH}] = \begin{bmatrix} 0.8 & -0.4 \\ 1 & -0.5 \end{bmatrix}$$

is 1. Hence, the system is not completely state controllable. The rank of the observability matrix

$$|\mathbf{C}^* \vdots \mathbf{G}^*\mathbf{C}^*] = \begin{bmatrix} 0 & 1 \\ 1 & -1.3 \end{bmatrix}$$

is 2. Hence, the system is observable.

The apparent difference in the controllability and observability of the same system is caused by the fact that the original system has a pole–zero cancellation in the pulse transfer function:

$$\frac{Y(z)}{U(z)} = \frac{z + 0.8}{z^2 + 1.3z + 0.4} = \frac{z + 0.8}{(z + 0.8)(z + 0.5)}$$

If a pole–zero cancellation occurs in the pulse transfer function, then the controllability and observability vary, depending on how the state variables are chosen.

Note that to be completely state controllable and completely observable the pulse-transfer-function system must not have any pole–zero cancellation.

Problem A–6–4

Consider the control system defined by

$$\mathbf{x}(k + 1) = \mathbf{G}\mathbf{x}(k) + \mathbf{H}u(k) \qquad (6\text{–}237)$$

$$y(k) = \mathbf{C}\mathbf{x}(k) + D \qquad (6\text{–}238)$$

where

$$\mathbf{x}(k) = \text{state vector } (n\text{-vector})$$

$$u(k) = \text{control signal (scalar)}$$

$$y(k) = \text{output signal (scalar)}$$

$$\mathbf{G} = n \times n \text{ matrix}$$

$$\mathbf{H} = n \times 1 \text{ matrix}$$

$$\mathbf{C} = 1 \times n \text{ matrix}$$

$$D = \text{scalar (constant)}$$

As stated by Equation (5–60), the pulse transfer function $F(z)$ can be given as follows:

$$F(z) = \mathbf{C}(z\mathbf{I} - \mathbf{G})^{-1}\mathbf{H} + D$$

Prove that, if the system is completely state controllable and completely observable, then there is no pole–zero cancellation in the pulse transfer function $F(z)$.

Solution   Suppose that there is a pole–zero cancellation in the pulse transfer function, even though the system is completely state controllable and completely observable. Consider the following identity equation:

$$\begin{bmatrix} \mathbf{I} & 0 \\ \mathbf{C}(z\mathbf{I} - \mathbf{G})^{-1} & 1 \end{bmatrix}\begin{bmatrix} z\mathbf{I} - \mathbf{G} & \mathbf{H} \\ -\mathbf{C} & D \end{bmatrix} = \begin{bmatrix} z\mathbf{I} - \mathbf{G} & \mathbf{H} \\ 0 & F(z) \end{bmatrix}$$

Taking the determinant of the left-hand side of the equation and equating it with the determinant of the right-hand side, we obtain

$$\begin{vmatrix} z\mathbf{I} - \mathbf{G} & \mathbf{H} \\ -\mathbf{C} & D \end{vmatrix} = |z\mathbf{I} - \mathbf{G}|F(z)$$

or

$$F(z) = \frac{\begin{vmatrix} z\mathbf{I} - \mathbf{G} & \mathbf{H} \\ -\mathbf{C} & D \end{vmatrix}}{|z\mathbf{I} - \mathbf{G}|}$$

The poles of $F(z)$ are the roots of $|z\mathbf{I} - \mathbf{G}| = 0$, and the zeros of $F(z)$ are the roots of

$$\begin{vmatrix} z\mathbf{I} - \mathbf{G} & \mathbf{H} \\ -\mathbf{C} & D \end{vmatrix} = 0 \qquad (6\text{–}239)$$

Now suppose that a pole–zero cancellation occurs. Let us assume that $z = z_1$ is a pole of $F(z)$ and is also a zero of $F(z)$, so cancellation occurs. Then $z = z_1$ is a root of $|z\mathbf{I} - \mathbf{G}| = 0$. Also, it is a root of the determinant equation given by Equation (6–239).

This means that there exists a vector

$$\left[\begin{array}{c} \mathbf{v} \\ \hline w \end{array}\right]$$

where $\mathbf{v}$ is an n-vector and w is a scalar, such that

$$\left[\begin{array}{c|c} z_1 \mathbf{I} - \mathbf{G} & \mathbf{H} \\ \hline -\mathbf{C} & D \end{array}\right]\left[\begin{array}{c} \mathbf{v} \\ \hline w \end{array}\right] = \left[\begin{array}{c} \mathbf{0} \\ 0 \end{array}\right] \tag{6-240}$$

If $w \neq 0$, then from Equation (6–240) we have

$$(z_1 \mathbf{I} - \mathbf{G})\mathbf{v} + \mathbf{H}w = \mathbf{0}$$

or

$$(\mathbf{G} - z_1 \mathbf{I})\mathbf{v} = \mathbf{H}w \tag{6-241}$$

Since $z = z_1$ is a root of the characteristic equation, the characteristic polynomial $\phi(z)$ of $\mathbf{G}$ can be written as follows:

$$\phi(z) = (z - z_1)\hat{\phi}(z)$$

or

$$\phi(\mathbf{G}) = (\mathbf{G} - z_1 \mathbf{I})\hat{\phi}(\mathbf{G}) = \hat{\phi}(\mathbf{G})(\mathbf{G} - z_1 \mathbf{I}) = \mathbf{0}$$

From Equation (6–241) we have

$$\phi(\mathbf{G})\mathbf{v} = \hat{\phi}(\mathbf{G})(\mathbf{G} - z_1 \mathbf{I})\mathbf{v} = \hat{\phi}(\mathbf{G})\mathbf{H}w = \mathbf{0}$$

Hence,

$$\hat{\phi}(\mathbf{G})\mathbf{H} = \mathbf{0}$$

Since $\hat{\phi}(z)$ is a polynomial of degree $n - 1$, the fact that $\hat{\phi}(\mathbf{G})\mathbf{H} = \mathbf{0}$ means that vector $\mathbf{G}^{n-1}\mathbf{H}$ can be written in terms of $\mathbf{H}, \mathbf{GH}, \ldots, \mathbf{G}^{n-2}\mathbf{H}$. Hence,

$$\text{rank}\,[\mathbf{H} \vdots \mathbf{GH} \vdots \cdots \vdots \mathbf{G}^{n-1}\mathbf{H}] < n$$

This contradicts the assumption that the system is completely state controllable. Thus, if the system is completely state controllable, then there is no pole–zero cancellation in the pulse transfer function.

Next, referring to Equation (6–240), if $w = 0$ and $\mathbf{v} \neq \mathbf{0}$, then we have

$$(z_1 \mathbf{I} - \mathbf{G})\mathbf{v} = \mathbf{0} \tag{6-242}$$

$$\mathbf{Cv} = \mathbf{0} \tag{6-243}$$

From Equation (6–243) we have

$$\mathbf{v}^*\mathbf{C}^* = \mathbf{0} \tag{6-244}$$

From Equation (6–242) we obtain

$$\mathbf{v}^*\mathbf{G}^* = z_1 \mathbf{v}^*$$

Hence,

$$\mathbf{v}^*\mathbf{G}^*\mathbf{C}^* = z_1 \mathbf{v}^*\mathbf{C}^* = \mathbf{0}$$

where we have used Equation (6–244). Similarly,

$$\mathbf{v}^*(\mathbf{G}^*)^2 \mathbf{C}^* = \mathbf{v}^*\mathbf{G}^*\mathbf{G}^*\mathbf{C}^* = z_1 \mathbf{v}^*\mathbf{G}^*\mathbf{C}^* = z_1^2 \mathbf{v}^*\mathbf{C}^* = \mathbf{0}$$

and

$$\mathbf{v}^*(\mathbf{G}^*)^{k-1}\mathbf{C}^* = z_1^{k-1}\mathbf{v}^*\mathbf{C}^* = 0, \qquad k = 1, 2, 3, \ldots, n$$

Hence,

$$\mathbf{v}^*[\mathbf{C}^* \vdots \mathbf{G}^*\mathbf{C}^* \vdots \cdots \vdots (\mathbf{G}^*)^{n-1}\mathbf{C}^*] = \mathbf{0}$$

or

$$\text{rank}\,[\mathbf{C}^* \vdots \mathbf{G}^*\mathbf{C}^* \vdots \cdots \vdots (\mathbf{G}^*)^{n-1}\mathbf{C}^*] < n$$

This contradicts the assumption that the system is completely observable. Thus, there is no pole–zero cancellation if the system is completely observable.

This completes the proof that if the system is completely state controllable and completely observable then there is no pole–zero cancellation in the pulse transfer function $F(z)$.

Problem A–6–5

Consider the completely state controllable system

$$\mathbf{x}(k + 1) = \mathbf{G}\mathbf{x}(k) + \mathbf{H}u(k)$$

Define the controllability matrix as $\mathbf{M}$:

$$\mathbf{M} = [\mathbf{H} \vdots \mathbf{GH} \vdots \cdots \vdots \mathbf{G}^{n-1}\mathbf{H}]$$

Show that

$$\mathbf{M}^{-1}\mathbf{GM} = \begin{bmatrix} 0 & 0 & \cdots & 0 & -a_n \\ 1 & 0 & \cdots & 0 & -a_{n-1} \\ 0 & 1 & \cdots & 0 & -a_{n-2} \\ \vdots & \vdots & & \vdots & \vdots \\ 0 & 0 & \cdots & 1 & -a_1 \end{bmatrix} \qquad (6\text{--}245)$$

where $a_1, a_2, \ldots, a_n$ are the coefficients of the characteristic polynomial

$$|z\mathbf{I} - \mathbf{G}| = z^n + a_1 z^{n-1} + \cdots + a_{n-1} z + a_n$$

Solution   Let us consider the case where $n = 3$. We shall show that

$$\mathbf{GM} = \mathbf{M}\begin{bmatrix} 0 & 0 & -a_3 \\ 1 & 0 & -a_2 \\ 0 & 1 & -a_1 \end{bmatrix} \qquad (6\text{--}246)$$

The left-hand side of Equation (6–246) is

$$\mathbf{GM} = \mathbf{G}[\mathbf{H} \vdots \mathbf{GH} \vdots \mathbf{G}^2\mathbf{H}] = [\mathbf{GH} \vdots \mathbf{G}^2\mathbf{H} \vdots \mathbf{G}^3\mathbf{H}]$$

The right-hand side of Equation (6–246) is

$$[\mathbf{H} \vdots \mathbf{GH} \vdots \mathbf{G}^2\mathbf{H}]\begin{bmatrix} 0 & 0 & -a_3 \\ 1 & 0 & -a_2 \\ 0 & 1 & -a_1 \end{bmatrix} = [\mathbf{GH} \vdots \mathbf{G}^2\mathbf{H} \vdots -a_3\mathbf{H} - a_2\mathbf{GH} - a_1\mathbf{G}^2\mathbf{H}] \qquad (6\text{--}247)$$

The Cayley–Hamilton theorem states that matrix $\mathbf{G}$ satisfies its own characteristic equation, or

$$\mathbf{G}^n + a_1\mathbf{G}^{n-1} + \cdots + a_{n-1}\mathbf{G} + a_n\mathbf{I} = \mathbf{0}$$

For $n = 3$, we have

$$\mathbf{G}^3 + a_1\mathbf{G}^2 + a_2\mathbf{G} + a_3\mathbf{I} = \mathbf{0} \qquad (6\text{--}248)$$

Using Equation (6–248), the third column of the right-hand side of Equation (6–247) becomes

$$-a_3 \mathbf{H} - a_2 \mathbf{GH} - a_1 \mathbf{G}^2 \mathbf{H} = \mathbf{G}^3 \mathbf{H}$$

Thus, Equation (6–247), the right-hand side of Equation (6–246), becomes

$$[\mathbf{H} \vdots \mathbf{GH} \vdots \mathbf{G}^2 \mathbf{H}] \begin{bmatrix} 0 & 0 & -a_3 \\ 1 & 0 & -a_2 \\ 0 & 1 & -a_1 \end{bmatrix} = [\mathbf{GH} \vdots \mathbf{G}^2 \mathbf{H} \vdots \mathbf{G}^3 \mathbf{H}]$$

Hence, we have shown that Equation (6–246) is true. thus,

$$\mathbf{M}^{-1} \mathbf{GM} = \begin{bmatrix} 0 & 0 & -a_3 \\ 1 & 0 & -a_2 \\ 0 & 1 & -a_1 \end{bmatrix}$$

The preceding derivation can easily be extended to the general case of any positive integer n.

Problem A–6–6

Consider the completely state controllable system

$$\mathbf{x}(k + 1) = \mathbf{Gx}(k) + \mathbf{H}u(k)$$

Define

$$\mathbf{M} = [\mathbf{H} \vdots \mathbf{GH} \vdots \cdots \vdots \mathbf{G}^{n-1} \mathbf{H}]$$

and

$$\mathbf{W} = \begin{bmatrix} a_{n-1} & a_{n-2} & \cdots & a_1 & 1 \\ a_{n-2} & a_{n-3} & \cdots & 1 & 0 \\ \vdots & \vdots & & \vdots & \vdots \\ a_1 & 1 & \cdots & 0 & 0 \\ 1 & 0 & \cdots & 0 & 0 \end{bmatrix}$$

where the a_i's are coefficients of the characteristic polynomial

$$|z\mathbf{I} - \mathbf{G}| = z^n + a_1 z^{n-1} + \cdots + a_{n-1} z + a_n$$

Define also

$$\mathbf{T} = \mathbf{MW}$$

Show that

$$\mathbf{T}^{-1} \mathbf{GT} = \begin{bmatrix} 0 & 1 & 0 & \cdots & 0 \\ 0 & 0 & 1 & \cdots & 0 \\ \vdots & \vdots & \vdots & & \vdots \\ 0 & 0 & 0 & \cdots & 1 \\ -a_n & -a_{n-1} & -a_{n-2} & \cdots & -a_1 \end{bmatrix}, \quad \mathbf{T}^{-1} \mathbf{H} = \begin{bmatrix} 0 \\ 0 \\ \vdots \\ 0 \\ 1 \end{bmatrix}$$

Solution   Let us consider the case where $n = 3$. We shall show that

$$\mathbf{T}^{-1} \mathbf{GT} = (\mathbf{MW})^{-1} \mathbf{G}(\mathbf{MW}) = \begin{bmatrix} 0 & 1 & 0 \\ 0 & 0 & 1 \\ -a_3 & -a_2 & -a_1 \end{bmatrix} \qquad (6\text{–}249)$$

Referring to Problem A–6–5, we have

$$(MW)^{-1} G(MW) = W^{-1}(M^{-1} GM)W = W^{-1} \begin{bmatrix} 0 & 0 & -a_3 \\ 1 & 0 & -a_2 \\ 0 & 1 & -a_1 \end{bmatrix} W$$

Hence, Equation (6–249) can be rewritten as follows:

$$W^{-1} \begin{bmatrix} 0 & 0 & -a_3 \\ 1 & 0 & -a_2 \\ 0 & 1 & -a_1 \end{bmatrix} W = \begin{bmatrix} 0 & 1 & 0 \\ 0 & 0 & 1 \\ -a_3 & -a_2 & -a_1 \end{bmatrix}$$

Consequently, we need to show that

$$\begin{bmatrix} 0 & 0 & -a_3 \\ 1 & 0 & -a_2 \\ 0 & 1 & -a_1 \end{bmatrix} W = W \begin{bmatrix} 0 & 1 & 0 \\ 0 & 0 & 1 \\ -a_3 & -a_2 & -a_1 \end{bmatrix} \qquad (6\text{–}250)$$

The left-hand side of Equation (6–250) is

$$\begin{bmatrix} 0 & 0 & -a_3 \\ 1 & 0 & -a_2 \\ 0 & 1 & -a_1 \end{bmatrix} \begin{bmatrix} a_2 & a_1 & 1 \\ a_1 & 1 & 0 \\ 1 & 0 & 0 \end{bmatrix} = \begin{bmatrix} -a_3 & 0 & 0 \\ 0 & a_1 & 1 \\ 0 & 1 & 0 \end{bmatrix}$$

The right-hand side of Equation (6–250) is

$$\begin{bmatrix} a_2 & a_1 & 1 \\ a_1 & 1 & 0 \\ 1 & 0 & 0 \end{bmatrix} \begin{bmatrix} 0 & 1 & 0 \\ 0 & 0 & 1 \\ -a_3 & -a_2 & -a_1 \end{bmatrix} = \begin{bmatrix} -a_3 & 0 & 0 \\ 0 & a_1 & 1 \\ 0 & 1 & 0 \end{bmatrix}$$

Clearly, Equation (6–250) holds true. Thus, we have shown that

$$T^{-1}GT = \begin{bmatrix} 0 & 1 & 0 \\ 0 & 0 & 1 \\ -a_3 & -a_2 & -a_1 \end{bmatrix}$$

Next, we shall show that

$$T^{-1}H = \begin{bmatrix} 0 \\ 0 \\ 1 \end{bmatrix} \qquad (6\text{–}251)$$

Note that Equation (6–251) can be written as follows:

$$H = T \begin{bmatrix} 0 \\ 0 \\ 1 \end{bmatrix} = MW \begin{bmatrix} 0 \\ 0 \\ 1 \end{bmatrix}$$

This last equation can easily be verified, since

$$T \begin{bmatrix} 0 \\ 0 \\ 1 \end{bmatrix} = MW \begin{bmatrix} 0 \\ 0 \\ 1 \end{bmatrix} = [H \vdots GH \vdots G^2 H] \begin{bmatrix} a_2 & a_1 & 1 \\ a_1 & 1 & 0 \\ 1 & 0 & 0 \end{bmatrix} \begin{bmatrix} 0 \\ 0 \\ 1 \end{bmatrix}$$

$$= [H \vdots GH \vdots G^2 H] \begin{bmatrix} 1 \\ 0 \\ 0 \end{bmatrix} = H$$

Hence,

$$\mathbf{T}^{-1}\mathbf{H} = \begin{bmatrix} 0 \\ 0 \\ 1 \end{bmatrix}$$

The derivation shown here can easily be extended to the general case of any positive integer n.

Problem A–6–7

Consider the following system:

$$\mathbf{x}(k + 1) = \mathbf{G}\mathbf{x}(k) + \mathbf{H}u(k)$$

where

$$\mathbf{G} = \begin{bmatrix} 0 & 1 & 0 \\ 0 & 0 & 1 \\ -a_3 & -a_2 & -a_1 \end{bmatrix}, \quad \mathbf{H} = \begin{bmatrix} 0 \\ 0 \\ 1 \end{bmatrix}$$

Notice that the system is in the controllable canonical form.
Define the transformation matrix $\mathbf{T}$ as follows:

$$\mathbf{T} = \mathbf{MW}$$

where

$$\mathbf{M} = [\mathbf{H} \vdots \mathbf{GH} \vdots \mathbf{G}^2\mathbf{H}]$$

and

$$\mathbf{W} = \begin{bmatrix} a_2 & a_1 & 1 \\ a_1 & 1 & 0 \\ 1 & 0 & 0 \end{bmatrix}$$

Show that if the system is in the controllable canonical form then $\mathbf{T} = \mathbf{I}$. Consequently, if the system is in the controllable canonical form, then

$$\mathbf{M}^{-1} = \mathbf{W} = \begin{bmatrix} a_2 & a_1 & 1 \\ a_1 & 1 & 0 \\ 1 & 0 & 0 \end{bmatrix}$$

Solution　Since

$$\mathbf{M} = \begin{bmatrix} 0 & 0 & 1 \\ 0 & 1 & -a_1 \\ 1 & -a_1 & -a_2 + a_1^2 \end{bmatrix}$$

we have

$$\mathbf{T} = \mathbf{MW} = \begin{bmatrix} 0 & 0 & 1 \\ 0 & 1 & -a_1 \\ 1 & -a_1 & -a_2 + a_1^2 \end{bmatrix}\begin{bmatrix} a_2 & a_1 & 1 \\ a_1 & 1 & 0 \\ 1 & 0 & 0 \end{bmatrix} = \begin{bmatrix} 1 & 0 & 0 \\ 0 & 1 & 0 \\ 0 & 0 & 1 \end{bmatrix} = \mathbf{I}$$

Hence,

$$\mathbf{M}^{-1} = \mathbf{W} = \begin{bmatrix} a_2 & a_1 & 1 \\ a_1 & 1 & 0 \\ 1 & 0 & 0 \end{bmatrix}$$

Problem A-6-8

Consider the completely observable system

$$\mathbf{x}(k + 1) = \mathbf{Gx}(k)$$

$$y(k) = \mathbf{Cx}(k)$$

Define the observability matrix as $\mathbf{N}$:

$$\mathbf{N} = [\mathbf{C}^* \vdots \mathbf{G}^*\mathbf{C}^* \vdots \cdots \vdots (\mathbf{G}^*)^{n-1}\mathbf{C}^*]$$

Show that

$$\mathbf{N}^*\mathbf{G}(\mathbf{N}^*)^{-1} = \begin{bmatrix} 0 & 1 & 0 & \cdots & 0 \\ 0 & 0 & 1 & \cdots & 0 \\ \vdots & \vdots & \vdots & & \vdots \\ 0 & 0 & 0 & \cdots & 1 \\ -a_n & -a_{n-1} & -a_{n-2} & \cdots & -a_1 \end{bmatrix} \tag{6-252}$$

where $a_1, a_2, \ldots, a_n$ are the coefficients of the characteristic polynomial

$$|z\mathbf{I} - \mathbf{G}| = z^n + a_1 z^{n-1} + \cdots + a_{n-1}z + a_n$$

Solution   Let us consider the case where $n = 3$. Then Equation (6-252) can be written as

$$\mathbf{N}^*\mathbf{G}(\mathbf{N}^*)^{-1} = \begin{bmatrix} 0 & 1 & 0 \\ 0 & 0 & 1 \\ -a_3 & -a_2 & -a_1 \end{bmatrix} \tag{6-253}$$

Equation (6-253) may be rewritten as

$$\mathbf{N}^*\mathbf{G} = \begin{bmatrix} 0 & 1 & 0 \\ 0 & 0 & 1 \\ -a_3 & -a_2 & -a_1 \end{bmatrix} \mathbf{N}^* \tag{6-254}$$

We shall show that Equation (6-254) holds true. The left-hand side of Equation (6-254) is

$$\mathbf{N}^*\mathbf{G} = \begin{bmatrix} \mathbf{C} \\ \mathbf{CG} \\ \mathbf{CG}^2 \end{bmatrix} \mathbf{G} = \begin{bmatrix} \mathbf{CG} \\ \mathbf{CG}^2 \\ \mathbf{CG}^3 \end{bmatrix}$$

The right-hand side of Equation (6-254) is

$$\begin{bmatrix} 0 & 1 & 0 \\ 0 & 0 & 1 \\ -a_3 & -a_2 & -a_1 \end{bmatrix} \mathbf{N}^* = \begin{bmatrix} 0 & 1 & 0 \\ 0 & 0 & 1 \\ -a_3 & -a_2 & -a_1 \end{bmatrix}\begin{bmatrix} \mathbf{C} \\ \mathbf{CG} \\ \mathbf{CG}^2 \end{bmatrix} = \begin{bmatrix} \mathbf{CG} \\ \mathbf{CG}^2 \\ -a_3\mathbf{C} - a_2\mathbf{CG} - a_1\mathbf{CG}^2 \end{bmatrix}$$

The Cayley-Hamilton theorem states that matrix $\mathbf{G}$ satisfies its own characteristic equation, or, for the case of $n = 3$,

$$\mathbf{G}^3 + a_1\mathbf{G}^2 + a_2\mathbf{G} + a_3\mathbf{I} = \mathbf{0}$$

Hence,

$$-a_1\mathbf{CG}^2 - a_2\mathbf{CG} - a_3\mathbf{C} = \mathbf{CG}^3$$

Consequently,

$$
\begin{bmatrix} 0 & 1 & 0 \\ 0 & 0 & 1 \\ -a_3 & -a_2 & -a_1 \end{bmatrix} \mathbf{N}^* = \begin{bmatrix} \mathbf{CG} \\ \mathbf{CG}^2 \\ \mathbf{CG}^3 \end{bmatrix}
$$

Thus, we have shown that Equation (6–254) holds true. Hence,

$$
\mathbf{N}^*\mathbf{G}(\mathbf{N}^*)^{-1} = \begin{bmatrix} 0 & 1 & 0 \\ 0 & 0 & 1 \\ -a_3 & -a_2 & -a_1 \end{bmatrix}
$$

The derivation presented here can be extended to the general case of any positive integer n.

Problem A–6–9

Consider the completely state controllable and completely observable system given by

$$
\mathbf{x}(k + 1) = \mathbf{Gx}(k) + \mathbf{H}u(k) \tag{6–255}
$$

$$
y(k) = \mathbf{Cx}(k) + Du(k) \tag{6–256}
$$

Define

$$
\mathbf{N} = [\mathbf{C}^* \vdots \mathbf{G}^*\mathbf{C}^* \vdots \cdots \vdots (\mathbf{G}^*)^{n-1}\mathbf{C}^*]
$$

and

$$
\mathbf{W} = \begin{bmatrix} a_{n-1} & a_{n-2} & \cdots & a_1 & 1 \\ a_{n-2} & a_{n-3} & \cdots & 1 & 0 \\ \vdots & \vdots & & \vdots & \vdots \\ a_1 & 1 & \cdots & 0 & 0 \\ 1 & 0 & \cdots & 0 & 0 \end{bmatrix}
$$

where the a_i's are coefficients of the characteristic polynomial

$$
|z\mathbf{I} - \mathbf{G}| = z^n + a_1 z^{n-1} + \cdots + a_{n-1}z + a_n
$$

Define also

$$
\mathbf{Q} = (\mathbf{WN}^*)^{-1}
$$

Show that

$$
\mathbf{Q}^{-1}\mathbf{GQ} = \begin{bmatrix} 0 & 0 & \cdots & 0 & -a_n \\ 1 & 0 & \cdots & 0 & -a_{n-1} \\ 0 & 1 & \cdots & 0 & -a_{n-2} \\ \vdots & \vdots & & \vdots & \vdots \\ 0 & 0 & \cdots & 1 & -a_1 \end{bmatrix}
$$

$$
\mathbf{CQ} = [0 \quad 0 \quad \cdots \quad 0 \quad 1]
$$

$$
\mathbf{Q}^{-1}\mathbf{H} = \begin{bmatrix} b_n - a_n b_0 \\ b_{n-1} - a_{n-1} b_0 \\ \vdots \\ b_1 - a_1 b_0 \end{bmatrix}
$$

where the b_k's ($k = 0, 1, 2, \ldots, n$) are those coefficients appearing in the numerator of the pulse transfer function when $\mathbf{C}(z\mathbf{I} - \mathbf{G})^{-1}\mathbf{H} + D$ is written as follows:

$$C(zI - G)^{-1}H + D = \frac{b_0 z^n + b_1 z^{n-1} + \cdots + b_{n-1}z + b_n}{z^n + a_1 z^{n-1} + \cdots + a_{n-1}z + a_n}$$

where $D = b_0$.

Solution   Let us consider the case where $n = 3$. We shall show that

$$Q^{-1}GQ = (WN^*)G(WN^*)^{-1} = \begin{bmatrix} 0 & 0 & -a_3 \\ 1 & 0 & -a_2 \\ 0 & 1 & -a_1 \end{bmatrix} \qquad (6\text{--}257)$$

Note that, by referring to Problem A–6–8, we have

$$(WN^*)G(WN^*)^{-1} = W[N^*G(N^*)^{-1}]W^{-1} = W\begin{bmatrix} 0 & 1 & 0 \\ 0 & 0 & 1 \\ -a_3 & -a_2 & -a_1 \end{bmatrix}W^{-1}$$

Hence, we need to show that

$$W\begin{bmatrix} 0 & 1 & 0 \\ 0 & 0 & 1 \\ -a_3 & -a_2 & -a_1 \end{bmatrix}W^{-1} = \begin{bmatrix} 0 & 0 & -a_3 \\ 1 & 0 & -a_2 \\ 0 & 1 & -a_1 \end{bmatrix}$$

or

$$W\begin{bmatrix} 0 & 1 & 0 \\ 0 & 0 & 1 \\ -a_3 & -a_2 & -a_1 \end{bmatrix} = \begin{bmatrix} 0 & 0 & -a_3 \\ 1 & 0 & -a_2 \\ 0 & 1 & -a_1 \end{bmatrix}W \qquad (6\text{--}258)$$

The left-hand side of Equation (6–258) is

$$W\begin{bmatrix} 0 & 1 & 0 \\ 0 & 0 & 1 \\ -a_3 & -a_2 & -a_1 \end{bmatrix} = \begin{bmatrix} a_2 & a_1 & 1 \\ a_1 & 1 & 0 \\ 1 & 0 & 0 \end{bmatrix}\begin{bmatrix} 0 & 1 & 0 \\ 0 & 0 & 1 \\ -a_3 & -a_2 & -a_1 \end{bmatrix} = \begin{bmatrix} -a_3 & 0 & 0 \\ 0 & a_1 & 1 \\ 0 & 1 & 0 \end{bmatrix}$$

The right-hand side of Equation (6–258) is

$$\begin{bmatrix} 0 & 0 & -a_3 \\ 1 & 0 & -a_2 \\ 0 & 1 & -a_1 \end{bmatrix}W = \begin{bmatrix} 0 & 0 & -a_3 \\ 1 & 0 & -a_2 \\ 0 & 1 & -a_1 \end{bmatrix}\begin{bmatrix} a_2 & a_1 & 1 \\ a_1 & 1 & 0 \\ 1 & 0 & 0 \end{bmatrix} = \begin{bmatrix} -a_3 & 0 & 0 \\ 0 & a_1 & 1 \\ 0 & 1 & 0 \end{bmatrix}$$

Thus, Equation (6–258) holds true. Hence, we have shown that Equation (6–257) holds true.

Next, we shall show that

$$CQ = [0 \quad 0 \quad 1]$$

or

$$C(WN^*)^{-1} = [0 \quad 0 \quad 1] \qquad (6\text{--}259)$$

Notice that

$$[0 \quad 0 \quad 1](WN^*) = [0 \quad 0 \quad 1]\begin{bmatrix} a_2 & a_1 & 1 \\ a_1 & 1 & 0 \\ 1 & 0 & 0 \end{bmatrix}\begin{bmatrix} C \\ CG \\ CG^2 \end{bmatrix} = [1 \quad 0 \quad 0]\begin{bmatrix} C \\ CG \\ CG^2 \end{bmatrix} = C$$

Hence, we have shown that

$$[0 \quad 0 \quad 1] = C(WN^*)^{-1}$$

which is Equation (6–259).

Next, define

$$\mathbf{x} = \mathbf{Q}\hat{\mathbf{x}}$$

Then Equation (6–255) becomes

$$\hat{\mathbf{x}}(k + 1) = \mathbf{Q}^{-1}\mathbf{GQ}\hat{\mathbf{x}}(k) + \mathbf{Q}^{-1}\mathbf{H}u(k) \tag{6–260}$$

and Equation (6–256) becomes

$$y(k) = \mathbf{CQ}\hat{\mathbf{x}}(k) + Du(k) \tag{6–261}$$

For the case of $n = 3$, Equation (6–260) becomes as follows:

$$\hat{\mathbf{x}}(k + 1) = \begin{bmatrix} 0 & 0 & -a_3 \\ 1 & 0 & -a_2 \\ 0 & 1 & -a_1 \end{bmatrix} \hat{\mathbf{x}}(k) + \begin{bmatrix} \gamma_3 \\ \gamma_2 \\ \gamma_1 \end{bmatrix} u(k)$$

where

$$\begin{bmatrix} \gamma_3 \\ \gamma_2 \\ \gamma_1 \end{bmatrix} = \mathbf{Q}^{-1}\mathbf{H}$$

The pulse transfer function $F(z)$ for the system defined by Equations (6–260) and (6–261) is

$$F(z) = (\mathbf{CQ})(z\mathbf{I} - \mathbf{Q}^{-1}\mathbf{GQ})^{-1}\mathbf{Q}^{-1}\mathbf{H} + D$$

Noting that

$$\mathbf{CQ} = [0 \quad 0 \quad 1]$$

we have

$$F(z) = [0 \quad 0 \quad 1] \begin{bmatrix} z & 0 & a_3 \\ -1 & z & a_2 \\ 0 & -1 & z + a_1 \end{bmatrix}^{-1} \begin{bmatrix} \gamma_3 \\ \gamma_2 \\ \gamma_1 \end{bmatrix} + D$$

Note that $D = b_0$. Since

$$\begin{bmatrix} z & 0 & a_3 \\ -1 & z & a_2 \\ 0 & -1 & z + a_1 \end{bmatrix}^{-1} = \frac{1}{z^3 + a_1 z^2 + a_2 z + a_3} \begin{bmatrix} z^2 + a_1 z + a_2 & -a_3 & -a_3 z \\ z + a_1 & z^2 + a_1 z & -a_2 z - a_3 \\ 1 & z & z^2 \end{bmatrix}$$

we have

$$\begin{aligned} F(z) &= \frac{1}{z^3 + a_1 z^2 + a_2 z + a_3} [1 \quad z \quad z^2] \begin{bmatrix} \gamma_3 \\ \gamma_2 \\ \gamma_1 \end{bmatrix} + D \\ &= \frac{\gamma_1 z^2 + \gamma_2 z + \gamma_3}{z^3 + a_1 z^2 + a_2 z + a_3} + b_0 \\ &= \frac{b_0 z^3 + (\gamma_1 + a_1 b_0)z^2 + (\gamma_2 + a_2 b_0)z + (\gamma_3 + a_3 b_0)}{z^3 + a_1 z^2 + a_2 z + a_3} \\ &= \frac{b_0 z^3 + b_1 z^2 + b_2 z + b_3}{z^3 + a_1 z^2 + a_2 z + a_3} \end{aligned}$$

Hence, $\gamma_1 = b_1 - a_1 b_0$, $\gamma_2 = b_2 - a_2 b_0$, and $\gamma_3 = b_3 - a_3 b_0$. Thus, we have shown that

$$\mathbf{Q}^{-1}\mathbf{H} = \begin{bmatrix} \gamma_3 \\ \gamma_2 \\ \gamma_1 \end{bmatrix} = \begin{bmatrix} b_3 - a_3 b_0 \\ b_2 - a_2 b_0 \\ b_1 - a_1 b_0 \end{bmatrix}$$

Note that what we have derived here can easily be extended to the case where n is any positive integer.

Problem A–6–10

Consider the system defined by

$$G(z) = \frac{z + 1}{z^2 + z + 0.16} \tag{6-262}$$

Referring to Section 6–4, obtain state-space representations for this system in the following three different forms:

1. Controllable canonical form
2. Observable canonical form
3. Diagonal canonical form

Solution

1. *Controllable canonical form.* By comparing Equation (6–262) with Equation (6–39), we obtain

$$a_1 = 1, \qquad a_2 = 0.16, \qquad b_0 = 0, \qquad b_1 = 1, \qquad b_2 = 1$$

Hence, referring to Equations (6–37) and (6–38), we obtain

$$\begin{bmatrix} x_1(k+1) \\ x_2(k+1) \end{bmatrix} = \begin{bmatrix} 0 & 1 \\ -0.16 & -1 \end{bmatrix} \begin{bmatrix} x_1(k) \\ x_2(k) \end{bmatrix} + \begin{bmatrix} 0 \\ 1 \end{bmatrix} u(k)$$

$$y(k) = \begin{bmatrix} 1 & 1 \end{bmatrix} \begin{bmatrix} x_1(k) \\ x_2(k) \end{bmatrix}$$

2. *Observable canonical form.* Since $a_1 = 1$, $a_2 = 0.16$, $b_0 = 0$, $b_1 = 1$, and $b_2 = 1$, referring to Equations (6–41) and (6–42), we obtain

$$\begin{bmatrix} x_1(k+1) \\ x_2(k+1) \end{bmatrix} = \begin{bmatrix} 0 & -0.16 \\ 1 & -1 \end{bmatrix} \begin{bmatrix} x_1(k) \\ x_2(k) \end{bmatrix} + \begin{bmatrix} 1 \\ 1 \end{bmatrix} u(k)$$

$$y(k) = \begin{bmatrix} 0 & 1 \end{bmatrix} \begin{bmatrix} x_1(k) \\ x_2(k) \end{bmatrix}$$

3. *Diagonal canonical form.* Notice that

$$G(z) = \frac{\frac{4}{3}}{z + 0.2} + \frac{-\frac{1}{3}}{z + 0.8}$$

By comparing this last equation with Equation (6–47), we obtain

$$\alpha_1 \beta_1 = \tfrac{4}{3}, \qquad \alpha_2 \beta_2 = -\tfrac{1}{3}, \qquad p_1 = -0.2, \qquad p_2 = -0.8, \qquad D = 0$$

Hence, by arbitrarily choosing $\alpha_1 = \alpha_2 = 1$ and referring to Equations (6–45) and (6–46), we obtain

$$\begin{bmatrix} x_1(k+1) \\ x_2(k+1) \end{bmatrix} = \begin{bmatrix} -0.2 & 0 \\ 0 & -0.8 \end{bmatrix} \begin{bmatrix} x_1(k) \\ x_2(k) \end{bmatrix} + \begin{bmatrix} 1 \\ 1 \end{bmatrix} u(k)$$

$$y(k) = \begin{bmatrix} \tfrac{4}{3} & -\tfrac{1}{3} \end{bmatrix} \begin{bmatrix} x_1(k) \\ x_2(k) \end{bmatrix}$$

Problem A–6–11

Consider the double-integrator system

$$\mathbf{x}((k + 1)T) = \mathbf{G}\mathbf{x}(kT) + \mathbf{H}u(kT)$$

where

$$\mathbf{G} = \begin{bmatrix} 1 & T \\ 0 & 1 \end{bmatrix}, \qquad \mathbf{H} = \begin{bmatrix} T^2/2 \\ T \end{bmatrix}$$

and where T is the sampling period. (See Problem A–5–16 for the derivation of this discrete-time state equation for the double integrator system.) Determine a state feedback gain matrix $\mathbf{K}$ such that the response to an arbitrary initial condition is deadbeat. For the initial state

$$\mathbf{x}(0) = \begin{bmatrix} 1 \\ 1 \end{bmatrix}$$

determine $u(0)$ and $u(T)$ for $T = 0.1$ sec, $T = 1$ sec, and $T = 10$ sec.

Solution   Let us define

$$\mathbf{K} = \begin{bmatrix} k_1 & k_2 \end{bmatrix}$$

Then

$$
|z\mathbf{I} - \mathbf{G} + \mathbf{H}\mathbf{K}| = \begin{vmatrix} z - 1 + \dfrac{T^2}{2}k_1 & -T + \dfrac{T^2}{2}k_2 \\ Tk_1 & z - 1 + Tk_2 \end{vmatrix}
$$

$$
= z^2 - \left(2 - \dfrac{T^2}{2}k_1 - Tk_2\right)z + 1 + \dfrac{T^2}{2}k_1 - Tk_2
$$

$$
= 0 \tag{6–263}
$$

The desired characteristic equation is

$$z^2 = 0 \tag{6–264}$$

Hence, by comparing Equations (6–263) and (6–264), we obtain

$$2 - \dfrac{T^2}{2}k_1 - Tk_2 = 0$$

$$1 + \dfrac{T^2}{2}k_1 - Tk_2 = 0$$

from which we get

$$k_1 = \dfrac{1}{T^2}, \qquad k_2 = \dfrac{3}{2T}$$

Hence,

$$\mathbf{K} = \begin{bmatrix} \dfrac{1}{T^2} & \dfrac{3}{2T} \end{bmatrix}$$

In what follows, we shall show that the response to initial conditions is deadbeat. Assume that the initial state is

$$\begin{bmatrix} x_1(0) \\ x_2(0) \end{bmatrix} = \begin{bmatrix} a \\ b \end{bmatrix}$$

The state feedback equation is

$$\mathbf{x}((k+1)T) = (\mathbf{G} - \mathbf{HK})\mathbf{x}(kT)$$

or

$$\begin{bmatrix} x_1((k+1)T) \\ x_2((k+1)T) \end{bmatrix} = \begin{bmatrix} \dfrac{1}{2} & \dfrac{T}{4} \\ -\dfrac{1}{T} & -\dfrac{1}{2} \end{bmatrix} \begin{bmatrix} x_1(kT) \\ x_2(kT) \end{bmatrix}$$

Notice that

$$\begin{bmatrix} x_1(T) \\ x_2(T) \end{bmatrix} = \begin{bmatrix} \dfrac{1}{2} & \dfrac{T}{4} \\ -\dfrac{1}{T} & -\dfrac{1}{2} \end{bmatrix} \begin{bmatrix} x_1(0) \\ x_2(0) \end{bmatrix} = \begin{bmatrix} \dfrac{1}{2} & \dfrac{T}{4} \\ -\dfrac{1}{T} & -\dfrac{1}{2} \end{bmatrix} \begin{bmatrix} a \\ b \end{bmatrix} = \begin{bmatrix} \dfrac{1}{2}a + \dfrac{T}{4}b \\ -\dfrac{1}{T}a - \dfrac{1}{2}b \end{bmatrix}$$

$$\begin{bmatrix} x_1(2T) \\ x_2(2T) \end{bmatrix} = \begin{bmatrix} \dfrac{1}{2} & \dfrac{T}{4} \\ -\dfrac{1}{T} & -\dfrac{1}{2} \end{bmatrix} \begin{bmatrix} \dfrac{1}{2}a + \dfrac{T}{4}b \\ -\dfrac{1}{T}a - \dfrac{1}{2}b \end{bmatrix} = \begin{bmatrix} 0 \\ 0 \end{bmatrix}$$

Thus, clearly, the response is deadbeat.

Now let us determine $u(0)$ and $u(T)$. Notice that

$$u(kT) = -\mathbf{K}\mathbf{x}(kT) = -\begin{bmatrix} \dfrac{1}{T^2} & \dfrac{3}{2T} \end{bmatrix} \mathbf{x}(kT)$$

Hence,

$$u(0) = -\begin{bmatrix} \dfrac{1}{T^2} & \dfrac{3}{2T} \end{bmatrix} \begin{bmatrix} a \\ b \end{bmatrix} = -\dfrac{1}{T^2}a - \dfrac{3}{2T}b$$

$$u(T) = -\begin{bmatrix} \dfrac{1}{T^2} & \dfrac{3}{2T} \end{bmatrix} \begin{bmatrix} \dfrac{1}{2}a + \dfrac{T}{4}b \\ -\dfrac{1}{T}a - \dfrac{1}{2}b \end{bmatrix} = \dfrac{1}{T^2}a + \dfrac{b}{2T}$$

For $a = 1$ and $b = 1$, we have

$$u(0) = -\dfrac{1}{T^2} - \dfrac{3}{2T}, \qquad u(T) = \dfrac{1}{T^2} + \dfrac{1}{2T}$$

In particular, for $T = 0.1$ sec,

$$u(0) = -115, \qquad u(T) = u(0.1) = 105$$

For $T = 1$ sec,

$$u(0) = -2.5, \qquad u(T) = u(1) = 1.5$$

For $T = 10$ sec,

$$u(0) = -0.16, \qquad u(T) = u(10) = 0.06$$

Notice that, for a small value of the sampling period T, $u(0)$ and $u(T)$ become large. Increasing the value of T reduces the magnitudes of $u(0)$ and $u(T)$ significantly.

Problem A–6–12

Consider the system defined by

$$\mathbf{x}(k + 1) = \mathbf{G}\mathbf{x}(k) + \mathbf{H}u(k)$$

where $\mathbf{x}(k)$ is a 3-vector. It is assumed that the system is completely state controllable. By use of the pole placement technique, we wish to design the system to have closed-loop poles at $z = \mu_1$, $z = \mu_2$, and $z = \mu_3$, where the μ_i's are distinct. That is, using the state feedback control

$$u(k) = -\mathbf{K}\mathbf{x}(k)$$

we wish to have

$$|z\mathbf{I} - \mathbf{G} + \mathbf{H}\mathbf{K}| = (z - \mu_1)(z - \mu_2)(z - \mu_3) = z^3 + \alpha_1 z^2 + \alpha_2 z + \alpha_3$$

Show that the desired state feedback gain matrix $\mathbf{K}$ can be given by

$$\mathbf{K} = [1 \quad 1 \quad 1][\boldsymbol{\xi}_1 \quad \boldsymbol{\xi}_2 \quad \boldsymbol{\xi}_3]^{-1} \tag{6–265}$$

where

$$\boldsymbol{\xi}_i = (\mathbf{G} - \mu_i \mathbf{I})^{-1}\mathbf{H}, \qquad i = 1, 2, 3 \tag{6–266}$$

Show also that the vectors $\boldsymbol{\xi}_i$ are eigenvectors of matrix $\mathbf{G} - \mathbf{H}\mathbf{K}$; that is, $\boldsymbol{\xi}_i$ satisfies the equation

$$(\mathbf{G} - \mathbf{H}\mathbf{K})\boldsymbol{\xi}_i = \mu_i \boldsymbol{\xi}_i, \qquad i = 1, 2, 3$$

Solution   Let us define

$$\tilde{\mathbf{G}} = \mathbf{G} - \mathbf{H}\mathbf{K}$$

By use of the Cayley–Hamilton theorem, $\tilde{\mathbf{G}}$ satisfies its own characteristic equation:

$$\tilde{\mathbf{G}}^3 + \alpha_1 \tilde{\mathbf{G}}^2 + \alpha_2 \tilde{\mathbf{G}} + \alpha_3 \mathbf{I} = \phi(\tilde{\mathbf{G}}) = \mathbf{0} \tag{6–267}$$

Consider the following identities:

$$\mathbf{I} = \mathbf{I}$$

$$\tilde{\mathbf{G}} = \mathbf{G} - \mathbf{H}\mathbf{K}$$

$$\tilde{\mathbf{G}}^2 = (\mathbf{G} - \mathbf{H}\mathbf{K})^2 = \mathbf{G}^2 - \mathbf{G}\mathbf{H}\mathbf{K} - \mathbf{H}\mathbf{K}\tilde{\mathbf{G}}$$

$$\tilde{\mathbf{G}}^3 = (\mathbf{G} - \mathbf{H}\mathbf{K})^3 = \mathbf{G}^3 - \mathbf{G}^2\mathbf{H}\mathbf{K} - \mathbf{G}\mathbf{H}\mathbf{K}\tilde{\mathbf{G}} - \mathbf{H}\mathbf{K}\tilde{\mathbf{G}}^2$$

Multiplying each preceding equation by α_3, α_2, α_1, and α_0 (where $\alpha_0 = 1$) in this order and adding the results, we obtain

$$\alpha_3 \mathbf{I} + \alpha_2 \tilde{\mathbf{G}} + \alpha_1 \tilde{\mathbf{G}}^2 + \tilde{\mathbf{G}}^3 = \alpha_3 \mathbf{I} + \alpha_2 \mathbf{G} + \alpha_1 \mathbf{G}^2 + \mathbf{G}^3 - \alpha_2 \mathbf{H}\mathbf{K}$$

$$- \alpha_1 \mathbf{G}\mathbf{H}\mathbf{K} - \alpha_1 \mathbf{H}\mathbf{K}\tilde{\mathbf{G}} - \mathbf{G}^2\mathbf{H}\mathbf{K} - \mathbf{G}\mathbf{H}\mathbf{K}\tilde{\mathbf{G}} - \mathbf{H}\mathbf{K}\tilde{\mathbf{G}}^2$$

Noting that the left-hand side of this last equation is $\mathbf{0}$, this last equation can be reduced to

$$0 = \phi(\mathbf{G}) - [\mathbf{H} \,\vdots\, \mathbf{G}\mathbf{H} \,\vdots\, \mathbf{G}^2\mathbf{H}] \begin{bmatrix} \alpha_2 \mathbf{K} + \alpha_1 \mathbf{K}\tilde{\mathbf{G}} + \mathbf{K}\tilde{\mathbf{G}}^2 \\ \alpha_1 \mathbf{K} + \mathbf{K}\tilde{\mathbf{G}} \\ \mathbf{K} \end{bmatrix}$$

Hence,

$$\begin{bmatrix} \alpha_2\,\mathbf{K} + \alpha_1\,\mathbf{K}\tilde{\mathbf{G}} + \mathbf{K}\tilde{\mathbf{G}}^2 \\ \alpha_1\,\mathbf{K} + \mathbf{K}\tilde{\mathbf{G}} \\ \mathbf{K} \end{bmatrix} = [\mathbf{H}\,\vdots\,\mathbf{G}\mathbf{H}\,\vdots\,\mathbf{G}^2\mathbf{H}]^{-1}\,\phi(\mathbf{G})$$

Premultiplying both sides of this last equation by $[0 \quad 0 \quad 1]$, we obtain

$$[0 \quad 0 \quad 1]\begin{bmatrix} \alpha_2\,\mathbf{K} + \alpha_1\,\mathbf{K}\tilde{\mathbf{G}} + \mathbf{K}\tilde{\mathbf{G}}^2 \\ \alpha_1\,\mathbf{K} + \mathbf{K}\tilde{\mathbf{G}} \\ \mathbf{K} \end{bmatrix} = [0 \quad 0 \quad 1][\mathbf{H}\,\vdots\,\mathbf{G}\mathbf{H}\,\vdots\,\mathbf{G}^2\mathbf{H}]^{-1}\,\phi(\mathbf{G})$$

or

$$\mathbf{K} = [0 \quad 0 \quad 1][\mathbf{H}\,\vdots\,\mathbf{G}\mathbf{H}\,\vdots\,\mathbf{G}^2\mathbf{H}]^{-1}\,\phi(\mathbf{G}) \tag{6-268}$$

which is Ackermann's formula. Noting that

$$\phi(\mathbf{G}) = \mathbf{G}^3 + \alpha_1\,\mathbf{G}^2 + \alpha_2\,\mathbf{G} + \alpha_3\,\mathbf{I}$$
$$= (\mathbf{G} - \mu_1\mathbf{I})(\mathbf{G} - \mu_2\mathbf{I})(\mathbf{G} - \mu_3\mathbf{I})$$

we have

$$\mathbf{K} = [0 \quad 0 \quad 1][\mathbf{H}\,\vdots\,\mathbf{G}\mathbf{H}\,\vdots\,\mathbf{G}^2\mathbf{H}]^{-1}(\mathbf{G} - \mu_1\mathbf{I})(\mathbf{G} - \mu_2\mathbf{I})(\mathbf{G} - \mu_3\mathbf{I})$$

By postmultiplying both sides of this last equation by $\boldsymbol{\xi}_1 = (\mathbf{G} - \mu_1\mathbf{I})^{-1}\mathbf{H}$, we obtain

$$\mathbf{K}\boldsymbol{\xi}_1 = [0 \quad 0 \quad 1][\mathbf{H}\,\vdots\,\mathbf{G}\mathbf{H}\,\vdots\,\mathbf{G}^2\mathbf{H}]^{-1}(\mathbf{G} - \mu_1\mathbf{I})(\mathbf{G} - \mu_2\mathbf{I})(\mathbf{G} - \mu_3\mathbf{I})(\mathbf{G} - \mu_1\mathbf{I})^{-1}\mathbf{H}$$
$$= [0 \quad 0 \quad 1][\mathbf{H}\,\vdots\,\mathbf{G}\mathbf{H}\,\vdots\,\mathbf{G}^2\mathbf{H}]^{-1}(\mathbf{G} - \mu_2\mathbf{I})(\mathbf{G} - \mu_3\mathbf{I})\mathbf{H} \tag{6-269}$$

Let us define

$$(\mathbf{G} - \mu_1\mathbf{I})(\mathbf{G} - \mu_2\mathbf{I}) = \mathbf{G}^2 + \beta_{12}\mathbf{G} + \beta_{13}\mathbf{I}$$
$$(\mathbf{G} - \mu_2\mathbf{I})(\mathbf{G} - \mu_3\mathbf{I}) = \mathbf{G}^2 + \beta_{22}\mathbf{G} + \beta_{23}\mathbf{I}$$
$$(\mathbf{G} - \mu_3\mathbf{I})(\mathbf{G} - \mu_1\mathbf{I}) = \mathbf{G}^2 + \beta_{32}\mathbf{G} + \beta_{33}\mathbf{I}$$

Then Equation (6–269) can be written as follows:

$$\mathbf{K}\boldsymbol{\xi}_1 = [0 \quad 0 \quad 1][\mathbf{H}\,\vdots\,\mathbf{G}\mathbf{H}\,\vdots\,\mathbf{G}^2\mathbf{H}]^{-1}(\mathbf{G}^2 + \beta_{22}\mathbf{G} + \beta_{23}\mathbf{I})\mathbf{H}$$

$$= [0 \quad 0 \quad 1][\mathbf{H}\,\vdots\,\mathbf{G}\mathbf{H}\,\vdots\,\mathbf{G}^2\mathbf{H}]^{-1}[\mathbf{H}\,\vdots\,\mathbf{G}\mathbf{H}\,\vdots\,\mathbf{G}^2\mathbf{H}]\begin{bmatrix} \beta_{23} \\ \beta_{22} \\ 1 \end{bmatrix}$$

$$= [0 \quad 0 \quad 1]\begin{bmatrix} \beta_{23} \\ \beta_{22} \\ 1 \end{bmatrix} = 1$$

Hence,

$$\mathbf{K}\boldsymbol{\xi}_1 = 1$$

Similarly, we obtain

$$\mathbf{K}\boldsymbol{\xi}_2 = 1, \qquad \mathbf{K}\boldsymbol{\xi}_3 = 1$$

Hence,

$$\mathbf{K}[\boldsymbol{\xi}_1 \quad \boldsymbol{\xi}_2 \quad \boldsymbol{\xi}_3] = [1 \quad 1 \quad 1]$$

or

$$\mathbf{K} = [1 \quad 1 \quad 1][\boldsymbol{\xi}_1 \quad \boldsymbol{\xi}_2 \quad \boldsymbol{\xi}_3]^{-1} \tag{6-270}$$

Equation (6–270) gives the desired state feedback gain matrix $\mathbf{K}$ in terms of ξ_1, ξ_2, and ξ_3.

To show that the ξ_i's are eigenvectors of matrix $\mathbf{G} - \mathbf{HK}$, notice that

$$(\mathbf{G} - \mathbf{HK})\xi_i = (\mathbf{G} - \mathbf{HK})(\mathbf{G} - \mu_i \mathbf{I})^{-1}\mathbf{H}$$

$$= (\mathbf{G} - \mu_i \mathbf{I} + \mu_i \mathbf{I} - \mathbf{HK})(\mathbf{G} - \mu_i \mathbf{I})^{-1}\mathbf{H}$$

$$= (\mathbf{G} - \mu_i \mathbf{I})(\mathbf{G} - \mu_i \mathbf{I})^{-1}\mathbf{H} + (\mu_i \mathbf{I} - \mathbf{HK})(\mathbf{G} - \mu_i \mathbf{I})^{-1}\mathbf{H}$$

$$= \mathbf{H} + (\mu_i \mathbf{I} - \mathbf{HK})\xi_i$$

$$= \mathbf{H} - \mathbf{HK}\xi_i + \mu_i \xi_i \qquad (6\text{–}271)$$

As we have shown earlier,

$$\mathbf{K}\xi_i = 1, \qquad i = 1, 2, 3$$

Hence Equation (6–271) can be simplified to

$$(\mathbf{G} - \mathbf{HK})\xi_i = \mu_i \xi_i, \qquad i = 1, 2, 3$$

Thus, vectors ξ_1, ξ_2, and ξ_3 are eigenvectors of matrix $\mathbf{G} - \mathbf{HK}$ corresponding to eigenvalues μ_1, μ_2, and μ_3, respectively.

Problem A–6–13

Consider the system defined by

$$x(k + 1) = \mathbf{G}x(k) + \mathbf{H}u(k)$$

where $x(k)$ is a 3-vector. It is assumed that the system is completely state controllable and that the deadbeat response to the initial state $x(0)$ is desired. (That is, the desired closed-loop poles must be at the origin so that $\mu_1 = \mu_2 = \mu_3 = 0$.)

Show that the desired state feedback gain matrix $\mathbf{K}$ can be given by

$$\mathbf{K} = [1 \quad 0 \quad 0][\xi_1 \quad \xi_2 \quad \xi_3]^{-1} \qquad (6\text{–}272)$$

where

$$\xi_1 = \mathbf{G}^{-1}\mathbf{H}$$

$$\xi_2 = \mathbf{G}^{-2}\mathbf{H}$$

$$\xi_3 = \mathbf{G}^{-3}\mathbf{H}$$

Show also that the vectors ξ_i are generalized eigenvectors of matrix $\mathbf{G} - \mathbf{HK}$; that is, ξ_i satisfies the equations

$$(\mathbf{G} - \mathbf{HK})\xi_1 = \mathbf{0}$$

$$(\mathbf{G} - \mathbf{HK})\xi_2 = \xi_1$$

$$(\mathbf{G} - \mathbf{HK})\xi_3 = \xi_2$$

Solution   Referring to Equation (6–268), we have

$$\mathbf{K} = [0 \quad 0 \quad 1][\mathbf{H} \vdots \mathbf{GH} \vdots \mathbf{G}^2\mathbf{H}]^{-1}\phi(\mathbf{G})$$

where

$$\phi(\mathbf{G}) = \mathbf{G}^3$$

Hence,

$$\mathbf{K} = [0 \quad 0 \quad 1][\mathbf{H} \vdots \mathbf{GH} \vdots \mathbf{G}^2\mathbf{H}]^{-1}\mathbf{G}^3 \qquad (6\text{–}273)$$

By postmultiplying both sides of Equation (6–273) by $\boldsymbol{\xi}_1 = \mathbf{G}^{-1}\mathbf{H}$, we obtain

$$\mathbf{K}\boldsymbol{\xi}_1 = [0 \quad 0 \quad 1][\mathbf{H} \vdots \mathbf{GH} \vdots \mathbf{G}^2\mathbf{H}]^{-1}\mathbf{G}^3\mathbf{G}^{-1}\mathbf{H}$$

$$= [0 \quad 0 \quad 1][\mathbf{H} \vdots \mathbf{GH} \vdots \mathbf{G}^2\mathbf{H}]^{-1}\mathbf{G}^2\mathbf{H}$$

$$= [0 \quad 0 \quad 1][\mathbf{H} \vdots \mathbf{GH} \vdots \mathbf{G}^2\mathbf{H}]^{-1}[\mathbf{H} \vdots \mathbf{GH} \vdots \mathbf{G}^2\mathbf{H}]\begin{bmatrix} 0 \\ 0 \\ 1 \end{bmatrix}$$

$$= [0 \quad 0 \quad 1]\begin{bmatrix} 0 \\ 0 \\ 1 \end{bmatrix} = 1$$

Hence,

$$\mathbf{K}\boldsymbol{\xi}_1 = 1$$

By postmultiplying both sides of Equation (6–273) by $\boldsymbol{\xi}_2 = \mathbf{G}^{-2}\mathbf{H}$, we obtain

$$\mathbf{K}\boldsymbol{\xi}_2 = [0 \quad 0 \quad 1][\mathbf{H} \vdots \mathbf{GH} \vdots \mathbf{G}^2\mathbf{H}]^{-1}\mathbf{G}^3\mathbf{G}^{-2}\mathbf{H}$$

$$= [0 \quad 0 \quad 1][\mathbf{H} \vdots \mathbf{GH} \vdots \mathbf{G}^2\mathbf{H}]^{-1}\mathbf{GH}$$

$$= [0 \quad 0 \quad 1][\mathbf{H} \vdots \mathbf{GH} \vdots \mathbf{G}^2\mathbf{H}]^{-1}[\mathbf{H} \vdots \mathbf{GH} \vdots \mathbf{G}^2\mathbf{H}]\begin{bmatrix} 0 \\ 1 \\ 0 \end{bmatrix} = 0$$

Hence,

$$\mathbf{K}\boldsymbol{\xi}_2 = 0$$

Similarly, by postmultiplying both sides of Equation (6–273) by $\boldsymbol{\xi}_3 = \mathbf{G}^{-3}\mathbf{H}$, we obtain

$$\mathbf{K}\boldsymbol{\xi}_3 = [0 \quad 0 \quad 1][\mathbf{H} \vdots \mathbf{GH} \vdots \mathbf{G}^2\mathbf{H}]^{-1}\mathbf{G}^3\mathbf{G}^{-3}\mathbf{H}$$

$$= [0 \quad 0 \quad 1][\mathbf{H} \vdots \mathbf{GH} \vdots \mathbf{G}^2\mathbf{H}]^{-1}\mathbf{H}$$

$$= [0 \quad 0 \quad 1][\mathbf{H} \vdots \mathbf{GH} \vdots \mathbf{G}^2\mathbf{H}]^{-1}[\mathbf{H} \vdots \mathbf{GH} \vdots \mathbf{G}^2\mathbf{H}]\begin{bmatrix} 1 \\ 0 \\ 0 \end{bmatrix} = 0$$

Hence,

$$\mathbf{K}\boldsymbol{\xi}_3 = 0$$

Consequently, we have

$$\mathbf{K}[\boldsymbol{\xi}_1 \quad \boldsymbol{\xi}_2 \quad \boldsymbol{\xi}_3] = [1 \quad 0 \quad 0]$$

Hence,

$$\mathbf{K} = [1 \quad 0 \quad 0][\boldsymbol{\xi}_1 \quad \boldsymbol{\xi}_2 \quad \boldsymbol{\xi}_3]^{-1}$$

which is Equation (6–272).

To show that $\boldsymbol{\xi}_1$ is an eigenvector of matrix $\mathbf{G} - \mathbf{HK}$, notice that

$$(\mathbf{G} - \mathbf{HK})\boldsymbol{\xi}_1 = (\mathbf{G} - \mathbf{HK})\mathbf{G}^{-1}\mathbf{H} = \mathbf{H} - \mathbf{HKG}^{-1}\mathbf{H} = \mathbf{H} - \mathbf{HK}\boldsymbol{\xi}_1$$

Since $\mathbf{K}\boldsymbol{\xi}_1 = 1$, we obtain

$$(\mathbf{G} - \mathbf{HK})\boldsymbol{\xi}_1 = \mathbf{0}$$

To show that $\xi_2 = G^{-2}H$ is a generalized eigenvector of matrix $G - HK$, notice that

$$(G - HK)\xi_2 = (G - HK)G^{-2}H = G^{-1}H - HKG^{-2}H = \xi_1 - HK\xi_2$$

Since $K\xi_2 = 0$, we obtain

$$(G - HK)\xi_2 = \xi_1$$

Similarly, to show that $\xi_3 = G^{-3}H$ is a generalized eigenvector of matrix $G - HK$, notice that

$$(G - HK)\xi_3 = (G - HK)G^{-3}H = G^{-2}H - HKG^{-3}H = \xi_2 - HK\xi_3$$

Since $K\xi_3 = 0$, we obtain

$$(G - HK)\xi_3 = \xi_2$$

Problem A–6–14

Consider the system defined by

$$\mathbf{x}(k + 1) = \mathbf{G}\mathbf{x}(k)$$

$$y(k) = \mathbf{C}\mathbf{x}(k)$$

where $\mathbf{x}(k)$ is a 3-vector and $y(k)$ is a scalar. It is assumed that the system is completely observable. It is desired to determine the observer feedback gain matrix $\mathbf{K}_e$ for a full-order prediction observer such that the error dynamics have characteristic roots at $z = \mu_1$, $z = \mu_2$, and $z = \mu_3$, or

$$|z\mathbf{I} - \mathbf{G} + \mathbf{K}_e\mathbf{C}| = (z - \mu_1)(z - \mu_2)(z - \mu_3)$$

$$= z^3 + \alpha_1 z^2 + \alpha_2 z + \alpha_3$$

Assume that the eigenvalues of $\mathbf{G}$ are λ_1, λ_2, and λ_3 and they are different from μ_1, μ_2, and μ_3. We also assume that μ_1, μ_2, and μ_3 are distinct.

Show that the matrix $\mathbf{K}_e$ can be given by

$$\mathbf{K}_e = \begin{bmatrix} \mathbf{f}_1 \\ \mathbf{f}_2 \\ \mathbf{f}_3 \end{bmatrix}^{-1} \begin{bmatrix} 1 \\ 1 \\ 1 \end{bmatrix}$$

where

$$\mathbf{f}_i = \mathbf{C}(\mathbf{G} - \mu_i\mathbf{I})^{-1}, \qquad i = 1, 2, 3$$

Show also that the $\mathbf{f}_i^*$'s are eigenvectors of matrix $(\mathbf{G} - \mathbf{K}_e\mathbf{C})^*$; that is, $\mathbf{f}_i^*$ satisfies the equation

$$(\mathbf{G} - \mathbf{K}_e\mathbf{C})^*\mathbf{f}_i^* = \bar{\mu}_i\mathbf{f}_i^*, \qquad i = 1, 2, 3$$

where $\bar{\mu}_i$ is the complex conjugate of μ_i. (Note that any complex eigenvalues occur as conjugate pairs.)

Solution Referring to Ackermann's formula as given by Equation (6–126) for the observer feedback gain matrix $\mathbf{K}_e$, we have

$$\mathbf{K}_e = \phi(\mathbf{G}) \begin{bmatrix} \mathbf{C} \\ \mathbf{C}\mathbf{G} \\ \mathbf{C}\mathbf{G}^2 \end{bmatrix}^{-1} \begin{bmatrix} 0 \\ 0 \\ 1 \end{bmatrix}$$

Noting that

$$\phi(\mathbf{G}) = \mathbf{G}^3 + \alpha_1\mathbf{G}^2 + \alpha_2\mathbf{G} + \alpha_3\mathbf{I} = (\mathbf{G} - \mu_1\mathbf{I})(\mathbf{G} - \mu_2\mathbf{I})(\mathbf{G} - \mu_3\mathbf{I})$$

we have

$$\mathbf{K}_e = (\mathbf{G} - \mu_1 \mathbf{I})(\mathbf{G} - \mu_2 \mathbf{I})(\mathbf{G} - \mu_3 \mathbf{I}) \begin{bmatrix} \mathbf{C} \\ \mathbf{CG} \\ \mathbf{CG}^2 \end{bmatrix}^{-1} \begin{bmatrix} 0 \\ 0 \\ 1 \end{bmatrix} \tag{6-274}$$

By premultiplying both sides of this last equation by $\mathbf{f}_1 = \mathbf{C}(\mathbf{G} - \mu_1 \mathbf{I})^{-1}$, we obtain

$$\mathbf{f}_1 \mathbf{K}_e = \mathbf{C}(\mathbf{G} - \mu_2 \mathbf{I})(\mathbf{G} - \mu_3 \mathbf{I}) \begin{bmatrix} \mathbf{C} \\ \mathbf{CG} \\ \mathbf{CG}^2 \end{bmatrix}^{-1} \begin{bmatrix} 0 \\ 0 \\ 1 \end{bmatrix} \tag{6-275}$$

Let us define

$$(\mathbf{G} - \mu_1 \mathbf{I})(\mathbf{G} - \mu_2 \mathbf{I}) = \mathbf{G}^2 + \beta_{12} \mathbf{G} + \beta_{13} \mathbf{I}$$

$$(\mathbf{G} - \mu_2 \mathbf{I})(\mathbf{G} - \mu_3 \mathbf{I}) = \mathbf{G}^2 + \beta_{22} \mathbf{G} + \beta_{23} \mathbf{I}$$

$$(\mathbf{G} - \mu_3 \mathbf{I})(\mathbf{G} - \mu_1 \mathbf{I}) = \mathbf{G}^2 + \beta_{32} \mathbf{G} + \beta_{33} \mathbf{I}$$

Then Equation (6–275) becomes

$$\mathbf{f}_1 \mathbf{K}_e = \mathbf{C}(\mathbf{G}^2 + \beta_{22} \mathbf{G} + \beta_{23} \mathbf{I}) \begin{bmatrix} \mathbf{C} \\ \mathbf{CG} \\ \mathbf{CG}^2 \end{bmatrix}^{-1} \begin{bmatrix} 0 \\ 0 \\ 1 \end{bmatrix}$$

$$= \begin{bmatrix} \beta_{23} & \beta_{22} & 1 \end{bmatrix} \begin{bmatrix} \mathbf{C} \\ \mathbf{CG} \\ \mathbf{CG}^2 \end{bmatrix} \begin{bmatrix} \mathbf{C} \\ \mathbf{CG} \\ \mathbf{CG}^2 \end{bmatrix}^{-1} \begin{bmatrix} 0 \\ 0 \\ 1 \end{bmatrix} = 1$$

Hence,

$$\mathbf{f}_1 \mathbf{K}_e = 1$$

Similarly, we obtain

$$\mathbf{f}_2 \mathbf{K}_e = 1, \qquad \mathbf{f}_3 \mathbf{K}_e = 1$$

Thus,

$$\begin{bmatrix} \mathbf{f}_1 \\ \mathbf{f}_2 \\ \mathbf{f}_3 \end{bmatrix} \mathbf{K}_e = \begin{bmatrix} 1 \\ 1 \\ 1 \end{bmatrix}$$

or

$$\mathbf{K}_e = \begin{bmatrix} \mathbf{f}_1 \\ \mathbf{f}_2 \\ \mathbf{f}_3 \end{bmatrix}^{-1} \begin{bmatrix} 1 \\ 1 \\ 1 \end{bmatrix} \tag{6-276}$$

Equation (6–276) gives the desired matrix $\mathbf{K}_e$ in terms of $\mathbf{f}_1$, $\mathbf{f}_2$, and $\mathbf{f}_3$, where

$$\mathbf{f}_i = \mathbf{C}(\mathbf{G} - \mu_i \mathbf{I})^{-1}, \qquad i = 1, 2, 3$$

To show that the $\mathbf{f}_i^*$'s are eigenvectors of matrix $(\mathbf{G} - \mathbf{K}_e \mathbf{C})^*$, notice that

$$(\mathbf{G} - \mathbf{K}_e \mathbf{C})^* \mathbf{f}_i^* = (\mathbf{G} - \mathbf{K}_e \mathbf{C})^* [\mathbf{C}(\mathbf{G} - \mu_i \mathbf{I})^{-1}]^*$$

$$= (\mathbf{G}^* - \mathbf{C}^* \mathbf{K}_e^*)(\mathbf{G}^* - \overline{\mu}_i \mathbf{I})^{-1} \mathbf{C}^*$$

$$= (\mathbf{G}^* - \overline{\mu}_i \mathbf{I} + \overline{\mu}_i \mathbf{I} - \mathbf{C}^* \mathbf{K}_e^*)(\mathbf{G}^* - \overline{\mu}_i \mathbf{I})^{-1} \mathbf{C}^*$$

$$= \mathbf{C}^* + (\overline{\mu}_i \mathbf{I} - \mathbf{C}^* \mathbf{K}_e^*) \mathbf{f}_i^*$$

$$= \mathbf{C}^* - \mathbf{C}^* \mathbf{K}_e^* \mathbf{f}_i^* + \overline{\mu}_i \mathbf{f}_i^* \tag{6-277}$$

As we have shown earlier,

$$\mathbf{f}_i \, \mathbf{K}_e = 1$$

Hence,

$$\mathbf{K}_e^* \, \mathbf{f}_i^* = 1$$

Thus, Equation (6–277) becomes

$$(\mathbf{G} - \mathbf{K}_e \, \mathbf{C})^* \mathbf{f}_i^* = \mathbf{C}^* - \mathbf{C}^* + \overline{\mu}_i \, \mathbf{f}_i^* = \overline{\mu}_i \, \mathbf{f}_i^*$$

We have thus shown that vectors $\mathbf{f}_1^*$, $\mathbf{f}_2^*$, and $\mathbf{f}_3^*$ are eigenvectors of matrix $(\mathbf{G} - \mathbf{K}_e \, \mathbf{C})^*$ corresponding to eigenvalues μ_1, μ_2, and μ_3, respectively.

Problem A–6–15

In Problem A–6–14 we obtained the observer feedback gain matrix $\mathbf{K}_e$ for the case where the eigenvalues μ_1, μ_2, and μ_3 of $\mathbf{G} - \mathbf{K}_e \, \mathbf{C}$ were distinct. Suppose that we desire the deadbeat response for the error vector. Then we require that $\mu_1 = \mu_2 = \mu_3 = 0$. Show that for this case matrix $\mathbf{K}_e$ can be given as follows:

$$\mathbf{K}_e = \begin{bmatrix} \mathbf{f}_3 \\ \mathbf{f}_2 \\ \mathbf{f}_1 \end{bmatrix}^{-1} \begin{bmatrix} 0 \\ 0 \\ 1 \end{bmatrix} \quad \text{or} \quad \mathbf{K}_e = \begin{bmatrix} \mathbf{f}_1 \\ \mathbf{f}_2 \\ \mathbf{f}_3 \end{bmatrix}^{-1} \begin{bmatrix} 1 \\ 0 \\ 0 \end{bmatrix}$$

where

$$\mathbf{f}_1 = \mathbf{CG}^{-1}, \qquad \mathbf{f}_2 = \mathbf{CG}^{-2}, \qquad \mathbf{f}_3 = \mathbf{CG}^{-3}$$

[Note that vectors $\mathbf{f}_1$, $\mathbf{f}_2$, and $\mathbf{f}_3$ given here are the eigenvectors or generalized eigenvectors of matrix $(\mathbf{G} - \mathbf{K}_e \, \mathbf{C})^*$.] The system is assumed to be completely observable.

Solution Referring to Equation (6–274), we have

$$\mathbf{K}_e = (\mathbf{G} - \mu_1 \mathbf{I})(\mathbf{G} - \mu_2 \mathbf{I})(\mathbf{G} - \mu_3 \mathbf{I}) \begin{bmatrix} \mathbf{C} \\ \mathbf{CG} \\ \mathbf{CG}^2 \end{bmatrix}^{-1} \begin{bmatrix} 0 \\ 0 \\ 1 \end{bmatrix}$$

By taking $\mu_1 = \mu_2 = \mu_3 = 0$ and substituting accordingly in this last equation, we obtain

$$\mathbf{K}_e = \mathbf{G}^3 \begin{bmatrix} \mathbf{C} \\ \mathbf{CG} \\ \mathbf{CG}^2 \end{bmatrix}^{-1} \begin{bmatrix} 0 \\ 0 \\ 1 \end{bmatrix}$$

which can be rewritten as follows:

$$\mathbf{K}_e = \begin{bmatrix} \mathbf{CG}^{-3} \\ \mathbf{CG}^{-2} \\ \mathbf{CG}^{-1} \end{bmatrix}^{-1} \begin{bmatrix} 0 \\ 0 \\ 1 \end{bmatrix} = \begin{bmatrix} \mathbf{f}_3 \\ \mathbf{f}_2 \\ \mathbf{f}_1 \end{bmatrix}^{-1} \begin{bmatrix} 0 \\ 0 \\ 1 \end{bmatrix} \tag{6–278}$$

Equation (6–278) gives the desired observer feedback gain matrix $\mathbf{K}_e$ when $\mu_1 = \mu_2 = \mu_3 = 0$.

Notice that Equation (6–278) can be modified to read

$$\begin{bmatrix} \mathbf{CG}^{-3} \\ \mathbf{CG}^{-2} \\ \mathbf{CG}^{-1} \end{bmatrix} \mathbf{K}_e = \begin{bmatrix} 0 \\ 0 \\ 1 \end{bmatrix}$$

which is equivalent to the following three equations:

$$\mathbf{CG}^{-3}\mathbf{K}_e = \mathbf{f}_3\mathbf{K}_e = 0$$

$$\mathbf{CG}^{-2}\mathbf{K}_e = \mathbf{f}_2\mathbf{K}_e = 0$$

$$\mathbf{CG}^{-1}\mathbf{K}_e = \mathbf{f}_1\mathbf{K}_e = 1$$

Hence, we obtain

$$
\begin{bmatrix} \mathbf{f}_1\,\mathbf{K}_e \\ \mathbf{f}_2\,\mathbf{K}_e \\ \mathbf{f}_3\,\mathbf{K}_e \end{bmatrix} = \begin{bmatrix} \mathbf{f}_1 \\ \mathbf{f}_2 \\ \mathbf{f}_3 \end{bmatrix}\mathbf{K}_e = \begin{bmatrix} 1 \\ 0 \\ 0 \end{bmatrix}
$$

or

$$
\mathbf{K}_e = \begin{bmatrix} \mathbf{f}_1 \\ \mathbf{f}_2 \\ \mathbf{f}_3 \end{bmatrix}^{-1}\begin{bmatrix} 1 \\ 0 \\ 0 \end{bmatrix}
$$

which also gives the desired observer feedback gain matrix when $\mu_1 = \mu_2 = \mu_3 = 0$.

Problem A–6–16

Consider the system

$$\mathbf{x}(k + 1) = \mathbf{Gx}(k) + \mathbf{H}u(k)$$

where

$$
\mathbf{G} = \begin{bmatrix} 0 & 1 \\ -0.16 & -1 \end{bmatrix}, \qquad \mathbf{H} = \begin{bmatrix} 0 \\ 1 \end{bmatrix}
$$

Assume that the following control scheme is used:

$$u = -\mathbf{Kx}$$

By use of MATLAB, determine the state feedback gain matrix $\mathbf{K}$ such that the system will have closed-loop poles at

$$z = 0.5 + j0.5, \qquad z = 0.5 - j0.5$$

Use Ackermann's formula given by Equation (6–68).

Solution We first construct matrix $\mathbf{J}$ whose eigenvalues are the desired closed-loop poles.

$$
\mathbf{J} = \begin{bmatrix} 0.5 + j0.5 & 0 \\ 0 & 0.5 - j0.5 \end{bmatrix}
$$

The command poly($\mathbf{J}$) gives the characteristic polynomial for $\mathbf{J}$.

```
p = poly(J)

p =

    1.0000   -1.0000   0.5000
```

This is the MATLAB expression for the characteristic polynomial for $\mathbf{J}$.

$$\text{poly}(\mathbf{J}) = \phi(\mathbf{J}) = \mathbf{J}^2 - \mathbf{J} + 0.5\mathbf{I}$$

where **I** is the identity matrix. For the matrix

$$\mathbf{G} = \begin{bmatrix} 0 & 1 \\ -0.16 & -1 \end{bmatrix}$$

the command polyvalm(poly(**J**), **G**) evaluates the following $\phi(\mathbf{G})$:

$$\phi(\mathbf{G}) = \mathbf{G}^2 - \mathbf{G} + 0.5\mathbf{I} = \begin{bmatrix} -0.16 & -1 \\ 0.16 & 0.84 \end{bmatrix} - \begin{bmatrix} 0 & 1 \\ -0.16 & -1 \end{bmatrix} + \begin{bmatrix} 0.5 & 0 \\ 0 & 0.5 \end{bmatrix}$$

$$= \begin{bmatrix} 0.34 & -2 \\ 0.32 & 2.34 \end{bmatrix}$$

See the following MATLAB output.

```
polyvalm(poly(J),G)

ans =

    0.3400   -2.0000
    0.3200    2.3400
```

Referring to the Ackermann's formula given by Equation (6–68), the desired matrix **K** is obtained from

$$\mathbf{K} = \begin{bmatrix} 0 & 1 \end{bmatrix}\begin{bmatrix} \mathbf{H} & \mathbf{GH} \end{bmatrix}^{-1}\phi(\mathbf{G})$$

$$= \begin{bmatrix} 0 & 1 \end{bmatrix}\mathbf{M}^{-1}\phi(\mathbf{G})$$

where $\mathbf{M} = \begin{bmatrix} \mathbf{H} & \mathbf{GH} \end{bmatrix}$. A MATLAB program for the determination of state feedback gain matrix **K** is given in MATLAB Program 6–4.

```
MATLAB Program 6–4

% --------- Pole placement in the z plane ---------

% ***** This program determines state feedback gain matrix K
% based on Ackermann's formula *****

% ***** Enter state matrix G and control matrix H *****

G = [0  1;-0.16 -1];
H = [0;1];

% ***** Enter the controllability matrix M and check its
% rank *****

M = [H  G*H];
rank(M)

ans =

   2

% ***** Since the rank of controllability matrix M is 2,
```

```
% arbitrary pole placement is possible *****

% ***** Enter the desired characteristic polynomial by
% defining the following matrix J and computing poly(J) *****

J = [0.5+0.5*i  0;0  0.5-0.5*i];

JJ = poly(J)

JJ =

    1.0000    -1.0000     0.5000

% ***** Enter characteristic polynomial Phi *****

Phi = polyvalm(poly(J),G);

% ***** State feedback gain matrix K can be given by *****

K = [0  1]*inv(M)*Phi

K =

    0.3400    -2.0000

k1 = K(1), k2 = K(2)

k1 =

    0.3400


k2 =

    -2
```

Problem A–6–17

Consider the system

$$\mathbf{x}(k + 1) = \mathbf{Gx}(k) + \mathbf{H}u(k)$$
$$y(k) = \mathbf{Cx}(k)$$

where

$$\mathbf{x}(k) = \text{state vector (3-vector)}$$
$$u(k) = \text{control signal (scalar)}$$
$$y(k) = \text{output signal (scalar)}$$

and

$$\mathbf{G} = \begin{bmatrix} 0 & 1 & 0 \\ 0 & 0 & 1 \\ -0.5 & -0.2 & 1.1 \end{bmatrix}, \quad \mathbf{H} = \begin{bmatrix} 0 \\ 0 \\ 1 \end{bmatrix}, \quad \mathbf{C} = \begin{bmatrix} 0 & 1 & 0 \end{bmatrix}$$

1. Determine the state feedback gain matrix **K** such that the system will exhibit a deadbeat response to any initial state. Assuming that the state is completely measurable so that the actual state $\mathbf{x}(k)$ can be fed back for control, or that

$$u(k) = -\mathbf{K}\mathbf{x}(k)$$

determine the response of the system to the initial state

$$x(0) = \begin{bmatrix} a \\ b \\ c \end{bmatrix}$$

where a, b, and c are arbitrary constants.

2. Assuming that only a portion of the state vector is measurable, that is, only the output $y(k)$ is measurable, design a minimum-order observer such that the response to the observer error is deadbeat. Assume that the system configuration is the same as that shown in Figure 6–11.

3. Assuming that the observed state is used for feedback, obtain the response of the system to

$$\mathbf{x}(0) = \begin{bmatrix} a \\ b \\ c \end{bmatrix}, \qquad \hat{\mathbf{e}}(0) = \begin{bmatrix} \alpha \\ \beta \end{bmatrix}$$

where $\hat{\mathbf{e}}(0)$ is the initial observer error for the minimum-order observer and a, b, c, α, and β are arbitrary constants.

4. Derive the pulse transfer function $G_D(z)$ of the observer regulator.

Solution Notice that the system is completely state controllable and observable.

1. The required state feedback gain matrix **K** for deadbeat response can be obtained easily, as follows. Let us define

$$\mathbf{K} = [k_1 \quad k_2 \quad k_3]$$

Then

$$|z\mathbf{I} - \mathbf{G} + \mathbf{HK}| = \begin{vmatrix} z & -1 & 0 \\ 0 & z & -1 \\ k_1 + 0.5 & k_2 + 0.2 & z + k_3 - 1.1 \end{vmatrix}$$

$$= z^3 + (k_3 - 1.1)z^2 + (k_2 + 0.2)z + k_1 + 0.5 = 0$$

By equating this characteristic equation with the desired characteristic equation (for deadbeat response),

$$z^3 = 0$$

we obtain

$$\mathbf{K} = [k_1 \quad k_2 \quad k_3] = [-0.5 \quad -0.2 \quad 1.1]$$

With this matrix **K**, the system equation becomes

$$\mathbf{x}(k + 1) = \mathbf{G}\mathbf{x}(k) + \mathbf{H}u(k) = (\mathbf{G} - \mathbf{HK})\mathbf{x}(k)$$

or

$$\begin{bmatrix} x_1(k + 1) \\ x_2(k + 1) \\ x_3(k + 1) \end{bmatrix} = \begin{bmatrix} 0 & 1 & 0 \\ 0 & 0 & 1 \\ 0 & 0 & 0 \end{bmatrix} \begin{bmatrix} x_1(k) \\ x_2(k) \\ x_3(k) \end{bmatrix}$$

The response of this system to an arbitrary initial state becomes as follows:

$$
\begin{bmatrix} x_1(1) \\ x_2(1) \\ x_3(1) \end{bmatrix} = \begin{bmatrix} 0 & 1 & 0 \\ 0 & 0 & 1 \\ 0 & 0 & 0 \end{bmatrix} \begin{bmatrix} a \\ b \\ c \end{bmatrix} = \begin{bmatrix} b \\ c \\ 0 \end{bmatrix}
$$

$$
\begin{bmatrix} x_1(2) \\ x_2(2) \\ x_3(2) \end{bmatrix} = \begin{bmatrix} 0 & 1 & 0 \\ 0 & 0 & 1 \\ 0 & 0 & 0 \end{bmatrix} \begin{bmatrix} b \\ c \\ 0 \end{bmatrix} = \begin{bmatrix} c \\ 0 \\ 0 \end{bmatrix}
$$

$$
\begin{bmatrix} x_1(3) \\ x_2(3) \\ x_3(3) \end{bmatrix} = \begin{bmatrix} 0 & 1 & 0 \\ 0 & 0 & 1 \\ 0 & 0 & 0 \end{bmatrix} \begin{bmatrix} c \\ 0 \\ 0 \end{bmatrix} = \begin{bmatrix} 0 \\ 0 \\ 0 \end{bmatrix}
$$

or

$$
\mathbf{x}(k) = \mathbf{0}, \qquad k = 3, 4, 5, \ldots
$$

Clearly, the response is deadbeat.

2. We shall now design a minimum-order observer assuming that only the output $y(k)$ is measurable. We shall first transform the state vector $\mathbf{x}(k)$ into a new state vector $\boldsymbol{\xi}(k)$ such that the output matrix $\mathbf{C}$ is transformed from $[0 \ \ 1 \ \ 0]$ to $[1 \ \ 0 \ \ 0]$. The following matrix $\mathbf{T}$ will accomplish the required transformation:

$$
\mathbf{T} = \begin{bmatrix} 0 & 1 & 0 \\ 1 & 0 & 0 \\ 0 & 0 & 1 \end{bmatrix}
$$

Thus, we define

$$
\mathbf{x}(k) = \mathbf{T}\boldsymbol{\xi}(k)
$$

Then the system equations become

$$
\boldsymbol{\xi}(k+1) = \mathbf{T}^{-1}\mathbf{G}\mathbf{T}\boldsymbol{\xi}(k) + \mathbf{T}^{-1}\mathbf{H}u(k) = \hat{\mathbf{G}}\boldsymbol{\xi}(k) + \hat{\mathbf{H}}u(k)
$$

$$
y(k) = \mathbf{C}\mathbf{T}\boldsymbol{\xi}(k) = \hat{\mathbf{C}}\boldsymbol{\xi}(k)
$$

where

$$
\hat{\mathbf{G}} = \mathbf{T}^{-1}\mathbf{G}\mathbf{T} = \begin{bmatrix} 0 & 1 & 0 \\ 1 & 0 & 0 \\ 0 & 0 & 1 \end{bmatrix} \begin{bmatrix} 0 & 1 & 0 \\ 0 & 0 & 1 \\ -0.5 & -0.2 & 1.1 \end{bmatrix} \begin{bmatrix} 0 & 1 & 0 \\ 1 & 0 & 0 \\ 0 & 0 & 1 \end{bmatrix}
$$

$$
= \begin{bmatrix} 0 & 0 & 1 \\ 1 & 0 & 0 \\ -0.2 & -0.5 & 1.1 \end{bmatrix} = \begin{bmatrix} \hat{G}_{aa} & \hat{G}_{ab} \\ \hat{G}_{ba} & \hat{G}_{bb} \end{bmatrix}
$$

$$
\hat{\mathbf{H}} = \mathbf{T}^{-1}\mathbf{H} = \begin{bmatrix} 0 & 1 & 0 \\ 1 & 0 & 0 \\ 0 & 0 & 1 \end{bmatrix} \begin{bmatrix} 0 \\ 0 \\ 1 \end{bmatrix} = \begin{bmatrix} 0 \\ 0 \\ 1 \end{bmatrix}
$$

$$
\hat{\mathbf{C}} = \mathbf{C}\mathbf{T} = [0 \ \ 1 \ \ 0] \begin{bmatrix} 0 & 1 & 0 \\ 1 & 0 & 0 \\ 0 & 0 & 1 \end{bmatrix} = [1 \ \ 0 \ \ 0]
$$

The transformed system is thus given by

$$
\begin{bmatrix} \xi_1(k+1) \\ \xi_2(k+1) \\ \xi_3(k+1) \end{bmatrix} = \left[\begin{array}{c|cc} 0 & 0 & 1 \\ \hline 1 & 0 & 0 \\ -0.2 & -0.5 & 1.1 \end{array} \right] \begin{bmatrix} \xi_1(k) \\ \xi_2(k) \\ \xi_3(k) \end{bmatrix} + \begin{bmatrix} 0 \\ 0 \\ 1 \end{bmatrix} u(k) \tag{6–279}
$$

$$
y(k) = [1 \ \vdots \ 0 \ \ 0] \begin{bmatrix} \xi_1(k) \\ \xi_2(k) \\ \xi_3(k) \end{bmatrix} \tag{6–280}
$$

Since only one state variable, $\xi_1(k)$, can be measured, we need to observe two state variables. Hence, the order of the minimum-order observer is 2. Since

$$\hat{\mathbf{G}}_{bb} - \mathbf{K}_e \hat{\mathbf{G}}_{ab} = \begin{bmatrix} 0 & 0 \\ -0.5 & 1.1 \end{bmatrix} - \begin{bmatrix} k_{e_1} \\ k_{e_2} \end{bmatrix} \begin{bmatrix} 0 & 1 \end{bmatrix} = \begin{bmatrix} 0 & -k_{e_1} \\ -0.5 & 1.1 - k_{e_2} \end{bmatrix}$$

the observer characteristic equation becomes

$$|z\mathbf{I} - \hat{\mathbf{G}}_{bb} + \mathbf{K}_e \hat{\mathbf{G}}_{ab}| = \begin{vmatrix} z & k_{e_1} \\ 0.5 & z - 1.1 + k_{e_2} \end{vmatrix} = z^2 + (k_{e_2} - 1.1)z - 0.5k_{e_1}$$

The desired characteristic equation (for deadbeat response) is

$$z^2 = 0$$

Hence, we obtain

$$k_{e_1} = 0, \qquad k_{e_2} = 1.1$$

or

$$\mathbf{K}_e = \begin{bmatrix} 0 \\ 1.1 \end{bmatrix}$$

3. The equation for the state feedback control system with a minimum-order observer is given by Equation (6–163):

$$\begin{bmatrix} \mathbf{x}(k+1) \\ \hline \mathbf{e}(k+1) \end{bmatrix} = \begin{bmatrix} \mathbf{G} - \mathbf{HK} & \vdots & \mathbf{HK\Gamma} \\ \hline \mathbf{0} & \vdots & \mathbf{G}_{bb} - \mathbf{K}_e \mathbf{G}_{ab} \end{bmatrix} \begin{bmatrix} \mathbf{x}(k) \\ \hline \mathbf{e}(k) \end{bmatrix} \qquad (6\text{–}281)$$

Let us rewrite Equation (6–281) in terms of the new state vector $\boldsymbol{\xi}(k)$ and error vector $\hat{\mathbf{e}}(k)$. Noting that the observed state is used for feedback, that is,

$$u(k) = -\hat{\mathbf{K}}\tilde{\boldsymbol{\xi}}(k)$$

we have

$$\boldsymbol{\xi}(k+1) = \hat{\mathbf{G}}\boldsymbol{\xi}(k) + \hat{\mathbf{H}}u(k)$$

$$= \hat{\mathbf{G}}\boldsymbol{\xi}(k) - \hat{\mathbf{H}}\hat{\mathbf{K}}\tilde{\boldsymbol{\xi}}(k)$$

$$= (\hat{\mathbf{G}} - \hat{\mathbf{H}}\hat{\mathbf{K}})\boldsymbol{\xi}(k) + \hat{\mathbf{H}}\hat{\mathbf{K}}[\boldsymbol{\xi}(k) - \tilde{\boldsymbol{\xi}}(k)]$$

$$= (\hat{\mathbf{G}} - \hat{\mathbf{H}}\hat{\mathbf{K}})\boldsymbol{\xi}(k) + \hat{\mathbf{H}}\hat{\mathbf{K}}\Gamma\hat{\mathbf{e}}(k)$$

where

$$\boldsymbol{\Gamma} = \begin{bmatrix} 0 & 0 \\ \hline 1 & 0 \\ 0 & 1 \end{bmatrix}$$

and

$$\hat{\mathbf{K}} = \mathbf{KT} = \begin{bmatrix} -0.5 & -0.2 & 1.1 \end{bmatrix} \begin{bmatrix} 0 & 1 & 0 \\ 1 & 0 & 0 \\ 0 & 0 & 1 \end{bmatrix} = \begin{bmatrix} -0.2 & -0.5 & 1.1 \end{bmatrix}$$

Hence, Equation (6–281) can be modified to read

$$\begin{bmatrix} \boldsymbol{\xi}(k+1) \\ \hline \hat{\mathbf{e}}(k+1) \end{bmatrix} = \begin{bmatrix} \hat{\mathbf{G}} - \hat{\mathbf{H}}\hat{\mathbf{K}} & \vdots & \hat{\mathbf{H}}\hat{\mathbf{K}}\Gamma \\ \hline \mathbf{0} & \vdots & \hat{\mathbf{G}}_{bb} - \mathbf{K}_e \hat{\mathbf{G}}_{ab} \end{bmatrix} \begin{bmatrix} \boldsymbol{\xi}(k) \\ \hline \hat{\mathbf{e}}(k) \end{bmatrix}$$

or

$$\begin{bmatrix} \xi_1(k+1) \\ \xi_2(k+1) \\ \xi_3(k+1) \\ \hat{e}_1(k+1) \\ \hat{e}_2(k+1) \end{bmatrix} = \begin{bmatrix} 0 & 0 & 1 & 0 & 0 \\ 1 & 0 & 0 & 0 & 0 \\ 0 & 0 & 0 & -0.5 & 1.1 \\ 0 & 0 & 0 & 0 & 0 \\ 0 & 0 & 0 & -0.5 & 0 \end{bmatrix} \begin{bmatrix} \xi_1(k) \\ \xi_2(k) \\ \xi_3(k) \\ \hat{e}_1(k) \\ \hat{e}_2(k) \end{bmatrix}$$

The response of this system to the given initial condition can be obtained as follows. First note that the assumed initial condition is

$$\begin{bmatrix} x_1(0) \\ x_2(0) \\ x_3(0) \\ \hat{e}_1(0) \\ \hat{e}_2(0) \end{bmatrix} = \begin{bmatrix} a \\ b \\ c \\ \alpha \\ \beta \end{bmatrix}$$

Hence,

$$\begin{bmatrix} \xi_1(0) \\ \xi_2(0) \\ \xi_3(0) \\ \hat{e}_1(0) \\ \hat{e}_2(0) \end{bmatrix} = \begin{bmatrix} b \\ a \\ c \\ \alpha \\ \beta \end{bmatrix}, \qquad \begin{bmatrix} \xi_1(1) \\ \xi_2(1) \\ \xi_3(1) \\ \hat{e}_1(1) \\ \hat{e}_2(1) \end{bmatrix} = \begin{bmatrix} c \\ b \\ -0.5\alpha + 1.1\beta \\ 0 \\ -0.5\alpha \end{bmatrix}$$

$$\begin{bmatrix} \xi_1(2) \\ \xi_2(2) \\ \xi_3(2) \\ \hat{e}_1(2) \\ \hat{e}_2(2) \end{bmatrix} = \begin{bmatrix} -0.5\alpha + 1.1\beta \\ c \\ -0.55\alpha \\ 0 \\ 0 \end{bmatrix}, \qquad \begin{bmatrix} \xi_1(3) \\ \xi_2(3) \\ \xi_3(3) \\ \hat{e}_1(3) \\ \hat{e}_2(3) \end{bmatrix} = \begin{bmatrix} -0.55\alpha \\ -0.5\alpha + 1.1\beta \\ 0 \\ 0 \\ 0 \end{bmatrix}$$

$$\begin{bmatrix} \xi_1(4) \\ \xi_2(4) \\ \xi_3(4) \\ \hat{e}_1(4) \\ \hat{e}_2(4) \end{bmatrix} = \begin{bmatrix} 0 \\ -0.55\alpha \\ 0 \\ 0 \\ 0 \end{bmatrix}, \qquad \begin{bmatrix} \xi_1(5) \\ \xi_2(5) \\ \xi_3(5) \\ \hat{e}_1(5) \\ \hat{e}_2(5) \end{bmatrix} = \begin{bmatrix} 0 \\ 0 \\ 0 \\ 0 \\ 0 \end{bmatrix}$$

The response is clearly deadbeat. For any initial condition, the settling time is at most five sampling periods. (This means that at most two sampling periods are needed for the error vector to become zero and, additionally, at most three sampling periods are needed for the state vector to become zero.)

   4. To derive the pulse transfer function $G_D(z)$ of the observer regulator, we refer to the state equation and output equation given by Equations (6–279) and (6–280), respectively. The equations for the minimum-order observer are given by Equations (6–148) and (6–149), rewritten thus:

$$\tilde{\xi}_b(k) - \mathbf{K}_e y(k) = \tilde{\eta}(k)$$

$$\tilde{\eta}(k+1) = (\hat{\mathbf{G}}_{bb} - \mathbf{K}_e \hat{\mathbf{G}}_{ab})\tilde{\eta}(k) + [(\hat{\mathbf{G}}_{bb} - \mathbf{K}_e \hat{\mathbf{G}}_{ab})\mathbf{K}_e$$
$$+ \hat{\mathbf{G}}_{ba} - \mathbf{K}_e \hat{\mathbf{G}}_{aa}]y(k) + (\hat{\mathbf{H}}_b - \mathbf{K}_e \hat{H}_a)u(k) \qquad (6\text{–}282)$$

For this problem,

$$\hat{\mathbf{G}}_{bb} - \mathbf{K}_e \hat{\mathbf{G}}_{ab} = \begin{bmatrix} 0 & 0 \\ -0.5 & 0 \end{bmatrix}$$

$$\hat{\mathbf{G}}_{ba} - \mathbf{K}_e \hat{\mathbf{G}}_{aa} = \begin{bmatrix} 1 \\ -0.2 \end{bmatrix}$$

$$\hat{\mathbf{H}}_b - \mathbf{K}_e \hat{\mathbf{H}}_a = \begin{bmatrix} 0 \\ 1 \end{bmatrix}$$

Hence, Equation (6–282) becomes

$$\bar{\boldsymbol{\eta}}(k + 1) = \begin{bmatrix} 0 & 0 \\ -0.5 & 0 \end{bmatrix} \bar{\boldsymbol{\eta}}(k) + \begin{bmatrix} 1 \\ -0.2 \end{bmatrix} y(k) + \begin{bmatrix} 0 \\ 1 \end{bmatrix} u(k)$$

Taking the z transform of this last equation, we obtain

$$z\bar{\boldsymbol{\eta}}(z) = \begin{bmatrix} 0 & 0 \\ -0.5 & 0 \end{bmatrix} \bar{\boldsymbol{\eta}}(z) + \begin{bmatrix} 1 \\ -0.2 \end{bmatrix} Y(z) + \begin{bmatrix} 0 \\ 1 \end{bmatrix} U(z)$$

or

$$\begin{bmatrix} z & 0 \\ 0.5 & z \end{bmatrix} \bar{\boldsymbol{\eta}}(z) = \begin{bmatrix} 1 \\ -0.2 \end{bmatrix} Y(z) + \begin{bmatrix} 0 \\ 1 \end{bmatrix} U(z)$$

Solving for $\bar{\boldsymbol{\eta}}(z)$, we obtain

$$\bar{\boldsymbol{\eta}}(z) = \begin{bmatrix} \dfrac{1}{z} \\ -\dfrac{0.5}{z^2} - \dfrac{0.2}{z} \end{bmatrix} Y(z) + \begin{bmatrix} 0 \\ \dfrac{1}{z} \end{bmatrix} U(z)$$

Equation (6–148) becomes

$$\bar{\boldsymbol{\xi}}_b(z) = \begin{bmatrix} 0 \\ 1.1 \end{bmatrix} Y(z) + \bar{\boldsymbol{\eta}}(z)$$

$$= \begin{bmatrix} \dfrac{1}{z} \\ 1.1 - \dfrac{0.5}{z^2} - \dfrac{0.2}{z} \end{bmatrix} Y(z) + \begin{bmatrix} 0 \\ \dfrac{1}{z} \end{bmatrix} U(z)$$

The control signal $u(k)$ is given by

$$u(k) = -\hat{\mathbf{K}}\bar{\boldsymbol{\xi}}(k) = -[-0.2 \quad -0.5 \quad 1.1]\bar{\boldsymbol{\xi}}(k)$$

$$= 0.2y(k) - [-0.5 \quad 1.1]\bar{\boldsymbol{\xi}}_b(k)$$

The z transform of this last equation becomes

$$U(z) = 0.2Y(z) - [-0.5 \quad 1.1]\begin{bmatrix} \dfrac{1}{z} \\ 1.1 - \dfrac{0.5}{z^2} - \dfrac{0.2}{z} \end{bmatrix} Y(z)$$

$$- [-0.5 \quad 1.1]\begin{bmatrix} 0 \\ \dfrac{1}{z} \end{bmatrix} U(z)$$

$$= \left[0.2 + \dfrac{0.5}{z} - 1.1\left(1.1 - \dfrac{0.5}{z^2} - \dfrac{0.2}{z} \right) \right] Y(z) - \dfrac{1.1}{z} U(z)$$

or

$$\left(1 + \frac{1.1}{z}\right)U(z) = \left(-1.01 + \frac{0.72}{z} + \frac{0.55}{z^2}\right)Y(z)$$

from which we get the pulse transfer function $G_D(z)$ of the observer regulator as follows:

$$G_D(z) = -\frac{U(z)}{Y(z)} = \frac{1.01z^2 - 0.72z - 0.55}{z^2 + 1.1z} \qquad (6\text{-}283)$$

The pulse transfer function of the plant can be obtained by use of Equation (5-60) as follows:

$$G_p(z) = \frac{Y(z)}{U(z)} = \mathbf{C}(z\mathbf{I} - \mathbf{G})^{-1}\mathbf{H} = \hat{\mathbf{C}}(z\mathbf{I} - \hat{\mathbf{G}})^{-1}\hat{\mathbf{H}}$$

$$= \frac{z}{z^3 - 1.1z^2 + 0.2z + 0.5}$$

A block diagram of the designed regulator system is shown in Figure 6–23.

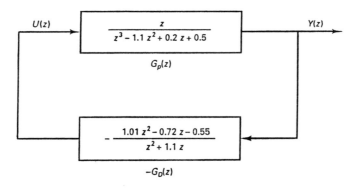

Figure 6–23   Block diagram of the regulator system designed in Problem A–6–17.

Problem A–6–18

Consider the servo system defined by Equation (6–185). The characteristic equation for the servo system is

$$\left| z\mathbf{I}_{n+m} - \begin{bmatrix} \mathbf{G} & \mathbf{H} \\ \mathbf{K}_2 - \mathbf{K}_2\mathbf{G} - \mathbf{K}_1\mathbf{CG} & \mathbf{I}_m - \mathbf{K}_2\mathbf{H} - \mathbf{K}_1\mathbf{CH} \end{bmatrix} \right| = 0 \qquad (6\text{-}284)$$

Rewriting Equation (6–284), we obtain

$$\left| \begin{array}{c|c} z\mathbf{I}_n - \mathbf{G} & -\mathbf{H} \\ \hline -\mathbf{K}_2 + \mathbf{K}_2\mathbf{G} + \mathbf{K}_1\mathbf{CG} & z\mathbf{I}_m - \mathbf{I}_m + \mathbf{K}_2\mathbf{H} + \mathbf{K}_1\mathbf{CH} \end{array} \right|$$

$$= \left| \begin{array}{c|c} \mathbf{I}_n & 0 \\ \hline -(\mathbf{K}_2 + \mathbf{K}_1\mathbf{C}) & \mathbf{I}_m \end{array} \right| \left| \begin{array}{c|c} z\mathbf{I}_n - \mathbf{G} + \mathbf{H}(\mathbf{K}_2 + \mathbf{K}_1\mathbf{C}) & -\mathbf{H} \\ \hline \mathbf{K}_1\mathbf{C} & z\mathbf{I}_m - \mathbf{I}_m \end{array} \right| \left| \begin{array}{c|c} \mathbf{I}_n & 0 \\ \hline \mathbf{K}_2 + \mathbf{K}_1\mathbf{C} & \mathbf{I}_m \end{array} \right|$$

$$= \left| \begin{array}{c|c} z\mathbf{I}_n - \mathbf{G} + \mathbf{H}(\mathbf{K}_2 + \mathbf{K}_1\mathbf{C}) & -\mathbf{H} \\ \hline \mathbf{K}_1\mathbf{C} & z\mathbf{I}_m - \mathbf{I}_m \end{array} \right|$$

$$= \left| \begin{array}{c|c} z\mathbf{I}_n - \mathbf{G} + \mathbf{HK}_2 + \mathbf{HK}_1\mathbf{C} + \mathbf{HK}_1\mathbf{C}(z\mathbf{I}_m - \mathbf{I}_m)^{-1} & -\mathbf{H} \\ \hline 0 & z\mathbf{I}_m - \mathbf{I}_m \end{array} \right|$$

$$\cdot \left[\begin{array}{c|c} \mathbf{I}_n & \mathbf{0} \\ \hline \mathbf{K}_1 \mathbf{C}(z\mathbf{I}_m - \mathbf{I}_m)^{-1} & \mathbf{I}_m \end{array} \right]$$

$$= \left| \begin{array}{c|c} z\mathbf{I}_n - \mathbf{G} + \mathbf{HK}_2 + \mathbf{HK}_1\mathbf{C} + \mathbf{HK}_1\mathbf{C}(z\mathbf{I}_m - \mathbf{I}_m)^{-1} & -\mathbf{H} \\ \hline \mathbf{0} & z\mathbf{I}_m - \mathbf{I}_m \end{array} \right|$$

$$= \left| z\mathbf{I}_n - \mathbf{G} + \mathbf{HK}_2 + \mathbf{HK}_1\mathbf{C} + \mathbf{HK}_1\mathbf{C}(z\mathbf{I}_m - \mathbf{I}_m)^{-1} \right| \left| z\mathbf{I}_m - \mathbf{I}_m \right| = 0 \qquad (6\text{–}285)$$

Equation (6–285) gives the characteristic equation for the system. We can determine matrices $\mathbf{K}_1$ and $\mathbf{K}_2$ so that the roots of this characteristic equation assume the desired values. For example, if deadbeat response to a step input is desired, then we determine $\mathbf{K}_1$ and $\mathbf{K}_2$ so that all roots of the characteristic equation are at the origin. [When the control $\mathbf{u}(k)$ is an m-vector (where $m > 1$), matrices $\mathbf{K}_1$ and $\mathbf{K}_2$ are not unique. That is, more than one set of $\mathbf{K}_1$ and $\mathbf{K}_2$ can be obtained.

Referring to the servo system design problem discussed in Example 6–13, consider first the problem of determining an integral gain constant K_1 and a state feedback gain matrix $\mathbf{K}_2$ by use of the characteristic equation given by Equation (6–284) or Equation (6–285) such that the unit-step response is deadbeat. Then consider a design of a full-order (third-order) prediction observer such that the response to the observer error is deadbeat. Defining the observer feedback gain matrix as $\mathbf{K}_e$, determine this matrix by equating the coefficients of the powers of z of

$$\left| z\mathbf{I} - \mathbf{G} + \mathbf{K}_e\mathbf{C} \right| = 0$$

and those of like powers of z in the desired characteristic equation, which is

$$z^3 = 0$$

Solution Let us define

$$\mathbf{K}_2 = [k_1 \quad k_2 \quad k_3]$$

Noting that

$$\mathbf{G} = \begin{bmatrix} 0 & 1 & 0 \\ 0 & 0 & 1 \\ -0.12 & -0.01 & 1 \end{bmatrix}, \quad \mathbf{H} = \begin{bmatrix} 0 \\ 0 \\ 1 \end{bmatrix}, \quad \mathbf{C} = [0.5 \quad 1 \quad 0]$$

Equation (6–285) can be written as follows:

$$\left| z\mathbf{I}_3 - \mathbf{G} + \mathbf{HK}_2 + \mathbf{HK}_1\mathbf{C} + \mathbf{HK}_1\mathbf{C}(z I_1 - I_1)^{-1} \right| \left| z I_1 - I_1 \right|$$

$$= \left| z\mathbf{I}_3 - \mathbf{G} + \mathbf{HK}_2 + \mathbf{HK}_1\mathbf{C}[1 + (z - 1)^{-1}] \right| \left| z - 1 \right|$$

$$= \left| \begin{bmatrix} z & 0 & 0 \\ 0 & z & 0 \\ 0 & 0 & z \end{bmatrix} - \begin{bmatrix} 0 & 1 & 0 \\ 0 & 0 & 1 \\ -0.12 & -0.01 & 1 \end{bmatrix} + \begin{bmatrix} 0 \\ 0 \\ 1 \end{bmatrix} [k_1 \quad k_2 \quad k_3] \right.$$

$$\left. + \begin{bmatrix} 0 \\ 0 \\ 1 \end{bmatrix} [K_1][0.5 \quad 1 \quad 0]\left(1 + \frac{1}{z - 1}\right) \right| \left| z - 1 \right|$$

$$= \left| \begin{array}{ccc} z & -1 & 0 \\ 0 & z & -1 \\ 0.12 + k_1 + \dfrac{0.5K_1 z}{z - 1} & 0.01 + k_2 + \dfrac{K_1 z}{z - 1} & z - 1 + k_3 \end{array} \right| \left| z - 1 \right|$$

$$= \left| \begin{array}{ccc} z & -1 & 0 \\ 0 & z & -1 \\ (0.12 + k_1)(z - 1) + 0.5K_1 z & (0.01 + k_2)(z - 1) + K_1 z & (z - 1)^2 + k_3(z - 1) \end{array} \right|$$

$$= z^4 + (-2 + k_3)z^3 + (1.01 + k_2 - k_3 + K_1)z^2$$
$$+ (0.11 + k_1 - k_2 + 0.5K_1)z - 0.12 - k_1 = 0$$

This characteristic equation must be equal to

$$z^4 = 0$$

Hence, we require

$$-2 + k_3 = 0$$
$$1.01 + k_2 - k_3 + K_1 = 0$$
$$0.11 + k_1 - k_2 + 0.5K_1 = 0$$
$$-0.12 - k_1 = 0$$

from which we get

$$K_1 = \tfrac{2}{3}, \qquad k_1 = -0.12, \qquad k_2 = \tfrac{0.97}{3}, \qquad k_3 = 2$$

or

$$K_1 = \tfrac{2}{3}, \qquad \mathbf{K} = [-0.12 \quad 0.3233 \quad 2]$$

[As a matter of course, these values agree with those given by Equations (6–199) and (6–200).]

Next, we shall design a full-order prediction observer. Define

$$\mathbf{K}_e = \begin{bmatrix} k_{e_1} \\ k_{e_2} \\ k_{e_3} \end{bmatrix}$$

Then

$$\mathbf{G} - \mathbf{K}_e\mathbf{C} = \begin{bmatrix} 0 & 1 & 0 \\ 0 & 0 & 1 \\ -0.12 & -0.01 & 1 \end{bmatrix} - \begin{bmatrix} k_{e_1} \\ k_{e_2} \\ k_{e_3} \end{bmatrix}[0.5 \quad 1 \quad 0]$$

$$= \begin{bmatrix} -0.5k_{e_1} & 1 - k_{e_1} & 0 \\ -0.5k_{e_2} & -k_{e_2} & 1 \\ -0.12 - 0.5k_{e_3} & -0.01 - k_{e_3} & 1 \end{bmatrix}$$

and we have

$$|z\mathbf{I} - \mathbf{G} + \mathbf{K}_e\mathbf{C}| = \begin{vmatrix} z + 0.5k_{e_1} & -1 + k_{e_1} & 0 \\ 0.5k_{e_2} & z + k_{e_2} & -1 \\ 0.12 + 0.5k_{e_3} & 0.01 + k_{e_3} & z - 1 \end{vmatrix}$$

$$= z^3 + (-1 + 0.5k_{e_1} + k_{e_2})z^2 + (0.01 - 0.5k_{e_1} - 0.5k_{e_2} + k_{e_3})z$$
$$+ 0.12 - 0.115k_{e_1} - 0.5k_{e_2} + 0.5k_{e_3} = 0$$

This characteristic equation must be equal to the desired characteristic equation

$$z^3 = 0$$

Hence, we require

$$-1 + 0.5k_{e_1} + k_{e_2} = 0$$
$$0.01 - 0.5k_{e_1} - 0.5k_{e_2} + k_{e_3} = 0$$
$$0.12 - 0.115k_{e_1} - 0.5k_{e_2} + 0.5k_{e_3} = 0$$

Solving these three simultaneous equations for k_{e_1}, k_{e_2}, and k_{e_3}, we obtain

$$\mathbf{K}_e = \begin{bmatrix} k_{e_1} \\ k_{e_2} \\ k_{e_3} \end{bmatrix} = \begin{bmatrix} 0.5192 \\ 0.7404 \\ 0.6198 \end{bmatrix}$$

This matrix gives the desired observer feedback gain matrix $\mathbf{K}_e$.

Remember that the design of the integral gain constant K_1 and the state feedback gain matrix $\mathbf{K}_2$ (a pole placement problem) and the design of the observer feedback gain matrix $\mathbf{K}_e$ (an observer problem) are independent problems. That is, matrix $\mathbf{K}_e$ does not depend on K_1 and $\mathbf{K}_2$, and vice versa.

PROBLEMS

Problem B-6-1

Consider the system defined by

$$\begin{bmatrix} x_1(k+1) \\ x_2(k+1) \end{bmatrix} = \begin{bmatrix} a & b \\ c & d \end{bmatrix} \begin{bmatrix} x_1(k) \\ x_2(k) \end{bmatrix} + \begin{bmatrix} 1 \\ 1 \end{bmatrix} u(k)$$

$$y(k) = \begin{bmatrix} 1 & 0 \end{bmatrix} \begin{bmatrix} x_1(k) \\ x_2(k) \end{bmatrix}$$

Determine the conditions on a, b, c, and d for complete state controllability and complete observability.

Problem B-6-2

The control system defined by

$$\begin{bmatrix} x_1(k+1) \\ x_2(k+1) \end{bmatrix} = \begin{bmatrix} 0 & 1 \\ -0.16 & -1 \end{bmatrix} \begin{bmatrix} x_1(k) \\ x_2(k) \end{bmatrix} + \begin{bmatrix} 1 \\ 0.5 \end{bmatrix} u(k)$$

$$\begin{bmatrix} x_1(0) \\ x_2(0) \end{bmatrix} = \begin{bmatrix} 1 \\ -1 \end{bmatrix}$$

is completely state controllable. Determine a sequence of control signals $u(0)$ and $u(1)$ such that the state $\mathbf{x}(2)$ becomes

$$\begin{bmatrix} x_1(2) \\ x_2(2) \end{bmatrix} = \begin{bmatrix} -1 \\ 2 \end{bmatrix}$$

Problem B-6-3

Consider the system

$$\begin{bmatrix} x_1(k+1) \\ x_2(k+1) \end{bmatrix} = \begin{bmatrix} 0 & 1 \\ -0.16 & -1 \end{bmatrix} \begin{bmatrix} x_1(k) \\ x_2(k) \end{bmatrix} + \begin{bmatrix} 1 \\ -0.8 \end{bmatrix} u(k)$$

$$\begin{bmatrix} x_1(0) \\ x_2(0) \end{bmatrix} = \begin{bmatrix} 1 \\ -1 \end{bmatrix}$$

Determine whether it is possible to bring the state to

1.

$$\begin{bmatrix} x_1(2) \\ x_2(2) \end{bmatrix} = \begin{bmatrix} 0 \\ -0.008 \end{bmatrix}$$

2.
$$\begin{bmatrix} x_1(2) \\ x_2(2) \end{bmatrix} = \begin{bmatrix} -1 \\ 2 \end{bmatrix}$$

Problem B–6–4

Consider the system

$$\begin{bmatrix} x_1(k + 1) \\ x_2(k + 1) \\ x_3(k + 1) \end{bmatrix} = \begin{bmatrix} 0 & 1 & 0 \\ 0 & 0 & 1 \\ a & b & -\dfrac{a}{b} \end{bmatrix} \begin{bmatrix} x_1(k) \\ x_2(k) \\ x_3(k) \end{bmatrix} + \begin{bmatrix} 0 \\ 1 \\ 0 \end{bmatrix} u(k)$$

Starting from the initial state

$$\mathbf{x}(0) = \begin{bmatrix} 1 \\ 1 \\ 1 \end{bmatrix}$$

determine whether or not the state $\mathbf{x}(3)$ can be brought to the origin. Also, determine whether or not the state can be brought to

$$\mathbf{x}(3) = \begin{bmatrix} 1 \\ 1 \\ 1 \end{bmatrix}$$

if the initial state is $\mathbf{x}(0) = \mathbf{0}$.

Problem B–6–5

For the system defined by

$$\begin{bmatrix} x_1(k + 1) \\ x_2(k + 1) \end{bmatrix} = \begin{bmatrix} 0 & 1 \\ -0.16 & -1 \end{bmatrix} \begin{bmatrix} x_1(k) \\ x_2(k) \end{bmatrix} + \begin{bmatrix} 0 \\ 1 \end{bmatrix} u(k)$$

$$y(k) = \begin{bmatrix} 1 & 0 \end{bmatrix} \begin{bmatrix} x_1(k) \\ x_2(k) \end{bmatrix}$$

assume that the following outputs are observed:

$$y(0) = 1, \qquad y(1) = 2$$

The control signals given are

$$u(0) = 2, \qquad u(1) = -1$$

Determine the initial state $\mathbf{x}(0)$. Also, determine states $\mathbf{x}(1)$ and $\mathbf{x}(2)$.

Problem B–6–6

Show that the system

$$\mathbf{x}(k + 1) = \mathbf{G}[\mathbf{x}(k) + \mathbf{C}^*u(k)]$$

$$y(k) = \mathbf{C}\mathbf{x}(k)$$

where

$$\mathbf{x}(k) = \text{state vector (4-vector)}$$

$$u(k) = \text{control signal (scalar)}$$

$$y(k) = \text{output signal (scalar)}$$

and

$$\mathbf{G} = \begin{bmatrix} 0 & 1 & 0 & 0 \\ 0 & 0 & 1 & 0 \\ 0 & 0 & 0 & 1 \\ 1 & 0 & 0 & 0 \end{bmatrix}, \qquad \mathbf{C} = [1 \quad 0 \quad 0 \quad 0]$$

is completely state controllable and completely observable.

Show also that given any initial state $\mathbf{x}(0)$ every state vector can be brought to the origin in at most four sampling periods if and only if the control signal is given by

$$u(k) = -\mathbf{C}\mathbf{x}(k)$$

Problem B–6–7

Consider the continuous-time control system

$$\begin{bmatrix} \dot{x}_1 \\ \dot{x}_2 \end{bmatrix} = \begin{bmatrix} 0 & 1 \\ -25 & -6 \end{bmatrix} \begin{bmatrix} x_1 \\ x_2 \end{bmatrix} + \begin{bmatrix} 0 \\ 1 \end{bmatrix} u$$

$$y = [3 \quad 1] \begin{bmatrix} x_1 \\ x_2 \end{bmatrix}$$

This system is completely state controllable and observable. Note that the eigenvalues of the state matrix are

$$\lambda_1 = -3 + j4, \qquad \lambda_2 = -3 - j4$$

Thus, this system involves complex poles.

As stated in Section 6–3, a system that is completely state controllable and completely observable in the absence of sampling remains completely state controllable and completely observable after the introduction of sampling if and only if, for every eigenvalue of the state matrix (root of the characteristic equation),

$$\mathrm{Re}\,\lambda_i = \mathrm{Re}\,\lambda_j$$

implies

$$\mathrm{Im}\,(\lambda_i - \lambda_j) \neq \frac{2\pi n}{T}$$

where T is the sampling period and $n = \pm 1, \pm 2, \dots$.

Consider the discretized version of this system. Show that for this system, if the sampling period T is equal to $\pi n/4$ (where $n = 1, 2, 3, \dots$), then the discretized system is uncontrollable and unobservable.

Problem B–6–8

Consider the pulse-transfer-function system

$$G(z) = \frac{z^{-1}(1 + z^{-1})}{(1 + 0.5z^{-1})(1 - 0.5z^{-1})}$$

Referring to Section 6–4, obtain the state-space representation of the system in the following forms:

1. Controllable canonical form
2. Observable canonical form
3. Diagonal canonical form

Problem B–6–9

Consider the pulse-transfer-function system

$$G(z) = \frac{1 + 0.8z^{-1}}{1 - z^{-1} + 0.5z^{-2}}$$

Obtain the state-space representation of the system in the following forms:

1. Controllable canonical form
2. Observable canonical form
3. Diagonal canonical form

Problem B–6–10

Consider the following system given in the controllable canonical form:

$$\begin{bmatrix} x_1(k+1) \\ x_2(k+1) \\ x_3(k+1) \end{bmatrix} = \begin{bmatrix} 0 & 1 & 0 \\ 0 & 0 & 1 \\ -a_3 & -a_2 & -a_1 \end{bmatrix} \begin{bmatrix} x_1(k) \\ x_2(k) \\ x_3(k) \end{bmatrix} + \begin{bmatrix} 0 \\ 0 \\ 1 \end{bmatrix} u(k)$$

$$y(k) = [b_3 - a_3 b_0 \vdots b_2 - a_2 b_0 \vdots b_1 - a_1 b_0] \begin{bmatrix} x_1(k) \\ x_2(k) \\ x_3(k) \end{bmatrix} + b_0 u(k)$$

It is desired to transform the system equations into the observable canonical form by means of the transformation of the state vector:

$$\mathbf{x} = \mathbf{Q}\hat{\mathbf{x}}$$

Determine a transformation matrix $\mathbf{Q}$ that will give the desired observable canonical form.

Problem B–6–11

Consider the double-integrator system

$$\mathbf{x}((k+1)T) = \mathbf{G}\mathbf{x}(kT) + \mathbf{H}u(kT)$$

where

$$\mathbf{G} = \begin{bmatrix} 1 & T \\ 0 & 1 \end{bmatrix}, \qquad \mathbf{H} = \begin{bmatrix} T^2/2 \\ T \end{bmatrix}$$

and T is the sampling period. (See Problem A–5–16 for the derivation of this discrete-time state equation for the double-integrator system.)

It is desired that the closed-loop poles be located at $z = \mu_1$ and $z = \mu_2$. Assuming that the state feedback control

$$u(kT) = -\mathbf{K}\mathbf{x}(kT)$$

is used, determine the state feedback gain matrix $\mathbf{K}$.

Problem B–6–12

Consider the system defined by

$$\begin{bmatrix} x_1(k+1) \\ x_2(k+1) \\ x_3(k+1) \end{bmatrix} = \begin{bmatrix} 0 & 1 & 0 \\ 0 & 0 & 1 \\ -0.16 & 0.84 & 0 \end{bmatrix} \begin{bmatrix} x_1(k) \\ x_2(k) \\ x_3(k) \end{bmatrix} + \begin{bmatrix} 1 \\ 1 \\ 1 \end{bmatrix} u(k)$$

Determine the state feedback gain matrix $\mathbf{K}$ such that when the control signal is given by

$$u(k) = -\mathbf{K}\mathbf{x}(k)$$

the closed-loop system will exhibit the deadbeat response to any initial state $\mathbf{x}(0)$.

Problem B–6–13

Consider the system

$$\mathbf{x}(k + 1) = \mathbf{G}\mathbf{x}(k) + \mathbf{H}u(k)$$
$$y(k) = \mathbf{C}\mathbf{x}(k)$$

where

$$\mathbf{x}(k) = \text{state vector (2-vector)}$$
$$u(k) = \text{control signal (scalar)}$$
$$y(k) = \text{output signal (scalar)}$$

and

$$\mathbf{G} = \begin{bmatrix} 0 & 1 \\ -0.16 & -1 \end{bmatrix}, \quad \mathbf{H} = \begin{bmatrix} 0 \\ 1 \end{bmatrix}, \quad \mathbf{C} = [1 \quad 1]$$

Design a current observer for the system. It is desired that the response to the initial observer error be deadbeat.

Problem B–6–14

Consider the system

$$\mathbf{x}(k + 1) = \mathbf{G}\mathbf{x}(k) + \mathbf{H}u(k)$$
$$y(k) = \mathbf{C}\mathbf{x}(k)$$

where

$$\mathbf{x}(k) = \text{state vector (3-vector)}$$
$$u(k) = \text{control signal (scalar)}$$
$$y(k) = \text{output signal (scalar)}$$

and

$$\mathbf{G} = \begin{bmatrix} 0 & 0 & -0.25 \\ 1 & 0 & 0 \\ 0 & 1 & 0.5 \end{bmatrix}, \quad \mathbf{H} = \begin{bmatrix} 1 \\ 0 \\ 1 \end{bmatrix}, \quad \mathbf{C} = [1 \quad 0 \quad 0]$$

Assuming that the output $y(k)$ is measurable, design a minimum-order observer such that the response to the initial observer error is deadbeat.

Problem B–6–15

Consider the system defined by

$$\begin{bmatrix} x_1(k + 1) \\ x_2(k + 1) \end{bmatrix} = \begin{bmatrix} 0 & 1 \\ -0.16 & -1 \end{bmatrix} \begin{bmatrix} x_1(k) \\ x_2(k) \end{bmatrix} + \begin{bmatrix} 0 \\ 1 \end{bmatrix} u(k)$$

Using MATLAB, determine the state feedback gain matrix $\mathbf{K}$ such that when the control signal is given by

$$u(k) = -\mathbf{Kx}(k)$$

the closed-loop system (regulator system) exhibits the deadbeat response to an initial state $\mathbf{x}(0)$. Write a MATLAB program for the determination of state feedback gain matrix $\mathbf{K}$.

Problem B–6–16

Consider the system defined by

$$\mathbf{x}(k + 1) = \mathbf{Gx}(k) + \mathbf{H}u(k)$$

$$y(k) = \mathbf{Cx}(k)$$

where

$$\mathbf{G} = \begin{bmatrix} 0 & -0.16 \\ 1 & -1 \end{bmatrix}, \qquad \mathbf{H} = \begin{bmatrix} 0 \\ 1 \end{bmatrix}, \qquad \mathbf{C} = [0 \quad 1]$$

Using MATLAB, determine the observer feedback gain matrix $\mathbf{K}_e$ such that the desired eigenvalues for the observer matrix are

$$\mu_1 = 0.5 + j0.5, \qquad \mu_2 = 0.5 - j0.5$$

Assume that the system configuration is identical to that shown in Figure 6–8. Using Ackermann's formula, write a MATLAB program.

Problem B–6–17

Figure 6–24 shows a servo system where the integral controller has a time delay of one sampling period. (Compare this system with the servo system shown in Figure 6–18.)

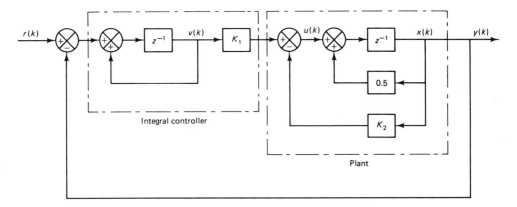

Figure 6–24 Servo system with state feedback and integral control involving a unit delay in the feedforward path.

Determine the feedforward gain K_1 and the feedback gain K_2 such that the response to the unit-step sequence input $r(k) = 1$ (where $k = 0, 1, 2, \ldots$) is deadbeat. Plot the response $y(k)$ versus k.

Problem B–6–18

Consider the servo system shown in Figure 6–25. (This system is similar to that shown in Figure 6–24, except that the integral controller has a unit delay element in the minor

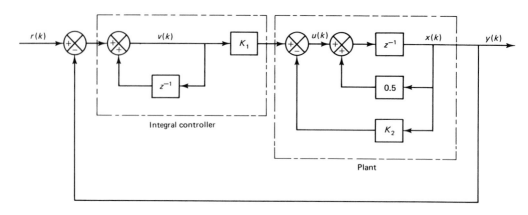

Figure 6–25   Servo system with state feedback and integral control involving a unit delay in the minor loop.

loop.) Determine the feedforward gain K_1 and the feedback gain K_2 such that the response to the unit-step sequence input $r(k) = 1$ (where $k = 0, 1, 2, \ldots$) is deadbeat. Plot the response $y(k)$ versus k.

7

Polynomial Equations Approach to Control Systems Design

7-1 INTRODUCTION

In Chapter 6 we designed state-feedback control systems using the pole placement technique. If some of the state variables were not directly measurable, we used observed states for the feedback purpose. The entire design was done in state space.

A different approach to the design of similar systems is available. It is called the polynomial equations approach. It is an alternative approach to the design via pole placement with a minimum-order state observer. (The polynomial equations approach can be applied to multiple-input–multiple-output systems. However, we shall consider only single-input–single-output systems in this chapter.)

This chapter presents an introductory account of the polynomial equations approach to control systems design. In this approach we solve Diophantine equations to determine polynomials in z that can be used in constructing necessary, physically realizable systems. This approach gives the mathematical solution to certain types of the design problem quickly.

The outline of this chapter is as follows: Section 7–1 has given an introductory remark. Section 7–2 discusses Diophantine equations and provides necessary mathematical preliminaries for the polynomial equations approach to the control systems design. Section 7–3 presents a simple example demonstrating the polynomial equations approach to the design of a regulator system having a desired characteristic polynomial. Section 7–4 discusses the polynomial equations approach to the design of control systems. Section 7–5 treats the design of a model matching control system. Here we design the system such that the response of the system to any input is the same as that of the specified mathematical model. To design such a system, we determine, based on the polynomial equations approach, physically realizable filters that will produce the desired system characteristics.

7–2 DIOPHANTINE EQUATION

In this section we shall discuss the Diophantine equation. Consider the system defined by the pulse transfer function

$$\frac{Y(z)}{U(z)} = \frac{B(z)}{A(z)} \tag{7-1}$$

where

$$A(z) = z^n + a_1 z^{n-1} + \cdots + a_{n-1} z + a_n$$

$$B(z) = b_0 z^n + b_1 z^{n-1} + \cdots + b_{n-1} z + b_n$$

Assume that this pulse transfer function system is completely state controllable and completely observable. That is, there is no pole–zero cancellation in the pulse transfer function, or $A(z)$ and $B(z)$ have no common factors. When polynomials $A(z)$ and $B(z)$ have no cancellation, these polynomials are called *coprime polynomials*. A polynomial in z is called *monic* if the coefficient of the highest-degree term is unity. Thus, polynomial $A(z)$ is monic.

Next, let us define a stable $(2n - 1)$th-degree polynomial $D(z)$ as follows:

$$D(z) = d_0 z^{2n-1} + d_1 z^{2n-2} + \cdots + d_{2n-2} z + d_{2n-1}$$

Then there exist unique $(n - 1)$th-degree polynomials $\alpha(z)$ and $\beta(z)$ such that

$$\alpha(z)A(z) + \beta(z)B(z) = D(z) \tag{7-2}$$

where

$$\alpha(z) = \alpha_0 z^{n-1} + \alpha_1 z^{n-2} + \cdots + \alpha_{n-2} z + \alpha_{n-1}$$

$$\beta(z) = \beta_0 z^{n-1} + \beta_1 z^{n-2} + \cdots + \beta_{n-2} z + \beta_{n-1}$$

Equation (7–2) is called a Diophantine equation, named after Diophantus of Alexandreia (A.D. 246?–330?). The Diophantine equation can be solved for $\alpha(z)$ and $\beta(z)$ by use of the following $2n \times 2n$ Sylvester matrix $\mathbf{E}$, which is defined in terms of the coefficients of coprime polynomials $A(z)$ and $B(z)$ as follows:

$$\mathbf{E} = \begin{bmatrix}
a_n & 0 & \cdots & 0 & b_n & 0 & \cdots & 0 \\
a_{n-1} & a_n & \cdots & 0 & b_{n-1} & b_n & \cdots & 0 \\
\vdots & a_{n-1} & \cdots & 0 & \vdots & b_{n-1} & \cdots & 0 \\
a_1 & \vdots & & \vdots & b_1 & \vdots & & \vdots \\
1 & a_1 & \cdots & a_{n-1} & b_0 & b_1 & \cdots & b_{n-1} \\
0 & 1 & \cdots & a_{n-2} & 0 & b_0 & \cdots & b_{n-2} \\
\vdots & \vdots & & \vdots & \vdots & \vdots & & \vdots \\
0 & 0 & \cdots & a_1 & 0 & 0 & \cdots & b_1 \\
0 & 0 & \cdots & 1 & 0 & 0 & \cdots & b_0
\end{bmatrix} \tag{7-3}$$

[To use Equation (7–3) polynomial $A(z)$ must be monic. Otherwise, we must modify Equation (7–3).] If $n = 4$, then this matrix becomes as follows:

$$
\mathbf{E} =
\begin{bmatrix}
a_4 & 0 & 0 & 0 & b_4 & 0 & 0 & 0 \\
a_3 & a_4 & 0 & 0 & b_3 & b_4 & 0 & 0 \\
a_2 & a_3 & a_4 & 0 & b_2 & b_3 & b_4 & 0 \\
a_1 & a_2 & a_3 & a_4 & b_1 & b_2 & b_3 & b_4 \\
1 & a_1 & a_2 & a_3 & b_0 & b_1 & b_2 & b_3 \\
0 & 1 & a_1 & a_2 & 0 & b_0 & b_1 & b_2 \\
0 & 0 & 1 & a_1 & 0 & 0 & b_0 & b_1 \\
0 & 0 & 0 & 1 & 0 & 0 & 0 & b_0
\end{bmatrix}
$$

The Sylvester matrix $\mathbf{E}$ is nonsingular if and only if $A(z)$ and $B(z)$ are coprime, or have no common factors. This fact may be seen from the following: Referring to the preceding 8×8 matrix $\mathbf{E}$, the determinant $|\mathbf{E}|$ becomes as follows:

$$
|\mathbf{E}| =
\begin{vmatrix}
a_4 & 0 & 0 & 0 & b_4 & 0 & 0 & 0 \\
a_3 & a_4 & 0 & 0 & b_3 & b_4 & 0 & 0 \\
a_2 & a_3 & a_4 & 0 & b_2 & b_3 & b_4 & 0 \\
a_1 & a_2 & a_3 & a_4 & b_1 & b_2 & b_3 & b_4 \\
1 & a_1 & a_2 & a_3 & b_0 & b_1 & b_2 & b_3 \\
0 & 1 & a_1 & a_2 & 0 & b_0 & b_1 & b_2 \\
0 & 0 & 1 & a_1 & 0 & 0 & b_0 & b_1 \\
0 & 0 & 0 & 1 & 0 & 0 & 0 & b_0
\end{vmatrix}
$$

$$
= b_0^4 (\lambda_1 - \lambda_5)(\lambda_1 - \lambda_6)(\lambda_1 - \lambda_7)(\lambda_1 - \lambda_8)
$$

$$
\cdot (\lambda_2 - \lambda_5)(\lambda_2 - \lambda_6)(\lambda_2 - \lambda_7)(\lambda_2 - \lambda_8)
$$

$$
\cdot (\lambda_3 - \lambda_5)(\lambda_3 - \lambda_6)(\lambda_3 - \lambda_7)(\lambda_3 - \lambda_8)
$$

$$
\cdot (\lambda_4 - \lambda_5)(\lambda_4 - \lambda_6)(\lambda_4 - \lambda_7)(\lambda_4 - \lambda_8) \tag{7-4}
$$

where $a_1, \ldots, a_4$ and $b_1, \ldots, b_4$ are coefficients of $A(z)$ and $B(z)$, respectively, and $\lambda_1, \ldots, \lambda_4$ and $\lambda_5, \ldots, \lambda_8$ are characteristic roots of $A(z)$ and $B(z)$, respectively:

$$
A(z) = z^4 + a_1 z^3 + a_2 z^2 + a_3 z + a_4 = (z - \lambda_1)(z - \lambda_2)(z - \lambda_3)(z - \lambda_4)
$$

$$
B(z) = b_0 z^4 + b_1 z^3 + b_2 z^2 + b_3 z + b_4 = b_0(z - \lambda_5)(z - \lambda_6)(z - \lambda_7)(z - \lambda_8)
$$

From Equation (7–4) it is clear that the determinant $|\mathbf{E}|$ is nonzero if and only if all multiplicative factors on the right-hand side of the equation are nonzero, that is, if and only if no cancellation occurs between $A(z)$ and $B(z)$. [For the derivation of Equation (7–4), refer to Problem A–7–1.]

Now define vectors $\mathbf{D}$ and $\mathbf{M}$ such that

$$
\mathbf{D} =
\begin{bmatrix}
d_{2n-1} \\
d_{2n-2} \\
\vdots \\
d_1 \\
d_0
\end{bmatrix},
\qquad
\mathbf{M} =
\begin{bmatrix}
\alpha_{n-1} \\
\alpha_{n-2} \\
\vdots \\
\alpha_0 \\
\beta_{n-1} \\
\beta_{n-2} \\
\vdots \\
\beta_0
\end{bmatrix}
$$

Then the coefficients $\alpha_0, \alpha_1, \ldots, \alpha_{n-1}$ and $\beta_0, \beta_1, \ldots, \beta_{n-1}$ can be determined from

$$\mathbf{M} = \mathbf{E}^{-1}\mathbf{D} \tag{7-5}$$

Equation (7–5) gives the solution to the Diophantine equation. [For the derivation of Equation (7–5), see Problem A–7–2.]

Example 7–1

Consider the following $A(z)$ (a monic polynomial of degree 2), $B(z)$ (a polynomial of degree 1), and $D(z)$ (a polynomial of degree 3):

$$A(z) = z^2 + z + 0.5$$

$$B(z) = z + 2$$

$$D(z) = z^3$$

[Clearly, there is no common factor between $A(z)$ and $B(z)$.] The problem here is to find unique polynomials $\alpha(z)$ and $\beta(z)$ such that

$$\alpha(z)A(z) + \beta(z)B(z) = D(z)$$

where

$$\alpha(z) = \alpha_0 z + \alpha_1$$

$$\beta(z) = \beta_0 z + \beta_1$$

or

$$(\alpha_0 z + \alpha_1)(z^2 + z + 0.5) + (\beta_0 z + \beta_1)(z + 2) = z^3 \tag{7-6}$$

Equation (7–6) is a Diophantine equation. To solve this equation for $\alpha(z)$ and $\beta(z)$, first note that

$$a_1 = 1, \qquad a_2 = 0.5$$

$$b_0 = 0, \qquad b_1 = 1, \qquad b_2 = 2$$

and then write the Sylvester matrix $\mathbf{E}$ as follows:

$$\mathbf{E} = \begin{bmatrix} 0.5 & 0 & 2 & 0 \\ 1 & 0.5 & 1 & 2 \\ 1 & 1 & 0 & 1 \\ 0 & 1 & 0 & 0 \end{bmatrix}$$

The inverse of such a matrix can be obtained easily if MATLAB is used. The MATLAB output for the inverse of matrix $\mathbf{E}$ is shown next.

```
E =

    0.5000         0    2.0000         0
    1.0000    0.5000    1.0000    2.0000
    1.0000    1.0000         0    1.0000
         0    1.0000         0         0

inv(E)
```

ans =

$$
\begin{array}{rrrr}
0.4000 & -0.8000 & 1.6000 & -1.2000 \\
0 & 0 & 0 & 1.0000 \\
0.4000 & 0.2000 & -0.4000 & 0.3000 \\
-0.4000 & 0.8000 & -0.6000 & 0.2000
\end{array}
$$

Since

$$D(z) = z^3$$

we have

$$d_0 = 1, \qquad d_1 = 0, \qquad d_2 = 0, \qquad d_3 = 0$$

Thus, matrix **D** becomes

$$
\mathbf{D} = \begin{bmatrix} d_3 \\ d_2 \\ d_1 \\ d_0 \end{bmatrix} = \begin{bmatrix} 0 \\ 0 \\ 0 \\ 1 \end{bmatrix}
$$

By defining matrix **M** as

$$
\mathbf{M} = \begin{bmatrix} \alpha_1 \\ \alpha_0 \\ \beta_1 \\ \beta_0 \end{bmatrix}
$$

the solution to the Diophantine equation is obtained from

$$\mathbf{M} = \mathbf{E}^{-1}\mathbf{D}$$

as follows:

M = (inv(E))*D

M =

$$
\begin{array}{r}
-1.2000 \\
1.0000 \\
0.3000 \\
0.2000
\end{array}
$$

From this MATLAB output, we obtain

$$\alpha_1 = -1.2, \qquad \alpha_0 = 1, \qquad \beta_1 = 0.3, \qquad \beta_0 = 0.2$$

or

$$\alpha(z) = \alpha_0 z + \alpha_1 = z - 1.2$$

$$\beta(z) = \beta_0 z + \beta_1 = 0.2z + 0.3$$

The polynomials $\alpha(z)$ and $\beta(z)$ thus determined will satisfy the Diophantine equation given by Equation (7–6). To verify, notice that

$$(z - 1.2)(z^2 + z + 0.5) + (0.2z + 0.3)(z + 2)$$
$$= z^3 - 1.2z^2 + z^2 - 1.2z + 0.5z - 0.6 + 0.2z^2 + 0.3z$$
$$\quad + 0.4z + 0.6$$
$$= z^3$$

7–3 ILLUSTRATIVE EXAMPLE

In Chapter 6 we discussed the pole placement approach to the control systems design. It was stated that some of the state variables might not be available for direct measurements, and in such case the pole placement approach required estimated or observed states for feedback.

Let us refer to Example 6–11, where we discussed the state feedback regulator system. In that system the desired characteristic equation was given, and one of the state variables was estimated by use of the deadbeat-type minimum-order observer. In this section we shall show that the same regulator system can be designed by use of the polynomial equations approach.

Regulator System Designed in Example 6–11. The regulator system designed in Example 6–11 is shown in Figure 7–1. The plant is completely state controllable and completely observable. (No cancellation occurs between the numerator polynomial and the denominator polynomial.) The sampling period was 0.2 sec, or $T = 0.2$. The controller was designed based on the pole placement approach by specifying the desired closed-loop poles at

$$z_1 = 0.6 + j0.4, \qquad z_2 = 0.6 - j0.4$$

and by incorporating a minimum-order observer to estimate one of the state variables for feedback. The minimum-order observer had the observer error equation of

$$\phi(z) = z$$

The regulator designed was

$$G_D(z) = 24\left(\frac{z - 0.6667}{z + 0.32}\right) \tag{7-7}$$

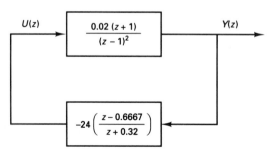

Figure 7–1 Regulator system designed in Example 6–11.

In what follows we shall present the polynomial equations approach to design the same regulator as given by Equation (7–7) by solving a Diophantine equation.

Polynomial Equations Approach to Design Regulator System.   Consider the block diagram shown in Figure 7–2. The feedback pulse transfer function $\beta(z)/\alpha(z)$ serves as a regulator. Let us determine $\alpha(z)$ and $\beta(z)$ by use of the polynomial equations approach. First note that the pulse transfer function of the plant is

$$\frac{Y(z)}{U(z)} = \frac{B(z)}{A(z)} = \frac{0.02(z+1)}{(z-1)^2}$$

[$A(z)$ is a monic polynomial of degree 2 and there is no cancellation between $A(z)$ and $B(z)$.] Then, although $R(z) = 0$, the closed-loop pulse transfer function for the system can be given by

$$\frac{Y(z)}{R(z)} = \frac{\alpha(z)B(z)}{\alpha(z)A(z) + \beta(z)B(z)} = \frac{0.02(z+1)\alpha(z)}{\alpha(z)(z-1)^2 + \beta(z)0.02(z+1)}$$

As stated earlier, in Example 6–11 we required the desired closed-loop poles for state feedback to be

$$z_1 = 0.6 + j0.4, \qquad z_2 = 0.6 - j0.4$$

or the desired characteristic polynomial was

$$H(z) = (z - 0.6 - j0.4)(z - 0.6 + j0.4)$$
$$= z^2 - 1.2z + 0.52$$

The desired minimum-order observer error polynomial was

$$F(z) = z$$

To determine $\alpha(z)$ and $\beta(z)$, we solve the following Diophantine equation:

$$\alpha(z)A(z) + \beta(z)B(z) = F(z)H(z) = D(z) \tag{7–8}$$

where

$$D(z) = F(z)H(z) = d_0 z^3 + d_1 z^2 + d_2 z + d_3 = z^3 - 1.2z^2 + 0.52z$$

Note that $D(z)$ is a stable, $(2n - 1)$th-degree polynomial in z (where $n = 2$ in the present case). Since

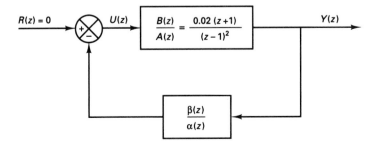

Figure 7–2 Block diagram of regulator system.

$$A(z) = z^2 - 2z + 1$$

$$B(z) = 0.02z + 0.02$$

we have

$$a_1 = -2, \qquad a_2 = 1, \qquad b_0 = 0, \qquad b_1 = 0.02, \qquad b_2 = 0.02$$

By substituting the polynomial expressions for $A(z)$, $B(z)$, and $D(z)$ into Equation (7–8), we obtain

$$\alpha(z)(z^2 - 2z + 1) + \beta(z)(0.02z + 0.02) = z^3 - 1.2z^2 + 0.52z$$

To solve this Diophantine equation for $\alpha(z)$ and $\beta(z)$, we first define $2n \times 2n$ (where $n = 2$) Sylvester matrix $\mathbf{E}$:

$$\mathbf{E} = \begin{bmatrix} 1 & 0 & 0.02 & 0 \\ -2 & 1 & 0.02 & 0.02 \\ 1 & -2 & 0 & 0.02 \\ 0 & 1 & 0 & 0 \end{bmatrix}$$

The inverse of matrix $\mathbf{E}$ can be obtained easily by use of MATLAB as follows:

$$\mathbf{E}^{-1} = \begin{bmatrix} 0.25 & -0.25 & 0.25 & 0.75 \\ 0 & 0 & 0 & 1 \\ 37.5 & 12.5 & -12.5 & -37.5 \\ -12.5 & 12.5 & 37.5 & 62.5 \end{bmatrix}$$

$\alpha(z)$ and $\beta(z)$ are polynomials of degree $n - 1 = 2 - 1 = 1$, or

$$\alpha(z) = \alpha_0 z + \alpha_1$$

$$\beta(z) = \beta_0 z + \beta_1$$

Define

$$\mathbf{D} = \begin{bmatrix} d_3 \\ d_2 \\ d_1 \\ d_0 \end{bmatrix} = \begin{bmatrix} 0 \\ 0.52 \\ -1.2 \\ 1 \end{bmatrix}, \qquad \mathbf{M} = \begin{bmatrix} \alpha_1 \\ \alpha_0 \\ \beta_1 \\ \beta_0 \end{bmatrix}$$

Then vector $\mathbf{M}$ is determined from

$$\mathbf{M} = \mathbf{E}^{-1}\mathbf{D} = \begin{bmatrix} 0.25 & -0.25 & 0.25 & 0.75 \\ 0 & 0 & 0 & 1 \\ 37.5 & 12.5 & -12.5 & -37.5 \\ -12.5 & 12.5 & 37.5 & 62.5 \end{bmatrix} \begin{bmatrix} 0 \\ 0.52 \\ -1.2 \\ 1 \end{bmatrix} = \begin{bmatrix} 0.32 \\ 1 \\ -16 \\ 24 \end{bmatrix}$$

Hence,

$$\alpha_1 = 0.32, \qquad \alpha_0 = 1, \qquad \beta_1 = -16, \qquad \beta_0 = 24$$

Therefore, $\alpha(z)$ and $\beta(z)$ are determined as

$$\alpha(z) = \alpha_0 z + \alpha_1 = z + 0.32$$

$$\beta(z) = \beta_0 z + \beta_1 = 24z - 16$$

and the feedback regulator is obtained as

$$\frac{\beta(z)}{\alpha(z)} = 24\left(\frac{z - 0.6667}{z + 0.32}\right)$$

which is identical to that designed in state space by the method based on pole placement combined with a minimum-order observer.

7–4 POLYNOMIAL EQUATIONS APPROACH TO CONTROL SYSTEMS DESIGN

In Section 7–3 we designed a regulator system by use of the polynomial equations approach. The block diagram of the regulator system designed is shown in Figure 7–3. Remember that $\alpha(z)$ and $\beta(z)$ were determined from the following Diophantine equation:

$$\alpha(z)A(z) + \beta(z)B(z) = H(z)F(z)$$

where $A(z)$ is a monic polynomial of degree n, $B(z)$ is a polynomial of degree m ($m \leq n$) [we assume that there are no common factors between $A(z)$ and $B(z)$], $H(z)$ is the desired characteristic polynomial for pole placement part, and $F(z)$ is the characteristic polynomial for the minimum-order observer. [Both polynomials $H(z)$ and $F(z)$ are stable polynomials.] The degree of polynomial $H(z)$ is n and the degree of polynomial $F(z)$ is $n - 1$. (We assume that the system output is the only measurable state variable. Therefore, the order of the minimum-order observer is $n - 1$.)

In the following we discuss the design of control systems based on the polynomial equations approach. We consider two different control system configurations.

Control System Configuration 1.   The regulator system shown in Figure 7–3 can be modified to a control system such that the output follows the reference input. A possible block diagram for the control system is shown in Figure 7–4. As a control system, it is necessary to have an adjustable gain K_0. This gain K_0 should be set such that the steady-state output $y(k)$ is equal to unity when the input $r(k)$ is a unit-step sequence.

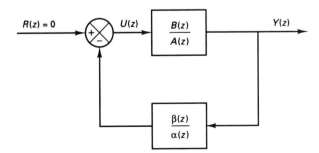

Figure 7–3 Block diagram of regulator system.

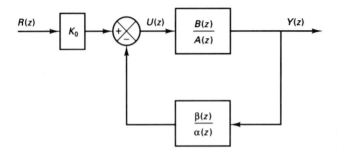

Figure 7–4 Block diagram of control system.

The closed-loop pulse transfer function $Y(z)/R(z)$ is

$$\frac{Y(z)}{R(z)} = K_0 \frac{\dfrac{B(z)}{A(z)}}{1 + \dfrac{B(z)\beta(z)}{A(z)\alpha(z)}}$$

$$= K_0 \frac{\alpha(z)B(z)}{\alpha(z)A(z) + \beta(z)B(z)}$$

$$= K_0 \frac{\alpha(z)B(z)}{H(z)F(z)} \tag{7-9}$$

Notice that the closed-loop system is of $(2n - 1)$th order, unless cancellation occurs between $\alpha(z)B(z)$ and $H(z)F(z)$. Notice also that the numerator dynamics has been changed from $B(z)$ to $K_0 \alpha(z)B(z)$.

To determine gain K_0, we set

$$\lim_{k \to \infty} y(k) = \lim_{z \to 1} (1 - z^{-1})Y(z)$$

$$= \lim_{z \to 1} \frac{z - 1}{z} K_0 \frac{\alpha(z)B(z)}{H(z)F(z)} \frac{z}{z - 1}$$

$$= K_0 \frac{\alpha(1)B(1)}{H(1)F(1)}$$

$$= 1$$

from which we get

$$K_0 = \frac{H(1)F(1)}{\alpha(1)B(1)}$$

Example 7–2

In the regulator system considered in Section 7–3,

$$A(z) = (z - 1)^2$$

$$B(z) = 0.02(z + 1)$$

$$H(z) = z^2 - 1.2z + 0.52$$

$$F(z) = z$$

$$\alpha(z) = z + 0.32$$

$$\beta(z) = 24z - 16$$

Hence, the closed-loop pulse transfer function $Y(z)/R(z)$ is obtained from Equation (7–9) as

$$\frac{Y(z)}{R(z)} = \frac{K_0(z + 0.32)(0.02)(z + 1)}{z^3 - 1.2z^2 + 0.52z}$$

Note that K_0 in this case is given by

$$K_0 = \frac{H(1)F(1)}{\alpha(1)B(1)} = \frac{0.32 \times 1}{1.32 \times 0.04} = 6.0606$$

Notice that the system is of third order. The unit-step response and the unit-ramp response of this system with $K_0 = 6.0606$ were shown in Figures 6–16 and 6–17, respectively.

Control System Configuration 2.   A control system with a different block diagram configuration may be designed by use of the polynomial equations approach. Consider the block diagram shown in Figure 7–5. (To figure out how such a block diagram came out, see Problem A–7–3.)

From Figure 7–5, we obtain the following equation:

$$U(z) = -\left[\frac{\alpha(z)}{F(z)}U(z) - U(z) + \frac{\beta(z)}{F(z)}Y(z)\right] + K_0 R(z)$$

which can be simplified to

$$\frac{\alpha(z)}{F(z)}U(z) = -\frac{\beta(z)}{F(z)}Y(z) + K_0 R(z) \qquad (7\text{–}10)$$

The pulse transfer function of the plant is

$$\frac{Y(z)}{U(z)} = \frac{B(z)}{A(z)}$$

where $A(z)$ is a monic polynomial of degree n and $B(z)$ is a stable polynomial of degree m ($m \le n$). Since

$$U(z) = \frac{A(z)}{B(z)}Y(z) \qquad (7\text{–}11)$$

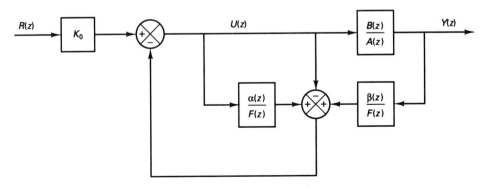

Figure 7–5 Block diagram of control system.

by substituting Equation (7–11) into Equation (7–10), we obtain

$$\left[\frac{\alpha(z)A(z)}{F(z)B(z)} + \frac{\beta(z)}{F(z)}\right]Y(z) = K_0 R(z)$$

Then

$$\frac{Y(z)}{R(z)} = \frac{K_0}{\dfrac{\alpha(z)A(z)}{F(z)B(z)} + \dfrac{\beta(z)}{F(z)}}$$

$$= \frac{K_0 F(z)B(z)}{\alpha(z)A(z) + \beta(z)B(z)}$$

Since

$$\alpha(z)A(z) + \beta(z)B(z) = H(z)F(z)$$

we obtain

$$\frac{Y(z)}{R(z)} = \frac{K_0 F(z)B(z)}{H(z)F(z)} = \frac{K_0 B(z)}{H(z)} \tag{7–12}$$

Notice that the observer polynomial $F(z)$ has been canceled [since $F(z)$ is a stable polynomial, cancellation of $F(z)$ is permissible], and the characteristic polynomial for the closed-loop system is given by $H(z)$. $H(z)$ is a desired, but in a sense "arbitrarily chosen," stable polynomial of degree n. Thus, the control system designed is of the nth order. (In the case of control system configuration 1, the order of the system is $2n - 1$, unless cancellations occur in the designed system, resulting in the reduction of the system order.) Notice also that the numerator dynamics of $Y(z)/R(z)$ has not been changed in the present approach. [The numerator is $B(z)$ times constant K_0.]

Example 7–3

Let us design a control system based on the block diagram shown in Figure 7–5. The plant we consider is given by

$$\frac{B(z)}{A(z)} = \frac{0.02(z + 1)}{(z - 1)^2}$$

(The sampling period T is 0.2 sec.) We shall use the same desired closed-loop poles as those used in Example 7–2, or

$$z_1 = 0.6 + j0.4, \qquad z_2 = 0.6 - j0.4$$

and use the same desired minimum-order observer polynomial, or

$$\phi(z) = z$$

Let us write the desired characteristic polynomial as $H(z)$,

$$H(z) = (z - 0.6 - j0.4)(z - 0.6 + j0.4)$$

$$= z^2 - 1.2z + 0.52$$

and the desired observer characteristic polynomial as $F(z)$,

$$F(z) = z$$

and solve the following Diophantine equation:

$$\alpha(z)A(z) + \beta(z)B(z) = H(z)F(z)$$

or

$$\alpha(z)(z - 1)^2 + \beta(z)(0.02)(z + 1) = z^3 - 1.2z^2 + 0.52z \qquad (7\text{–}13)$$

Equation (7–13) was solved in Section 7–3 and $\alpha(z)$ and $\beta(z)$ were obtained as follows:

$$\alpha(z) = z + 0.32$$

$$\beta(z) = 24z - 16$$

Using these $\alpha(z)$ and $\beta(z)$ and referring to Equation (7–12), the closed-loop pulse transfer function $Y(z)/R(z)$ can be written as follows:

$$\frac{Y(z)}{R(z)} = \frac{K_0 B(z)}{H(z)} = \frac{K_0(0.02z + 0.02)}{z^2 - 1.2z + 0.52}$$

To determine constant K_0, we require $y(\infty)$ in the unit-step response to be unity.

$$\lim_{k \to \infty} y(k) = \lim_{z \to 1} (1 - z^{-1})Y(z)$$

$$= \lim_{z \to 1} \frac{z - 1}{z} \frac{K_0(0.02z + 0.02)}{z^2 - 1.2z + 0.52} \frac{z}{z - 1}$$

$$= \frac{K_0}{8} = 1$$

Hence, K_0 is determined as

$$K_0 = 8$$

Then the closed-loop pulse transfer function becomes

$$\frac{Y(z)}{R(z)} = \frac{0.16z + 0.16}{z^2 - 1.2z + 0.52}$$

Clearly, the system designed is of second order. A block diagram for the designed system is shown in Figure 7–6(a). Figure 7–6(b) shows a simplified block diagram.

Next, we shall examine the unit-step response and the unit-ramp response of the system just designed. MATLAB Program 7–1 is used to obtain the unit-step response. The resulting unit-step response is shown in Figure 7–7. MATLAB Program 7–2 gives the unit-ramp response. The resulting response is shown in Figure 7–8.

The steady-state error $e(\infty)$ in following the unit-ramp input is obtained as follows: Since

$$\frac{Y(z)}{R(z)} = \frac{8(0.02z + 0.02)}{z^2 - 1.2z + 0.52}$$

we have

$$E(z) = R(z) - Y(z) = \left[1 - \frac{Y(z)}{R(z)} \right] R(z)$$

$$= \frac{(z - 1)(z - 0.36)}{z^2 - 1.2z + 0.52} R(z)$$

where

$$R(z) = \frac{0.2z}{(z - 1)^2}$$

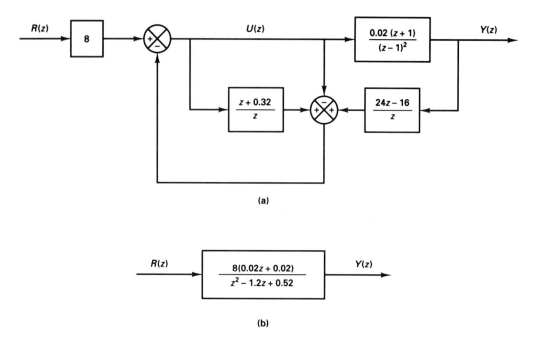

Figure 7–6 (a) Block diagram of the control system designed by use of polynomial equations approach; (b) simplified block diagram.

```
MATLAB Program 7–1

num = [0  0.16  0.16];
den = [1  -1.2  0.52];
r = ones(1,41);
k = 0: 40;
y = filter(num,den,r);
plot(k,y,'o')
v = [0  40  0  1.6];
axis(v);
grid
title('Unit-Step Response')
xlabel('k')
ylabel('y(k)')
```

Hence,

$$\lim_{k \to \infty} e(k) = \lim_{z \to 1} \frac{z-1}{z} \frac{(z-1)(z-0.36)}{z^2 - 1.2z + 0.52} \frac{0.2z}{(z-1)^2}$$

$$= 0.4$$

The steady-state error in following the unit-ramp input is 0.4.

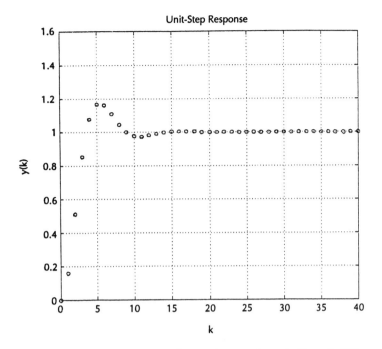

Figure 7-7 Unit-step response of control system shown in Figure 7-6(b).

```
MATLAB Program 7-2

num = [0  0.16  0.16];
den = [1  -1.2  0.52];
k = 0:20;
r = [0.2*k];
y = filter(num,den,r);
plot(k,y,'o',k,y,'-',k,0.2*k,'--')
v = [0  20  0  4];
axis(v);
grid
title('Unit-Ramp Response')
xlabel('k')
ylabel('y(k)')
```

In comparing the unit-step responses of the systems under control system configuration 1 and control system configuration 2, they are about the same. In comparing the unit-ramp responses of the two systems, the system under configuration 2 has approximately 10% smaller steady-state error in following the unit-ramp input than the system under configuration 1.

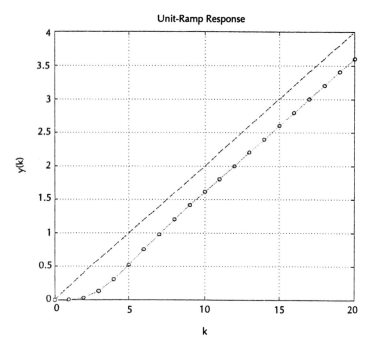

Figure 7–8   Unit-ramp response of control system shown in Figure 7–6(b).

7–5 DESIGN OF MODEL MATCHING CONTROL SYSTEMS

In the design technique presented in Section 7–4 (under control system configuration 2), the observer polynomial $F(z)$ has been canceled between the numerator and denominator of the closed-loop pulse transfer function. [See Equation (7–12).] The characteristic equation of the designed system was $H(z)$, a stable polynomial of degree n. [$H(z)$ was a desirable, but in a sense "arbitrary," stable polynomial of degree n.]

Suppose that the pulse transfer function of the plant is

$$\frac{Y(z)}{U(z)} = \frac{B(z)}{A(z)}$$

where $A(z)$ is an nth-degree monic polynomial in z and $B(z)$ is an mth-degree polynomial in z ($m \leq n$), where we assume that there are no common factors between $A(z)$ and $B(z)$. If $B(z)$ is a stable polynomial (meaning that all zeros lie in the unit circle in the z plane), it may be possible to choose $H(z)$ such that it includes polynomial $B(z)$, or

$$H(z) = B(z)H_1(z)$$

Then, referring to Equation (7–12), we can have

$$\frac{Y(z)}{R(z)} = \frac{K_0 B(z)}{H(z)} = \frac{K_0 B(z)}{B(z)H_1(z)} = \frac{K_0}{H_1(z)}$$

Thus, we eliminated the zeros of the numerator polynomial, which means that we can eliminate the zeros of the plant if we so desire.

Suppose that we wish to have desired zeros in the numerator and desired poles in the denominator. That is, we would like to have the system to possess desired poles and zeros like a "model system," or

$$\frac{Y(z)}{R(z)} = G_{\text{model}} = \frac{B_m(z)}{A_m(z)}$$

Under certain conditions, it is possible to design such a system by use of the polynomial equations approach. Since we force the pulse transfer function of the control system exactly like the model, we call such a control system a model matching control system.

In the design process discussed in Section 7-4, we chose $H(z)$ as the desired characteristic polynomial of degree n. [$H(z)$ is a stable nth-degree polynomial, but is not unique, but rather arbitrary, provided the response of the system is acceptable.] Let us choose a stable polynomial of degree $n - m$ as $H_1(z)$. [$H_1(z)$ must be a stable polynomial, but in a sense arbitrary, provided the response of the resulting system is acceptable.] Now we define the product of $B(z)$ and $H_1(z)$ as $H(z)$, or

$$H(z) = B(z)H_1(z)$$

Model Matching Control System.   We first refer to the block diagram of Figure 7-9. We assume that the plant $B(z)/A(z)$ is completely state controllable and completely observable; that is, there are no common factors between $A(z)$ and $B(z)$. We determine $\alpha(z)$ and $\beta(z)$ by solving the following Diophantine equation:

$$\alpha(z)A(z) + \beta(z)B(z) = F(z)B(z)H_1(z)$$

where $F(z)$ is a stable polynomial of $(n - 1)$th degree. [Note that $\alpha(z)$ and $\beta(z)$ are polynomials of $(n - 1)$th degree.] Then, from the block diagram of Figure 7-9, we have

$$U(z) = -\left[\frac{\alpha(z)}{F(z)}U(z) - U(z) + \frac{\beta(z)}{F(z)}Y(z)\right] + V(z)$$

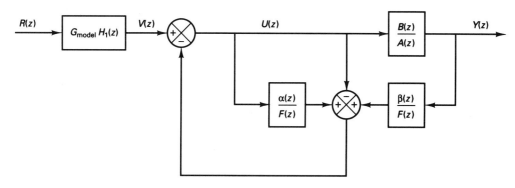

Figure 7-9   Block diagram of model matching control system.

or

$$\frac{\alpha(z)}{F(z)} U(z) + \frac{\beta(z)}{F(z)} Y(z) = V(z)$$

Since

$$U(z) = \frac{A(z)}{B(z)} Y(z)$$

we have

$$\frac{\alpha(z)}{F(z)} \frac{A(z)}{B(z)} Y(z) + \frac{\beta(z)}{F(z)} Y(z) = V(z)$$

or

$$\frac{Y(z)}{V(z)} = \frac{F(z)B(z)}{\alpha(z)A(z) + \beta(z)B(z)} = \frac{F(z)B(z)}{F(z)B(z)H_1(z)} = \frac{1}{H_1(z)}$$

Also,

$$V(z) = G_{\text{model}} H_1(z) R(z)$$

Hence,

$$\frac{Y(z)}{R(z)} = \frac{Y(z)}{V(z)} \frac{V(z)}{R(z)} = \frac{G_{\text{model}} H_1(z)}{H_1(z)} = G_{\text{model}}$$

We thus have seen that if we set

$$\frac{V(z)}{R(z)} = G_{\text{model}} H_1(z)$$

then the pulse transfer function between the output $Y(z)$ and the input $R(z)$ becomes G_{model}. Hence, we achieved model matching control.

Comments.   In applying the present approach to the design of model matching control systems, it is important to remember the following:

1. To make the pulse transfer function $G_{\text{model}} H_1(z)$ physically realizable, the degree of the numerator polynomial of $G_{\text{model}} H_1(z)$ must be equal to or less than the degree of the denominator polynomial of $G_{\text{model}} H_1(z)$. Otherwise, the present approach does not apply.
2. As noted earlier, the numerator polynomial $B(z)$ of the plant must be a stable polynomial, because the cancellation of $B(z)$ takes place between the numerator and denominator of $Y(z)/V(z)$. [If $B(z)$ were not a stable polynomial, that is, $B(z)$ possesses a zero or zeros on or outside the unit circle, then cancellation of $B(z)$ in $Y(z)/V(z)$ will generate unstable response and the designed system will become unstable.]

Example 7–4

Consider the plant defined by

$$\frac{Y(z)}{U(z)} = \frac{0.3679z + 0.2642}{(z - 0.3679)(z - 1)}$$

Assume that the sampling period T is 1 sec. It is desired to design a control system such that the closed-loop system will behave like

$$\frac{Y_m(z)}{R_m(z)} = \frac{0.62z - 0.3}{z^2 - 1.2z + 0.52}$$

Let us call this pulse transfer function the model pulse transfer function, or G_{model}, and

$$G_{\text{model}} = \frac{Y_m(z)}{R_m(z)} = \frac{0.62z - 0.3}{z^2 - 1.2z + 0.52} \tag{7-14}$$

Let us assume that we use the system configuration given in Figure 7–9. Note that for the given plant,

$$A(z) = z^2 - 1.3679z + 0.3679$$

$$B(z) = 0.3679z + 0.2642$$

Hence,

$$a_1 = -1.3679, \qquad a_2 = 0.3679$$

$$b_0 = 0, \qquad b_1 = 0.3679, \qquad b_2 = 0.2642$$

Clearly, the numerator $B(z)$ is a stable polynomial.

Since the pulse transfer function of the plant is of second order (or $n = 2$), we choose $H_1(z)$ as a stable, first-degree $[(n - 1)$ degree] polynomial. For example, we may choose $H_1(z)$ as

$$H_1(z) = z + 0.5$$

[Choice of $H_1(z)$ is, in a sense, arbitrary as long as it is a stable polynomial.] Now define

$$H(z) = B(z)H_1(z) = (0.3679z + 0.2642)(z + 0.5)$$

Next, we choose

$$F(z) = z$$

[$F(z)$ can be any stable $(n - 1)$th-degree polynomial.] Then

$$D(z) = F(z)H(z) = F(z)B(z)H_1(z) = z(0.3679z + 0.2642)(z + 0.5)$$

$$= 0.3679z^3 + 0.4482z^2 + 0.1321z$$

Hence,

$$d_0 = 0.3679, \qquad d_1 = 0.4482, \qquad d_2 = 0.1321, \qquad d_3 = 0$$

Now we need to solve the following Diophantine equation:

$$\alpha(z)A(z) + \beta(z)B(z) = F(z)B(z)H_1(z)$$

or

$$\alpha(z)(z^2 - 1.3679z + 0.3679) + \beta(z)(0.3679z + 0.2642)$$

$$= 0.3679z^3 + 0.4482z^2 + 0.1321z$$

where $\alpha(z)$ and $\beta(z)$ are first-degree polynomials in z, respectively. The 4×4 Sylvester matrix $\mathbf{E}$ for this problem becomes as follows:

$$\mathbf{E} = \begin{bmatrix} a_2 & 0 & b_2 & 0 \\ a_1 & a_2 & b_1 & b_2 \\ 1 & a_1 & b_0 & b_1 \\ 0 & 1 & 0 & b_0 \end{bmatrix} = \begin{bmatrix} 0.3679 & 0 & 0.2642 & 0 \\ -1.3679 & 0.3679 & 0.3679 & 0.2642 \\ 1 & -1.3679 & 0 & 0.3679 \\ 0 & 1 & 0 & 0 \end{bmatrix}$$

Then, by use of MATLAB, $\mathbf{E}^{-1}$ can be obtained as shown next.

```
E =

    0.3679        0      0.2642        0
   -1.3679    0.3679     0.3679     0.2642
    1.0000   -1.3679        0       0.3679
       0      1.0000        0          0

inv(E)

ans =

    0.5359   -0.3849     0.2764     0.5197
       0         0          0       1.0000
    3.0387    0.5359    -0.3849    -0.7236
   -1.4567    1.0461     1.9669     2.3056
```

Define

$$\mathbf{D} = \begin{bmatrix} d_3 \\ d_2 \\ d_1 \\ d_0 \end{bmatrix} = \begin{bmatrix} 0 \\ 0.1321 \\ 0.4482 \\ 0.3679 \end{bmatrix}, \qquad \mathbf{M} = \begin{bmatrix} \alpha_1 \\ \alpha_0 \\ \beta_1 \\ \beta_0 \end{bmatrix}$$

Then matrix $\mathbf{M}$ is obtained as follows:

$$\mathbf{M} = \mathbf{E}^{-1}\mathbf{D} = \begin{bmatrix} 0.2642 \\ 0.3679 \\ -0.3679 \\ 1.8680 \end{bmatrix}$$

Hence,

$$\alpha(z) = \alpha_0 z + \alpha_1 = 0.3679z + 0.2642$$

$$\beta(z) = \beta_0 z + \beta_1 = 1.8680z - 0.3679$$

Using $\alpha(z)$ and $\beta(z)$ thus determined, $Y(z)/V(z)$ becomes as follows:

$$\frac{Y(z)}{V(z)} = \frac{F(z)B(z)}{F(z)B(z)H_1(z)} = \frac{1}{H_1(z)} = \frac{1}{z + 0.5}$$

Since $V(z)/R(z)$ is

$$\frac{V(z)}{R(z)} = G_{\text{model}} H_1(z) = \frac{(0.62z - 0.3)(z + 0.5)}{z^2 - 1.2z + 0.52}$$

the pulse transfer function $Y(z)/R(z)$ becomes

$$\frac{Y(z)}{R(z)} = \frac{0.62z - 0.3}{z^2 - 1.2z + 0.52} = G_{\text{model}}$$

The designed model matching control system has the block diagram as shown in Figure 7–10(a). This block diagram can be simplified to those shown in Figures 7–10(b) and (c).

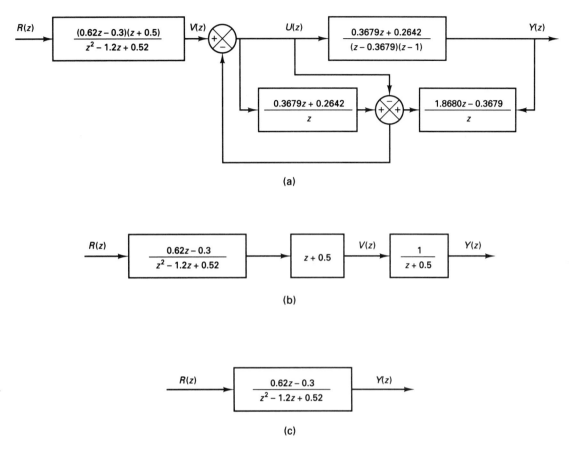

Figure 7–10 (a) Block diagram of model matching control system; (b) and (c) simplified block diagrams.

The unit-step response and unit-ramp response of the model system are shown in Figures 7–11 and 7–12, respectively. The unit-step response exhibits approximately 30% overshoot in the response, and the error in following the unit-ramp input is approximately 0.55.

Comments.   It is important to note that the present approach is different from multiplying the following filter (pulse transfer function)

$$\frac{(z - 0.3679)(z - 1)}{z^2 - 1.2z + 0.52} \frac{0.62z - 0.3}{0.3679z + 0.2642}$$

to the plant. Although, mathematically, the product becomes

$$\frac{(z - 0.3679)(z - 1)}{z^2 - 1.2z + 0.52} \frac{0.62z - 0.3}{0.3679z + 0.2642} \frac{0.3679z + 0.2642}{(z - 0.3679)(z - 1)} = \frac{0.62z - 0.3}{z^2 - 1.2z + 0.52}$$

and the resulting system has the pulse transfer function of the model, in this case cancellation takes place between a critically stable pole at $z = 1$ and zero at $z = 1$,

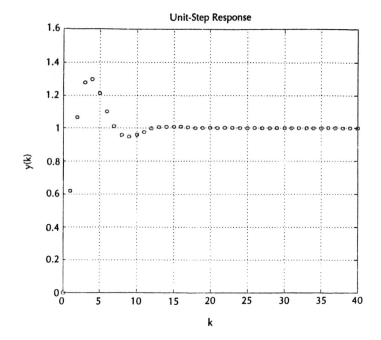

Figure 7–11   Unit-step response of model G_{model} given by Equation (7–14).

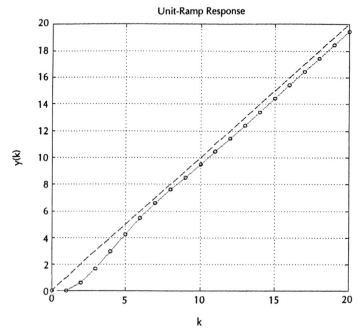

Figure 7–12   Unit-ramp response of model G_{model} given by Equation (7–14).

and the system will become unstable. [Remember that we should never cancel an unstable (or critically stable) pole and zero.] In the present polynomial equations approach, no cancellations take place between unstable (or critically stable) poles and zeros in the entire design process and, therefore, the resulting system is always stable.

If no steady-state error in following the ramp input is desired, then we need to change the model. For example, if we use the following pulse transfer function as the pulse transfer function of the revised model

$$G'_{\text{model}} = \frac{0.8z - 0.48}{z^2 - 1.2z + 0.52} \tag{7–15}$$

then the steady-state error in following the ramp input becomes zero. However, the maximum overshoot in the unit-step response becomes approximately 45%. The unit-step response and unit-ramp response of the revised model are shown in Figures 7–13 and 7–14, respectively.

Note that changing the model does not change the block diagram between $Y(z)$ and $V(z)$, because $Y(z)/V(z)$ is independent of the model pulse transfer function. Therefore, if a change of the model is desired, all we need is to change the pulse transfer function of the first block from G_{model} to G'_{model}.

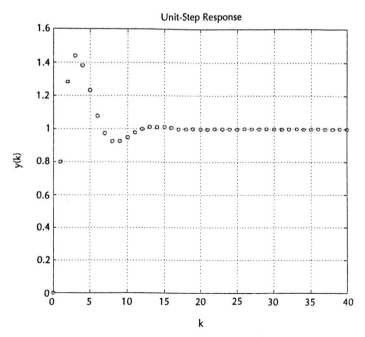

Figure 7–13   Unit-step response of model G'_{model} given by Equation (7–15).

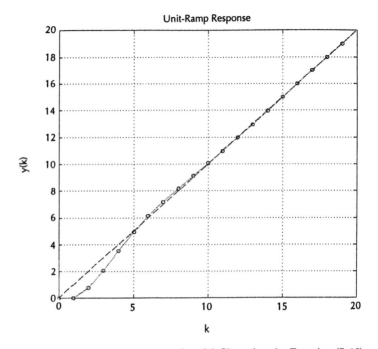

Figure 7–14   Unit-ramp response of model G'_{model} given by Equation (7–15).

EXAMPLE PROBLEMS AND SOLUTIONS

Problem A–7–1

Consider polynomials

$$A(z) = z^2 + a_1 z + a_2$$

$$B(z) = b_0 z^2 + b_1 z + b_2$$

The Sylvester matrix $\mathbf{E}$ is defined by

$$\mathbf{E} = \begin{bmatrix} a_2 & 0 & b_2 & 0 \\ a_1 & a_2 & b_1 & b_2 \\ 1 & a_1 & b_0 & b_1 \\ 0 & 1 & 0 & b_0 \end{bmatrix}$$

Show that matrix $\mathbf{E}$ is nonsingular if and only if there is no cancellation between polynomials $A(z)$ and $B(z)$.

Solution   Let us assume that the roots of $A(z) = 0$ are λ_1 and λ_2 and those of $B(z) = 0$ are λ_3 and λ_4. Thus,

$$A(z) = z^2 + a_1 z + a_2 = (z - \lambda_1)(z - \lambda_2)$$

$$= z^2 - (\lambda_1 + \lambda_2)z + \lambda_1 \lambda_2$$

$$B(z) = b_0 z^2 + b_1 z + b_2 = b_0(z - \lambda_3)(z - \lambda_4)$$

$$= b_0 z^2 - b_0(\lambda_3 + \lambda_4)z + b_0 \lambda_3 \lambda_4$$

Hence,

$$a_1 = -(\lambda_1 + \lambda_2), \qquad a_2 = \lambda_1 \lambda_2$$

$$b_1 = -b_0(\lambda_3 + \lambda_4), \qquad b_2 = b_0 \lambda_3 \lambda_4$$

Let us write matrix $\mathbf{E}$ in terms of λ's.

$$\mathbf{E} = \begin{bmatrix} \lambda_1 \lambda_2 & 0 & b_0 \lambda_3 \lambda_4 & 0 \\ -(\lambda_1 + \lambda_2) & \lambda_1 \lambda_2 & -b_0(\lambda_3 + \lambda_4) & b_0 \lambda_3 \lambda_4 \\ 1 & -(\lambda_1 + \lambda_2) & b_0 & -b_0(\lambda_3 + \lambda_4) \\ 0 & 1 & 0 & b_0 \end{bmatrix}$$

The determinant of matrix $\mathbf{E}$ can be calculated by use of Laplace's expansion by minors.

$$|E| = \begin{vmatrix} \lambda_1 \lambda_2 & 0 \\ -(\lambda_1 + \lambda_2) & \lambda_1 \lambda_2 \end{vmatrix} \begin{vmatrix} b_0 - b_0 (\lambda_3 + \lambda_4) \\ 0 & b_0 \end{vmatrix} - \begin{vmatrix} \lambda_1 \lambda_2 & 0 \\ 1 & -(\lambda_1 + \lambda_2) \end{vmatrix} \begin{vmatrix} -b_0(\lambda_3 + \lambda_4) & b_0 \lambda_3 \lambda_4 \\ 0 & b_0 \end{vmatrix}$$

$$+ \begin{vmatrix} \lambda_1 \lambda_2 & 0 \\ 0 & 1 \end{vmatrix} \begin{vmatrix} -b_0(\lambda_3 + \lambda_4) & b_0 \lambda_3 \lambda_4 \\ b_0 & -b_0(\lambda_3 + \lambda_4) \end{vmatrix} + \begin{vmatrix} -(\lambda_1 + \lambda_2) & \lambda_1 \lambda_2 \\ 1 & -(\lambda_1 + \lambda_2) \end{vmatrix} \begin{vmatrix} b_0 \lambda_3 \lambda_4 & 0 \\ 0 & b_0 \end{vmatrix}$$

$$- \begin{vmatrix} -(\lambda_1 + \lambda_2) & \lambda_1 \lambda_2 \\ 0 & 1 \end{vmatrix} \begin{vmatrix} b_0 \lambda_3 \lambda_4 & 0 \\ b_0 & -b_0(\lambda_3 + \lambda_4) \end{vmatrix} + \begin{vmatrix} 1 & -(\lambda_1 + \lambda_2) \\ 0 & 1 \end{vmatrix} \begin{vmatrix} b_0 \lambda_3 \lambda_4 & 0 \\ -b_0(\lambda_3 + \lambda_4) & b_0 \lambda_3 \lambda_4 \end{vmatrix}$$

$$= \lambda_1^2 \lambda_2^2 b_0^2 - (\lambda_1^2 \lambda_2 + \lambda_1 \lambda_2^2) b_0^2 (\lambda_3 + \lambda_4) + \lambda_1 \lambda_2 b_0^2 (\lambda_3^2 + 2\lambda_3 \lambda_4 + \lambda_4^2) - \lambda_1 \lambda_2 b_0^2 \lambda_3 \lambda_4$$

$$+ (\lambda_1^2 + \lambda_1 \lambda_2 + \lambda_2^2) b_0^2 \lambda_3 \lambda_4 - b_0^2 (\lambda_1 + \lambda_2)(\lambda_3 + \lambda_4)\lambda_3 \lambda_4 + b_0^2 \lambda_3^2 \lambda_4^2$$

$$= b_0^2 (\lambda_1^2 \lambda_2^2 - \lambda_1^2 \lambda_2 \lambda_3 - \lambda_1 \lambda_2^2 \lambda_3 - \lambda_1^2 \lambda_2 \lambda_4 - \lambda_1 \lambda_2^2 \lambda_4$$

$$+ \lambda_1 \lambda_2 \lambda_3^2 + \lambda_1 \lambda_2 \lambda_3 \lambda_4 + \lambda_1 \lambda_2 \lambda_4^2 + \lambda_1^2 \lambda_3 \lambda_4 + \lambda_1 \lambda_2 \lambda_3 \lambda_4$$

$$+ \lambda_2^2 \lambda_3 \lambda_4 - \lambda_1 \lambda_3^2 \lambda_4 - \lambda_2 \lambda_3^2 \lambda_4 - \lambda_1 \lambda_3 \lambda_4^2 - \lambda_2 \lambda_3 \lambda_4^2 + \lambda_3^2 \lambda_4^2)$$

Rearranging the terms on the right-hand side of this last equation, we obtain

$$|E| = b_0^2 (\lambda_1^2 \lambda_2^2 - \lambda_1 \lambda_2^2 \lambda_3 - \lambda_1 \lambda_2^2 \lambda_4 + \lambda_2^2 \lambda_3 \lambda_4 - \lambda_1^2 \lambda_2 \lambda_3$$

$$+ \lambda_1 \lambda_2 \lambda_3^2 + \lambda_1 \lambda_2 \lambda_3 \lambda_4 - \lambda_2 \lambda_3^2 \lambda_4 - \lambda_1^2 \lambda_2 \lambda_4$$

$$+ \lambda_1 \lambda_2 \lambda_3 \lambda_4 + \lambda_1 \lambda_2 \lambda_4^2 - \lambda_2 \lambda_3 \lambda_4^2 + \lambda_1^2 \lambda_3 \lambda_4$$

$$- \lambda_1 \lambda_3^2 \lambda_4 - \lambda_1 \lambda_3 \lambda_4^2 + \lambda_3^2 \lambda_4^2)$$

$$= b_0^2 (\lambda_1^2 - \lambda_1 \lambda_3 - \lambda_1 \lambda_4 + \lambda_3 \lambda_4)(\lambda_2^2 - \lambda_2 \lambda_3 - \lambda_2 \lambda_4 + \lambda_3 \lambda_4)$$

$$= b_0^2 (\lambda_1 - \lambda_3)(\lambda_1 - \lambda_4)(\lambda_2 - \lambda_3)(\lambda_2 - \lambda_4)$$

The determinant $|E|$ is nonzero unless at least $\lambda_1 = \lambda_3$ or $\lambda_1 = \lambda_4$ or $\lambda_2 = \lambda_3$ or $\lambda_2 = \lambda_4$. Hence, $\mathbf{E}$ is nonsingular if and only if no cancellation occurs between $A(z)$ and $B(z)$. That is, if $A(z)$ and $B(z)$ are coprime polynomials, then $|E|$ is nonzero and $\mathbf{E}^{-1}$ exists.

Problem A–7–2

Assume that polynomials $A(z)$ and $B(z)$ have no cancellations and are given by

$$A(z) = z^2 + a_1 z + a_2$$

$$B(z) = b_0 z^2 + b_1 z + b_2$$

Then the Sylvester matrix $\mathbf{E}$ is given by

$$\mathbf{E} = \begin{bmatrix} a_2 & 0 & b_2 & 0 \\ a_1 & a_2 & b_1 & b_2 \\ 1 & a_1 & b_0 & b_1 \\ 0 & 1 & 0 & b_0 \end{bmatrix}$$

Define

$$\alpha(z) = \alpha_0 z + \alpha_1$$

$$\beta(z) = \beta_0 z + \beta_1$$

$$D(z) = d_0 z^3 + d_1 z^2 + d_2 z + d_3$$

Show that if

$$\alpha(z)A(z) + \beta(z)B(z) = D(z)$$

then α_0, α_1, β_0, and β_1 can be determined by computing

$$\mathbf{M} = \mathbf{E}^{-1}\mathbf{D}$$

where

$$\mathbf{D} = \begin{bmatrix} d_3 \\ d_2 \\ d_1 \\ d_0 \end{bmatrix}, \qquad \mathbf{M} = \begin{bmatrix} \alpha_1 \\ \alpha_0 \\ \beta_1 \\ \beta_0 \end{bmatrix}$$

Solution   First, notice that

$$\alpha(z)A(z) + \beta(z)B(z) = (\alpha_0 z + \alpha_1)(z^2 + a_1 z + a_2) + (\beta_0 z + \beta_1)(b_0 z^2 + b_1 z + b_2)$$

$$= (\alpha_0 + \beta_0 b_0)z^3 + (\alpha_1 + \alpha_0 a_1 + \beta_1 b_0 + \beta_0 b_1)z^2$$

$$+ (\alpha_1 a_1 + \alpha_0 a_2 + \beta_1 b_1 + \beta_0 b_2)z + \alpha_1 a_2 + \beta_1 b_2$$

$$= D(z) = d_0 z^3 + d_1 z^2 + d_2 z + d_3$$

Hence,

$$d_0 = \alpha_0 + \beta_0 b_0$$

$$d_1 = \alpha_1 + \alpha_0 a_1 + \beta_1 b_0 + \beta_0 b_1$$

$$d_2 = \alpha_1 a_1 + \alpha_0 a_2 + \beta_1 b_1 + \beta_0 b_2$$

$$d_3 = \alpha_1 a_2 + \beta_1 b_2$$

Now compute **EM**. Since

$$\mathbf{EM} = \begin{bmatrix} a_2 & 0 & b_2 & 0 \\ a_1 & a_2 & b_1 & b_2 \\ 1 & a_1 & b_0 & b_1 \\ 0 & 1 & 0 & b_0 \end{bmatrix} \begin{bmatrix} \alpha_1 \\ \alpha_0 \\ \beta_1 \\ \beta_0 \end{bmatrix}$$

$$= \begin{bmatrix} \alpha_1 a_2 + \beta_1 b_2 \\ \alpha_1 a_1 + \alpha_0 a_2 + \beta_1 b_1 + \beta_0 b_2 \\ \alpha_1 + \alpha_0 a_1 + \beta_1 b_0 + \beta_1 b_1 \\ \alpha_0 + \beta_0 b_0 \end{bmatrix}$$

comparing each element of the right-hand side of this last equation with d_3, d_2, d_1, and d_0, respectively, we have

$$\mathbf{EM} = \begin{bmatrix} d_3 \\ d_2 \\ d_1 \\ d_0 \end{bmatrix} = \mathbf{D}$$

Hence,

$$\mathbf{M} = \mathbf{E}^{-1}\mathbf{D}$$

or the coefficients of polynomials $\alpha(z)$ and $\beta(z)$ can be determined by multiplying $\mathbf{E}^{-1}$ by $\mathbf{D}$.

Problem A–7–3

In Chapter 6 we designed regulator systems and control systems using the observed-state feedback scheme. In single-input–single-output systems, the output is always measurable. Hence, in such systems we may need minimum-order observers, rather than full-order observers.

Consider the regulator system designed by the pole placement approach combined with the minimum-order observer. The plant equations are

$$\mathbf{x}(k + 1) = \mathbf{G}\mathbf{x}(k) + \mathbf{H}u(k)$$

$$y(k) = \mathbf{C}\mathbf{x}(k)$$

where $\mathbf{x}$ is an n-vector, $u(k)$ is a scalar, and $y(k)$ is also a scalar. We assume that the plant is completely state controllable and completely observable. The plant equations can be rewritten as

$$\begin{bmatrix} x_a(k + 1) \\ \mathbf{x}_b(k + 1) \end{bmatrix} = \begin{bmatrix} G_{aa} & \mathbf{G}_{ab} \\ \mathbf{G}_{ba} & \mathbf{G}_{bb} \end{bmatrix}\begin{bmatrix} x_a(k) \\ \mathbf{x}_b(k) \end{bmatrix} + \begin{bmatrix} H_a \\ \mathbf{H}_b \end{bmatrix}u(k)$$

$$y(k) = \begin{bmatrix} 1 & \mathbf{0} \end{bmatrix}\begin{bmatrix} x_a(k) \\ \mathbf{x}_b(k) \end{bmatrix}$$

where $y(k) = x_a(k)$ is the measurable state variable and $\mathbf{x}_b(k)$ consists of unmeasurable state variables.

The equation for the observer regulator (controller) can be derived from Equations (6-148), (6-149), and (6-159), rewritten thus:

$$\tilde{\mathbf{x}}_b(k) - \mathbf{K}_e y(k) = \tilde{\mathbf{\eta}}(k) \tag{7–16}$$

$$\begin{aligned} \tilde{\mathbf{\eta}}(k + 1) = {}& (\mathbf{G}_{bb} - \mathbf{K}_e\,\mathbf{G}_{ab})\tilde{\mathbf{\eta}}(k) \\ & + [(\mathbf{G}_{bb} - \mathbf{K}_e\,\mathbf{G}_{ab})\mathbf{K}_e + \mathbf{G}_{ba} - \mathbf{K}_e\,\mathbf{G}_{aa}]y(k) \\ & + (\mathbf{H}_b - \mathbf{K}_e\,H_a)u(k) \end{aligned} \tag{7–17}$$

$$u(k) = -\mathbf{K}\tilde{\mathbf{x}}(k) \tag{7–18}$$

where $\mathbf{K}$ is the state feedback gain matrix and $\mathbf{K}_e$ is the observer gain matrix. Define

$$\tilde{\mathbf{x}}(k) = \begin{bmatrix} x_1(k) \\ \tilde{\mathbf{x}}_b(k) \end{bmatrix} = \begin{bmatrix} y(k) \\ \tilde{\mathbf{x}}_b(k) \end{bmatrix}$$

where $\tilde{\mathbf{x}}(k)$ is an n-vector and $\tilde{\mathbf{x}}_b(k)$ is an $(n - 1)$-vector consisting of $(n - 1)$ observed state variables. Define also

$$\mathbf{K} = \begin{bmatrix} k_1 & \mathbf{k}_b \end{bmatrix}$$

Then Equation (7–18) becomes

$$u(k) = -\begin{bmatrix} k_1 & \mathbf{k}_b \end{bmatrix}\begin{bmatrix} y(k) \\ \tilde{\mathbf{x}}_b(k) \end{bmatrix} = -k_1 y(k) - \mathbf{k}_b\,\tilde{\mathbf{x}}_b(k) \tag{7–19}$$

Show that the regulator (or controller) equation can be given by

$$G_D(z) = -\frac{U(z)}{Y(z)} = \frac{\beta(z)}{\alpha(z)}$$

where

$$\alpha(z) = \mathbf{k}_b \mathbf{W}(z)\mathbf{Q} + F(z)$$

$$\beta(z) = (k_1 + \mathbf{k}_b \mathbf{K}_e)F(z) + \mathbf{k}_b \mathbf{W}(z)\mathbf{P}$$

$$\mathbf{P} = (\mathbf{G}_{bb} - \mathbf{K}_e \mathbf{G}_{ab})\mathbf{K}_e + \mathbf{G}_{ba} - \mathbf{K}_e \mathbf{G}_{aa}$$

$$\mathbf{Q} = \mathbf{H}_b - \mathbf{K}_e \mathbf{H}_a$$

$F(z) = |z\mathbf{I} - \mathbf{G}_{bb} + \mathbf{K}_e \mathbf{G}_{ab}| =$ characteristic equation for the minimum-order observer (stable polynomial of degree $n - 1$)

$$\mathbf{W}(z) = (z\mathbf{I} - \mathbf{G}_{bb} + \mathbf{K}_e \mathbf{G}_{ab})^{-1} F(z)$$

Show also that the block diagram for the regulator system can be given as shown in Figure 7–15.

Solution   Taking the z transform of Equation (7–17), assuming zero initial conditions, we obtain

$$z\tilde{\mathbf{\eta}}(z) = (\mathbf{G}_{bb} - \mathbf{K}_e \mathbf{G}_{ab})\tilde{\mathbf{\eta}}(z) + [(\mathbf{G}_{bb} - \mathbf{K}_e \mathbf{G}_{ab})\mathbf{K}_e$$

$$+ \mathbf{G}_{ba} - \mathbf{K}_e \mathbf{G}_{aa}]Y(z) + (\mathbf{H}_b - \mathbf{K}_e \mathbf{H}_a)U(z)$$

$$= (\mathbf{G}_{bb} - \mathbf{K}_e \mathbf{G}_{ab})\tilde{\mathbf{\eta}}(z) + \mathbf{P}Y(z) + \mathbf{Q}U(z)$$

which can be written as

$$(z\mathbf{I} - \mathbf{G}_{bb} + \mathbf{K}_e \mathbf{G}_{ab})\tilde{\mathbf{\eta}}(z) = \mathbf{P}Y(z) + \mathbf{Q}U(z)$$

Solving this last equation for $\tilde{\mathbf{\eta}}(z)$, we have

$$\tilde{\mathbf{\eta}}(z) = (z\mathbf{I} - \mathbf{G}_{bb} + \mathbf{K}_e \mathbf{G}_{ab})^{-1}[\mathbf{P}Y(z) + \mathbf{Q}U(z)] \qquad (7\text{–}20)$$

Since

$$(z\mathbf{I} - \mathbf{G}_{bb} + \mathbf{K}_e \mathbf{G}_{ab})^{-1} = \frac{\mathbf{W}(z)}{|z\mathbf{I} - \mathbf{G}_{bb} + \mathbf{K}_e \mathbf{G}_{ab}|} = \frac{\mathbf{W}(z)}{F(z)}$$

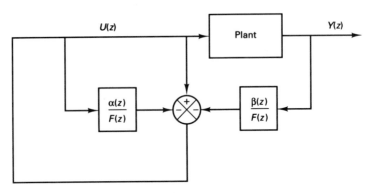

Figure 7–15   Block diagram of the regulator system.

Equation (7–20) becomes as follows:

$$\hat{\eta}(z) = \frac{\mathbf{W}(z)}{F(z)}[\mathbf{P}Y(z) + \mathbf{Q}U(z)] \tag{7–21}$$

By substituting Equation (7–16) into Equation (7–19), we obtain

$$u(k) = -k_1 y(k) - \mathbf{k}_b[\mathbf{K}_e y(k) + \hat{\eta}(k)]$$

The z transform of this last equation gives

$$U(z) = -k_1 Y(z) - \mathbf{k}_b[\mathbf{K}_e Y(z) + \hat{\eta}(z)] \tag{7–22}$$

Substituting Equation (7–21) into Equation (7–22) gives

$$U(z) = -k_1 Y(z) - \mathbf{k}_b \mathbf{K}_e Y(z) - \mathbf{k}_b \frac{\mathbf{W}(z)}{F(z)}[\mathbf{P}Y(z) + \mathbf{Q}U(z)]$$

$$= -\mathbf{k}_b \frac{\mathbf{W}(z)}{F(z)}\mathbf{Q}U(z) - \left[k_1 + \mathbf{k}_b \mathbf{K}_e + \mathbf{k}_b \frac{\mathbf{W}(z)}{F(z)}\mathbf{P}\right]Y(z)$$

$$= -\left[\frac{\mathbf{k}_b \mathbf{W}(z)\mathbf{Q} + F(z)}{F(z)} - 1\right]U(z)$$

$$\quad -\left[k_1 + \mathbf{k}_b \mathbf{K}_e + \mathbf{k}_b \frac{\mathbf{W}(z)}{F(z)}\mathbf{P}\right]Y(z) \tag{7–23}$$

Using $\alpha(z)$ and $\beta(z)$ defined in the problem statement, Equation (7–23) can be written as follows:

$$U(z) = -\frac{\alpha(z)}{F(z)}U(z) + U(z) - \frac{\beta(z)}{F(z)}Y(z) \tag{7–24}$$

from which we get

$$\frac{\alpha(z)}{F(z)}U(z) = -\frac{\beta(z)}{F(z)}Y(z)$$

or

$$-\frac{U(z)}{Y(z)} = \frac{\beta(z)}{\alpha(z)} \tag{7–25}$$

[Note that $F(z)$ is a stable polynomial. Hence, two $F(z)$'s can be canceled.] Equation (7–25) is the equation for the regulator (or controller). The block diagram representation of Equation (7–24) becomes as shown in Figure 7–15. (This corresponds to a regulator system.) If this block diagram is incorporated into the control system, we obtain the block diagram of Figure 7–5.

Problem A–7–4

Referring to the example system discussed in Section 7–3 (the regulator system designed in Example 6–11) and Problem A–7–3, verify that

$$\alpha(z) = \mathbf{k}_b W(z)\mathbf{Q} + F(z) = z + 0.32$$

$$\beta(z) = (k_1 + \mathbf{k}_b \mathbf{K}_e)F(z) + \mathbf{k}_b W(z)\mathbf{P} = 24(z - 0.6667)$$

Solution   Referring to Example 6–11, we have

$$\mathbf{x}(k + 1) = \mathbf{G}\mathbf{x}(k) + \mathbf{H}u(k)$$

$$y(k) = \mathbf{C}\mathbf{x}(k)$$

where

$$\mathbf{G} = \begin{bmatrix} 1 & 0.2 \\ 0 & 1 \end{bmatrix}, \quad \mathbf{H} = \begin{bmatrix} 0.02 \\ 0.2 \end{bmatrix}, \quad \mathbf{C} = [1 \quad 0]$$

Hence,

$$G_{aa} = 1, \quad G_{ab} = 0.2, \quad G_{ba} = 0, \quad G_{bb} = 1,$$

$$H_a = 0.02, \quad H_b = 0.2$$

For the pole placement part, the desired characteristic equation was

$$|z\mathbf{I} - \mathbf{G} + \mathbf{HK}| = z^2 - 1.2z + 0.52 = 0$$

The state feedback gain matrix $\mathbf{K}$ for the desired characteristic equation was obtained as

$$\mathbf{K} = [8 \quad 3.2]$$

Hence,

$$k_1 = 8, \quad k_b = 3.2$$

For the minimum-order observer part, the observer characteristic equation was

$$\phi(z) = z = 0$$

The observer gain K_e was obtained as

$$K_e = 5$$

Hence, P, Q, $F(z)$, and $W(z)$ are obtained as follows:

$$P = (G_{bb} - K_e G_{ab})K_e + G_{ba} - K_e G_{aa}$$

$$= (1 - 5 \times 0.2) \times 5 + 0 - 5 \times 1 = -5$$

$$Q = H_b - K_e H_a = 0.2 - 5 \times 0.02 = 0.1$$

$$F(z) = z$$

$$W(z) = (z - G_{bb} + K_e G_{ab})^{-1} F(z) = (z - 1 + 5 \times 0.2)^{-1} z = 1$$

Hence,

$$\alpha(z) = k_b W(z)Q + F(z) = 3.2 \times 1 \times 0.1 + z$$

$$= z + 0.32$$

$$\beta(z) = (k_1 + k_b K_e)F(z) + k_b W(z)P$$

$$= (8 + 3.2 \times 5)z + 3.2 \times 1 \times (-5)$$

$$= 24z - 16 = 24(z - 0.6667)$$

Problem A-7-5

Show that the block diagram of Figure 7-5 can be modified to that shown in Figure 7-16.

Solution The block diagram of Figure 7-5 can be modified to that shown in Figure 7-17. Since the pulse transfer function of the minor loop is

$$\frac{U(z)}{X(z)} = \frac{1}{1 + \dfrac{\alpha(z)}{F(z)} - 1} = \frac{F(z)}{\alpha(z)}$$

by eliminating the minor loop we obtain the block diagram of Figure 7-16.

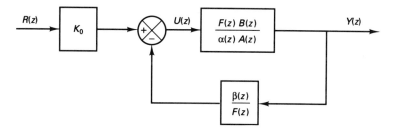

Figure 7–16   Modified block diagram.

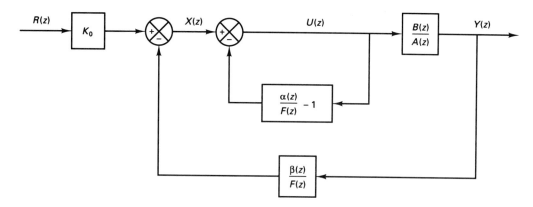

Figure 7–17   Block diagram of Figure 7–5 modified to that having a minor loop.

Problem A–7–6

Consider a plant defined by

$$\frac{Y(z)}{U(z)} = \frac{1}{z^2 + z + 0.16}$$

Using the polynomial equations approach, design a control system for this plant based on the block diagram shown in Figure 7–5. Assume that the desired characteristic equation is

$$H(z) = (z - 0.6 - j0.4)(z - 0.6 + j0.4)$$

$$= z^2 - 1.2z + 0.52$$

and $F(z)$ (minimum-order observer polynomial) is

$$F(z) = z$$

Solution   Referring to Figure 7–5, $\alpha(z)$ and $\beta(z)$ are determined from the following Diophantine equation:

$$\alpha(z)A(z) + \beta(z)B(z) = H(z)F(z) = D(z)$$

where

$$A(z) = z^2 + z + 0.16$$

$$B(z) = 1$$

$$D(z) = z^3 - 1.2z^2 + 0.52z$$

The Sylvester matrix $\mathbf{E}$ for this case is

$$\mathbf{E} = \begin{bmatrix} 0.16 & 0 & 1 & 0 \\ 1 & 0.16 & 0 & 1 \\ 1 & 1 & 0 & 0 \\ 0 & 1 & 0 & 0 \end{bmatrix}$$

The inverse of matrix $\mathbf{E}$ can be obtained as follows:

$$\mathbf{E}^{-1} = \begin{bmatrix} 0 & 0 & 1 & -1 \\ 0 & 0 & 0 & 1 \\ 1 & 0 & -0.16 & 0.16 \\ 0 & 1 & -1 & 0.84 \end{bmatrix}$$

Note that $\alpha(z)$ and $\beta(z)$ are polynomials of degree $n - 1 = 2 - 1 = 1$, respectively, or

$$\alpha(z) = \alpha_0 z + \alpha_1$$
$$\beta(z) = \beta_0 z + \beta_1$$

Define

$$\mathbf{D} = \begin{bmatrix} d_3 \\ d_2 \\ d_1 \\ d_0 \end{bmatrix} = \begin{bmatrix} 0 \\ 0.52 \\ -1.2 \\ 1 \end{bmatrix}, \qquad \mathbf{M} = \begin{bmatrix} \alpha_1 \\ \alpha_0 \\ \beta_1 \\ \beta_0 \end{bmatrix}$$

Then vector $\mathbf{M}$ is determined from $\mathbf{M} = \mathbf{E}^{-1}\mathbf{D}$ as follows:

$$\mathbf{M} = \mathbf{E}^{-1}\mathbf{D} = \begin{bmatrix} -2.2 \\ 1 \\ 0.352 \\ 2.56 \end{bmatrix}$$

Hence $\alpha(z)$ and $\beta(z)$ are determined as

$$\alpha(z) = z - 2.2$$
$$\beta(z) = 2.56z + 0.352$$

Referring to Equation (7–12), the designed system has the closed-loop pulse transfer function $Y(z)/R(z)$ as follows:

$$\frac{Y(z)}{R(z)} = \frac{K_0 B(z)}{H(z)} = K_0 \frac{1}{z^2 - 1.2z + 0.52}$$

Next we need to determine gain K_0 in Figure 7–5. We require the steady-state output $y(\infty)$ to a unit-step input to be unity, or

$$\lim_{k \to \infty} y(k) = \lim_{z \to 1} \frac{z - 1}{z} \frac{K_0}{z^2 - 1.2z + 0.52} \frac{z}{z - 1} = 1$$

from which we obtain

$$K_0 = 0.32$$

Hence, the pulse transfer function of the designed system becomes as follows:

$$\frac{Y(z)}{R(z)} = \frac{0.32}{z^2 - 1.2z + 0.52}$$

A block diagram for the designed system is shown in Figure 7–18. A unit-step response of the designed system is shown in Figure 7–19.

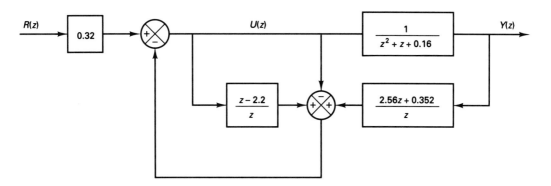

Figure 7–18   Block diagram of the system designed in Problem A–7–6.

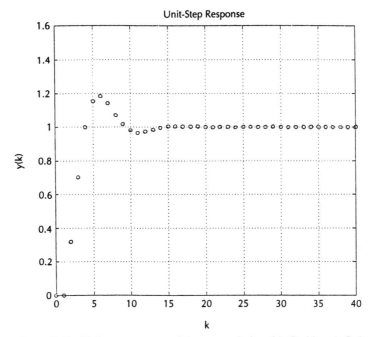

Figure 7–19   Unit-step response of the system designed in Problem A–7–6.

Problem A–7–7

In Problem A–6–17 we considered the design of the regulator system where the plant was defined by

$$\mathbf{x}(k + 1) = \mathbf{Gx}(k) + \mathbf{H}u(k) \tag{7–26}$$

$$y(k) = \mathbf{Cx}(k) \tag{7–27}$$

where

$$\mathbf{G} = \begin{bmatrix} 0 & 1 & 0 \\ 0 & 0 & 1 \\ -0.5 & -0.2 & 1.1 \end{bmatrix}, \quad \mathbf{H} = \begin{bmatrix} 0 \\ 0 \\ 1 \end{bmatrix}, \quad \mathbf{C} = [0 \quad 1 \quad 0]$$

We used the pole placement approach to determine the state feedback gain matrix $\mathbf{K}$ such that the regulator system would exhibit a deadbeat response to any initial state. We used a minimum-order observer such that the response to observer error was deadbeat.

Using the polynomial equations approach, design an equivalent control system for this plant such that the block diagram configuration is the same as that of Figure 7–4. Obtain the unit-step response and unit-ramp response of the designed system. The sampling period T of the system is 0.2 sec.

Solution   Referring to Problem A–6–17, the desired characteristic equation was

$$|z\mathbf{I} - \mathbf{G} + \mathbf{HK}| = z^3 = 0$$

Hence, we define

$$H(z) = z^3$$

The observer error characteristic equation was

$$|z\mathbf{I} - \hat{\mathbf{G}}_{bb} + \mathbf{K}_e\,\hat{\mathbf{G}}_{ab}| = z^2 = 0$$

Therefore, we define

$$F(z) = z^2$$

Next, we find the pulse transfer function $G_p(z)$ of the plant.

$$G_p(z) = \mathbf{C}(z\mathbf{I} - \mathbf{G})^{-1}\mathbf{H}$$

$$= \begin{bmatrix} 0 & 1 & 0 \end{bmatrix} \begin{bmatrix} z & -1 & 0 \\ 0 & z & -1 \\ 0.5 & 0.2 & z - 1.1 \end{bmatrix}^{-1} \begin{bmatrix} 0 \\ 0 \\ 1 \end{bmatrix}$$

$$= \begin{bmatrix} 0 & 1 & 0 \end{bmatrix} \frac{1}{z^3 - 1.1z^2 + 0.2z + 0.5} \begin{bmatrix} 1 \\ z \\ z^2 \end{bmatrix}$$

$$= \frac{z}{z^3 - 1.1z^2 + 0.2z + 0.5} = \frac{B(z)}{A(z)}$$

where

$$A(z) = z^3 - 1.1z^2 + 0.2z + 0.5$$

$$B(z) = z$$

Hence,

$$a_1 = -1.1, \qquad a_2 = 0.2, \qquad a_3 = 0.5$$

$$b_0 = 0, \qquad b_1 = 0, \qquad b_2 = 1, \qquad b_3 = 0$$

Referring to Figure 7–4, the block diagram for the present system can be drawn as shown in Figure 7–20.

The characteristic polynomial for the entire system (observed-state feedback system) is

$$D(z) = H(z)F(z) = z^3 \cdot z^2 = z^5$$

Hence,

$$d_0 = 1, \qquad d_1 = 0, \qquad d_2 = 0, \qquad d_3 = 0, \qquad d_4 = 0, \qquad d_5 = 0$$

From the block diagram of Figure 7–20, the characteristic equation for the system is

$$\alpha(z)A(z) + \beta(z)B(z) = 0$$

Hence, in the polynomial equations approach we set

$$\alpha(z)A(z) + \beta(z)B(z) = H(z)F(z)$$

or

$$\alpha(z)(z^3 - 1.1z^2 + 0.2z + 0.5) + \beta(z)z = z^5$$

To determine the pulse transfer function $\beta(z)/\alpha(z)$ of the controller, we solve this Diophantine equation.

For the present problem, $n = 3$ and the $2n \times 2n$ Sylvester matrix $\mathbf{E}$ becomes a 6×6 matrix as follows:

$$\mathbf{E} = \begin{bmatrix} 0.5 & 0 & 0 & 0 & 0 & 0 \\ 0.2 & 0.5 & 0 & 1 & 0 & 0 \\ -1.1 & 0.2 & 0.5 & 0 & 1 & 0 \\ 1 & -1.1 & 0.2 & 0 & 0 & 1 \\ 0 & 1 & -1.1 & 0 & 0 & 0 \\ 0 & 0 & 1 & 0 & 0 & 0 \end{bmatrix}$$

We can easily obtain $\mathbf{E}^{-1}$ by use of MATLAB, as shown next.

```
E =

     0.5000          0          0          0          0          0
     0.2000     0.5000          0     1.0000          0          0
    -1.1000     0.2000     0.5000          0     1.0000          0
     1.0000    -1.1000     0.2000          0          0     1.0000
          0     1.0000    -1.1000          0          0          0
          0          0     1.0000          0          0          0

inv(E)

ans =

     2.0000          0          0          0          0          0
          0          0          0          0     1.0000     1.1000
          0          0          0          0          0     1.0000
    -0.4000     1.0000     0.0000          0    -0.5000    -0.5500
     2.2000          0     1.0000          0    -0.2000    -0.7200
    -2.0000          0          0     1.0000     1.1000     1.0100
```

To determine $\alpha(z)$ and $\beta(z)$, we first define

$$\mathbf{D} = \begin{bmatrix} d_5 \\ d_4 \\ d_3 \\ d_2 \\ d_1 \\ d_0 \end{bmatrix} = \begin{bmatrix} 0 \\ 0 \\ 0 \\ 0 \\ 0 \\ 1 \end{bmatrix}, \qquad \mathbf{M} = \begin{bmatrix} \alpha_2 \\ \alpha_1 \\ \alpha_0 \\ \beta_2 \\ \beta_1 \\ \beta_0 \end{bmatrix}$$

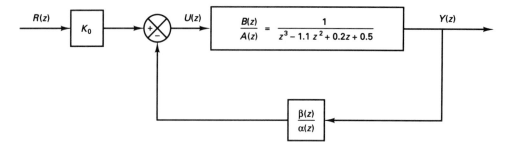

Figure 7–20   Block diagram of the control system considered in Problem A–7–7.

and then determine **M** by

$$\mathbf{M} = \mathbf{E}^{-1}\mathbf{D}$$

which can be calculated easily by MATLAB as follows:

```
D = [0;0;0;0;0;1];
M = (inv(E))*D

M =

           0
      1.1000
      1.0000
     -0.5500
     -0.7200
      1.0100
```

We can now determine $\alpha(z)$ and $\beta(z)$ as follows:

$$\alpha(z) = \alpha_0 z^2 + \alpha_1 z + \alpha_2 = z^2 + 1.1z$$

$$\beta(z) = \beta_0 z^2 + \beta_1 z + \beta_2 = 1.01z^2 - 0.72z - 0.55$$

Hence, the controller designed is

$$\frac{\beta(z)}{\alpha(z)} = \frac{1.01z^2 - 0.72z - 0.55}{z^2 + 1.1z}$$

Notice that this pulse transfer function of the controller is the same as Equation (6–283), the pulse transfer function of the observer controller derived in Problem A–6–17.

     Referring to Figure 7–20, the closed-loop pulse transfer function becomes as follows:

$$\frac{Y(z)}{R(z)} = \frac{K_0 \dfrac{B(z)}{A(z)}}{1 + \dfrac{\beta(z)}{\alpha(z)}\dfrac{B(z)}{A(z)}} = \frac{K_0 \alpha(z)B(z)}{\alpha(z)A(z) + \beta(z)B(z)}$$

$$= \frac{K_0 \alpha(z)B(z)}{H(z)F(z)} = \frac{K_0(z^2 + 1.1z)z}{z^5} = \frac{K_0(z + 1.1)}{z^3}$$

Notice that cancellation of z^2 occurred between $\alpha(z)B(z)$ and $H(z)F(z)$ and the system is of the third order. [The system would have been of the $(2n - 1)$th order (or fifth order) if no cancellation occurred.]

To determine gain K_0, we impose the condition that the steady-state output $y(\infty)$ to the unit-step input is unity, or

$$\lim_{k \to \infty} y(k) = \lim_{z \to 1} \frac{z - 1}{z} \frac{K_0(z + 1.1)}{z^3} \frac{z}{z - 1} = 2.1K_0 = 1$$

Hence,

$$K_0 = 0.4762$$

and the closed-loop pulse transfer function becomes

$$\frac{Y(z)}{R(z)} = \frac{0.4762(z + 1.1)}{z^3} = \frac{0.4762z + 0.5238}{z^3}$$

The system is of third order. The unit-step response of the designed system is shown in Figure 7–21. Notice that in the unit-step response the output reaches unity in three sampling periods.

The unit-ramp response of the designed system is shown in Figure 7–22. The error in following the unit-ramp input can be calculated as follows:

$$E(z) = R(z) - Y(z) = \left[1 - \frac{Y(z)}{R(z)} \right] R(z)$$

$$= \left(1 - \frac{0.4762z + 0.5238}{z^3} \right) R(z)$$

$$= \frac{(z - 1)(z^2 + z + 0.5238)}{z^3} R(z)$$

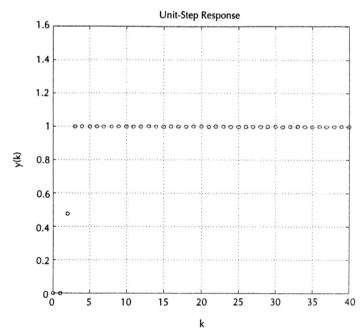

Figure 7–21   Unit-step response of the system designed in Problem A–7–7.

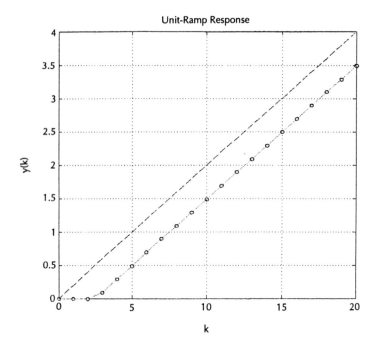

Figure 7–22 Unit-ramp response of the system designed in Problem A–7–7.

The unit-ramp input $R(z)$ is given by

$$R(z) = \frac{Tz^{-1}}{(1 - z^{-1})^2} = \frac{0.2z}{(z - 1)^2}$$

Hence,

$$\lim_{k \to \infty} e(k) = \lim_{z \to 1} \frac{z - 1}{z} \frac{(z - 1)(z^2 + z + 0.5238)}{z^3} \frac{0.2z}{(z - 1)^2}$$

$$= 0.5048$$

Thus, the error in following the unit-ramp input is 0.5048, as can be seen in Figure 7–22.

Problem A–7–8

Referring to Problem A–7–7, consider the same plant as given by Equations (7–26) and (7–27). (Refer also to Problem A–6–17.) Assume the same polynomials $H(z)$ and $F(z)$ as used in Problem A–7–7.

Using the polynomial equations approach, design a control system for the plant such that the block diagram configuration is the same as that of Figure 7–5. Then obtain the unit-step response and unit-ramp response of the designed system. The sampling period T of the system is 0.2 sec.

Solution For this problem

$$H(z) = z^3, \qquad F(z) = z^2$$

Referring to Problem A–7–7, the pulse transfer function $G_p(z)$ of the plant is

$$G_p(z) = \frac{z}{z^3 - 1.1z^2 + 0.2z + 0.5} = \frac{B(z)}{A(z)}$$

The block diagram for the present system is shown in Figure 7–23. The Diophantine equation for this problem is

$$\alpha(z)A(z) + \beta(z)B(z) = H(z)F(z)$$

or

$$\alpha(z)(z^3 - 1.1z^2 + 0.2z + 0.5) + \beta(z)z = z^5$$

This Diophantine equation was solved in Problem A–7–7 and the result was

$$\alpha(z) = z^2 + 1.1z$$

$$\beta(z) = 1.01z^2 - 0.72z - 0.55$$

Hence, referring to Equation (7–12), the closed-loop pulse transfer function $Y(z)/R(z)$ is given by

$$\frac{Y(z)}{R(z)} = \frac{K_0 B(z)}{H(z)}$$

or

$$\frac{Y(z)}{R(z)} = \frac{K_0 z}{z^3} = \frac{K_0}{z^2}$$

To determine gain K_0, we impose the condition that the steady-state output $y(\infty)$ to the unit-step input is unity, or

$$\lim_{k \to \infty} y(k) = \lim_{z \to 1} \frac{z-1}{z} \frac{K_0}{z^2} \frac{z}{z-1} = K_0 = 1$$

Hence,

$$K_0 = 1$$

Thus, $Y(z)/R(z)$ becomes as follows:

$$\frac{Y(z)}{R(z)} = \frac{1}{z^2}$$

The system designed is of second order. The unit-step response of the designed system is shown in Figure 7–24. Notice that in the unit-step response the output reaches unity in two sampling periods.

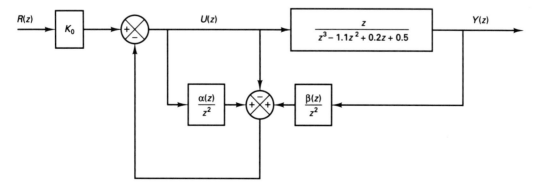

Figure 7–23 Block diagram of the control system considered in Problem A–7–8.

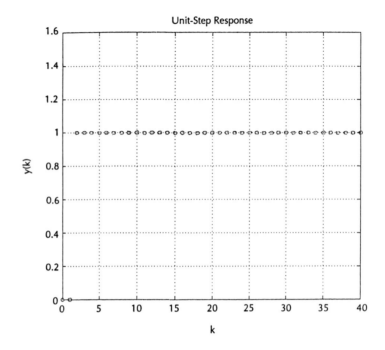

Figure 7–24 Unit-step response of the system designed in Problem A–7–8.

The unit-ramp response of the system is shown in Figure 7–25. The error in following the unit-ramp input can be calculated as follows:

$$E(z) = R(z) - Y(z) = \left[1 - \frac{Y(z)}{R(z)}\right] R(z)$$

$$= \left(1 - \frac{1}{z^2}\right) R(z) = \frac{(z-1)(z+1)}{z^2} R(z)$$

The unit-ramp input $R(z)$ is

$$R(z) = \frac{Tz^{-1}}{(1-z^{-1})^2} = \frac{0.2z}{(z-1)^2}$$

Hence,

$$\lim_{k \to \infty} e(k) = \lim_{z \to 1} \frac{z-1}{z} \frac{(z-1)(z+1)}{z^2} \frac{0.2z}{(z-1)^2} = 0.4$$

Thus, the error in following the unit-ramp input is 0.4, as can be seen in Figure 7–25. Comparing the systems designed in Problems A–7–7 and A–7–8, the latter system that uses the block diagram configuration of Figure 7–5 exhibits superior behavior.

Problem A–7–9

Consider a plant defined by

$$G(z) = \frac{z + 0.5}{z^3 - z^2 + 0.01z + 0.12}$$

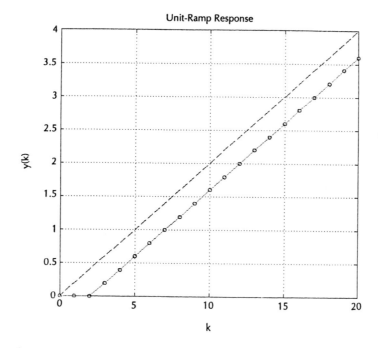

Figure 7–25   Unit-ramp response of the system designed in Problem A–7–8.

The sampling period T is 1 sec. Assume that in the present problem it is important to have zero following error to a ramp input.

It is desired to design a controller such that the control system will behave like the model system, whose pulse transfer function is

$$G_{\text{model}} = \frac{0.64z - 0.512}{(z^2 - 1.2z + 0.52)(z - 0.6)}$$

$$= \frac{0.64z - 0.512}{z^3 - 1.8z^2 + 1.24z - 0.312} \qquad (7\text{–}28)$$

(The sampling period for the model is also 1 sec.) The unit-step response and unit-ramp response of the model system G_{model} are shown in Figures 7–26 and 7–27, respectively. (The steady-state error in following the ramp input is zero, and the maximum overshoot in the unit-step response is approximately 70%.)

Solution   Let us assume that the block diagram of the system is the same as that of Figure 7–9. For the given plant,

$$A(z) = z^3 - z^2 + 0.01z + 0.12$$

$$B(z) = z + 0.5$$

Thus,

$$a_1 = -1, \qquad a_2 = 0.01, \qquad a_3 = 0.12$$

$$b_0 = 0, \qquad b_1 = 0, \qquad b_2 = 1, \qquad b_3 = 0.5$$

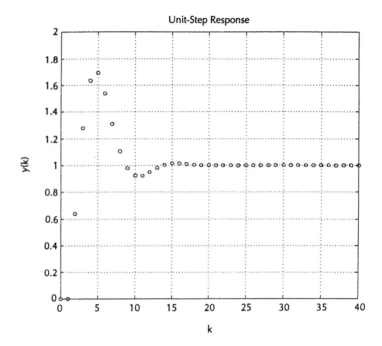

Figure 7–26 Unit-step response of the model system G_{model} given by Equation (7–28).

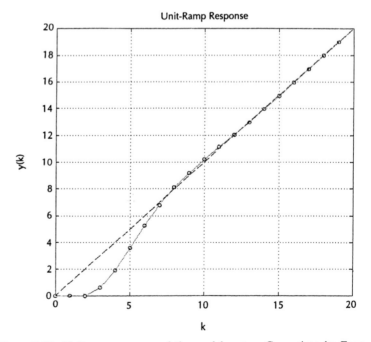

Figure 7–27 Unit-ramp response of the model system G_{model} given by Equation (7–28).

Clearly, there is no common factor between $A(z)$ and $B(z)$. In this problem we may choose $H_1(z)$ as

$$H_1(z) = z^2 - 1.2z + 0.52$$

Here we chose $H_1(z)$ to cancel a part of the denominator of G_{model}. [But this is not necessary. Requirement on $H_1(z)$ is that it must be a stable polynomial of degree $n - 1$ ($n = 3$ in this problem). Infinitely many choices of $H_1(z)$ are possible.] Define

$$H(z) = B(z)H_1(z)$$
$$= (z + 0.5)(z^2 - 1.2z + 0.52)$$
$$= z^3 - 0.7z^2 - 0.08z + 0.26$$

Next we choose $F(z)$ as

$$F(z) = z^2$$

[$F(z)$ must be a stable polynomial of degree $n - 1$. In this case, again, infinitely many choices are possible.] Then $D(z)$ becomes as follows:

$$D(z) = F(z)H(z) = F(z)B(z)H_1(z)$$
$$= z^5 - 0.7z^4 - 0.08z^3 + 0.26z^2$$

Hence,

$$d_0 = 1, \quad d_1 = -0.7, \quad d_2 = -0.08,$$
$$d_3 = 0.26, \quad d_4 = 0, \quad d_5 = 0$$

Now we need to solve the following Diophantine equation:

$$\alpha(z)A(z) + \beta(z)B(z) = F(z)B(z)H_1(z)$$

or

$$\alpha(z)(z^3 - z^2 + 0.01z + 0.12) + \beta(z)(z + 0.5) = z^5 - 0.7z^4 - 0.08z^3 + 0.26z^2$$

The 6×6 Sylvester matrix $\mathbf{E}$ for this problem is

$$\mathbf{E} = \begin{bmatrix} a_3 & 0 & 0 & b_3 & 0 & 0 \\ a_2 & a_3 & 0 & b_2 & b_3 & 0 \\ a_1 & a_2 & a_3 & b_1 & b_2 & b_3 \\ 1 & a_1 & a_2 & b_0 & b_1 & b_2 \\ 0 & 1 & a_1 & 0 & b_0 & b_1 \\ 0 & 0 & 1 & 0 & 0 & b_0 \end{bmatrix}$$

$$= \begin{bmatrix} 0.12 & 0 & 0 & 0.5 & 0 & 0 \\ 0.01 & 0.12 & 0 & 1 & 0.5 & 0 \\ -1 & 0.01 & 0.12 & 0 & 1 & 0.5 \\ 1 & -1 & 0.01 & 0 & 0 & 1 \\ 0 & 1 & -1 & 0 & 0 & 0 \\ 0 & 0 & 1 & 0 & 0 & 0 \end{bmatrix}$$

By use of MATLAB, $\mathbf{E}^{-1}$ can be obtained as shown next.

```
E =

     0.1200         0         0    0.5000         0         0
     0.0100    0.1200         0    1.0000    0.5000         0
    -1.0000    0.0100    0.1200         0    1.0000    0.5000
     1.0000   -1.0000    0.0100         0         0    1.0000
          0    1.0000   -1.0000         0         0         0
          0         0    1.0000         0         0         0

inv(E)

ans =

    -3.8462    1.9231   -0.9615    0.4808    0.2596    0.3702
          0         0         0         0    1.0000    1.0000
          0         0         0         0         0    1.0000
     2.9231   -0.4615    0.2308   -0.1154   -0.0623   -0.0888
    -5.7692    2.8846   -0.4423    0.2212   -0.1206   -0.0697
     3.8462   -1.9231    0.9615    0.5192    0.7404    0.6198
```

Define

$$
\mathbf{D} = \begin{bmatrix} d_5 \\ d_4 \\ d_3 \\ d_2 \\ d_1 \\ d_0 \end{bmatrix} = \begin{bmatrix} 0 \\ 0 \\ 0.26 \\ -0.08 \\ -0.7 \\ 1 \end{bmatrix}, \quad \mathbf{M} = \begin{bmatrix} \alpha_2 \\ \alpha_1 \\ \alpha_0 \\ \beta_2 \\ \beta_1 \\ \beta_0 \end{bmatrix}
$$

Then vector $\mathbf{M}$ can be obtained from

$$\mathbf{M} = \mathbf{E}^{-1}\mathbf{D}$$

MATLAB computation of this equation is shown next.

```
D = [0;0;0.26;-0.08;-0.7;1];
M = (inv(E))*D
M =
    -0.1000
     0.3000
     1.0000
     0.0240
    -0.1180
     0.3100
```

From vector $\mathbf{M}$ we get values of α's and β's. $\alpha(z)$ and $\beta(z)$ are determined as

$$\alpha(z) = \alpha_0 z^2 + \alpha_1 z + \alpha_2 = z^2 + 0.3z - 0.1$$

$$\beta(z) = \beta_0 z^2 + \beta_1 z + \beta_2 = 0.31z^2 - 0.118z + 0.024$$

Using $\alpha(z)$ and $\beta(z)$ thus determined, $Y(z)/V(z)$ becomes as follows:

$$\frac{Y(z)}{V(z)} = \frac{F(z)B(z)}{F(z)B(z)H_1(z)} = \frac{1}{H_1(z)} = \frac{1}{z^2 - 1.2z + 0.52}$$

Since $V(z)/R(z)$ is

$$\frac{V(z)}{R(z)} = G_{\text{model}} H_1(z) = \frac{(0.64z - 0.512)(z^2 - 1.2z + 0.52)}{(z^2 - 1.2z + 0.52)(z - 0.6)}$$

$$= \frac{0.64z - 0.512}{z - 0.6}$$

The pulse transfer function $Y(z)/R(z)$ becomes

$$\frac{Y(z)}{R(z)} = \frac{1}{z^2 - 1.2z + 0.52} \frac{0.64z - 0.512}{z - 0.6} = G_{\text{model}}$$

The designed model matching control system has the block diagram as shown in Figure 7–28(a). This block diagram can be simplified to those shown in Figures 7–28(b) and (c).

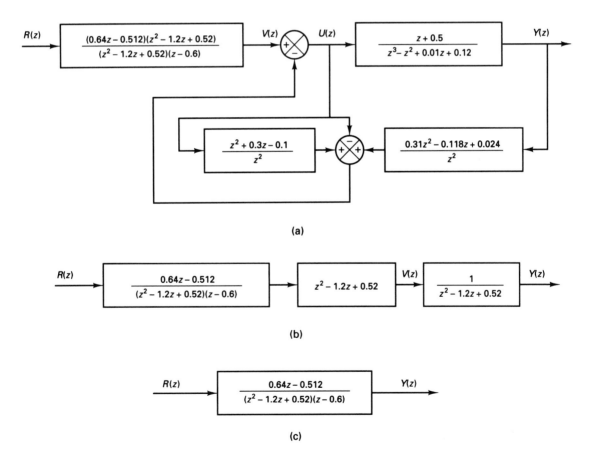

(a)

(b)

(c)

Figure 7–28 (a) Block diagram of the designed model matching control system; (b) and (c) simplified block diagrams.

As stated earlier, $G_{\text{model}} H_1(z)$ must be physically realizable. The degree of the numerator polynomial $G_{\text{model}} H_1(z)$ must be equal to or less than the degree of the denominator polynomial of $G_{\text{model}} H_1(z)$. This means that for

$$\frac{Y(z)}{U(z)} = \frac{B(z)}{A(z)} = \frac{m\text{th-degree polynomial in } z}{n\text{th-degree monic polynomial in } z}$$

where $A(z)$ and $B(z)$ have no common factors, and

$$G_{\text{model}} = \frac{B_m(z)}{A_m(z)} = \frac{m'\text{th-degree polynomial in } z}{n'\text{th-degree polynomial in } z}$$

we must have

$$n' - m' \geq n - m$$

If this condition is not satisfied, the best approach is to modify G_{model} so that this condition can be satisfied and the present approach can be used.

PROBLEMS

Problem B–7–1

Consider polynomials $A(z)$ and $B(z)$ defined by

$$A(z) = z^2 + a_1 z + a_2 = (z - \lambda_1)(z - \lambda_2) = z^2 - (\lambda_1 + \lambda_2)z + \lambda_1 \lambda_2$$

$$B(z) = b_1 z + b_2 = b(z - \lambda_3) = bz - b\lambda_3$$

where $b = b_1$. Define the Sylvester matrix $\mathbf{E}$ by

$$\mathbf{E} = \begin{bmatrix} a_2 & 0 & b_2 & 0 \\ a_1 & a_2 & b_1 & b_2 \\ 1 & a_1 & 0 & b_1 \\ 0 & 1 & 0 & 0 \end{bmatrix}$$

Show that the determinant of $\mathbf{E}$ can be given by

$$|\mathbf{E}| = b^2(\lambda_1 - \lambda_3)(\lambda_2 - \lambda_3)$$

Problem B–7–2

Consider the following Diophantine equation:

$$\alpha(z)A(z) + \beta(z)B(z) = 1$$

where

$$A(z) = z^2 - 0.7z + 0.1$$

$$B(z) = z^2 + 0.2z - 0.24$$

$$\alpha(z) = \alpha_0 z + \alpha_1$$

$$\beta(z) = \beta_0 z + \beta_1$$

Solve this Diophantine equation for $\alpha(z)$ and $\beta(z)$ and determine coefficients α_0, α_1, β_0, and β_1.

Problem B–7–3

Consider the plant $Y(z)/U(z)$, where

$$\frac{Y(z)}{U(z)} = \frac{B(z)}{A(z)}$$

Assume that $A(z)$ is an nth-degree monic polynomial in z and $B(z)$ is an mth-degree polynomial in z. Assume also that there are no common factors between $A(z)$ and $B(z)$; that is, the plant is completely state controllable and completely observable.

Consider the following Diophantine equation:

$$\gamma(z)A(z) + \beta(z)B(z) = F(z)[H(z) - A(z)]$$

where $H(z)$ is the desired characteristic polynomial for the pole placement part and $F(z)$ is the desired characteristic polynomial for the minimum-order observer. $[H(z)$ is an nth-degree monic polynomial and $F(z)$ is an $(n - 1)$th-degree polynomial.]

Show that if this Diophantine equation is solved for $\beta(z)$ and $\gamma(z)$, then the use of

$$U(z) = -\frac{\gamma(z)}{F(z)}U(z) - \frac{\beta(z)}{F(z)}Y(z)$$

will accomplish the desired observed-state feedback regulator system.

Problem B–7–4

In Example 7–3 a control system was designed such that the desired characteristic equation for the pole placement part was

$$H(z) = (z - 0.6 - j0.4)(z - 0.6 + j0.4) = z^2 - 1.2z + 0.52$$

and the desired characteristic polynomial for the minimum-order observer was

$$F(z) = z$$

The Diophantine equation given by Equation (7–13) was solved. $\alpha(z)$ and $\beta(z)$ were determined as follows:

$$\alpha(z) = z + 0.32$$

$$\beta(z) = 24z - 16$$

The constant K_0 was determined as 8. Figure 7–6(a) shows the designed system.

Show that the control signal $u(k)$ can be given by

$$u(k) = -0.32u(k - 1) - 24y(k) + 16y(k - 1) + 8r(k), \qquad k = 1, 2, 3, \ldots$$

$$u(0) = -24y(0) + 8r(0)$$

Plot $u(k)$ versus k when the input $r(k)$ is a unit-step sequence.

Problem B–7–5

Consider a plant defined by

$$x(k + 1) = Gx(k) + Hu(k)$$

$$y(k) = Cx(k)$$

where

$$G = \begin{bmatrix} 0 & 1 & 0 \\ 0 & 0 & 1 \\ -0.16 & 0.84 & 0 \end{bmatrix}, \qquad H = \begin{bmatrix} 0 \\ 0 \\ 1 \end{bmatrix}, \qquad C = \begin{bmatrix} 1 & 0 & 0 \end{bmatrix}$$

The pulse transfer function for the plant can be written as

$$\frac{Y(z)}{U(z)} = \frac{B(z)}{A(z)}$$

Determine polynomials $A(z)$ and $B(z)$.

Using the polynomial equations approach, design a control system for the plant. It is desired that the block diagram configuration of the designed system is the same as that of Figure 7–4. In solving the Diophantine equation

$$\alpha(z)A(z) + \beta(z)B(z) = F(z)H(z)$$

assume that $H(z)$ and $F(z)$ are, respectively, as follows:

$$H(z) = z^3, \qquad F(z) = z^2$$

Obtain the unit-step response and unit-ramp response of the designed control system. The sampling period T is 1 sec.

Problem B–7–6

Consider the same plant as given in Problem B–7–5. Using the polynomial equations approach, design a control system for the plant. Use the block diagram configuration shown in Figure 7–5. Assume the following $H(z)$ and $F(z)$:

$$H(z) = z^3, \qquad F(z) = z^2$$

Obtain the unit-step response and unit-ramp response of the designed control system. Assume the sampling period T to be 1 sec.

Problem B–7–7

Consider the plant defined by

$$x(k + 1) = Gx(k) + Hu(k)$$

$$y(k) = Cx(k)$$

where

$$G = \begin{bmatrix} 0 & 0 & -0.25 \\ 1 & 0 & 0 \\ 0 & 1 & 0.5 \end{bmatrix}, \qquad H = \begin{bmatrix} 1 \\ 0 \\ 1 \end{bmatrix}, \qquad C = [1 \ \ 0 \ \ 0]$$

Design a control system for the plant. For the pole placement part, we want to have three closed-loop poles at the origin, or

$$H(z) = z^3$$

and for the characteristic equation for the minimum-order observer, we want to have

$$F(z) = z^2$$

Use the polynomial equations approach to the design.

Problem B–7–8

Consider the plant defined by

$$\frac{Y(z)}{U(z)} = \frac{0.6z + 0.5}{(z - 1)^2}$$

Using the polynomial equations approach, design a control system such that the system will behave like the following model, G_{model}:

$$G_{model} = \frac{2z - 1}{z^2}$$

Obtain the unit-step response and unit-ramp response of the designed system (which is the same as G_{model}). The sampling period T is 1 sec.

Problem B–7–9

Consider the plant defined by

$$\frac{Y(z)}{U(z)} = \frac{0.01873(z + 0.9356)}{(z - 1)(z - 0.8187)}$$

Using the polynomial equations approach, design a control system such that the system will behave like the following model, G_{model}:

$$G_{model} = \frac{0.32}{z^2 - 1.2z + 0.52}$$

Obtain the unit-step response and unit-ramp response of the control system (which is the same as the model, G_{model}). The sampling period T is 0.2 sec.

8

Quadratic Optimal Control Systems

8-1 INTRODUCTION

Problems of optimal control have received a great deal of attention from control engineers. An optimal control system—a system whose design "optimizes" (minimizes or maximizes, as the case may be) the value of a function chosen as the *performance index*—differs from an ideal one in that the former is the best attainable in the presence of physical constraints, whereas the latter may well be an unattainable goal.

Performance Indexes. In designing an optimal control system or optimal regulator system, we need to find a rule for determining the present control decision, subject to certain constraints, so as to minimize some measure of the deviation from ideal behavior. That measure is usually provided by the chosen performance index, which is a function whose value we consider to be an indication of how well the actual performance of the system matches the desired performance. In most cases, the behavior of a system is optimized by choosing the control vector $\mathbf{u}(k)$ in such a way that the performance index is minimized (or maximized, depending on the nature of the performance index chosen). The selection of an appropriate performance index is important, because, to a large degree, it determines the nature of the resulting optimal control system. That is, whether the resulting control system will be linear, nonlinear, stationary, or time-varying will depend on the form of the performance index. The control engineer thus formulates this index on the basis of the requirements the system must meet and takes it into account in determining the nature of the resulting system. The requirements of the design usually include not only performance specifications, but also, to ensure physical realizability, restrictions on the form of control to be used.

The optimization process not only should provide optimal control laws and parameter configurations, but should also predict the degradation in performance due to any departure of the performance index function from its minimum (or maximum) value that results when nonoptimal control laws are applied.

Choosing the most appropriate performance index for a given problem is very difficult, especially in complex systems. To a considerable degree, the use of optimization theory in system design has been hampered by the conflict between analytical feasibility and practical utility in the selection of the performance index. It is desirable that the criteria for optimal control originate not from a mathematical but from an application point of view. In general, however, the choice of a performance index involves a compromise between a meaningful evaluation of system performance and a tractable mathematical problem.

Formulation of Optimization Problems.   The problem of optimization of a control system may be formulated if the following information is given:

1. System equations
2. Class of allowable control vectors
3. Constraints on the problem
4. Performance index
5. System parameters

The solution of an optimal control problem is to determine the optimal control vector $\mathbf{u}(k)$ within the class of allowable control vectors. This vector $\mathbf{u}(k)$ depends on

1. The nature of the performance index
2. The nature of the constraints
3. The initial state or initial output
4. The desired state or desired output

Except for special cases, the optimal control problem may be so complicated for an analytical solution that a computational solution has to be obtained.

Questions Concerning the Existence of Solutions to Optimal Control Problems. It has been stated that the optimal control problem, given any initial state $\mathbf{x}(0)$, consists of finding an allowable control vector $\mathbf{u}(k)$ that transfers the state to the desired region of the state space and for which the performance index is minimized.

It is important to mention that in some cases a particular combination of plant, desired state, performance index, and constraints makes optimal control impossible. This is a matter of requiring performance beyond the physical capabilities of the system.

Questions regarding the existence of an optimal control vector are important, since they serve to inform the designer whether or not optimal control is possible for a given system and given set of constraints. Two of the most important among these questions are those of controllability and observability, which were presented in Chapter 6.

Comments on Optimal Control Systems.   The system whose design minimizes (or maximizes) the selected performance index is, by definition, optimal. It is evident that the performance index, in reality, determines the configuration of the system. It is very important to point out that a control system that is optimal under one performance index is, in general, not optimal under other performance indexes. In addition, hardware realization of a particular optimal control law may be quite difficult and expensive. Hence, it may be pointless to devote too much expense to implementing an optimal controller that is the best only in some narrow, individualistic sense. A control system is seldom designed to perform a single task that is completely specified beforehand. Instead, it is designed to perform a task selected at random from a complete repertoire of possible tasks. In practical systems, then, it may be more sensible to seek approximate optimal control laws that are not rigidly tied to a single performance index.

Strictly speaking, we should realize that a mathematically obtained optimal control system gives, in most practical situations, the highest possible performance under the given performance index and is more a measuring stick than a practical goal. Therefore, before we decide whether to build an optimal control system or something inferior but simpler, we should carefully evaluate a measure of the degree to which the performance of the complex, optimal control system exceeds that of a simpler, suboptimal one. Unless it can be justified, we should not build an extremely complicated and expensive optimal control system.

Once the ultimate degree of performance is found by use of optimal control theory, we should make efforts to design a simple system that is close to optimal. Keeping this in mind, we build a prototype physical system, test it, and modify it until a satisfactory system is obtained that has performance characteristics close to the optimal control system we have worked out in theory.

Analytically solvable optimal control problems provide good insight into optimal structures and algorithms that may be applied to practical cases. An example of analytically solvable optimal control problems is the problem of optimal control of linear systems based on quadratic performance indexes. Quadratic performance indexes have been used very frequently in practical control systems as measures of system performance.

Quadratic Optimal Control.   Let us consider the control system defined by

$$\mathbf{x}(k + 1) = \mathbf{G}\mathbf{x}(k) + \mathbf{H}\mathbf{u}(k)$$

where

$$\mathbf{x}(k) = \text{state vector } (n\text{-vector})$$
$$\mathbf{u}(k) = \text{control vector } (r\text{-vector})$$
$$\mathbf{G} = n \times n \text{ matrix}$$
$$\mathbf{H} = n \times r \text{ matrix}$$

In the quadratic optimal control problem we desire to determine a law for the control vector $\mathbf{u}(k)$ such that a given quadratic performance index is minimized.

An example of a quadratic performance index is

$$J = \frac{1}{2}\mathbf{x}^*(N)\mathbf{S}\mathbf{x}(N) + \frac{1}{2}\sum_{k=0}^{N-1}[\mathbf{x}^*(k)\mathbf{Q}\mathbf{x}(k) + \mathbf{u}^*(k)\mathbf{R}\mathbf{u}(k)]$$

where matrices **S** and **Q** are positive definite or positive semidefinite Hermitian matrices and **R** is a positive definite Hermitian matrix. The first term on the right-hand side of this last equation accounts for the importance of the final state. The first term in the summation brackets accounts for the relative importance of the error during the control process, and the second term accounts for the expenditure of the energy of the control signals. We assume that the control vector $\mathbf{u}(k)$ is unconstrained.

It will be shown in Section 8–2 that the optimal control law is given by

$$\mathbf{u}(k) = -\mathbf{K}(k)\mathbf{x}(k)$$

where $\mathbf{K}(k)$ is a time-varying $r \times n$ matrix. If $N = \infty$, then $\mathbf{K}(k)$ becomes a constant $r \times n$ matrix. The design of optimal control systems based on such quadratic performance indexes boils down to the determination of matrix $\mathbf{K}(k)$.

The major characteristic of the optimal control law based on a quadratic performance index is that it is a linear function of the state vector $\mathbf{x}(k)$. Such a state feedback requires that all state variables be available for feedback. It is advantageous, therefore, to represent the system in terms of measurable state variables. If not all state variables can be measured, we need to estimate or observe the unmeasurable state variables. We then use the measured and observed state variables to generate optimal control signals.

The advantage of using the quadratic optimal control scheme is that the system designed will be asymptotically stable, except very special academic cases. (See Problems A–8–6 and A–8–7.)

There are many different approaches to the solution of quadratic optimal control problems. In this chapter, we present a commonly used approach based on the minimization technique using Lagrange multipliers. For the steady-state quadratic optimal control problem, we also present the Liapunov approach. It will be shown in Section 8–3 that there is a direct relationship between Liapunov functions and quadratic performance indexes.

Note that when an optimal control system is designed in state space it is important to check the frequency-response characteristics. Sometimes a specific compensation for noise effects may be needed. Then it may become necessary to modify the optimal configuration and accept a suboptimal configuration, or it may become necessary to modify the performance index.

Outline of the Chapter.   Section 8–1 has presented introductory material. Section 8–2 presents a basic quadratic optimal control problem and its solution. Section 8–3 treats the steady-state quadratic optimal control problem. Here we include the Liapunov approach to the solution of the quadratic optimal control problem. Section 8–4 discusses the quadratic optimal control of a servo system.

8–2 QUADRATIC OPTIMAL CONTROL

Quadratic optimal control problems can be solved by many different approaches. In this section we shall solve the basic quadratic optimal control problem by the conventional minimization method using Lagrange multipliers.

Quadratic Optimal Control Problem.   The quadratic optimal control problem may be stated as follows. Given a linear discrete-time control system

$$\mathbf{x}(k + 1) = \mathbf{Gx}(k) + \mathbf{Hu}(k), \qquad \mathbf{x}(0) = \mathbf{c} \tag{8-1}$$

where it is assumed to be completely state controllable and where

$$\mathbf{x}(k) = \text{state vector } (n\text{-vector})$$

$$\mathbf{u}(k) = \text{control vector } (r\text{-vector})$$

$$\mathbf{G} = n \times n \text{ nonsingular matrix}$$

$$\mathbf{H} = n \times r \text{ matrix}$$

find the optimal control sequence $\mathbf{u}(0), \mathbf{u}(1), \mathbf{u}(2), \ldots, \mathbf{u}(N - 1)$ that minimizes a quadratic performance index. An example of the quadratic performance indexes for a finite time process $(0 \le k \le N)$ is

$$J = \frac{1}{2}\mathbf{x}^*(N)\mathbf{Sx}(N) + \frac{1}{2}\sum_{k=0}^{N-1}[\mathbf{x}^*(k)\mathbf{Qx}(k) + \mathbf{u}^*(k)\mathbf{Ru}(k)] \tag{8-2}$$

where

$$\mathbf{Q} = n \times n \text{ positive definite or positive semidefinite}$$
$$\text{Hermitian matrix (or real symmetric matrix)}$$

$$\mathbf{R} = r \times r \text{ positive definite Hermitian matrix (or real}$$
$$\text{symmetric matrix)}$$

$$\mathbf{S} = n \times n \text{ positive definite or positive semidefinite}$$
$$\text{Hermitian matrix (or real symmetric matrix)}$$

Matrices $\mathbf{Q}$, $\mathbf{R}$, and $\mathbf{S}$ are selected to weigh the relative importance of the performance measures caused by the state vector $\mathbf{x}(k)$ $(k = 0, 1, 2, \ldots, N - 1)$, the control vector $\mathbf{u}(k)$ $(k = 0, 1, 2, \ldots, N - 1)$, and the final state $\mathbf{x}(N)$, respectively.

The initial state of the system is at some arbitrary state $\mathbf{x}(0) = \mathbf{c}$. The final state $\mathbf{x}(N)$ may be fixed, in which case the term $\frac{1}{2}\mathbf{x}^*(N)\mathbf{Sx}(N)$ is removed from the performance index of Equation (8–2) and instead the terminal condition $\mathbf{x}(N) = \mathbf{x}_f$ is imposed, where $\mathbf{x}_f$ is the fixed terminal state. If the final state $\mathbf{x}(N)$ is not fixed, then the first term in Equation (8–2) represents the weight of the performance measure due to the final state. Note that in the minimization problem the inclusion of the term $\frac{1}{2}\mathbf{x}^*(N)\mathbf{Sx}(N)$ in the performance index $\mathbf{J}$ implies that we desire the final state $\mathbf{x}(N)$ to be as close to the origin as possible.

Solution by the Conventional Minimization Method Using Lagrange Multipliers.
The quadratic optimal control problem is a minimization problem involving a function of several variables. Thus, it can be solved by the conventional minimization method. The minimization problem subjected to equality constraints may be solved by adjoining the constraints to the function to be minimized by use of Lagrange multipliers.

In the present optimization problem, we minimize J as given by Equation (8–2), repeated here,

$$J = \frac{1}{2}\mathbf{x}^*(N)\mathbf{Sx}(N) + \frac{1}{2}\sum_{k=0}^{N-1}[\mathbf{x}^*(k)\mathbf{Qx}(k) + \mathbf{u}^*(k)\mathbf{Ru}(k)] \tag{8-3}$$

when it is subjected to the constraint equation specified by Equation (8–1),

$$\mathbf{x}(k + 1) = \mathbf{G}\mathbf{x}(k) + \mathbf{H}\mathbf{u}(k) \tag{8-4}$$

where $k = 0, 1, 2, \ldots, N - 1$, and where the initial condition on the state vector is specified as

$$\mathbf{x}(0) = \mathbf{c} \tag{8-5}$$

Now, by using a set of Lagrange multipliers $\boldsymbol{\lambda}(1), \boldsymbol{\lambda}(2), \ldots, \boldsymbol{\lambda}(N)$, we define a new performance index L as follows:

$$
\begin{aligned}
L = \frac{1}{2}\mathbf{x}^*(N)\mathbf{S}\mathbf{x}(N) + \frac{1}{2}\sum_{k=0}^{N-1} & \{[\mathbf{x}^*(k)\mathbf{Q}\mathbf{x}(k) + \mathbf{u}^*(k)\mathbf{R}\mathbf{u}(k)] \\
& + \boldsymbol{\lambda}^*(k + 1)[\mathbf{G}\mathbf{x}(k) + \mathbf{H}\mathbf{u}(k) - \mathbf{x}(k + 1)] \\
& + [\mathbf{G}\mathbf{x}(k) + \mathbf{H}\mathbf{u}(k) - \mathbf{x}(k + 1)]^*\boldsymbol{\lambda}(k + 1)\}
\end{aligned}
\tag{8-6}
$$

The reason for writing the terms involving the Lagrange multiplier in the form shown in Equation (8–6) is to ensure that $L = L^*$. (L is a real scalar quantity.) Note that

$$\boldsymbol{\lambda}^*(0)[\mathbf{c} - \mathbf{x}(0)] + [\mathbf{c} - \mathbf{x}(0)]^*\boldsymbol{\lambda}(0)$$

may be added to the performance index L. However, we shall not do so, to simplify the presentation. It is a well-known fact that minimization of the function L defined by Equation (8–6) is equivalent to minimization of J as defined by Equation (8–3) when it is subjected to the equality constraint defined by Equation (8–4).

To minimize the function L, we need to differentiate L with respect to each component of vectors $\mathbf{x}(k)$, $\mathbf{u}(k)$, and $\boldsymbol{\lambda}(k)$ and set the results equal to zero. From the computational viewpoint, however, it is convenient to differentiate L with respect to $\bar{x}_i(k)$, $\bar{u}_i(k)$, and $\bar{\lambda}_i(k)$, where $\bar{x}_i(k)$, $\bar{u}_i(k)$, and $\bar{\lambda}_i(k)$ are, respectively, the complex conjugates of $x_i(k)$, $u_i(k)$, and $\lambda_i(k)$. (Note that the signal and its complex conjugate contain the same mathematical information.) Thus, we set

$$\frac{\partial L}{\partial \bar{x}_i(k)} = 0, \quad i = 1, 2, \ldots, n; \, k = 1, 2, \ldots, N$$

$$\frac{\partial L}{\partial \bar{u}_i(k)} = 0, \quad i = 1, 2, \ldots, r; \, k = 0, 1, \ldots, N - 1$$

$$\frac{\partial L}{\partial \bar{\lambda}_i(k)} = 0, \quad i = 1, 2, \ldots, n; \, k = 1, 2, \ldots, N$$

These equations are necessary conditions for L to have a minimum. Note that the simplified expressions for the preceding partial derivative equations are

$$\frac{\partial L}{\partial \bar{\mathbf{x}}(k)} = \mathbf{0}, \quad k = 1, 2, \ldots, N \tag{8-7}$$

$$\frac{\partial L}{\partial \bar{\mathbf{u}}(k)} = \mathbf{0}, \quad k = 0, 1, \ldots, N - 1 \tag{8-8}$$

$$\frac{\partial L}{\partial \bar{\boldsymbol{\lambda}}(k)} = \mathbf{0}, \quad k = 1, 2, \ldots, N \tag{8-9}$$

Referring to Appendix A (see Problems A–7 and A–8) for partial differentiation of complex quadratic and bilinear forms with respect to vector variables, we have

$$\frac{\partial}{\partial \overline{\mathbf{x}}} \mathbf{x}^* \mathbf{A}\mathbf{x} = \mathbf{A}\mathbf{x} \quad \text{and} \quad \frac{\partial}{\partial \overline{\mathbf{x}}} \mathbf{x}^* \mathbf{A}\mathbf{y} = \mathbf{A}\mathbf{y}$$

Then, Equations (8–7), (8–8), and (8–9) may be obtained as follows:

$$\frac{\partial L}{\partial \overline{\mathbf{x}}(k)} = \mathbf{0}: \quad \mathbf{Q}\mathbf{x}(k) + \mathbf{G}^*\boldsymbol{\lambda}(k+1) - \boldsymbol{\lambda}(k) = \mathbf{0}, \quad k = 1, 2, \ldots, N - 1 \quad (8\text{–}10)$$

$$\frac{\partial L}{\partial \overline{\mathbf{x}}(N)} = \mathbf{0}: \quad \mathbf{S}\mathbf{x}(N) - \boldsymbol{\lambda}(N) = \mathbf{0}, \quad (8\text{–}11)$$

$$\frac{\partial L}{\partial \overline{\mathbf{u}}(k)} = \mathbf{0}: \quad \mathbf{R}\mathbf{u}(k) + \mathbf{H}^*\boldsymbol{\lambda}(k+1) = \mathbf{0}, \quad k = 0, 1, \ldots, N - 1 \quad (8\text{–}12)$$

$$\frac{\partial L}{\partial \overline{\boldsymbol{\lambda}}(k)} = \mathbf{0}: \quad \mathbf{G}\mathbf{x}(k-1) + \mathbf{H}\mathbf{u}(k-1) - \mathbf{x}(k) = \mathbf{0}, \quad k = 1, 2, \ldots, N \quad (8\text{–}13)$$

Equation (8–13) is simply the system state equation. Equation (8–11) specifies the final value of the Lagrange multiplier. Note that the Lagrange multiplier $\boldsymbol{\lambda}(k)$ is often called a *covector* or *adjoint vector*.

Now we shall simplify the equations just obtained. From Equation (8–10) we have

$$\boldsymbol{\lambda}(k) = \mathbf{Q}\mathbf{x}(k) + \mathbf{G}^*\boldsymbol{\lambda}(k+1), \quad k = 1, 2, 3, \ldots, N - 1 \quad (8\text{–}14)$$

with the final condition $\boldsymbol{\lambda}(N) = \mathbf{S}\mathbf{x}(N)$. By solving Equation (8–12) for $\mathbf{u}(k)$ and noting that $\mathbf{R}^{-1}$ exists, we obtain

$$\mathbf{u}(k) = -\mathbf{R}^{-1}\mathbf{H}^*\boldsymbol{\lambda}(k+1), \quad k = 0, 1, 2, \ldots, N - 1 \quad (8\text{–}15)$$

Equation (8–13) can be rewritten as

$$\mathbf{x}(k+1) = \mathbf{G}\mathbf{x}(k) + \mathbf{H}\mathbf{u}(k), \quad k = 0, 1, 2, \ldots, N - 1 \quad (8\text{–}16)$$

which is simply the state equation. Substitution of Equation (8–15) into Equation (8–16) results in

$$\mathbf{x}(k+1) = \mathbf{G}\mathbf{x}(k) - \mathbf{H}\mathbf{R}^{-1}\mathbf{H}^*\boldsymbol{\lambda}(k+1) \quad (8\text{–}17)$$

with the initial condition $\mathbf{x}(0) = \mathbf{c}$.

To obtain the solution to the minimization problem, we need to solve Equations (8–14) and (8–17) simultaneously. Notice that for the system equation, Equation (8–16), the initial condition $\mathbf{x}(0)$ is specified, while for the Lagrange multiplier equation, Equation (8–14), the final condition $\boldsymbol{\lambda}(N)$ is specified. Thus, the problem here becomes a two-point boundary-value problem.

If the two-point boundary-value problem is solved, then the optimal values for the state vector and Lagrange multiplier vector may be determined and the optimal control vector $\mathbf{u}(k)$ may be obtained in the open-loop form. However, if we employ the Riccati transformation, the optimal control vector $\mathbf{u}(k)$ can be obtained in the following closed-loop, or feedback, form:

$$\mathbf{u}(k) = -\mathbf{K}(k)\mathbf{x}(k)$$

where $\mathbf{K}(k)$ is the $r \times n$ feedback matrix.

In what follows, we shall obtain the optimal control vector $\mathbf{u}(k)$ in the closed-loop form by first obtaining the Riccati equation. Assume that $\mathbf{\lambda}(k)$ can be written in the following form:

$$\mathbf{\lambda}(k) = \mathbf{P}(k)\mathbf{x}(k) \tag{8–18}$$

where $\mathbf{P}(k)$ is an $n \times n$ Hermitian matrix (or an $n \times n$ real symmetric matrix). Substitution of Equation (8–18) into Equation (8–14) results in

$$\mathbf{P}(k)\mathbf{x}(k) = \mathbf{Q}\mathbf{x}(k) + \mathbf{G}*\mathbf{P}(k + 1)\mathbf{x}(k + 1) \tag{8–19}$$

and substitution of Equation (8–18) into Equation (8–17) gives

$$\mathbf{x}(k + 1) = \mathbf{G}\mathbf{x}(k) - \mathbf{H}\mathbf{R}^{-1}\mathbf{H}*\mathbf{P}(k + 1)\mathbf{x}(k + 1) \tag{8–20}$$

Notice that Equations (8–19) and (8–20) do not involve $\mathbf{\lambda}(k)$ and thus we have eliminated $\mathbf{\lambda}(k)$. The transformation process employed here is called the *Riccati transformation*. It is of extreme importance in solving such a two-point boundary-value problem.

From Equation (8–20) we have

$$[\mathbf{I} + \mathbf{H}\mathbf{R}^{-1}\mathbf{H}*\mathbf{P}(k + 1)]\mathbf{x}(k + 1) = \mathbf{G}\mathbf{x}(k) \tag{8–21}$$

For completely state controllable systems, it can be shown that $\mathbf{P}(k + 1)$ is positive definite or positive semidefinite. For at least a positive semidefinite matrix $\mathbf{P}(k + 1)$, we have

$$|\mathbf{I}_n + \mathbf{H}\mathbf{R}^{-1}\mathbf{H}*\mathbf{P}(k + 1)| = |\mathbf{I}_r + \mathbf{H}*\mathbf{P}(k + 1)\mathbf{H}\mathbf{R}^{-1}| = |\mathbf{I}_r + \mathbf{R}^{-1}\mathbf{H}*\mathbf{P}(k + 1)\mathbf{H}|$$

$$= |\mathbf{R}^{-1}||\mathbf{R} + \mathbf{H}*\mathbf{P}(k + 1)\mathbf{H}| \neq 0$$

where we have used the relationship

$$|\mathbf{I}_n + \mathbf{A}\mathbf{B}| = |\mathbf{I}_r + \mathbf{B}\mathbf{A}|, \qquad \mathbf{A} = n \times r \text{ matrix}, \mathbf{B} = r \times n \text{ matrix}$$

(See Appendix A.) Hence, the inverse of $\mathbf{I} + \mathbf{H}\mathbf{R}^{-1}\mathbf{H}*\mathbf{P}(k + 1)$ exists. Consequently, Equation (8–21) can be written as follows:

$$\mathbf{x}(k + 1) = [\mathbf{I} + \mathbf{H}\mathbf{R}^{-1}\mathbf{H}*\mathbf{P}(k + 1)]^{-1}\mathbf{G}\mathbf{x}(k) \tag{8–22}$$

By substituting Equation (8–22) into Equation (8–19), we obtain

$$\mathbf{P}(k)\mathbf{x}(k) = \mathbf{Q}\mathbf{x}(k) + \mathbf{G}*\mathbf{P}(k + 1)[\mathbf{I} + \mathbf{H}\mathbf{R}^{-1}\mathbf{H}*\mathbf{P}(k + 1)]^{-1}\mathbf{G}\mathbf{x}(k)$$

or

$$\{\mathbf{P}(k) - \mathbf{Q} - \mathbf{G}*\mathbf{P}(k + 1)[\mathbf{I} + \mathbf{H}\mathbf{R}^{-1}\mathbf{H}*\mathbf{P}(k + 1)]^{-1}\mathbf{G}\}\mathbf{x}(k) = \mathbf{0}$$

This last equation must hold for all $\mathbf{x}(k)$. Hence, we must have

$$\mathbf{P}(k) = \mathbf{Q} + \mathbf{G}*\mathbf{P}(k + 1)[\mathbf{I} + \mathbf{H}\mathbf{R}^{-1}\mathbf{H}*\mathbf{P}(k + 1)]^{-1}\mathbf{G} \tag{8–23}$$

Equation (8–23) may be modified. By using the matrix inversion lemma

$$(\mathbf{A} + \mathbf{B}\mathbf{D})^{-1} = \mathbf{A}^{-1} - \mathbf{A}^{-1}\mathbf{B}(\mathbf{I} + \mathbf{D}\mathbf{A}^{-1}\mathbf{B})^{-1}\mathbf{D}\mathbf{A}^{-1}$$

and making the substitutions

$$\mathbf{A} = \mathbf{I}, \quad \mathbf{B} = \mathbf{HR}^{-1}, \quad \mathbf{D} = \mathbf{H}^*\mathbf{P}(k + 1)$$

we obtain

$$[\mathbf{I} + \mathbf{HR}^{-1}\mathbf{H}^*\mathbf{P}(k + 1)]^{-1} = \mathbf{I} - \mathbf{HR}^{-1}[\mathbf{I} + \mathbf{H}^*\mathbf{P}(k + 1)\mathbf{HR}^{-1}]^{-1}\mathbf{H}^*\mathbf{P}(k + 1)$$

$$= \mathbf{I} - \mathbf{H}[\mathbf{R} + \mathbf{H}^*\mathbf{P}(k + 1)\mathbf{H}]^{-1}\mathbf{H}^*\mathbf{P}(k + 1)$$

Hence, Equation (8–23) can be modified to

$$\mathbf{P}(k) = \mathbf{Q} + \mathbf{G}^*\mathbf{P}(k + 1)\mathbf{G}$$

$$- \mathbf{G}^*\mathbf{P}(k + 1)\mathbf{H}[\mathbf{R} + \mathbf{H}^*\mathbf{P}(k + 1)\mathbf{H}]^{-1}\mathbf{H}^*\mathbf{P}(k + 1)\mathbf{G} \quad (8\text{–}24)$$

Equation (8–24) or its equivalent [such as Equation (8–23)] is called the Riccati equation. Referring to Equations (8–11) and (8–18), notice that at $k = N$ we have

$$\mathbf{P}(N)\mathbf{x}(N) = \boldsymbol{\lambda}(N) = \mathbf{Sx}(N)$$

or

$$\mathbf{P}(N) = \mathbf{S} \quad (8\text{–}25)$$

Hence, Equation (8–23) or (8–24) can be solved uniquely backward from $k = N$ to $k = 0$. That is, we can obtain $\mathbf{P}(N), \mathbf{P}(N - 1), \ldots, \mathbf{P}(0)$ starting from $\mathbf{P}(N)$, which is known.

By referring to Equations (8–14) and (8–18), the optimal control vector $\mathbf{u}(k)$, given by Equation (8–15), now becomes

$$\mathbf{u}(k) = -\mathbf{R}^{-1}\mathbf{H}^*\boldsymbol{\lambda}(k + 1) = -\mathbf{R}^{-1}\mathbf{H}^*(\mathbf{G}^*)^{-1}[\boldsymbol{\lambda}(k) - \mathbf{Qx}(k)]$$

$$= -\mathbf{R}^{-1}\mathbf{H}^*(\mathbf{G}^*)^{-1}[\mathbf{P}(k) - \mathbf{Q}]\mathbf{x}(k) = -\mathbf{K}(k)\mathbf{x}(k) \quad (8\text{–}26)$$

where

$$\mathbf{K}(k) = \mathbf{R}^{-1}\mathbf{H}^*(\mathbf{G}^*)^{-1}[\mathbf{P}(k) - \mathbf{Q}] \quad (8\text{–}27)$$

Equation (8–26) gives the closed-loop form, or feedback form, for the optimal control vector $\mathbf{u}(k)$. Notice that the optimal control vector is proportional to the state vector.

Note that the optimal control vector $\mathbf{u}(k)$ can be given in a few different forms. Referring to Equations (8–18) and (8–22), $\mathbf{u}(k)$ may be given by

$$\mathbf{u}(k) = -\mathbf{R}^{-1}\mathbf{H}^*\boldsymbol{\lambda}(k + 1) = -\mathbf{R}^{-1}\mathbf{H}^*\mathbf{P}(k + 1)\mathbf{x}(k + 1)$$

$$= -\mathbf{R}^{-1}\mathbf{H}^*\mathbf{P}(k + 1)[\mathbf{I} + \mathbf{HR}^{-1}\mathbf{H}^*\mathbf{P}(k + 1)]^{-1}\mathbf{Gx}(k)$$

$$= -\mathbf{R}^{-1}\mathbf{H}^*[\mathbf{P}^{-1}(k + 1) + \mathbf{HR}^{-1}\mathbf{H}^*]^{-1}\mathbf{Gx}(k)$$

$$= -\mathbf{K}(k)\mathbf{x}(k) \quad (8\text{–}28)$$

where

$$\mathbf{K}(k) = \mathbf{R}^{-1}\mathbf{H}^*[\mathbf{P}^{-1}(k + 1) + \mathbf{HR}^{-1}\mathbf{H}^*]^{-1}\mathbf{G} \quad (8\text{–}29)$$

A slightly different form of the optimal control vector $\mathbf{u}(k)$ can be given by

$$\mathbf{u}(k) = -[\mathbf{R} + \mathbf{H}^*\mathbf{P}(k + 1)\mathbf{H}]^{-1}\mathbf{H}^*\mathbf{P}(k + 1)\mathbf{Gx}(k)$$

$$= -\mathbf{K}(k)\mathbf{x}(k) \tag{8–30}$$

where

$$\mathbf{K}(k) = [\mathbf{R} + \mathbf{H}^*\mathbf{P}(k + 1)\mathbf{H}]^{-1}\mathbf{H}^*\mathbf{P}(k + 1)\mathbf{G} \tag{8–31}$$

The equivalence of the expressions for the optimal control vector $\mathbf{u}(k)$ given by Equations (8–26), (8–28), and (8–30) can be shown easily; see Problem A–8–1.

Equation (8–26), (8–28), or (8–30) clearly indicates that the optimal control law requires feedback of the state vector with time-varying gain $\mathbf{K}(k)$. Figure 8–1 shows the optimal control scheme of the regulator system based on the quadratic performance index. It is important to point out that a time-varying gain $\mathbf{K}(k)$ can be computed before the process begins, once the system state matrix $\mathbf{G}$, control matrix $\mathbf{H}$, and weighting matrices $\mathbf{Q}$, $\mathbf{R}$, and $\mathbf{S}$ are given. Consequently, $\mathbf{K}(k)$ can be precomputed off-line and stored for future use. Note that the initial state $\mathbf{x}(0)$ does not enter the computation for $\mathbf{K}(k)$. The optimal control vector $\mathbf{u}(k)$ at each stage can be determined immediately by premultiplying the state vector $\mathbf{x}(k)$ by $-\mathbf{K}(k)$.

Note that a property of the feedback gain matrix $\mathbf{K}(k)$ is that it is almost constant, except near the end of the process at $k = N$. (See Example 8–1, which follows shortly, and Problem A–8–3.)

Evaluation of the Minimum Performance Index.   We shall next evaluate the minimum value of the performance index:

$$\min J = \min\left\{\frac{1}{2}\mathbf{x}^*(N)\mathbf{Sx}(N) + \frac{1}{2}\sum_{k=0}^{N-1}[\mathbf{x}^*(k)\mathbf{Qx}(k) + \mathbf{u}^*(k)\mathbf{Ru}(k)]\right\}$$

Premultiplying both sides of Equation (8–19) by $\mathbf{x}^*(k)$, we have

$$\mathbf{x}^*(k)\mathbf{P}(k)\mathbf{x}(k) = \mathbf{x}^*(k)\mathbf{Qx}(k) + \mathbf{x}^*(k)\mathbf{G}^*\mathbf{P}(k + 1)\mathbf{x}(k + 1)$$

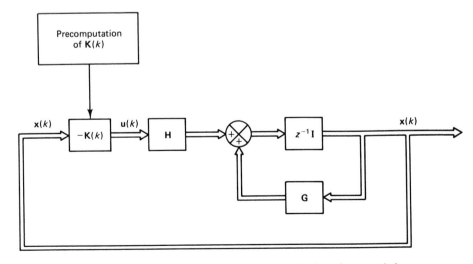

Figure 8–1   Optimal regulator system based on a quadratic performance index.

Substituting Equation (8–21) into this last equation, we obtain

$$\mathbf{x}^*(k)\mathbf{P}(k)\mathbf{x}(k) = \mathbf{x}^*(k)\mathbf{Q}\mathbf{x}(k) + \mathbf{x}^*(k+1)[\mathbf{I} + \mathbf{H}\mathbf{R}^{-1}\mathbf{H}^*\mathbf{P}(k+1)]^*\mathbf{P}(k+1)\mathbf{x}(k+1)$$

$$= \mathbf{x}^*(k)\mathbf{Q}\mathbf{x}(k) + \mathbf{x}^*(k+1)[\mathbf{I} + \mathbf{P}(k+1)\mathbf{H}\mathbf{R}^{-1}\mathbf{H}^*]\mathbf{P}(k+1)\mathbf{x}(k+1)$$

Hence,

$$\mathbf{x}^*(k)\mathbf{Q}\mathbf{x}(k) = \mathbf{x}^*(k)\mathbf{P}(k)\mathbf{x}(k) - \mathbf{x}^*(k+1)\mathbf{P}(k+1)\mathbf{x}(k+1)$$

$$- \mathbf{x}^*(k+1)\mathbf{P}(k+1)\mathbf{H}\mathbf{R}^{-1}\mathbf{H}^*\mathbf{P}(k+1)\mathbf{x}(k+1) \quad (8\text{–}32)$$

Also, from Equations (8–15) and (8–18) we have

$$\mathbf{u}(k) = -\mathbf{R}^{-1}\mathbf{H}^*\mathbf{P}(k+1)\mathbf{x}(k+1)$$

Hence,

$$\mathbf{u}^*(k)\mathbf{R}\mathbf{u}(k) = [-\mathbf{x}^*(k+1)\mathbf{P}(k+1)\mathbf{H}\mathbf{R}^{-1}]\mathbf{R}[-\mathbf{R}^{-1}\mathbf{H}^*\mathbf{P}(k+1)\mathbf{x}(k+1)]$$

$$= \mathbf{x}^*(k+1)\mathbf{P}(k+1)\mathbf{H}\mathbf{R}^{-1}\mathbf{H}^*\mathbf{P}(k+1)\mathbf{x}(k+1) \quad (8\text{–}33)$$

By adding Equations (8–32) and (8–33), we have

$$\mathbf{x}^*(k)\mathbf{Q}\mathbf{x}(k) + \mathbf{u}^*(k)\mathbf{R}\mathbf{u}(k) = \mathbf{x}^*(k)\mathbf{P}(k)\mathbf{x}(k) - \mathbf{x}^*(k+1)\mathbf{P}(k+1)\mathbf{x}(k+1) \quad (8\text{–}34)$$

By substituting Equation (8–34) into Equation (8–3), we obtain

$$J_{\min} = \frac{1}{2}\mathbf{x}^*(N)\mathbf{S}\mathbf{x}(N) + \frac{1}{2}\sum_{k=0}^{N-1}[\mathbf{x}^*(k)\mathbf{P}(k)\mathbf{x}(k) - \mathbf{x}^*(k+1)\mathbf{P}(k+1)\mathbf{x}(k+1)]$$

$$= \frac{1}{2}\mathbf{x}^*(N)\mathbf{S}\mathbf{x}(N) + \frac{1}{2}[\mathbf{x}^*(0)\mathbf{P}(0)\mathbf{x}(0) - \mathbf{x}^*(1)\mathbf{P}(1)\mathbf{x}(1) + \mathbf{x}^*(1)\mathbf{P}(1)\mathbf{x}(1)$$

$$- \mathbf{x}^*(2)\mathbf{P}(2)\mathbf{x}(2) + \cdots + \mathbf{x}^*(N-1)\mathbf{P}(N-1)\mathbf{x}(N-1) - \mathbf{x}^*(N)\mathbf{P}(N)\mathbf{x}(N)]$$

$$= \frac{1}{2}\mathbf{x}^*(N)\mathbf{S}\mathbf{x}(N) + \frac{1}{2}\mathbf{x}^*(0)\mathbf{P}(0)\mathbf{x}(0) - \frac{1}{2}\mathbf{x}^*(N)\mathbf{P}(N)\mathbf{x}(N) \quad (8\text{–}35)$$

Notice that from Equation (8–25) we have $\mathbf{P}(N) = \mathbf{S}$. Hence, Equation (8–35) becomes

$$J_{\min} = \frac{1}{2}\mathbf{x}^*(0)\mathbf{P}(0)\mathbf{x}(0) \quad (8\text{–}36)$$

Thus, the minimum value of the performance index J is given by Equation (8–36). It is a function of $\mathbf{P}(0)$ and the initial state $\mathbf{x}(0)$.

Example 8–1

Consider the discrete-time control system defined by

$$x(k+1) = 0.3679x(k) + 0.6321u(k), \qquad x(0) = 1$$

Determine the optimal control law to minimize the following performance index:

$$J = \frac{1}{2}[x(10)]^2 + \frac{1}{2}\sum_{k=0}^{9}[x^2(k) + u^2(k)]$$

Note that in this example $S = 1$, $Q = 1$, and $R = 1$. Also, determine the minimum value of the performance index J.

Referring to Equation (8–23), we obtain $P(k)$ as follows:

$$P(k) = 1 + (0.3679)P(k + 1)[1 + (0.6321)(1)(0.6321)P(k + 1)]^{-1}(0.3679)$$

which can be simplified to

$$P(k) = 1 + 0.1354P(k + 1)[1 + 0.3996P(k + 1)]^{-1}$$

The boundary condition for $P(k)$ is specified by Equation (8–25), and in this example

$$P(N) = P(10) = S = 1$$

We now compute $P(k)$ backward from $k = 9$ to $k = 0$:

$$P(9) = 1 + 0.1354 \times 1(1 + 0.3996 \times 1)^{-1} = 1.0967$$

$$P(8) = 1 + 0.1354 \times 1.0967(1 + 0.3996 \times 1.0967)^{-1} = 1.1032$$

$$P(7) = 1 + 0.1354 \times 1.1032(1 + 0.3996 \times 1.1032)^{-1} = 1.1036$$

$$P(6) = 1 + 0.1354 \times 1.1036(1 + 0.3996 \times 1.1037)^{-1} = 1.1037$$

$$P(k) = 1.1037, \qquad k = 5, 4, 3, 2, 1, 0$$

Notice that the values of $P(k)$ rapidly approach the steady-state value. The steady-state value P_{ss} can be obtained from

$$P_{ss} = 1 + 0.1354P_{ss}(1 + 0.3996P_{ss})^{-1}$$

or

$$0.3996P_{ss}^2 + 0.4650P_{ss} - 1 = 0$$

Solving this last equation for P_{ss}, we have

$$P_{ss} = 1.1037 \qquad \text{or} \qquad -2.2674$$

Since $P(k)$ must be positive, we find the steady-state value for $P(k)$ to be 1.1037.

The feedback gain $K(k)$ can be computed from Equation (8–27):

$$K(k) = (1)(0.6321)(0.3679)^{-1}[P(k) - 1] = 1.7181[P(k) - 1]$$

By substituting the values of $P(k)$ we have obtained, we get

$$K(10) = 1.7181(1 - 1) = 0$$

$$K(9) = 1.7181(1.0967 - 1) = 0.1662$$

$$K(8) = 1.7181(1.1032 - 1) = 0.1773$$

$$K(7) = 1.7181(1.1036 - 1) = 0.1781$$

$$K(6) = K(5) = \cdots = K(0) = 0.1781$$

The optimal control law is given by

$$u(k) = -K(k)x(k)$$

Since

$$x(k + 1) = 0.3679x(k) + 0.6321u(k) = [0.3679 - 0.6321K(k)]x(k)$$

we obtain

$$x(1) = [0.3679 - 0.6321K(0)]x(0)$$
$$= (0.3679 - 0.6321 \times 0.1781) \times 1 = 0.2553$$
$$x(2) = (0.3679 - 0.6321 \times 0.1781) \times 0.2553 = 0.0652$$
$$x(3) = (0.3679 - 0.6321 \times 0.1781) \times 0.0652 = 0.0166$$
$$x(4) = (0.3679 - 0.6321 \times 0.1781) \times 0.0166 = 0.00424$$

The values of $x(k)$ for $k = 5, 6, \ldots, 10$ approach zero rapidly.

The optimal control sequence $u(k)$ is now obtained as follows:

$$u(0) = -K(0)x(0) = -0.1781 \times 1 = -0.1781$$
$$u(1) = -K(1)x(1) = -0.1781 \times 0.2553 = -0.0455$$
$$u(2) = -K(2)x(2) = -0.1781 \times 0.0652 = -0.0116$$
$$u(3) = -K(3)x(3) = -0.1781 \times 0.0166 = -0.00296$$
$$u(4) = -K(4)x(4) = -0.1781 \times 0.00424 = -0.000756$$
$$u(k) \doteq 0, \quad k = 5, 6, \ldots, 10$$

The values of $P(k)$, $K(k)$, $x(k)$, and $u(k)$ are plotted in Figure 8–2. Notice that the values of $P(k)$ and $K(k)$ are constant except for the final few stages.

Finally, the minimum value of the performance index J can be obtained from Equation (8–36):

$$J_{\min} = \frac{1}{2}x^*(0)P(0)x(0) = \frac{1}{2}(1 \times 1.1037 \times 1) = 0.5518$$

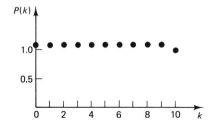
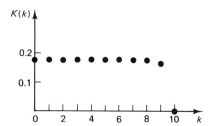
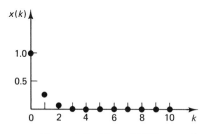
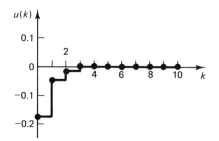

Figure 8–2　Plots of $P(k)$ versus k, $x(k)$ versus k, $K(k)$ versus k, and $u(k)$ versus k for the system considered in Example 8–1.

MATLAB approach to the solution of this example problem. A MATLAB program can be easily written for the solution of this example problem. MATLAB Program 8–1 shows a possible program. [Here, we used Equation (8–24) for the computation of P(k).]

MATLAB Program 8–1

```
G = 0.3679; H = 0.6321; Q = 1; R = 1; S = 1; x0 = 1;
N = 11; p(N) = S; x(1) = 1; Pnext = S;

for i = N−1:−1:1,
  P = Q+G'*Pnext*G−G'*Pnext*H*inv(R+H'*Pnext*H)*H'*Pnext*G;
  p(i) = P; Pnext = P;
end
for i = N:−1:1,
  K = inv(R)*H'*inv(G')*(p(i)−Q);
  k(i) = K;
end
for i = 1:N−1,
  xnext = (G−H*k(i))*x(i);
  x(i+1) =xnext;
end
for i = 1:N,
  u(i) = −k(i)*x(i);
end
```

To print out the values of P0, P1, ..., P10 [which correspond to p(1), p(2), ..., p(11)], K0, K1, ..., K10 [which correspond to k(1), k(2), ..., k(11)], x0, x1, ..., x10 [which correspond to x(1), x(2), ..., x(11)], and u0, u1, ..., u10 [which correspond to u(1), u(2), ..., u(11)], we enter the printout statement as follows.

```
% ***** Printout P, K, x, and u *****

M = [p'  k'  x'  u']

M =

    1.1037    0.1781    1.0000   −0.1781
    1.1037    0.1781    0.2553   −0.0455
    1.1037    0.1781    0.0652   −0.0116
    1.1037    0.1781    0.0166   −0.0030
    1.1037    0.1781    0.0042   −0.0008
    1.1037    0.1781    0.0011   −0.0002
    1.1037    0.1781    0.0003   −0.0000
    1.1036    0.1781    0.0001   −0.0000
    1.1032    0.1773    0.0000   −0.0000
    1.0967    0.1662    0.0000   −0.0000
    1.0000         0    0.0000         0
```

The first column of matrix M gives from top to bottom the values of P0, P1, ..., P10. Similarly, the second, third, and fourth columns give K0, K1, ..., K10; x0, x1, ..., x10; and u0, u1, ..., u10; respectively.

Discretized Quadratic Optimal Control Problem.   We shall next consider the quadratic optimal control of a discretized control system. Consider the continuous-time control system

$$\dot{\mathbf{x}} = \mathbf{A}\mathbf{x} + \mathbf{B}\mathbf{u} \tag{8–37}$$

where

$$\mathbf{u}(t) = \mathbf{u}(kT), \qquad kT \le t < (k + 1)T$$

and the performance index to be minimized is

$$J = \frac{1}{2}\mathbf{x}^*(t_f)\mathbf{S}\mathbf{x}(t_f) + \frac{1}{2}\int_0^{t_f} [\mathbf{x}^*(t)\mathbf{Q}\mathbf{x}(t) + \mathbf{u}^*(t)\mathbf{R}\mathbf{u}(t)]\, dt \tag{8–38}$$

Suppose the continuous-time control system is approximated by its discrete equivalent. The discretized system equation is

$$\mathbf{x}((k + 1)T) = \mathbf{G}(T)\mathbf{x}(kT) + \mathbf{H}(T)\mathbf{u}(kT)$$

and the discretized performance index when $t_f = NT$ will become as follows:

$$J = \frac{1}{2}\mathbf{x}^*(NT)\mathbf{S}\mathbf{x}^*(NT)$$

$$+ \frac{1}{2}\sum_{k=0}^{N-1} [\mathbf{x}^*(kT)\mathbf{Q}_1\,\mathbf{x}(kT) + 2\mathbf{x}^*(kT)\mathbf{M}_1\,\mathbf{u}(kT) + \mathbf{u}^*(kT)\mathbf{R}_1\,\mathbf{u}(kT)] \tag{8–39}$$

It is noted that the integral term in Equation (8–38) is not replaced by

$$\frac{1}{2}\sum_{k=0}^{N-1} [\mathbf{x}^*(kT)\mathbf{Q}\mathbf{x}(kT) + \mathbf{u}^*(kT)\mathbf{R}\mathbf{u}(kT)]$$

but is modified to include a cross term involving $\mathbf{x}(kT)$ and $\mathbf{u}(kT)$. Also, matrices $\mathbf{Q}$ and $\mathbf{R}$ are modified. In what follows, we shall consider the discretized quadratic optimal control problem by use of a simple example.

Consider the continuous-time system defined by

$$\dot{x}(t) = ax(t) + bu(t) \tag{8–40}$$

where a and b are constants and

$$u(t) = u(kT), \qquad kT \le t < (k + 1)T$$

The performance index to be minimized is

$$J = \frac{1}{2}x^2(NT) + \frac{1}{2}\int_0^{NT} [Qx^2(t) + Ru^2(t)]\, dt \tag{8–41}$$

Let us discretize the system equation and the performance index and formulate the discretized quadratic optimal control problem.

Equation (8–40) may be discretized as follows:

$$x((k + 1)T) = G(T)x(kT) + H(T)u(kT)$$

where

$$G(T) = e^{aT}$$

$$H(T) = \int_0^T e^{a\tau}b\,d\tau = \frac{b}{a}(e^{aT} - 1)$$

or

$$x((k + 1)T) = e^{aT}x(kT) + \frac{b}{a}(e^{aT} - 1)u(kT) \tag{8–42}$$

The performance index J given by Equation (8–41) may be discretized. First, rewrite J as

$$J_1 = \frac{1}{2}x^2(NT) + \frac{1}{2}\sum_{k=0}^{N-1}\int_{kT}^{(k+1)T}[Qx^2(t) + Ru^2(t)]\,dt$$

Noting that the solution $x(t)$ for $kT \le t < (k + 1)T$ can be written as

$$x(t) = e^{a(t-kT)}x(kT) + \int_{kT}^t e^{a(t-\tau)}bu(\tau)\,d\tau$$

$$= \xi(t - kT)x(kT) + \eta(t - kT)u(kT)$$

where

$$\xi(t - kT) = e^{a(t-kT)}$$

$$\eta(t - kT) = \int_{kT}^t \xi(t - \tau)b\,d\tau = \frac{b}{a}[e^{a(t-kT)} - 1]$$

the performance index J_1 can be written as follows:

$$J_1 = \frac{1}{2}x^2(NT) + \frac{1}{2}\sum_{k=0}^{N-1}\int_{kT}^{(k+1)T}\{Q[\xi(t - kT)x(kT) + \eta(t - kT)u(kT)]^2$$

$$+ Ru^2(kT)\}\,dt$$

$$= \frac{1}{2}x^2(NT) + \frac{1}{2}\sum_{k=0}^{N-1}\int_{kT}^{(k+1)T}[Q\xi^2(t - kT)x^2(kT)$$

$$+ 2Q\xi(t - kT)\eta(t - kT)x(kT)u(kT)$$

$$+ Q\eta^2(t - kT)u^2(kT) + Ru^2(kT)]\,dt$$

$$= \frac{1}{2}x^2(NT) + \frac{1}{2}\sum_{k=0}^{N-1}[Q_1x^2(kT) + 2M_1x(kT)u(kT) + R_1u^2(kT)] \tag{8–43}$$

where

$$Q_1 = \int_{kT}^{(k+1)T} Q\xi^2(t - kT)\,dt$$

$$M_1 = \int_{kT}^{(k+1)T} Q\xi(t - kT)\eta(t - kT)\,dt$$

$$R_1 = \int_{kT}^{(k+1)T} [Q\eta^2(t - kT) + R]\,dt$$

Notice that Q_1, M_1, and R_1 may be simplified as follows:

$$Q_1 = \int_{kT}^{(k+1)T} Qe^{2a(t-kT)}\,dt = \frac{Q}{2a}(e^{2aT} - 1) \tag{8-44}$$

$$M_1 = \int_{kT}^{(k+1)T} Qe^{a(t-kT)}\frac{b}{a}[e^{a(t-kT)} - 1]\,dt = \frac{bQ}{2a^2}(e^{aT} - 1)^2 \tag{8-45}$$

$$R_1 = \int_{kT}^{(k+1)T} \left[Q\left\{\frac{b}{a}[e^{a(t-kT)} - 1]\right\}^2 + R\right]dt$$

$$= \frac{b^2 Q}{2a^3}[(e^{aT} - 3)(e^{aT} - 1) + 2aT] + RT \tag{8-46}$$

Summarizing, the present discretized quadratic optimal control problem may be stated as follows. Given the discretized system equation

$$x((k + 1)T) = G(T)x(kT) + H(T)u(kT)$$

where

$$G(T) = e^{aT} \quad \text{and} \quad H(T) = \frac{b}{a}(e^{aT} - 1)$$

find the optimal control sequence $u(0), u(T), \ldots, u((N - 1)T)$ such that the following performance index is minimized:

$$J_1 = \frac{1}{2}x^2(NT) + \frac{1}{2}\sum_{k=0}^{N-1} [Q_1 x^2(kT) + 2M_1 x(kT)u(kT) + R_1 u^2(kT)]$$

Such a performance index including a cross term involving $x(kT)$ and $u(kT)$ can be modified to a form that does not include a cross term, and the solution to the discretized quadratic optimal control problem can then be obtained in a manner similar to that for the quadratic optimal control problem presented earlier in this section. This subject is presented in the following.

Performance Index Including a Cross Term Involving x(k) and u(k).   Next, we shall consider the quadratic optimal control problem where the system is as given by Equation (8–1), which was

$$\mathbf{x}(k + 1) = \mathbf{Gx}(k) + \mathbf{Hu}(k), \qquad \mathbf{x}(0) = \mathbf{c}$$

and the performance index is given by

$$J = \frac{1}{2}\mathbf{x}^*(N)\mathbf{Sx}(N) + \frac{1}{2}\sum_{k=0}^{N-1} [\mathbf{x}^*(k)\mathbf{Qx}(k) + 2\mathbf{x}^*(k)\mathbf{Mu}(k) + \mathbf{u}^*(k)\mathbf{Ru}(k)] \tag{8-47}$$

where $\mathbf{Q}$ and $\mathbf{S}$ are $n \times n$ positive definite or positive semidefinite Hermitian matrices, $\mathbf{R}$ is an $r \times r$ positive definite Hermitian matrix, and $\mathbf{M}$ is an $n \times r$ matrix such that matrix

$$\begin{bmatrix} \mathbf{Q} & \mathbf{M} \\ \mathbf{M}^* & \mathbf{R} \end{bmatrix}$$

is positive definite. This means that

$$[\mathbf{x}^*(k) \quad \mathbf{u}^*(k)]\begin{bmatrix} \mathbf{Q} & \mathbf{M} \\ \mathbf{M}^* & \mathbf{R} \end{bmatrix}\begin{bmatrix} \mathbf{x}(k) \\ \mathbf{u}(k) \end{bmatrix}$$

$$= \mathbf{x}^*(k)\mathbf{Q}\mathbf{x}(k) + \mathbf{x}^*(k)\mathbf{M}\mathbf{u}(k) + \mathbf{u}^*(k)\mathbf{M}^*\mathbf{x}(k) + \mathbf{u}^*(k)\mathbf{R}\mathbf{u}(k)$$

$$= \mathbf{x}^*(k)\mathbf{Q}\mathbf{x}(k) + 2\mathbf{x}^*(k)\mathbf{M}\mathbf{u}(k) + \mathbf{u}^*(k)\mathbf{R}\mathbf{u}(k)$$

is positive definite. Note that the performance index J given by Equation (8–47) includes a cross term involving $\mathbf{x}(k)$ and $\mathbf{u}(k)$.

To obtain the optimal control vector $\mathbf{u}(k)$, let us define

$$\hat{\mathbf{Q}} = \mathbf{Q} - \mathbf{M}\mathbf{R}^{-1}\mathbf{M}^* \tag{8–48}$$

and eliminate $\mathbf{Q}$ from the performance index J. Then Equation (8–47) becomes

$$J = \frac{1}{2}\mathbf{x}^*(N)\mathbf{S}\mathbf{x}(N) + \frac{1}{2}\sum_{k=0}^{N-1}\{\mathbf{x}^*(k)[\hat{\mathbf{Q}} + \mathbf{M}\mathbf{R}^{-1}\mathbf{M}^*]\mathbf{x}(k)$$

$$+ 2\mathbf{x}^*(k)\mathbf{M}\mathbf{u}(k) + \mathbf{u}^*(k)\mathbf{R}\mathbf{u}(k)\}$$

$$= \frac{1}{2}\mathbf{x}^*(N)\mathbf{S}\mathbf{x}(N) + \frac{1}{2}\sum_{k=0}^{N-1}[\mathbf{x}^*(k)\hat{\mathbf{Q}}\mathbf{x}(k) + \mathbf{x}^*(k)\mathbf{M}\mathbf{R}^{-1}\mathbf{M}^*\mathbf{x}(k)$$

$$+ 2\mathbf{x}^*(k)\mathbf{M}\mathbf{u}(k) + \mathbf{u}^*(k)\mathbf{R}\mathbf{u}(k)]$$

$$= \frac{1}{2}\mathbf{x}^*(N)\mathbf{S}\mathbf{x}(N) + \frac{1}{2}\sum_{k=0}^{N-1}\{\mathbf{x}^*(k)\hat{\mathbf{Q}}\mathbf{x}(k)$$

$$+ [\mathbf{x}^*(k)\mathbf{M}\mathbf{R}^{-1} + \mathbf{u}^*(k)]\mathbf{R}[\mathbf{R}^{-1}\mathbf{M}^*\mathbf{x}(k) + \mathbf{u}(k)]\} \tag{8–49}$$

Define

$$\mathbf{v}(k) = \mathbf{R}^{-1}\mathbf{M}^*\mathbf{x}(k) + \mathbf{u}(k) \tag{8–50}$$

Then Equation (8–49) can be written as follows:

$$J = \frac{1}{2}\mathbf{x}^*(N)\mathbf{S}\mathbf{x}(N) + \frac{1}{2}\sum_{k=0}^{N-1}[\mathbf{x}^*(k)\hat{\mathbf{Q}}\mathbf{x}(k) + \mathbf{v}^*(k)\mathbf{R}\mathbf{v}(k)] \tag{8–51}$$

Notice that Equation (8–51) no longer involves the cross term. We have effectively eliminated the cross term involving $\mathbf{x}(k)$ and $\mathbf{u}(k)$.

By substituting Equation (8–50) into the system equation, Equation (8–1), we obtain

$$\mathbf{x}(k + 1) = \mathbf{G}\mathbf{x}(k) + \mathbf{H}[\mathbf{v}(k) - \mathbf{R}^{-1}\mathbf{M}^*\mathbf{x}(k)]$$

$$= (\mathbf{G} - \mathbf{H}\mathbf{R}^{-1}\mathbf{M}^*)\mathbf{x}(k) + \mathbf{H}\mathbf{v}(k)$$

$$= \hat{\mathbf{G}}\mathbf{x}(k) + \mathbf{H}\mathbf{v}(k) \tag{8–52}$$

where

$$\hat{\mathbf{G}} = \mathbf{G} - \mathbf{H}\mathbf{R}^{-1}\mathbf{M}^* \tag{8-53}$$

Note that the quadratic optimal control of the system given by Equation (8–1) with the performance index given by Equation (8–47) is equivalent to the quadratic optimal control of the system given by Equation (8–52) with the performance index given by Equation (8–51). Hence, the optimal control vector $\mathbf{v}(k)$ that minimizes the performance index given by Equation (8–51) can be given as follows. Referring to Equation (8–26), (8–28), or (8–30), we have

$$\mathbf{v}(k) = -\mathbf{R}^{-1}\mathbf{H}^*(\hat{\mathbf{G}}^*)^{-1}[\hat{\mathbf{P}}(k) - \hat{\mathbf{Q}}]\mathbf{x}(k) \tag{8-54}$$

or

$$\mathbf{v}(k) = -\mathbf{R}^{-1}\mathbf{H}^*[\hat{\mathbf{P}}^{-1}(k + 1) + \mathbf{H}\mathbf{R}^{-1}\mathbf{H}^*]^{-1}\hat{\mathbf{G}}\mathbf{x}(k) \tag{8-55}$$

or

$$\mathbf{v}(k) = -[\mathbf{R} + \mathbf{H}^*\hat{\mathbf{P}}(k + 1)\mathbf{H}]^{-1}\mathbf{H}^*\hat{\mathbf{P}}(k + 1)\hat{\mathbf{G}}\mathbf{x}(k) \tag{8-56}$$

where $\hat{\mathbf{P}}(k)$ is a modified version of Equation (8–23), or

$$\hat{\mathbf{P}}(k) = \hat{\mathbf{Q}} + \hat{\mathbf{G}}^*\hat{\mathbf{P}}(k + 1)[\mathbf{I} + \mathbf{H}\mathbf{R}^{-1}\mathbf{H}^*\hat{\mathbf{P}}(k + 1)]^{-1}\hat{\mathbf{G}}, \qquad \hat{\mathbf{P}}(N) = \mathbf{S} \tag{8-57}$$

The optimal control vector $\mathbf{u}(k)$ can then be given by

$$\mathbf{u}(k) = \mathbf{v}(k) - \mathbf{R}^{-1}\mathbf{M}^*\mathbf{x}(k) \tag{8-58}$$

where $\mathbf{v}(k)$ is given by Equation (8–54), (8–55), or (8–56). Whichever expression for $\mathbf{v}(k)$ is used, Equation (8–58) may be reduced to the following form:

$$\mathbf{u}(k) = -[\mathbf{R} + \mathbf{H}^*\hat{\mathbf{P}}(k + 1)\mathbf{H}]^{-1}[\mathbf{H}^*\hat{\mathbf{P}}(k + 1)\mathbf{G} + \mathbf{M}^*]\mathbf{x}(k) \tag{8-59}$$

(See Problem A–8–2 for details.)

Example 8–2

Consider the continuous-time control system

$$\dot{x}(t) = -x(t) + u(t), \qquad x(0) = 1 \tag{8-60}$$

where

$$u(t) = u(kT), \qquad kT \le t < (k + 1)T$$

and the performance index

$$J = \frac{1}{2}x^2(NT) + \frac{1}{2}\int_0^{NT} [x^2(t) + u^2(t)]\,dt \tag{8-61}$$

where $T = 1$ sec and $N = 10$. Discretize the system equation and the performance index. Then determine the optimal control sequence $u(kT)$ for $k = 0, 1, 2, \ldots, 9$; this will be the control sequence for which the performance index is minimum. Also, obtain the minimum value of J.

Referring to Equations (8–40) and (8–42), the discretized system equation is

$$x((k + 1)T) = e^{aT}x(kT) + \frac{b}{a}(e^{aT} - 1)u(kT)$$

where $a = -1$, $b = 1$, and $T = 1$. Thus, the system equation becomes

$$x(k + 1) = 0.3679x(k) + 0.6321u(k), \qquad x(0) = 1 \tag{8-62}$$

The discretized performance index becomes

$$J_1 = \frac{1}{2}x^2(N) + \frac{1}{2}\sum_{k=0}^{N-1}[Q_1 x^2(k) + 2M_1 x(k)u(k) + R_1 u^2(k)] \tag{8-63}$$

where Q_1, M_1, and R_1 are given by substituting $Q = 1$ and $R = 1$ into Equations (8–44), (8–45), and (8–46), respectively, as follows:

$$Q_1 = \frac{1}{2a}(e^{2aT} - 1) = \frac{1}{-2}(e^{-2} - 1) = 0.4323$$

$$M_1 = \frac{b}{2a^2}(e^{aT} - 1)^2 = \frac{1}{2}(e^{-1} - 1)^2 = 0.1998$$

$$R_1 = \frac{b^2}{2a^3}[(e^{aT} - 3)(e^{aT} - 1) + 2aT] + T$$

$$= \frac{1}{2(-1)^3}[(e^{-1} - 3)(e^{-1} - 1) - 2] + 1 = 1.1681$$

Thus, the performance index given by Equation (8–63) can be written as follows:

$$J_1 = \frac{1}{2}x^2(10) + \frac{1}{2}\sum_{k=0}^{9}[0.4323x^2(k) + 0.3996x(k)u(k) + 1.1681u^2(k)] \tag{8-64}$$

Therefore, our problem becomes as follows. Given the system equation, Equation (8–62), find the optimal control sequence $u(k)$, where $k = 0, 1, 2, \ldots, 9$, such that the performance index given by Equation (8–64) is minimum.

Now, comparing Equations (8–47) and (8–64), we have

$$S = 1, \qquad Q = 0.4323, \qquad M = 0.1998, \qquad R = 1.1681$$

Notice that

$$\begin{bmatrix} Q & M \\ M^* & R \end{bmatrix} = \begin{bmatrix} 0.4323 & 0.1998 \\ 0.1998 & 1.1681 \end{bmatrix}$$

is positive definite. The next step is to modify J_1 as given by Equation (8–64) into the form given by Equation (8–51). Since $\hat{Q} = Q - MR^{-1}M^* = 0.3981$,

$$J_1 = \frac{1}{2}x^2(10) + \frac{1}{2}\sum_{k=0}^{9}[0.3981x^2(k) + 1.1681v^2(k)] \tag{8-65}$$

The optimal control signal $u(k)$ can be found from Equation (8–58):

$$u(k) = v(k) - R^{-1}M^*x(k)$$

which can be written in the form given by Equation (8–59):

$$u(k) = -[R + H^*\hat{P}(k + 1)H]^{-1}[H^*\hat{P}(k + 1)G + M^*]x(k) \tag{8-66}$$

where

$$G = 0.3679 \qquad \text{and} \qquad H = 0.6321$$

Equation (8–66) can be rewritten as follows:

$$u(k) = -[1.1681 + 0.3996\hat{P}(k+1)]^{-1}[0.2325\hat{P}(k+1) + 0.1998]x(k)$$

$$= -\frac{0.2325\hat{P}(k+1) + 0.1998}{1.1681 + 0.3996\hat{P}(k+1)}x(k) = -K(k)x(k) \tag{8-67}$$

where

$$K(k) = \frac{0.2325\hat{P}(k+1) + 0.1998}{1.1681 + 0.3996\hat{P}(k+1)} \tag{8-68}$$

Note that $\hat{P}(k)$ is as given by Equation (8–57), or

$$\hat{P}(k) = \hat{Q} + \hat{G}^*\hat{P}(k+1)[1 + HR^{-1}H^*\hat{P}(k+1)]^{-1}\hat{G} \tag{8-69}$$

where $\hat{P}(N) = \hat{P}(10) = 1$ and

$$\hat{Q} = Q - MR^{-1}M^* = 0.3981$$

$$\hat{G} = G - HR^{-1}M^* = 0.3679 - 0.1081 = 0.2598$$

Equation (8–69) can be simplified into the following form:

$$\hat{P}(k) = 0.3981 + \frac{0.06750\hat{P}(k+1)}{1 + 0.3421\hat{P}(k+1)} \tag{8-70}$$

We shall now compute $\hat{P}(k)$ with the boundary condition $\hat{P}(10) = 1$. Using Equation (8–70), we find $\hat{P}(k)$ backward from $k = 9$ to $k = 0$. The results are tabulated in Table 8–1. Using the values of $\hat{P}(k)$ just obtained, we compute $K(k)$ from Equation (8–68). The results are also shown in Table 8–1. Next, we compute $x(k)$. By substituting Equation (8–67) into Equation (8–62) and eliminating $u(k)$ from these two equations, we obtain

$$x(k+1) = \frac{0.3035}{1.1681 + 0.3996\hat{P}(k+1)}x(k), \qquad x(0) = 1 \tag{8-71}$$

TABLE 8–1　VALUES OF $\hat{P}(k)$, $K(k)$, $x(k)$, AND $u(k)$ FOR THE SYSTEM CONSIDERED IN EXAMPLE 8–2

k	$\hat{P}(k)$	$K(k)$	$x(k)$	$u(k)$
0	0.4230	0.2230	1.0000	−0.2230
1	0.4230	0.2230	0.2270	−0.05062
2	0.4230	0.2230	0.05152	−0.01149
3	0.4230	0.2230	0.01169	−0.002607
4	0.4230	0.2230	0.002653	−0.0005916
5	0.4230	0.2230	0.000602	−0.0001342
6	0.4230	0.2230	0.0001366	−0.0000304
7	0.4231	0.2231	$\doteq 0$	$\doteq 0$
8	0.4243	0.2257	$\doteq 0$	$\doteq 0$
9	0.4484	0.2758	$\doteq 0$	$\doteq 0$

Starting with $x(0) = 1$, the values of $x(k)$ can be computed from Equation (8–71) by using the values of $\hat{P}(k)$ already obtained. The computed results are shown in Table 8–1. Once we get the values of $K(k)$ and $x(k)$, the optimal control signal $u(k)$ can be obtained from Equation (8–67), or

$$u(k) = -K(k)x(k)$$

The results are also shown in Table 8–1.

Finally, referring to Equation (8–36), the minimum value of J_1 can be obtained as follows:

$$J_{1,\,min} = \frac{1}{2}\hat{P}(0)x^2(0) = \frac{1}{2} \times 0.4230 \times 1^2 = 0.2115$$

Comments.   The approach presented here is useful in solving a problem of finite-time quadratic optimal control of continuous-time systems by means of computer simulation. Unless the given continuous-time quadratic performance index is rigidly specified for some reason, however, it is better to define a new discrete-time quadratic performance index after the system equations are discretized. (See Section 8–4.)

Note that in most cases matrices **S**, **Q**, and **R** in the performance index are not rigidly fixed, but are "in a sense" arbitrarily chosen positive definite (or positive semidefinite as the case may be) matrices; minimization of an arbitrarily chosen performance index does not have much meaning. As noted earlier, the main reason that the quadratic optimal control scheme is useful and frequently used is that it produces an asymptotically stable control system, except in very special academic cases. (See Problems A–8–6 and A–8–7.)

8–3 STEADY-STATE QUADRATIC OPTIMAL CONTROL

We have seen that when the control process is finite (when N is finite) the feedback gain matrix **K**(k) becomes a time-varying matrix.

Let us now consider the quadratic optimal control problem where the process continues without bound, or where $N = \infty$ (that is, where the process is an infinite-stage process). As N approaches infinity, the optimal control solution becomes a steady-state solution, and the time-varying gain matrix **K**(k) becomes a constant gain matrix. Such a constant gain matrix **K**(k) is called a steady-state gain matrix and is written as **K**.

In what follows, we shall consider steady-state quadratic optimal control of a regulator system. The plant equation is given by

$$\mathbf{x}(k + 1) = \mathbf{G}\mathbf{x}(k) + \mathbf{H}\mathbf{u}(k) \tag{8–72}$$

For $N = \infty$, the performance index may be modified to

$$J = \frac{1}{2}\sum_{k=0}^{\infty} [\mathbf{x}^*(k)\mathbf{Q}\mathbf{x}(k) + \mathbf{u}^*(k)\mathbf{R}\mathbf{u}(k)] \tag{8–73}$$

The term $\frac{1}{2}\mathbf{x}^*(N)\mathbf{S}\mathbf{x}(N)$, which appeared in Equation (8–2), is not included in this representation of J. This is because, if the optimal regulator system is stable so that the value of J converges to a constant, $\mathbf{x}(\infty)$ becomes zero and $\frac{1}{2}\mathbf{x}^*(\infty)\mathbf{S}\mathbf{x}(\infty) = 0$.

Let us now define the steady-state matrix $\mathbf{P}(k)$ as $\mathbf{P}$. Referring to Equation (8–23), matrix $\mathbf{P}$ can be determined as follows:

$$\mathbf{P} = \mathbf{Q} + \mathbf{G}^*\mathbf{P}(\mathbf{I} + \mathbf{H}\mathbf{R}^{-1}\mathbf{H}^*\mathbf{P})^{-1}\mathbf{G}$$

$$= \mathbf{Q} + \mathbf{G}^*(\mathbf{P}^{-1} + \mathbf{H}\mathbf{R}^{-1}\mathbf{H}^*)^{-1}\mathbf{G} \tag{8–74}$$

Clearly, matrix $\mathbf{P}$ is determined by matrices $\mathbf{G}$, $\mathbf{H}$, $\mathbf{Q}$, and $\mathbf{R}$. A slightly different expression for $\mathbf{P}$ can be derived from Equation (8–24):

$$\mathbf{P} = \mathbf{Q} + \mathbf{G}^*\mathbf{P}\mathbf{G} - \mathbf{G}^*\mathbf{P}\mathbf{H}(\mathbf{R} + \mathbf{H}^*\mathbf{P}\mathbf{H})^{-1}\mathbf{H}^*\mathbf{P}\mathbf{G} \tag{8–75}$$

The steady-state gain matrix $\mathbf{K}$ can be obtained in terms of $\mathbf{P}$ as follows. From Equation (8–27),

$$\mathbf{K} = \mathbf{R}^{-1}\mathbf{H}^*(\mathbf{G}^*)^{-1}(\mathbf{P} - \mathbf{Q}) \tag{8–76}$$

From Equation (8–29),

$$\mathbf{K} = \mathbf{R}^{-1}\mathbf{H}^*(\mathbf{P}^{-1} + \mathbf{H}\mathbf{R}^{-1}\mathbf{H}^*)^{-1}\mathbf{G} \tag{8–77}$$

Still another expression for $\mathbf{K}$ is possible. From Equation (8–31),

$$\mathbf{K} = (\mathbf{R} + \mathbf{H}^*\mathbf{P}\mathbf{H})^{-1}\mathbf{H}^*\mathbf{P}\mathbf{G} \tag{8–78}$$

The optimal control law for steady-state operation is given by

$$\mathbf{u}(k) = -\mathbf{K}\mathbf{x}(k)$$

If, for example, Equation (8–78) is substituted into this last equation, we obtain

$$\mathbf{u}(k) = -(\mathbf{R} + \mathbf{H}^*\mathbf{P}\mathbf{H})^{-1}\mathbf{H}^*\mathbf{P}\mathbf{G}\mathbf{x}(k) \tag{8–79}$$

and the control system becomes an optimal regulator system:

$$\mathbf{x}(k + 1) = [\mathbf{G} - \mathbf{H}(\mathbf{R} + \mathbf{H}^*\mathbf{P}\mathbf{H})^{-1}\mathbf{H}^*\mathbf{P}\mathbf{G}]\mathbf{x}(k)$$

$$= (\mathbf{I} + \mathbf{H}\mathbf{R}^{-1}\mathbf{H}^*\mathbf{P})^{-1}\mathbf{G}\mathbf{x}(k) \tag{8–80}$$

where we have used the matrix inversion lemma,

$$(\mathbf{A} + \mathbf{B}\mathbf{C})^{-1} = \mathbf{A}^{-1} - \mathbf{A}^{-1}\mathbf{B}(\mathbf{I} + \mathbf{C}\mathbf{A}^{-1}\mathbf{B})^{-1}\mathbf{C}\mathbf{A}^{-1}$$

with $\mathbf{A} = \mathbf{I}$, $\mathbf{B} = \mathbf{H}$, and $\mathbf{C} = \mathbf{R}^{-1}\mathbf{H}^*\mathbf{P}$. (Refer to Appendix A.)

The performance index J associated with the steady-state optimal control law can be obtained from Equation (8–36) by substituting $\mathbf{P}$ for $\mathbf{P}(0)$:

$$J_{\min} = \frac{1}{2}\mathbf{x}^*(0)\mathbf{P}\mathbf{x}(0) \tag{8–81}$$

In many practical systems, instead of using a time-varying gain matrix $\mathbf{K}(k)$, we approximate such a gain matrix by the constant gain matrix $\mathbf{K}$. Deviations from the optimal performance due to the approximation will appear only near the end of the control process.

Steady-State Riccati Equation.   In implementing the steady-state (or time-invariant) optimal controller, we require the steady-state solution of the Riccati equation. There are several ways to obtain the steady-state solution.

One way to solve the steady-state Riccati equation given by Equation (8–75),

$$\mathbf{P} = \mathbf{Q} + \mathbf{G}^*\mathbf{PG} - \mathbf{G}^*\mathbf{PH}(\mathbf{R} + \mathbf{H}^*\mathbf{PH})^{-1}\mathbf{H}^*\mathbf{PG}$$

is to start with the following non-steady-state Riccati equation, which was given by Equation (8–24):

$$\mathbf{P}(k) = \mathbf{Q} + \mathbf{G}^*\mathbf{P}(k + 1)\mathbf{G}$$
$$- \mathbf{G}^*\mathbf{P}(k + 1)\mathbf{H}[\mathbf{R} + \mathbf{H}^*\mathbf{P}(k + 1)\mathbf{H}]^{-1}\mathbf{H}^*\mathbf{P}(k + 1)\mathbf{G} \qquad (8\text{–}82)$$

By reversing the direction of time, we may modify Equation (8–82) to read

$$\mathbf{P}(k + 1) = \mathbf{Q} + \mathbf{G}^*\mathbf{P}(k)\mathbf{G} - \mathbf{G}^*\mathbf{P}(k)\mathbf{H}[\mathbf{R} + \mathbf{H}^*\mathbf{P}(k)\mathbf{H}]^{-1}\mathbf{H}^*\mathbf{P}(k)\mathbf{G} \qquad (8\text{–}83)$$

and begin the solution with $\mathbf{P}(0) = \mathbf{0}$ and iterate the equation until a stationary solution is obtained. In computing the numerical solution, it is important to note that matrix $\mathbf{P}$ is either a Hermitian or a real symmetric matrix and is positive definite.

In what follows, we shall first present a MATLAB approach to the solution of steady-state quadratic optimal control problem. Then we discuss another approach based on the Liapunov method.

MATLAB Approach to the Solution of Steady-State Quadratic Optimal Control Problem.   In what follows we consider an example problem of steady-state quadratic optimal control of a regulator system and solve it with MATLAB.

Consider the system

$$\mathbf{x}(k + 1) = \mathbf{Gx}(k) + \mathbf{H}u(k)$$

where

$$\mathbf{G} = \begin{bmatrix} 0.2 & 0 \\ 0 & 0.4 \end{bmatrix}, \qquad \mathbf{H} = \begin{bmatrix} 1 \\ 1 \end{bmatrix}$$

The performance index J is given by

$$J = \frac{1}{2}\sum_{k=0}^{\infty}[\mathbf{x}'(k)\mathbf{Qx}(k) + u'(k)Ru(k)]$$

where

$$\mathbf{Q} = \begin{bmatrix} 1 & 0 \\ 0 & 0.5 \end{bmatrix}, \qquad R = 1$$

The control law that minimizes J can be given by

$$u(k) = -\mathbf{Kx}(k)$$

Determine the steady-state gain matrix $\mathbf{K}$.

MATLAB Program 8-2 solves this problem. Note that MATLAB carries out the iterative solution of

$$\mathbf{P} = \mathbf{Q} + \mathbf{G}'\mathbf{PG} - \mathbf{G}'\mathbf{PH}(\mathbf{R} + \mathbf{H}'\mathbf{PH})^{-1}\mathbf{H}'\mathbf{PG}$$

MATLAB Program 8–2

```
% ---------- Steady-state quadratic optimal control ----------

% ***** Solving steady-state Riccati equation and finding
% optimal feedback gain matrix K *****

% ***** Enter matrices G, H, Q, and R *****

G = [0.2  0;0  0.4];
H = [1;1];
Q = [1  0;0  0.5];
R = [1];

% ***** Start with the solution of steady-state Riccati equation
% with P = [0  0;0  0] *****

P = [0  0;0  0];
P = Q + G'*P*G - G'*P*H*inv(R+H'*P*H)*H'*P*G;

% ***** Check solution P every 10 or 20 steps of iteration.
% Stop iteration when P stays constant *****

for i = 1:10,
  P = Q + G'*P*G - G'*P*H*inv(R+H'*P*H)*H'*P*G;
end
P

P =

     1.0252    -0.0189
    -0.0189     0.5724

for i = 1:10,
  P = Q + G'*P*G - G'*P*H*inv(R+H'*P*H)*H'*P*G;
end
P

P =

     1.0252    -0.0189
    -0.0189     0.5724

% ***** P matrix stays constant.  Thus steady state has been
% reached.  The steady-state P matrix is *****

P

P =

     1.0252    -0.0189
    -0.0189     0.5724
```

```
% ***** Optimal feedback gain matrix K is obtained from *****

K = inv(R + H'*P*H)*H'*P*G

K =

    0.0786      0.0865
```

with the boundary condition

$$\mathbf{P} = \begin{bmatrix} 0 & 0 \\ 0 & 0 \end{bmatrix}$$

until the solution reaches steady state. Then MATLAB Program 8–2 computes the optimal feedback gain matrix **K** using the following equation:

$$\mathbf{K} = (\mathbf{R} + \mathbf{H'PH})^{-1}\mathbf{H'PG}$$

Liapunov Approach to the Solution of the Steady-State Quadratic Optimal Regulator Problem.   In what follows we shall present the Liapunov approach to the solution of the parameter optimization problem and the steady-state quadratic optimal regulator problem. As we shall see, there is a direct relationship between Liapunov functions and quadratic performance indexes.

Let us consider the system

$$\mathbf{x}(k + 1) = \mathbf{Gx}(k) \tag{8–84}$$

where matrix **G** involves one or more adjustable parameters and all eigenvalues of **G** lie inside the unit circle, or the origin $\mathbf{x} = \mathbf{0}$ is asymptotically stable. Let us assume that we desire to minimize the following performance index by adjusting the parameter (or parameters):

$$J = \frac{1}{2}\sum_{k=0}^{\infty} \mathbf{x}^*(k)\mathbf{Qx}(k) \tag{8–85}$$

where **Q** is a positive definite or positive semidefinite Hermitian (or real symmetric) matrix. We shall show that a Liapunov function can be utilized for solving this problem.

For the system of Equation (8–84), a Liapunov function may be given by

$$V(\mathbf{x}(k)) = \mathbf{x}^*(k)\mathbf{Px}(k)$$

where **P** is a positive definite Hermitian (or real symmetric) matrix and

$$\Delta V(\mathbf{x}(k)) = V(\mathbf{x}(k + 1)) - V(\mathbf{x}(k))$$

$$= \mathbf{x}^*(k + 1)\mathbf{Px}(k + 1) - \mathbf{x}^*(k)\mathbf{Px}(k)$$

Let us set

$$\mathbf{x}^*(k)\mathbf{Qx}(k) = -[\mathbf{x}^*(k + 1)\mathbf{Px}(k + 1) - \mathbf{x}^*(k)\mathbf{Px}(k)] \tag{8–86}$$

Notice that Equation (8–86) can be rewritten as follows:

$$\mathbf{x}^*(k)\mathbf{Q}\mathbf{x}(k) = -\{[\mathbf{G}\mathbf{x}(k)]^*\mathbf{P}[\mathbf{G}\mathbf{x}(k)] - \mathbf{x}^*(k)\mathbf{P}\mathbf{x}(k)\}$$

$$= -\mathbf{x}^*(k)[\mathbf{G}^*\mathbf{P}\mathbf{G} - \mathbf{P}]\mathbf{x}(k)$$

By the second method of Liapunov, we know that for a given matrix $\mathbf{Q}$ there exists a positive definite matrix $\mathbf{P}$, since matrix $\mathbf{G}$ is stable, such that

$$\mathbf{G}^*\mathbf{P}\mathbf{G} - \mathbf{P} = -\mathbf{Q} \qquad (8\text{-}87)$$

Hence, we can determine the elements of $\mathbf{P}$ from this equation.

The performance index J can be evaluated as follows:

$$J = \frac{1}{2}\sum_{k=0}^{\infty} \mathbf{x}^*(k)\mathbf{Q}\mathbf{x}(k) = \frac{1}{2}\sum_{k=0}^{\infty} [\mathbf{x}^*(k)\mathbf{P}\mathbf{x}(k) - \mathbf{x}^*(k+1)\mathbf{P}\mathbf{x}(k+1)]$$

$$= \frac{1}{2}\mathbf{x}^*(0)\mathbf{P}\mathbf{x}(0) \qquad (8\text{-}88)$$

where $\mathbf{P}$ is a function of the adjustable parameter(s). In obtaining Equation (8–88) we used the condition that $\mathbf{x}(\infty) \to \mathbf{0}$, since all eigenvalues of $\mathbf{G}$ lie inside the unit circle. Thus, the performance index J can be obtained in terms of the initial state $\mathbf{x}(0)$ and matrix $\mathbf{P}$, which is related to matrices $\mathbf{G}$ and $\mathbf{Q}$ by Equation (8–87). Minimization of the performance index J can be accomplished by minimizing $\mathbf{x}^*(0)\mathbf{P}\mathbf{x}(0)$ with respect to the parameter in question.

It is important to note that the optimal value of the parameter depends, in general, upon the initial condition $\mathbf{x}(0)$. However, if $\mathbf{x}(0)$ involves only one nonzero component, for example, if $x_1(0) \neq 0$ and the other initial conditions are zero, then the optimal value of the parameter does not depend on the numerical value of $x_1(0)$.

Liapunov Approach to the Solution of the Steady-State Quadratic Optimal Control Problem. We shall now consider the optimal control problem where, given the plant equation

$$\mathbf{x}(k+1) = \mathbf{G}\mathbf{x}(k) + \mathbf{H}\mathbf{u}(k) \qquad (8\text{-}89)$$

we wish to determine the matrix $\mathbf{K}$ of the optimal control law

$$\mathbf{u}(k) = -\mathbf{K}\mathbf{x}(k) \qquad (8\text{-}90)$$

such that the performance index

$$J = \frac{1}{2}\sum_{k=0}^{\infty} [\mathbf{x}^*(k)\mathbf{Q}\mathbf{x}(k) + \mathbf{u}^*(k)\mathbf{R}\mathbf{u}(k)] \qquad (8\text{-}91)$$

is minimized, where $\mathbf{Q}$ is a positive definite or positive semidefinite Hermitian (or real symmetric) matrix and $\mathbf{R}$ is a positive definite Hermitian (or real symmetric) matrix.

Substituting Equation (8–90) into Equation (8–89), we obtain

$$\mathbf{x}(k+1) = \mathbf{G}\mathbf{x}(k) - \mathbf{H}\mathbf{K}\mathbf{x}(k) = (\mathbf{G} - \mathbf{H}\mathbf{K})\mathbf{x}(k) \qquad (8\text{-}92)$$

Substituting Equation (8–90) into Equation (8–91) yields

$$J = \frac{1}{2}\sum_{k=0}^{\infty}[\mathbf{x}^*(k)\mathbf{Q}\mathbf{x}(k) + \mathbf{x}^*(k)\mathbf{K}^*\mathbf{R}\mathbf{K}\mathbf{x}(k)]$$

$$= \frac{1}{2}\sum_{k=0}^{\infty}\mathbf{x}^*(k)(\mathbf{Q} + \mathbf{K}^*\mathbf{R}\mathbf{K})\mathbf{x}(k) \tag{8–93}$$

In the following analysis, we assume that the matrix $\mathbf{G} - \mathbf{HK}$ is stable, or that the eigenvalues of $\mathbf{G} - \mathbf{HK}$ lie inside the unit circle. (If the system is completely state controllable and observable, it can be proved that $\mathbf{G} - \mathbf{HK}$ is a stable matrix. Refer to Problem B–8–6.) Then a Liapunov function exists that is positive definite and whose derivative is negative definite. Following the discussion given in solving the parameter optimization problem, we set

$$\mathbf{x}^*(k)(\mathbf{Q} + \mathbf{K}^*\mathbf{R}\mathbf{K})\mathbf{x}(k) = -[\mathbf{x}^*(k + 1)\mathbf{P}\mathbf{x}(k + 1) - \mathbf{x}^*(k)\mathbf{P}\mathbf{x}(k)] \tag{8–94}$$

By referring to Equation (8–92), Equation (8–94) can be modified to

$$\mathbf{x}^*(k)(\mathbf{Q} + \mathbf{K}^*\mathbf{R}\mathbf{K})\mathbf{x}(k) = -[(\mathbf{G} - \mathbf{HK})\mathbf{x}(k)]^*\mathbf{P}[(\mathbf{G} - \mathbf{HK})\mathbf{x}(k)] + \mathbf{x}^*(k)\mathbf{P}\mathbf{x}(k)$$

$$= -\mathbf{x}^*(k)[(\mathbf{G} - \mathbf{HK})^*\mathbf{P}(\mathbf{G} - \mathbf{HK}) - \mathbf{P}]\mathbf{x}(k) \tag{8–95}$$

Comparing the two sides of Equation (8–95) and noting that this equation must hold true for any $\mathbf{x}(k)$, we require that

$$\mathbf{Q} + \mathbf{K}^*\mathbf{R}\mathbf{K} = -(\mathbf{G} - \mathbf{HK})^*\mathbf{P}(\mathbf{G} - \mathbf{HK}) + \mathbf{P} \tag{8–96}$$

Note that by the second method of Liapunov, for a stable matrix $\mathbf{G} - \mathbf{HK}$, there exists a positive definite matrix $\mathbf{P}$ that satisfies Equation (8–96).

Equation (8–96) can be modified as follows:

$$\mathbf{Q} + \mathbf{K}^*\mathbf{R}\mathbf{K} + (\mathbf{G}^* - \mathbf{K}^*\mathbf{H}^*)\mathbf{P}(\mathbf{G} - \mathbf{HK}) - \mathbf{P} = 0$$

or

$$\mathbf{Q} + \mathbf{G}^*\mathbf{P}\mathbf{G} - \mathbf{P} + \mathbf{K}^*(\mathbf{R} + \mathbf{H}^*\mathbf{P}\mathbf{H})\mathbf{K} - (\mathbf{K}^*\mathbf{H}^*\mathbf{P}\mathbf{G} + \mathbf{G}^*\mathbf{P}\mathbf{H}\mathbf{K}) = 0$$

This last equation can further be modified as follows:

$$\mathbf{Q} + \mathbf{G}^*\mathbf{P}\mathbf{G} - \mathbf{P} + [(\mathbf{R} + \mathbf{H}^*\mathbf{P}\mathbf{H})^{1/2}\mathbf{K} - (\mathbf{R} + \mathbf{H}^*\mathbf{P}\mathbf{H})^{-1/2}\mathbf{H}^*\mathbf{P}\mathbf{G}]^* \cdot [(\mathbf{R} + \mathbf{H}^*\mathbf{P}\mathbf{H})^{1/2}\mathbf{K}$$

$$- (\mathbf{R} + \mathbf{H}^*\mathbf{P}\mathbf{H})^{-1/2}\mathbf{H}^*\mathbf{P}\mathbf{G}] - \mathbf{G}^*\mathbf{P}\mathbf{H}(\mathbf{R} + \mathbf{H}^*\mathbf{P}\mathbf{H})^{-1}\mathbf{H}^*\mathbf{P}\mathbf{G} = 0 \tag{8–97}$$

Matrix $\mathbf{K}$ that minimizes J can be obtained by minimizing the left-hand side of Equation (8–97) with respect to $\mathbf{K}$. (See Problem A–8–5.) Since

$$[(\mathbf{R} + \mathbf{H}^*\mathbf{P}\mathbf{H})^{1/2}\mathbf{K} - (\mathbf{R} + \mathbf{H}^*\mathbf{P}\mathbf{H})^{-1/2}\mathbf{H}^*\mathbf{P}\mathbf{G}]^*[(\mathbf{R} + \mathbf{H}^*\mathbf{P}\mathbf{H})^{1/2}\mathbf{K}$$

$$- (\mathbf{R} + \mathbf{H}^*\mathbf{P}\mathbf{H})^{-1/2}\mathbf{H}^*\mathbf{P}\mathbf{G}]$$

is nonnegative, the minimum occurs when it is zero, or when

$$(\mathbf{R} + \mathbf{H}^*\mathbf{P}\mathbf{H})^{1/2}\mathbf{K} = (\mathbf{R} + \mathbf{H}^*\mathbf{P}\mathbf{H})^{-1/2}\mathbf{H}^*\mathbf{P}\mathbf{G}$$

Hence, we obtain

$$\mathbf{K} = (\mathbf{R} + \mathbf{H}^*\mathbf{P}\mathbf{H})^{-1}\mathbf{H}^*\mathbf{P}\mathbf{G} \tag{8–98}$$

Substitution of Equation (8–98) into Equation (8–97) gives

$$\mathbf{P} = \mathbf{Q} + \mathbf{G^*PG} - \mathbf{G^*PH}(\mathbf{R} + \mathbf{H^*PH})^{-1}\mathbf{H^*PG} \tag{8–99}$$

Matrix $\mathbf{P}$ must satisfy Equation (8–99), which is the same as Equation (8–75). Equation (8–99) can be modified to read

$$\mathbf{P} = \mathbf{Q} + \mathbf{G^*P}[\mathbf{I} - \mathbf{H}(\mathbf{I} + \mathbf{R}^{-1}\mathbf{H^*PH})^{-1}\mathbf{R}^{-1}\mathbf{H^*P}]\mathbf{G} \tag{8–100}$$

By use of the matrix inversion lemma

$$(\mathbf{I} + \mathbf{HR}^{-1}\mathbf{H^*P})^{-1} = \mathbf{I} - \mathbf{H}(\mathbf{I} + \mathbf{R}^{-1}\mathbf{H^*PH})^{-1}\mathbf{R}^{-1}\mathbf{H^*P}$$

Equation (8–100) may be modified to

$$\mathbf{P} = \mathbf{Q} + \mathbf{G^*P}(\mathbf{I} + \mathbf{HR}^{-1}\mathbf{H^*P})^{-1}\mathbf{G} \tag{8–101}$$

Matrix $\mathbf{P}$ may be determined from Equation (8–101).

Finally, the minimum value of J can be obtained as follows. Referring to Equations (8–93) and (8–94) and noting that $\mathbf{x}(\infty) = \mathbf{0}$, we obtain the minimum value of the performance index J as follows:

$$
\begin{aligned}
J_{\min} &= \frac{1}{2}\sum_{k=0}^{\infty}\mathbf{x}^*(k)(\mathbf{Q} + \mathbf{K^*RK})\mathbf{x}(k) \\
&= \frac{1}{2}\sum_{k=0}^{\infty}[\mathbf{x}^*(k)\mathbf{Px}(k) - \mathbf{x}^*(k+1)\mathbf{Px}(k+1)] \\
&= \frac{1}{2}\mathbf{x}^*(0)\mathbf{Px}(0)
\end{aligned}
$$

Example 8–3

Consider the system

$$
\begin{bmatrix} x_1(k+1) \\ x_2(k+1) \end{bmatrix} = \begin{bmatrix} 1 & 1 \\ a & -1 \end{bmatrix}\begin{bmatrix} x_1(k) \\ x_2(k) \end{bmatrix}, \qquad \begin{bmatrix} x_1(0) \\ x_2(0) \end{bmatrix} = \begin{bmatrix} 1 \\ 0 \end{bmatrix}
$$

where $-0.25 \le a < 0$. We desire to determine an optimal value of a that will minimize the following performance index:

$$J = \frac{1}{2}\sum_{k=0}^{\infty}\mathbf{x}^*(k)\mathbf{Qx}(k)$$

where $\mathbf{Q} = \mathbf{I}$.

From Equation (8–88), the performance index J in terms of the system parameter a is given by

$$J = \frac{1}{2}\mathbf{x}^*(0)\mathbf{Px}(0)$$

where $\mathbf{P}$ involves the system parameter a. We now determine $\mathbf{P}$ from Equation (8–87):

$$\mathbf{G^*PG} - \mathbf{P} = -\mathbf{Q}$$

or

$$
\begin{bmatrix} 1 & a \\ 1 & -1 \end{bmatrix}\begin{bmatrix} p_{11} & p_{12} \\ p_{12} & p_{22} \end{bmatrix}\begin{bmatrix} 1 & 1 \\ a & -1 \end{bmatrix} - \begin{bmatrix} p_{11} & p_{12} \\ p_{12} & p_{22} \end{bmatrix} = -\begin{bmatrix} 1 & 0 \\ 0 & 1 \end{bmatrix}
$$

which can be simplified to

$$\begin{bmatrix} 2ap_{12} + a^2 p_{22} & p_{11} + (a-2)p_{12} - ap_{22} \\ p_{11} + (a-2)p_{12} - ap_{22} & p_{11} - 2p_{12} \end{bmatrix} = \begin{bmatrix} -1 & 0 \\ 0 & -1 \end{bmatrix}$$

This last equation results in the following three equations:

$$2ap_{12} + a^2 p_{22} = -1$$

$$p_{11} + (a-2)p_{12} - ap_{22} = 0$$

$$p_{11} - 2p_{12} = -1$$

Solving these three equations for the p_{ij}'s, we obtain

$$P = \begin{bmatrix} -\dfrac{1 + 0.5a^2}{a(1 + 0.5a)} & \dfrac{0.5(a-1)}{a(1 + 0.5a)} \\ \dfrac{0.5(a-1)}{a(1 + 0.5a)} & -\dfrac{1.5}{a(1 + 0.5a)} \end{bmatrix}$$

Since $-0.25 \le a < 0$, **P** is positive definite.

The performance index J becomes

$$J = \frac{1}{2}\mathbf{x}^*(0)\mathbf{P}\mathbf{x}(0) = \frac{1}{2}[1 \quad 0]\begin{bmatrix} p_{11} & p_{12} \\ p_{12} & p_{22} \end{bmatrix}\begin{bmatrix} 1 \\ 0 \end{bmatrix} = \frac{1}{2}p_{11}$$

$$= -\frac{1 + 0.5a^2}{2a(1 + 0.5a)}$$

The minimum of J occurs at the end point $a = -0.25$. Thus, the minimum value of J is found to be

$$J_{min} = -\frac{1 + 0.5(-0.25)^2}{2(-0.25)(1 - 0.5 \times 0.25)} = 2.3571$$

Example 8–4

Consider the system

$$\begin{bmatrix} x_1(k+1) \\ x_2(k+1) \end{bmatrix} = \begin{bmatrix} 1 & 1 \\ 1 & 0 \end{bmatrix}\begin{bmatrix} x_1(k) \\ x_2(k) \end{bmatrix} + \begin{bmatrix} 1 \\ 0 \end{bmatrix}[u(k)], \qquad \begin{bmatrix} x_1(0) \\ x_2(0) \end{bmatrix} = \begin{bmatrix} 1 \\ 0 \end{bmatrix}$$

and the performance index

$$J = \frac{1}{2}\sum_{k=0}^{\infty}[\mathbf{x}^*(k)\mathbf{Q}\mathbf{x}(k) + u^*(k)Ru(k)]$$

where $\mathbf{Q} = \mathbf{I}$ and $R = 1$. From Equation (8–98), the optimal control law that minimizes the performance index J is given by

$$u(k) = -\mathbf{K}\mathbf{x}(k) = -(R + \mathbf{H}^*\mathbf{P}\mathbf{H})^{-1}\mathbf{H}^*\mathbf{P}\mathbf{G}\mathbf{x}(k)$$

where matrix **P** may be obtained from Equation (8–101):

$$\mathbf{P} = \mathbf{Q} + \mathbf{G}^*\mathbf{P}(\mathbf{I} + \mathbf{H}R^{-1}\mathbf{H}^*\mathbf{P})^{-1}\mathbf{G} \tag{8–102}$$

Let us determine matrix **P** for the present system.

Noting that

$$\mathbf{G} = \begin{bmatrix} 1 & 1 \\ 1 & 0 \end{bmatrix}$$

which is nonsingular, we may modify Equation (8–102) to read

$$(\mathbf{P} - \mathbf{Q})\mathbf{G}^{-1}(\mathbf{I} + \mathbf{H}\mathbf{R}^{-1}\mathbf{H}*\mathbf{P}) = \mathbf{G}*\mathbf{P} \qquad (8\text{–}103)$$

Matrix **P** in this problem is a real symmetric matrix. Hence, Equation (8–103) can be written as follows:

$$\left\{ \begin{bmatrix} p_{11} & p_{12} \\ p_{12} & p_{22} \end{bmatrix} - \begin{bmatrix} 1 & 0 \\ 0 & 1 \end{bmatrix} \right\} \begin{bmatrix} 0 & 1 \\ 1 & -1 \end{bmatrix} \left\{ \begin{bmatrix} 1 & 0 \\ 0 & 1 \end{bmatrix} + \begin{bmatrix} 1 \\ 0 \end{bmatrix} [1][1 \ 0] \begin{bmatrix} p_{11} & p_{12} \\ p_{12} & p_{22} \end{bmatrix} \right\} = \begin{bmatrix} 1 & 1 \\ 1 & 0 \end{bmatrix} \begin{bmatrix} p_{11} & p_{12} \\ p_{12} & p_{22} \end{bmatrix}$$

or

$$\begin{bmatrix} p_{12}(1 + p_{11}) & p_{12}^2 + p_{11} - 1 - p_{12} \\ (p_{22} - 1)(1 + p_{11}) & (p_{22} - 1)p_{12} + p_{12} - p_{22} + 1 \end{bmatrix} = \begin{bmatrix} p_{11} + p_{12} & p_{12} + p_{22} \\ p_{11} & p_{12} \end{bmatrix}$$

This last equation yields four scalar equations. Note, however, that only three of them are linearly independent. The four scalar equations are as follows:

$$p_{12}(1 + p_{11}) = p_{11} + p_{12}$$

$$p_{12}^2 + p_{11} - 1 - p_{12} = p_{12} + p_{22}$$

$$(p_{22} - 1)(1 + p_{11}) = p_{11}$$

$$(p_{22} - 1)p_{12} + p_{12} - p_{22} + 1 = p_{12}$$

From the first of these four equations, we obtain

$$p_{12} = 1 \qquad (8\text{–}104)$$

From the second of the four equations, we have

$$p_{22} = p_{11} - 2 \qquad (8\text{–}105)$$

Then the third of the four equations gives

$$p_{11}^2 - 3p_{11} - 3 = 0 \qquad (8\text{–}106)$$

[By substituting Equations (8–104) and (8–105) into the last of the four equations, we find that it is always satisfied.] By solving Equation (8–106), we find

$$p_{11} = 3.7913 \qquad \text{or} \qquad -0.7913$$

Since matrix **P** must be positive definite, we choose $p_{11} = 3.7913$. Then

$$p_{22} = p_{11} - 2 = 3.7913 - 2 = 1.7913$$

Consequently, matrix **P** is found to be as follows:

$$\mathbf{P} = \begin{bmatrix} 3.7913 & 1.0000 \\ 1.0000 & 1.7913 \end{bmatrix}$$

8–4 QUADRATIC OPTIMAL CONTROL OF A SERVO SYSTEM

In this section we shall discuss quadratic optimal control of a servo system. The plant we consider here is the inverted pendulum system shown in Figure 8–3. We shall design a digital control scheme for this inverted pendulum control system.

The inverted pendulum is unstable in that it may fall over any time in any direction unless a suitable control force is applied. Here we consider only a two-dimensional problem in which the pendulum moves only in the plane of the page.

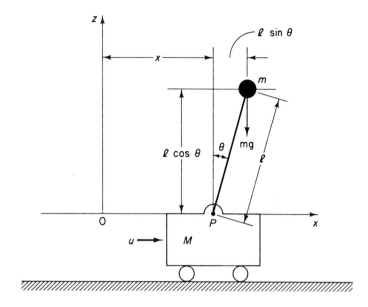

Figure 8-3   Inverted pendulum system.

Assume that the pendulum mass is concentrated at the end of the rod, as shown in the figure. (The rod is massless.) The control force u is applied to the cart.

It is desired to keep the inverted pendulum upright as much as possible and yet control the position of the cart, for instance, by moving the cart in a step fashion. To control the position of the cart, we need to build a type 1 servo system. The inverted-pendulum system mounted on a cart does not have an integrator. Therefore, we feed the position signal y (which indicates the position of the cart) back to the input and insert an integrator in the feed-forward path, as shown in Figure 8-4. (Note that other configurations are possible.) We choose the sampling period T to be 0.1 sec.

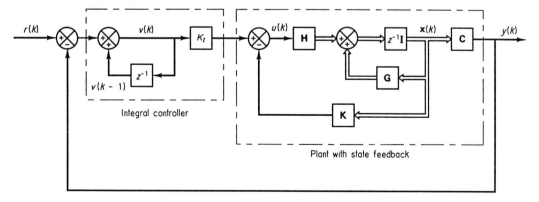

Figure 8-4   Block diagram of servo system.

The control system to be designed here will include state feedback and an integrator in the closed loop. Design variables are gain constant K_I and feedback gain matrix $\mathbf{K}$. Therefore, designing the digital control scheme means determining constant K_I and matrix $\mathbf{K}$. We shall solve this design problem with MATLAB. We shall obtain the unit-step response of the designed system with MATLAB.

In solving this design problem, we shall define state variables x_1, x_2, x_3, and x_4 as follows:

$$x_1 = \theta$$

$$x_2 = \dot{\theta}$$

$$x_3 = x$$

$$x_4 = \dot{x}$$

In this system we would like to keep angle θ as small as possible when the cart is moved in the step fashion. We consider the displacement of the cart as the output of the system. Thus, the output equation becomes

$$y = [0 \quad 0 \quad 1 \quad 0] \begin{bmatrix} x_1 \\ x_2 \\ x_3 \\ x_4 \end{bmatrix}$$

We shall assume the following numerical values for M, m, and l.

$$M = 2 \text{ kg}, \qquad m = 0.1 \text{ kg}, \qquad l = 0.5 \text{ m}$$

In designing the control system, we shall use the discretized model. The discretized state and output equations for the plant can be derived as shown next. (Refer to Problem A–8–10 for the derivation of these equations.)

$$\mathbf{x}(k + 1) = \mathbf{G}\mathbf{x}(k) + \mathbf{H}u(k)$$

$$y(k) = \mathbf{C}\mathbf{x}(k) + Du(k)$$

where

$$\mathbf{G} = \begin{bmatrix} 1.1048 & 0.1035 & 0 & 0 \\ 2.1316 & 1.1048 & 0 & 0 \\ -0.0025 & -0.0001 & 1 & 0.1 \\ -0.0508 & -0.0025 & 0 & 1 \end{bmatrix}, \quad \mathbf{H} = \begin{bmatrix} -0.0051 \\ -0.1035 \\ 0.0025 \\ 0.0501 \end{bmatrix}, \quad \mathbf{C} = [0 \ 0 \ 1 \ 0], \quad D = [0]$$

From Figure 8–4 the state-space representation for the entire control system is given by

$$\mathbf{x}(k + 1) = \mathbf{G}\mathbf{x}(k) + \mathbf{H}u(k)$$

$$y(k) = \mathbf{C}\mathbf{x}(k)$$

$$v(k) = v(k - 1) + r(k) - y(k)$$

$$u(k) = -\mathbf{K}\mathbf{x}(k) + K_I v(k)$$

where

$$\mathbf{K} = [k_1 \quad k_2 \quad k_3 \quad k_4]$$

Since

$$v(k + 1) = v(k) + r(k + 1) - y(k + 1)$$
$$= v(k) + r(k + 1) - \mathbf{C}[\mathbf{G}\mathbf{x}(k) + \mathbf{H}u(k)]$$
$$= -\mathbf{C}\mathbf{G}\mathbf{x}(k) + v(k) - \mathbf{C}\mathbf{H}u(k) + r(k + 1)$$

we have

$$\begin{bmatrix} \mathbf{x}(k + 1) \\ v(k + 1) \end{bmatrix} = \begin{bmatrix} \mathbf{G} & \mathbf{0} \\ -\mathbf{C}\mathbf{G} & 1 \end{bmatrix} \begin{bmatrix} \mathbf{x}(k) \\ v(k) \end{bmatrix} + \begin{bmatrix} \mathbf{H} \\ -\mathbf{C}\mathbf{H} \end{bmatrix} u(k) + \begin{bmatrix} \mathbf{0} \\ 1 \end{bmatrix} r(k + 1)$$

Let us assume that the input r is a step function, or

$$r(k) = r(k + 1) = r$$

Then, as k approaches infinity,

$$\begin{bmatrix} \mathbf{x}(\infty) \\ v(\infty) \end{bmatrix} = \begin{bmatrix} \mathbf{G} & \mathbf{0} \\ -\mathbf{C}\mathbf{G} & 1 \end{bmatrix} \begin{bmatrix} \mathbf{x}(\infty) \\ v(\infty) \end{bmatrix} + \begin{bmatrix} \mathbf{H} \\ -\mathbf{C}\mathbf{H} \end{bmatrix} u(\infty) + \begin{bmatrix} \mathbf{0} \\ 1 \end{bmatrix} r(\infty)$$

Define

$$\mathbf{x}_e(k) = \mathbf{x}(k) - \mathbf{x}(\infty)$$
$$v_e(k) = v(k) - v(\infty)$$

Then the error equation becomes

$$\begin{bmatrix} \mathbf{x}_e(k + 1) \\ v_e(k + 1) \end{bmatrix} = \begin{bmatrix} \mathbf{G} & \mathbf{0} \\ -\mathbf{C}\mathbf{G} & 1 \end{bmatrix} \begin{bmatrix} \mathbf{x}_e(k) \\ v_e(k) \end{bmatrix} + \begin{bmatrix} \mathbf{H} \\ -\mathbf{C}\mathbf{H} \end{bmatrix} u_e(k)$$

Note that

$$u_e(k) = -\mathbf{K}\mathbf{x}_e(k) + K_I v_e(k) = -[\mathbf{K} \quad -K_I] \begin{bmatrix} \mathbf{x}_e(k) \\ v_e(k) \end{bmatrix}$$

Now define

$$\hat{\mathbf{G}} = \begin{bmatrix} \mathbf{G} & \mathbf{0} \\ -\mathbf{C}\mathbf{G} & 1 \end{bmatrix}, \quad \hat{\mathbf{H}} = \begin{bmatrix} \mathbf{H} \\ -\mathbf{C}\mathbf{H} \end{bmatrix}, \quad \hat{\mathbf{K}} = [\mathbf{K} \quad -K_I], \quad w(k) = u_e(k)$$

$$\boldsymbol{\xi}(k) = \begin{bmatrix} \mathbf{x}_e(k) \\ v_e(k) \end{bmatrix} = \begin{bmatrix} x_{1e}(k) \\ x_{2e}(k) \\ x_{3e}(k) \\ x_{4e}(k) \\ x_{5e}(k) \end{bmatrix}$$

where $x_{5e}(k) = v_e(k)$. Then we have

$$\boldsymbol{\xi}(k + 1) = \hat{\mathbf{G}}\boldsymbol{\xi}(k) + \hat{\mathbf{H}}w(k)$$
$$w(k) = -\hat{\mathbf{K}}\boldsymbol{\xi}(k)$$

It is noted that since the system is continuous time, a continuous-time quadratic performance index could be considered. However, discretization of the continuous-time quadratic performance index generates a cross term involving $\boldsymbol{\xi}$ and w. (See

Section 8–3.) To simplify the design process, it is better to define a discrete-time quadratic performance index. Our problem, therefore, becomes that of determining matrix $\hat{\mathbf{K}}$ such that the following quadratic performance index is minimized:

$$J = \frac{1}{2}\sum_{k=0}^{\infty}[\xi'\mathbf{Q}\xi + w'Rw]$$

where $\mathbf{Q}$ and R should be chosen properly so that the response of the system is acceptable. (The purpose of using the quadratic performance index is to assure the stability of the system.)

Let us choose $\mathbf{Q}$ and R as follows:

$$\mathbf{Q} = \begin{bmatrix} 10 & 0 & 0 & 0 & 0 \\ 0 & 1 & 0 & 0 & 0 \\ 0 & 0 & 100 & 0 & 0 \\ 0 & 0 & 0 & 1 & 0 \\ 0 & 0 & 0 & 0 & 1 \end{bmatrix}, \qquad R = [1]$$

Our emphasis is on state variables x_{3e} and x_{1e}. (Note that many different sets of $\mathbf{Q}$ R may be used.) In the MATLAB program for solving this problem, we shall use the notations

$$G1 = \hat{\mathbf{G}}, \qquad H1 = \hat{\mathbf{H}}, \qquad KK = \hat{\mathbf{K}}$$

MATLAB Program 8–3 yields the solution $\mathbf{P}$ of the steady-state Riccati equation, feedback gain matrix $\mathbf{K}$, and integral gain constant K_I.

MATLAB Program 8–3

```
% ---------- Design of an inverted pendulum control system
% based on minimization of a quadratic performance index ----------

% ***** The following program solves steady-state Riccati
% equation and gives optimal feedback gain matrix K *****

% ***** Enter matrices G, H, C, and D *****

G = [1.1048    0.1035    0    0
     2.1316    1.1048    0    0
    -0.0025   -0.0001    1    0.1
    -0.0508   -0.0025    0    1];
H = [-0.0051
     -0.1035
      0.0025
      0.0501];
C = [0  0  1  0];
D = [0];

% ***** Enter matrices G1, H1, Q and R *****
```

```
G1 = [G   zeros(4,1);-C*G  1];
H1 = [H;-C*H];
Q = [10  0    0  0  0
      0  1    0  0  0
      0  0  100  0  0
      0  0    1  0  0
      0  0    0  0  1];
R = [1];

% ***** Start solving steady-state Riccati equation
% for P with P = diag(0,4) *****

P = diag(0,4);
P = Q + G1'*P*G1 - G1'*P*H1*inv(R+H1'*P*H1)*H1'*P*G1;

% ***** Check solution P every 20 steps of iteration.
% Stop iteration when P stays constant *****

for i = 1:20,
 P = Q + G1'*P*G1 - G1'*P*H1*inv(R+H1'*P*H1)*H1'*P*G1;
end
P


P =


  1.0e + 003 *


    9.6887    2.1677    3.4319    2.4341   -0.1741
    2.1677    0.4876    0.7743    0.5490   -0.0393
    3.4319    0.7743    2.2988    1.1788   -0.1312
    2.4341    0.5490    1.1788    0.7824   -0.0625
   -0.1741   -0.0393   -0.1312   -0.0625    0.0185

for i = 1:20,
 P = Q + G1'*P*G1 - G1'*P*H1*inv(R+H1'*P*H1)*H1'*P*G1;
end
P


P =


  1.0e + 004 *


    1.0707    0.2397    0.3996    0.2772   -0.0220
    0.2397    0.0539    0.0902    0.0625   -0.0050
    0.3996    0.0902    0.2617    0.1367   -0.0158
    0.2772    0.0625    0.1367    0.0895   -0.0078
   -0.0220   -0.0050   -0.0158   -0.0078    0.0021

for i = 1:20,
 P = Q + G1'*P*G1 - G1'*P*H1*inv(R+H1'*P*H1)*H1'*P*G1;
end
P
```

```
P =

   1.0e + 004 *

   1.0724    0.2401    0.4006    0.2778   -0.0221
   0.2401    0.0540    0.0904    0.0627   -0.0050
   0.4006    0.0904    0.2623    0.1371   -0.0158
   0.2778    0.0627    0.1371    0.0897   -0.0078
  -0.0221   -0.0050   -0.0158   -0.0078    0.0021

for i = 1:20,
 P = Q + G1'*P*G1 - G1'*P*H1*inv(R+H1'*P*H1)*H1'*P*G1;
end
P

P =

   1.0e + 004 *

   1.0724    0.2401    0.4006    0.2778   -0.0221
   0.2401    0.0540    0.0904    0.0627   -0.0050
   0.4006    0.0904    0.2623    0.1371   -0.0158
   0.2778    0.0627    0.1371    0.0897   -0.0078
  -0.0221   -0.0050   -0.0158   -0.0078    0.0021

% ***** P matrix stays constant. Thus steady state has
% been reached.  The steady-state P matrix is *****

P

P =

   1.0e + 004 *

   1.0724    0.2401    0.4006    0.2778   -0.0221
   0.2401    0.0540    0.0904    0.0627   -0.0050
   0.4006    0.0904    0.2623    0.1371   -0.0158
   0.2778    0.0627    0.1371    0.0897   -0.0078
  -0.0221   -0.0050   -0.0158   -0.0078    0.0021

% ***** Optimal feedback gain matrix KK is obtained from *****

KK = inv(R + H1'*P*H1)*H1'*P*G1

KK =

  -64.9346  -14.4819  -10.8475   -9.2871   0.5189

K = [KK(1)  KK(2)  KK(3)  KK(4)]

K =

  -64.9346   -14.4819   -10.8475   -9.2871
```

```
KI = -KK(5)

KI =

    -0.5189
```

We thus determined feedback gain matrix **K** and integral gain constant K_I as follows:

$$\mathbf{K} = [-64.9346 \quad -14.4819 \quad -10.8475 \quad -9.2871], \qquad K_I = -0.5189$$

Our mathematical design is completed.

Unit-Step Response of the Designed System.   To obtain the unit-step response, we proceed as follows: Since

$$\mathbf{x}(k + 1) = \mathbf{G}\mathbf{x}(k) + \mathbf{H}[-\mathbf{K}\mathbf{x}(k) + K_I v(k)]$$

$$= (\mathbf{G} - \mathbf{H}\mathbf{K})\mathbf{x}(k) + \mathbf{H}K_I v(k)$$

$$v(k + 1) = v(k) + r(k + 1) - y(k + 1)$$

$$= v(k) + r(k + 1) - \mathbf{C}[\mathbf{G}\mathbf{x}(k) + \mathbf{H}u(k)]$$

$$= (-\mathbf{C}\mathbf{G} + \mathbf{C}\mathbf{H}\mathbf{K})\mathbf{x}(k) + (1 - \mathbf{C}\mathbf{H}K_I)v(k) + r$$

we obtain

$$\begin{bmatrix} \mathbf{x}(k + 1) \\ v(k + 1) \end{bmatrix} = \begin{bmatrix} \mathbf{G} - \mathbf{H}\mathbf{K} & \mathbf{H}K_I \\ -\mathbf{C}\mathbf{G} + \mathbf{C}\mathbf{H}\mathbf{K} & 1 - \mathbf{C}\mathbf{H}K_I \end{bmatrix} \begin{bmatrix} \mathbf{x}(k) \\ v(k) \end{bmatrix} + \begin{bmatrix} \mathbf{0} \\ 1 \end{bmatrix} r \qquad (8\text{–}107)$$

$$y(k) = [\mathbf{C} \quad 0] \begin{bmatrix} \mathbf{x}(k) \\ v(k) \end{bmatrix} + [0]r \qquad (8\text{–}108)$$

where $r = 1$. To determine the unit-step response $y(k)$ (the cart position), first define

$$\mathbf{GG} = \begin{bmatrix} \mathbf{G} - \mathbf{H}\mathbf{K} & \mathbf{H}K_I \\ -\mathbf{C}\mathbf{G} + \mathbf{C}\mathbf{H}\mathbf{K} & 1 - \mathbf{C}\mathbf{H}K_I \end{bmatrix}$$

$$\mathbf{HH} = \begin{bmatrix} \mathbf{0} \\ 1 \end{bmatrix}$$

$$\mathbf{CC} = [\mathbf{C} \quad 0] = [0 \quad 0 \quad 1 \quad 0 \quad 0]$$

$$DD = [D] = [0]$$

and then convert the state-space equations [Equations (8–107) and (8–108)] into the pulse transfer function $Y(z)/R(z)$ by using the following MATLAB command:

$$[\text{num,den}] = \text{ss2tf(GG,HH,CC,DD)}$$

Then use the *filter* command:

$$y = \text{filter(num,den,r)}$$

where r = unit step input.

To obtain the response $x_1(k)$, we note that

$$x_1(k) = [1 \quad 0 \quad 0 \quad 0 \quad 0]\begin{bmatrix} \mathbf{x}(k) \\ v(k) \end{bmatrix}$$

Define

$$\mathbf{FF} = [1 \quad 0 \quad 0 \quad 0 \quad 0]$$

Then

$$x_1(k) = \mathbf{FF}\begin{bmatrix} \mathbf{x}(k) \\ v(k) \end{bmatrix} \tag{8-109}$$

Convert the state-space equations [Equations (8–107) and (8–109)] into the pulse transfer function $X_1(z)/R(z)$ using the command

$$[\text{num1,den1}] = \text{ss2tf(GG,HH,FF,DD)}$$

Then use the filter command

$$\text{x1} = \text{filter(num1,den1,r)}$$

Similarly, to obtain the response $x_2(k)$, we note that

$$x_2(k) = [0 \quad 1 \quad 0 \quad 0 \quad 0]\begin{bmatrix} \mathbf{x}(k) \\ v(k) \end{bmatrix} \tag{8-110}$$

Define

$$\mathbf{JJ} = [0 \quad 1 \quad 0 \quad 0 \quad 0]$$

Then convert the state-space equations [Equations (8–107) and (8–110)] into the pulse transfer function $X_2(z)/R(z)$ using the command

$$[\text{num2,den2}] = \text{ss2tf(GG,HH,JJ,DD)}$$

Then use the filter command

$$\text{x2} = \text{filter(num2,den2,r)}$$

Similarly, by defining

$$\mathbf{LL} = [0 \quad 0 \quad 0 \quad 1 \quad 0]$$
$$\mathbf{MM} = [0 \quad 0 \quad 0 \quad 0 \quad 1]$$

the responses $x_4(k)$ and $x_5(k) = v(k)$ can be obtained using the commands

$$[\text{num4,den4}] = \text{ss2tf(GG,HH,LL,DD)}$$
$$\text{x4} = \text{filter(num4,den4,r)}$$

and

$$[num5,den5] = ss2tf(GG,HH,MM,DD)$$

$$x5 = filter(num5,den5,r)$$

MATLAB Program 8–4 yields $y(k)$, $x_1(k)$, $x_2(k)$, $x_4(k)$, and $x_5(k)$ when the unit-step input ($r = 1$) is given.

```
MATLAB Program 8–4

% ---------- Step response of the designed system ----------

% ***** This program calculates the response of the system
% when subjected to a unit-step input. The values that are
% used for K and KI are computed in MATLAB Program 8–3. The
% response is obtained using the method to convert the discrete-
% time state-space equations into pulse transfer-function form. The
% response is then found with the conventional 'filter' command *****

% ***** Enter matrices K, KI, GG, HH, CC, FF, JJ, LL, MM, DD *****

K = [-64.9346  -14.4819  -10.8475  -9.2871];
KI = -0.5189;
GG = [G-H*K  H*KI;-C*G+C*H*K  1-C*H*KI];
HH = [0;0;0;0;1];
CC = [0  0  1  0  0];
FF = [1  0  0  0  0];
JJ  = [0  1  0  0  0];
LL  = [0  0  0  1  0];
MM = [0  0  0  0  1];
DD = [0];

% ***** To obtain y(k) convert state-space equations into pulse
% transfer function X3(z)/R(z) *****

[num,den] = ss2tf(GG,HH,CC,DD);

% ***** Enter command to obtain unit-step response *****

r = ones(1,101);
k = 0:100;
y = filter(num,den,r);
plot(k,y,'o',k,y,'-')
axis([0  100  -0.2  1.2]);
grid
title('Position of Cart : y(k) = x3(k)')
xlabel('k')
ylabel('y(k) = x3(k)')

% ***** To obtain x1(k) convert state-space equations into pulse
% transfer function X1(z)/R(z) *****

[num1,den1] = ss2tf(GG,HH,FF,DD);
```

```
% ***** Enter command to obtain unit-step response *****

x1 = filter(num1,den1,r);
plot(k,x1,'o',k,x1,'-')
axis([0  100  -0.1  0.2]);
grid
title('Angular Displacement Theta : x1(k)')
xlabel('k')
ylabel('x1(k)')

% ***** To obtain x2(k) convert state-space equations into pulse
% transfer function X2(z)/R(z) *****

[num2,den2] = ss2tf(GG,HH,JJ,DD);

% ***** Enter command to obtain unit-step response *****

x2 = filter(num2,den2,r);
plot(k,x2,'o',k,x2,'-')
axis([0  100  -0.5  0.5]);
grid
title('Angular Velocity Theta Dot : x2(k)')
xlabel('k')
ylabel('x2(k)')

% ***** To obtain x4(k) convert state-space equations into pulse
% transfer function X4(z)/R(z) *****

[num4,den4] = ss2tf(GG,HH,LL,DD);

% ***** Enter command to obtain unit-step response *****

x4 = filter(num4,den4,r);
plot(k,x4,'o',k,x4,'-')
axis([0  100  -0.5  1]);
grid
title('Velocity of Cart : x4(k)')
xlabel('k')
ylabel('x4(k)')

% ***** To obtain x5(k) convert state-space equations into pulse
% transfer function X5(z)/R(z) *****

[num5,den5] = ss2tf(GG,HH,MM,DD);

% ***** Enter command to obtain unit-step response *****

x5 = filter(num5,den5,r);
plot(k,x5,'o',k,x5,'-')
axis([0  100  -5  30]);
grid
title('Output of Integrator : x5(k) = v(k)')
xlabel('k')
ylabel('x5(k) = v(k)')
```

Based on MATLAB computations, the position of the cart [$y(k)$ versus k] can be obtained as shown in Figure 8–5. (Notice that the initial move of the cart is in the negative direction.) Figure 8–6 depicts the angular displacement of the pendu-

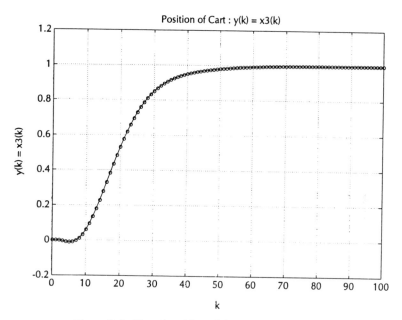

Figure 8–5   Plot of position of the cart $y(k)$ versus k.

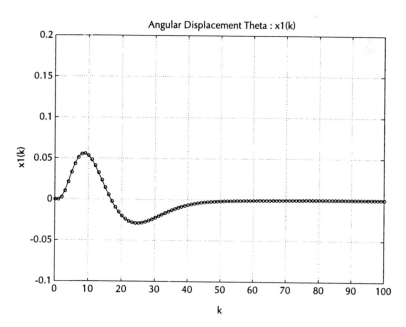

Figure 8–6   Plot of angular displacement $x_1(k)$ versus k.

lum, $x_1(k) = \theta(k)$, plotted versus k. Figure 8–7 shows the angular velocity of the pendulum, $x_2(k)$, plotted versus k. Figure 8–8 shows the velocity of the cart, $x_4(k)$, plotted versus k. The output of the integrator, $v(k)$ versus k, is shown in Figure 8–9.

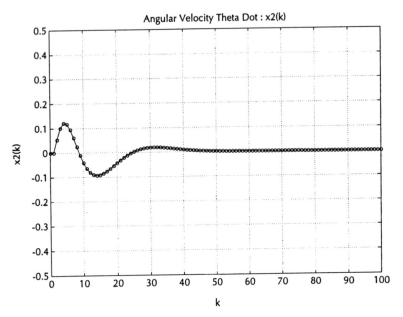

Figure 8–7   Plot of angular velocity $x_2(k)$ versus k.

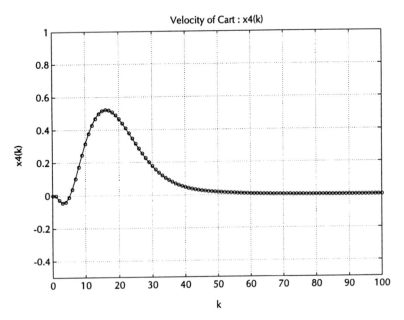

Figure 8–8   Plot of velocity of cart $x_4(k)$ versus k.

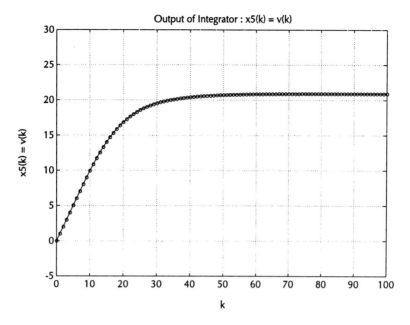

Figure 8–9   Plot of output of integrator $v(k)$ versus k.

Since in the present system the sampling period T is 0.1 sec, it takes approximately 6 sec to reach steady state.

EXAMPLE PROBLEMS AND SOLUTIONS

Problem A–8–1

Consider the discrete-time control system

$$\mathbf{x}(k + 1) = \mathbf{Gx}(k) + \mathbf{Hu}(k)$$

where

$$\mathbf{x}(k) = \text{state vector } (n\text{-vector})$$

$$\mathbf{u}(k) = \text{control vector } (r\text{-vector})$$

$$\mathbf{G} = n \times n \text{ nonsingular matrix}$$

$$\mathbf{H} = n \times r \text{ matrix}$$

We wish to find the optimal control vector that will minimize the following performance index:

$$J = \frac{1}{2}\mathbf{x}^*(N)\mathbf{Sx}(N) + \frac{1}{2}\sum_{k=0}^{N-1}[\mathbf{x}^*(k)\mathbf{Qx}(k) + \mathbf{u}^*(k)\mathbf{Ru}(k)]$$

where $\mathbf{Q}$ and $\mathbf{S}$ are $n \times n$ positive definite or positive semidefinite Hermitian matrices and $\mathbf{R}$ is an $r \times r$ positive definite Hermitian matrix.

In Section 8–2 we obtained the optimal control vector $\mathbf{u}(k)$ in the form given by Equation (8–26):

$$\mathbf{u}(k) = -\mathbf{R}^{-1}\mathbf{H}*(\mathbf{G}*)^{-1}[\mathbf{P}(k) - \mathbf{Q}]\mathbf{x}(k) \tag{8–111}$$

where $\mathbf{P}(k)$ is given by Equations (8–23) and (8–25):

$$\mathbf{P}(k) = \mathbf{Q} + \mathbf{G}*\mathbf{P}(k + 1)[\mathbf{I} + \mathbf{H}\mathbf{R}^{-1}\mathbf{H}*\mathbf{P}(k + 1)]^{-1}\mathbf{G}, \qquad \mathbf{P}(N) = \mathbf{S} \tag{8–112}$$

1. Show that the optimal control vector $\mathbf{u}(k)$ can be modified to read

$$\mathbf{u}(k) = -\mathbf{R}^{-1}\mathbf{H}*[\mathbf{P}^{-1}(k + 1) + \mathbf{H}\mathbf{R}^{-1}\mathbf{H}*]^{-1}\mathbf{G}\mathbf{x}(k) \tag{8–113}$$

where

$$\mathbf{P}(k) = \mathbf{Q} + \mathbf{G}*[\mathbf{P}^{-1}(k + 1) + \mathbf{H}\mathbf{R}^{-1}\mathbf{H}*]^{-1}\mathbf{G}, \qquad \mathbf{P}(N) = \mathbf{S} \tag{8–114}$$

2. Show that the optimal control vector $\mathbf{u}(k)$ can also be given by

$$\mathbf{u}(k) = -[\mathbf{R} + \mathbf{H}*\mathbf{P}(k + 1)\mathbf{H}]^{-1}\mathbf{H}*\mathbf{P}(k + 1)\mathbf{G}\mathbf{x}(k) \tag{8–115}$$

where

$$\mathbf{P}(k) = \mathbf{Q} + \mathbf{G}*\mathbf{P}(k + 1)\mathbf{G}$$
$$- \mathbf{G}*\mathbf{P}(k + 1)\mathbf{H}[\mathbf{R} + \mathbf{H}*\mathbf{P}(k + 1)\mathbf{H}]^{-1}\mathbf{H}*\mathbf{P}(k + 1)\mathbf{G}, \qquad \mathbf{P}(N) = \mathbf{S} \tag{8–116}$$

3. Show that the three different expressions for $\mathbf{P}(k)$ given by Equations (8–112), (8–114), and (8–116) are equivalent.

Solution

1. We shall first show that Equations (8–111) and (8–113) are equivalent. Referring to Equation (8–23),

$$(\mathbf{G}*)^{-1}[\mathbf{P}(k) - \mathbf{Q}] = (\mathbf{G}*)^{-1}\mathbf{G}*\mathbf{P}(k + 1)[\mathbf{I} + \mathbf{H}\mathbf{R}^{-1}\mathbf{H}*\mathbf{P}(k + 1)]^{-1}\mathbf{G}$$
$$= \mathbf{P}(k + 1)[\mathbf{I} + \mathbf{H}\mathbf{R}^{-1}\mathbf{H}*\mathbf{P}(k + 1)]^{-1}\mathbf{G}$$
$$= [\mathbf{P}^{-1}(k + 1) + \mathbf{H}\mathbf{R}^{-1}\mathbf{H}*]^{-1}\mathbf{G}$$

Hence,

$$\mathbf{u}(k) = -\mathbf{R}^{-1}\mathbf{H}*(\mathbf{G}*)^{-1}[\mathbf{P}(k) - \mathbf{Q}]\mathbf{x}(k)$$
$$= -\mathbf{R}^{-1}\mathbf{H}*[\mathbf{P}^{-1}(k + 1) + \mathbf{H}\mathbf{R}^{-1}\mathbf{H}*]^{-1}\mathbf{G}\mathbf{x}(k)$$

and we have shown that Equations (8–111) and (8–113) are equivalent.

2. To show that Equations (8–113) and (8–115) are equivalent, note that

$$[\mathbf{R} + \mathbf{H}*\mathbf{P}(k + 1)\mathbf{H}]^{-1}\mathbf{H}*\mathbf{P}(k + 1)[\mathbf{P}^{-1}(k + 1) + \mathbf{H}\mathbf{R}^{-1}\mathbf{H}*]$$
$$= [\mathbf{R} + \mathbf{H}*\mathbf{P}(k + 1)\mathbf{H}]^{-1}\mathbf{H}*[\mathbf{I} + \mathbf{P}(k + 1)\mathbf{H}\mathbf{R}^{-1}\mathbf{H}*]$$
$$= [\mathbf{R} + \mathbf{H}*\mathbf{P}(k + 1)\mathbf{H}]^{-1}[\mathbf{R} + \mathbf{H}*\mathbf{P}(k + 1)\mathbf{H}]\mathbf{R}^{-1}\mathbf{H}*$$
$$= \mathbf{R}^{-1}\mathbf{H}*$$

Hence,

$$\mathbf{R}^{-1}\mathbf{H}*[\mathbf{P}^{-1}(k + 1) + \mathbf{H}\mathbf{R}^{-1}\mathbf{H}*]^{-1} = [\mathbf{R} + \mathbf{H}*\mathbf{P}(k + 1)\mathbf{H}]^{-1}\mathbf{H}*\mathbf{P}(k + 1)$$

and consequently

$$\mathbf{u}(k) = -\mathbf{R}^{-1}\mathbf{H}^*[\mathbf{P}^{-1}(k+1) + \mathbf{H}\mathbf{R}^{-1}\mathbf{H}^*]^{-1}\mathbf{G}\mathbf{x}(k)$$

$$= -[\mathbf{R} + \mathbf{H}^*\mathbf{P}(k+1)\mathbf{H}]^{-1}\mathbf{H}^*\mathbf{P}(k+1)\mathbf{G}\mathbf{x}(k)$$

We have thus shown that Equations (8–113) and (8–115) are equivalent.

3. Next, we shall prove that Equations (8–112) and (8–114) are equivalent. If we note that

$$\mathbf{P}(k+1)[\mathbf{I} + \mathbf{H}\mathbf{R}^{-1}\mathbf{H}^*\mathbf{P}(k+1)]^{-1} = [\mathbf{P}^{-1}(k+1) + \mathbf{H}\mathbf{R}^{-1}\mathbf{H}^*]^{-1}$$

then the equivalence of Equations (8–112) and (8–114) is apparent.

To show that Equations (8–114) and (8–116) are equivalent, notice that

$$[\mathbf{P}^{-1}(k+1) + \mathbf{H}\mathbf{R}^{-1}\mathbf{H}^*]\{\mathbf{P}(k+1) - \mathbf{P}(k+1)\mathbf{H}[\mathbf{R} + \mathbf{H}^*\mathbf{P}(k+1)\mathbf{H}]^{-1}\mathbf{H}^*\mathbf{P}(k+1)\}$$

$$= \mathbf{I} + \mathbf{H}\mathbf{R}^{-1}\mathbf{H}^*\mathbf{P}(k+1) - \mathbf{H}[\mathbf{R} + \mathbf{H}^*\mathbf{P}(k+1)\mathbf{H}]^{-1}\mathbf{H}^*\mathbf{P}(k+1)$$

$$- \mathbf{H}\mathbf{R}^{-1}\mathbf{H}^*\mathbf{P}(k+1)\mathbf{H}[\mathbf{R} + \mathbf{H}^*\mathbf{P}(k+1)\mathbf{H}]^{-1}\mathbf{H}^*\mathbf{P}(k+1)$$

$$= \mathbf{I} - \mathbf{H}\{-\mathbf{R}^{-1} + [\mathbf{R} + \mathbf{H}^*\mathbf{P}(k+1)\mathbf{H}]^{-1}$$

$$+ \mathbf{R}^{-1}\mathbf{H}^*\mathbf{P}(k+1)\mathbf{H}[\mathbf{R} + \mathbf{H}^*\mathbf{P}(k+1)\mathbf{H}]^{-1}\}\mathbf{H}^*\mathbf{P}(k+1)$$

$$= \mathbf{I} - \mathbf{H}\{[\mathbf{I} + \mathbf{R}^{-1}\mathbf{H}^*\mathbf{P}(k+1)\mathbf{H}][\mathbf{R} + \mathbf{H}^*\mathbf{P}(k+1)\mathbf{H}]^{-1} - \mathbf{R}^{-1}\}\mathbf{H}^*\mathbf{P}(k+1)$$

$$= \mathbf{I} - \mathbf{H}\{\mathbf{R}^{-1}[\mathbf{R} + \mathbf{H}^*\mathbf{P}(k+1)\mathbf{H}][\mathbf{R} + \mathbf{H}^*\mathbf{P}(k+1)\mathbf{H}]^{-1} - \mathbf{R}^{-1}\}\mathbf{H}^*\mathbf{P}(k+1)$$

$$= \mathbf{I} - \mathbf{H}[\mathbf{R}^{-1} - \mathbf{R}^{-1}]\mathbf{H}^*\mathbf{P}(k+1) = \mathbf{I}$$

Hence,

$$[\mathbf{P}^{-1}(k+1) + \mathbf{H}\mathbf{R}^{-1}\mathbf{H}^*]^{-1}$$

$$= \mathbf{P}(k+1) - \mathbf{P}(k+1)\mathbf{H}[\mathbf{R} + \mathbf{H}^*\mathbf{P}(k+1)\mathbf{H}]^{-1}\mathbf{H}^*\mathbf{P}(k+1)$$

and we have

$$\mathbf{P}(k) = \mathbf{Q} + \mathbf{G}^*[\mathbf{P}^{-1}(k+1) + \mathbf{H}\mathbf{R}^{-1}\mathbf{H}^*]^{-1}\mathbf{G}$$

$$= \mathbf{Q} + \mathbf{G}^*\mathbf{P}(k+1)\mathbf{G} - \mathbf{G}^*\mathbf{P}(k+1)\mathbf{H}[\mathbf{R} + \mathbf{H}^*\mathbf{P}(k+1)\mathbf{H}]^{-1}\mathbf{H}^*\mathbf{P}(k+1)\mathbf{G}$$

Thus, we have shown that Equations (8–114) and (8–116) are equivalent.

Problem A–8–2

For the quadratic optimal control problem where the system is as given by Equation (8–1) and the performance index is as given by Equation (8–47), we have found in Section 8–2 that the optimal control vector $\mathbf{u}(k)$ can be given by the equation

$$\mathbf{u}(k) = \mathbf{v}(k) - \mathbf{R}^{-1}\mathbf{M}^*\mathbf{x}(k) \tag{8–117}$$

where $\mathbf{v}(k)$ is given by Equation (8–54), (8–55), or (8–56) as follows:

$$\mathbf{v}(k) = -\mathbf{R}^{-1}\mathbf{H}^*(\hat{\mathbf{G}}^*)^{-1}[\hat{\mathbf{P}}(k) - \hat{\mathbf{Q}}]\mathbf{x}(k)$$

or

$$\mathbf{v}(k) = -\mathbf{R}^{-1}\mathbf{H}^*[\hat{\mathbf{P}}^{-1}(k+1) + \mathbf{H}\mathbf{R}^{-1}\mathbf{H}^*]^{-1}\hat{\mathbf{G}}\mathbf{x}(k)$$

or

$$\mathbf{v}(k) = -[\mathbf{R} + \mathbf{H}^*\hat{\mathbf{P}}(k+1)\mathbf{H}]^{-1}\mathbf{H}^*\hat{\mathbf{P}}(k+1)\hat{\mathbf{G}}\mathbf{x}(k) \tag{8–118}$$

where

$$\hat{\mathbf{P}}(k) = \hat{\mathbf{Q}} + \hat{\mathbf{G}}^*\hat{\mathbf{P}}(k+1)[\mathbf{I} + \mathbf{H}\mathbf{R}^{-1}\mathbf{H}^*\hat{\mathbf{P}}(k+1)]^{-1}\hat{\mathbf{G}}, \qquad \hat{\mathbf{P}}(N) = \mathbf{S}$$

and

$$\hat{G} = G - HR^{-1}M^* \quad \text{and} \quad \hat{Q} = Q - MR^{-1}M^*$$

Show that the optimal control vector $u(k)$ can be expressed as follows:

$$u(k) = -[R + H^*\hat{P}(k + 1)H]^{-1}[H^*\hat{P}(k + 1)G + M^*]x(k) \qquad (8\text{–}119)$$

Solution   The equivalence of the right-hand sides of the three expressions for $v(k)$ was shown in Problem A–8–1. Hence, we may derive Equation (8–119) using, for example, Equation (8–118).

From Equations (8–117) and (8–118), we have

$$u(k) = v(k) - R^{-1}M^*x(k)$$

$$= -[R + H^*\hat{P}(k + 1)H]^{-1}H^*\hat{P}(k + 1)\hat{G}x(k) - R^{-1}M^*x(k)$$

$$= -\{[R + H^*\hat{P}(k + 1)H]^{-1}H^*\hat{P}(k + 1)[G - HR^{-1}M^*]$$

$$\quad + [R + H^*\hat{P}(k + 1)H]^{-1}[R + H^*\hat{P}(k + 1)H]R^{-1}M^*\}x(k)$$

$$= -[R + H^*\hat{P}(k + 1)H]^{-1}[H^*\hat{P}(k + 1)G - H^*\hat{P}(k + 1)HR^{-1}M^*$$

$$\quad + M^* + H^*\hat{P}(k + 1)HR^{-1}M^*]x(k)$$

$$= -[R + H^*\hat{P}(k + 1)H]^{-1}[H^*\hat{P}(k + 1)G + M^*]x(k)$$

which is Equation (8–119).

Problem A–8–3

Consider the discrete-time control system defined by

$$x(k + 1) = Gx(k) + Hu(k)$$

where

$$G = \begin{bmatrix} 1 & 1 \\ 1 & 0 \end{bmatrix}, \qquad H = \begin{bmatrix} 1 \\ 0 \end{bmatrix}, \qquad x(0) = \begin{bmatrix} 1 \\ 0 \end{bmatrix}$$

Determine the optimal control sequence $u(k)$ that will minimize the following performance index:

$$J = \frac{1}{2}x^*(8)Sx(8) + \frac{1}{2}\sum_{k=0}^{7}[x^*(k)Qx(k) + u^*(k)Ru(k)]$$

where

$$Q = \begin{bmatrix} 1 & 0 \\ 0 & 1 \end{bmatrix}, \qquad R = 1, \qquad S = \begin{bmatrix} 1 & 0 \\ 0 & 1 \end{bmatrix}$$

Solution   Referring to Equation (8–23), we have

$$P(k) = Q + G^*P(k + 1)[I + HR^{-1}H^*P(k + 1)]^{-1}G$$

$$= \begin{bmatrix} 1 & 0 \\ 0 & 1 \end{bmatrix} + \begin{bmatrix} 1 & 1 \\ 1 & 0 \end{bmatrix}\begin{bmatrix} p_{11}(k + 1) & p_{12}(k + 1) \\ p_{12}(k + 1) & p_{22}(k + 1) \end{bmatrix}$$

$$\times \left\{\begin{bmatrix} 1 & 0 \\ 0 & 1 \end{bmatrix} + \begin{bmatrix} 1 & 0 \\ 0 & 0 \end{bmatrix}\begin{bmatrix} p_{11}(k + 1) & p_{12}(k + 1) \\ p_{12}(k + 1) & p_{22}(k + 1) \end{bmatrix}\right\}^{-1}\begin{bmatrix} 1 & 1 \\ 1 & 0 \end{bmatrix}$$

The boundary condition for $P(k)$ is specified by Equation (8–25) and is given by

$$P(N) = P(8) = S = \begin{bmatrix} 1 & 0 \\ 0 & 1 \end{bmatrix}$$

Now we compute $\mathbf{P}(k)$ backward from $\mathbf{P}(7)$ to $\mathbf{P}(0)$:

$$\mathbf{P}(7) = \begin{bmatrix} 1 & 0 \\ 0 & 1 \end{bmatrix} + \begin{bmatrix} 1 & 1 \\ 1 & 0 \end{bmatrix} \begin{bmatrix} 1 & 0 \\ 0 & 1 \end{bmatrix} \left\{ \begin{bmatrix} 1 & 0 \\ 0 & 1 \end{bmatrix} \right.$$

$$+ \begin{bmatrix} 1 & 0 \\ 0 & 0 \end{bmatrix} \left. \begin{bmatrix} 1 & 0 \\ 0 & 1 \end{bmatrix} \right\}^{-1} \begin{bmatrix} 1 & 1 \\ 1 & 0 \end{bmatrix}$$

$$= \begin{bmatrix} \frac{5}{2} & \frac{1}{2} \\ \frac{1}{2} & \frac{3}{2} \end{bmatrix} = \begin{bmatrix} 2.5 & 0.5 \\ 0.5 & 1.5 \end{bmatrix}$$

$$\mathbf{P}(6) = \begin{bmatrix} 1 & 0 \\ 0 & 1 \end{bmatrix} + \begin{bmatrix} 1 & 1 \\ 1 & 0 \end{bmatrix} \begin{bmatrix} 2.5 & 0.5 \\ 0.5 & 1.5 \end{bmatrix} \left\{ \begin{bmatrix} 1 & 0 \\ 0 & 1 \end{bmatrix} \right.$$

$$+ \begin{bmatrix} 1 & 0 \\ 0 & 0 \end{bmatrix} \left. \begin{bmatrix} 2.5 & 0.5 \\ 0.5 & 1.5 \end{bmatrix} \right\}^{-1} \begin{bmatrix} 1 & 1 \\ 1 & 0 \end{bmatrix}$$

$$= \begin{bmatrix} \frac{24}{7} & \frac{6}{7} \\ \frac{6}{7} & \frac{12}{7} \end{bmatrix} = \begin{bmatrix} 3.4286 & 0.8571 \\ 0.8571 & 1.7143 \end{bmatrix}$$

Similarly, $\mathbf{P}(5), \mathbf{P}(4), \ldots, \mathbf{P}(0)$ can be computed as shown in Table 8–2.

Next, we shall determine the feedback gain matrix $\mathbf{K}(k)$. Referring to Equation (8–27), matrix $\mathbf{K}(k)$ can be given as follows:

$$\mathbf{K}(k) = R^{-1}\mathbf{H}^*(\mathbf{G}^*)^{-1}[\mathbf{P}(k) - \mathbf{Q}]$$

$$= [1][1 \quad 0] \begin{bmatrix} 1 & 1 \\ 1 & 0 \end{bmatrix}^{-1} [\mathbf{P}(k) - \mathbf{Q}]$$

TABLE 8–2 TABLE SHOWING $\mathbf{P}(k)$, $\mathbf{K}(k)$, $\mathbf{x}(k)$, AND $u(k)$ FOR $k = 0, 1, 2, \ldots, 8$, RESPECTIVELY, FOR THE SYSTEM CONSIDERED IN PROBLEM A–8–3

k	$\mathbf{P}(k)$	$\mathbf{K}(k)$	$\mathbf{x}(k)$	$u(k)$
0	$\begin{bmatrix} 3.7913 & 1.0000 \\ 1.0000 & 1.7913 \end{bmatrix}$	$[1.0000 \quad 0.7913]$	$\begin{bmatrix} 1.0000 \\ 0.0000 \end{bmatrix}$	-1.0000
1	$\begin{bmatrix} 3.7911 & 0.9999 \\ 0.9999 & 1.7913 \end{bmatrix}$	$[0.9999 \quad 0.7913]$	$\begin{bmatrix} 0.0000 \\ 1.0000 \end{bmatrix}$	-0.7913
2	$\begin{bmatrix} 3.7905 & 0.9997 \\ 0.9997 & 1.7911 \end{bmatrix}$	$[0.9997 \quad 0.7911]$	$\begin{bmatrix} 0.2087 \\ 0.0000 \end{bmatrix}$	-0.2087
3	$\begin{bmatrix} 3.7877 & 0.9986 \\ 0.9986 & 1.7905 \end{bmatrix}$	$[0.9986 \quad 0.7905]$	$\begin{bmatrix} 0.0001 \\ 0.2087 \end{bmatrix}$	-0.1651
4	$\begin{bmatrix} 3.7740 & 0.9932 \\ 0.9932 & 1.7877 \end{bmatrix}$	$[0.9932 \quad 0.7877]$	$\begin{bmatrix} 0.0437 \\ 0.0001 \end{bmatrix}$	-0.0435
5	$\begin{bmatrix} 3.7097 & 0.9677 \\ 0.9677 & 1.7742 \end{bmatrix}$	$[0.9677 \quad 0.7742]$	$\begin{bmatrix} 0.0003 \\ 0.0437 \end{bmatrix}$	-0.0342
6	$\begin{bmatrix} 3.4286 & 0.8571 \\ 0.8571 & 1.7143 \end{bmatrix}$	$[0.8571 \quad 0.7143]$	$\begin{bmatrix} 0.0099 \\ 0.0003 \end{bmatrix}$	-0.0087
7	$\begin{bmatrix} 2.5000 & 0.5000 \\ 0.5000 & 1.5000 \end{bmatrix}$	$[0.5000 \quad 0.5000]$	$\begin{bmatrix} 0.0015 \\ 0.0099 \end{bmatrix}$	-0.0057
8	$\begin{bmatrix} 1.0000 & 0.0000 \\ 0.0000 & 1.0000 \end{bmatrix}$	$[0.0000 \quad 0.0000]$	$\begin{bmatrix} 0.0057 \\ 0.0015 \end{bmatrix}$	0.0000

$$= [0 \quad 1]\begin{bmatrix} p_{11}(k) - 1 & p_{12}(k) \\ p_{12}(k) & p_{22}(k) - 1 \end{bmatrix}$$

$$= [p_{12}(k) \quad p_{22}(k) - 1]$$

Thus,

$$\mathbf{K}(8) = [p_{12}(8) \quad p_{22}(8) - 1] = [0.0000 \quad 0.0000]$$

$$\mathbf{K}(7) = [p_{12}(7) \quad p_{22}(7) - 1] = [0.5000 \quad 0.5000]$$

Similarly, $\mathbf{K}(6), \mathbf{K}(5), \ldots, \mathbf{K}(0)$ can be computed to give the values shown in Table 8–2. Next, we shall compute $\mathbf{x}(k)$. Let us write

$$\mathbf{K}(k) = [k_1(k) \quad k_2(k)]$$

Then

$$\mathbf{x}(k + 1) = \mathbf{Gx}(k) + \mathbf{H}u(k)$$

$$= [\mathbf{G} - \mathbf{HK}(k)]\mathbf{x}(k)$$

$$= \begin{bmatrix} 1 - k_1(k) & 1 - k_2(k) \\ 1 & 0 \end{bmatrix}\begin{bmatrix} x_1(k) \\ x_2(k) \end{bmatrix}$$

Since the initial state is

$$\mathbf{x}(0) = \begin{bmatrix} 1 \\ 0 \end{bmatrix}$$

$\mathbf{x}(k)$, where $k = 1, 2, \ldots, 8$, can be obtained as follows:

$$\mathbf{x}(1) = \begin{bmatrix} 1 - 1 & 1 - 0.7913 \\ 1 & 0 \end{bmatrix}\begin{bmatrix} 1 \\ 0 \end{bmatrix} = \begin{bmatrix} 0.0000 \\ 1.0000 \end{bmatrix}$$

$$\mathbf{x}(2) = \begin{bmatrix} 1 - 0.9999 & 1 - 0.7913 \\ 1 & 0 \end{bmatrix}\begin{bmatrix} 0.0000 \\ 1.0000 \end{bmatrix} = \begin{bmatrix} 0.2087 \\ 0.0000 \end{bmatrix}$$

Similarly, $\mathbf{x}(3), \mathbf{x}(4), \ldots, \mathbf{x}(8)$ can be computed. The results are shown in Table 8–2.

Finally, the optimal control sequence $u(k)$ can be obtained from Equation (8–28):

$$u(k) = -\mathbf{K}(k)\mathbf{x}(k)$$

That is,

$$u(0) = -\mathbf{K}(0)\mathbf{x}(0) = -[1 \quad 0.7913]\begin{bmatrix} 1 \\ 0 \end{bmatrix} = -1.0000$$

$$u(1) = -\mathbf{K}(1)\mathbf{x}(1) = -[0.9999 \quad 0.7913]\begin{bmatrix} 0 \\ 1 \end{bmatrix} = -0.7913$$

Similarly, $u(2), u(3), \ldots, u(8)$ can be computed to give the values shown in Table 8–2.

As mentioned earlier, the feedback gain matrix $\mathbf{K}(k)$ is constant except for the last several values of k. This means that if the number of stages is not 8 but 100 then $\mathbf{K}(0), \mathbf{K}(1), \ldots, \mathbf{K}(93)$ will be constant matrices and $\mathbf{K}(94), \mathbf{K}(95), \ldots, \mathbf{K}(100)$ will vary. This fact is important, because if the number of stages N is sufficiently large, then the feedback gain matrix becomes a constant matrix and so the designer is able to use a constant feedback gain matrix to approximate the time-varying optimal gain matrix.

The minimum value of J is obtained from Equation (8–36), as follows:

$$J_{\min} = \frac{1}{2}\mathbf{x}^*(0)\mathbf{P}(0)\mathbf{x}(0) = \frac{1}{2}[1 \quad 0]\begin{bmatrix} 3.7913 & 1.0000 \\ 1.0000 & 1.7913 \end{bmatrix}\begin{bmatrix} 1 \\ 0 \end{bmatrix}$$

$$= 1.8956$$

Problem A–8–4

Referring to Problem A–8–3, solve that problem with MATLAB. Write a MATLAB program for finding $P(k)$, $K(k)$, $x(k)$, and $u(k)$. Print out $P(k)$, $K(k)$, $x(k)$, and $u(k)$.

Solution   MATLAB Program 8–5 shows a possible program for solving the problem.

```
MATLAB Program 8–5

% ---------- Quadratic optimal control ----------

% ***** Solving Riccati equation and finding optimal feedback
% gain matrix K *****

% ***** Enter matrices G, H, S, Q, and R *****

G = [1  1;1  0];
H = [1;0];
S = [1  0;0  1];
Q = [1  0;0  1];
R = [1];

% ***** Enter x0 = [1;0], N = 9, p11(N) = 1, p12(N) = 0,
% p22(N) = 1, x1(1) = 1, x2(1) = 0, Pnext = S *****

x0 = [1;0]; N = 9;
p11(N) = 1; p12(N) = 0; p22(N) = 1; x1(1) = 1; x2(1) = 0; Pnext = S;

% ***** Start with the solution of Riccati equation *****

for i = N−1:−1:1,
  P = Q+G'*Pnext*inv(eye(2)+H*inv(R)*H'*Pnext)*G;
  p11(i) = P(1,1); p12(i) = P(1,2); p22(i) = P(2,2); Pnext = P;
end

% ***** Optimal feedback gain matrix K is obtained from *****

for i = N:−1:1,
  K = inv(R)*H'*inv(G')*([p11(i)   p12(i);p12(i)   p22(i)]−Q);
  k1(i) = K(1); k2(i) = K(2);
end

% ***** Optimal  control u(i) is obtained from *****

for i = 1:N−1,
  xnext = (G−H*[k1(i)   k2(i)])*[x1(i);x2(i)];
  x1(i+1) = xnext(1); x2(i+1) = xnext(2);
end
for i = 1:N,
  u(i) = −[k1(i)   k2(i)]*[x1(i);x2(i)];
end
```

Using this program, matrix P, matrix K, vector x, and vector u can be obtained as shown next.

```
% ***** Printout P, K, x, and u *****

P = [p11;p12;p12;p22]

P =

Columns 1 through 7

    3.7913    3.7911    3.7905    3.7877    3.7740    3.7097    3.4286
    1.0000    0.9999    0.9997    0.9986    0.9932    0.9677    0.8571
    1.0000    0.9999    0.9997    0.9986    0.9932    0.9677    0.8571
    1.7913    1.7913    1.7911    1.7905    1.7877    1.7742    1.7143

Columns 8 through 9

    2.5000    1.0000
    0.5000         0
    0.5000         0
    1.5000    1.0000

K = [k1;k2]'

K =

    1.0000    0.7913
    0.9999    0.7913
    0.9997    0.7911
    0.9986    0.7905
    0.9932    0.7877
    0.9677    0.7742
    0.8571    0.7143
    0.5000    0.5000
         0         0

x = [x1;x2]

x =

Columns 1 through 7

    1.0000    0.0000    0.2087    0.0001    0.0437    0.0003    0.0099
         0    1.0000    0.0000    0.2087    0.0001    0.0437    0.0003
```

Columns 8 through 9

   0.0015   0.0057
   0.0099   0.0015

u = u'

u =

  −1.0000
  −0.7913
  −0.2087
  −0.1651
  −0.0435
  −0.0342
  −0.0087
  −0.0057
      0

In this printout, P0, P1, . . . , P8 are given as column vectors. The first column of matrix P gives P0, the second column gives P1, and so forth. In each column the first row gives p11, the second and third row give p12, and the fourth row gives p22. K0, K1, . . . , K8 are given as row vectors in matrix K. The first row corresponds to K0 and the last row corresponds to K8. x0, x1, . . . , x8 are given as columns of matrix x. The first column corresponds to x0 and the last column corresponds to x8. u0, u1, . . . , u8 are given as the first, second, . . . , ninth row of vector u.

Problem A–8–5

Consider the scalar control system

$$x(k + 1) = gx(k) + hu(k) \qquad (8\text{–}120)$$

and the performance index

$$J = \frac{1}{2} \sum_{k=0}^{\infty} [qx^2(k) + ru^2(k)] \qquad (8\text{–}121)$$

where $q > 0$ and $r > 0$. It was shown in Section 8–3 that the optimal control law that will minimize the performance index J can be given by

$$u(k) = -Kx(k) \qquad (8\text{–}122)$$

Substituting Equation (8–122) into Equation (8–120), we obtain

$$x(k + 1) = (g - hK)x(k) \qquad (8\text{–}123)$$

By substituting Equation (8–122) into Equation (8–121), we have

$$J = \frac{1}{2} \sum_{k=0}^{\infty} (q + rK^2)x^2(k)$$

Using the Liapunov approach and referring to Equation (8–94), we set

$$(q + rK^2)x^2(k) = -[px^2(k + 1) - px^2(k)] \qquad (8\text{–}124)$$

By substituting Equation (8–123) into Equation (8–124), we obtain

$$(q + rK^2)x^2(k) = [-p(g - hK)^2 + p]x^2(k)$$

or

$$[q + rK^2 + p(g - hK)^2 - p]x^2(k) = 0$$

This last equation must hold true for any $x(k)$. Hence, we require that

$$q + rK^2 + p(g - hK)^2 - p = 0 \tag{8–125}$$

Show that the optimal control law can be given by

$$u(k) = -Kx(k) = -ghp(r + ph^2)^{-1}x(k)$$

or

$$K = ghp(r + ph^2)^{-1} \tag{8–126}$$

Also show that p can be determined as a positive root of the following equation:

$$q - p + g^2 rp(r + ph^2)^{-1} = 0 \tag{8–127}$$

Solution   By referring to Equation (8–88), the performance index J can be given as follows:

$$J = \tfrac{1}{2} px^2(0)$$

To minimize this value of J for a given $x(0)$ with respect to K, we set

$$\frac{\partial p}{\partial K} = 0 \tag{8–128}$$

where p is as given by Equation (8–125). Notice that in Equation (8–125) $q + rK^2 > 0$. Hence, $1 - (g - hK)^2 \neq 0$. Therefore, p can be given as follows:

$$p = \frac{q + rK^2}{1 - (g - hK)^2} \tag{8–129}$$

By differentiating p with respect to K and equating the result to zero, we obtain

$$\frac{\partial p}{\partial K} = \frac{2rK[1 - (g - hK)^2] - (q + rK^2)[2(g - hK)h]}{[1 - (g - hK)^2]^2} = 0$$

which yields

$$rK[1 - (g - hK)^2] - (q + rK^2)(g - hK)h = 0$$

Hence, we obtain

$$\frac{q + rK^2}{1 - (g - hK)^2} = \frac{rK}{h(g - hK)} \tag{8–130}$$

From Equations (8–129) and (8–130), we get

$$p = \frac{rK}{h(g - hK)} \tag{8–131}$$

Solving Equation (8–131) for K and noting that $r + ph^2 > 0$, we have

$$K = \frac{ghp}{r + ph^2} = ghp(r + ph^2)^{-1} \tag{8–132}$$

which is Equation (8–126).

By substituting Equation (8–132) into Equation (8–125),

$$q + \frac{g^2h^2p^2r}{(r + ph^2)^2} + p\left(\frac{gr}{r + ph^2}\right)^2 - p = 0$$

which can be simplified to

$$q - p + g^2rp(r + ph^2)^{-1} = 0$$

which is Equation (8–127).

The same results can also be obtained in the following way. First note that Equation (8–125) can be modified as follows:

$$q + (r + ph^2)K^2 - 2ghpK + pg^2 - p = 0$$

or

$$q + pg^2 - p + \left(\sqrt{r + ph^2}K - \frac{ghp}{\sqrt{r + ph^2}}\right)^2 - \frac{g^2h^2p^2}{r + ph^2} = 0 \qquad (8–133)$$

Then, considering this last equation as a function of K, the minimum of the left-hand side of this last equation with respect to K occurs when

$$\sqrt{r + ph^2}K - \frac{ghp}{\sqrt{1 + ph^2}} = 0$$

or

$$K = ghp(r + ph^2)^{-1} \qquad (8–134)$$

which is Equation (8–126).

By substituting Equation (8–134) into Equation (8–133), we obtain

$$q + pg^2 - p - \frac{g^2h^2p^2}{r + ph^2} = 0$$

which can be simplified as follows:

$$q - p + g^2rp(r + ph^2)^{-1} = 0$$

which is Equation (8–127).

Problem A–8–6

Consider the system defined by

$$\begin{bmatrix} x_1(k + 1) \\ x_2(k + 1) \end{bmatrix} = \begin{bmatrix} -0.5 & -0.5 \\ 0 & 1.5 \end{bmatrix}\begin{bmatrix} x_1(k) \\ x_2(k) \end{bmatrix} + \begin{bmatrix} 1 \\ 0 \end{bmatrix}u$$

Show that this system cannot be stabilized by the state feedback control scheme:

$$u(k) = -\mathbf{K}x(k)$$

whatever matrix **K** is chosen.

Solution   Define

$$\mathbf{K} = [k_1 \quad k_2]$$

Then

$$\mathbf{G} - \mathbf{HK} = \begin{bmatrix} -0.5 & -0.5 \\ 0 & 1.5 \end{bmatrix} - \begin{bmatrix} 1 \\ 0 \end{bmatrix}[k_1 \quad k_2] = \begin{bmatrix} -0.5 - k_1 & -0.5 - 0.5k_2 \\ 0 & 1.5 \end{bmatrix}$$

Hence, the characteristic equation becomes

$$|z\mathbf{I} - \mathbf{G} + \mathbf{H}\mathbf{K}| = \begin{vmatrix} z + 0.5 + k_1 & 0.5 + 0.5k_2 \\ 0 & z - 1.5 \end{vmatrix}$$

$$= (z + 0.5 + k_1)(z - 1.5) = 0$$

The closed-loop poles are located at

$$z = -0.5 - k_1, \qquad z = 1.5$$

Since the pole at $z = 1.5$ is located outside the unit circle, the system is unstable, whatever $\mathbf{K}$ matrix is chosen. Hence, the quadratic optimal control technique cannot be applied to this system. (The solution to the quadratic optimal control problem does not exist.)

Problem A–8–7

Consider the system

$$\begin{bmatrix} x_1(k + 1) \\ x_2(k + 1) \end{bmatrix} = \begin{bmatrix} 0 & 0 \\ 1 & 1 \end{bmatrix}\begin{bmatrix} x_1(k) \\ x_2(k) \end{bmatrix} + \begin{bmatrix} 1 \\ 0 \end{bmatrix}u(k), \qquad \begin{bmatrix} x_1(0) \\ x_2(0) \end{bmatrix} = \begin{bmatrix} 1 \\ 1 \end{bmatrix} \qquad (8\text{–}135)$$

and the performance index

$$J = \frac{1}{2}\sum_{k=0}^{\infty}[\mathbf{x}^*(k)\mathbf{Q}\mathbf{x}(k) + u^*(k)Ru(k)] \qquad (8\text{–}136)$$

where

$$\mathbf{Q} = \begin{bmatrix} 1 & 0 \\ 0 & 0 \end{bmatrix}, \qquad R = 1$$

Determine the optimal control law to minimize the performance index. Also, determine the minimum value of J.

Solution From Equation (8–135) we have

$$\mathbf{G} = \begin{bmatrix} 0 & 0 \\ 1 & 1 \end{bmatrix}, \qquad \mathbf{H} = \begin{bmatrix} 1 \\ 0 \end{bmatrix}$$

Matrix $\mathbf{P}$ can be determined from Equation (8–101), or

$$\mathbf{P} = \mathbf{Q} + \mathbf{G}^*\mathbf{P}(\mathbf{I} + \mathbf{H}R^{-1}\mathbf{H}^*\mathbf{P})^{-1}\mathbf{G} \qquad (8\text{–}137)$$

Since matrices $\mathbf{Q}$, $\mathbf{G}$, $\mathbf{H}$, and R are real, matrix $\mathbf{P}$ is a real symmetric matrix. By substituting given matrices $\mathbf{Q}$, $\mathbf{G}$, $\mathbf{H}$, and R into Equation (8–137), we obtain

$$\begin{bmatrix} p_{11} & p_{12} \\ p_{12} & p_{22} \end{bmatrix} = \begin{bmatrix} 1 & 0 \\ 0 & 0 \end{bmatrix} + \begin{bmatrix} 0 & 1 \\ 0 & 1 \end{bmatrix}\begin{bmatrix} p_{11} & p_{12} \\ p_{12} & p_{22} \end{bmatrix}\left(\begin{bmatrix} 1 & 0 \\ 0 & 1 \end{bmatrix}\right.$$

$$+ \begin{bmatrix} 1 \\ 0 \end{bmatrix}[1 \quad 0]\begin{bmatrix} p_{11} & p_{12} \\ p_{12} & p_{22} \end{bmatrix}\left.\right)^{-1}\begin{bmatrix} 0 & 0 \\ 1 & 1 \end{bmatrix}$$

Simplifying this last equation, we get

$$\begin{bmatrix} p_{11} & p_{12} \\ p_{12} & p_{22} \end{bmatrix} = \begin{bmatrix} 1 & 0 \\ 0 & 0 \end{bmatrix} + \frac{1}{1 + p_{11}}\begin{bmatrix} p_{12} & p_{22} \\ p_{12} & p_{22} \end{bmatrix}\begin{bmatrix} -p_{12} & -p_{12} \\ 1 + p_{11} & 1 + p_{11} \end{bmatrix}$$

or

$$\begin{bmatrix} p_{11}(1 + p_{11}) & p_{12}(1 + p_{11}) \\ p_{12}(1 + p_{11}) & p_{22}(1 + p_{11}) \end{bmatrix} = \begin{bmatrix} 1 + p_{11} & 0 \\ 0 & 0 \end{bmatrix}$$

$$+ \begin{bmatrix} -p_{12}^2 + p_{22}(1 + p_{11}) & -p_{12}^2 + p_{22}(1 + p_{11}) \\ -p_{12}^2 + p_{22}(1 + p_{11}) & -p_{12}^2 + p_{22}(1 + p_{11}) \end{bmatrix}$$

This last equation is equivalent to the following three equations:

$$p_{11}(1 + p_{11}) = 1 + p_{11} - p_{12}^2 + p_{22}(1 + p_{11})$$

$$p_{12}(1 + p_{11}) = -p_{12}^2 + p_{22}(1 + p_{11})$$

$$p_{22}(1 + p_{11}) = -p_{12}^2 + p_{22}(1 + p_{11})$$

Solving these three equations for p_{11}, p_{12}, and p_{22}, requiring that $p_{11} > 0$, we obtain

$$p_{11} = 1, \qquad p_{12} = 0, \qquad p_{22} = 0$$

Hence

$$\mathbf{P} = \begin{bmatrix} 1 & 0 \\ 0 & 0 \end{bmatrix} \tag{8–138}$$

Equation (8–138) gives the required solution of the steady-state Riccati equation.
Referring to Equation (8–79), we have

$$u(k) = -(R + \mathbf{H^*PH})^{-1} \mathbf{H^*PG} x(k)$$

$$= -(1 + 1)^{-1} [1 \quad 0] \begin{bmatrix} 1 & 0 \\ 0 & 0 \end{bmatrix} \begin{bmatrix} 0 & 0 \\ 1 & 1 \end{bmatrix} x(k)$$

$$= -2^{-1} [0 \quad 0] x(k) = 0 \tag{8–139}$$

Equation (8–139) gives the optimal control law.
The closed-loop system now becomes

$$x(k + 1) = \mathbf{G}x(k) + \mathbf{H}u(k) = \begin{bmatrix} 0 & 0 \\ 1 & 1 \end{bmatrix} x(k) \tag{8–140}$$

Equation (8–140) gives the optimal closed-loop operation for the system. The closed-loop poles are at $\mu_1 = 1$ and $\mu_2 = 0$. The closed-loop system is not asymptotically stable.

The minimum value of J is obtained from Equation (8–81), as follows:

$$J_{\min} = \frac{1}{2} x^*(0) \mathbf{P} x(0) = \frac{1}{2} [1 \quad 1] \begin{bmatrix} 1 & 0 \\ 0 & 0 \end{bmatrix} \begin{bmatrix} 1 \\ 1 \end{bmatrix} = \frac{1}{2}$$

Although the system is not asymptotically stable, the performance index becomes finite and is minimum. In fact, since $u(k) = 0$ for $k = 0, 1, 2, \ldots$, the system equation becomes

$$x_1(k + 1) = 0$$

$$x_2(k + 1) = x_1(k) + x_2(k)$$

or

$$x_1(0) = 1, \qquad x_1(k) = 0, \qquad k = 1, 2, 3, \ldots$$

$$x_2(0) = 1, \qquad x_2(k) = 2, \qquad k = 1, 2, 3, \ldots$$

Notice that the performance index becomes finite, because it involves $x_1(k)$, but does not include $x_2(k)$.

This example problem has shown that in an academic but not practical case, the quadratic optimal control does not yield an asymptotically stable system.

Problem A–8–8

If an nth-order linear single-input–single-output discrete-time control system is completely state controllable, we need at most n sampling periods to bring an arbitrary initial state to the desired final state, provided the control vector is not constrained. Hence, if we allow N (where $N > n$) sampling periods, then we have extra freedom to satisfy additional constraints.

The amount of control energy needed depends on the time period (number of sampling periods) allowed for control. If the number of sampling periods allowed is n, the order of the system, then the time-optimal control sequence $u(0), u(1), \ldots, u(n - 1)$ is unique. However, if N sampling periods ($N > n$) are allowed, then more than one control sequence is possible. Each possible control sequence requires a certain amount of control energy. In many industrial applications, if many control sequences are possible, it is desirable to accomplish control tasks using the minimum amount of control energy.

In this problem, we treat the problem of transferring the state from an arbitrary initial state to the desired final state (which we assume to be the origin of the state space) in N sampling periods and at the same time using the minimum control energy. Consider the discrete-time control system defined by

$$\mathbf{x}(k + 1) = \mathbf{G}\mathbf{x}(k) + \mathbf{H}u(k) \tag{8–141}$$

where

$$\mathbf{x}(k) = \text{state vector } (n\text{-vector}) \text{ at } k\text{th sampling instant}$$

$$u(k) = \text{control signal (scalar) at } k\text{th sampling instant}$$

$$\mathbf{G} = n \times n \text{ nonsingular matrix}$$

$$\mathbf{H} = n \times 1 \text{ matrix}$$

Determine the control law that will bring the system state from an arbitrary initial state to the origin in N sampling periods (where $N > n$) using a minimum amount of control energy, where the control energy is measured by

$$\frac{1}{2} \sum_{k=0}^{N-1} u^2(k)$$

Assume that the system is completely state controllable.

Solution Referring to Equation (5–30), the state $\mathbf{x}(N)$ of Equation (8–141) can be given by

$$\mathbf{x}(N) = \mathbf{G}^N\mathbf{x}(0) + \mathbf{G}^{N-1}\mathbf{H}u(0) + \mathbf{G}^{N-2}\mathbf{H}u(1) + \cdots + \mathbf{G}\mathbf{H}u(N - 2) + \mathbf{H}u(N - 1)$$

Substituting $\mathbf{0}$ for $\mathbf{x}(N)$ in this last equation yields

$$\mathbf{x}(0) = -\mathbf{G}^{-1}\mathbf{H}u(0) - \mathbf{G}^{-2}\mathbf{H}u(1) - \cdots - \mathbf{G}^{-N+1}\mathbf{H}u(N - 2) - \mathbf{G}^{-N}\mathbf{H}u(N - 1) \tag{8–142}$$

Define

$$\mathbf{f}_i = \mathbf{G}^{-i}\mathbf{H} \tag{8–143}$$

Then Equation (8–142) becomes

$$\mathbf{x}(0) = -\mathbf{f}_1 u(0) - \mathbf{f}_2 u(1) - \cdots - \mathbf{f}_{N-1} u(N-2) - \mathbf{f}_N u(N-1) \qquad (8\text{–}144)$$

Since the system is completely state controllable, the vectors $\mathbf{f}_1, \mathbf{f}_2, \ldots, \mathbf{f}_n$ are linearly independent. (The remaining $N - n$ vectors can be expressed as linear combinations of these n linearly independent vectors.) Equation (8–144) can be rewritten as

$$\mathbf{x}(0) = -\mathbf{FU} \qquad (8\text{–}145)$$

where

$$\mathbf{F} = [\mathbf{f}_1 \vdots \mathbf{f}_2 \vdots \cdots \vdots \mathbf{f}_N], \qquad \mathbf{U} = \begin{bmatrix} u(0) \\ u(1) \\ \vdots \\ u(N-1) \end{bmatrix}$$

We shall now find the control sequence that satisfies Equation (8–145) and at the same time minimizes the total control energy. Note that matrix $\mathbf{F}$ is an $n \times N$ matrix and has rank n. Since $\mathbf{F}$ is not a square matrix, the inverse of matrix $\mathbf{F}$ is not defined. Notice that since $N > n$ the number of unknown control signals $u(0), u(1), \ldots, u(N-1)$ in Equation (8–145) is greater than the number n of component scalar equations. A set of scalar equations in such a situation is said to be *underdetermined* and possesses an indefinite number of solutions. However, in the present case we have a constraint that a set of N unknown variables $u(0), u(1), \ldots, u(N-1)$ gives a minimum norm:

$$\frac{1}{2} \sum_{k=0}^{N-1} u^2(k) = \text{minimum}$$

Then, as seen in Appendix A (Section A–8), there is a unique solution. Such a unique solution gives the control sequence that brings an arbitrary initial state $\mathbf{x}(0)$ to the origin in N sampling periods and in so doing minimizes the total energy of control.

The minimizing solution in such a problem, where the number of unknown variables is greater than the number of equations, can be obtained in terms of the right pseudoinverse (refer to Appendix A). The right pseudoinverse is defined as follows:

$$\mathbf{F}^{RM} = \mathbf{F}^*(\mathbf{FF}^*)^{-1} \qquad (8\text{–}146)$$

By using the right pseudoinverse, the minimum-energy control sequence $u(0)$, $u(1), \ldots, u(N-1)$ that transfers an arbitrary initial state $\mathbf{x}(0)$ to the origin can be given by

$$\mathbf{U} = -\mathbf{F}^{RM}\mathbf{x}(0) = -\mathbf{F}^*(\mathbf{FF}^*)^{-1}\mathbf{x}(0) \qquad (8\text{–}147)$$

Note that $\mathbf{F}^*(\mathbf{FF}^*)^{-1}$ is an $N \times n$ matrix. Hence, $\mathbf{F}^*(\mathbf{FF}^*)^{-1}$ postmultiplied by $\mathbf{x}(0)$ is an $N \times 1$ matrix. Equation (8–147) can be rewritten as follows:

$$\begin{bmatrix} u(0) \\ u(1) \\ \vdots \\ u(N-1) \end{bmatrix} = -\mathbf{F}^*(\mathbf{FF}^*)^{-1}\mathbf{x}(0) \qquad (8\text{–}148)$$

The control sequence given by Equation (8–148) will bring an arbitrary initial state to the origin in N sampling periods and will require the minimum control energy among all possible control sequences requiring N sampling periods.

Problem A–8–9

Consider the system

$$\mathbf{x}(k + 1) = \mathbf{G}\mathbf{x}(k) + \mathbf{H}u(k) \qquad (8\text{–}149)$$

where

$$\mathbf{G} = \begin{bmatrix} 1 & 0.6321 \\ 0 & 0.3679 \end{bmatrix}, \quad \mathbf{H} = \begin{bmatrix} 0.3679 \\ 0.6321 \end{bmatrix}, \quad \begin{bmatrix} x_1(0) \\ x_2(0) \end{bmatrix} = \begin{bmatrix} 5 \\ -5 \end{bmatrix}$$

It is desired to bring the initial state to the origin in three sampling periods. (The sampling period is assumed to be 1 sec.) Among infinitely many possible choices for the control sequence, determine the optimal control sequence that will minimize the control energy, or will minimize the following performance index:

$$J = \frac{1}{2} \sum_{k=0}^{2} u^2(k)$$

Solution From Equation (8–144), the initial state $\mathbf{x}(0)$ can be written as follows:

$$\mathbf{x}(0) = -\mathbf{f}_1 u(0) - \mathbf{f}_2 u(1) - \mathbf{f}_3 u(2)$$

where

$$\mathbf{f}_1 = \mathbf{G}^{-1}\mathbf{H} = \begin{bmatrix} -0.7181 \\ 1.7181 \end{bmatrix}, \quad \mathbf{f}_2 = \mathbf{G}^{-2}\mathbf{H} = \begin{bmatrix} -3.6701 \\ 4.6701 \end{bmatrix}, \quad \mathbf{f}_3 = \mathbf{G}^{-3}\mathbf{H} = \begin{bmatrix} -11.6939 \\ 12.6939 \end{bmatrix}$$

Hence,

$$\begin{bmatrix} x_1(0) \\ x_2(0) \end{bmatrix} = -\begin{bmatrix} -0.7181 \\ 1.7181 \end{bmatrix}u(0) - \begin{bmatrix} -3.6701 \\ 4.6701 \end{bmatrix}u(1) - \begin{bmatrix} -11.6939 \\ 12.6939 \end{bmatrix}u(2)$$

or

$$\begin{bmatrix} x_1(0) \\ x_2(0) \end{bmatrix} = -\begin{bmatrix} -0.7181 & -3.6701 & -11.6939 \\ 1.7181 & 4.6701 & 12.6939 \end{bmatrix}\begin{bmatrix} u(0) \\ u(1) \\ u(2) \end{bmatrix} \qquad (8\text{–}150)$$

By use of the right pseudoinverse, we can give the minimum norm solution to Equation (8–150) as

$$\begin{bmatrix} u(0) \\ u(1) \\ u(2) \end{bmatrix} = -\mathbf{F}^{RM}\mathbf{x}(0) = -\mathbf{F}^*(\mathbf{F}\mathbf{F}^*)^{-1}\begin{bmatrix} x_1(0) \\ x_2(0) \end{bmatrix}$$

where

$$\mathbf{F} = \begin{bmatrix} -0.7181 & -3.6701 & -11.6939 \\ 1.7181 & 4.6701 & 12.6939 \end{bmatrix}$$

The right pseudoinverse $\mathbf{F}^{RM}$ is determined as follows:

$$\mathbf{F}^{RM} = \mathbf{F}^*(\mathbf{F}\mathbf{F}^*)^{-1} = \begin{bmatrix} -0.7181 & 1.7181 \\ -3.6701 & 4.6701 \\ -11.6939 & 12.6939 \end{bmatrix}\begin{bmatrix} 150.7326 & -166.8147 \\ -166.8147 & 185.8968 \end{bmatrix}^{-1}$$

$$= \begin{bmatrix} 0.7910 & 0.7191 \\ 0.5000 & 0.4738 \\ -0.2910 & -0.1929 \end{bmatrix}$$

Hence,

$$\begin{bmatrix} u(0) \\ u(1) \\ u(2) \end{bmatrix} = -\begin{bmatrix} 0.7910 & 0.7191 \\ 0.5000 & 0.4738 \\ -0.2910 & -0.1929 \end{bmatrix}\begin{bmatrix} 5 \\ -5 \end{bmatrix} = \begin{bmatrix} -0.3598 \\ -0.1310 \\ 0.4908 \end{bmatrix} \qquad (8\text{–}151)$$

The control sequence given by Equation (8–151) will bring the state to the origin in three sampling periods and will also minimize the total control energy.

By using the optimal control sequence given by Equation (8–151), the state can be transferred as follows:

$$\begin{bmatrix} x_1(1) \\ x_2(1) \end{bmatrix} = \begin{bmatrix} 1 & 0.6321 \\ 0 & 0.3679 \end{bmatrix}\begin{bmatrix} 5 \\ -5 \end{bmatrix} + \begin{bmatrix} 0.3679 \\ 0.6321 \end{bmatrix}[-0.3598] = \begin{bmatrix} 1.7071 \\ -2.0669 \end{bmatrix}$$

$$\begin{bmatrix} x_1(2) \\ x_2(2) \end{bmatrix} = \begin{bmatrix} 1 & 0.6321 \\ 0 & 0.3679 \end{bmatrix}\begin{bmatrix} 1.7071 \\ -2.0669 \end{bmatrix} + \begin{bmatrix} 0.3679 \\ 0.6321 \end{bmatrix}[-0.1310] = \begin{bmatrix} 0.3524 \\ -0.8432 \end{bmatrix}$$

$$\begin{bmatrix} x_1(3) \\ x_2(3) \end{bmatrix} = \begin{bmatrix} 1 & 0.6321 \\ 0 & 0.3679 \end{bmatrix}\begin{bmatrix} 0.3524 \\ -0.8432 \end{bmatrix} + \begin{bmatrix} 0.3679 \\ 0.6321 \end{bmatrix}[0.4908] = \begin{bmatrix} 0 \\ 0 \end{bmatrix}$$

The minimum energy required for this control is

$$J_{min} = \frac{1}{2}\sum_{k=0}^{2} u^2(k) = \frac{1}{2}[u^2(0) + u^2(1) + u^2(2)] = \frac{1}{2}[(-0.3598)^2 + (-0.1310)^2 + (0.4908)^2]$$

$$= 0.1937$$

It is interesting to compare the minimum energy obtained here with the energy required for time-optimal control of this system. The time-optimal control requires two sampling periods. In Problem A–6–2, the time-optimal control sequence $u(0)$ and $u(1)$, where the sampling period was 1 sec, was found to be

$$u(0) = -1.5820x_1(0) - 1.2433x_2(0)$$

$$u(1) = 0.5820x_1(0) + 0.2433x_2(0)$$

[Refer to Equation (6–231).] By taking $x_1(0) = 5$ and $x_2(0) = -5$ and substituting these values in these two equations, we obtain $u(0) = -1.6935$ and $u(1) = 1.6935$. Hence, for time-optimal control the total energy required is

$$\frac{1}{2}[u^2(0) + u^2(1)] = \frac{1}{2}[(-1.6935)^2 + (1.6935)^2] = 2.8679$$

Notice that by allowing the control duration to be three sampling periods (3 sec), rather than two sampling periods (2 sec), the energy required can be reduced remarkably.

Problem A–8–10

Consider the inverted pendulum system shown in Figure 8–10, where an inverted pendulum is mounted on a motor-driven cart. Here we consider only the two-dimensional problem in which the pendulum moves only in the plane of the paper. The inverted pendulum is unstable in that it may fall over anytime unless a suitable control force is applied. Assume that the pendulum mass is concentrated at the end of the rod as shown in the figure. (We assume that the rod is massless.) The control force u is applied to the cart.

In the diagram, θ is the angle of the rod from the vertical line. We assume that angle θ is small so that we may approximate $\sin\theta$ by θ, $\cos\theta$ by 1, and also assume that $\dot{\theta}$ is small so that $\theta\dot{\theta}^2 \doteq 0$. (Under these conditions, the system's nonlinear equations can be linearized.)

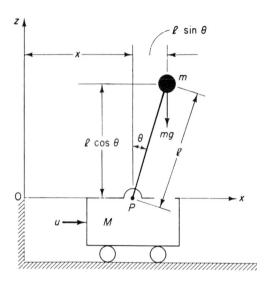

Figure 8–10   Inverted pendulum system.

It is desired to keep the pendulum upright in response to step changes in the cart position. (The control force u is the force applied to the cart.) Derive first the continuous-time state-space model. Then discretize the continuous-time state-space model and obtain the discrete-time model. Assume that the sampling period T is 0.1 sec. Assume the following numerical values for M, m, and l:

$$M = 2 \text{ kg}, \qquad m = 0.1 \text{ kg}, \qquad l = 0.5 \text{ m}$$

(In Section 8–4 we designed a digital controller for this inverted pendulum system.)

Solution   Define the angle of the rod from the vertical line as θ. (Since we want to keep the inverted pendulum vertical, angle θ is assumed to be small.) Define also the (x, z) coordinates of the center of gravity of the mass as (x_G, z_G). Then

$$x_G = x + l \sin \theta$$

$$z_G = l \cos \theta$$

Applying Newton's second law to the x direction of motion yields

$$M \frac{d^2 x}{dt^2} + m \frac{d^2 x_G}{dt^2} = u$$

or

$$M \frac{d^2 x}{dt^2} + m \frac{d^2}{dt^2} (x + l \sin \theta) = u \qquad (8\text{--}152)$$

Noting that

$$\frac{d}{dt} \sin \theta = (\cos \theta) \dot{\theta}$$

$$\frac{d^2}{dt^2} \sin \theta = -(\sin \theta) \dot{\theta}^2 + (\cos \theta) \ddot{\theta}$$

$$\frac{d}{dt} \cos \theta = -(\sin \theta) \dot{\theta}$$

$$\frac{d^2}{dt^2} \cos \theta = -(\cos \theta) \dot{\theta}^2 - (\sin \theta) \ddot{\theta}$$

Equation (8–152) can be written as

$$(M + m)\ddot{x} - ml(\sin \theta)\dot{\theta}^2 + ml(\cos \theta)\ddot{\theta} = u \qquad (8\text{–}153)$$

The equation of motion of the mass m in the z direction cannot be written without considering the motion of the mass m in the x direction. Therefore, instead of considering the motion of the mass m in the z direction, we consider the rotational motion of the mass m around point P. Applying Newton's second law to the rotational motion, we obtain

$$m\frac{d^2 x_G}{dt^2} l \cos \theta - m\frac{d^2 z_G}{dt^2} l \sin \theta = mgl \sin \theta$$

or

$$\left[m\frac{d^2}{dt^2}(x + l \sin \theta) \right] l \cos \theta - \left[m\frac{d^2}{dt^2}(l \cos \theta) \right] l \sin \theta = mgl \sin \theta$$

which can be simplified as follows:

$$m[\ddot{x} - l(\sin \theta)\dot{\theta}^2 + l(\cos \theta)\ddot{\theta}]l \cos \theta - m[-l(\cos \theta)\dot{\theta}^2 - l(\sin \theta)\ddot{\theta}]l \sin \theta = mgl \sin \theta$$

Further simplification results in

$$m\ddot{x} \cos \theta + ml\ddot{\theta} = mg \sin \theta \qquad (8\text{–}154)$$

By substituting $\sin \theta \doteq \theta$, $\cos \theta \doteq 1$, and $\theta\dot{\theta}^2 \doteq 0$, Equations (8–153) and (8–154) can be linearized as follows:

$$(M + m)\ddot{x} + ml\ddot{\theta} = u \qquad (8\text{–}155)$$

$$m\ddot{x} + ml\ddot{\theta} = mg\theta \qquad (8\text{–}156)$$

These linearized equations are valid as long as θ and $\dot{\theta}$ are small. Equations (8–155) and (8–156) define a mathematical model of the inverted pendulum system.

The linearized system equations, Equations (8–155) and (8–156), can be modified to

$$Ml\ddot{\theta} = (M + m)g\theta - u \qquad (8\text{–}157)$$

$$M\ddot{x} = u - mg\theta \qquad (8\text{–}158)$$

Equation (8–157) was obtained by eliminating $\ddot{x}$ from Equations (8–155) and (8–156). Equation (8–158) was obtained by eliminating $\ddot{\theta}$ from Equations (8–155) and (8–156). Define state variables x_1, x_2, x_3, and x_4 by

$$x_1 = \theta$$

$$x_2 = \dot{\theta}$$

$$x_3 = x$$

$$x_4 = \dot{x}$$

Note that angle θ indicates the rotation of the pendulum rod about point P, and x is the location of the cart. We consider x as the output of the system, or

$$y = x = x_3$$

Then, from the definition of the state variables and Equations (8–157) and (8–158), we obtain

$$\dot{x}_1 = x_2$$

$$\dot{x}_2 = \frac{M + m}{Ml}gx_1 - \frac{1}{Ml}u$$

$$\dot{x}_3 = x_4$$

$$\dot{x}_4 = -\frac{m}{M}gx_1 + \frac{1}{M}u$$

In terms of vector-matrix equations, we have

$$\begin{bmatrix} \dot{x}_1 \\ \dot{x}_2 \\ \dot{x}_3 \\ \dot{x}_4 \end{bmatrix} = \begin{bmatrix} 0 & 1 & 0 & 0 \\ \dfrac{M+m}{Ml}g & 0 & 0 & 0 \\ 0 & 0 & 0 & 1 \\ -\dfrac{m}{M}g & 0 & 0 & 0 \end{bmatrix} \begin{bmatrix} x_1 \\ x_2 \\ x_3 \\ x_4 \end{bmatrix} + \begin{bmatrix} 0 \\ -\dfrac{1}{Ml} \\ 0 \\ \dfrac{1}{M} \end{bmatrix} u \qquad (8\text{--}159)$$

$$y = \begin{bmatrix} 0 & 0 & 1 & 0 \end{bmatrix} \begin{bmatrix} x_1 \\ x_2 \\ x_3 \\ x_4 \end{bmatrix} \qquad (8\text{--}160)$$

Equations (8–159) and (8–160) give a state-space representation of the inverted pendulum system. (Note that the state-space representation of the system is not unique. There are infinitely many such representations.)

By substituting the given numerical values for M, m, and l, we obtain

$$\frac{M + m}{Ml}g = 20.601, \qquad \frac{m}{M}g = 0.4905, \qquad \frac{1}{Ml} = 1, \qquad \frac{1}{M} = 0.5$$

Then the state equation and output equation for the inverted pendulum with cart become as follows:

$$\dot{\mathbf{x}} = \mathbf{Ax} + \mathbf{B}u \qquad (8\text{--}161)$$

$$y = \mathbf{Cx} + Du \qquad (8\text{--}162)$$

where

$$\mathbf{A} = \begin{bmatrix} 0 & 1 & 0 & 0 \\ 20.601 & 0 & 0 & 0 \\ 0 & 0 & 0 & 1 \\ -0.4905 & 0 & 0 & 0 \end{bmatrix}, \qquad \mathbf{B} = \begin{bmatrix} 0 \\ -1 \\ 0 \\ 0.5 \end{bmatrix}, \qquad \mathbf{C} = \begin{bmatrix} 0 & 0 & 1 & 0 \end{bmatrix}, \qquad D = 0$$

Next, we discretize the state equation, Equation (8–161). The discretization can be accomplished by using the following MATLAB command:

$$[G,H] = c2d(A,B,T)$$

where T is the sampling period involved in the discrete-time control system. In this problem T = 0.1 sec. Then the command

$$[G,H] = c2d(A,B,0.1)$$

will transform the continuous-time state-space equation into the discrete-time state-space equation. See the following MATLAB command and output:

$$
\begin{aligned}
A = [0 &&& 1 & 0 & 0 \\
20.601 &&& 0 & 0 & 0 \\
0 &&& 0 & 0 & 1 \\
-0.4905 &&& 0 & 0 & 0];
\end{aligned}
$$

$B = [0; -1; 0; 0.5];$

$[G,H] = c2d(A,B,0.1)$

$G =$

$$
\begin{array}{cccc}
1.1048 & 0.1035 & 0 & 0 \\
2.1316 & 1.1048 & 0 & 0 \\
-0.0025 & -0.0001 & 1.0000 & 0.1000 \\
-0.0508 & -0.0025 & 0 & 1.0000
\end{array}
$$

$H =$

$$
\begin{array}{c}
-0.0051 \\
-0.1035 \\
0.0025 \\
0.0501
\end{array}
$$

Thus, the discretized state-space model is given as follows:

$$
\mathbf{x}(k + 1) = \mathbf{G}\mathbf{x}(k) + \mathbf{H}u(k)
$$

$$
y(k) = \mathbf{C}\mathbf{x}(k) + Du(k)
$$

where

$$
\mathbf{G} = \begin{bmatrix} 1.1048 & 0.1035 & 0 & 0 \\ 2.1316 & 1.1048 & 0 & 0 \\ -0.0025 & -0.0001 & 1 & 0.1 \\ -0.0508 & -0.0025 & 0 & 1 \end{bmatrix}, \quad \mathbf{H} = \begin{bmatrix} -0.0051 \\ -0.1035 \\ 0.0025 \\ 0.0501 \end{bmatrix}
$$

$$
\mathbf{C} = [0 \quad 0 \quad 1 \quad 0], \qquad D = 0
$$

PROBLEMS

Problem B–8–1

Consider the discrete-time system

$$
\mathbf{x}(k + 1) = \mathbf{G}\mathbf{x}(k) + \mathbf{H}u(k)
$$

where

$$
\mathbf{G} = \begin{bmatrix} 0 & 1 \\ -0.5 & 1 \end{bmatrix}, \qquad \mathbf{H} = \begin{bmatrix} 1 \\ 1 \end{bmatrix}, \qquad \mathbf{x}(0) = \begin{bmatrix} 2 \\ 2 \end{bmatrix}
$$

Determine the optimal control sequence $u(k)$ that will minimize the following performance index:

$$J = \frac{1}{2}\mathbf{x}^*(8)\mathbf{S}\mathbf{x}(8) + \frac{1}{2}\sum_{k=0}^{7}[\mathbf{x}^*(k)\mathbf{Q}\mathbf{x}(k) + u^*(k)Ru(k)]$$

where

$$\mathbf{Q} = \begin{bmatrix} 1 & 0 \\ 0 & 1 \end{bmatrix}, \qquad R = 1, \qquad \mathbf{S} = \begin{bmatrix} 1 & 0 \\ 0 & 1 \end{bmatrix}$$

Problem B–8–2

Consider the system

$$\mathbf{x}(k + 1) = \mathbf{G}\mathbf{x}(k) + \mathbf{H}u(k)$$

where

$$\mathbf{G} = \begin{bmatrix} 0 & 0 \\ -0.5 & 1 \end{bmatrix}, \qquad \mathbf{H} = \begin{bmatrix} 1 \\ 0 \end{bmatrix}, \qquad \mathbf{x}(0) = \begin{bmatrix} 2 \\ 2 \end{bmatrix}$$

and the performance index

$$J = \frac{1}{2}\sum_{k=0}^{\infty}[\mathbf{x}^*(k)\mathbf{Q}\mathbf{x}(k) + u^*(k)Ru(k)]$$

where

$$\mathbf{Q} = \begin{bmatrix} 1 & 0 \\ 0 & 0.5 \end{bmatrix}, \qquad R = 1$$

Determine the optimal control law to minimize the performance index. Also, determine the minimum value of J.

Problem B–8–3

Consider the system defined by

$$\begin{bmatrix} x_1(k + 1) \\ x_2(k + 1) \end{bmatrix} = \begin{bmatrix} 1 & 1 \\ a & -1 \end{bmatrix}\begin{bmatrix} x_1(k) \\ x_2(k) \end{bmatrix}, \qquad \begin{bmatrix} x_1(0) \\ x_2(0) \end{bmatrix} = \begin{bmatrix} 1 \\ 1 \end{bmatrix}$$

where $-1 \le a < 0$. Determine the value of a such that the performance index

$$J = \frac{1}{2}\sum_{k=0}^{\infty}\mathbf{x}^*(k)\mathbf{Q}\mathbf{x}(k)$$

where

$$\mathbf{Q} = \begin{bmatrix} 1 & 0 \\ 0 & 0.5 \end{bmatrix}$$

is minimized.

Problem B–8–4

A discrete-time control system is described by the equation

$$x(k + 1) = 0.3679x(k) + 0.6321u(k)$$

Determine the optimal control law to minimize the following performance index:

$$J = \frac{1}{2}\sum_{k=0}^{\infty}[x^2(k) + u^2(k)]$$

Also, determine the minimum value of the performance index J.

Problem B–8–5

Consider the same system as discussed in Problem A–8–6. Is it possible to determine a positive definite matrix **P** for this system? Use Equation (8–101) for the determination of matrix **P**.

Problem B–8–6

Consider the system defined by the equations

$$\mathbf{x}(k + 1) = \mathbf{Gx}(k) + \mathbf{Hu}(k)$$

$$\mathbf{y}(k) = \mathbf{Cx}(k)$$

where $\mathbf{x}(k)$ is an n-vector, $\mathbf{u}(k)$ is an r-vector, $\mathbf{y}(k)$ is an m-vector, **G** is an $n \times n$ matrix, **H** is an $n \times r$ matrix, and **C** is an $m \times n$ matrix. The performance index is

$$J = \frac{1}{2} \sum_{k=0}^{\infty} [\mathbf{x}^*(k)\mathbf{Qx}(k) + \mathbf{u}^*(k)\mathbf{Ru}(k)]$$

where **Q** is an $n \times n$ positive definite Hermitian matrix and **R** is an $r \times r$ positive definite Hermitian matrix. Let us define the optimal control law that minimizes the performance index as $\mathbf{u}(k) = -\mathbf{Kx}(k)$.

Show that if the system is completely state controllable and observable then the algebraic Riccati equation

$$\mathbf{P} = \mathbf{Q} + \mathbf{GPG}^* - \mathbf{GPC}^*(\mathbf{R} + \mathbf{CPC}^*)^{-1}\mathbf{CPG}^*$$

has a unique positive definite solution. Show also that the optimal closed-loop system is stable, or $\mathbf{G} - \mathbf{HK}$ is a stable matrix.

Problem B–8–7

Referring to Problem A–8–9, solve the same problem with MATLAB. Determine the optimal control sequence $u(0)$, $u(1)$, and $u(2)$.

Problem B–8–8

Consider the system

$$\begin{bmatrix} x_1(k + 1) \\ x_2(k + 1) \end{bmatrix} = \begin{bmatrix} 1 & 1 \\ 1 & 0 \end{bmatrix}\begin{bmatrix} x_1(k) \\ x_2(k) \end{bmatrix} + \begin{bmatrix} 1 \\ 1 \end{bmatrix}u(k), \qquad \begin{bmatrix} x_1(0) \\ x_2(0) \end{bmatrix} = \begin{bmatrix} 1 \\ 1 \end{bmatrix}$$

It is desired to bring the initial state to the origin in n sampling periods. Determine the optimal control law to minimize the control energy measured by

$$J = \frac{1}{2} \sum_{k=0}^{n} u^2(k)$$

Consider three values of n: $n = 2$, $n = 3$, and $n = 4$.

Problem B–8–9

Consider the design of the servo system shown in Figure 8–11. The plant does not involve an integrator and, therefore, an integral controller is included in the loop. The sampling period T is 0.1 sec.

Show that the system equations may be given by the following state-space equations:

$$\mathbf{x}(k + 1) = \mathbf{Gx}(k) + \mathbf{H}w(k)$$

$$w(k) = -\mathbf{Kx}(k)$$

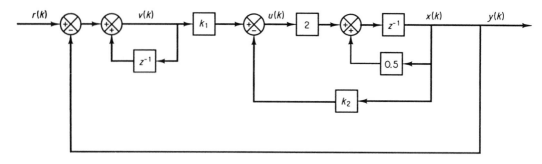

Figure 8–11 A servo system.

where

$$\mathbf{x}(k) = \begin{bmatrix} x_1(k) \\ x_2(k) \end{bmatrix} = \begin{bmatrix} x(k) - x(\infty) \\ v(k) - v(\infty) \end{bmatrix}, \qquad w(k) = u(k) - u(\infty)$$

$$\mathbf{G} = \begin{bmatrix} 0.5 & 0 \\ -0.5 & 1 \end{bmatrix}, \qquad \mathbf{H} = \begin{bmatrix} 2 \\ -2 \end{bmatrix}, \qquad \mathbf{K} = [k_2 \quad -k_1]$$

Problem B–8–10

Referring to Problem B–8–9, it is desired to design the state feedback gain matrix $\mathbf{K} = [k_2 \quad -k_1]$ such that the system has reasonable step-response characteristics. Let us assume that we use the quadratic optimal control scheme.

Let us assume the following performance index:

$$J = \frac{1}{2} \sum_{k=0}^{\infty} [\mathbf{x}(k)^*\mathbf{Q}\mathbf{x}(k) + w(k)^*Rw(k)]$$

If $\mathbf{Q}$ and R are chosen to be positive definite, the resulting system is stable. For this problem, we choose

$$\mathbf{Q} = \begin{bmatrix} 100 & 0 \\ 0 & 1 \end{bmatrix}, \qquad R = 1$$

Note that the present $\mathbf{Q}$ and R are only one possible set. (Other positive definite $\mathbf{Q}$ and R may be chosen. The resulting system is stable but different for each different set of $\mathbf{Q}$ and R.)

Using the state-space representation shown in Problem B–8–9, determine matrix $\mathbf{K}$ with MATLAB. Write a MATLAB program. Using matrix $\mathbf{K}$ thus determined, obtain the unit-step response of the designed system with MATLAB. Plot $y(k)$ versus k and $v(k)$ versus k.

Appendix A

Vector-Matrix Analysis

A–1 DEFINITIONS

Matrices that we frequently encounter in the study of modern control theory are the symmetric matrix, skew-symmetric matrix, orthogonal matrix, Hermitian matrix, skew-Hermitian matrix, unitary matrix, and normal matrix. The following equations define these matrices:

$$\mathbf{A}^T = \mathbf{A} \qquad\qquad\qquad \mathbf{A} \text{ is symmetric}$$

$$\mathbf{A}^T = -\mathbf{A} \qquad\qquad\qquad \mathbf{A} \text{ is skew-symmetric}$$

$$\mathbf{A}\mathbf{A}^T = \mathbf{A}^T\mathbf{A} = \mathbf{I} \qquad\qquad \mathbf{A} \text{ is orthogonal}$$

$$\mathbf{A}^* = \mathbf{A} \qquad\qquad\qquad \mathbf{A} \text{ is Hermitian}$$

$$\mathbf{A}^* = -\mathbf{A} \qquad\qquad\qquad \mathbf{A} \text{ is skew-Hermitian}$$

$$\mathbf{A}\mathbf{A}^* = \mathbf{A}^*\mathbf{A} = \mathbf{I} \qquad\qquad \mathbf{A} \text{ is unitary}$$

$$\mathbf{A}\mathbf{A}^* = \mathbf{A}^*\mathbf{A} \quad \text{or} \quad \mathbf{A}\mathbf{A}^T = \mathbf{A}^T\mathbf{A} \quad \mathbf{A} \text{ is normal}$$

where the superscript * denotes the conjugate transpose and superscript T signifies the transpose.

A–2 DETERMINANTS

Determinants of a 2 × 2 Matrix, a 3 × 3 Matrix, and a 4 × 4 Matrix. For a 2 × 2 matrix **A**, we have

$$|\mathbf{A}| = \begin{vmatrix} a_1 & a_2 \\ b_1 & b_2 \end{vmatrix} = a_1 b_2 - b_1 a_2$$

For a 3×3 matrix $\mathbf{A}$,

$$|\mathbf{A}| = \begin{vmatrix} a_1 & a_2 & a_3 \\ b_1 & b_2 & b_3 \\ c_1 & c_2 & c_3 \end{vmatrix} = a_1 b_2 c_3 + b_1 c_2 a_3 + c_1 a_2 b_3 - c_1 b_2 a_3 - b_1 a_2 c_3 - a_1 b_3 c_2$$

For a 4×4 matrix $\mathbf{A}$,

$$|\mathbf{A}| = \begin{vmatrix} a_1 & a_2 & a_3 & a_4 \\ b_1 & b_2 & b_3 & b_4 \\ c_1 & c_2 & c_3 & c_4 \\ d_1 & d_2 & d_3 & d_4 \end{vmatrix}$$

$$= \begin{vmatrix} a_1 & a_2 \\ b_1 & b_2 \end{vmatrix} \begin{vmatrix} c_3 & c_4 \\ d_3 & d_4 \end{vmatrix} - \begin{vmatrix} a_1 & a_2 \\ c_1 & c_2 \end{vmatrix} \begin{vmatrix} b_3 & b_4 \\ d_3 & d_4 \end{vmatrix}$$

$$+ \begin{vmatrix} a_1 & a_2 \\ d_1 & d_2 \end{vmatrix} \begin{vmatrix} b_3 & b_4 \\ c_3 & c_4 \end{vmatrix} + \begin{vmatrix} b_1 & b_2 \\ c_1 & c_2 \end{vmatrix} \begin{vmatrix} a_3 & a_4 \\ d_3 & d_4 \end{vmatrix}$$

$$- \begin{vmatrix} b_1 & b_2 \\ d_1 & d_2 \end{vmatrix} \begin{vmatrix} a_3 & a_4 \\ c_3 & c_4 \end{vmatrix} + \begin{vmatrix} c_1 & c_2 \\ d_1 & d_2 \end{vmatrix} \begin{vmatrix} a_3 & a_4 \\ b_3 & b_4 \end{vmatrix} \qquad \text{(A–1)}$$

(This expansion is called Laplace's expansion by the minors.)

Properties of the Determinant. The determinant of an $n \times n$ matrix has the following properties:

1. If two rows (or two columns) of the determinant are interchanged, only the sign of the determinant is changed.
2. The determinant is invariant under the addition of a scalar multiple of a row (or a column) to another row (or column).
3. If an $n \times n$ matrix has two identical rows (or columns), then the determinant is zero.
4. For an $n \times n$ matrix $\mathbf{A}$,

$$|\mathbf{A}^T| = |\mathbf{A}|, \qquad |\mathbf{A}^*| = \overline{|\mathbf{A}|}$$

5. The determinant of a product of two $n \times n$ matrices $\mathbf{A}$ and $\mathbf{B}$ is the product of their determinants:

$$|\mathbf{AB}| = |\mathbf{A}|\,|\mathbf{B}| = |\mathbf{BA}|$$

6. If a row (or a column) is multiplied by a scalar k, then the determinant is multiplied by k.
7. If all elements of an $n \times n$ matrix are multiplied by k, then the determinant is multiplied by k^n; that is,

$$|k\mathbf{A}| = k^n|\mathbf{A}|$$

8. If the eigenvalues of $\mathbf{A}$ are λ_i ($i = 1, 2, \ldots, n$), then

$$|\mathbf{A}| = \lambda_1 \lambda_2 \ldots \lambda_n$$

Hence, $|\mathbf{A}| \neq 0$ implies $\lambda_i \neq 0$ for $i = 1, 2, \ldots, n$. (For details of the eigenvalue, see Section A–6.)

9. If matrices **A**, **B**, **C**, and **D** are an $n \times n$, an $n \times m$, an $m \times n$, and an $m \times m$ matrix, respectively, then

$$\begin{vmatrix} \mathbf{A} & \mathbf{B} \\ \mathbf{0} & \mathbf{D} \end{vmatrix} = \begin{vmatrix} \mathbf{A} & \mathbf{0} \\ \mathbf{C} & \mathbf{D} \end{vmatrix} = |\mathbf{A}|\,|\mathbf{D}|, \qquad \text{if } |\mathbf{A}| \neq 0 \text{ and } |\mathbf{D}| \neq 0 \tag{A-2}$$

$$\begin{vmatrix} \mathbf{A} & \mathbf{B} \\ \mathbf{0} & \mathbf{D} \end{vmatrix} = \begin{vmatrix} \mathbf{A} & \mathbf{0} \\ \mathbf{C} & \mathbf{D} \end{vmatrix} = 0, \qquad \text{if } |\mathbf{A}| = 0 \text{ or } |\mathbf{D}| = 0 \text{ or } |\mathbf{A}| = |\mathbf{D}| = 0$$

Also,

$$\begin{vmatrix} \mathbf{A} & \mathbf{B} \\ \mathbf{C} & \mathbf{D} \end{vmatrix} = \begin{cases} |\mathbf{A}|\,|\mathbf{D} - \mathbf{C}\mathbf{A}^{-1}\mathbf{B}|, & \text{if } |\mathbf{A}| \neq 0 & \text{(A-3)} \\ |\mathbf{D}|\,|\mathbf{A} - \mathbf{B}\mathbf{D}^{-1}\mathbf{C}|, & \text{if } |\mathbf{D}| \neq 0 & \text{(A-4)} \end{cases}$$

[For the derivation of Equation (A–2), see Problem A–1. For derivations of Equations (A–3) and (A–4), refer to Problem A–2.]

10. For an $n \times m$ matrix **A** and an $m \times n$ matrix **B**,

$$|\mathbf{I}_n + \mathbf{A}\mathbf{B}| = |\mathbf{I}_m + \mathbf{B}\mathbf{A}| \tag{A-5}$$

(For the proof, see Problem A–3.) In particular, for $m = 1$, that is, for an $n \times 1$ matrix **A** and a $1 \times n$ matrix **B**, we have

$$|\mathbf{I}_n + \mathbf{A}\mathbf{B}| = 1 + \mathbf{B}\mathbf{A} \tag{A-6}$$

Equations (A–2) through (A–6) are useful in computing the determinants of matrices of large order.

A–3 INVERSION OF MATRICES

Nonsingular Matrix and Singular Matrix.   A square matrix **A** is called a nonsingular matrix if a matrix **B** exists such that $\mathbf{BA} = \mathbf{AB} = \mathbf{I}$. If such a matrix **B** exists, then it is denoted by $\mathbf{A}^{-1}$. $\mathbf{A}^{-1}$ is called the *inverse* of **A**. The inverse matrix $\mathbf{A}^{-1}$ exists if $|\mathbf{A}|$ is nonzero. If $\mathbf{A}^{-1}$ does not exist, **A** is said to be *singular*.

If **A** and **B** are nonsingular matrices, then the product **AB** is a nonsingular matrix and

$$(\mathbf{AB})^{-1} = \mathbf{B}^{-1}\mathbf{A}^{-1}$$

Also,

$$(\mathbf{A}^T)^{-1} = (\mathbf{A}^{-1})^T$$

and

$$(\mathbf{A}^*)^{-1} = (\mathbf{A}^{-1})^*$$

Properties of the Inverse Matrix.   The inverse of a matrix has the following properties.

1. If k is a nonzero scalar and **A** is an $n \times n$ nonsingular matrix, then

$$(k\mathbf{A})^{-1} = \frac{1}{k}\mathbf{A}^{-1}$$

2. The determinant of $\mathbf{A}^{-1}$ is the inverse of the determinant of $\mathbf{A}$, or

$$|\mathbf{A}^{-1}| = \frac{1}{|\mathbf{A}|}$$

This can be verified easily as follows:

$$|\mathbf{A}\mathbf{A}^{-1}| = |\mathbf{A}| \, |\mathbf{A}^{-1}| = 1$$

Useful Formulas for Finding the Inverse of a Matrix

1. For a 2×2 matrix $\mathbf{A}$, where

$$\mathbf{A} = \begin{bmatrix} a & b \\ c & d \end{bmatrix}, \qquad ad - bc \neq 0$$

the inverse matrix is given by

$$\mathbf{A}^{-1} = \frac{1}{ad - bc} \begin{bmatrix} d & -b \\ -c & a \end{bmatrix}$$

2. For a 3×3 matrix $\mathbf{A}$, where

$$\mathbf{A} = \begin{bmatrix} a & b & c \\ d & e & f \\ g & h & i \end{bmatrix}, \qquad |\mathbf{A}| \neq 0$$

the inverse matrix is given by

$$\mathbf{A}^{-1} = \frac{1}{|\mathbf{A}|} \begin{bmatrix} \begin{vmatrix} e & f \\ h & i \end{vmatrix} & -\begin{vmatrix} b & c \\ h & i \end{vmatrix} & \begin{vmatrix} b & c \\ e & f \end{vmatrix} \\[6pt] -\begin{vmatrix} d & f \\ g & i \end{vmatrix} & \begin{vmatrix} a & c \\ g & i \end{vmatrix} & -\begin{vmatrix} a & c \\ d & f \end{vmatrix} \\[6pt] \begin{vmatrix} d & e \\ g & h \end{vmatrix} & -\begin{vmatrix} a & b \\ g & h \end{vmatrix} & \begin{vmatrix} a & b \\ d & e \end{vmatrix} \end{bmatrix}$$

3. If $\mathbf{A}$, $\mathbf{B}$, $\mathbf{C}$, and $\mathbf{D}$ are, respectively, an $n \times n$, an $n \times m$, an $m \times n$, and an $m \times m$ matrix, then

$$(\mathbf{A} + \mathbf{BDC})^{-1} = \mathbf{A}^{-1} - \mathbf{A}^{-1}\mathbf{B}(\mathbf{D}^{-1} + \mathbf{CA}^{-1}\mathbf{B})^{-1}\mathbf{CA}^{-1} \qquad \text{(A–7)}$$

provided the indicated inverses exist. Equation (A–7) is commonly referred to as the *matrix inversion lemma*. (For the proof, see Problem A–4.)

If $\mathbf{D} = \mathbf{I}_m$, then Equation (A–7) simplifies to

$$(\mathbf{A} + \mathbf{BC})^{-1} = \mathbf{A}^{-1} - \mathbf{A}^{-1}\mathbf{B}(\mathbf{I}_m + \mathbf{CA}^{-1}\mathbf{B})^{-1}\mathbf{CA}^{-1}$$

In this last equation, if $\mathbf{B}$ and $\mathbf{C}$ are an $n \times 1$ matrix and a $1 \times n$ matrix, respectively, then

$$(\mathbf{A} + \mathbf{BC})^{-1} = \mathbf{A}^{-1} - \frac{\mathbf{A}^{-1}\mathbf{BCA}^{-1}}{1 + \mathbf{CA}^{-1}\mathbf{B}} \qquad \text{(A–8)}$$

Equation (A–8) is useful in that if an $n \times n$ matrix $\mathbf{X}$ can be written as $\mathbf{A} + \mathbf{BC}$, where $\mathbf{A}$ is an $n \times n$ matrix whose inverse is known and $\mathbf{BC}$ is a product of a column vector and a row vector, then $\mathbf{X}^{-1}$ can be obtained easily in terms of the known $\mathbf{A}^{-1}$, $\mathbf{B}$, and $\mathbf{C}$.

4. If $\mathbf{A}$, $\mathbf{B}$, $\mathbf{C}$, and $\mathbf{D}$ are, respectively, an $n \times n$, an $n \times m$, an $m \times n$, and an $m \times m$ matrix, then

$$\begin{bmatrix} \mathbf{A} & \mathbf{B} \\ \mathbf{C} & \mathbf{D} \end{bmatrix}^{-1} = \begin{bmatrix} \mathbf{A}^{-1} + \mathbf{A}^{-1}\mathbf{B}(\mathbf{D} - \mathbf{CA}^{-1}\mathbf{B})^{-1}\mathbf{CA}^{-1} & -\mathbf{A}^{-1}\mathbf{B}(\mathbf{D} - \mathbf{CA}^{-1}\mathbf{B})^{-1} \\ -(\mathbf{D} - \mathbf{CA}^{-1}\mathbf{B})^{-1}\mathbf{CA}^{-1} & (\mathbf{D} - \mathbf{CA}^{-1}\mathbf{B})^{-1} \end{bmatrix} \tag{A–9}$$

provided $|\mathbf{A}| \neq 0$ and $|\mathbf{D} - \mathbf{CA}^{-1}\mathbf{B}| \neq 0$, or

$$\begin{bmatrix} \mathbf{A} & \mathbf{B} \\ \mathbf{C} & \mathbf{D} \end{bmatrix}^{-1} = \begin{bmatrix} (\mathbf{A} - \mathbf{BD}^{-1}\mathbf{C})^{-1} & -(\mathbf{A} - \mathbf{BD}^{-1}\mathbf{C})^{-1}\mathbf{BD}^{-1} \\ -\mathbf{D}^{-1}\mathbf{C}(\mathbf{A} - \mathbf{BD}^{-1}\mathbf{C})^{-1} & \mathbf{D}^{-1}\mathbf{C}(\mathbf{A} - \mathbf{BD}^{-1}\mathbf{C})^{-1}\mathbf{BD}^{-1} + \mathbf{D}^{-1} \end{bmatrix} \tag{A–10}$$

provided $|\mathbf{D}| \neq 0$ and $|\mathbf{A} - \mathbf{BD}^{-1}\mathbf{C}| \neq 0$. In particular, if $\mathbf{C} = \mathbf{0}$ or $\mathbf{B} = \mathbf{0}$, then Equations (A–9) and (A–10) can be simplified as follows:

$$\begin{bmatrix} \mathbf{A} & \mathbf{B} \\ \mathbf{0} & \mathbf{D} \end{bmatrix}^{-1} = \begin{bmatrix} \mathbf{A}^{-1} & -\mathbf{A}^{-1}\mathbf{BD}^{-1} \\ \mathbf{0} & \mathbf{D}^{-1} \end{bmatrix} \tag{A–11}$$

or

$$\begin{bmatrix} \mathbf{A} & \mathbf{0} \\ \mathbf{C} & \mathbf{D} \end{bmatrix}^{-1} = \begin{bmatrix} \mathbf{A}^{-1} & \mathbf{0} \\ -\mathbf{D}^{-1}\mathbf{CA}^{-1} & \mathbf{D}^{-1} \end{bmatrix} \tag{A–12}$$

[For the derivation of Equations (A–9) through (A–12), refer to Problems A–5 and A–6.]

A–4 RULES OF MATRIX OPERATIONS

In this section we shall review some of the rules of algebraic operations with matrices and then give definitions of the derivative and the integral of matrices. Then the rules of differentiation of matrices will be presented.

Note that matrix algebra differs from ordinary number algebra in that matrix multiplication is not commutative and cancellation of matrices is not valid.

Multiplication of a Matrix by a Scalar. The product of a matrix and a scalar is a matrix in which each element is multiplied by the scalar. That is,

$$k\mathbf{A} = \begin{bmatrix} ka_{11} & ka_{12} & \cdots & ka_{1m} \\ ka_{21} & ka_{22} & \cdots & ka_{2m} \\ \vdots & \vdots & & \vdots \\ ka_{n1} & ka_{n2} & \cdots & ka_{nm} \end{bmatrix}$$

Multiplication of a Matrix by a Matrix. Multiplication of a matrix by a matrix is possible between matrices in which the number of columns in the first matrix is equal to the number of rows in the second. Otherwise, multiplication is not defined.

Consider the product of an $n \times m$ matrix $\mathbf{A}$ and an $m \times r$ matrix $\mathbf{B}$:

$$\mathbf{AB} = \begin{bmatrix} a_{11} & a_{12} & \cdots & a_{1m} \\ a_{21} & a_{22} & \cdots & a_{2m} \\ \vdots & \vdots & & \vdots \\ a_{n1} & a_{n2} & \cdots & a_{nm} \end{bmatrix} \begin{bmatrix} b_{11} & b_{12} & \cdots & b_{1r} \\ b_{21} & b_{22} & \cdots & b_{2r} \\ \vdots & \vdots & & \vdots \\ b_{m1} & b_{m2} & \cdots & b_{mr} \end{bmatrix}$$

$$= \begin{bmatrix} c_{11} & c_{12} & \cdots & c_{1r} \\ c_{21} & c_{22} & \cdots & c_{2r} \\ \vdots & \vdots & & \vdots \\ c_{n1} & c_{n2} & \cdots & c_{nr} \end{bmatrix}$$

where

$$c_{ik} = \sum_{j=1}^{m} a_{ij} b_{jk}$$

Thus, multiplication of an $n \times m$ matrix by an $m \times r$ matrix yields an $n \times r$ matrix. It should be noted that, in general, matrix multiplication is not commutative; that is

$$\mathbf{AB} \neq \mathbf{BA} \qquad \text{in general}$$

For example,

$$\mathbf{AB} = \begin{bmatrix} a_{11} & a_{12} \\ a_{21} & a_{22} \end{bmatrix} \begin{bmatrix} b_{11} & b_{12} \\ b_{21} & b_{22} \end{bmatrix} = \begin{bmatrix} a_{11}b_{11} + a_{12}b_{21} & a_{11}b_{12} + a_{12}b_{22} \\ a_{21}b_{11} + a_{22}b_{21} & a_{21}b_{12} + a_{22}b_{22} \end{bmatrix}$$

and

$$\mathbf{BA} = \begin{bmatrix} b_{11} & b_{12} \\ b_{21} & b_{22} \end{bmatrix} \begin{bmatrix} a_{11} & a_{12} \\ a_{21} & a_{22} \end{bmatrix} = \begin{bmatrix} b_{11}a_{11} + b_{12}a_{21} & b_{11}a_{12} + b_{12}a_{22} \\ b_{21}a_{11} + b_{22}a_{21} & b_{21}a_{12} + b_{22}a_{22} \end{bmatrix}$$

Thus, in general, $\mathbf{AB} \neq \mathbf{BA}$. Hence, the order of multiplication is significant and must be preserved. If $\mathbf{AB} = \mathbf{BA}$, matrices $\mathbf{A}$ and $\mathbf{B}$ are said to commute. In the preceding matrices $\mathbf{A}$ and $\mathbf{B}$, if, for example, $a_{12} = a_{21} = b_{12} = b_{21} = 0$, then $\mathbf{A}$ and $\mathbf{B}$ commute.

For $n \times n$ diagonal matrices $\mathbf{A}$ and $\mathbf{B}$,

$$\mathbf{AB} = [a_{ij}\,\delta_{ij}][b_{ij}\,\delta_{ij}] = \begin{bmatrix} a_{11}b_{11} & & & 0 \\ & a_{22}b_{22} & & \\ & & \cdot\cdot\cdot & \\ 0 & & & a_{nn}b_{nn} \end{bmatrix}$$

If $\mathbf{A}$, $\mathbf{B}$, and $\mathbf{C}$ are an $n \times m$ matrix, an $m \times r$ matrix, and an $r \times p$ matrix, respectively, then the following associativity law holds true:

$$(\mathbf{AB})\mathbf{C} = \mathbf{A}(\mathbf{BC})$$

This may be proved as follows:

$$(i, k)\text{th element of } \mathbf{AB} = \sum_{j=1}^{m} a_{ij} b_{jk}$$

$$(j, h)\text{th element of } \mathbf{BC} = \sum_{k=1}^{r} b_{jk} c_{kh}$$

$$(i,h)\text{th element of } (\mathbf{AB})\mathbf{C} = \sum_{k=1}^{r}\left(\sum_{j=1}^{m} a_{ij}b_{jk}\right)c_{kh} = \sum_{j=1}^{m}\sum_{k=1}^{r}(a_{ij}b_{jk})c_{kh}$$

$$= \sum_{j=1}^{m}\sum_{k=1}^{r} a_{ij}(b_{jk}c_{kh}) = \sum_{j=1}^{m} a_{ij}\left[\sum_{k=1}^{r} b_{jk}c_{kh}\right]$$

$$= (i,h)\text{th element of } \mathbf{A}(\mathbf{BC})$$

Since the associativity of multiplication of matrices holds true, we have

$$\mathbf{ABCD} = (\mathbf{AB})(\mathbf{CD}) = \mathbf{A}(\mathbf{BCD}) = (\mathbf{ABC})\mathbf{D}$$

$$\mathbf{A}^{m+n} = \mathbf{A}^m \mathbf{A}^n, \qquad m, n = 1, 2, 3, \dots$$

If **A** and **B** are $n \times m$ matrices and **C** and **D** are $m \times r$ matrices, then the following distributivity law holds true:

$$(\mathbf{A} + \mathbf{B})(\mathbf{C} + \mathbf{D}) = \mathbf{AC} + \mathbf{AD} + \mathbf{BC} + \mathbf{BD}$$

This can be proved by comparing the (i,j)th element of $(\mathbf{A} + \mathbf{B})(\mathbf{C} + \mathbf{D})$ and the (i,j)th element of $(\mathbf{AC} + \mathbf{AD} + \mathbf{BC} + \mathbf{BD})$.

Remarks on Cancellation of Matrices.   Cancellation of matrices is not valid in matrix algebra. Consider the product of two singular matrices **A** and **B**. Take, for example,

$$\mathbf{A} = \begin{bmatrix} 2 & 1 \\ 6 & 3 \end{bmatrix} \neq \mathbf{0}, \qquad \mathbf{B} = \begin{bmatrix} 1 & -2 \\ -2 & 4 \end{bmatrix} \neq \mathbf{0}$$

Then

$$\mathbf{AB} = \begin{bmatrix} 2 & 1 \\ 6 & 3 \end{bmatrix}\begin{bmatrix} 1 & -2 \\ -2 & 4 \end{bmatrix} = \begin{bmatrix} 0 & 0 \\ 0 & 0 \end{bmatrix} = \mathbf{0}$$

Clearly, $\mathbf{AB} = \mathbf{0}$ implies neither $\mathbf{A} = \mathbf{0}$ nor $\mathbf{B} = \mathbf{0}$. In fact, $\mathbf{AB} = \mathbf{0}$ implies one of the following three:

1. $\mathbf{A} = \mathbf{0}$.
2. $\mathbf{B} = \mathbf{0}$.
3. Both **A** and **B** are singular.

It can easily be proved that, if both **A** and **B** are nonzero matrices and $\mathbf{AB} = \mathbf{0}$, then both **A** and **B** must be singular. Assume that **B** is nonzero and **A** is not singular. Then $|\mathbf{A}| \neq \mathbf{0}$ and $\mathbf{A}^{-1}$ exists. Then we obtain

$$\mathbf{A}^{-1}\mathbf{AB} = \mathbf{B} = \mathbf{0}$$

which contradicts the assumption that **B** is nonzero. In this way we can prove that both **A** and **B** must be singular if $\mathbf{A} \neq \mathbf{0}$ and $\mathbf{B} \neq \mathbf{0}$.

Similarly, notice that if **A** is singular then neither $\mathbf{AB} = \mathbf{AC}$ nor $\mathbf{BA} = \mathbf{CA}$ implies $\mathbf{B} = \mathbf{C}$. If, however, **A** is a nonsingular matrix, then $\mathbf{AB} = \mathbf{AC}$ implies $\mathbf{B} = \mathbf{C}$ and $\mathbf{BA} = \mathbf{CA}$ also implies $\mathbf{B} = \mathbf{C}$.

Derivative and Integral of a Matrix. The derivative of an $n \times m$ matrix $\mathbf{A}(t)$ is defined by the matrix whose (i, j)th element is the derivative of the (i, j)th element of the original matrix, provided that all the elements $a_{ij}(t)$ have derivatives with respect to t:

$$\frac{d}{dt}\mathbf{A}(t) = \begin{bmatrix} \dfrac{d}{dt}a_{11}(t) & \cdots & \dfrac{d}{dt}a_{1m}(t) \\ \vdots & & \vdots \\ \dfrac{d}{dt}a_{n1}(t) & \cdots & \dfrac{d}{dt}a_{nm}(t) \end{bmatrix}$$

In the case of an n-dimensional vector $\mathbf{x}(t)$,

$$\frac{d}{dt}\mathbf{x}(t) = \begin{bmatrix} \dfrac{d}{dt}x_1(t) \\ \vdots \\ \dfrac{d}{dt}x_n(t) \end{bmatrix}$$

Similarly, the integral of an $n \times m$ matrix $\mathbf{A}(t)$ with respect to t is defined by the matrix whose (i, j)th element is the integral of the (i, j)th element of the original matrix, or

$$\int \mathbf{A}(t)\, dt = \begin{bmatrix} \displaystyle\int a_{11}(t)\, dt & \cdots & \displaystyle\int a_{1m}(t)\, dt \\ \vdots & & \vdots \\ \displaystyle\int a_{n1}(t)\, dt & \cdots & \displaystyle\int a_{nm}(t)\, dt \end{bmatrix}$$

provided that the $a_{ij}(t)$'s are integrable as functions of t.

Differentiation of a Matrix. If the elements of matrices $\mathbf{A}$ and $\mathbf{B}$ are functions of t, then

$$\frac{d}{dt}(\mathbf{A} + \mathbf{B}) = \frac{d}{dt}\mathbf{A} + \frac{d}{dt}\mathbf{B} \tag{A-13}$$

$$\frac{d}{dt}(\mathbf{AB}) = \frac{d\mathbf{A}}{dt}\mathbf{B} + \mathbf{A}\frac{d\mathbf{B}}{dt} \tag{A-14}$$

If $k(t)$ is a scalar and is a function of t, then

$$\frac{d}{dt}[\mathbf{A}k(t)] = \frac{d\mathbf{A}}{dt}k(t) + \mathbf{A}\frac{dk(t)}{dt} \tag{A-15}$$

Also,

$$\int_a^b \frac{d\mathbf{A}}{dt}\mathbf{B}\, dt = \mathbf{AB}\Big|_a^b - \int_a^b \mathbf{A}\frac{d\mathbf{B}}{dt}\, dt \tag{A-16}$$

It is important to note that the derivative of $\mathbf{A}^{-1}$ is given by

$$\frac{d}{dt}\mathbf{A}^{-1} = -\mathbf{A}^{-1}\frac{d\mathbf{A}}{dt}\mathbf{A}^{-1} \tag{A-17}$$

Equation (A–17) can be derived easily by differentiating $\mathbf{A}\mathbf{A}^{-1}$ with respect to t. Since

$$\frac{d}{dt}\mathbf{A}\mathbf{A}^{-1} = \frac{d\mathbf{A}}{dt}\mathbf{A}^{-1} + \mathbf{A}\frac{d\mathbf{A}^{-1}}{dt}$$

and also

$$\frac{d}{dt}\mathbf{A}\mathbf{A}^{-1} = \frac{d}{dt}\mathbf{I} = \mathbf{0}$$

we obtain

$$\mathbf{A}\frac{d\mathbf{A}^{-1}}{dt} = -\frac{d\mathbf{A}}{dt}\mathbf{A}^{-1}$$

or

$$\mathbf{A}^{-1}\mathbf{A}\frac{d\mathbf{A}^{-1}}{dt} = \frac{d\mathbf{A}^{-1}}{dt} = -\mathbf{A}^{-1}\frac{d\mathbf{A}}{dt}\mathbf{A}^{-1}$$

which is the desired result.

Derivatives of a Scalar Function with Respect to a Vector.   If $J(\mathbf{x})$ is a scalar function of a vector $\mathbf{x}$, then

$$\frac{\partial J}{\partial \mathbf{x}} = \begin{bmatrix} \dfrac{\partial J}{\partial x_1} \\ \vdots \\ \dfrac{\partial J}{\partial x_n} \end{bmatrix}, \qquad \frac{\partial^2 J}{\partial \mathbf{x}^2} = \begin{bmatrix} \dfrac{\partial^2 J}{\partial^2 x_1} & \dfrac{\partial^2 J}{\partial x_1\,\partial x_2} & \cdots & \dfrac{\partial^2 J}{\partial x_1\,\partial x_n} \\ \vdots & \vdots & & \vdots \\ \dfrac{\partial^2 J}{\partial x_n\,\partial x_1} & \dfrac{\partial^2 J}{\partial x_n\,\partial x_2} & \cdots & \dfrac{\partial^2 J}{\partial x_n^2} \end{bmatrix}$$

Also, for a scalar function $V(\mathbf{x}(t))$, we have

$$\frac{d}{dt}V(\mathbf{x}(t)) = \left(\frac{\partial V}{\partial \mathbf{x}}\right)^T \frac{d\mathbf{x}}{dt}$$

Jacobian.   If an $m \times 1$ matrix $\mathbf{f}(\mathbf{x})$ is a vector function of an n-vector $\mathbf{x}$ (*note:* an n-vector is meant as an n-dimensional vector), then

$$\frac{\partial \mathbf{f}}{\partial \mathbf{x}} = \begin{bmatrix} \dfrac{\partial f_1}{\partial x_1} & \dfrac{\partial f_2}{\partial x_1} & \cdots & \dfrac{\partial f_m}{\partial x_1} \\ \dfrac{\partial f_1}{\partial x_2} & \dfrac{\partial f_2}{\partial x_2} & \cdots & \dfrac{\partial f_m}{\partial x_2} \\ \vdots & \vdots & & \vdots \\ \dfrac{\partial f_1}{\partial x_n} & \dfrac{\partial f_2}{\partial x_n} & \cdots & \dfrac{\partial f_m}{\partial x_n} \end{bmatrix} \qquad (A\text{–}18)$$

Such an $n \times m$ matrix is called a *Jacobian*.

Notice that, by using this definition of the Jacobian, we have

$$\frac{\partial}{\partial \mathbf{x}}\mathbf{A}\mathbf{x} = \mathbf{A}^T \qquad (A\text{–}19)$$

The fact that Equation (A–19) holds true can be easily seen from the following example. If $\mathbf{A}$ and $\mathbf{x}$ are given by

$$\mathbf{A} = \begin{bmatrix} a_{11} & a_{12} & a_{13} \\ a_{21} & a_{22} & a_{23} \end{bmatrix}, \qquad \mathbf{x} = \begin{bmatrix} x_1 \\ x_2 \\ x_3 \end{bmatrix}$$

then

$$\mathbf{Ax} = \begin{bmatrix} a_{11} & a_{12} & a_{13} \\ a_{21} & a_{22} & a_{23} \end{bmatrix} \begin{bmatrix} x_1 \\ x_2 \\ x_3 \end{bmatrix} = \begin{bmatrix} a_{11}x_1 + a_{12}x_2 + a_{13}x_3 \\ a_{21}x_1 + a_{22}x_2 + a_{23}x_3 \end{bmatrix} = \begin{bmatrix} f_1 \\ f_2 \end{bmatrix}$$

and

$$\frac{\partial}{\partial \mathbf{x}}\mathbf{Ax} = \begin{bmatrix} \dfrac{\partial f_1}{\partial x_1} & \dfrac{\partial f_2}{\partial x_1} \\ \dfrac{\partial f_1}{\partial x_2} & \dfrac{\partial f_2}{\partial x_2} \\ \dfrac{\partial f_1}{\partial x_3} & \dfrac{\partial f_2}{\partial x_3} \end{bmatrix} = \begin{bmatrix} a_{11} & a_{21} \\ a_{12} & a_{22} \\ a_{13} & a_{23} \end{bmatrix} = \mathbf{A}^T$$

Also, we have the following useful formula. For an $n \times n$ real matrix $\mathbf{A}$ and a real n-vector $\mathbf{x}$,

$$\frac{\partial}{\partial \mathbf{x}}\mathbf{x}^T \mathbf{Ax} = \mathbf{Ax} + \mathbf{A}^T \mathbf{x} \qquad\qquad (A\text{–}20)$$

In addition, if matrix $\mathbf{A}$ is a real symmetric matrix, then

$$\frac{\partial}{\partial \mathbf{x}}\mathbf{x}^T \mathbf{Ax} = 2\mathbf{Ax}$$

Note that if $\mathbf{A}$ is an $n \times n$ Hermitian matrix and $\mathbf{x}$ is a complex n-vector then

$$\frac{\partial}{\partial \overline{\mathbf{x}}}\mathbf{x}^*\mathbf{Ax} = \mathbf{Ax} \qquad\qquad (A\text{–}21)$$

[For derivations of Equations (A–20) and (A–21), see Problem A–7.]

For an $n \times m$ real matrix $\mathbf{A}$, a real n-vector $\mathbf{x}$, and a real m-vector $\mathbf{y}$, we have

$$\frac{\partial}{\partial \mathbf{x}}\mathbf{x}^T \mathbf{Ay} = \mathbf{Ay} \qquad\qquad (A\text{–}22)$$

$$\frac{\partial}{\partial \mathbf{y}}\mathbf{x}^T \mathbf{Ay} = \mathbf{A}^T \mathbf{x} \qquad\qquad (A\text{–}23)$$

Similarly, for an $n \times m$ complex matrix $\mathbf{A}$, a complex n-vector $\mathbf{x}$, and a complex m-vector $\mathbf{y}$, we have

$$\frac{\partial}{\partial \overline{\mathbf{x}}}\mathbf{x}^*\mathbf{Ay} = \mathbf{Ay} \qquad\qquad (A\text{–}24)$$

$$\frac{\partial}{\partial \mathbf{y}}\mathbf{x}^*\mathbf{Ay} = \mathbf{A}^T \overline{\mathbf{x}} \qquad\qquad (A\text{–}25)$$

[For derivations of Equations (A–22) through (A–25), refer to Problem A–8.] Note that Equation (A–25) is equivalent to the following equation:

$$\frac{\overline{\partial}}{\partial \mathbf{y}} \mathbf{x}^* \mathbf{A} \mathbf{y} = \mathbf{A}^* \mathbf{x}$$

A–5 VECTORS AND VECTOR ANALYSIS

Linear Dependence and Independence of Vectors.   Vectors $\mathbf{x}_1, \mathbf{x}_2, \ldots, \mathbf{x}_n$ are said to be *linearly independent* if the equation

$$c_1 \mathbf{x}_1 + c_2 \mathbf{x}_2 + \cdots + c_n \mathbf{x}_n = \mathbf{0}$$

where $c_1, c_2, \ldots, c_n$ are constants, implies that $c_1 = c_2 = \cdots = c_n = 0$. Conversely, vectors $\mathbf{x}_1, \mathbf{x}_2, \ldots, \mathbf{x}_n$ are said to be *linearly dependent* if and only if $\mathbf{x}_i$ can be expressed as a linear combination of $\mathbf{x}_j$ ($j = 1, 2, \ldots, n; j \neq i$).

It is important to note that if vectors $\mathbf{x}_1, \mathbf{x}_2, \ldots, \mathbf{x}_n$ are linearly independent and vectors $\mathbf{x}_1, \mathbf{x}_2, \ldots, \mathbf{x}_n, \mathbf{x}_{n+1}$ are linearly dependent, then $\mathbf{x}_{n+1}$ can be expressed as a unique linear combination of $\mathbf{x}_1, \mathbf{x}_2, \ldots, \mathbf{x}_n$.

Necessary and Sufficient Conditions for Linear Independence of Vectors.   It can be proved that the necessary and sufficient conditions for n-vectors $\mathbf{x}_i$ ($i = 1, 2, \ldots, m$) to be linearly independent are that

1. $m \leq n$.
2. There exists at least one nonzero m-column determinant of the $n \times m$ matrix whose columns consist of $\mathbf{x}_1, \mathbf{x}_2, \ldots, \mathbf{x}_m$.

Hence, for n vectors $\mathbf{x}_1, \mathbf{x}_2, \ldots, \mathbf{x}_n$ the necessary and sufficient condition for linear independence is

$$|\mathbf{A}| \neq 0$$

where $\mathbf{A}$ is the $n \times n$ matrix whose ith column is made up of the components of $\mathbf{x}_i$ ($i = 1, 2, \ldots, n$).

Inner Product.   Any rule that assigns to each pair of vectors $\mathbf{x}$ and $\mathbf{y}$ in a vector space a scalar quantity is called an *inner product* or *scalar product* and is given the symbol $\langle \mathbf{x}, \mathbf{y} \rangle$, provided that the following four axioms are satisfied:

1.
$$\langle \mathbf{y}, \mathbf{x} \rangle = \overline{\langle \mathbf{x}, \mathbf{y} \rangle}$$

where the bar denotes the conjugate of a complex number

2.
$$\langle c\mathbf{x}, \mathbf{y} \rangle = \bar{c} \langle \mathbf{x}, \mathbf{y} \rangle = \langle \mathbf{x}, \bar{c}\mathbf{y} \rangle$$

where c is a complex number

3.
$$\langle \mathbf{x} + \mathbf{y}, \mathbf{z} + \mathbf{w} \rangle = \langle \mathbf{x}, \mathbf{z} \rangle + \langle \mathbf{y}, \mathbf{z} \rangle + \langle \mathbf{x}, \mathbf{w} \rangle + \langle \mathbf{y}, \mathbf{w} \rangle$$

4.
$$\langle \mathbf{x}, \mathbf{x} \rangle > 0, \quad \text{for } \mathbf{x} \neq \mathbf{0}$$

In any finite-dimensional vector space, there are many different definitions of the inner product, all satisfying the four axioms.

In this book, unless the contrary is stated, we shall adopt the following definition of the inner product: The inner product of a pair of n-vectors $\mathbf{x}$ and $\mathbf{y}$ in a vector space V is given by

$$\langle \mathbf{x}, \mathbf{y} \rangle = \bar{x}_1 y_1 + \bar{x}_2 y_2 + \cdots + \bar{x}_n y_n = \sum_{i=1}^{n} \bar{x}_i y_i \qquad \text{(A–26)}$$

where the summation is a complex number and where the $\bar{x}_i$'s are the complex conjugates of the x_i's. This definition clearly satisfies the four axioms. The inner product can then be expressed as follows:

$$\langle \mathbf{x}, \mathbf{y} \rangle = \mathbf{x}^* \mathbf{y}$$

where $\mathbf{x}^*$ denotes the conjugate transpose of $\mathbf{x}$. Also,

$$\langle \mathbf{x}, \mathbf{y} \rangle = \overline{\langle \mathbf{y}, \mathbf{x} \rangle} = \overline{\mathbf{y}^* \mathbf{x}} = \mathbf{y}^T \bar{\mathbf{x}} = \mathbf{x}^* \mathbf{y} \qquad \text{(A–27)}$$

The inner product of two n-vectors $\mathbf{x}$ and $\mathbf{y}$ with real components is therefore given by

$$\langle \mathbf{x}, \mathbf{y} \rangle = x_1 y_1 + x_2 y_2 + \cdots + x_n y_n = \sum_{i=1}^{n} x_i y_i \qquad \text{(A–28)}$$

In this case, clearly we have

$$\langle \mathbf{x}, \mathbf{y} \rangle = \mathbf{x}^T \mathbf{y} = \mathbf{y}^T \mathbf{x}, \qquad \text{for real vectors } \mathbf{x} \text{ and } \mathbf{y}$$

It is noted that the real or complex vector $\mathbf{x}$ is said to be *normalized* if $\langle \mathbf{x}, \mathbf{x} \rangle = 1$. It is also noted that, for an n-vector $\mathbf{x}$, $\mathbf{x}^* \mathbf{x}$ is a nonnegative scalar, but $\mathbf{x}\mathbf{x}^*$ is an $n \times n$ matrix. That is,

$$\mathbf{x}^* \mathbf{x} = \langle \mathbf{x}, \mathbf{x} \rangle = \bar{x}_1 x_1 + \bar{x}_2 x_2 + \cdots + \bar{x}_n x_n$$
$$= |x_1|^2 + |x_2|^2 + \cdots + |x_n|^2$$

and

$$\mathbf{x}\mathbf{x}^* = \begin{bmatrix} x_1 \bar{x}_1 & x_1 \bar{x}_2 & \cdots & x_1 \bar{x}_n \\ x_2 \bar{x}_1 & x_2 \bar{x}_2 & \cdots & x_2 \bar{x}_n \\ \vdots & \vdots & & \vdots \\ x_n \bar{x}_1 & x_n \bar{x}_2 & \cdots & x_n \bar{x}_n \end{bmatrix}$$

Notice that, for an $n \times n$ complex matrix $\mathbf{A}$ and complex n-vectors $\mathbf{x}$ and $\mathbf{y}$, the inner product of $\mathbf{x}$ and $\mathbf{A}\mathbf{y}$ and that of $\mathbf{A}^*\mathbf{x}$ and $\mathbf{y}$ are the same, or

$$\langle \mathbf{x}, \mathbf{A}\mathbf{y} \rangle = \mathbf{x}^* \mathbf{A} \mathbf{y}, \qquad \langle \mathbf{A}^*\mathbf{x}, \mathbf{y} \rangle = \mathbf{x}^* \mathbf{A} \mathbf{y}$$

Similarly, for an $n \times n$ real matrix $\mathbf{A}$ and real n-vectors $\mathbf{x}$ and $\mathbf{y}$, the inner product of $\mathbf{x}$ and $\mathbf{A}\mathbf{y}$ and that of $\mathbf{A}^T\mathbf{x}$ and $\mathbf{y}$ are the same, or

$$\langle \mathbf{x}, \mathbf{A}\mathbf{y} \rangle = \mathbf{x}^T \mathbf{A} \mathbf{y}, \qquad \langle \mathbf{A}^T\mathbf{x}, \mathbf{y} \rangle = \mathbf{x}^T \mathbf{A} \mathbf{y}$$

Unitary Transformation.   If $\mathbf{A}$ is a unitary matrix (that is, if $\mathbf{A}^{-1} = \mathbf{A}^*$), then the inner product $\langle \mathbf{x}, \mathbf{x} \rangle$ is invariant under the linear transformation $\mathbf{x} = \mathbf{Ay}$, because

$$\langle \mathbf{x}, \mathbf{x} \rangle = \langle \mathbf{Ay}, \mathbf{Ay} \rangle = \langle \mathbf{y}, \mathbf{A}^*\mathbf{Ay} \rangle = \langle \mathbf{y}, \mathbf{A}^{-1}\mathbf{Ay} \rangle = \langle \mathbf{y}, \mathbf{y} \rangle$$

Such a transformation $\mathbf{x} = \mathbf{Ay}$, where $\mathbf{A}$ is a unitary matrix, which transforms $\sum_{i=1}^{n} \bar{x}_i x_i$ into $\sum_{i=1}^{n} \bar{y}_i y_i$, is called a *unitary transformation*.

Orthogonal Transformation.   If $\mathbf{A}$ is an orthogonal matrix (that is, if $\mathbf{A}^{-1} = \mathbf{A}^T$), then the inner product $\langle \mathbf{x}, \mathbf{x} \rangle$ is invariant under the linear transformation $\mathbf{x} = \mathbf{Ay}$, because

$$\langle \mathbf{x}, \mathbf{x} \rangle = \langle \mathbf{Ay}, \mathbf{Ay} \rangle = \langle \mathbf{y}, \mathbf{A}^T\mathbf{Ay} \rangle = \langle \mathbf{y}, \mathbf{A}^{-1}\mathbf{Ay} \rangle = \langle \mathbf{y}, \mathbf{y} \rangle$$

Such a transformation $\mathbf{x} = \mathbf{Ay}$, which transforms $\sum_{i=1}^{n} x_i^2$ into $\sum_{i=1}^{n} y_i^2$, is called an *orthogonal transformation*.

Norms of a Vector.   Once we define the inner product, we can use this inner product to define norms of a vector $\mathbf{x}$. The concept of a norm is somewhat similar to that of the absolute value. A norm is a function that assigns to every vector $\mathbf{x}$ in a given vector space a real number denoted by $\|\mathbf{x}\|$ such that

1. $$\|\mathbf{x}\| > 0, \quad \text{for } \mathbf{x} \neq \mathbf{0}$$

2. $$\|\mathbf{x}\| = 0, \quad \text{if and only if } \mathbf{x} = \mathbf{0}$$

3. $$\|k\mathbf{x}\| = |k|\,\|\mathbf{x}\|,$$

   where k is a scalar and $|k|$ is the absolute value of k

4. $$\|\mathbf{x} + \mathbf{y}\| \leq \|\mathbf{x}\| + \|\mathbf{y}\|, \quad \text{for all } \mathbf{x} \text{ and } \mathbf{y}$$

5. $$|\langle \mathbf{x}, \mathbf{y} \rangle| \leq \|\mathbf{x}\|\,\|\mathbf{y}\| \quad \text{(Schwarz inequality)}$$

Several different definitions of norms are commonly used in the literature. However, the following definition is widely used. A norm of a vector is defined as the nonnegative square root of $\langle \mathbf{x}, \mathbf{x} \rangle$:

$$\|\mathbf{x}\| = \langle \mathbf{x}, \mathbf{x} \rangle^{1/2} = (\mathbf{x}^*\mathbf{x})^{1/2} = \sqrt{|x_1|^2 + |x_2|^2 + \cdots + |x_n|^2} \qquad \text{(A–29)}$$

If $\mathbf{x}$ is a real vector, the quantity $\|\mathbf{x}\|^2$ can be interpreted geometrically as the square of the distance from the origin to the point represented by the vector $\mathbf{x}$. Note that

$$\|\mathbf{x} - \mathbf{y}\| = \langle \mathbf{x} - \mathbf{y}, \mathbf{x} - \mathbf{y} \rangle^{1/2} = \sqrt{(x_1 - y_1)^2 + (x_2 - y_2)^2 + \cdots + (x_n - y_n)^2}$$

The five properties of norms listed earlier may be obvious, except perhaps the last two inequalities. These two inequalities may be proved as follows. From the definitions of the inner product and the norm, we have

$$\|\lambda\mathbf{x} + \mathbf{y}\|^2 = \langle \lambda\mathbf{x} + \mathbf{y}, \lambda\mathbf{x} + \mathbf{y} \rangle = \langle \lambda\mathbf{x}, \lambda\mathbf{x} \rangle + \langle \mathbf{y}, \lambda\mathbf{x} \rangle + \langle \lambda\mathbf{x}, \mathbf{y} \rangle + \langle \mathbf{y}, \mathbf{y} \rangle$$

$$= \bar{\lambda}\lambda\|\mathbf{x}\|^2 + \lambda\langle \mathbf{y}, \mathbf{x} \rangle + \bar{\lambda}\langle \mathbf{x}, \mathbf{y} \rangle + \|\mathbf{y}\|^2$$

$$= \bar{\lambda}(\lambda\|\mathbf{x}\|^2 + \langle \mathbf{x}, \mathbf{y} \rangle) + \lambda\overline{\langle \mathbf{x}, \mathbf{y} \rangle} + \|\mathbf{y}\|^2 \geq 0$$

If we choose

$$\lambda = -\frac{\langle \mathbf{x}, \mathbf{y} \rangle}{\|\mathbf{x}\|^2}, \qquad \text{for } \mathbf{x} \neq \mathbf{0}$$

then

$$\lambda \overline{\langle \mathbf{x}, \mathbf{y} \rangle} + \|\mathbf{y}\|^2 = -\frac{\langle \mathbf{x}, \mathbf{y} \rangle \overline{\langle \mathbf{x}, \mathbf{y} \rangle}}{\|\mathbf{x}\|^2} + \|\mathbf{y}\|^2 \geq 0$$

and

$$\|\mathbf{x}\|^2 \|\mathbf{y}\|^2 \geq \langle \mathbf{x}, \mathbf{y} \rangle \overline{\langle \mathbf{x}, \mathbf{y} \rangle} = |\langle \mathbf{x}, \mathbf{y} \rangle|^2, \qquad \text{for } \mathbf{x} \neq \mathbf{0}$$

For $\mathbf{x} = \mathbf{0}$, clearly,

$$\|\mathbf{x}\|^2 \|\mathbf{y}\|^2 = |\langle \mathbf{x}, \mathbf{y} \rangle|^2$$

Therefore, we obtain the Schwarz inequality,

$$|\langle \mathbf{x}, \mathbf{y} \rangle| \leq \|\mathbf{x}\| \, \|\mathbf{y}\| \tag{A–30}$$

By use of the Schwarz inequality, we obtain the following inequality:

$$\|\mathbf{x} + \mathbf{y}\| \leq \|\mathbf{x}\| + \|\mathbf{y}\| \tag{A–31}$$

This can be proved easily, since

$$
\begin{aligned}
\|\mathbf{x} + \mathbf{y}\|^2 &= \langle \mathbf{x} + \mathbf{y}, \mathbf{x} + \mathbf{y} \rangle \\
&= \langle \mathbf{x}, \mathbf{x} \rangle + \langle \mathbf{x}, \mathbf{y} \rangle + \langle \mathbf{y}, \mathbf{x} \rangle + \langle \mathbf{y}, \mathbf{y} \rangle \\
&= \|\mathbf{x}\|^2 + \langle \mathbf{x}, \mathbf{y} \rangle + \overline{\langle \mathbf{x}, \mathbf{y} \rangle} + \|\mathbf{y}\|^2 \\
&= \|\mathbf{x}\|^2 + \|\mathbf{y}\|^2 + 2 \operatorname{Re} \langle \mathbf{x}, \mathbf{y} \rangle \\
&\leq \|\mathbf{x}\|^2 + \|\mathbf{y}\|^2 + 2|\langle \mathbf{x}, \mathbf{y} \rangle| \\
&\leq \|\mathbf{x}\|^2 + \|\mathbf{y}\|^2 + 2\|\mathbf{x}\| \, \|\mathbf{y}\| \\
&= (\|\mathbf{x}\| + \|\mathbf{y}\|)^2
\end{aligned}
$$

Equations (A–26) through (A–31) are useful in modern control theory.

As stated earlier, different definitions of norms are used in the literature. Three such definitions of norms follow.

1. A norm $\|\mathbf{x}\|$ may be defined as follows:

$$\|\mathbf{x}\| = [(\mathbf{Tx})*(\mathbf{Tx})]^{1/2} = (\mathbf{x}*\mathbf{T}*\mathbf{Tx})^{1/2} = (\mathbf{x}*\mathbf{Qx})^{1/2}$$

$$= \left[\sum_{i=1}^{n} \sum_{j=1}^{n} q_{ij} \bar{x}_i x_j \right]^{1/2} \geq 0$$

The matrix $\mathbf{Q} = \mathbf{T}*\mathbf{T}$ is Hermitian, since $\mathbf{Q}* = \mathbf{T}*\mathbf{T} = \mathbf{Q}$. The norm $\|\mathbf{x}\| = (\mathbf{x}*\mathbf{Qx})^{1/2}$ is a generalized form of $(\mathbf{x}*\mathbf{x})^{1/2}$, which can be written as $(\mathbf{x}*\mathbf{Ix})^{1/2}$.

2. A norm may be defined as the sum of the magnitudes of all the components x_i:

$$\|\mathbf{x}\| = \sum_{i=1}^{n} |x_i|$$

3. A norm may be defined as the maximum of the magnitudes of all the components x_i:

$$\|\mathbf{x}\| = \max_i \{|x_i|\}$$

It can be shown that the various norms just defined are equivalent. Among these definitions of norms, norm $(\mathbf{x}^*\mathbf{x})^{1/2}$ is most commonly used in explicit calculations.

Norms of a Matrix.   The concept of norms of a vector can be extended to matrices. There are several different definitions of norms of a matrix. Some of them follow.

1. A norm $\|\mathbf{A}\|$ of an $n \times n$ matrix $\mathbf{A}$ may be defined by

$$\|\mathbf{A}\| = \min k$$

such that

$$\|\mathbf{Ax}\| \le k\|\mathbf{x}\|$$

For the norm $(\mathbf{x}^*\mathbf{x})^{1/2}$, this definition is equivalent to

$$\|\mathbf{A}\|^2 = \max_{\mathbf{x}} \{\mathbf{x}^*\mathbf{A}^*\mathbf{Ax}; \mathbf{x}^*\mathbf{x} = 1\}$$

which means that $\|\mathbf{A}\|^2$ is the maximum of the "absolute value" of the vector $\mathbf{Ax}$ when $\mathbf{x}^*\mathbf{x} = 1$.

2. A norm of an $n \times n$ matrix $\mathbf{A}$ may be defined by

$$\|\mathbf{A}\| = \sum_{i=1}^{n} \sum_{j=1}^{n} |a_{ij}|$$

where $|a_{ij}|$ is the absolute value of a_{ij}.

3. A norm may be defined by

$$\|\mathbf{A}\| = \left(\sum_{i=1}^{n} \sum_{j=1}^{n} |a_{ij}|^2 \right)^{1/2}$$

4. Another definition of a norm is given by

$$\|\mathbf{A}\| = \max_i \left(\sum_{j=1}^{n} |a_{ij}| \right)$$

Note that all definitions of norms of an $n \times n$ matrix $\mathbf{A}$ have the following properties:

1. $$\|\mathbf{A}\| = \|\mathbf{A}^*\| \quad \text{or} \quad \|\mathbf{A}\| = \|\mathbf{A}^T\|$$

2. $$\|\mathbf{A} + \mathbf{B}\| \le \|\mathbf{A}\| + \|\mathbf{B}\|$$

3. $$\|\mathbf{AB}\| \le \|\mathbf{A}\| \|\mathbf{B}\|$$

4. $$\|\mathbf{Ax}\| \le \|\mathbf{A}\| \|\mathbf{x}\|$$

Orthogonality of Vectors.   If the inner product of two vectors $\mathbf{x}$ and $\mathbf{y}$ is zero, or $\langle \mathbf{x}, \mathbf{y} \rangle = 0$, then vectors $\mathbf{x}$ and $\mathbf{y}$ are said to be *orthogonal to each other*. For example, vectors

$$\mathbf{x}_1 = \begin{bmatrix} 1 \\ 1 \\ 0 \end{bmatrix}, \qquad \mathbf{x}_2 = \begin{bmatrix} 0 \\ 0 \\ 1 \end{bmatrix}, \qquad \mathbf{x}_3 = \begin{bmatrix} 1 \\ -1 \\ 0 \end{bmatrix}$$

are orthogonal in pairs and thus form an orthogonal set.

In an n-dimensional vector space, vectors $\mathbf{x}_1, \mathbf{x}_2, \ldots, \mathbf{x}_n$ defined by

$$\mathbf{x}_1 = \begin{bmatrix} 1 \\ 0 \\ \vdots \\ 0 \end{bmatrix}, \qquad \mathbf{x}_2 = \begin{bmatrix} 0 \\ 1 \\ \vdots \\ 0 \end{bmatrix}, \qquad \ldots, \qquad \mathbf{x}_n = \begin{bmatrix} 0 \\ 0 \\ \vdots \\ 1 \end{bmatrix}$$

satisfy the conditions $\langle \mathbf{x}_i, \mathbf{x}_j \rangle = \delta_{ij}$, or

$$\langle \mathbf{x}_i, \mathbf{x}_i \rangle = 1$$

$$\langle \mathbf{x}_i, \mathbf{x}_j \rangle = 0, \qquad i \neq j$$

where $i, j = 1, 2, \ldots, n$. Such a set of vectors is said to be *orthonormal*, since the vectors are orthogonal to each other and each vector is normalized.

A nonzero vector $\mathbf{x}$ can be normalized by dividing $\mathbf{x}$ by $\|\mathbf{x}\|$. The normalized vector $\mathbf{x}/\|\mathbf{x}\|$ is a unit vector. Unit vectors $\mathbf{x}_1, \mathbf{x}_2, \ldots, \mathbf{x}_n$ form an orthonormal set if they are orthogonal in pairs.

Consider a unitary matrix $\mathbf{A}$. By partitioning $\mathbf{A}$ into column vectors $\mathbf{A}_1$, $\mathbf{A}_2, \ldots, \mathbf{A}_n$, we have

$$\mathbf{A}^*\mathbf{A} = \begin{bmatrix} \mathbf{A}_1^* \\ \hline \mathbf{A}_2^* \\ \hline \vdots \\ \hline \mathbf{A}_n^* \end{bmatrix} [\mathbf{A}_1 \vdots \mathbf{A}_2 \vdots \cdots \vdots \mathbf{A}_n]$$

$$= \begin{bmatrix} \mathbf{A}_1^* \mathbf{A}_1 & \mathbf{A}_1^* \mathbf{A}_2 & \cdots & \mathbf{A}_1^* \mathbf{A}_n \\ \mathbf{A}_2^* \mathbf{A}_1 & \mathbf{A}_2^* \mathbf{A}_2 & \cdots & \mathbf{A}_2^* \mathbf{A}_n \\ \vdots & \vdots & & \vdots \\ \mathbf{A}_n^* \mathbf{A}_1 & \mathbf{A}_n^* \mathbf{A}_2 & \cdots & \mathbf{A}_n^* \mathbf{A}_n \end{bmatrix}$$

$$= \begin{bmatrix} 1 & 0 & \cdots & 0 \\ 0 & 1 & \cdots & 0 \\ \vdots & \vdots & & \vdots \\ 0 & 0 & \cdots & 1 \end{bmatrix}$$

it follows that

$$\mathbf{A}_i^* \mathbf{A}_i = \langle \mathbf{A}_i, \mathbf{A}_i \rangle = 1$$

$$\mathbf{A}_i^* \mathbf{A}_j = \langle \mathbf{A}_i, \mathbf{A}_j \rangle = 0, \qquad i \neq j$$

Thus, we see that the column vectors (or row vectors) of a unitary matrix $\mathbf{A}$ are orthonormal. The same is true for orthogonal matrices, since they are unitary.

A–6 EIGENVALUES, EIGENVECTORS, AND SIMILARITY TRANSFORMATION

In this section we shall first review important properties of the rank of a matrix and then give definitions of eigenvalues and eigenvectors. Finally, we shall discuss Jordan canonical forms, similarity transformation, and the trace of an $n \times n$ matrix.

Rank of a Matrix.   A matrix **A** is called of rank m if the maximum number of linearly independent rows (or columns) is m. Hence, if there exists an $m \times m$ submatrix **M** of **A** such that $|\mathbf{M}| \neq 0$ and the determinant of every $r \times r$ submatrix (where $r \geq m + 1$) of **A** is zero, then the rank of **A** is m. [Note that, if the determinant of every $(m + 1) \times (m + 1)$ submatrix of **A** is zero, then any determinant of order s (where $s > m + 1$) is zero, since any determinant of order $s > m + 1$ can be expressed as a linear sum of determinants of order $m + 1$.]

Properties of Rank of a Matrix.   We shall list important properties of the rank of a matrix in the following.

1. The rank of a matrix is invariant under the interchange of two rows (or columns), or the addition of a scalar multiple of a row (or column) to another row (or column), or the multiplication of any row (or column) by a nonzero scalar.

2. For an $n \times m$ matrix **A**,

$$\text{rank } \mathbf{A} \leq \min(n, m)$$

3. For an $n \times n$ matrix **A**, a necessary and sufficient condition for rank $\mathbf{A} = n$ is that $|\mathbf{A}| \neq 0$.

4. For an $n \times m$ matrix **A**,

$$\text{rank } \mathbf{A}^* = \text{rank } \mathbf{A} \qquad \text{or} \qquad \text{rank } \mathbf{A}^T = \text{rank } \mathbf{A}$$

5. The rank of a product of two matrices **AB** cannot exceed the rank of **A** or the rank of **B**; that is,

$$\text{rank } \mathbf{AB} \leq \min(\text{rank } \mathbf{A}, \text{rank } \mathbf{B})$$

Hence, if **A** is an $n \times 1$ matrix and **B** is a $1 \times m$ matrix, then rank $\mathbf{AB} = 1$ unless $\mathbf{AB} = \mathbf{0}$. If a matrix has rank 1, then this matrix can be expressed as a product of a column vector and a row vector.

6. For an $n \times n$ matrix **A** (where $|\mathbf{A}| \neq 0$) and an $n \times m$ matrix **B**,

$$\text{rank } \mathbf{AB} = \text{rank } \mathbf{B}$$

Similarly, for an $m \times m$ matrix **A** (where $|\mathbf{A}| \neq 0$) and an $n \times m$ matrix **B**,

$$\text{rank } \mathbf{BA} = \text{rank } \mathbf{B}$$

Eigenvalues of a Square Matrix.   For an $n \times n$ matrix **A**, the determinant

$$|\lambda \mathbf{I} - \mathbf{A}|$$

is called the *characteristic polynomial* of **A**. It is an nth-degree polynomial in λ. The characteristic equation is given by

$$|\lambda \mathbf{I} - \mathbf{A}| = 0$$

If the determinant $|\lambda\mathbf{I} - \mathbf{A}|$ is expanded, the characteristic equation becomes

$$|\lambda\mathbf{I} - \mathbf{A}| = \begin{vmatrix} \lambda - a_{11} & -a_{12} & \cdots & -a_{1n} \\ -a_{21} & \lambda - a_{22} & \cdots & -a_{2n} \\ \vdots & \vdots & & \vdots \\ -a_{n1} & -a_{n2} & \cdots & \lambda - a_{nn} \end{vmatrix}$$

$$= \lambda^n + a_1\lambda^{n-1} + \cdots + a_{n-1}\lambda + a_n = 0$$

The n roots of the characteristic equation are called the *eigenvalues* of $\mathbf{A}$. They are also called the *characteristic roots*.

It is noted that an $n \times n$ real matrix $\mathbf{A}$ does not necessarily possess real eigenvalues. However, for an $n \times n$ real matrix $\mathbf{A}$, the characteristic equation $|\lambda\mathbf{I} - \mathbf{A}| = 0$ is a polynomial with real coefficients, and therefore any complex eigenvalues must occur in conjugate pairs; that is, if $\alpha + j\beta$ is an eigenvalue of $\mathbf{A}$, then $\alpha - j\beta$ is also an eigenvalue of $\mathbf{A}$.

There is an important relationship between the eigenvalues of an $n \times n$ matrix $\mathbf{A}$ and those of $\mathbf{A}^{-1}$. If we assume the eigenvalues of $\mathbf{A}$ to be λ_i and those of $\mathbf{A}^{-1}$ to be μ_i, then

$$\mu_i = \lambda_i^{-1}, \qquad i = 1, 2, \ldots, n$$

That is, if λ_i is an eigenvalue of $\mathbf{A}$, then λ_i^{-1} is an eigenvalue of $\mathbf{A}^{-1}$. To prove this, notice that the characteristic equation for matrix $\mathbf{A}$ can be written as

$$|\lambda\mathbf{I} - \mathbf{A}| = |\lambda\mathbf{A}^{-1} - \mathbf{I}||\mathbf{A}| = |\lambda||\mathbf{A}^{-1} - \lambda^{-1}\mathbf{I}||\mathbf{A}| = 0$$

or

$$|\lambda^{-1}\mathbf{I} - \mathbf{A}^{-1}| = 0$$

By assumption, the characteristic equation for the inverse matrix $\mathbf{A}^{-1}$ is

$$|\mu\mathbf{I} - \mathbf{A}^{-1}| = 0$$

By comparing the last two equations, we see that

$$\mu = \lambda^{-1}$$

Hence, if λ is an eigenvalue of $\mathbf{A}$, then $\mu = \lambda^{-1}$ is an eigenvalue of $\mathbf{A}^{-1}$.

Finally, note that it is possible to prove that, for two square matrices $\mathbf{A}$ and $\mathbf{B}$,

$$|\lambda\mathbf{I} - \mathbf{AB}| = |\lambda\mathbf{I} - \mathbf{BA}|$$

(For the proof, see Problem A–9.)

Eigenvectors of an $n \times n$ Matrix. Any nonzero vector $\mathbf{x}_i$ such that

$$\mathbf{A}\mathbf{x}_i = \lambda_i\mathbf{x}_i$$

is said to be an *eigenvector* associated with an eigenvalue λ_i of $\mathbf{A}$, where $\mathbf{A}$ is an $n \times n$ matrix. Since the components of $\mathbf{x}_i$ are determined from n linear homogeneous algebraic equations within a constant factor, if $\mathbf{x}_i$ is an eigenvector, then for any scalar $\alpha \neq 0$, $\alpha\mathbf{x}_i$ is also an eigenvector. The eigenvector is said to be a *normalized* eigenvector if its length or absolute value is unity.

Similar Matrices.   The $n \times n$ matrices **A** and **B** are said to be *similar* if a nonsingular matrix **P** exists such that

$$\mathbf{P}^{-1}\mathbf{A}\mathbf{P} = \mathbf{B}$$

The matrix **B** is said to be obtained from **A** by a *similarity transformation*, in which **P** is the transformation matrix. Notice that **A** can be obtained from **B** by a similarity transformation with a transformation matrix $\mathbf{P}^{-1}$, since

$$\mathbf{A} = \mathbf{P}\mathbf{B}\mathbf{P}^{-1} = (\mathbf{P}^{-1})^{-1}\mathbf{B}(\mathbf{P}^{-1})$$

Diagonalization of Matrices.   If an $n \times n$ matrix **A** has n distinct eigenvalues, then there are n linearly independent eigenvectors. If matrix **A** has a multiple eigenvalue of multiplicity k, then there are at least one and not more than k linearly independent eigenvectors associated with this eigenvalue.

If an $n \times n$ matrix has n linearly independent eigenvectors, it can be diagonalized by a similarity transformation. However, a matrix that does not have a complete set of n linearly independent eigenvectors cannot be diagonalized. Such a matrix can be transformed into a Jordan canonical form.

Jordan Canonical Form.   A $k \times k$ matrix **J** is said to be in the Jordan canonical form if

$$\mathbf{J} = \begin{bmatrix} \mathbf{J}_{p_1} & & & \mathbf{0} \\ & \mathbf{J}_{p_2} & & \\ & & \cdot \cdot \cdot & \\ \mathbf{0} & & & \mathbf{J}_{p_s} \end{bmatrix}$$

where the $\mathbf{J}_{p_i}$'s are $p_i \times p_i$ matrices of the form

$$\mathbf{J}_{p_i} = \begin{bmatrix} \lambda & 1 & 0 & \cdots & 0 & 0 \\ 0 & \lambda & 1 & \cdots & 0 & 0 \\ \vdots & \vdots & \vdots & & \vdots & \vdots \\ 0 & 0 & 0 & \cdots & \lambda & 1 \\ 0 & 0 & 0 & \cdots & 0 & \lambda \end{bmatrix}$$

The matrices $\mathbf{J}_{p_i}$ are called p_ith-order Jordan blocks. Note that the λ in $\mathbf{J}_{p_i}$ and that in $\mathbf{J}_{p_j}$ may or may not be the same, and that

$$p_1 + p_2 + \cdots + p_s = k$$

For example, in a 7×7 matrix **J**, if $p_1 = 3, p_2 = 2, p_3 = 1, p_4 = 1$, and the eigenvalues of **J** are $\lambda_1, \lambda_1, \lambda_1, \lambda_1, \lambda_1, \lambda_6, \lambda_7$, then the Jordan canonical form may be given by

$$\mathbf{J} = \begin{bmatrix} \mathbf{J}_3(\lambda_1) & & & \mathbf{0} \\ & \mathbf{J}_2(\lambda_1) & & \\ & & \mathbf{J}_1(\lambda_6) & \\ \mathbf{0} & & & \mathbf{J}_1(\lambda_7) \end{bmatrix} = \begin{bmatrix} \lambda_1 & 1 & 0 & & & & \mathbf{0} \\ 0 & \lambda_1 & 1 & & & & \\ 0 & 0 & \lambda_1 & & & & \\ & & & \lambda_1 & 1 & & \\ & & & 0 & \lambda_1 & & \\ & & & & & \lambda_6 & \\ \mathbf{0} & & & & & & \lambda_7 \end{bmatrix}$$

Notice that a diagonal matrix is a special case of the Jordan canonical form.

Jordan canonical forms have the properties that the elements on the main diagonal of the matrix are the eigenvalues of **A** and that the elements immediately above (or below) the main diagonal are either 1 or 0 and all other elements are zeros.

The determination of the exact form of the Jordan block may not be simple. To illustrate some possible structures, consider a 3×3 matrix having a triple eigenvalue of λ_1. Then any one of the following Jordan canonical forms is possible:

$$\begin{bmatrix} \lambda_1 & 1 & 0 \\ 0 & \lambda_1 & 1 \\ 0 & 0 & \lambda_1 \end{bmatrix}, \qquad \begin{bmatrix} \lambda_1 & 1 & 0 \\ 0 & \lambda_1 & 0 \\ 0 & 0 & \lambda_1 \end{bmatrix}, \qquad \begin{bmatrix} \lambda_1 & 0 & 0 \\ 0 & \lambda_1 & 0 \\ 0 & 0 & \lambda_1 \end{bmatrix}$$

Each of the three preceding matrices has the same characteristic equation $(\lambda - \lambda_1)^3 = 0$. The first one corresponds to the case where there exists only one linearly independent eigenvector, since by denoting the first matrix by **A** and solving the following equation for **x**,

$$(\mathbf{A} - \lambda_1 \mathbf{I})\mathbf{x} = \mathbf{0}$$

we obtain only one eigenvector:

$$\mathbf{x} = \begin{bmatrix} a \\ 0 \\ 0 \end{bmatrix}, \qquad a = \text{nonzero constant}$$

The second and third of these matrices have, respectively, two and three linearly independent eigenvectors. (Notice that only the diagonal matrix has three linearly independent eigenvectors.)

As we have seen, if a $k \times k$ matrix **A** has a k-multiple eigenvalue, then the following can be shown:

1. If the rank of $\lambda \mathbf{I} - \mathbf{A}$ is $k - s$ (where $1 \le s \le k$), then there exist s linearly independent eigenvectors associated with λ.
2. There are s Jordan blocks corresponding to the s eigenvectors.
3. The sum of the orders p_i of the Jordan blocks equals the multiplicity k.

Therefore, as demonstrated in the preceding three 3×3 matrices, even if the multiplicity of the eigenvalue is the same, the number of Jordan blocks and their orders may be different depending on the structure of matrix **A**.

Similarity Transformation When an $n \times n$ Matrix Has Distinct Eigenvalues. If n eigenvalues of **A** are distinct, there exists one eigenvector associated with each eigenvalue λ_i. It can be proved that such n eigenvectors $\mathbf{x}_1, \mathbf{x}_2, \dots, \mathbf{x}_n$ are linearly independent.

Let us define an $n \times n$ matrix **P** such that

$$\mathbf{P} = [\mathbf{P}_1 \vdots \mathbf{P}_2 \vdots \cdots \vdots \mathbf{P}_n] = [\mathbf{x}_1 \vdots \mathbf{x}_2 \vdots \cdots \vdots \mathbf{x}_n]$$

where column vector $\mathbf{P}_i$ is equal to column vector $\mathbf{x}_i$, or

$$\mathbf{P}_i = \mathbf{x}_i, \qquad i = 1, 2, \dots, n$$

Matrix **P** defined in this way is nonsingular, and $\mathbf{P}^{-1}$ exists. Noting that eigenvectors $\mathbf{x}_1, \mathbf{x}_2, \ldots, \mathbf{x}_n$ satisfy the equations

$$\mathbf{Ax}_1 = \lambda_1 \mathbf{x}_1$$

$$\mathbf{Ax}_2 = \lambda_2 \mathbf{x}_2$$

$$\vdots$$

$$\mathbf{Ax}_n = \lambda_n \mathbf{x}_n$$

we may combine these n equations into one, as follows:

$$\mathbf{A}[\mathbf{x}_1 \vdots \mathbf{x}_2 \vdots \cdots \vdots \mathbf{x}_n] = [\mathbf{x}_1 \vdots \mathbf{x}_2 \vdots \cdots \vdots \mathbf{x}_n] \begin{bmatrix} \lambda_1 & & & 0 \\ & \lambda_2 & & \\ & & \ddots & \\ 0 & & & \lambda_n \end{bmatrix}$$

or, in terms of matrix **P**,

$$\mathbf{AP} = \mathbf{P} \begin{bmatrix} \lambda_1 & & & 0 \\ & \lambda_2 & & \\ & & \ddots & \\ 0 & & & \lambda_n \end{bmatrix}$$

By premultiplying this last equation by $\mathbf{P}^{-1}$, we obtain

$$\mathbf{P}^{-1}\mathbf{AP} = \begin{bmatrix} \lambda_1 & & & 0 \\ & \lambda_2 & & \\ & & \ddots & \\ 0 & & & \lambda_n \end{bmatrix} = \mathrm{diag}\,(\lambda_1, \lambda_2, \ldots, \lambda_n)$$

Thus, matrix **A** is transformed into a diagonal matrix by a similarity transformation.

The process that transforms matrix **A** into a diagonal matrix is called the *diagonalization* of matrix **A**.

As noted earlier, a scalar multiple of eigenvector $\mathbf{x}_i$ is also an eigenvector, since $\alpha\mathbf{x}_i$ satisfies the following equation:

$$\mathbf{A}(\alpha\mathbf{x}_i) = \lambda_i(\alpha\mathbf{x}_i)$$

Consequently, we may choose an α such that the transformation matrix **P** becomes as simple as possible.

To summarize, if the eigenvalues of an $n \times n$ matrix **A** are distinct, then there are exactly n eigenvectors and they are linearly independent. A transformation matrix **P** that transforms **A** into a diagonal matrix can be constructed from such n linearly independent eigenvectors.

Similarity Transformation When an $n \times n$ Matrix Has Multiple Eigenvalues. Let us assume that an $n \times n$ matrix **A** involves a k-multiple eigenvalue λ_1 and other eigenvalues $\lambda_{k+1}, \lambda_{k+2}, \ldots, \lambda_n$ that are all distinct and different from λ_1. That is, the eigenvalues of **A** are

$$\lambda_1, \lambda_1, \ldots, \lambda_1, \lambda_{k+1}, \lambda_{k+2}, \ldots, \lambda_n$$

We shall first consider the case where the rank of $\lambda_1 I - A$ is $n - 1$. For such a case there exists only one Jordan block for the multiple eigenvalue λ_1, and there is only one eigenvector associated with this multiple eigenvalue. The order of the Jordan block is k, which is the same as the order of multiplicity of the eigenvalue λ_1.

Note that, when an $n \times n$ matrix A does not possess n linearly independent eigenvectors, it cannot be diagonalized, but can be reduced to a Jordan canonical form.

In the present case, only one linearly independent eigenvector exists for λ_1. We shall now investigate whether it is possible to find $k - 1$ vectors that are somehow associated with this eigenvalue and that are linearly independent of the eigenvectors. Without proof, we shall show that this is possible. First, note that the eigenvector x_1 is a vector that satisfies the equation

$$(A - \lambda_1 I)x_1 = 0$$

so that x_1 is annihilated by $A - \lambda_1 I$. Since we do not have enough vectors that are annihilated by $A - \lambda_1 I$, we seek vectors that are annihilated by $(A - \lambda_1 I)^2$, $(A - \lambda_1 I)^3$, and so on, until we obtain $k - 1$ vectors. The $k - 1$ vectors determined in this way are called *generalized eigenvectors*.

Let us define the desired $k - 1$ generalized eigenvectors as $x_2, x_3, \ldots, x_k$. Then these $k - 1$ generalized eigenvectors can be determined from the equations

$$(A - \lambda_1 I)x_1 = 0$$
$$(A - \lambda_1 I)^2 x_2 = 0$$
$$\vdots$$
$$(A - \lambda_1 I)^k x_k = 0 \tag{A-32}$$

which can be rewritten as

$$(A - \lambda_1 I)x_1 = 0$$
$$(A - \lambda_1 I)x_2 = x_1$$
$$\vdots$$
$$(A - \lambda_1 I)x_k = x_{k-1}$$

Notice that

$$(A - \lambda_1 I)^{k-1} x_k = (A - \lambda_1 I)^{k-2} x_{k-1} = \cdots = (A - \lambda_1 I)x_2 = x_1$$

or

$$(A - \lambda_1 I)^{k-1} x_k = x_1 \tag{A-33}$$

The eigenvector x_1 and the $k - 1$ generalized eigenvectors $x_2, x_3, \ldots, x_k$ determined in this way form a set of k linearly independent vectors.

A proper way to determine the generalized eigenvectors is to start with x_k. That is, we first determine the x_k that will satisfy Equation (A-32) and at the same time will yield a nonzero vector $(A - \lambda_1 I)^{k-1} x_k$. Any such nonzero vector can be considered as a possible eigenvector x_1. Therefore, to find eigenvector x_1, we apply a row reduction process to $(\mathbf{A} - \lambda_1 \mathbf{I})^k$ and find k linearly independent vectors satisfying Equation (A–32). Then these vectors are tested to find one that yields a nonzero vector on the right-hand side of Equation (A–33). (Note that if we start with $\mathbf{x}_1$ then we must make arbitrary choices at each step along the way to determine $\mathbf{x}_2$, $\mathbf{x}_3, \ldots, \mathbf{x}_k$. This is time consuming and inconvenient. For this reason, this approach is not recommended.)

To summarize what we have discussed so far, the eigenvector $\mathbf{x}_1$ and the generalized eigenvectors $\mathbf{x}_2, \mathbf{x}_3, \ldots, \mathbf{x}_k$ satisfy the following equations:

$$\mathbf{Ax}_1 = \lambda_1 \mathbf{x}_1$$

$$\mathbf{Ax}_2 = \mathbf{x}_1 + \lambda_1 \mathbf{x}_2$$

$$\vdots$$

$$\mathbf{Ax}_k = \mathbf{x}_{k-1} + \lambda_1 \mathbf{x}_k$$

The eigenvectors $\mathbf{x}_{k+1}, \mathbf{x}_{k+2}, \ldots, \mathbf{x}_n$ associated with distinct eigenvalues λ_{k+1}, $\lambda_{k+2}, \ldots, \lambda_n$, respectively, can be determined from

$$\mathbf{Ax}_{k+1} = \lambda_{k+1} \mathbf{x}_{k+1}$$

$$\mathbf{Ax}_{k+2} = \lambda_{k+2} \mathbf{x}_{k+2}$$

$$\vdots$$

$$\mathbf{Ax}_n = \lambda_n \mathbf{x}_n$$

Now define

$$\mathbf{S} = [\mathbf{S}_1 \vdots \mathbf{S}_2 \vdots \cdots \vdots \mathbf{S}_n] = [\mathbf{x}_1 \vdots \mathbf{x}_2 \vdots \cdots \vdots \mathbf{x}_n]$$

where the n column vectors of $\mathbf{S}$ are linearly independent. Thus, matrix $\mathbf{S}$ is nonsingular. Then, combining the preceding eigenvector equations and generalized eigenvector equations into one, we obtain

$$\mathbf{A}[\mathbf{x}_1 \vdots \mathbf{x}_2 \vdots \cdots \vdots \mathbf{x}_k \vdots \mathbf{x}_{k+1} \vdots \cdots \vdots \mathbf{x}_n]$$

$$= [\mathbf{x}_1 \vdots \mathbf{x}_2 \vdots \cdots \vdots \mathbf{x}_k \vdots \mathbf{x}_{k+1} \vdots \cdots \vdots \mathbf{x}_n]
\begin{bmatrix}
\lambda_1 & 1 & & & 0 & & & 0 \\
 & \lambda_1 & 1 & & & & & \\
 & & \ddots & \ddots & & & & \\
 & & & & 1 & & & \\
0 & & & & \lambda_1 & 0 & & \\
\hline
 & & & & 0 & \lambda_{k+1} & & 0 \\
 & & & & & & \ddots & \\
0 & & & & & 0 & & \lambda_n
\end{bmatrix}$$

Hence,

$$\mathbf{AS} = \mathbf{S}
\begin{bmatrix}
\mathbf{J}_k(\lambda_1) & & 0 \\
\hline
 & \lambda_{k+1} & \\
 & & \ddots \\
0 & & \lambda_n
\end{bmatrix}$$

By premultiplying this last equation by $\mathbf{S}^{-1}$, we obtain

$$
\mathbf{S}^{-1}\mathbf{A}\mathbf{S} = \begin{bmatrix} \mathbf{J}_k(\lambda_1) & & & 0 \\ \hline & \lambda_{k+1} & & \\ & & \ddots & \\ 0 & & & \lambda_n \end{bmatrix}
$$

In the preceding discussion we considered the case where the rank of $\lambda_1\mathbf{I} - \mathbf{A}$ was $n - 1$. Next we shall consider the case where the rank of $\lambda_1\mathbf{I} - \mathbf{A}$ is $n - s$ (where $2 \le s \le n$). Since we assumed that matrix $\mathbf{A}$ involves the k-multiple eigenvalue λ_1 and other eigenvalues $\lambda_{k+1}, \lambda_{k+2}, \ldots, \lambda_n$ that are all distinct and different from λ_1, we have s linearly independent eigenvectors associated with eigenvalue λ_1. Hence, there are s Jordan blocks corresponding to eigenvalue λ_1.

For notational convenience, let us define the s linearly independent eigenvectors associated with eigenvalue λ_1 as $\mathbf{v}_{11}, \mathbf{v}_{21}, \ldots, \mathbf{v}_{s1}$. We shall define the generalized eigenvectors associated with $\mathbf{v}_{i1}$ as $\mathbf{v}_{i2}, \mathbf{v}_{i3}, \ldots, \mathbf{v}_{ip_i}$, where $i = 1, 2, \ldots, s$. Then there are altogether k such vectors (eigenvectors and generalized eigenvectors), which are

$$
\mathbf{v}_{11}, \mathbf{v}_{12}, \ldots, \mathbf{v}_{1p_1}, \mathbf{v}_{21}, \mathbf{v}_{22}, \ldots, \mathbf{v}_{2p_2}, \ldots, \mathbf{v}_{s1}, \mathbf{v}_{s2}, \ldots, \mathbf{v}_{sp_s}
$$

The generalized eigenvectors are determined from

$$
\begin{aligned}
(\mathbf{A} - \lambda_1\mathbf{I})\mathbf{v}_{11} &= \mathbf{0}, & \cdots & & (\mathbf{A} - \lambda_1\mathbf{I})\mathbf{v}_{s1} &= \mathbf{0} \\
(\mathbf{A} - \lambda_1\mathbf{I})\mathbf{v}_{12} &= \mathbf{v}_{11}, & \cdots & & (\mathbf{A} - \lambda_1\mathbf{I})\mathbf{v}_{s2} &= \mathbf{v}_{s1} \\
&\;\;\vdots & & & &\;\;\vdots \\
(\mathbf{A} - \lambda_1\mathbf{I})\mathbf{v}_{1p_1} &= \mathbf{v}_{1p_1-1}, & \cdots & & (\mathbf{A} - \lambda_1\mathbf{I})\mathbf{v}_{sp_s} &= \mathbf{v}_{sp_s-1}
\end{aligned}
$$

where the s eigenvectors $\mathbf{v}_{11}, \mathbf{v}_{21}, \ldots, \mathbf{v}_{s1}$ are linearly independent and

$$
p_1 + p_2 + \cdots + p_s = k
$$

Note that $p_1, p_2, \ldots, p_s$ represent the order of each of the s Jordan blocks. (For the determination of the generalized eigenvectors, we follow the method discussed earlier. For an example showing the details of such a determination, see Problem A–11.)

Let us define an $n \times k$ matrix consisting of $\mathbf{v}_{11}, \mathbf{v}_{12}, \ldots, \mathbf{v}_{sp_s}$ as

$$
\begin{aligned}
\mathbf{S}(\lambda_1) &= [\mathbf{v}_{11} \,\vdots\, \mathbf{v}_{12} \,\vdots\, \cdots \,\vdots\, \mathbf{v}_{1p_1} \,\vdots\, \cdots \,\vdots\, \mathbf{v}_{s1} \,\vdots\, \mathbf{v}_{s2} \,\vdots\, \cdots \,\vdots\, \mathbf{v}_{sp_s}] \\
&= [\mathbf{x}_1 \,\vdots\, \mathbf{x}_2 \,\vdots\, \cdots \,\vdots\, \mathbf{x}_{p_1} \,\vdots\, \cdots \,\vdots\, \mathbf{x}_k] \\
&= [\mathbf{S}_1 \,\vdots\, \mathbf{S}_2 \,\vdots\, \cdots \,\vdots\, \mathbf{S}_k]
\end{aligned}
$$

and define

$$
\begin{aligned}
\mathbf{S} &= [\mathbf{S}(\lambda_1) \,\vdots\, \mathbf{S}_{k+1} \,\vdots\, \mathbf{S}_{k+2} \,\vdots\, \cdots \,\vdots\, \mathbf{S}_n] \\
&= [\mathbf{S}_1 \,\vdots\, \mathbf{S}_2 \,\vdots\, \cdots \,\vdots\, \mathbf{S}_n]
\end{aligned}
$$

where

$$
\mathbf{S}_{k+1} = \mathbf{x}_{k+1}, \qquad \mathbf{S}_{k+2} = \mathbf{x}_{k+2}, \qquad \ldots, \qquad \mathbf{S}_n = \mathbf{x}_n
$$

Note that $\mathbf{x}_{k+1}, \mathbf{x}_{k+2}, \ldots, \mathbf{x}_n$ are eigenvectors associated with eigenvalues λ_{k+1}, $\lambda_{k+2}, \ldots, \lambda_n$, respectively. Matrix $\mathbf{S}$ defined in this way is nonsingular. Now we obtain

$$
\mathbf{AS} = \mathbf{S}
\begin{bmatrix}
\mathbf{J}_{p_1}(\lambda_1) & & & & 0 & & & 0 \\
& \mathbf{J}_{p_2}(\lambda_1) & & & & & & \\
& & \ddots & & & & & \\
0 & & & \mathbf{J}_{p_s}(\lambda_1) & 0 & & & \\
\hline
& & & 0 & \lambda_{k+1} & & & 0 \\
& & & & & \ddots & & \\
0 & & & & 0 & & & \lambda_n
\end{bmatrix}
$$

where $\mathbf{J}_{p_i}(\lambda_1)$ is in the form

$$
\mathbf{J}_{p_i}(\lambda_1) =
\begin{bmatrix}
\lambda_1 & 1 & & & 0 \\
& \lambda_1 & 1 & & \\
& & \ddots & \ddots & \\
& & & \ddots & 1 \\
0 & & & & \lambda_1
\end{bmatrix}
$$

which is a $p_i \times p_i$ matrix. Hence,

$$
\mathbf{S}^{-1}\mathbf{AS} =
\begin{bmatrix}
\mathbf{J}_{p_1}(\lambda_1) & & & & 0 & & & 0 \\
& \mathbf{J}_{p_2}(\lambda_1) & & & & & & \\
& & \ddots & & & & & \\
0 & & & \mathbf{J}_{p_s}(\lambda_1) & 0 & & & \\
\hline
& & & 0 & \lambda_{k+1} & & & 0 \\
& & & & & \ddots & & \\
0 & & & & 0 & & & \lambda_n
\end{bmatrix}
$$

Thus, as we have shown, by using a set of n linearly independent vectors (eigenvectors and generalized eigenvectors), any $n \times n$ matrix can be reduced to a Jordan canonical form by a similarity transformation.

Similarity Transformation When an $n \times n$ Matrix Is Normal. First, recall that a matrix is normal if it is a real symmetric, a Hermitian, a real skew-symmetric, a skew-Hermitian, an orthogonal, or a unitary matrix.

Assume that an $n \times n$ normal matrix has a k-multiple eigenvalue λ_1 and that its other $n - k$ eigenvalues are distinct and different from λ_1. Then the rank of $\mathbf{A} - \lambda_1 \mathbf{I}$ becomes $n - k$. (Refer to Problem A–12 for the proof.) If the rank of $\mathbf{A} - \lambda_1 \mathbf{I}$ is $n - k$, there are k linearly independent eigenvectors $\mathbf{x}_1, \mathbf{x}_2, \ldots, \mathbf{x}_k$ that satisfy the equation

$$(\mathbf{A} - \lambda_1 \mathbf{I})\mathbf{x}_i = \mathbf{0}, \qquad i = 1, 2, \ldots, k$$

Therefore, there exist k Jordan blocks for eigenvalue λ_1. Since the number of Jordan blocks is the same as the multiplicity number of eigenvalue λ_1, all k Jordan blocks become first order. Since the remaining $n - k$ eigenvalues are distinct, the eigenvectors associated with these eigenvalues are linearly independent. Hence, the $n \times n$ normal matrix possesses altogether n linearly independent eigenvectors, and the Jordan canonical form of the normal matrix becomes a diagonal matrix.

It can be proved that if $\mathbf{A}$ is an $n \times n$ normal matrix, then, regardless of whether or not the eigenvalues include multiple eigenvalues, there exists an $n \times n$ unitary matrix $\mathbf{U}$ such that

$$\mathbf{U}^{-1}\mathbf{A}\mathbf{U} = \mathbf{U}^*\mathbf{A}\mathbf{U} = \mathbf{D} = \text{diag}\,(\lambda_1, \lambda_2, \ldots, \lambda_n)$$

where $\mathbf{D}$ is a diagonal matrix with n eigenvalues as diagonal elements.

Trace of an $n \times n$ Matrix.   The trace of an $n \times n$ matrix $\mathbf{A}$ is defined as follows:

$$\text{trace of } \mathbf{A} = \text{tr}\,\mathbf{A} = \sum_{i=1}^{n} a_{ii}$$

The trace of an $n \times n$ matrix $\mathbf{A}$ has the following properties:

1. $$\text{tr}\,\mathbf{A}^T = \text{tr}\,\mathbf{A}$$

2. For $n \times n$ matrices $\mathbf{A}$ and $\mathbf{B}$,

$$\text{tr}\,(\mathbf{A} + \mathbf{B}) = \text{tr}\,\mathbf{A} + \text{tr}\,\mathbf{B}$$

3. If the eigenvalues of $\mathbf{A}$ are denoted by $\lambda_1, \lambda_2, \ldots, \lambda_n$, then

$$\text{tr}\,\mathbf{A} = \lambda_1 + \lambda_2 + \cdots + \lambda_n \tag{A–34}$$

4. For an $n \times m$ matrix $\mathbf{A}$ and an $m \times n$ matrix $\mathbf{B}$, regardless of whether $\mathbf{AB} = \mathbf{BA}$ or $\mathbf{AB} \neq \mathbf{BA}$, we have

$$\text{tr}\,\mathbf{AB} = \text{tr}\,\mathbf{BA} = \sum_{i=1}^{n}\sum_{j=1}^{m} a_{ij}b_{ji}$$

If $m = 1$, then by writing $\mathbf{A}$ and $\mathbf{B}$ as $\mathbf{a}$ and $\mathbf{b}$, respectively, we have

$$\text{tr}\,\mathbf{ab} = \mathbf{ba}$$

Hence, for an $n \times m$ matrix $\mathbf{C}$, we have

$$\mathbf{a}^T\mathbf{C}\mathbf{a} = \text{tr}\,\mathbf{a}\mathbf{a}^T\mathbf{C}$$

Note that Equation (A–34) may be proved as follows. By use of a similarity transformation, we have

$$\mathbf{P}^{-1}\mathbf{A}\mathbf{P} = \mathbf{D} = \text{diagonal matrix}$$

or

$$\mathbf{S}^{-1}\mathbf{A}\mathbf{S} = \mathbf{J} = \text{Jordan canonical form}$$

That is,

$$\mathbf{A} = \mathbf{P}\mathbf{D}\mathbf{P}^{-1} \quad \text{or} \quad \mathbf{A} = \mathbf{S}\mathbf{J}\mathbf{S}^{-1}$$

Hence, by using property 4 listed here, we have

$$\text{tr}\,\mathbf{A} = \text{tr}\,\mathbf{P}\mathbf{D}\mathbf{P}^{-1} = \text{tr}\,\mathbf{P}^{-1}\mathbf{P}\mathbf{D} = \text{tr}\,\mathbf{D} = \lambda_1 + \lambda_2 + \cdots + \lambda_n$$

Similarly,

$$\text{tr}\,\mathbf{A} = \text{tr}\,\mathbf{S}\mathbf{J}\mathbf{S}^{-1} = \text{tr}\,\mathbf{S}^{-1}\mathbf{S}\mathbf{J} = \text{tr}\,\mathbf{J} = \lambda_1 + \lambda_2 + \cdots + \lambda_n$$

Invariant Properties Under Similarity Transformation.   If an $n \times n$ matrix $\mathbf{A}$ can be reduced to a similar matrix that has a simple form, then important properties of $\mathbf{A}$ can be readily observed. A property of a matrix is said to be invariant if it is possessed by all similar matrices. For example, the determinant and the characteristic polynomial are invariant under a similarity transformation, as shown in the following. Suppose that $\mathbf{P}^{-1}\mathbf{AP} = \mathbf{B}$. Then

$$|\mathbf{B}| = |\mathbf{P}^{-1}\mathbf{AP}| = |\mathbf{P}^{-1}||\mathbf{A}||\mathbf{P}| = |\mathbf{A}||\mathbf{P}^{-1}||\mathbf{P}| = |\mathbf{A}||\mathbf{P}^{-1}\mathbf{P}|$$

$$= |\mathbf{A}||\mathbf{I}| = |\mathbf{A}|$$

and

$$|\lambda\mathbf{I} - \mathbf{B}| = |\lambda\mathbf{I} - \mathbf{P}^{-1}\mathbf{AP}| = |\mathbf{P}^{-1}(\lambda\mathbf{I})\mathbf{P} - \mathbf{P}^{-1}\mathbf{AP}|$$

$$= |\mathbf{P}^{-1}(\lambda\mathbf{I} - \mathbf{A})\mathbf{P}| = |\mathbf{P}^{-1}||\lambda\mathbf{I} - \mathbf{A}||\mathbf{P}|$$

$$= |\lambda\mathbf{I} - \mathbf{A}||\mathbf{P}^{-1}||\mathbf{P}| = |\lambda\mathbf{I} - \mathbf{A}|$$

Notice that the trace of a matrix is also invariant under similarity transformation, as was shown earlier:

$$\text{tr}\,\mathbf{A} = \text{tr}\,\mathbf{P}^{-1}\mathbf{AP}$$

The property of symmetry of a matrix, however, is not invariant.

Notice that only invariant properties of matrices present intrinsic characteristics of the class of similar matrices. To determine the invariant properties of a matrix $\mathbf{A}$, we examine the Jordan canonical form of $\mathbf{A}$, since the similarity of two matrices can be defined in terms of the Jordan canonical form: The necessary and sufficient condition for $n \times n$ matrices $\mathbf{A}$ and $\mathbf{B}$ to be similar is that the Jordan canonical form of $\mathbf{A}$ and that of $\mathbf{B}$ be identical.

A-7 QUADRATIC FORMS

Quadratic Forms.   For an $n \times n$ real symmetric matrix $\mathbf{A}$ and a real n-vector $\mathbf{x}$, the form

$$\mathbf{x}^T\mathbf{Ax} = \sum_{i=1}^{n}\sum_{j=1}^{n} a_{ij}x_i x_j, \qquad a_{ji} = a_{ij}$$

is called a *real quadratic form* in x_i. Frequently, a real quadratic form is called simply a *quadratic form*. Note that $\mathbf{x}^T\mathbf{Ax}$ is a real scalar quantity.

Any real quadratic form can always be written as $\mathbf{x}^T\mathbf{Ax}$. For example,

$$x_1^2 - 2x_1 x_2 + 4x_1 x_3 + x_2^2 + 8x_3^2 = [x_1 \quad x_2 \quad x_3]\begin{bmatrix} 1 & -1 & 2 \\ -1 & 1 & 0 \\ 2 & 0 & 8 \end{bmatrix}\begin{bmatrix} x_1 \\ x_2 \\ x_3 \end{bmatrix}$$

It is worthwhile to mention that, for an $n \times n$ real matrix $\mathbf{A}$, if we define

$$\mathbf{B} = \tfrac{1}{2}(\mathbf{A} + \mathbf{A}^T) \qquad \text{and} \qquad \mathbf{C} = \tfrac{1}{2}(\mathbf{A} - \mathbf{A}^T)$$

then

$$\mathbf{A} = \mathbf{B} + \mathbf{C}$$

Notice that

$$\mathbf{B}^T = \mathbf{B} \quad \text{and} \quad \mathbf{C}^T = -\mathbf{C}$$

Hence, an $n \times n$ real matrix $\mathbf{A}$ can be expressed as a sum of a real symmetric and a real skew-symmetric matrix. Since $\mathbf{x}^T \mathbf{C} \mathbf{x}$ is a real scalar quantity, we have

$$\mathbf{x}^T \mathbf{C} \mathbf{x} = (\mathbf{x}^T \mathbf{C} \mathbf{x})^T = \mathbf{x}^T \mathbf{C}^T \mathbf{x} = -\mathbf{x}^T \mathbf{C} \mathbf{x}$$

Consequently, we have

$$\mathbf{x}^T \mathbf{C} \mathbf{x} = 0$$

This means that a quadratic form for a real skew-symmetric matrix is zero. Hence,

$$\mathbf{x}^T \mathbf{A} \mathbf{x} = \mathbf{x}^T (\mathbf{B} + \mathbf{C}) \mathbf{x} = \mathbf{x}^T \mathbf{B} \mathbf{x}$$

and we see that the real quadratic form $\mathbf{x}^T \mathbf{A} \mathbf{x}$ involves only the symmetric component $\mathbf{x}^T \mathbf{B} \mathbf{x}$. This is the reason why the real quadratic form is defined only for a real symmetric matrix.

For a Hermitian matrix $\mathbf{A}$ and a complex n-vector $\mathbf{x}$, the form

$$\mathbf{x}^* \mathbf{A} \mathbf{x} = \sum_{i=1}^{n} \sum_{j=1}^{n} a_{ij} \bar{x}_i x_j, \qquad a_{ji} = \bar{a}_{ij}$$

is called a *complex quadratic form*, or Hermitian form. Notice that the scalar quantity $\mathbf{x}^* \mathbf{A} \mathbf{x}$ is real, because

$$\overline{\mathbf{x}^* \mathbf{A} \mathbf{x}} = \mathbf{x}^T \overline{\mathbf{A}} \overline{\mathbf{x}} = (\mathbf{x}^T \overline{\mathbf{A}} \overline{\mathbf{x}})^T = \overline{\mathbf{x}}^T \overline{\mathbf{A}}^T \mathbf{x} = \mathbf{x}^* \mathbf{A} \mathbf{x}$$

Bilinear Forms.   For an $n \times m$ real matrix $\mathbf{A}$, a real n-vector $\mathbf{x}$, and a real m-vector $\mathbf{y}$, the form

$$\mathbf{x}^T \mathbf{A} \mathbf{y} = \sum_{i=1}^{n} \sum_{j=1}^{m} a_{ij} x_i y_j$$

is called a *real bilinear form* in x_i and y_j. $\mathbf{x}^T \mathbf{A} \mathbf{y}$ is a real scalar quantity.

For an $n \times m$ complex matrix $\mathbf{A}$, a complex n-vector $\mathbf{x}$, and a complex m-vector $\mathbf{y}$, the form

$$\mathbf{x}^* \mathbf{A} \mathbf{y} = \sum_{i=1}^{n} \sum_{j=1}^{m} a_{ij} \bar{x}_i y_j$$

is called a *complex bilinear form*. $\mathbf{x}^* \mathbf{A} \mathbf{y}$ is a complex scalar quantity.

Definiteness and Semidefiniteness.   A quadratic form $\mathbf{x}^T \mathbf{A} \mathbf{x}$, where $\mathbf{A}$ is a real symmetric matrix (or a Hermitian form $\mathbf{x}^* \mathbf{A} \mathbf{x}$, where $\mathbf{A}$ is a Hermitian matrix), is said to be positive definite if

$$\mathbf{x}^T \mathbf{A} \mathbf{x} > 0 \quad (\text{or } \mathbf{x}^* \mathbf{A} \mathbf{x} > 0), \quad \text{for } \mathbf{x} \neq \mathbf{0}$$

$$\mathbf{x}^T \mathbf{A} \mathbf{x} = 0 \quad (\text{or } \mathbf{x}^* \mathbf{A} \mathbf{x} = 0), \quad \text{for } \mathbf{x} = \mathbf{0}$$

$\mathbf{x}^T \mathbf{Ax}$ (or $\mathbf{x}^*\mathbf{Ax}$) is said to be positive semidefinite if

$$\mathbf{x}^T \mathbf{Ax} \geq 0 \quad (\text{or } \mathbf{x}^*\mathbf{Ax} \geq 0), \quad \text{for } \mathbf{x} \neq \mathbf{0}$$

$$\mathbf{x}^T \mathbf{Ax} = 0 \quad (\text{or } \mathbf{x}^*\mathbf{Ax} = 0), \quad \text{for } \mathbf{x} = \mathbf{0}$$

$\mathbf{x}^T \mathbf{Ax}$ (or $\mathbf{x}^*\mathbf{Ax}$) is said to be negative definite if

$$\mathbf{x}^T \mathbf{Ax} < 0 \quad (\text{or } \mathbf{x}^*\mathbf{Ax} < 0), \quad \text{for } \mathbf{x} \neq \mathbf{0}$$

$$\mathbf{x}^T \mathbf{Ax} = 0 \quad (\text{or } \mathbf{x}^*\mathbf{Ax} = 0), \quad \text{for } \mathbf{x} = \mathbf{0}$$

$\mathbf{x}^T \mathbf{Ax}$ (or $\mathbf{x}^*\mathbf{Ax}$) is said to be negative semidefinite if

$$\mathbf{x}^T \mathbf{Ax} \leq 0 \quad (\text{or } \mathbf{x}^*\mathbf{Ax} \leq 0), \quad \text{for } \mathbf{x} \neq \mathbf{0}$$

$$\mathbf{x}^T \mathbf{Ax} = 0 \quad (\text{or } \mathbf{x}^*\mathbf{Ax} = 0), \quad \text{for } \mathbf{x} = \mathbf{0}$$

If $\mathbf{x}^T \mathbf{Ax}$ (or $\mathbf{x}^*\mathbf{Ax}$) can be of either sign, then $\mathbf{x}^T \mathbf{Ax}$ (or $\mathbf{x}^*\mathbf{Ax}$) is said to be indefinite.

Note that if $\mathbf{x}^T \mathbf{Ax}$ or $\mathbf{x}^*\mathbf{Ax}$ is positive (or negative) definite we say that $\mathbf{A}$ is a positive (or negative) definite matrix. Similarly, matrix $\mathbf{A}$ is called a positive (or negative) semidefinite matrix if $\mathbf{x}^T \mathbf{Ax}$ or $\mathbf{x}^*\mathbf{Ax}$ is positive (or negative) semidefinite; matrix $\mathbf{A}$ is called an indefinite matrix if $\mathbf{x}^T \mathbf{Ax}$ or $\mathbf{x}^*\mathbf{Ax}$ is indefinite.

Note also that the eigenvalues of an $n \times n$ real symmetric or Hermitian matrix are real. (For the proof, see Problem A–13.) It can be shown that an $n \times n$ real symmetric or Hermitian matrix $\mathbf{A}$ is a positive definite matrix if all eigenvalues λ_i $(i = 1, 2, \ldots, n)$ are positive. Matrix $\mathbf{A}$ is positive semidefinite if all eigenvalues are nonnegative, or $\lambda_i \geq 0$ $(i = 1, 2, \ldots, n)$, and at least one of them is zero.

Notice that if $\mathbf{A}$ is a positive definite matrix then $|\mathbf{A}| \neq 0$, because all eigenvalues are positive. Hence, the inverse matrix always exists for a positive definite matrix.

In the process of determining the stability of an equilibrium state, we frequently encounter a scalar function $\mathbf{V}(\mathbf{x})$. A scalar function $\mathbf{V}(\mathbf{x})$, which is a function of $x_1, x_2, \ldots, x_n$, is said to be positive definite if

$$\mathbf{V}(\mathbf{x}) > 0, \quad \text{for } \mathbf{x} \neq \mathbf{0}$$

$$\mathbf{V}(\mathbf{0}) = 0$$

$\mathbf{V}(\mathbf{x})$ is said to be positive semidefinite if

$$\mathbf{V}(\mathbf{x}) \geq 0, \quad \text{for } \mathbf{x} \neq \mathbf{0}$$

$$\mathbf{V}(\mathbf{0}) = 0$$

If $-\mathbf{V}(\mathbf{x})$ is positive definite (or positive semidefinite), then $\mathbf{V}(\mathbf{x})$ is said to be negative definite (or negative semidefinite).

Necessary and sufficient conditions for the quadratic form $\mathbf{x}^T \mathbf{Ax}$ (or the Hermitian form $\mathbf{x}^*\mathbf{Ax}$) to be positive definite, negative definite, positive semidefinite, or negative semidefinite have been given by J. J. Sylvester. Sylvester's criteria follow.

Sylvester's Criterion for Positive Definiteness of a Quadratic Form or Hermitian Form. A necessary and sufficient condition for a quadratic form $\mathbf{x}^T \mathbf{Ax}$ (or a Hermitian form $\mathbf{x}^*\mathbf{Ax}$), where $\mathbf{A}$ is an $n \times n$ real symmetric matrix (or Hermitian matrix), to be positive definite is that the determinant of $\mathbf{A}$ be positive and the successive principal minors of the determinant of $\mathbf{A}$ (the determinants of the $k \times k$ matrices in the top-left corner of matrix $\mathbf{A}$, where $k = 1, 2, \ldots, n - 1$) be positive; that is, we must have

$$a_{11} > 0, \qquad \begin{vmatrix} a_{11} & a_{12} \\ a_{21} & a_{22} \end{vmatrix} > 0, \qquad \begin{vmatrix} a_{11} & a_{12} & a_{13} \\ a_{21} & a_{22} & a_{23} \\ a_{31} & a_{32} & a_{33} \end{vmatrix} > 0, \qquad \ldots, \qquad |\mathbf{A}| > 0$$

where

$$a_{ij} = a_{ji}, \qquad \text{for real symmetric matrix } \mathbf{A}$$

$$a_{ij} = \bar{a}_{ji}, \qquad \text{for Hermitian matrix } \mathbf{A}$$

Sylvester's Criterion for Negative Definiteness of a Quadratic Form or Hermitian Form. A necessary and sufficient condition for a quadratic form $\mathbf{x}^T \mathbf{A} \mathbf{x}$ (or a Hermitian form $\mathbf{x}^* \mathbf{A} \mathbf{x}$), where $\mathbf{A}$ is an $n \times n$ real symmetric matrix (or Hermitian matrix), to be negative definite is that the determinant of $\mathbf{A}$ be positive if n is even and negative if n is odd, and that the successive principal minors of even order be positive and the successive principal minors of odd order be negative; that is, we must have

$$a_{11} < 0, \qquad \begin{vmatrix} a_{11} & a_{12} \\ a_{21} & a_{22} \end{vmatrix} > 0, \qquad \begin{vmatrix} a_{11} & a_{12} & a_{13} \\ a_{21} & a_{22} & a_{23} \\ a_{31} & a_{32} & a_{33} \end{vmatrix} < 0, \qquad \ldots$$

$$|\mathbf{A}| > 0 \qquad (n \text{ even})$$

$$|\mathbf{A}| < 0 \qquad (n \text{ odd})$$

where

$$a_{ij} = a_{ji}, \qquad \text{for real symmetric matrix } \mathbf{A}$$

$$a_{ij} = \bar{a}_{ji}, \qquad \text{for Hermitian matrix } \mathbf{A}$$

[This condition can be derived by requiring that $\mathbf{x}^T(-\mathbf{A})\mathbf{x}$ be positive definite.]

Sylvester's Criterion for Positive Semidefiniteness of a Quadratic Form or Hermitian Form. A necessary and sufficient condition for a quadratic form $\mathbf{x}^T \mathbf{A} \mathbf{x}$ (or a Hermitian form $\mathbf{x}^* \mathbf{A} \mathbf{x}$), where $\mathbf{A}$ is a real symmetric matrix (or a Hermitian matrix), to be positive semidefinite is that $\mathbf{A}$ be singular ($|\mathbf{A}| = 0$) and all the principal minors be nonnegative:

$$a_{ii} \geq 0, \qquad \begin{vmatrix} a_{ii} & a_{ij} \\ a_{ji} & a_{jj} \end{vmatrix} \geq 0, \qquad \begin{vmatrix} a_{ii} & a_{ij} & a_{ik} \\ a_{ji} & a_{jj} & a_{jk} \\ a_{ki} & a_{kj} & a_{kk} \end{vmatrix} \geq 0, \qquad \ldots, \qquad |\mathbf{A}| = 0$$

where $i < j < k$ and

$$a_{ij} = a_{ji}, \qquad \text{for real symmetric matrix } \mathbf{A}$$

$$a_{ij} = \bar{a}_{ji}, \qquad \text{for Hermitian matrix } \mathbf{A}$$

(It is important to point out that in the positive semidefiniteness test or negative semidefiniteness test we must check the signs of all the principal minors, not just successive principal minors. See Problem A–15.)

Sylvester's Criterion for Negative Semidefiniteness of a Quadratic Form or a Hermitian Form.   A necessary and sufficient condition for a quadratic form $\mathbf{x}^T \mathbf{A} \mathbf{x}$ (or a Hermitian form $\mathbf{x}^* \mathbf{A} \mathbf{x}$), where $\mathbf{A}$ is an $n \times n$ real symmetric matrix (or Hermitian matrix), to be negative semidefinite is that $\mathbf{A}$ be singular ($|\mathbf{A}| = 0$) and that all the principal minors of even order be nonnegative and those of odd order be nonpositive:

$$a_{ii} \leq 0, \qquad \begin{vmatrix} a_{ii} & a_{ij} \\ a_{ji} & a_{jj} \end{vmatrix} \geq 0, \qquad \begin{vmatrix} a_{ii} & a_{ij} & a_{ik} \\ a_{ji} & a_{jj} & a_{jk} \\ a_{ki} & a_{kj} & a_{kk} \end{vmatrix} \leq 0, \qquad \dots, \qquad |\mathbf{A}| = 0$$

where $i < j < k$ and

$$a_{ij} = a_{ji}, \qquad \text{for real symmetric matrix } \mathbf{A}$$

$$a_{ij} = \bar{a}_{ji}, \qquad \text{for Hermitian matrix } \mathbf{A}$$

A–8 PSEUDOINVERSES

The concept of pseudoinverses of a matrix is a generalization of the notion of an inverse. It is useful for finding a "solution" to a set of algebraic equations in which the number of unknown variables and the number of independent linear equations are not equal.

In what follows, we shall consider pseudoinverses that enable us to determine minimum norm solutions.

Minimum Norm Solution That Minimizes $\|\mathbf{x}\|$.   Consider a linear algebraic equation

$$x_1 + 5x_2 = 1$$

Since we have two variables and only one equation, no unique solution exists. Instead, there exist an infinite number of solutions. Graphically, any point on line $x_1 + 5x_2 = 1$, as shown in Figure A–1, is a possible solution. However, if we decide to pick the point that is closest to the origin, the solution becomes unique.

Consider the vector-matrix equation

$$\mathbf{A}\mathbf{x} = \mathbf{b} \tag{A–35}$$

where $\mathbf{A}$ is an $n \times m$ matrix, $\mathbf{x}$ is an m-vector, and $\mathbf{b}$ is an n-vector. We assume that $m > n$ (that is, the number of unknown variables is greater than the number of equations) and that the equation has an infinite number of solutions. Let us find the unique solution $\mathbf{x}$ that is located closest to the origin or that has the minimum norm $\|\mathbf{x}\|$.

Let us define the minimum norm solution as $\mathbf{x}^\circ$. That is, $\mathbf{x}^\circ$ satisfies the condition that $\mathbf{A}\mathbf{x}^\circ = \mathbf{b}$ and $\|\mathbf{x}^\circ\| \leq \|\mathbf{x}\|$ for all $\mathbf{x}$ that satisfy $\mathbf{A}\mathbf{x} = \mathbf{b}$. This means that

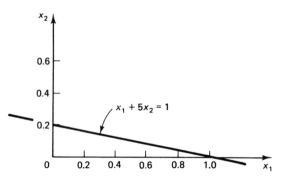

Figure A–1   Line $x_1 + 5x_2 = 1$ on the x_1x_2 plane.

the solution point $\mathbf{x}°$ is nearest to the origin of the m-dimensional space among all possible solutions of Equation (A–35). We shall obtain such a minimum norm solution in the following.

Right Pseudoinverse Matrix.   For a vector-matrix equation

$$\mathbf{Ax} = \mathbf{b}$$

where $\mathbf{A}$ is an $n \times m$ matrix having rank n, $\mathbf{x}$ is an m-vector, and $\mathbf{b}$ is an n-vector, the solution that minimizes the norm $\|\mathbf{x}\|$ is given by

$$\mathbf{x}° = \mathbf{A}^{RM}\mathbf{b}$$

where $\mathbf{A}^{RM} = \mathbf{A}^T(\mathbf{AA}^T)^{-1}$.

This can be proved as follows. First, note that norm $\|\mathbf{x}\|$ can be written as follows:

$$\|\mathbf{x}\| = \|\mathbf{x} - \mathbf{x}° + \mathbf{x}°\| = \|\mathbf{x}°\| + \|\mathbf{x} - \mathbf{x}°\| + 2(\mathbf{x}°)^T(\mathbf{x} - \mathbf{x}°)$$

The last term, $2(\mathbf{x}°)^T(\mathbf{x} - \mathbf{x}°)$, can be shown to be zero, since

$$\begin{aligned}
(\mathbf{x}°)^T(\mathbf{x} - \mathbf{x}°) &= [\mathbf{A}^T(\mathbf{AA}^T)^{-1}\mathbf{b}]^T[\mathbf{x} - \mathbf{A}^T(\mathbf{AA}^T)^{-1}\mathbf{b}] \\
&= \mathbf{b}^T(\mathbf{AA}^T)^{-1}\mathbf{A}[\mathbf{x} - \mathbf{A}^T(\mathbf{AA}^T)^{-1}\mathbf{b}] \\
&= \mathbf{b}^T(\mathbf{AA}^T)^{-1}[\mathbf{Ax} - (\mathbf{AA}^T)(\mathbf{AA}^T)^{-1}\mathbf{b}] \\
&= \mathbf{b}^T(\mathbf{AA}^T)^{-1}(\mathbf{b} - \mathbf{b}) \\
&= 0
\end{aligned}$$

Hence,

$$\|\mathbf{x}\| = \|\mathbf{x}°\| + \|\mathbf{x} - \mathbf{x}°\|$$

which can be rewritten as

$$\|\mathbf{x}\| - \|\mathbf{x}°\| = \|\mathbf{x} - \mathbf{x}°\|$$

Since $\|\mathbf{x} - \mathbf{x}°\| \geq 0$, we obtain

$$\|\mathbf{x}\| \geq \|\mathbf{x}°\|$$

Thus, we have shown that $\mathbf{x}°$ is the solution that gives the minimum norm $\|\mathbf{x}\|$..

The matrix $\mathbf{A}^{RM} = \mathbf{A}^T(\mathbf{A}\mathbf{A}^T)^{-1}$ that yields the minimum norm solution ($\|\mathbf{x}^\circ\|$ = minimum) is called the *right pseudoinverse* or *minimal right inverse* of $\mathbf{A}$.

Summary On the Right Pseudoinverse Matrix.   The right pseudoinverse $\mathbf{A}^{RM}$ gives the solution $\mathbf{x}^\circ = \mathbf{A}^{RM}\mathbf{b}$ that minimizes the norm, or gives $\|\mathbf{x}^\circ\|$ = minimum. Note that the right pseudoinverse $\mathbf{A}^{RM}$ is an $m \times n$ matrix, since $\mathbf{A}$ is an $n \times m$ matrix and

$$\mathbf{A}^{RM} = \mathbf{A}^T(\mathbf{A}\mathbf{A}^T)^{-1}$$

$$= (m \times n \text{ matrix})(n \times n \text{ matrix})^{-1}$$

$$= m \times n \text{ matrix}, \quad m > n$$

Notice that the dimension of $\mathbf{A}\mathbf{A}^T$ is smaller than the dimension of vector $\mathbf{x}$, which is m. Notice also that the right pseudoinverse $\mathbf{A}^{RM}$ possesses the property that it is indeed an "inverse" matrix if premultiplied by $\mathbf{A}$:

$$\mathbf{A}\mathbf{A}^{RM} = \mathbf{A}[\mathbf{A}^T(\mathbf{A}\mathbf{A}^T)^{-1}] = \mathbf{A}\mathbf{A}^T(\mathbf{A}\mathbf{A}^T)^{-1} = \mathbf{I}_n$$

Solution That Minimizes $\|\mathbf{A}\mathbf{x} - \mathbf{b}\|$.   Consider a vector-matrix equation

$$\mathbf{A}\mathbf{x} = \mathbf{b} \tag{A–36}$$

where $\mathbf{A}$ is an $n \times m$ matrix, $\mathbf{x}$ is an m-vector, and $\mathbf{b}$ is an n-vector. Here we assume that $n > m$. That is, the number of unknown variables is smaller than the number of equations. In the classical sense, there may or may not exist any solution.

If no solution exists, we may wish to find a unique "solution" that minimizes the norm $\|\mathbf{A}\mathbf{x} - \mathbf{b}\|$. Let us define a "solution" to Equation (A–36) that will minimize $\|\mathbf{A}\mathbf{x} - \mathbf{b}\|$ as $\mathbf{x}^\circ$. In other words, $\mathbf{x}^\circ$ satisfies the condition

$$\|\mathbf{A}\mathbf{x} - \mathbf{b}\| \geq \|\mathbf{A}\mathbf{x}^\circ - \mathbf{b}\|, \quad \text{for all } \mathbf{x}$$

Note that $\mathbf{x}^\circ$ is not a solution in the classical sense, since it does not satisfy the original vector-matrix equation $\mathbf{A}\mathbf{x} = \mathbf{b}$. Therefore, we may call $\mathbf{x}^\circ$ an "approximate solution," in that it minimizes norm $\|\mathbf{A}\mathbf{x} - \mathbf{b}\|$. We shall obtain such an approximate solution in the following.

Left Pseudoinverse Matrix.   For a vector-matrix equation

$$\mathbf{A}\mathbf{x} = \mathbf{b}$$

where $\mathbf{A}$ is an $n \times m$ matrix having rank m, $\mathbf{x}$ is an m-vector, and $\mathbf{b}$ is an n-vector, the vector $\mathbf{x}^\circ$ that minimizes the norm $\|\mathbf{A}\mathbf{x} - \mathbf{b}\|$ is given by

$$\mathbf{x}^\circ = \mathbf{A}^{LM}\mathbf{b} = (\mathbf{A}^T\mathbf{A})^{-1}\mathbf{A}^T\mathbf{b}$$

where $\mathbf{A}^{LM} = (\mathbf{A}^T\mathbf{A})^{-1}\mathbf{A}^T$.

To verify this, first note that

$$\|\mathbf{A}\mathbf{x} - \mathbf{b}\| = \|\mathbf{A}(\mathbf{x} - \mathbf{x}^\circ) + \mathbf{A}\mathbf{x}^\circ - \mathbf{b}\|$$

$$= \|\mathbf{A}(\mathbf{x} - \mathbf{x}^\circ)\| + \|\mathbf{A}\mathbf{x}^\circ - \mathbf{b}\| + 2[\mathbf{A}(\mathbf{x} - \mathbf{x}^\circ)]^T(\mathbf{A}\mathbf{x}^\circ - \mathbf{b})$$

The last term can be shown to be zero as follows:

$$[\mathbf{A}(\mathbf{x} - \mathbf{x}^\circ)]^T(\mathbf{A}\mathbf{x}^\circ - \mathbf{b}) = (\mathbf{x} - \mathbf{x}^\circ)^T\mathbf{A}^T[\mathbf{A}(\mathbf{A}^T\mathbf{A})^{-1}\mathbf{A}^T - \mathbf{I}_n]\mathbf{b}$$
$$= (\mathbf{x} - \mathbf{x}^\circ)^T[(\mathbf{A}^T\mathbf{A})(\mathbf{A}^T\mathbf{A})^{-1}\mathbf{A}^T - \mathbf{A}^T]\mathbf{b}$$
$$\mathbf{)} = (\mathbf{x} - \mathbf{x}^\circ)^T(\mathbf{A}^T - \mathbf{A}^T)\mathbf{b}$$
$$= 0$$

Hence,

$$\|\mathbf{A}\mathbf{x} - \mathbf{b}\| = \|\mathbf{A}(\mathbf{x} - \mathbf{x}^\circ)\| + \|\mathbf{A}\mathbf{x}^\circ - \mathbf{b}\|$$

Noting that $\|\mathbf{A}(\mathbf{x} - \mathbf{x}^\circ)\| \geq 0$, we obtain

$$\|\mathbf{A}\mathbf{x} - \mathbf{b}\| - \|\mathbf{A}\mathbf{x}^\circ - \mathbf{b}\| = \|\mathbf{A}(\mathbf{x} - \mathbf{x}^\circ)\| \geq 0$$

or

$$\|\mathbf{A}\mathbf{x} - \mathbf{b}\| \geq \|\mathbf{A}\mathbf{x}^\circ - \mathbf{b}\|$$

Thus,

$$\mathbf{x}^\circ = \mathbf{A}^{LM}\mathbf{b} = (\mathbf{A}^T\mathbf{A})^{-1}\mathbf{A}^T\mathbf{b}$$

minimizes $\|\mathbf{A}\mathbf{x} - \mathbf{b}\|$.

The matrix $\mathbf{A}^{LM} = (\mathbf{A}^T\mathbf{A})^{-1}\mathbf{A}^T$ is called the *left pseudoinverse* or *minimal left inverse* of matrix $\mathbf{A}$. Note that $\mathbf{A}^{LM}$ is indeed the inverse matrix of $\mathbf{A}$, in that if postmultiplied by $\mathbf{A}$ it will give an identity matrix $\mathbf{I}_m$:

$$\mathbf{A}^{LM}\mathbf{A} = (\mathbf{A}^T\mathbf{A})^{-1}\mathbf{A}^T\mathbf{A} = (\mathbf{A}^T\mathbf{A})^{-1}(\mathbf{A}^T\mathbf{A}) = \mathbf{I}_m$$

EXAMPLE PROBLEMS AND SOLUTIONS

Problem A–1

Show that if matrices $\mathbf{A}$, $\mathbf{B}$, $\mathbf{C}$, and $\mathbf{D}$ are an $n \times n$, an $n \times m$, an $m \times n$, and an $m \times m$ matrix, respectively, and if $|\mathbf{A}| \neq 0$ and $|\mathbf{D}| \neq 0$, then

$$\begin{vmatrix} \mathbf{A} & \mathbf{B} \\ \mathbf{0} & \mathbf{D} \end{vmatrix} = \begin{vmatrix} \mathbf{A} & \mathbf{0} \\ \mathbf{C} & \mathbf{D} \end{vmatrix} = |\mathbf{A}|\,|\mathbf{D}| \neq 0, \qquad \text{if } |\mathbf{A}| \neq 0 \text{ and } |\mathbf{D}| \neq 0$$

Solution   Since matrix $\mathbf{A}$ is nonsingular, we have

$$\begin{bmatrix} \mathbf{A} & \mathbf{B} \\ \mathbf{0} & \mathbf{D} \end{bmatrix} = \begin{bmatrix} \mathbf{A} & \mathbf{0} \\ \mathbf{0} & \mathbf{I} \end{bmatrix}\begin{bmatrix} \mathbf{I} & \mathbf{0} \\ \mathbf{0} & \mathbf{D} \end{bmatrix}\begin{bmatrix} \mathbf{I} & \mathbf{A}^{-1}\mathbf{B} \\ \mathbf{0} & \mathbf{I} \end{bmatrix}$$

Hence,

$$\begin{vmatrix} \mathbf{A} & \mathbf{B} \\ \mathbf{0} & \mathbf{D} \end{vmatrix} = \begin{vmatrix} \bar{\mathbf{A}} & \mathbf{0} \\ \mathbf{0} & \mathbf{I} \end{vmatrix}\begin{vmatrix} \mathbf{I} & \mathbf{0} \\ \mathbf{0} & \mathbf{D} \end{vmatrix}\begin{vmatrix} \mathbf{I} & \mathbf{A}^{-1}\mathbf{B} \\ \mathbf{0} & \mathbf{I} \end{vmatrix} = |\mathbf{A}|\,|\mathbf{D}|$$

Similarly, since $\mathbf{D}$ is nonsingular, we get

$$\begin{vmatrix} \mathbf{A} & \mathbf{0} \\ \mathbf{C} & \mathbf{D} \end{vmatrix} = \begin{vmatrix} \mathbf{A} & \mathbf{0} \\ \mathbf{0} & \mathbf{I} \end{vmatrix}\begin{vmatrix} \mathbf{I} & \mathbf{0} \\ \mathbf{0} & \mathbf{D} \end{vmatrix}\begin{vmatrix} \mathbf{I} & \mathbf{0} \\ \mathbf{D}^{-1}\mathbf{C} & \mathbf{I} \end{vmatrix} = |\mathbf{A}|\,|\mathbf{D}|$$

Problem A–2

Show that if matrices $\mathbf{A}, \mathbf{B}, \mathbf{C}$, and $\mathbf{D}$ are an $n \times n$, an $n \times m$, an $m \times n$, and an $m \times m$ matrix, respectively, then

$$\begin{vmatrix} \mathbf{A} & \mathbf{B} \\ \mathbf{C} & \mathbf{D} \end{vmatrix} = \begin{cases} |\mathbf{A}| \, |\mathbf{D} - \mathbf{C}\mathbf{A}^{-1}\mathbf{B}|, & \text{if } |\mathbf{A}| \neq 0 \\ |\mathbf{D}| \, |\mathbf{A} - \mathbf{B}\mathbf{D}^{-1}\mathbf{C}|, & \text{if } |\mathbf{D}| \neq 0 \end{cases}$$

Solution   if $|\mathbf{A}| \neq 0$, the matrix

$$\begin{bmatrix} \mathbf{A} & \mathbf{B} \\ \mathbf{C} & \mathbf{D} \end{bmatrix}$$

can be written as a product of two matrices:

$$\begin{bmatrix} \mathbf{A} & \mathbf{0} \\ \mathbf{C} & \mathbf{I}_m \end{bmatrix} \quad \text{and} \quad \begin{bmatrix} \mathbf{I}_n & \mathbf{A}^{-1}\mathbf{B} \\ \mathbf{0} & \mathbf{D} - \mathbf{C}\mathbf{A}^{-1}\mathbf{B} \end{bmatrix}$$

or

$$\begin{bmatrix} \mathbf{A} & \mathbf{B} \\ \mathbf{C} & \mathbf{D} \end{bmatrix} = \begin{bmatrix} \mathbf{A} & \mathbf{0} \\ \mathbf{C} & \mathbf{I}_m \end{bmatrix} \begin{bmatrix} \mathbf{I}_n & \mathbf{A}^{-1}\mathbf{B} \\ \mathbf{0} & \mathbf{D} - \mathbf{C}\mathbf{A}^{-1}\mathbf{B} \end{bmatrix}$$

Hence,

$$\begin{vmatrix} \mathbf{A} & \mathbf{B} \\ \mathbf{C} & \mathbf{D} \end{vmatrix} = \begin{vmatrix} \mathbf{A} & \mathbf{0} \\ \mathbf{C} & \mathbf{I}_m \end{vmatrix} \begin{vmatrix} \mathbf{I}_n & \mathbf{A}^{-1}\mathbf{B} \\ \mathbf{0} & \mathbf{D} - \mathbf{C}\mathbf{A}^{-1}\mathbf{B} \end{vmatrix}$$

$$= |\mathbf{A}| \, |\mathbf{I}_m| \, |\mathbf{I}_n| \, |\mathbf{D} - \mathbf{C}\mathbf{A}^{-1}\mathbf{B}|$$

$$= |\mathbf{A}| \, |\mathbf{D} - \mathbf{C}\mathbf{A}^{-1}\mathbf{B}|$$

Similarly, if $|\mathbf{D}| \neq 0$, then

$$\begin{bmatrix} \mathbf{A} & \mathbf{B} \\ \mathbf{C} & \mathbf{D} \end{bmatrix} = \begin{bmatrix} \mathbf{I}_n & \mathbf{B} \\ \mathbf{0} & \mathbf{D} \end{bmatrix} \begin{bmatrix} \mathbf{A} - \mathbf{B}\mathbf{D}^{-1}\mathbf{C} & \mathbf{0} \\ \mathbf{D}^{-1}\mathbf{C} & \mathbf{I}_m \end{bmatrix}$$

and therefore

$$\begin{vmatrix} \mathbf{A} & \mathbf{B} \\ \mathbf{C} & \mathbf{D} \end{vmatrix} = \begin{vmatrix} \mathbf{I}_n & \mathbf{B} \\ \mathbf{0} & \mathbf{D} \end{vmatrix} \begin{vmatrix} \mathbf{A} - \mathbf{B}\mathbf{D}^{-1}\mathbf{C} & \mathbf{0} \\ \mathbf{D}^{-1}\mathbf{C} & \mathbf{I}_m \end{vmatrix}$$

$$= |\mathbf{I}_n| \, |\mathbf{D}| \, |\mathbf{A} - \mathbf{B}\mathbf{D}^{-1}\mathbf{C}| \, |\mathbf{I}_m|$$

$$= |\mathbf{D}| \, |\mathbf{A} - \mathbf{B}\mathbf{D}^{-1}\mathbf{C}|$$

Problem A–3

For an $n \times m$ matrix $\mathbf{A}$ and an $m \times n$ matrix $\mathbf{B}$, show that

$$|\mathbf{I}_n + \mathbf{A}\mathbf{B}| = |\mathbf{I}_m + \mathbf{B}\mathbf{A}|$$

Solution   Consider the following matrix:

$$\begin{bmatrix} \mathbf{I}_n & -\mathbf{A} \\ \mathbf{B} & \mathbf{I}_m \end{bmatrix}$$

Referring to Problem A–2,

$$\begin{vmatrix} \mathbf{A} & \mathbf{B} \\ \mathbf{C} & \mathbf{D} \end{vmatrix} = \begin{cases} |\mathbf{A}| \, |\mathbf{D} - \mathbf{C}\mathbf{A}^{-1}\mathbf{B}|, & \text{if } |\mathbf{A}| \neq 0 \\ |\mathbf{D}| \, |\mathbf{A} - \mathbf{B}\mathbf{D}^{-1}\mathbf{C}|, & \text{if } |\mathbf{D}| \neq 0 \end{cases}$$

Hence,

$$\begin{vmatrix} \mathbf{I}_n & -\mathbf{A} \\ \mathbf{B} & \mathbf{I}_m \end{vmatrix} = \begin{cases} |\mathbf{I}_n| |\mathbf{I}_m + \mathbf{BA}| = |\mathbf{I}_m + \mathbf{BA}| \\ |\mathbf{I}_m| |\mathbf{I}_n + \mathbf{AB}| = |\mathbf{I}_n + \mathbf{AB}| \end{cases}$$

and we have

$$|\mathbf{I}_n + \mathbf{AB}| = |\mathbf{I}_m + \mathbf{BA}|$$

Problem A–4

If $\mathbf{A}$, $\mathbf{B}$, $\mathbf{C}$, and $\mathbf{D}$ are, respectively, an $n \times n$, an $n \times m$, an $m \times n$, and an $m \times m$ matrix, then we have the following matrix inversion lemma:

$$(\mathbf{A} + \mathbf{BDC})^{-1} = \mathbf{A}^{-1} - \mathbf{A}^{-1}\mathbf{B}(\mathbf{D}^{-1} + \mathbf{CA}^{-1}\mathbf{B})^{-1}\mathbf{CA}^{-1}$$

where we assume the indicated inverses to exist. Prove this matrix inversion lemma.

Solution Let us premultiply both sides of the equation by $(\mathbf{A} + \mathbf{BDC})$:

$$(\mathbf{A} + \mathbf{BDC})(\mathbf{A} + \mathbf{BDC})^{-1} = (\mathbf{A} + \mathbf{BDC})[\mathbf{A}^{-1} - \mathbf{A}^{-1}\mathbf{B}(\mathbf{D}^{-1} + \mathbf{CA}^{-1}\mathbf{B})^{-1}\mathbf{CA}^{-1}]$$

or

$$\mathbf{I} = \mathbf{I} + \mathbf{BDCA}^{-1} - \mathbf{B}(\mathbf{D}^{-1} + \mathbf{CA}^{-1}\mathbf{B})^{-1}\mathbf{CA}^{-1} - \mathbf{BDCA}^{-1}\mathbf{B}(\mathbf{D}^{-1} + \mathbf{CA}^{-1}\mathbf{B})^{-1}\mathbf{CA}^{-1}$$

$$= \mathbf{I} + \mathbf{BDCA}^{-1} - (\mathbf{B} + \mathbf{BDCA}^{-1}\mathbf{B})(\mathbf{D}^{-1} + \mathbf{CA}^{-1}\mathbf{B})^{-1}\mathbf{CA}^{-1}$$

$$= \mathbf{I} + \mathbf{BDCA}^{-1} - \mathbf{BD}(\mathbf{D}^{-1} + \mathbf{CA}^{-1}\mathbf{B})(\mathbf{D}^{-1} + \mathbf{CA}^{-1}\mathbf{B})^{-1}\mathbf{CA}^{-1}$$

$$= \mathbf{I} + \mathbf{BDCA}^{-1} - \mathbf{BDCA}^{-1}$$

$$= \mathbf{I}$$

Hence, we have proved the matrix inversion lemma.

Problem A–5

Prove that if $\mathbf{A}$, $\mathbf{B}$, $\mathbf{C}$, and $\mathbf{D}$ are, respectively, an $n \times n$, an $n \times m$, an $m \times n$, and an $m \times m$ matrix, then

$$\begin{bmatrix} \mathbf{A} & \mathbf{B} \\ \mathbf{0} & \mathbf{D} \end{bmatrix}^{-1} = \begin{bmatrix} \mathbf{A}^{-1} & -\mathbf{A}^{-1}\mathbf{BD}^{-1} \\ \mathbf{0} & \mathbf{D}^{-1} \end{bmatrix} \tag{A–37}$$

provided $|\mathbf{A}| \neq 0$ and $|\mathbf{D}| \neq 0$.
  Prove also that

$$\begin{bmatrix} \mathbf{A} & \mathbf{0} \\ \mathbf{C} & \mathbf{D} \end{bmatrix}^{-1} = \begin{bmatrix} \mathbf{A}^{-1} & \mathbf{0} \\ -\mathbf{D}^{-1}\mathbf{CA}^{-1} & \mathbf{D}^{-1} \end{bmatrix} \tag{A–38}$$

provided $|\mathbf{A}| \neq 0$ and $|\mathbf{D}| \neq 0$.

Solution Note that

$$\begin{bmatrix} \mathbf{A}^{-1} & -\mathbf{A}^{-1}\mathbf{BD}^{-1} \\ \mathbf{0} & \mathbf{D}^{-1} \end{bmatrix}\begin{bmatrix} \mathbf{A} & \mathbf{B} \\ \mathbf{0} & \mathbf{D} \end{bmatrix} = \begin{bmatrix} \mathbf{I}_n & \mathbf{A}^{-1}\mathbf{B} - \mathbf{A}^{-1}\mathbf{B} \\ \mathbf{0} & \mathbf{I}_m \end{bmatrix} = \begin{bmatrix} \mathbf{I}_n & \mathbf{0} \\ \mathbf{0} & \mathbf{I}_m \end{bmatrix}$$

Hence, Equation (A–37) is proved. Similarly,

$$\begin{bmatrix} \mathbf{A}^{-1} & \mathbf{0} \\ -\mathbf{D}^{-1}\mathbf{CA}^{-1} & \mathbf{D}^{-1} \end{bmatrix}\begin{bmatrix} \mathbf{A} & \mathbf{0} \\ \mathbf{C} & \mathbf{D} \end{bmatrix} = \begin{bmatrix} \mathbf{I}_n & \mathbf{0} \\ -\mathbf{D}^{-1}\mathbf{C} + \mathbf{D}^{-1}\mathbf{C} & \mathbf{I}_m \end{bmatrix} = \begin{bmatrix} \mathbf{I}_n & \mathbf{0} \\ \mathbf{0} & \mathbf{I}_m \end{bmatrix}$$

Hence, we have proved Equation (A–38).

Problem A–6

Prove that if **A**, **B**, **C**, and **D** are, respectively, an $n \times n$, an $n \times m$, an $m \times n$, and an $m \times m$ matrix, then

$$\begin{bmatrix} \mathbf{A} & \mathbf{B} \\ \mathbf{C} & \mathbf{D} \end{bmatrix}^{-1} = \begin{bmatrix} \mathbf{A}^{-1} + \mathbf{A}^{-1}\mathbf{B}(\mathbf{D} - \mathbf{C}\mathbf{A}^{-1}\mathbf{B})^{-1}\mathbf{C}\mathbf{A}^{-1} & -\mathbf{A}^{-1}\mathbf{B}(\mathbf{D} - \mathbf{C}\mathbf{A}^{-1}\mathbf{B})^{-1} \\ -(\mathbf{D} - \mathbf{C}\mathbf{A}^{-1}\mathbf{B})^{-1}\mathbf{C}\mathbf{A}^{-1} & (\mathbf{D} - \mathbf{C}\mathbf{A}^{-1}\mathbf{B})^{-1} \end{bmatrix}$$

provided $|\mathbf{A}| \neq 0$ and $|\mathbf{D} - \mathbf{C}\mathbf{A}^{-1}\mathbf{B}| \neq 0$.

Prove also that

$$\begin{bmatrix} \mathbf{A} & \mathbf{B} \\ \mathbf{C} & \mathbf{D} \end{bmatrix}^{-1} = \begin{bmatrix} (\mathbf{A} - \mathbf{B}\mathbf{D}^{-1}\mathbf{C})^{-1} & -(\mathbf{A} - \mathbf{B}\mathbf{D}^{-1}\mathbf{C})^{-1}\mathbf{B}\mathbf{D}^{-1} \\ -\mathbf{D}^{-1}\mathbf{C}(\mathbf{A} - \mathbf{B}\mathbf{D}^{-1}\mathbf{C})^{-1} & \mathbf{D}^{-1}\mathbf{C}(\mathbf{A} - \mathbf{B}\mathbf{D}^{-1}\mathbf{C})^{-1}\mathbf{B}\mathbf{D}^{-1} + \mathbf{D}^{-1} \end{bmatrix}$$

provided $|\mathbf{D}| \neq 0$ and $|\mathbf{A} - \mathbf{B}\mathbf{D}^{-1}\mathbf{C}| \neq 0$.

Solution   First, note that

$$\begin{bmatrix} \mathbf{A} & \mathbf{B} \\ \mathbf{C} & \mathbf{D} \end{bmatrix} = \begin{bmatrix} \mathbf{A} & \mathbf{0} \\ \mathbf{C} & \mathbf{I}_m \end{bmatrix}\begin{bmatrix} \mathbf{I}_n & \mathbf{A}^{-1}\mathbf{B} \\ \mathbf{0} & \mathbf{D} - \mathbf{C}\mathbf{A}^{-1}\mathbf{B} \end{bmatrix} \tag{A–39}$$

By taking the inverse of both sides of Equation (A–39), we obtain

$$\begin{bmatrix} \mathbf{A} & \mathbf{B} \\ \mathbf{C} & \mathbf{D} \end{bmatrix}^{-1} = \begin{bmatrix} \mathbf{I}_n & \mathbf{A}^{-1}\mathbf{B} \\ \mathbf{0} & \mathbf{D} - \mathbf{C}\mathbf{A}^{-1}\mathbf{B} \end{bmatrix}^{-1}\begin{bmatrix} \mathbf{A} & \mathbf{0} \\ \mathbf{C} & \mathbf{I}_m \end{bmatrix}^{-1}$$

By referring to Problem A–5, we find

$$\begin{bmatrix} \mathbf{I}_n & \mathbf{A}^{-1}\mathbf{B} \\ \mathbf{0} & \mathbf{D} - \mathbf{C}\mathbf{A}^{-1}\mathbf{B} \end{bmatrix}^{-1} = \begin{bmatrix} \mathbf{I}_n & -\mathbf{A}^{-1}\mathbf{B}(\mathbf{D} - \mathbf{C}\mathbf{A}^{-1}\mathbf{B})^{-1} \\ \mathbf{0} & (\mathbf{D} - \mathbf{C}\mathbf{A}^{-1}\mathbf{B})^{-1} \end{bmatrix}$$

and

$$\begin{bmatrix} \mathbf{A} & \mathbf{0} \\ \mathbf{C} & \mathbf{I}_m \end{bmatrix}^{-1} = \begin{bmatrix} \mathbf{A}^{-1} & \mathbf{0} \\ -\mathbf{C}\mathbf{A}^{-1} & \mathbf{I}_m \end{bmatrix}$$

Hence,

$$\begin{aligned} \begin{bmatrix} \mathbf{A} & \mathbf{B} \\ \mathbf{C} & \mathbf{D} \end{bmatrix}^{-1} &= \begin{bmatrix} \mathbf{I}_n & \mathbf{A}^{-1}\mathbf{B} \\ \mathbf{0} & \mathbf{D} - \mathbf{C}\mathbf{A}^{-1}\mathbf{B} \end{bmatrix}^{-1}\begin{bmatrix} \mathbf{A} & \mathbf{0} \\ \mathbf{C} & \mathbf{I}_m \end{bmatrix}^{-1} \\ &= \begin{bmatrix} \mathbf{I}_n & -\mathbf{A}^{-1}\mathbf{B}(\mathbf{D} - \mathbf{C}\mathbf{A}^{-1}\mathbf{B})^{-1} \\ \mathbf{0} & (\mathbf{D} - \mathbf{C}\mathbf{A}^{-1}\mathbf{B})^{-1} \end{bmatrix}\begin{bmatrix} \mathbf{A}^{-1} & \mathbf{0} \\ -\mathbf{C}\mathbf{A}^{-1} & \mathbf{I}_m \end{bmatrix} \\ &= \begin{bmatrix} \mathbf{A}^{-1} + \mathbf{A}^{-1}\mathbf{B}(\mathbf{D} - \mathbf{C}\mathbf{A}^{-1}\mathbf{B})^{-1}\mathbf{C}\mathbf{A}^{-1} & -\mathbf{A}^{-1}\mathbf{B}(\mathbf{D} - \mathbf{C}\mathbf{A}^{-1}\mathbf{B})^{-1} \\ -(\mathbf{D} - \mathbf{C}\mathbf{A}^{-1}\mathbf{B})^{-1}\mathbf{C}\mathbf{A}^{-1} & (\mathbf{D} - \mathbf{C}\mathbf{A}^{-1}\mathbf{B})^{-1} \end{bmatrix} \end{aligned}$$

provided $|\mathbf{A}| \neq 0$ and $|\mathbf{D} - \mathbf{C}\mathbf{A}^{-1}\mathbf{B}| \neq 0$.

Similarly, notice that

$$\begin{bmatrix} \mathbf{A} & \mathbf{B} \\ \mathbf{C} & \mathbf{D} \end{bmatrix} = \begin{bmatrix} \mathbf{I}_n & \mathbf{B} \\ \mathbf{0} & \mathbf{D} \end{bmatrix}\begin{bmatrix} \mathbf{A} - \mathbf{B}\mathbf{D}^{-1}\mathbf{C} & \mathbf{0} \\ \mathbf{D}^{-1}\mathbf{C} & \mathbf{I}_m \end{bmatrix} \tag{A–40}$$

By taking the inverse of both sides of Equation (A–40) and referring to Problem A–5, we obtain

$$\begin{aligned} \begin{bmatrix} \mathbf{A} & \mathbf{B} \\ \mathbf{C} & \mathbf{D} \end{bmatrix}^{-1} &= \begin{bmatrix} \mathbf{A} - \mathbf{B}\mathbf{D}^{-1}\mathbf{C} & \mathbf{0} \\ \mathbf{D}^{-1}\mathbf{C} & \mathbf{I}_m \end{bmatrix}^{-1}\begin{bmatrix} \mathbf{I}_n & \mathbf{B} \\ \mathbf{0} & \mathbf{D} \end{bmatrix}^{-1} \\ &= \begin{bmatrix} (\mathbf{A} - \mathbf{B}\mathbf{D}^{-1}\mathbf{C})^{-1} & \mathbf{0} \\ -\mathbf{D}^{-1}\mathbf{C}(\mathbf{A} - \mathbf{B}\mathbf{D}^{-1}\mathbf{C})^{-1} & \mathbf{I}_m \end{bmatrix}\begin{bmatrix} \mathbf{I}_n & -\mathbf{B}\mathbf{D}^{-1} \\ \mathbf{0} & \mathbf{D}^{-1} \end{bmatrix} \\ &= \begin{bmatrix} (\mathbf{A} - \mathbf{B}\mathbf{D}^{-1}\mathbf{C})^{-1} & -(\mathbf{A} - \mathbf{B}\mathbf{D}^{-1}\mathbf{C})^{-1}\mathbf{B}\mathbf{D}^{-1} \\ -\mathbf{D}^{-1}\mathbf{C}(\mathbf{A} - \mathbf{B}\mathbf{D}^{-1}\mathbf{C})^{-1} & \mathbf{D}^{-1}\mathbf{C}(\mathbf{A} - \mathbf{B}\mathbf{D}^{-1}\mathbf{C})^{-1}\mathbf{B}\mathbf{D}^{-1} + \mathbf{D}^{-1} \end{bmatrix} \end{aligned}$$

provided $|\mathbf{D}| \neq 0$ and $|\mathbf{A} - \mathbf{B}\mathbf{D}^{-1}\mathbf{C}| \neq 0$.

Problem A–7

For an $n \times n$ real matrix $\mathbf{A}$ and real n-vectors $\mathbf{x}$ and $\mathbf{y}$, show that

(a)
$$\frac{\partial}{\partial \mathbf{x}} \mathbf{y}^T \mathbf{x} = \mathbf{y}$$

(b)
$$\frac{\partial}{\partial \mathbf{x}} \mathbf{x}^T \mathbf{A} \mathbf{x} = \mathbf{A}\mathbf{x} + \mathbf{A}^T \mathbf{x}$$

For an $n \times n$ Hermitian matrix $\mathbf{A}$ and a complex n-vector $\mathbf{x}$, show that

(c)
$$\frac{\partial}{\partial \overline{\mathbf{x}}} \mathbf{x}^* \mathbf{A} \mathbf{x} = \mathbf{A}\mathbf{x}$$

Solution

(a) Note that
$$\mathbf{y}^T \mathbf{x} = y_1 x_1 + y_2 x_2 + \cdots + y_n x_n$$

which is a scalar quantity. Hence,

$$\frac{\partial}{\partial \mathbf{x}} \mathbf{y}^T \mathbf{x} = \begin{bmatrix} \dfrac{\partial}{\partial x_1} \mathbf{y}^T \mathbf{x} \\ \vdots \\ \dfrac{\partial}{\partial x_n} \mathbf{y}^T \mathbf{x} \end{bmatrix} = \begin{bmatrix} y_1 \\ \vdots \\ y_n \end{bmatrix} = \mathbf{y}$$

(b) Notice that
$$\mathbf{x}^T \mathbf{A} \mathbf{x} = \sum_{i=1}^{n} \sum_{j=1}^{n} a_{ij} x_i x_j$$

which is a scalar quantity. Hence,

$$\frac{\partial}{\partial \mathbf{x}} \mathbf{x}^T \mathbf{A} \mathbf{x} = \begin{bmatrix} \dfrac{\partial}{\partial x_1}\left(\sum_{i=1}^{n} \sum_{j=1}^{n} a_{ij} x_i x_j \right) \\ \vdots \\ \dfrac{\partial}{\partial x_n}\left(\sum_{i=1}^{n} \sum_{j=1}^{n} a_{ij} x_i x_j \right) \end{bmatrix} = \begin{bmatrix} \displaystyle\sum_{j=1}^{n} a_{1j} x_j + \sum_{i=1}^{n} a_{i1} x_i \\ \vdots \\ \displaystyle\sum_{j=1}^{n} a_{nj} x_j + \sum_{i=1}^{n} a_{in} x_i \end{bmatrix}$$

$$= \mathbf{A}\mathbf{x} + \mathbf{A}^T \mathbf{x}$$

which is Equation (A–20)

If matrix $\mathbf{A}$ is a real symmetric matrix, then

$$\frac{\partial}{\partial \mathbf{x}} \mathbf{x}^T \mathbf{A} \mathbf{x} = 2\mathbf{A}\mathbf{x}, \qquad \text{if } \mathbf{A} = \mathbf{A}^T$$

(c) For a Hermitian matrix $\mathbf{A}$, we have

$$\mathbf{x}^* \mathbf{A} \mathbf{x} = \sum_{i=1}^{n} \sum_{j=1}^{n} a_{ij} \overline{x}_i x_j$$

and

$$\frac{\partial}{\partial \overline{\mathbf{x}}} \mathbf{x}^* \mathbf{A} \mathbf{x} = \begin{bmatrix} \dfrac{\partial}{\partial \overline{x}_1}\left(\sum_{i=1}^{n} \sum_{j=1}^{n} a_{ij} \overline{x}_i x_j \right) \\ \vdots \\ \dfrac{\partial}{\partial \overline{x}_n}\left(\sum_{i=1}^{n} \sum_{j=1}^{n} a_{ij} \overline{x}_i x_j \right) \end{bmatrix} = \begin{bmatrix} \displaystyle\sum_{j=1}^{n} a_{1j} x_j \\ \vdots \\ \displaystyle\sum_{j=1}^{n} a_{nj} x_j \end{bmatrix} = \mathbf{A}\mathbf{x}$$

which is Equation (A–21).

Note that

$$
\frac{\partial}{\partial \mathbf{x}} \mathbf{x}^* \mathbf{A} \mathbf{x} =
\begin{bmatrix}
\frac{\partial}{\partial x_1}\left(\sum_{i=1}^{n}\sum_{j=1}^{n} a_{ij}\bar{x}_i x_j\right) \\
\vdots \\
\frac{\partial}{\partial x_n}\left(\sum_{i=1}^{n}\sum_{j=1}^{n} a_{ij}\bar{x}_i x_j\right)
\end{bmatrix}
=
\begin{bmatrix}
\sum_{i=1}^{n} a_{i1}\bar{x}_i \\
\vdots \\
\sum_{i=1}^{n} a_{in}\bar{x}_i
\end{bmatrix}
= \mathbf{A}^T \bar{\mathbf{x}}
$$

Therefore,

$$
\frac{\overline{\partial}}{\partial \mathbf{x}} \mathbf{x}^* \mathbf{A} \mathbf{x} = \mathbf{A}^* \mathbf{x} = \mathbf{A} \mathbf{x}
$$

Problem A–8

For an $n \times m$ complex matrix $\mathbf{A}$, a complex n-vector $\mathbf{x}$, and a complex m-vector $\mathbf{y}$, show that

(a)
$$
\frac{\partial}{\partial \bar{\mathbf{x}}} \mathbf{x}^* \mathbf{A} \mathbf{y} = \mathbf{A} \mathbf{y}
$$

(b)
$$
\frac{\partial}{\partial \mathbf{y}} \mathbf{x}^* \mathbf{A} \mathbf{y} = \mathbf{A}^T \bar{\mathbf{x}}
$$

Solution

(a) Notice that

$$
\mathbf{x}^* \mathbf{A} \mathbf{y} = \sum_{i=1}^{n}\sum_{j=1}^{m} a_{ij}\bar{x}_i y_j
$$

Hence,

$$
\frac{\partial}{\partial \bar{\mathbf{x}}} \mathbf{x}^* \mathbf{A} \mathbf{y} =
\begin{bmatrix}
\frac{\partial}{\partial \bar{x}_1}\left(\sum_{i=1}^{n}\sum_{j=1}^{m} a_{ij}\bar{x}_i y_j\right) \\
\vdots \\
\frac{\partial}{\partial \bar{x}_n}\left(\sum_{i=1}^{n}\sum_{j=1}^{m} a_{ij}\bar{x}_i y_j\right)
\end{bmatrix}
=
\begin{bmatrix}
\sum_{j=1}^{m} a_{1j} y_j \\
\vdots \\
\sum_{j=1}^{m} a_{nj} y_j
\end{bmatrix}
= \mathbf{A} \mathbf{y}
$$

which is Equation (A–24).

(b) Notice that

$$
\frac{\partial}{\partial \mathbf{y}} \mathbf{x}^* \mathbf{A} \mathbf{y} =
\begin{bmatrix}
\frac{\partial}{\partial y_1}\left(\sum_{i=1}^{n}\sum_{j=1}^{m} a_{ij}\bar{x}_i y_j\right) \\
\vdots \\
\frac{\partial}{\partial y_m}\left(\sum_{i=1}^{n}\sum_{j=1}^{m} a_{ij}\bar{x}_i y_j\right)
\end{bmatrix}
=
\begin{bmatrix}
\sum_{i=1}^{n} a_{i1}\bar{x}_i \\
\vdots \\
\sum_{i=1}^{n} a_{im}\bar{x}_i
\end{bmatrix}
= \mathbf{A}^T \bar{\mathbf{x}}
$$

which is Equation (A–25).

Similarly, for an $n \times m$ real matrix $\mathbf{A}$, a real n-vector $\mathbf{x}$, and a real m-vector $\mathbf{y}$, we have

$$
\frac{\partial}{\partial \mathbf{x}} \mathbf{x}^T \mathbf{A} \mathbf{y} = \mathbf{A} \mathbf{y}, \qquad \frac{\partial}{\partial \mathbf{y}} \mathbf{x}^T \mathbf{A} \mathbf{y} = \mathbf{A}^T \mathbf{x}
$$

which are Equations (A–22) and (A–23), respectively.

Problem A–9

Given two $n \times n$ matrices $\mathbf{A}$ and $\mathbf{B}$, prove that the eigenvalues of $\mathbf{A}\mathbf{B}$ and those of $\mathbf{B}\mathbf{A}$ are the same, even if $\mathbf{A}\mathbf{B} \neq \mathbf{B}\mathbf{A}$.

Solution   First, we shall consider the case where $\mathbf{A}$ (or $\mathbf{B}$) is nonsingular. In this case,

$$|\lambda\mathbf{I} - \mathbf{BA}| = |\lambda\mathbf{I} - \mathbf{A}^{-1}(\mathbf{AB})\mathbf{A}| = |\mathbf{A}^{-1}(\lambda\mathbf{I} - \mathbf{AB})\mathbf{A}| = |\mathbf{A}^{-1}||\lambda\mathbf{I} - \mathbf{AB}||\mathbf{A}| = |\lambda\mathbf{I} - \mathbf{AB}|$$

Next we shall consider the case where both $\mathbf{A}$ and $\mathbf{B}$ are singular. There exist $n \times n$ nonsingular matrices $\mathbf{P}$ and $\mathbf{Q}$ such that

$$\mathbf{PAQ} = \begin{bmatrix} \mathbf{I}_r & \mathbf{0} \\ \mathbf{0} & \mathbf{0} \end{bmatrix}$$

where $\mathbf{I}_r$ is the $r \times r$ identity matrix and r is the rank of $\mathbf{A}$, $r < n$. We have

$$|\lambda\mathbf{I} - \mathbf{BA}| = |\lambda\mathbf{I} - \mathbf{Q}^{-1}\mathbf{BAQ}| = |\lambda\mathbf{I} - \mathbf{Q}^{-1}\mathbf{BP}^{-1}\mathbf{PAQ}|$$

$$= \left|\lambda\mathbf{I} - \begin{bmatrix} \mathbf{G}_{11} & \mathbf{G}_{12} \\ \mathbf{G}_{21} & \mathbf{G}_{22} \end{bmatrix}\begin{bmatrix} \mathbf{I}_r & \mathbf{0} \\ \mathbf{0} & \mathbf{0} \end{bmatrix}\right|$$

where

$$\mathbf{Q}^{-1}\mathbf{BP}^{-1} = \begin{bmatrix} \mathbf{G}_{11} & \mathbf{G}_{12} \\ \mathbf{G}_{21} & \mathbf{G}_{22} \end{bmatrix}$$

Then

$$|\lambda\mathbf{I} - \mathbf{BA}| = \left|\lambda\mathbf{I} - \begin{bmatrix} \mathbf{G}_{11} & \mathbf{0} \\ \mathbf{G}_{21} & \mathbf{0} \end{bmatrix}\right| = \begin{vmatrix} \lambda\mathbf{I}_r - \mathbf{G}_{11} & \mathbf{0} \\ -\mathbf{G}_{21} & \lambda\mathbf{I}_{n-r} \end{vmatrix}$$

$$= |\lambda\mathbf{I}_r - \mathbf{G}_{11}||\lambda\mathbf{I}_{n-r}|$$

Also,

$$|\lambda\mathbf{I} - \mathbf{AB}| = |\lambda\mathbf{I} - \mathbf{PABP}^{-1}| = |\lambda\mathbf{I} - \mathbf{PAQQ}^{-1}\mathbf{BP}^{-1}|$$

$$= \left|\lambda\mathbf{I} - \begin{bmatrix} \mathbf{I}_r & \mathbf{0} \\ \mathbf{0} & \mathbf{0} \end{bmatrix}\begin{bmatrix} \mathbf{G}_{11} & \mathbf{G}_{12} \\ \mathbf{G}_{21} & \mathbf{G}_{22} \end{bmatrix}\right|$$

$$= \left|\lambda\mathbf{I} - \begin{bmatrix} \mathbf{G}_{11} & \mathbf{G}_{12} \\ \mathbf{0} & \mathbf{0} \end{bmatrix}\right|$$

$$= \begin{vmatrix} \lambda\mathbf{I}_r - \mathbf{G}_{11} & -\mathbf{G}_{12} \\ \mathbf{0} & \lambda\mathbf{I}_{n-r} \end{vmatrix}$$

$$= |\lambda\mathbf{I}_r - \mathbf{G}_{11}||\lambda\mathbf{I}_{n-r}|$$

Hence, we have proved that

$$|\lambda\mathbf{I} - \mathbf{BA}| = |\lambda\mathbf{I} - \mathbf{AB}|$$

or that the eigenvalues of $\mathbf{AB}$ and $\mathbf{BA}$ are the same regardless of whether $\mathbf{AB} = \mathbf{BA}$ or $\mathbf{AB} \neq \mathbf{BA}$.

Problem A–10

Show that the following 2×2 matrix $\mathbf{A}$ has two distinct eigenvalues and that the eigenvectors are linearly independent of each other:

$$\mathbf{A} = \begin{bmatrix} 1 & 1 \\ 0 & 2 \end{bmatrix}$$

Then normalize the eigenvectors.

Solution   The eigenvalues are obtained from

$$|\lambda\mathbf{I} - \mathbf{A}| = \begin{vmatrix} \lambda - 1 & -1 \\ 0 & \lambda - 2 \end{vmatrix} = (\lambda - 1)(\lambda - 2) = 0$$

as

$$\lambda_1 = 1 \quad \text{and} \quad \lambda_2 = 2$$

Thus, matrix **A** has two distinct eigenvalues.

There are two eigenvectors $\mathbf{x}_1$ and $\mathbf{x}_2$ associated with λ_1 and λ_2, respectively. If we define

$$\mathbf{x}_1 = \begin{bmatrix} x_{11} \\ x_{21} \end{bmatrix}, \quad \mathbf{x}_2 = \begin{bmatrix} x_{12} \\ x_{22} \end{bmatrix}$$

then the eigenvector $\mathbf{x}_1$ can be found from

$$\mathbf{A}\mathbf{x}_1 = \lambda_1 \mathbf{x}_1$$

or

$$(\lambda_1 \mathbf{I} - \mathbf{A})\mathbf{x}_1 = \mathbf{0}$$

Noting that $\lambda_1 = 1$, we have

$$\begin{bmatrix} 1 - 1 & -1 \\ 0 & 1 - 2 \end{bmatrix} \begin{bmatrix} x_{11} \\ x_{21} \end{bmatrix} = \begin{bmatrix} 0 \\ 0 \end{bmatrix}$$

which gives

$$x_{11} = \text{arbitrary constant} \quad \text{and} \quad x_{21} = 0$$

Hence, eigenvector $\mathbf{x}_1$ may be written as

$$\mathbf{x}_1 = \begin{bmatrix} x_{11} \\ x_{21} \end{bmatrix} = \begin{bmatrix} c_1 \\ 0 \end{bmatrix}$$

where $c_1 \neq 0$ is an arbitrary constant.

Similarly, for the eigenvector $\mathbf{x}_2$, we have

$$\mathbf{A}\mathbf{x}_2 = \lambda_2 \mathbf{x}_2$$

or

$$(\lambda_2 \mathbf{I} - \mathbf{A})\mathbf{x}_2 = \mathbf{0}$$

Noting that $\lambda_2 = 2$, we obtain

$$\begin{bmatrix} 2 - 1 & -1 \\ 0 & 2 - 2 \end{bmatrix} \begin{bmatrix} x_{12} \\ x_{22} \end{bmatrix} = \begin{bmatrix} 0 \\ 0 \end{bmatrix}$$

from which we get

$$x_{12} - x_{22} = 0$$

Hence, the eigenvector associated with $\lambda_2 = 2$ may be selected as

$$\mathbf{x}_2 = \begin{bmatrix} x_{12} \\ x_{22} \end{bmatrix} = \begin{bmatrix} c_2 \\ c_2 \end{bmatrix}$$

where $c_2 \neq 0$ is an arbitrary constant.

The two eigenvectors are therefore given by

$$\mathbf{x}_1 = \begin{bmatrix} c_1 \\ 0 \end{bmatrix} \quad \text{and} \quad \mathbf{x}_2 = \begin{bmatrix} c_2 \\ c_2 \end{bmatrix}$$

The fact that eigenvectors $\mathbf{x}_1$ and $\mathbf{x}_2$ are linearly independent can be seen from the fact that the determinant of the matrix $[\mathbf{x}_1 \ \mathbf{x}_2]$ is nonzero:

$$\begin{vmatrix} c_1 & c_2 \\ 0 & c_2 \end{vmatrix} \neq 0$$

To normalize the eigenvectors, we choose $c_1 = 1$ and $c_2 = 1/\sqrt{2}$, or

$$\mathbf{x}_1 = \begin{bmatrix} 1 \\ 0 \end{bmatrix}, \qquad \mathbf{x}_2 = \begin{bmatrix} \dfrac{1}{\sqrt{2}} \\ \dfrac{1}{\sqrt{2}} \end{bmatrix}$$

Clearly, the absolute value of each eigenvector becomes unity and therefore the eigenvectors are normalized.

Problem A–11

Obtain a transformation matrix $\mathbf{T}$ that transforms the matrix

$$\mathbf{A} = \begin{bmatrix} 0 & 1 & 0 & 3 \\ 0 & -1 & 1 & 1 \\ 0 & 0 & 0 & 1 \\ 0 & 0 & -1 & -2 \end{bmatrix}$$

into a Jordan canonical form.

Solution The characteristic equation is

$$|\lambda \mathbf{I} - \mathbf{A}| = \begin{vmatrix} \lambda & -1 & 0 & -3 \\ 0 & \lambda+1 & -1 & -1 \\ 0 & 0 & \lambda & -1 \\ 0 & 0 & 1 & \lambda+2 \end{vmatrix} = \begin{vmatrix} \lambda & -1 \\ 0 & \lambda+1 \end{vmatrix} \begin{vmatrix} \lambda & -1 \\ 1 & \lambda+2 \end{vmatrix}$$

$$= (\lambda + 1)^3 \lambda = 0$$

Hence, matrix $\mathbf{A}$ involves eigenvalues

$$\lambda_1 = -1, \qquad \lambda_2 = -1, \qquad \lambda_3 = -1, \qquad \lambda_4 = 0$$

For the multiple eigenvalue -1, we have

$$\lambda_1 \mathbf{I} - \mathbf{A} = \begin{bmatrix} -1 & -1 & 0 & -3 \\ 0 & 0 & -1 & -1 \\ 0 & 0 & -1 & -1 \\ 0 & 0 & 1 & 1 \end{bmatrix}$$

which is of rank 2, or rank $(4 - 2)$. From the rank condition we see that there must be two Jordan blocks for eigenvalue -1, that is, one $p_1 \times p_1$ Jordan block and one $p_2 \times p_2$ Jordan block, where $p_1 + p_2 = 3$. Notice that for $p_1 + p_2 = 3$ there is only one combination (2 and 1) for the orders of p_1 and p_2. Let us choose

$$p_1 = 2 \qquad \text{and} \qquad p_2 = 1$$

Then there are one eigenvector and one generalized eigenvector for Jordan block $\mathbf{J}_{p1}$ and one eigenvector for Jordan block $\mathbf{J}_{p2}$.

Let us define an eigenvector and a generalized eigenvector for Jordan block $\mathbf{J}_{p1}$ as $\mathbf{v}_{11}$ and $\mathbf{v}_{12}$, respectively, and an eigenvector for Jordan block $\mathbf{J}_{p2}$ as $\mathbf{v}_{21}$. Then there must be vectors $\mathbf{v}_{11}$, $\mathbf{v}_{12}$, and $\mathbf{v}_{21}$ that satisfy the following equations:

$$(\mathbf{A} - \lambda_1 \mathbf{I})\mathbf{v}_{11} = \mathbf{0}, \qquad (\mathbf{A} - \lambda_1 \mathbf{I})\mathbf{v}_{21} = \mathbf{0}$$

$$(\mathbf{A} - \lambda_1 \mathbf{I})\mathbf{v}_{12} = \mathbf{v}_{11}$$

For $\lambda_1 = -1$, $A - \lambda_1 I$ can be given as follows:

$$A - \lambda_1 I = \begin{bmatrix} 1 & 1 & 0 & 3 \\ 0 & 0 & 1 & 1 \\ 0 & 0 & 1 & 1 \\ 0 & 0 & -1 & -1 \end{bmatrix}$$

Noting that

$$(A - \lambda_1 I)^2 = \begin{bmatrix} 1 & 1 & -2 & 1 \\ 0 & 0 & 0 & 0 \\ 0 & 0 & 0 & 0 \\ 0 & 0 & 0 & 0 \end{bmatrix}$$

we determine vector v_{12} to be such that it will satisfy the equation

$$(A - \lambda_1 I)^2 v_{12} = 0$$

and at the same time will make $(A - \lambda_1 I)v_{12}$ nonzero. An example of such a generalized eigenvector v_{12} can be found to be

$$v_{12} = \begin{bmatrix} -a \\ 0 \\ 0 \\ a \end{bmatrix}, \quad a = \text{arbitrary nonzero constant}$$

The eigenvector v_{11} is then found to be a nonzero vector $(A - \lambda_1 I)v_{12}$:

$$v_{11} = (A - \lambda_1 I)v_{12} = \begin{bmatrix} 2a \\ a \\ a \\ -a \end{bmatrix}$$

Since a is an arbitrary nonzero constant, let us choose $a = 1$. Then we have

$$v_{11} = \begin{bmatrix} 2 \\ 1 \\ 1 \\ -1 \end{bmatrix} \quad \text{and} \quad v_{12} = \begin{bmatrix} -1 \\ 0 \\ 0 \\ 1 \end{bmatrix}$$

Next, we determine v_{21} so that v_{21} and v_{11} are linearly independent. For v_{21} we may choose

$$v_{21} = \begin{bmatrix} b + 3c \\ -b \\ c \\ -c \end{bmatrix}$$

where b and c are arbitrary constants. Let us choose, for example, $b = 1$ and $c = 0$. Then

$$v_{21} = \begin{bmatrix} 1 \\ -1 \\ 0 \\ 0 \end{bmatrix}$$

Clearly, v_{11}, v_{12}, and v_{21} are linearly independent. Let us define

$$v_{11} = x_1, \qquad v_{12} = x_2, \qquad v_{21} = x_3$$

and

$$\mathbf{T}(\lambda_1) = [\mathbf{v}_{11} \vdots \mathbf{v}_{12} \vdots \mathbf{v}_{21}] = [\mathbf{x}_1 \vdots \mathbf{x}_2 \vdots \mathbf{x}_3] = \begin{bmatrix} 2 & -1 & 1 \\ 1 & 0 & -1 \\ 1 & 0 & 0 \\ -1 & 1 & 0 \end{bmatrix}$$

For the distinct eigenvalue $\lambda_4 = 0$, the eigenvector $\mathbf{x}_4$ can be determined from

$$(\mathbf{A} - \lambda_4 \mathbf{I})\mathbf{x}_4 = \mathbf{0}$$

Noting that

$$\mathbf{A} - \lambda_4 \mathbf{I} = \mathbf{A} = \begin{bmatrix} 0 & 1 & 0 & 3 \\ 0 & -1 & 1 & 1 \\ 0 & 0 & 0 & 1 \\ 0 & 0 & -1 & -2 \end{bmatrix}$$

we find

$$\mathbf{x}_4 = \begin{bmatrix} d \\ 0 \\ 0 \\ 0 \end{bmatrix}$$

where $d \neq 0$ is an arbitrary constant. By choosing $d = 1$, we have

$$\mathbf{T}(\lambda_4) = \mathbf{x}_4 = \begin{bmatrix} 1 \\ 0 \\ 0 \\ 0 \end{bmatrix}$$

Thus, the transformation matrix $\mathbf{T}$ can be written as

$$\mathbf{T} = [\mathbf{T}(\lambda_1) \vdots \mathbf{T}(\lambda_4)] = \begin{bmatrix} 2 & -1 & 1 & 1 \\ 1 & 0 & -1 & 0 \\ 1 & 0 & 0 & 0 \\ -1 & 1 & 0 & 0 \end{bmatrix}$$

Then

$$\mathbf{T}^{-1}\mathbf{A}\mathbf{T} = \begin{bmatrix} 0 & 0 & 1 & 0 \\ 0 & 0 & 1 & 1 \\ 0 & -1 & 1 & 0 \\ 1 & 1 & -2 & 1 \end{bmatrix}\begin{bmatrix} 0 & 1 & 0 & 3 \\ 0 & -1 & 1 & 1 \\ 0 & 0 & 0 & 1 \\ 0 & 0 & -1 & -2 \end{bmatrix}\begin{bmatrix} 2 & -1 & 1 & 1 \\ 1 & 0 & -1 & 0 \\ 1 & 0 & 0 & 0 \\ -1 & 1 & 0 & 0 \end{bmatrix}$$

$$= \begin{bmatrix} -1 & 1 & 0 & 0 \\ 0 & -1 & 0 & 0 \\ 0 & 0 & -1 & 0 \\ 0 & 0 & 0 & 0 \end{bmatrix} = \text{diag}\,[\mathbf{J}_2(-1), \mathbf{J}_1(-1), \mathbf{J}_1(0)]$$

Problem A–12

Assume that an $n \times n$ normal matrix $\mathbf{A}$ has a k-multiple eigenvalue λ_1. Prove that the rank of $\mathbf{A} - \lambda_1 \mathbf{I}$ is $n - k$.

Solution Suppose that the rank of $\mathbf{A} - \lambda_1 \mathbf{I}$ is $n - m$. Then the equation

$$(\mathbf{A} - \lambda_1 \mathbf{I})\mathbf{x} = \mathbf{0} \tag{A–41}$$

will have m linearly independent vector solutions. Let us choose m such vectors so that they are orthogonal to each other and normalized. That is, vectors $x_1, x_2, \ldots, x_m$ will satisfy Equation (A–41) and will be orthonormal.

Let us consider $n - m$ vectors $x_{m+1}, x_{m+2}, \ldots, x_n$ such that all n vectors

$$x_1, x_2, \ldots, x_n$$

will be orthonormal to each other. Then matrix U, defined by

$$U = [x_1 \vdots x_2 \vdots \cdots \vdots x_n]$$

is a unitary matrix.

Since for $1 \le i \le m$, we have

$$Ax_i = \lambda_1 x_i$$

and therefore we can write

$$AU = U \begin{bmatrix} \lambda_1 I_m & B \\ 0 & C \end{bmatrix}$$

or

$$U^*AU = \begin{bmatrix} \lambda_1 I_m & B \\ 0 & C \end{bmatrix}$$

Noting that

$$\begin{aligned}
\|Ax_i - \lambda x_i\|^2 &= \langle (A - \lambda I)x_i, (A - \lambda I)x_i \rangle \\
&= \langle (A^* - \bar{\lambda} I)(A - \lambda I)x_i, x_i \rangle \\
&= \langle (A - \lambda I)(A^* - \bar{\lambda} I)x_i, x_i \rangle \\
&= \langle (A^* - \bar{\lambda} I)x_i, (A^* - \bar{\lambda} I)x_i \rangle \\
&= \|A^*x_i - \bar{\lambda} x_i\|^2 \\
&= 0
\end{aligned}$$

we have

$$A^*x_i = \bar{\lambda} x_i$$

Therefore, we can write

$$A^*U = U \begin{bmatrix} \bar{\lambda}_1 I_m & B_1 \\ 0 & C_1 \end{bmatrix}$$

or

$$U^*A^*U = \begin{bmatrix} \bar{\lambda}_1 I_m & B_1 \\ 0 & C_1 \end{bmatrix}$$

Hence,

$$\begin{bmatrix} \lambda_1 I_m & B \\ 0 & C \end{bmatrix} = U^*AU = (U^*A^*U)^* = \begin{bmatrix} \bar{\lambda}_1 I_m & B_1 \\ 0 & C_1 \end{bmatrix}^* = \begin{bmatrix} \lambda_1 I_m & 0 \\ B_1^* & C_1^* \end{bmatrix}$$

Comparing the left and right sides of this last equation, we obtain

$$B = 0$$

Hence, we get

$$A = U \begin{bmatrix} \lambda_1 I_m & 0 \\ 0 & C \end{bmatrix} U^*$$

Then

$$A - \lambda I = U \begin{bmatrix} (\lambda_1 - \lambda)I_m & 0 \\ 0 & C - \lambda I_{n-m} \end{bmatrix} U^*$$

The determinant of this last equation is

$$|A - \lambda I| = (\lambda_1 - \lambda)^m |C - \lambda I_{n-m}| \tag{A-42}$$

On the other hand, we have

$$\text{rank}(A - \lambda_1 I) = n - m = \text{rank}\left\{ U \begin{bmatrix} 0 & 0 \\ 0 & C - \lambda_1 I_{n-m} \end{bmatrix} U^* \right\}$$

$$= \text{rank} \begin{bmatrix} 0 & 0 \\ 0 & C - \lambda_1 I_{n-m} \end{bmatrix} = \text{rank}(C - \lambda_1 I_{n-m})$$

Hence, we conclude that the rank of $C - \lambda_1 I_{n-m}$ is $n - m$. Consequently,

$$|C - \lambda_1 I_{n-m}| \neq 0$$

and from Equation (A–42), λ_1 is shown to be the m-multiple eigenvalue of $|A - \lambda I| = 0$. Since λ_1 is the k-multiple eigenvalue of A, we must have $m = k$. Therefore, the rank of $A - \lambda_1 I$ is $n - k$.

Note that, since the rank of $A - \lambda_1 I$ is $n - k$, the equation

$$(A - \lambda_1 I)x_i = 0$$

will have k linearly independent eigenvectors $x_1, x_2, \ldots, x_k$.

Problem A–13

Prove that the eigenvalues of an $n \times n$ Hermitian matrix and of an $n \times n$ real symmetric matrix are real. Prove also that the eigenvalues of a skew-Hermitian matrix and of a real skew-symmetric matrix are either zero or purely imaginary.

Solution   Let us define any eigenvalue of an $n \times n$ Hermitian matrix A by $\lambda = \alpha + j\beta$. There exists a vector $x \neq 0$ such that

$$Ax = (\alpha + j\beta)x$$

The conjugate transpose of this last equation is

$$x^*A^* = (\alpha - j\beta)x^*$$

Since A is Hermitian $A^* = A$. Therefore, we obtain

$$x^*Ax = (\alpha - j\beta)x^*x$$

On the other hand, since $Ax = (\alpha + j\beta)x$, we have

$$x^*Ax = (\alpha + j\beta)x^*x$$

Hence, we obtain

$$[(\alpha - j\beta) - (\alpha + j\beta)]x^*x = 0$$

or

$$-2j\beta x^*x = 0$$

Since $x^*x \neq 0$ (for $x \neq 0$), we conclude that

$$\beta = 0$$

This proves that any eigenvalue of an $n \times n$ Hermitian matrix A is real. It follows that the eigenvalues of a real symmetric matrix are also real, since it is Hermitian.

To prove the second half of the problem, notice that if B is skew-Hermitian, then jB is Hermitian. Hence, the eigenvalues of jB are real, which implies that the eigenvalues of B are either zero or purely imaginary.

The eigenvalues of a real skew-symmetric matrix are also either zero or purely imaginary, since a real skew-symmetric matrix is skew-Hermitian.

Note that, in the real skew-symmetric matrix, purely imaginary eigenvalues always occur in conjugate pairs, since the coefficients of the characteristic equation are real. Note also that an $n \times n$ real skew-symmetric matrix is singular if n is odd, since such a matrix must include at least one zero eigenvalue.

Problem A–14

Examine whether or not the following 3×3 matrix A is positive definite:

$$A = \begin{bmatrix} 2 & 2 & -1 \\ 2 & 6 & 0 \\ -1 & 0 & 1 \end{bmatrix}$$

Solution   We shall demonstrate three different ways to test the positive definiteness of matrix A.

1. We may first apply Sylvester's criterion for positive definiteness of a quadratic form $x^T A x$. For the given matrix A, we have

$$2 > 0, \quad \begin{vmatrix} 2 & 2 \\ 2 & 6 \end{vmatrix} > 0, \quad \begin{vmatrix} 2 & 2 & -1 \\ 2 & 6 & 0 \\ -1 & 0 & 1 \end{vmatrix} > 0$$

Thus, the successive principal minors are all positive. Hence, matrix A is positive definite.

2. We may examine the positive definiteness of $x^T A x$. Since

$$x^T A x = [x_1 x_2 x_3] \begin{bmatrix} 2 & 2 & -1 \\ 2 & 6 & 0 \\ -1 & 0 & 1 \end{bmatrix} \begin{bmatrix} x_1 \\ x_2 \\ x_2 \end{bmatrix}$$

$$= 2x_1^2 + 4x_1 x_2 - 2x_1 x_3 + 6x_2^2 + x_3^2$$

$$= (x_1 - x_3)^2 + (x_1 + 2x_2)^2 + 2x_2^2$$

we find that $x^T A x$ is positive except at the origin ($x = 0$). Hence, we conclude that matrix A is positive definite.

3. We may examine the eigenvalues of matrix A. Note that

$$|\lambda I - A| = \lambda^3 - 9\lambda^2 + 15\lambda - 2$$

$$= (\lambda - 2)(\lambda - 0.1459)(\lambda - 6.8541)$$

Hence,

$$\lambda_1 = 2, \quad \lambda_2 = 0.1459, \quad \lambda_3 = 6.8541$$

Since all eigenvalues are positive, we conclude that A is a positive definite matrix.

Problem A–15

Examine whether the following matrix A is positive semidefinite:

$$A = \begin{bmatrix} 1 & 2 & 1 \\ 2 & 4 & 2 \\ 1 & 2 & 0 \end{bmatrix}$$

Solution In the positive semidefiniteness test, we need to examine the signs of all principal minors in addition to the sign of the determinant of the given matrix, which must be zero; that is, $|\mathbf{A}|$ must be equal to 0.

For the 3×3 matrix

$$\begin{bmatrix} a_{11} & a_{12} & a_{13} \\ a_{21} & a_{22} & a_{23} \\ a_{31} & a_{32} & a_{33} \end{bmatrix}$$

there are six principal minors:

$$a_{11}, \quad a_{22}, \quad a_{33}, \quad \begin{vmatrix} a_{11} & a_{12} \\ a_{21} & a_{22} \end{vmatrix}, \quad \begin{vmatrix} a_{22} & a_{23} \\ a_{32} & a_{33} \end{vmatrix}, \quad \begin{vmatrix} a_{11} & a_{13} \\ a_{31} & a_{33} \end{vmatrix}$$

We need to examine the signs of all six principal minors and the sign of $|\mathbf{A}|$.

For the given matrix $\mathbf{A}$,

$$a_{11} = 1 > 0$$

$$a_{22} = 4 > 0$$

$$a_{33} = 0$$

$$\begin{vmatrix} a_{11} & a_{12} \\ a_{21} & a_{22} \end{vmatrix} = \begin{vmatrix} 1 & 2 \\ 2 & 4 \end{vmatrix} = 0$$

$$\begin{vmatrix} a_{22} & a_{23} \\ a_{32} & a_{33} \end{vmatrix} = \begin{vmatrix} 4 & 2 \\ 2 & 0 \end{vmatrix} = -4 < 0$$

$$\begin{vmatrix} a_{11} & a_{13} \\ a_{31} & a_{33} \end{vmatrix} = \begin{vmatrix} 1 & 1 \\ 1 & 0 \end{vmatrix} = -1 < 0$$

$$\begin{vmatrix} a_{11} & a_{12} & a_{13} \\ a_{21} & a_{22} & a_{23} \\ a_{31} & a_{32} & a_{33} \end{vmatrix} = \begin{vmatrix} 1 & 2 & 1 \\ 2 & 4 & 2 \\ 1 & 2 & 0 \end{vmatrix} = 0$$

Clearly, two of the principal minors are negative. Hence, we conclude that matrix $\mathbf{A}$ is not positive semidefinite.

It is important to note that, had we tested the signs of only the successive principal minors and the determinant of $\mathbf{A}$,

$$1 > 0, \quad \begin{vmatrix} 1 & 2 \\ 2 & 4 \end{vmatrix} = 0, \quad |\mathbf{A}| = \begin{vmatrix} 1 & 2 & 1 \\ 2 & 4 & 2 \\ 1 & 2 & 0 \end{vmatrix} = 0$$

we would have reached the wrong conclusion that matrix $\mathbf{A}$ is positive semidefinite.

In fact, for the given matrix $\mathbf{A}$,

$$|\lambda \mathbf{I} - \mathbf{A}| = \begin{vmatrix} \lambda - 1 & -2 & -1 \\ -2 & \lambda - 4 & -2 \\ -1 & -2 & \lambda \end{vmatrix} = (\lambda^2 - 5\lambda - 5)\lambda$$

$$= (\lambda - 5.8541)\lambda(\lambda + 0.8541)$$

and so the eigenvalues are

$$\lambda_1 = 5.8541, \quad \lambda_2 = 0, \quad \lambda_3 = -0.8541$$

For matrix $\mathbf{A}$ to be positive semidefinite, all eigenvalues must be nonnegative and at least one of them must be zero. Clearly, matrix $\mathbf{A}$ is an indefinite matrix.

Appendix B

z Transform Theory

B–1 INTRODUCTION

This appendix first presents useful theorems of the z transform theory that were not treated in Chapter 2. Then we discuss details of the inversion integral method for finding the inverse z transform. Finally, we present the modified z transform method. At the end of this appendix (in the Example Problems and Solutions section), we discuss some of the interesting problems dealing with the z transformation, not treated in Chapter 2.

B–2 USEFUL THEOREMS OF THE z TRANSFORM THEORY

In this section we present some of the useful theorems of the z transform theory that were not discussed in Chapter 2.

Complex Differentiation. In the region of convergence a power series in z may be differentiated with respect to z any number of times to get a convergent series. The derivatives of $X(z)$ converge in the same region as $X(z)$.

Consider

$$X(z) = \sum_{k=0}^{\infty} x(k)z^{-k}$$

which converges in a certain region in the z plane. Differentiating $X(z)$ with respect to z, we obtain

$$\frac{d}{dz}X(z) = \sum_{k=0}^{\infty} (-k)x(k)z^{-k-1}$$

681

Multiplying both sides of this last equation by $-z$ gives

$$-z\frac{d}{dz}X(z) = \sum_{k=0}^{\infty} kx(k)z^{-k} \tag{B-1}$$

Thus, we have

$$\mathscr{Z}[kx(k)] = -z\frac{d}{dz}X(z) \tag{B-2}$$

Similarly, by differentiating both sides of Equation (B–1) with respect to z, we have

$$\frac{d}{dz}\left[-z\frac{d}{dz}X(z)\right] = \sum_{k=0}^{\infty} (-k^2)x(k)z^{-k-1}$$

Multiplying both sides of this last equation by $-z$, we obtain

$$-z\frac{d}{dz}\left[-z\frac{d}{dz}X(z)\right] = \sum_{k=0}^{\infty} k^2 x(k)z^{-k}$$

or

$$\mathscr{Z}[k^2 x(k)] = \left(-z\frac{d}{dz}\right)^2 X(z)$$

The operation $\left(-z\dfrac{d}{dz}\right)^2$ implies that we apply the operator $-z\dfrac{d}{dz}$ twice. Similarly, by repeating this process we have

$$\mathscr{Z}[k^m x(k)] = \left(-z\frac{d}{dz}\right)^m X(z) \tag{B-3}$$

Such complex differentiation enables us to obtain new z transform pairs from the known z transform pairs.

Example B–1

The z transform of the unit-step sequence $1(k)$ is given by

$$\mathscr{Z}[1(k)] = \frac{1}{1-z^{-1}}$$

Obtain the z transform of the unit-ramp sequence $x(k)$, where

$$x(k) = k$$

by using the complex differentiation theorem.

$$\mathscr{Z}[x(k)] = \mathscr{Z}[k] = \mathscr{Z}[k \cdot 1(k)] = -z\frac{d}{dz}\left(\frac{1}{1-z^{-1}}\right) = \frac{z^{-1}}{(1-z^{-1})^2}$$

Complex Integration. Consider the sequence

$$g(k) = \frac{x(k)}{k}$$

where $x(k)/k$ is finite for $k = 0$. The z transform of $x(k)/k$ is given by

$$\mathscr{Z}\left[\frac{x(k)}{k}\right] = \int_z^\infty \frac{X(z_1)}{z_1} dz_1 + \lim_{k\to0} \frac{x(k)}{k} \tag{B–4}$$

where $\mathscr{Z}[x(k)] = X(z)$.

To prove Equation (B–4), note that

$$\mathscr{Z}\left[\frac{x(k)}{k}\right] = G(z) = \sum_{k=0}^\infty \frac{x(k)}{k} z^{-k}$$

Differentiating this last equation with respect to z yields

$$\frac{d}{dz} G(z) = -\sum_{k=0}^\infty x(k) z^{-k-1} = -z^{-1} \sum_{k=0}^\infty x(k) z^{-k} = -\frac{X(z)}{z}$$

Integrating both sides of this last equation with respect to z from z to ∞ gives

$$\int_z^\infty \frac{d}{dz} G(z)\, dz = G(\infty) - G(z) = -\int_z^\infty \frac{X(z_1)}{z_1} dz_1$$

or

$$G(z) = \int_z^\infty \frac{X(z_1)}{z_1} dz_1 + G(\infty)$$

Noting that $G(\infty)$ is given by

$$G(\infty) = \lim_{z\to\infty} G(z) = g(0) = \lim_{k\to0} \frac{x(k)}{k}$$

we have

$$\mathscr{Z}\left[\frac{x(k)}{k}\right] = \int_z^\infty \frac{X(z_1)}{z_1} dz_1 + \lim_{k\to0} \frac{x(k)}{k}$$

Partial Differentiation Theorem.   Consider a function $x(t, a)$ or $x(kT, a)$ that is z-transformable. Here a is a constant or an independent variable. Define the z transform of $x(t, a)$ or $x(kT, a)$ as $X(z, a)$. Thus,

$$\mathscr{Z}[x(t, a)] = \mathscr{Z}[x(kT, a)] = X(z, a)$$

The z transform of the partial derivative of $x(t, a)$ or $x(kT, a)$ with respect to a can be given by

$$\mathscr{Z}\left[\frac{\partial}{\partial a} x(t, a)\right] = \mathscr{Z}\left[\frac{\partial}{\partial a} x(kT, a)\right] = \frac{\partial}{\partial a} X(z, a) \tag{B–5}$$

This equation is called the partial differentiation theorem.

To prove this theorem, note that

$$\mathscr{Z}\left[\frac{\partial}{\partial a} x(t, a)\right] = \mathscr{Z}\left[\frac{\partial}{\partial a} x(kT, a)\right] = \sum_{k=0}^\infty \frac{\partial}{\partial a} x(kT, a) z^{-k}$$

$$= \frac{\partial}{\partial a} \sum_{k=0}^\infty x(kT, a) z^{-k} = \frac{\partial}{\partial a} X(z, a)$$

Example B–2

Consider

$$x(t, a) = t^2 e^{-at}$$

Obtain the z transform of this function $x(t, a)$ by use of the partial differentiation theorem.

Notice that

$$\frac{\partial}{\partial a}(-te^{-at}) = t^2 e^{-at}$$

and

$$\mathscr{Z}[te^{-at}] = \frac{Te^{-aT}z^{-1}}{(1 - e^{-aT}z^{-1})^2}$$

Then we have

$$\mathscr{Z}[x(t, a)] = \mathscr{Z}[t^2 e^{-at}] = \mathscr{Z}\left[\frac{\partial}{\partial a}(-te^{-at})\right]$$

$$= \frac{\partial}{\partial a}\left[-\frac{Te^{-aT}z^{-1}}{(1 - e^{-aT}z^{-1})^2}\right]$$

$$= \frac{T^2 e^{-aT}(1 + e^{-aT}z^{-1})z^{-1}}{(1 - e^{-aT}z^{-1})^3}$$

Real Convolution Theorem.   Consider the functions $x_1(t)$ and $x_2(t)$, where

$$x_1(t) = 0, \quad \text{for } t < 0$$

$$x_2(t) = 0, \quad \text{for } t < 0$$

Assume that $x_1(t)$ and $x_2(t)$ are z-transformable and their z transforms are $X_1(z)$ and $X_2(z)$, respectively. Then

$$X_1(z)X_2(z) = \mathscr{Z}\left[\sum_{h=0}^{k} x_1(hT)x_2(kT - hT)\right] \tag{B-6}$$

This equation is called the real convolution theorem.

To prove this theorem, notice that

$$\mathscr{Z}\left[\sum_{h=0}^{k} x_1(hT)x_2(kT - hT)\right] = \sum_{k=0}^{\infty}\sum_{h=0}^{k} x_1(hT)x_2(kT - hT)z^{-k}$$

$$= \sum_{k=0}^{\infty}\sum_{h=0}^{\infty} x_1(hT)x_2(kT - hT)z^{-k}$$

where we used the condition that $x_2(kT - hT) = 0$ for $h > k$. Now define $m = k - h$. Then

$$\mathscr{Z}\left[\sum_{h=0}^{k} x_1(hT)x_2(kT - hT)\right] = \sum_{h=0}^{\infty} x_1(hT)z^{-h} \sum_{m=-h}^{\infty} x_2(mT)z^{-m}$$

Since $x_2(mT) = 0$ for $m < 0$, this last equation becomes

$$\mathscr{Z}\left[\sum_{h=0}^{k} x_1(hT)x_2(kT - hT)\right] = \sum_{h=0}^{\infty} x_1(hT)z^{-h} \sum_{m=0}^{\infty} x_2(mT)z^{-m} = X_1(z)X_2(z)$$

Complex Convolution Theorem.   The following, known as the complex convolution theorem, is useful in obtaining the z transform of the product of two sequences $x_1(k)$ and $x_2(k)$.

Suppose both $x_1(k)$ and $x_2(k)$ are zero for $k < 0$. Assume that

$$X_1(z) = \mathcal{Z}[x_1(k)], \qquad |z| > R_1$$

$$X_2(z) = \mathcal{Z}[x_2(k)], \qquad |z| > R_2$$

where R_1 and R_2 are the radii of absolute convergence for $x_1(k)$ and $x_2(k)$, respectively. Then the z transform of the product of $x_1(k)$ and $x_2(k)$ can be given by

$$\mathcal{Z}[x_1(k)x_2(k)] = \frac{1}{2\pi j}\oint_C \zeta^{-1} X_2(\zeta)X_1(\zeta^{-1}z)\,d\zeta \tag{B–7}$$

where $R_2 < |\zeta| < |z|/R_1$.

To prove this theorem, let us take the z transform of $x_1(k)x_2(k)$:

$$\mathcal{Z}[x_1(k)x_2(k)] = \sum_{k=0}^{\infty} x_1(k)x_2(k)z^{-k} \tag{B–8}$$

The series on the right-hand side of Equation (B–8) converges for $|z| > R$, where R is the radius of absolute convergence for $x_1(k)x_2(k)$. From Equation (2–23), we have

$$x_2(k) = \frac{1}{2\pi j}\oint_C X_2(z)z^{k-1}\,dz$$

$$= \frac{1}{2\pi j}\oint_C X_2(\zeta)\zeta^{k-1}\,d\zeta \tag{B–9}$$

Substituting Equation (B–9) into Equation (B–8), we obtain

$$\mathcal{Z}[x_1(k)x_2(k)] = \frac{1}{2\pi j}\sum_{k=0}^{\infty}\oint_C X_2(\zeta)\zeta^{k-1}x_1(k)z^{-k}\,d\zeta$$

Noting that Equation (B–8) converges uniformly for the region $|z| > R$, we may interchange the order of summation and integration. Then

$$\mathcal{Z}[x_1(k)z_2(k)] = \frac{1}{2\pi j}\oint_C \zeta^{-1} X_2(\zeta)\sum_{k=0}^{\infty} x_1(k)(\zeta^{-1}z)^{-k}\,d\zeta$$

Since

$$\sum_{k=0}^{\infty} x_1(k)(\zeta^{-1}z)^{-k} = X_1(\zeta^{-1}z)$$

we have

$$\mathcal{Z}[x_1(k)x_2(k)] = \frac{1}{2\pi j}\oint_C \zeta^{-1} X_2(\zeta)X_1(\zeta^{-1}z)\,d\zeta \tag{B–10}$$

where C is a contour (a circle with its center at the origin), which lies in the region given by $|\zeta| > R_2$ and $|\zeta^{-1}z| > R_1$, or

$$R_2 < |\zeta| < \frac{|z|}{R_1} \tag{B–11}$$

Parseval's Theorem. Suppose the z transforms of sequences $x_1(k)$ and $x_2(k)$ are such that

$$X_1(z) = \mathcal{Z}[x_1(k)], \qquad |z| > R_1 \text{ (where } R_1 < 1)$$

$$X_2(z) = \mathcal{Z}[x_2(k)], \qquad |z| > R_2$$

and inequality (B–11) is satisfied for $|z| = 1$, or

$$R_2 < |\zeta| < \frac{1}{R_1}$$

Then, by substituting $|z| = 1$ into Equation (8–10), we obtain the following equation:

$$\mathcal{Z}[x_1(k)x_2(k)]|_{|z|=1} = \sum_{k=0}^{\infty} x_1(k)x_2(k) = \frac{1}{2\pi j} \oint_C \zeta^{-1} X_2(\zeta) X_1(\zeta^{-1}) \, d\zeta$$

If we set $x_1(k) = x_2(k) = x(k)$ in this last equation, we get

$$\sum_{k=0}^{\infty} x^2(k) = \frac{1}{2\pi j} \oint_C \zeta^{-1} X(\zeta) X(\zeta^{-1}) \, d\zeta$$

$$= \frac{1}{2\pi j} \oint_C z^{-1} X(z) X(z^{-1}) \, dz \qquad \text{(B–12)}$$

Equation (B–12) is Parseval's theorem. This theorem is useful for obtaining the summation of $x^2(k)$.

B–3 INVERSE z TRANSFORMATION AND INVERSION INTEGRAL METHOD

If $X(z)$ is expanded into a power series in z^{-1},

$$X(z) = \sum_{k=0}^{\infty} x(kT)z^{-k} = x(0) + x(T)z^{-1} + x(2T)z^{-2} + \cdots + x(kT)z^{-k} + \cdots$$

or

$$X(z) = \sum_{k=0}^{\infty} x(k)z^{-k} = x(0) + x(1)z^{-1} + x(2)z^{-2} + \cdots + x(k)z^{-k} + \cdots$$

then the values of $x(kT)$ or $x(k)$ give the inverse z transform. If $X(z)$ is given in the form of a rational function, the expansion into an infinite power series in increasing powers of z^{-1} can be accomplished by simply dividing the numerator by the denominator. If the resulting series is convergent, the coefficients of the z^{-k} term in the series are the values $x(kT)$ of the time sequence. However, it is usually difficult to get the closed-form expressions.

The following formulas are sometimes useful in recognizing the closed-form expressions for finite or infinite series in z^{-1}:

$$(1 - az^{-1})^3 = 1 - 3az^{-1} + 3a^2 z^{-2} - a^3 z^{-3}$$

$$(1 - az^{-1})^4 = 1 - 4az^{-1} + 6a^2 z^{-2} - 4a^3 z^{-3} + a^4 z^{-4}$$

$$(1 - az^{-1})^{-1} = 1 + az^{-1} + a^2 z^{-2} + a^3 z^{-3} + a^4 z^{-4} + a^5 z^{-5} + \cdots \qquad |z| > 1$$

$$(1 - az^{-1})^{-2} = 1 + 2az^{-1} + 3a^2 z^{-2} + 4a^3 z^{-3} + 5a^4 z^{-4} + 6a^5 z^{-5} + \cdots, \qquad |z| > 1$$

$$(1 - az^{-1})^{-3} = 1 + 3az^{-1} + 6a^2 z^{-2} + 10a^3 z^{-3} + 15a^4 z^{-4}$$
$$+ 21a^5 z^{-5} + 28a^6 z^{-6} + \cdots \qquad |z| > 1$$

$$(1 - az^{-1})^{-4} = 1 + 4az^{-1} + 10a^2 z^{-2} + 20a^3 z^{-3} + 35a^4 z^{-4}$$
$$+ 56a^5 z^{-5} + 84a^6 z^{-6} + 120a^7 z^{-7} + \cdots \qquad |z| > 1$$

For a given z transform $X(z)$, if a closed-form expression for $x(k)$ is desired, we may use the partial-fraction-expansion method or the inversion integral method discussed in what follows.

Inversion Integral Method.   The inversion integral method, based on the inversion integral, is the most general method for obtaining the inverse z transform. It is based on complex variable theory. (For a rigorous and complete derivation of the inversion integral, refer to a book on complex variable theory.) In presenting the inversion integral formula for the z transform, we need to review the residue theorem and its associated background material.

Review of Background Material in Deriving the Inversion Integral Formula.
Suppose z_0 is an isolated singular point (pole) of $F(z)$. It can be seen that a positive number r_1 exists such that the function $F(z)$ is analytic at every point z for which $0 < |z - z_0| \le r_1$. Let us denote the circle with center at $z = z_0$ and radius r_1 as Γ_1. Define Γ_2 as any circle with center at $z = z_0$ and radius $|z - z_0| = r_2$ for which $r_2 \le r_1$. Circles Γ_1 and Γ_2 are shown in Figure B–1. Then the Laurent series expansion of $F(z)$ about pole $z = z_0$ may be given by

$$F(z) = \sum_{n=0}^{\infty} a_n (z - z_0)^n + \sum_{n=1}^{\infty} \frac{b_n}{(z - z_0)^n}$$

where coefficients a_n and b_n are given by

$$a_n = \frac{1}{2\pi j} \oint_{\Gamma_1} \frac{F(z)}{(z - z_0)^{n+1}} dz, \qquad n = 0, 1, 2, \ldots$$

$$b_n = \frac{1}{2\pi j} \oint_{\Gamma_2} \frac{F(z)}{(z - z_0)^{-n+1}} dz, \qquad n = 1, 2, 3, \ldots$$

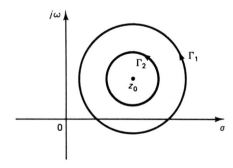

Figure B–1 Analytic region for function $F(z)$.

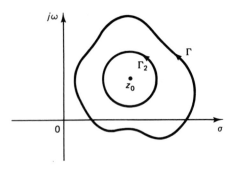

Figure B–2 Analytic region for function $F(z)$ as bonded by closed curve Γ.

Notice that the coefficient b_1 is given by

$$b_1 = \frac{1}{2\pi j} \oint_{\Gamma_2} F(z)\, dz \tag{B-13}$$

It can be proved that the value of the integral of Equation (B–13) is unchanged if Γ_1 is replaced by any closed curve Γ around z_0 such that $F(z)$ is analytic on and inside Γ except at pole $z = z_0$ (see Figure B–2). The closed curve Γ may extend outside the circle Γ_1. Then, by referring to the Cauchy–Goursat theorem, we have

$$\oint_{\Gamma} F(z)\, dz - \oint_{\Gamma_2} F(z)\, dz = 0$$

Thus, Equation (B–13) can be written as

$$b_1 = \frac{1}{2\pi j} \oint_{\Gamma} F(z)\, dz$$

The coefficient b_1 is called the *residue* of $F(z)$ at the pole z_0.

Next, let us assume that the closed curve Γ enclosed m isolated poles $z_1, z_2, \ldots, z_m$, as shown in Figure B–3. Notice that the function $F(z)$ is analytic in

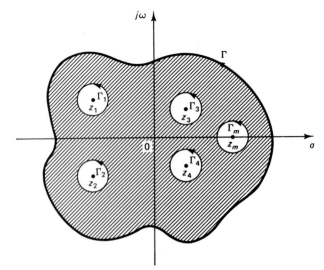

Figure B–3 Closed curve Γ enclosing m isolated poles $z_1, z_2, \ldots, z_m$.

the shaded region. According to the Cauchy–Goursat theorem, the integral of $F(z)$ over the shaded region is zero. The integral over the total shaded region is

$$\oint_\Gamma F(z)\,dz - \oint_{\Gamma_1} F(z)\,dz - \oint_{\Gamma_2} F(z)\,dz - \cdots - \oint_{\Gamma_m} F(z)\,dz = 0$$

where $\Gamma_1, \Gamma_2, \ldots, \Gamma_m$ are closed curves around the poles $z_1, z_2, \ldots, z_m$, respectively. Hence,

$$\oint_\Gamma F(z)\,dz = \oint_{\Gamma_1} F(z)\,dz + \oint_{\Gamma_2} F(z)\,dz + \cdots + \oint_{\Gamma_m} F(z)\,dz$$

$$= 2\pi j(b_{1_1} + b_{1_2} + \cdots + b_{1_m})$$

$$= 2\pi j(K_1 + K_2 + \cdots + K_m) \tag{B–14}$$

where $K_1 = b_{1_1}, K_2 = b_{1_2}, \ldots, K_m = b_{1_m}$ are residues of $F(z)$ at poles $z_1, z_2, \ldots, z_m$, respectively.

Equation (B–14) is known as the *residue theorem*. It states that if a function $F(z)$ is analytic within and on a closed curve Γ, except at a finite number of poles $z_1, z_2, \ldots, z_m$ inside Γ, then the integral of $F(z)$ taken counterclockwise around Γ is equal to $2\pi j$ times the sum of the residues at poles $z_1, z_2, \ldots, z_m$.

Inversion Integral for the z Transform. We shall now use the Cauchy–Goursat theorem and the residue theorem to derive the inversion integral for the z transform.

From the definition of the z transform, we have

$$X(z) = \sum_{k=0}^{\infty} x(kT)z^{-k} = x(0) + x(T)z^{-1} + z(2T)z^{-2} + \cdots + x(kT)z^{-k} + \cdots$$

By multiplying both sides of this last equation by z^{k-1}, we obtain

$$X(z)z^{k-1} = x(0)z^{k-1} + x(T)z^{k-2} + x(2T)z^{k-3} + \cdots + x(kT)z^{-1} + \cdots \tag{B–15}$$

Notice that Equation (B–15) is the Laurent series expansion of $X(z)z^{k-1}$ around point $z = 0$.

Consider a circle C with its center at the origin of the z plane such that all poles of $X(z)z^{k-1}$ are inside it. Noting that the coefficient $x(kT)$ associated with the term z^{-1} in Equation (B–15) is the residue, we obtain

$$x(kT) = \frac{1}{2\pi j}\oint_C X(z)z^{k-1}\,dz \tag{B–16}$$

Equation (B–16) is the inversion integral for the z transform. The evaluation of the inversion integral can be done as presented next.

Let us define the poles of $X(z)z^{k-1}$ as $z_1, z_2, \ldots, z_m$. Since the closed curve C encloses all poles $z_1, z_2, \ldots, z_m$, then referring to Equation (B–14) we have

$$\oint_C X(z)z^{k-1}\,dz = \oint_{C_1} X(z)z^{k-1}\,dz + \oint_{C_2} X(z)z^{k-1}\,dz + \cdots + \oint_{C_m} X(z)z^{k-1}\,dz$$

$$= 2\pi j(K_1 + K_2 + \cdots + K_m) \tag{B–17}$$

where $K_1, K_2, \ldots, K_m$ denote the residues of $X(z)z^{k-1}$ at poles $z_1, z_2, \ldots, z_m$, respectively, and $C_1, C_2, \ldots, C_m$ are small closed curves around the isolated poles $z_1, z_2, \ldots, z_m$, respectively.

Now we combine Equations (B–16) and (B–17) to obtain a very useful result. Since $X(z)z^{k-1}$ has m poles, that is, $z_1, z_2, \ldots, z_m$,

$$x(k) = x(kT) = K_1 + K_2 + \cdots + K_m$$

$$= \sum_{i=1}^{m} [\text{residue of } X(z)z^{k-1} \text{ at pole } z = z_i \text{ of } X(z)z^{k-1}] \qquad \text{(B–18)}$$

In evaluating residues, note that if the denominator of $X(z)z^{k-1}$ contains a simple pole $z = z_i$ then the corresponding residue K is

$$K = \lim_{z \to z_i} [(z - z_i)X(z)z^{k-1}]$$

If $X(z)z^{k-1}$ contains a multiple pole z_j of order q, then the residue K is given by

$$K = \frac{1}{(q-1)!} \lim_{z \to z_j} \frac{d^{q-1}}{dz^{q-1}}[(z - z_j)^q X(z)z^{k-1}]$$

Note that in this book we treat only one-sided z transforms. This implies that $x(k) = 0$ for $k < 0$. Hence, we restrict the values of k in Equation (B–17) to the nonnegative integer values.

If $X(z)$ has a zero of order r at the origin, then $X(z)z^{k-1}$ in Equations (B–17) will involve a zero of order $r + k - 1$ at the origin. If $r \geq 1$, then $r + k - 1 \geq 0$ for $k \geq 0$, and there is no pole at $z = 0$ in $X(z)z^{k-1}$. However, if $r \leq 0$, then there will be a pole at $z = 0$ for one or more nonnegative values of k. In such a case, separate inversion of Equation (B–17) is necessary for each of such values of k.

It should be noted that the inversion integral method, when evaluated by residues, is a very simple technique for obtaining the inverse z transform, provided that $X(z)z^{k-1}$ has no poles at the origin, $z = 0$. If, however, $X(z)z^{k-1}$ has a simple pole or a multiple pole at $z = 0$, then calculations may become cumbersome and the partial-fraction-expansion method may prove to be simpler to apply.

Comments on Calculating Residues.   In obtaining the residues of a function $X(z)$, note that, regardless of the way we calculate the residues, the final result is the same. Therefore, we may use any method that is convenient for a given situation. As an example, consider the following function $X(z)$:

$$X(z) = \frac{2z^2 + 5z + 6}{(z + 1)^3} + \frac{4z}{(z + 1)^2} + \frac{5}{z + 1}$$

We shall demonstrate three methods for calculating the residue of this function $X(z)$.

Method 1.   The residue of this function may be obtained as the sum of the residues of the respective terms:

[Residue K of $X(z)$ at pole $z = -1$]

$$= \frac{1}{(3-1)!} \lim_{z \to -1} \frac{d^2}{dz^2}\left[(z + 1)^3 \frac{2z^2 + 5z + 6}{(z + 1)^3}\right]$$

$$+ \frac{1}{(2-1)!} \lim_{z \to -1} \frac{d}{dz}\left[(z + 1)^2 \frac{4z}{(z + 1)^2}\right] + \lim_{z \to -1}\left[(z + 1)\frac{5}{z + 1}\right]$$

$$= \tfrac{1}{2} \lim_{z \to -1} (4) + \lim_{z \to -1} (4) + \lim_{z \to -1} (5) = 2 + 4 + 5$$

$$= 11$$

Method 2.   If the three terms of $X(z)$ are combined into one as shown next,

$$X(z) = \frac{2z^2 + 5z + 6}{(z + 1)^3} + \frac{4z}{(z + 1)^2} + \frac{5}{z + 1} = \frac{11z^2 + 19z + 11}{(z + 1)^3}$$

then the residue can be calculated as follows:

[Residue K of $X(z)$ at pole $z = -1$]

$$= \frac{1}{(3 - 1)!} \lim_{z \to -1} \frac{d^2}{dz^2}\left[(z + 1)^3 \frac{11z^2 + 19z + 11}{(z + 1)^3}\right]$$

$$= \tfrac{1}{2} \lim_{z \to -1} (22)$$

$$= 11$$

Method 3.   If $X(z)$ is expanded into usual partial fractions as shown next,

$$X(z) = \frac{11z^2 + 19z + 11}{(z + 1)^3} = \frac{3}{(z + 1)^3} - \frac{3}{(z + 1)^2} + \frac{11}{z + 1}$$

then the residue of $X(z)$ is the coefficient of the term $1/(z + 1)$. Thus,

[Residue K of $X(z)$ at pole $z = -1$] $= 11$

B–4 MODIFIED z TRANSFORM METHOD

The modified z transform method is a modification of the z transform method. It is based on inserting a fictitious delay time at the output of the system, in addition to the insertion of the fictitious output sampler, and varying the amount of the fictitious delay time so that the output at any time between two consecutive sampling instants can be obtained.

The modified z transform method is useful not only in obtaining the response between two consecutive sampling instants, but also in obtaining the z transform of the process with pure delay or transportation lag. In addition, the modified z transform method is applicable to most sampling schemes.

Consider the system shown in Figure B–4(a). In this system a fictitious delay of $(1 - m)T$ seconds, where $0 \le m \le 1$ and T is the sampling period, is inserted at the output of the system. By varying m between 0 and 1, the output $y(t)$ at $t = kT - (1 - m)T$ (where $k = 1, 2, 3, \ldots$) may be obtained. Noting that $G^*(s)$ is given by

$$G^*(s) = \mathcal{L}[g(t)\delta_T(t)]$$

we define the modified pulse transfer function $G(z, m)$ by

$$\mathcal{Z}_m[G(s)] = G(z, m) = G^*(s, m)\big|_{s=(1/T)\ln z}$$

$$= \mathcal{L}[g(t - (1 - m)T)\delta_T(t)]\big|_{s=(1/T)\ln z} \qquad (B-19)$$

where the notation $\mathcal{Z}_m$ signifies the modified z transform.

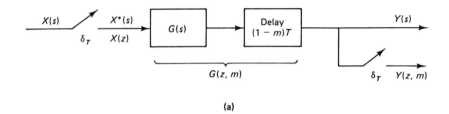

(a)

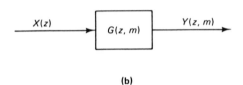

(b)

Figure B–4   (a) System with a fictitious delay time of $(1 - m)T$ sec; (b) modified pulse-transfer-function system with input $X(z)$ and output $Y(z, m)$.

Noting that

$$\mathcal{L}[g(t - (1 - m)T)\delta_T(t)] = \mathcal{L}[g(t - T + mT)\delta_T(t)]$$
$$= e^{-Ts}\mathcal{L}[g(t + mT)\delta_T(t)]$$

we have

$$G^*(s, m) = e^{-Ts}\mathcal{L}[g(t + mT)\delta_T(t)] \tag{B–20}$$

Since $\mathcal{L}[g(t + mT)\delta_T(t)]$ is the Laplace transform of the product of two time functions, by referring to Equation (3–19) it can be obtained as follows:

$$\mathcal{L}[g(t + mT)\delta_T(t)] = \frac{1}{2\pi j}\int_{c-j\infty}^{c+j\infty} G(p)\frac{e^{mTp}}{1 - e^{-T(s-p)}}\,dp \tag{B–21}$$

The integration on the right-hand side of Equation (B–21) can be carried out in a way similar to that discussed in Section 3–3; that is, the convolution integral can be integrated in either the left half-plane or the right half-plane.

Let us consider the contour integration along the infinite semicircle in the left half-plane. Then

$$\mathcal{L}[g(t + mT)\delta_T(t)] = \sum\left[\text{residue of }\frac{G(s)e^{mTs}\,z}{z - e^{Ts}}\text{ at pole of }G(s)\right] \tag{B–22}$$

Hence, from Equations (B–19), (B–20), and (B–22), we obtain the modified z transform of $G(s)$ as follows:

$$G(z, m) = z^{-1}\sum\left[\text{residue of }\frac{G(s)e^{mTs}\,z}{z - e^{Ts}}\text{ at pole of }G(s)\right] \tag{B–23}$$

Note that the modified z transform $G(z, m)$ and the z transform $G(z)$ are related as follows:

$$G(z) = \lim_{m \to 0} zG(z, m) \tag{B–24}$$

Referring to Figure B–4(b), the output $Y(z, m)$ is obtained as follows:

$$Y(z, m) = G(z, m)X(z) \tag{B–25}$$

As in the case of the z transform, the modified z transform $Y(z, m)$ can be expanded into an infinite series in z^{-1}, as follows:

$$Y(z, m) = y_0(m)z^{-1} + y_1(m)z^{-2} + y_2(m)z^{-3} + \cdots \tag{B–26}$$

By multiplying both sides of Equation (B–26) by z, we have

$$zY(z, m) = y_0(m) + y_1(m)z^{-1} + y_2(m)z^{-2} + \cdots \tag{B–27}$$

where $y_k(m)$ represents the value of $y(t)$ between $t = kT$ and $t = (k + 1)T$ $(k = 0, 1, 2, \ldots)$, or

$$y_k(m) = y((k + m)T) \tag{B–28}$$

Note that if $y(k)$ is continuous then

$$\lim_{m \to 1} y_{k-1}(m) = \lim_{m \to 0} y_k(m) \tag{B–29}$$

The left-hand side of Equation (B–29) gives the values $y(0-), y(T-), y(2T-), \ldots,$ and the right-hand side gives the values $y(0+), y(T+), y(2T+), \ldots.$ If the output $y(kT)$ is continuous, then $y(kT-) = y(kT+)$.

Example B–3

Obtain the modified z transform of $G(s)$, where

$$G(s) = \frac{1}{s + a}$$

Referring to Equation (B–23), we obtain the modified z transform of $G(s)$ as follows:

$$G(z, m) = z^{-1}\left[\text{residue of } \frac{1}{s + a} \frac{e^{mTs} z}{z - e^{Ts}} \text{ at pole } s = -a\right]$$

$$= z^{-1}\left\{\lim_{s \to -a}\left[(s + a)\frac{1}{s + a} \frac{e^{mTs} z}{z - e^{Ts}}\right]\right\}$$

$$= z^{-1}\frac{e^{-maT} z}{z - e^{-aT}} = \frac{e^{-maT} z^{-1}}{1 - e^{-aT} z^{-1}}$$

Example B–4

Consider the systems shown in Figures B–5(a) and (b). Obtain the output $Y(z, m)$ of each system.

For the system shown in Figure B–5(a), we have

$$Y(z, m) = \mathscr{Z}_m[Y(s)] = G_2(z, m)G_1(z)X(z)$$

Note that

$$Y(z) = \mathscr{Z}[Y(s)] = G_2(z)G_1(z)X(z)$$

For the system shown in Figure B–5(b), we have

$$Y(z, m) = \mathscr{Z}_m[Y(s)] = G_1 G_2(z, m)X(z)$$

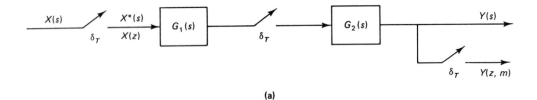

(a)

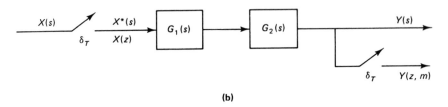

(b)

Figure B–5 (a) System with a sampler between $G_1(s)$ and $G_2(s)$; (b) system with no sampler between $G_1(s)$ and $G_2(s)$.

where

$$G_1 G_2(z, m) = \mathcal{Z}_m[G_1(s)G_2(s)]$$

Note that

$$Y(z) = G_1 G_2(z)X(z)$$

Example B–5

Consider the system shown in Figure B–6. Obtain the modified z transform of $C(s)$. The output $C(z)$ is given by

$$C(z) = \frac{G(z)}{1 + GH(z)} R(z)$$

The modified z transform of $C(z)$ is given by

$$C(z, m) = \frac{G(z, m)}{1 + GH(z)} R(z) \qquad \text{(B–30)}$$

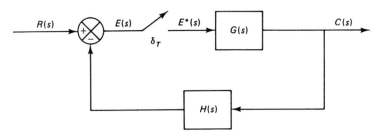

Figure B–6 Closed-loop control system.

Example B–6

Consider the system shown in Figure B–7. The sampling period T is 1 sec. or $T = 1$. Suppose that the system is subjected to a unit-step input. Obtain $c_k(m)$ for $m = 0.5$ and $k = 0, 1, 2, \ldots, 9$. Also, verify that Equation (B–24) holds true. The modified z transform of $G(s)$ is obtained from Equation (B–23) as follows:

$$G(z, m) = z^{-1} \sum \left[\text{residue of } \frac{G(s)e^{ms} z}{z - e^s} \text{ at pole of } G(s) \right]$$

$$= z^{-1}(1 - z^{-1}) \left\{ \left[\text{residue of } \frac{1}{s^2(s + 1)} \frac{e^{ms} z}{z - e^s} \text{ at double pole } s = 0 \right] \right.$$

$$\left. + \left[\text{residue of } \frac{1}{s^2(s + 1)} \frac{e^{ms} z}{z - e^s} \text{ at simple pole } s = -1 \right] \right\}$$

$$= z^{-1}(1 - z^{-1}) \left\{ \frac{1}{(2 - 1)!} \lim_{s \to 0} \frac{d}{ds} \left[s^2 \frac{1}{s^2(s + 1)} \frac{e^{ms} z}{z - e^s} \right] \right.$$

$$\left. + \lim_{s \to -1} \left[(s + 1) \frac{1}{s^2(s + 1)} \frac{e^{ms} z}{z - e^s} \right] \right\}$$

$$= z^{-1}(1 - z^{-1}) \left[\frac{mz^2 - mz - z^2 + 2z}{(z - 1)^2} + \frac{e^{-m} z}{z - e^{-1}} \right]$$

$$= \frac{(m - 1)z^{-1} + (2 - m)z^{-2}}{1 - z^{-1}} + \frac{e^{-m} z^{-1}(1 - z^{-1})}{1 - e^{-1} z^{-1}}$$

$$= \frac{(m - 1 + e^{-m})z^{-1} + (2.3679 - 1.3679m - 2e^{-m})z^{-2}}{+ [-0.3679(2 - m) + e^{-m}]z^{-3}}{(1 - z^{-1})(1 - 0.3679z^{-1})}$$

Referring to Equation (B–30) and noting that $R(z) = 1/(1 - z^{-1})$, we have

$$C(z, m) = \frac{G(z, m)}{1 + G(z)} \frac{1}{1 - z^{-1}}$$

$$= \frac{(m - 1 + e^{-m})z^{-1} + (2.3679 - 1.3679m - 2e^{-m})z^{-2}}{+ (-0.7358 + 0.3679m + e^{-m})z^{-3}}{1 - 2z^{-1} + 1.6321z^{-2} - 0.6321z^{-3}}$$

Hence, for $m = 0.5$ we have

$$C(z, 0.5) = \frac{0.1065z^{-1} + 0.4709z^{-2} + 0.05468z^{-3}}{1 - 2z^{-1} + 1.6321z^{-2} - 0.6321z^{-3}} \tag{B–31}$$

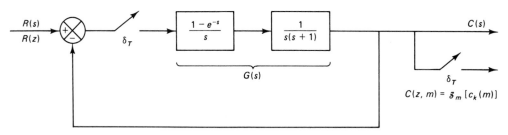

Figure B–7 Closed-loop control system.

By referring to Equation (B–27), Equation (B–31) can be expanded into an infinite series in z^{-1} as follows:

$$C(z, 0.5) = c_0(0.5)z^{-1} + c_1(0.5)z^{-2} + c_2(0.5)z^{-3} + \cdots$$

or

$$zC(z, 0.5) = c_0(0.5) + c_1(0.5)z^{-1} + c_2(0.5)z^{-2} + \cdots$$

where $c_k(0.5) = c((k + 0.5)T) = c(k + 0.5)$ and $k = 0, 1, 2, \ldots$. The values of $c_k(0.5)$ can easily be obtained with a digital computer. The computer solution for $k = 0, 1, 2, \ldots, 9$ is as follows:

$$c_0(0.5) = c(0.5) = 0.1065$$

$$c_1(0.5) = c(1.5) = 0.6839$$

$$c_2(0.5) = c(2.5) = 1.2487$$

$$c_3(0.5) = c(3.5) = 1.4485$$

$$c_4(0.5) = c(4.5) = 1.2913$$

$$c_5(0.5) = c(5.5) = 1.0078$$

$$c_6(0.5) = c(6.5) = 0.8236$$

$$c_7(0.5) = c(7.5) = 0.8187$$

$$c_8(0.5) = c(8.5) = 0.9302$$

$$c_9(0.5) = c(9.5) = 1.0447$$

These values give the response at the midpoints between pairs of consecutive sampling points. Note that by varying the value of m between 0 and 1 it is possible to find the response at any point between two consecutive sampling points, such as $c(1.2)$ and $c(2.8)$.

Finally, note that

$$G(z) = \mathscr{Z}[G(s)] = \mathscr{Z}\left[\frac{1 - e^{-s}}{s} \frac{1}{s(s + 1)}\right]$$

$$= \frac{(T - 1 + e^{-T})z^{-1} + (1 - e^{-T} - Te^{-T})z^{-2}}{(1 - z^{-1})(1 - e^{-T}z^{-1})}$$

$$= \frac{0.3679z^{-1} + 0.2642z^{-2}}{(1 - z^{-1})(1 - 0.3679z^{-1})}$$

and

$$\lim_{m \to 0} zG(z, m) = \frac{0.3679z^{-1} + 0.2642z^{-2}}{(1 - z^{-1})(1 - 0.3679z^{-1})}$$

Hence,

$$G(z) = \lim_{m \to 0} zG(z, m)$$

Clearly, Equation (B–24) holds true.

Summary.   The main purpose of this section has been to present the modified *z* transform method for finding the response for any time between two consecutive sampling instants. It is noted that the modified *z* transform method can be used not only for such a purpose, but also for dealing with multirate sampling schemes.

EXAMPLE PROBLEMS AND SOLUTIONS

Problem B–1

Obtain the z transform of $1/k!$

Solution

$$Z\left[\frac{1}{k!}\right] = \sum_{k=0}^{\infty} \frac{1}{k!} z^{-k}$$

$$= 1 + z^{-1} + \frac{1}{2!} z^{-2} + \frac{1}{3!} z^{-3} + \frac{1}{4!} z^{-4} + \cdots$$

$$= \exp(z^{-1})$$

Problem B–2

Obtain

$$\sum_{k=1}^{\infty} \left(\frac{1}{k}\right) z^{-k}$$

(This series looks like the z transform of $1/k$, but the k sequence begins here with 1 instead of 0.)

Solution Since

$$\sum_{k=0}^{\infty} z^{-k} = 1 + z^{-1} + z^{-2} + \cdots = \frac{1}{1 - z^{-1}}, \qquad |z| > 1$$

by multiplying both sides of this last equation by z^{-2}, we have

$$\sum_{k=0}^{\infty} z^{-k-2} = \frac{z^{-2}}{1 - z^{-1}}$$

Integrating this last equation with respect to z, we have

$$\int \sum_{k=0}^{\infty} z^{-k-2} \, dz = \int \frac{z^{-2}}{1 - z^{-1}} \, dz$$

or

$$\sum_{k=0}^{\infty} \frac{z^{-k-1}}{-k - 1} = \ln(1 - z^{-1}) + \text{constant} \qquad\qquad \text{(B–32)}$$

where the constant in Equation (B–32) is zero. [To verify this, substitute ∞ for z in both sides of Equation (B–32).] Equation (B–32) can thus be rewritten as follows:

$$\sum_{k=1}^{\infty} \frac{z^{-k}}{-k} = \ln(1 - z^{-1}), \qquad |z| > 1$$

or

$$\sum_{k=1}^{\infty} \left(\frac{1}{k}\right) z^{-k} = -\ln(1 - z^{-1}), \qquad |z| > 1$$

Problem B–3

The first backward difference between $x(k)$ and $x(k - 1)$ is defined by

$$\nabla x(k) = x(k) - x(k - 1)$$

The second backward difference is defined by

$$\nabla^2 x(k) = \nabla[\nabla x(k)] = \nabla[x(k) - x(k-1)]$$
$$= \nabla x(k) - \nabla x(k-1)$$

and the third backward difference is defined by

$$\nabla^3 x(k) = \nabla^2 x(k) - \nabla^2 x(k-1)$$

Similarly, the mth backward difference is given by

$$\nabla^m x(k) = \nabla^{m-1} x(k) - \nabla^{m-1} x(k-1)$$

Obtain the z transforms of $\nabla x(k)$, $\nabla^2 x(k)$, $\nabla^3 x(k)$, and $\nabla^m x(k)$.

Solution   The z transform of the first backward difference is obtained as follows:

$$\mathscr{Z}[\nabla x(k)] = \mathscr{Z}[x(k)] - \mathscr{Z}[x(k-1)]$$
$$= X(z) - z^{-1} X(z)$$
$$= (1 - z^{-1})X(z) \tag{B–33}$$

Since

$$\nabla^2 x(k) = [x(k) - x(k-1)] - [x(k-1) - x(k-2)]$$
$$= x(k) - 2x(k-1) + x(k-2)$$

the z transform of $\nabla^2 x(k)$ is

$$\mathscr{Z}[\nabla^2 x(k)] = \mathscr{Z}[x(k)] - 2\,\mathscr{Z}[x(k-1)] + \mathscr{Z}[x(k-2)]$$
$$= X(z) - 2z^{-1} X(z) + z^{-2} X(z)$$
$$= (1 - z^{-1})^2 X(z) \tag{B–34}$$

In this way we obtain

$$\mathscr{Z}[\nabla^3 x(k)] = (1 - z^{-1})^3 X(z)$$

Notice that the operation of taking the backward difference corresponds to multiplying $X(z)$ by $(1 - z^{-1})$. Thus, for the mth backward difference,

$$\nabla^m x(k) = \nabla^{m-1} x(k) - \nabla^{m-1} x(k-1)$$

we have

$$\mathscr{Z}[\nabla^m x(k)] = (1 - z^{-1})^m X(z) \tag{B–35}$$

Problem B–4

The first forward difference between $x(k+1)$ and $x(k)$ is defined by

$$\Delta x(k) = x(k+1) - x(k)$$

The second forward difference is defined by

$$\Delta^2 x(k) = \Delta[\Delta x(k)] = \Delta[x(k+1) - x(k)]$$
$$= \Delta x(k+1) - \Delta x(k)$$

The third forward difference is defined by

$$\Delta^3 x(k) = \Delta^2 x(k+1) - \Delta^2 x(k)$$

and the mth forward difference is given by

$$\Delta^m x(k) = \Delta^{m-1} x(k+1) - \Delta^{m-1} x(k)$$

Obtain the z transforms of $\Delta x(k)$, $\Delta^2 x(k)$, $\Delta^3 x(k)$, and $\Delta^m x(k)$.

Solution   The z transform of the first forward difference is given by

$$\mathcal{Z}[\Delta x(k)] = \mathcal{Z}[x(k+1)] - \mathcal{Z}[x(k)]$$
$$= zX(z) - zx(0) - X(z)$$
$$= (z-1)X(z) - zx(0) \tag{B-36}$$

Since

$$\Delta^2 x(k) = [x(k+2) - x(k+1)] - [x(k+1) - x(k)]$$
$$= x(k+2) - 2x(k+1) + x(k)$$

the z transform of $\Delta^2 x(k)$ is

$$\mathcal{Z}[\Delta^2 x(k)] = \mathcal{Z}[x(k+2) - 2x(k+1) + x(k)]$$
$$= z^2 X(z) - z^2 x(0) - zx(1) - 2[zX(z) - zx(0)] + X(z)$$
$$= (z-1)^2 X(z) - z(z-1)x(0) - z\Delta x(0) \tag{B-37}$$

where $\Delta x(0) = x(1) - x(0)$. The z transform of $\Delta^3 x(k)$ becomes

$$\mathcal{Z}[\Delta^3 x(k)] = \mathcal{Z}[x(k+3) - 3x(k+2) + 3x(k+1) - x(k)]$$
$$= (z-1)^3 X(z) - z(z-1)^2 x(0) - z(z-1)\Delta x(0) - z\Delta^2 x(0)$$

where $\Delta x(0) = x(1) - x(0)$ and $\Delta^2 x(0) = x(2) - 2x(1) + x(0)$. Similarly, for the mth forward difference

$$\Delta^m x(k) = \Delta^{m-1} x(k+1) - \Delta^{m-1} x(k)$$

we have

$$\mathcal{Z}[\Delta^m x(k)] = (z-1)^m X(z) - z \sum_{j=0}^{m-1} (z-1)^{m-j-1} \Delta^j x(0) \tag{B-38}$$

Problem B–5

Solve the following difference equation:

$$(k+1)x(k+1) - x(k) = 0$$

where $x(k) = 0$ for $k < 0$ and $x(0) = 1$. Notice that this difference equation is of the time-varying kind. The solution of this type of difference equation may be obtained by use of the z transform. (It should be cautioned that, in general, the z transform approach to the solution of time-varying difference equations may not be successful.)

Solution   First, note that

$$\mathcal{Z}[kx(k)] = -z \frac{d}{dz} X(z)$$

Since the original difference equation can be written as

$$kx(k) - x(k-1) = 0$$

the z transform of this last equation can be obtained as follows:

$$-z\frac{d}{dz}X(z) - z^{-1}X(z) = 0$$

or

$$z^2\frac{d}{dz}X(z) + X(z) = 0$$

from which we have

$$\frac{dX(z)}{X(z)} = -\frac{dz}{z^2}$$

or

$$\ln X(z) = \frac{1}{z} + \ln K$$

where K is a constant. Then $X(z)$ can be found from

$$X(z) = K\exp z^{-1}$$

Since $\exp z^{-1}$ may be expanded into the series

$$\exp z^{-1} = 1 + z^{-1} + \frac{1}{2!}z^{-2} + \frac{1}{3!}z^{-3} + \cdots, \qquad |z| > 0$$

we have

$$X(z) = K\left(1 + z^{-1} + \frac{1}{2!}z^{-2} + \frac{1}{3!}z^{-3} + \cdots\right)$$

from which we find the inverse z transform of $X(z)$ to be

$$x(k) = K\frac{1}{k!}, \qquad k = 0, 1, 2, \ldots$$

Since $x(0)$ is given as 1, we have

$$x(0) = K = 1$$

Thus, we have determined the unknown constant K. Hence, the solution to the given difference equation is

$$x(k) = \frac{1}{k!}, \qquad k = 0, 1, 2, \ldots$$

Problem B–6

Solve the following difference equation:

$$(k + 1)x(k + 1) - kx(k) = k + 1$$

where $x(k) = 0$ for $k \leq 0$.

Solution   First note that by substituting $k = 0$ into the given difference equation, we have

$$x(1) = 1$$

Now define

$$y(k) = kx(k)$$

Then the given difference equation can be written as

$$y(k + 1) - y(k) = k + 1$$

Taking the z transform of this last equation, we have

$$zY(z) - zy(0) - Y(z) = \frac{z^{-1}}{(1 - z^{-1})^2} + \frac{1}{1 - z^{-1}}$$

Since $y(0) = 0$, we have

$$Y(z) = \frac{z^{-2}}{(1 - z^{-1})^3} + \frac{z^{-1}}{(1 - z^{-1})^2}$$

Referring to Problem A–2–8, we have

$$\mathcal{Z}^{-1}\left[\frac{z^{-2}}{(1 - z^{-1})^3}\right] = \frac{1}{2}(k^2 - k)$$

Hence, the inverse z transform of $Y(z)$ can be given by

$$y(k) = \tfrac{1}{2}(k^2 - k) + k = \tfrac{1}{2}(k^2 + k)$$

Then, $x(k)$ for $k = 1, 2, 3, \ldots$ is determined from

$$kx(k) = y(k) = \tfrac{1}{2}(k^2 + k)$$

as follows:

$$x(k) = \tfrac{1}{2}(k + 1), \qquad k = 1, 2, 3, \ldots$$

Problem B–7

Consider the system shown in Figure B–8. The sampling period is 2 sec, or $T = 2$. The input $x(t)$ is a Kronecker delta function $\delta_0(t)$; that is,

$$\delta_0(k) = \begin{cases} 1, & k = 0 \\ 0, & k \neq 0 \end{cases}$$

Obtain the response every 0.5 sec by using the modified z transform method.

Solution   Since the input $x(t)$ is a Kronecker delta function, we have

$$X(z) = 1$$

The modified pulse transfer function $G(z, m)$ is obtained as follows. Referring to Equation (B–23),

$$G(z, m) = z^{-1}\left[\text{residue of } \frac{1}{s + 0.6931} \frac{e^{mTs} z}{z - e^{Ts}} \text{ at pole } s = -0.6931\right]$$

Noting that $T = 2$, we obtain

$$G(z, m) = z^{-1}\left\{ \lim_{s \to -0.6931} \left[(s + 0.6931)\frac{1}{s + 0.6931} \frac{e^{2ms} z}{z - e^{2s}} \right] \right\}$$

$$= z^{-1}\frac{(e^{-1.3862})^m z}{z - e^{-1.3862}} = \frac{4^{-m}}{z - 0.25}$$

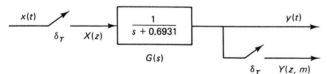

Figure B–8   Impulse-sampled system.

Hence, the output $Y(z, m)$ can be obtained as follows:

$$Y(z, m) = G(z, m)X(z) = \frac{4^{-m}}{z - 0.25}$$

Referring to Equation (B–27), we have

$$zY(z, m) = y_0(m) + y_1(m)z^{-1} + y_2(m)z^{-2} + \cdots$$

where $y_k(m) = y((k + m)T) = y(2k + 2m)$. In this problem $zY(z, m)$ can be expanded into an infinite series in z^{-1} as follows:

$$zY(z, m) = \frac{4^{-m}}{1 - 0.25z^{-1}}$$

$$= 4^{-m} + 4^{-m-1}z^{-1} + 4^{-m-2}z^{-2} + 4^{-m-3}z^{-3} + \cdots$$

Hence,

$$y_0(m) = 4^{-m}$$

$$y_1(m) = 4^{-m-1}$$

$$y_2(m) = 4^{-m-2}$$

$$y_3(m) = 4^{-m-3}$$
$$\vdots$$

To obtain the system output every 0.5 sec, we set $m = 0$, 0.25, 0.50, and 0.75. For $m = 0.25$, we obtain

$$y_0(0.25) = y(0.5) = 4^{-0.25} = 0.7071$$

$$y_1(0.25) = y(2.5) = 4^{-1.25} = 0.1768$$

$$y_2(0.25) = y(4.5) = 4^{-2.25} = 0.04419$$

$$y_3(0.25) = y(6.5) = 4^{-3.25} = 0.01105$$
$$\vdots$$

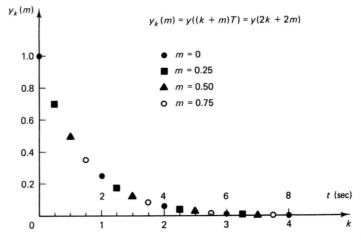

Figure B–9 Plot of $y_k(m)$ versus k for the system considered in Problem B–7.

Similarly, the values of $y_k(m)$ for $m = 0, 0.5$, and 0.75 can be calculated. The result is shown in Figure B–9 as a plot of $y_k(m)$ versus k.

Problem B–8

Obtain $C(z, m)$, the modified z transform of the output, of the system shown in Figure B–10.

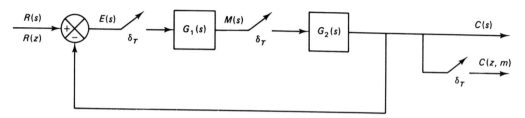

Figure B–10   Closed-loop discrete-time control system.

Solution   From Figure B–10 we have

$$E(s) = R(s) - C(s)$$
$$M(s) = G_1(s)E^*(s)$$
$$C(s) = G_2(s)M^*(s)$$

Hence,

$$M^*(s) = G_1^*(s)E^*(s)$$

or

$$M(z) = G_1(z)E(z)$$

Also,

$$E^*(s) = R^*(s) - C^*(s) = R^*(s) - G_2^*(s)M^*(s)$$

or

$$E(z) = R(z) - G_2(z)M(z)$$

Therefore,

$$M(z) = G_1(z)[R(z) - G_2(z)M(z)]$$

from which we obtain

$$M(z) = \frac{G_1(z)R(z)}{1 + G_1(z)G_2(z)}$$

Since $C(z, m)$ can be given by $G_2(z, m)M(z)$, we have

$$C(z, m) = G_2(z, m)M(z) = \frac{G_1(z)G_2(z, m)}{1 + G_1(z)G_2(z)}R(z)$$

Appendix C

Pole Placement Design with Vector Control

C-1 INTRODUCTION

In Chapter 6 we presented the pole placement technique and state observer design when the control signal $u(k)$ was a scalar. If the control signal is a vector quantity (r-vector), however, we can expect to improve the system's response characteristics, because we have more freedom to choose control signals $u_1(k), u_2(k), \ldots, u_r(k)$. For example, in the case of the nth-order system with a scalar control, the deadbeat response can be achieved in at most n sampling periods. In the case of the vector control $\mathbf{u}(k)$, the deadbeat response can be achieved in less than n sampling periods.

It is noted that with the vector control it is possible to choose freely more than n parameters; that is, in addition to being able to place n closed-loop poles properly, we have the freedom to satisfy other requirements, if any, of the closed-loop system.

In the case of the vector control, however, the determination of the state feedback gain matrix $\mathbf{K}$ becomes more complex, as we shall see in this appendix.

C-2 PRELIMINARY DISCUSSIONS

Consider the system

$$\mathbf{x}(k + 1) = \mathbf{G}\mathbf{x}(k) + \mathbf{H}\mathbf{u}(k) \tag{C-1}$$

where

$\mathbf{x}(k)$ = state vector (n-vector) at kth sampling instant

$\mathbf{u}(k)$ = control vector (r-vector) at kth sampling instant

$\mathbf{G}$ = $n \times n$ matrix

$\mathbf{H}$ = $n \times r$ matrix

We assume that the magnitudes of the r components of $\mathbf{u}(k)$ are unconstrained. As in the case of the system with a scalar control signal, it can be proved that a necessary and sufficient condition for arbitrary pole placement for the system defined by Equation (C–1) is that the system be completely state controllable.

Let us assume that the system defined by Equation (C–1) is completely state controllable. In the state feedback control scheme, the control vector $\mathbf{u}(k)$ is chosen as

$$\mathbf{u}(k) = -\mathbf{K}\mathbf{x}(k) \tag{C–2}$$

where $\mathbf{K}$ is the state feedback gain matrix. It is an $r \times n$ matrix. With state feedback the system becomes a closed-loop system and its state equation becomes

$$\mathbf{x}(k + 1) = (\mathbf{G} - \mathbf{HK})\mathbf{x}(k)$$

where we choose matrix $\mathbf{K}$ so that the eigenvalues of $\mathbf{G} - \mathbf{HK}$ are the desired closed-loop poles $\mu_1, \mu_2, \ldots, \mu_n$.

Transforming State Equation Into Controllable Canonical Form.   Consider the system defined by

$$\mathbf{x}(k + 1) = \mathbf{G}\mathbf{x}(k) + \mathbf{H}_1 u(k) \tag{C–3}$$

where

$$\mathbf{x}(k) = \text{state vector } (n\text{-vector})$$

$$u(k) = \text{control signal (scalar)}$$

$$\mathbf{G} = n \times n \text{ matrix}$$

$$\mathbf{H}_1 = n \times 1 \text{ matrix}$$

Assume that the system is completely state controllable. Then the controllability matrix has its inverse. Define

$$[\mathbf{H}_1 \vdots \mathbf{G}\mathbf{H}_1 \vdots \cdots \vdots \mathbf{G}^{n-1}\mathbf{H}_1]^{-1} = \begin{bmatrix} \mathbf{f}_1 \\ \mathbf{f}_2 \\ \vdots \\ \mathbf{f}_n \end{bmatrix}$$

where the $\mathbf{f}_i$'s are the row vectors. Then construct a transformation matrix $\mathbf{T}_1$ as follows:

$$\mathbf{T}_1 = \begin{bmatrix} \mathbf{f}_n \\ \mathbf{f}_n\mathbf{G} \\ \vdots \\ \mathbf{f}_n\mathbf{G}^{n-1} \end{bmatrix}^{-1} \tag{C–4}$$

where the $\mathbf{f}_n\mathbf{G}^k$ are row vectors ($k = 0, 1, 2, \ldots, n - 1$). Then it can be shown that

$$\mathbf{T}_1^{-1}\mathbf{G}\mathbf{T}_1 = \begin{bmatrix} \mathbf{f}_n \\ \mathbf{f}_n\mathbf{G} \\ \vdots \\ \mathbf{f}_n\mathbf{G}^{n-1} \end{bmatrix} \mathbf{G} \begin{bmatrix} \mathbf{f}_n \\ \mathbf{f}_n\mathbf{G} \\ \vdots \\ \mathbf{f}_n\mathbf{G}^{n-1} \end{bmatrix}^{-1}$$

$$= \begin{bmatrix} 0 & 1 & 0 & \cdots & 0 \\ 0 & 0 & 1 & \cdots & 0 \\ \vdots & \vdots & \vdots & & \vdots \\ 0 & 0 & 0 & \cdots & 1 \\ -a_n & -a_{n-1} & -a_{n-2} & \cdots & -a_1 \end{bmatrix} \qquad (\text{C--5})$$

and

$$\mathbf{T}_1^{-1}\mathbf{H}_1 = \begin{bmatrix} 0 \\ 0 \\ \vdots \\ 0 \\ 1 \end{bmatrix} \qquad (\text{C--6})$$

[See Problem C–1 for the derivation of Equations (C–5) and (C–6).]
Now if we define

$$\mathbf{x}(k) = \mathbf{T}_1 \hat{\mathbf{x}}(k)$$

then Equation (C–3) becomes

$$\hat{\mathbf{x}}(k+1) = \mathbf{T}_1^{-1}\mathbf{G}\mathbf{T}_1\hat{\mathbf{x}}(k) + \mathbf{T}_1^{-1}\mathbf{H}_1 u(k)$$

or

$$\begin{bmatrix} \hat{x}_1(k+1) \\ \hat{x}_2(k+1) \\ \vdots \\ \hat{x}_{n-1}(k+1) \\ \hat{x}_n(k+1) \end{bmatrix} = \begin{bmatrix} 0 & 1 & 0 & \cdots & 0 \\ 0 & 0 & 1 & \cdots & 0 \\ \vdots & \vdots & \vdots & & \vdots \\ 0 & 0 & 0 & \cdots & 1 \\ -a_n & -a_{n-1} & -a_{n-2} & \cdots & -a_1 \end{bmatrix} \begin{bmatrix} \hat{x}_1(k) \\ \hat{x}_2(k) \\ \vdots \\ \hat{x}_{n-1}(k) \\ \hat{x}_n(k) \end{bmatrix} + \begin{bmatrix} 0 \\ 0 \\ \vdots \\ 0 \\ 1 \end{bmatrix} u(k) \qquad (\text{C--7})$$

We have thus shown that the state equation, Equation (C–3), can be transformed into the controllable canonical form by use of the transformation matrix $\mathbf{T}_1$ defined by Equation (C–4).

Design Steps.   In what follows we shall discuss the procedure for determining a state feedback gain matrix $\mathbf{K}$ such that the eigenvalues of $\mathbf{G} - \mathbf{H}\mathbf{K}$ are the desired values $\mu_1, \mu_2, \ldots, \mu_n$.

The state equation to be considered in the following was given by Equation (C–1):

$$\mathbf{x}(k+1) = \mathbf{G}\mathbf{x}(k) + \mathbf{H}u(k)$$

We assume that the rank of the $n \times r$ matrix $\mathbf{H}$ is r. This last equation is equivalent to

$$\mathbf{x}(k+1) = \mathbf{G}\mathbf{x}(k) + [\mathbf{H}_1 \vdots \mathbf{H}_2 \vdots \cdots \vdots \mathbf{H}_r]\mathbf{u}(k)$$

where

$$[\mathbf{H}_1 \vdots \mathbf{H}_2 \vdots \cdots \vdots \mathbf{H}_r] = \mathbf{H}, \qquad \mathbf{H}_i = \begin{bmatrix} h_{1i} \\ h_{2i} \\ \vdots \\ h_{ni} \end{bmatrix}, \qquad i = 1, 2, \ldots, r$$

The procedure for designing the state feedback gain matrix **K** involves the following two steps:

Step 1.   Extend the transformation process [the process that transforms the state equation given by Equation (C–3) into the state equation in the controllable canonical form given by Equation (C–7)] to the case where matrix **H** is an $n \times r$ matrix. That is, we transform the given state equation into a controllable canonical form by use of a transformation matrix **T**, the exact form of which will be given later. By defining

$$\mathbf{x}(k) = \mathbf{T}\hat{\mathbf{x}}(k)$$

the original state equation, Equation (C–1), can be transformed into

$$\hat{\mathbf{x}}(k + 1) = \mathbf{T}^{-1}\mathbf{GT}\hat{\mathbf{x}}(k) + \mathbf{T}^{-1}\mathbf{Hu}(k) = \hat{\mathbf{G}}\hat{\mathbf{x}}(k) + \hat{\mathbf{H}}\mathbf{u}(k) \qquad (C-8)$$

where $\hat{\mathbf{G}} = \mathbf{T}^{-1}\mathbf{GT}$ is in a controllable canonical form and $\hat{\mathbf{H}} = \mathbf{T}^{-1}\mathbf{H}$. (This controllable canonical form is slightly different from the usual form, as we shall see later.)

Step 2.   By use of a state feedback gain matrix **K**, the control vector can be given by

$$\mathbf{u}(k) = -\mathbf{Kx}(k) = -\mathbf{KT}\hat{\mathbf{x}}(k)$$

and the system state equation becomes

$$\hat{\mathbf{x}}(k + 1) = (\hat{\mathbf{G}} - \hat{\mathbf{H}}\mathbf{KT})\hat{\mathbf{x}}(k)$$

We choose matrix **K** so that matrix $\hat{\mathbf{G}} - \hat{\mathbf{H}}\mathbf{KT}$ will have the desired eigenvalues $\mu_1, \mu_2, \ldots, \mu_n$.

C–3 POLE PLACEMENT DESIGN

We shall first discuss the determination of a necessary transformation matrix **T** and then determine the state feedback gain matrix **K**.

Consider the completely state controllable system defined by

$$\mathbf{x}(k + 1) = \mathbf{Gx}(k) + \mathbf{Hu}(k) \qquad (C-9)$$

where

$$\mathbf{x}(k) = \text{state vector } (n\text{-vector})$$

$$\mathbf{u}(k) = \text{control vector } (r\text{-vector})$$

$$\mathbf{G} = n \times n \text{ matrix}$$

$$\mathbf{H} = [\mathbf{H}_1 \vdots \mathbf{H}_2 \vdots \cdots \vdots \mathbf{H}_r] = n \times r \text{ matrix}$$

We assume that the rank of matrix **H** is r. Thus, the component vectors $\mathbf{H}_1$, $\mathbf{H}_2, \ldots, \mathbf{H}_r$ of matrix **H** are linearly independent of each other. Since the system is assumed to be completely state controllable, the rank of the $n \times nr$ controllability matrix

$$[\mathbf{H} \vdots \mathbf{GH} \vdots \cdots \vdots \mathbf{G}^{n-1}\mathbf{H}]$$

is n. The controllability matrix can be written in an expanded form as follows:

$$[\mathbf{H}_1 \vdots \mathbf{H}_2 \vdots \cdots \vdots \mathbf{H}_r \vdots \mathbf{GH}_1 \vdots \mathbf{GH}_2 \vdots \cdots \vdots \mathbf{GH}_r \vdots \cdots \vdots \mathbf{G}^{n-1}\mathbf{H}_1 \vdots \mathbf{G}^{n-1}\mathbf{H}_2 \vdots \cdots \vdots \mathbf{G}^{n-1}\mathbf{H}_r]$$

Let us choose n linearly independent vectors from this $n \times nr$ matrix. Let us begin from the left-hand side of this matrix. Since the first r vectors $\mathbf{H}_1, \mathbf{H}_2, \ldots, \mathbf{H}_r$ are linearly independent of each other, we choose these r vectors first. Then we examine $\mathbf{GH}_1$ if it is linearly independent of the r vectors already chosen. If it is, we have chosen $r + 1$ linearly independent vectors. Next, we examine $\mathbf{GH}_2, \mathbf{GH}_3, \ldots,$ $\mathbf{GH}_r, \ldots$ in the order shown in the expanded controllability matrix until we find altogether n linearly independent vectors. (Since the rank of the controllability matrix is n, there always exist n linearly independent vectors.)

Once we have chosen n linearly independent vectors, we rearrange these vectors in the following way:

$$\mathbf{F} = [\mathbf{H}_1 \vdots \mathbf{GH}_1 \vdots \cdots \vdots \mathbf{G}^{n_1-1}\mathbf{H}_1 \vdots \mathbf{H}_2 \vdots \mathbf{GH}_2 \vdots \cdots \vdots$$

$$\mathbf{G}^{n_2-1}\mathbf{H}_2 \vdots \cdots \vdots \mathbf{H}_r \vdots \mathbf{GH}_r \vdots \cdots \vdots \mathbf{G}^{n_r-1}\mathbf{H}_r] \qquad \text{(C–10)}$$

The numbers n_i are said to be *Kronecker invariant* and satisfy the equation

$$n_1 + n_2 + \cdots + n_r = n$$

We shall define the maximum of $n_1, n_2, \ldots, n_r$ as $n_{\min}$:

$$n_{\min} = \max(n_1, n_2, \ldots, n_r) \qquad \text{(C–11)}$$

We shall refer to this equation later in the discussion of deadbeat response. Next, we compute $\mathbf{F}^{-1}$ and define the η_i th row vector as $\mathbf{f}_i$, where

$$\eta_i = n_1 + n_2 + \cdots + n_i, \qquad i = 1, 2, \ldots, r$$

Then the required transformation matrix $\mathbf{T}$ can be given by

$$\mathbf{T} = \begin{bmatrix} \mathbf{S}_1 \\ \mathbf{S}_2 \\ \vdots \\ \mathbf{S}_r \end{bmatrix}^{-1} \qquad \text{(C–12)}$$

where

$$\mathbf{S}_i = \begin{bmatrix} \mathbf{f}_i \\ \mathbf{f}_i\mathbf{G} \\ \vdots \\ \mathbf{f}_i\mathbf{G}^{n_i-1} \end{bmatrix}$$

Notice that the transformation matrix $\mathbf{T}$ given by Equation (C–12) is an extension of the transformation matrix given by Equation (C–4).

To simplify the presentation, in what follows we shall consider a simple case where $n = 4$ and $r = 2$. (In this case, only n_1 and n_2 are involved.) (Extension to more general cases is straightforward.) Then the transformation matrix $\mathbf{T}$ becomes a 4×4 matrix. The transformation matrix $\mathbf{T}$ given by Equation (C–12) becomes

$$\mathbf{T} = \begin{bmatrix} \mathbf{S}_1 \\ \hline \mathbf{S}_2 \end{bmatrix}^{-1}$$

where

$$
S_1 = \begin{bmatrix} f_1 \\ \vdots \\ f_1 G^{n_1-1} \end{bmatrix}, \qquad S_2 = \begin{bmatrix} f_2 \\ \vdots \\ f_2 G^{n_2-1} \end{bmatrix}
$$

(Note that in the case of $n = 4$ there are three possibilities for the combinations of n_1 and n_2: $n_1 = 1, n_2 = 3$; $n_1 = 2, n_2 = 2$; and $n_1 = 3, n_2 = 1$.) For example, if $n_1 = 2$ and $n_2 = 2$, then matrices $\hat{G}$ and $\hat{H}$ become, respectively, as

$$
\hat{G} = T^{-1} GT = \left[\begin{array}{cc|cc} 0 & 1 & 0 & 0 \\ -a_{11} & -a_{12} & -a_{13} & -a_{14} \\ \hline 0 & 0 & 0 & 1 \\ -a_{21} & -a_{22} & -a_{23} & -a_{24} \end{array} \right], \qquad \text{if } n_1 = 2, n_2 = 2 \qquad \text{(C–13)}
$$

and

$$
\hat{H} = T^{-1} H = \left[\begin{array}{cc} 0 & 0 \\ 1 & b_{12} \\ \hline 0 & 0 \\ 0 & 1 \end{array} \right], \qquad \begin{array}{l} \text{if } n_1 = 2, n_2 = 2 \\ \text{(Note: } b_{12} = f_1 GH_2 = 0 \text{ in this case)} \end{array} \qquad \text{(C–14)}
$$

(see Problem C–2). As another example, if $n_1 = 3$ and $n_2 = 1$, then

$$
\hat{G} = T^{-1} GT = \left[\begin{array}{ccc|c} 0 & 1 & 0 & 0 \\ 0 & 0 & 1 & 0 \\ -a_{11} & -a_{12} & -a_{13} & -a_{14} \\ \hline -a_{21} & -a_{22} & -a_{23} & -a_{24} \end{array} \right], \qquad \text{if } n_1 = 3, n_2 = 1 \qquad \text{(C–15)}
$$

and

$$
\hat{H} = T^{-1} H = \left[\begin{array}{cc} 0 & 0 \\ 0 & 0 \\ 1 & b_{12} \\ \hline 0 & 1 \end{array} \right], \qquad \begin{array}{l} \text{if } n_1 = 3, n_2 = 1 \\ \text{(Note: } b_{12} = f_1 G^2 H_2 \text{ may or} \\ \text{may not be zero)} \end{array} \qquad \text{(C–16)}
$$

(see Problem C–4). In what follows, we shall focus on the case where $n_1 = 2$ and $n_2 = 2$. (Other cases can be handled similarly. For example, for the case where $n_1 = 3$ and $n_2 = 1$, see Problems C–3, C–4, and C–5.) For the case where $n_1 = 2$ and $n_2 = 2$, matrix $\hat{G} = T^{-1} GT$ can be given by Equation (C–13), and the characteristic equation is

$$
|zI - \hat{G}| = \begin{vmatrix} z & -1 & 0 & 0 \\ a_{11} & z + a_{12} & a_{13} & a_{14} \\ 0 & 0 & z & -1 \\ a_{21} & a_{22} & a_{23} & z + a_{24} \end{vmatrix}
$$

$$
= \begin{vmatrix} z & -1 \\ a_{11} & z + a_{12} \end{vmatrix} \begin{vmatrix} z & -1 \\ a_{23} & z + a_{24} \end{vmatrix} + \begin{vmatrix} z & -1 \\ a_{21} & a_{22} \end{vmatrix} \begin{vmatrix} a_{13} & a_{14} \\ z & -1 \end{vmatrix}
$$

$$
= (z^2 + a_{12} z + a_{11})(z^2 + a_{24} z + a_{23}) - (a_{22} z + a_{21})(a_{14} z + a_{13})
$$

$$
= 0 \qquad \text{(C–17)}
$$

where we have used Laplace's expansion by the minors. (See Appendix A for the details.) From Equation (C–17) the characteristic equation $|z\mathbf{I} - \hat{\mathbf{G}}| = 0$ becomes

$$|z\mathbf{I} - \hat{\mathbf{G}}| = \begin{vmatrix} z^2 + a_{12}z + a_{11} & a_{14}z + a_{13} \\ \hline a_{22}z + a_{21} & z^2 + a_{24}z + a_{23} \end{vmatrix} = 0 \qquad \text{(C–18)}$$

The eigenvalues of $\hat{\mathbf{G}}$ can be determined by solving this characteristic equation.

Next, we shall determine the state feedback gain matrix $\mathbf{K}$ so that the eigenvalues of $\hat{\mathbf{G}} - \hat{\mathbf{H}}\mathbf{K}$ are $\mu_1, \mu_2, \ldots, \mu_n$, the desired values. Let us define a 2×2 matrix $\mathbf{B}$ such that

$$\mathbf{B} = \begin{bmatrix} 1 & b_{12} \\ 0 & 1 \end{bmatrix}^{-1}$$

(Note that b_{12} is a constant appearing in $\hat{\mathbf{H}}$ matrix.) In the particular case where $n_1 = 2$ and $n_2 = 2$, the value of b_{12} is equal to 0. Thus, $\mathbf{B} = \mathbf{I}$. For more general cases, matrix $\mathbf{B}$ may not be the identity matrix.

Also, define a 2×4 matrix $\boldsymbol{\Delta}$ such that

$$\boldsymbol{\Delta} = \begin{bmatrix} \delta_{11} & \delta_{12} & \delta_{13} & \delta_{14} \\ \delta_{21} & \delta_{22} & \delta_{23} & \delta_{24} \end{bmatrix} \qquad \text{(C–19)}$$

Then it will be seen that matrix $\mathbf{K}$ can be given by

$$\mathbf{K} = \mathbf{B}\boldsymbol{\Delta}\mathbf{T}^{-1}$$

and the control vector $\mathbf{u}(k)$ can be given by

$$\mathbf{u}(k) = -\mathbf{B}\boldsymbol{\Delta}\mathbf{T}^{-1}\mathbf{x}(k) = -\mathbf{B}\boldsymbol{\Delta}\hat{\mathbf{x}}(k)$$

Thus, the system state equation given by Equation (C–8) becomes

$$\hat{\mathbf{x}}(k + 1) = \hat{\mathbf{G}}\hat{\mathbf{x}}(k) - \hat{\mathbf{H}}\mathbf{B}\boldsymbol{\Delta}\hat{\mathbf{x}}(k) = (\hat{\mathbf{G}} - \hat{\mathbf{H}}\mathbf{B}\boldsymbol{\Delta})\hat{\mathbf{x}}(k)$$

For the present case, matrix $\hat{\mathbf{H}}\mathbf{B}\boldsymbol{\Delta}$ becomes as follows:

$$\hat{\mathbf{H}}\mathbf{B}\boldsymbol{\Delta} = \begin{bmatrix} 0 & 0 \\ 1 & 0 \\ 0 & 0 \\ 0 & 1 \end{bmatrix} \begin{bmatrix} 1 & 0 \\ 0 & 1 \end{bmatrix}^{-1} \begin{bmatrix} \delta_{11} & \delta_{12} & \delta_{13} & \delta_{14} \\ \delta_{21} & \delta_{22} & \delta_{23} & \delta_{24} \end{bmatrix}$$

$$= \begin{bmatrix} 0 & 0 & 0 & 0 \\ \delta_{11} & \delta_{12} & \delta_{13} & \delta_{14} \\ 0 & 0 & 0 & 0 \\ \delta_{21} & \delta_{22} & \delta_{23} & \delta_{24} \end{bmatrix}$$

Hence,

$$\hat{\mathbf{G}} - \hat{\mathbf{H}}\mathbf{B}\boldsymbol{\Delta} = \begin{bmatrix} 0 & 1 & 0 & 0 \\ -a_{11} - \delta_{11} & -a_{12} - \delta_{12} & -a_{13} - \delta_{13} & -a_{14} - \delta_{14} \\ 0 & 0 & 0 & 1 \\ -a_{21} - \delta_{21} & -a_{22} - \delta_{22} & -a_{23} - \delta_{23} & -a_{24} - \delta_{24} \end{bmatrix}$$

Then, referring to Equation (C–18), the characteristic equation $|z\mathbf{I} - \hat{\mathbf{G}} + \hat{\mathbf{H}}\mathbf{B}\boldsymbol{\Delta}|$ becomes

$$|z\mathbf{I} - \hat{\mathbf{G}} + \hat{\mathbf{H}}\mathbf{B}\boldsymbol{\Delta}| = \begin{vmatrix} z^2 + (a_{12} + \delta_{12})z + a_{11} + \delta_{11} & (a_{14} + \delta_{14})z + a_{13} + \delta_{13} \\ (a_{22} + \delta_{22})z + a_{21} + \delta_{21} & z^2 + (a_{24} + \delta_{24})z + a_{23} + \delta_{23} \end{vmatrix}$$

$$= [z^2 + (a_{12} + \delta_{12})z + a_{11} + \delta_{11}][z^2 + (a_{24} + \delta_{24})z + a_{23} + \delta_{23}]$$

$$- [(a_{14} + \delta_{14})z + a_{13} + \delta_{13}][(a_{22} + \delta_{22})z + a_{21} + \delta_{21}]$$

$$= 0 \tag{C–20}$$

We desire the eigenvalues of $\hat{\mathbf{G}} - \hat{\mathbf{H}}\mathbf{B}\boldsymbol{\Delta}$ to be μ_1, μ_2, μ_3, and μ_4, or the desired characteristic equation to be

$$(z - \mu_1)(z - \mu_2)(z - \mu_3)(z - \mu_4) = z^4 + \alpha_1 z^3 + \alpha_2 z^2 + \alpha_3 z + \alpha_4 = 0 \tag{C–21}$$

If we equate the coefficients of equal powers of z of the two characteristic equations, Equations (C–20) and (C–21), we obtain the following equations:

$$a_{12} + \delta_{12} + a_{24} + \delta_{24} = \alpha_1$$

$$a_{11} + \delta_{11} + (a_{12} + \delta_{12})(a_{24} + \delta_{24}) + a_{23} + \delta_{23} - (a_{14} + \delta_{14})(a_{22} + \delta_{22}) = \alpha_2$$

$$(a_{11} + \delta_{11})(a_{24} + \delta_{24}) + (a_{12} + \delta_{12})(a_{23} + \delta_{23})$$

$$- (a_{13} + \delta_{13})(a_{22} + \delta_{22}) - (a_{21} + \delta_{21})(a_{14} + \delta_{14}) = \alpha_3$$

$$(a_{11} + \delta_{11})(a_{23} + \delta_{23}) - (a_{13} + \delta_{13})(a_{21} + \delta_{21}) = \alpha_4$$

Notice that we have eight δ variables and four equations. Hence, the values of δ_{11}, δ_{12}, δ_{13}, δ_{14}, δ_{21}, δ_{22}, δ_{23}, and δ_{24} cannot be determined uniquely. There are many possible sets of values $\delta_{11}, \delta_{12}, \ldots, \delta_{24}$ and thus matrix $\boldsymbol{\Delta}$ is not unique. Any matrix $\boldsymbol{\Delta}$ whose elements satisfy the foregoing four equations is acceptable.

Once matrix $\boldsymbol{\Delta}$ is chosen, the required state feedback gain matrix $\mathbf{K}$ is given by

$$\mathbf{K} = \mathbf{B}\boldsymbol{\Delta}\mathbf{T}^{-1}$$

and the state feedback control vector is

$$\mathbf{u}(k) = -\mathbf{B}\boldsymbol{\Delta}\mathbf{T}^{-1}\mathbf{x}(k)$$

and the state equation given by Equation (C–9) becomes

$$\mathbf{x}(k + 1) = \mathbf{G}\mathbf{x}(k) - \mathbf{H}\mathbf{B}\boldsymbol{\Delta}\mathbf{T}^{-1}\mathbf{x}(k) = (\mathbf{G} - \mathbf{H}\mathbf{B}\boldsymbol{\Delta}\mathbf{T}^{-1})\mathbf{x}(k)$$

As a matter of course, note that

$$|\mathbf{G} - \mathbf{H}\mathbf{B}\boldsymbol{\Delta}\mathbf{T}^{-1}| = |\mathbf{T}^{-1}||\mathbf{G} - \mathbf{H}\mathbf{B}\boldsymbol{\Delta}\mathbf{T}^{-1}||\mathbf{T}| = |\mathbf{T}^{-1}\mathbf{G}\mathbf{T} - \mathbf{T}^{-1}\mathbf{H}\mathbf{B}\boldsymbol{\Delta}| = |\hat{\mathbf{G}} - \hat{\mathbf{H}}\mathbf{B}\boldsymbol{\Delta}|$$

For a given set of desired eigenvalues $\mu_1, \mu_2, \ldots, \mu_n$, we have the corresponding coefficients $\alpha_1, \alpha_2, \ldots, \alpha_n$ in the characteristic equation $|z\mathbf{I} - \hat{\mathbf{G}} + \hat{\mathbf{H}}\mathbf{B}\boldsymbol{\Delta}| = 0$. For the given $\alpha_1, \alpha_2, \ldots, \alpha_n$, it is possible to choose a matrix $\boldsymbol{\Delta}$ that is not unique. (This means that we have some freedom to satisfy other requirements, if any.)

If the deadbeat response is desired, we require $\mu_1 = \mu_2 = \mu_3 = \mu_4 = 0$. The desired characteristic equation given by Equation (C–21) becomes

$$z^4 = 0$$

Notice that if we choose, for example,

$$\mathbf{\Delta} = \begin{bmatrix} -a_{11} & -a_{12} & -a_{13} & -a_{14} \\ * & * & -a_{23} & -a_{24} \end{bmatrix} \tag{C–22}$$

where the elements indicated by asterisks are arbitrary constants, then $\mathbf{\hat{G}} - \mathbf{\hat{H}B\Delta}$ becomes

$$\mathbf{\hat{G}} - \mathbf{\hat{H}B\Delta} = \begin{bmatrix} 0 & 1 & 0 & 0 \\ 0 & 0 & 0 & 0 \\ 0 & 0 & 0 & 1 \\ ** & ** & 0 & 0 \end{bmatrix}$$

where the elements indicated by the double asterisks are arbitrary constants.

$$(\mathbf{\hat{G}} - \mathbf{\hat{H}B\Delta})^2 = \begin{bmatrix} 0 & 0 & 0 & 0 \\ 0 & 0 & 0 & 0 \\ ** & 0 & 0 & 0 \\ 0 & ** & 0 & 0 \end{bmatrix}$$

$$(\mathbf{\hat{G}} - \mathbf{\hat{H}B\Delta})^3 = \begin{bmatrix} 0 & 0 & 0 & 0 \\ 0 & 0 & 0 & 0 \\ 0 & ** & 0 & 0 \\ 0 & 0 & 0 & 0 \end{bmatrix}$$

and

$$(\mathbf{\hat{G}} - \mathbf{\hat{H}B\Delta})^4 = \begin{bmatrix} 0 & 0 & 0 & 0 \\ 0 & 0 & 0 & 0 \\ 0 & 0 & 0 & 0 \\ 0 & 0 & 0 & 0 \end{bmatrix}$$

Thus, the deadbeat response is obtained. Matrix $\mathbf{\Delta}$ given by Equation (C–22) is not unique because different choices of elements can yield the deadbeat response. Hence, more than one state feedback gain matrix $\mathbf{K}$ exists that will yield the deadbeat response. This is expected, since we have two control signals $u_1(k)$ and $u_2(k)$ available, instead of just one control signal.

It is important to note that if we choose

$$\mathbf{\Delta} = \begin{bmatrix} -a_{11} & -a_{12} & -a_{13} & -a_{14} \\ -a_{21} & -a_{22} & -a_{23} & -a_{24} \end{bmatrix} \tag{C–23}$$

then

$$\mathbf{\hat{G}} - \mathbf{\hat{H}B\Delta} = \begin{bmatrix} 0 & 1 & 0 & 0 \\ 0 & 0 & 0 & 0 \\ 0 & 0 & 0 & 1 \\ 0 & 0 & 0 & 0 \end{bmatrix}$$

and

$$(\hat{\mathbf{G}} - \hat{\mathbf{H}}\mathbf{B}\mathbf{\Delta})^2 = \mathbf{0}$$

Thus, $(\hat{\mathbf{G}} - \hat{\mathbf{H}}\mathbf{B}\mathbf{\Delta})^k$ becomes zero for $k = 2, 3, 4, \ldots$. The deadbeat response is achieved in two sampling periods. In fact, in general, by choosing the elements of $\mathbf{\Delta}$ in the manner given by Equation (C–23), the deadbeat response can be achieved in n_{min} steps rather than n steps, where

$$n_{min} = \max(n_1, n_2, \ldots, n_r)$$

Since $n_1 + n_2 + \cdots + n_r = n$, we note that n_{min} is always less than n.

Extension to the More General Case.   Thus far, we have given detailed discussions for the case where $n = 4$ ($n_1 = n_2 = 2$) and $r = 2$. Extension of the preceding discussions to the more general case is straightforward. For example, consider the case where $n = 6$ and $r = 3$. For this case,

$$n_1 + n_2 + n_3 = 6$$

and we have several possible combinations of n_1, n_2, and n_3.

Now consider the case where $n_1 = 3$, $n_2 = 2$, and $n_3 = 1$. The modified 6×6 controllability matrix $\mathbf{F}$ for this case is

$$\mathbf{F} = [\mathbf{H}_1 \,\vdots\, \mathbf{GH}_1 \,\vdots\, \mathbf{G}^2\mathbf{H}_1 \,\vdots\, \mathbf{H}_2 \,\vdots\, \mathbf{GH}_2 \,\vdots\, \mathbf{H}_3]$$

Define

$$\mathbf{F}^{-1} = \begin{bmatrix} *** \\ \hline *** \\ \hline \mathbf{f}_1 \\ \hline *** \\ \hline \mathbf{f}_2 \\ \hline \mathbf{f}_3 \end{bmatrix} \begin{matrix} \\ \left.\vphantom{\begin{matrix}*\\ *\\ *\end{matrix}}\right\} n_1 = 3 \\ \\ \left.\vphantom{\begin{matrix}*\\ *\end{matrix}}\right\} n_2 = 2 \\ \left.\vphantom{*}\right\} n_3 = 1 \end{matrix}$$

where a row of asterisks denotes a row vector. Then the transformation matrix $\mathbf{T}$ can be formed as follows:

$$\mathbf{T} = \begin{bmatrix} \mathbf{S}_1 \\ \hline \mathbf{S}_2 \\ \hline \mathbf{S}_3 \end{bmatrix}^{-1}$$

where

$$\mathbf{S}_1 = \begin{bmatrix} \mathbf{f}_1 \\ \hline \mathbf{f}_1\mathbf{G} \\ \hline \mathbf{f}_1\mathbf{G}^2 \end{bmatrix}, \qquad \mathbf{S}_2 = \begin{bmatrix} \mathbf{f}_2 \\ \hline \mathbf{f}_2\mathbf{G} \end{bmatrix}, \qquad \mathbf{S}_3 = \mathbf{f}_3$$

Then the matrices $\hat{\mathbf{G}}$ and $\hat{\mathbf{H}}$ will have the following forms:

$$\hat{\mathbf{G}} = \left[\begin{array}{ccc:ccc} 0 & 1 & 0 & 0 & 0 & 0 \\ 0 & 0 & 1 & 0 & 0 & 0 \\ -a_{11} & -a_{12} & -a_{13} & -a_{14} & -a_{15} & -a_{16} \\ \hdashline 0 & 0 & 0 & 0 & 1 & 0 \\ -a_{21} & -a_{22} & -a_{23} & -a_{24} & -a_{25} & -a_{26} \\ \hdashline -a_{31} & -a_{32} & -a_{33} & -a_{34} & -a_{35} & -a_{36} \end{array}\right]$$

$$
\hat{\mathbf{H}} = \left[\begin{array}{ccc}
0 & 0 & 0 \\
0 & 0 & 0 \\
\hline
1 & b_{12} & b_{13} \\
\hline
0 & 0 & 0 \\
0 & 1 & b_{23} \\
\hline
0 & 0 & 1
\end{array}\right]
$$

where $b_{12} = \mathbf{f}_1 \mathbf{G}^2 \mathbf{H}_2$, $b_{13} = \mathbf{f}_1 \mathbf{G}^2 \mathbf{H}_3$, and $b_{23} = \mathbf{f}_2 \mathbf{GH}_3$. These values may or may not be zero. (Notice that in matrix $\hat{\mathbf{G}}$ the principal minors are in the controllable canonical form.) The state feedback gain matrix $\mathbf{K}$ is given as follows:

$$
\mathbf{K} = \mathbf{B\Delta T}^{-1}
$$

where

$$
\mathbf{B} = \left[\begin{array}{ccc}
1 & b_{12} & b_{13} \\
0 & 1 & b_{23} \\
0 & 0 & 1
\end{array}\right]^{-1}
$$

and

$$
\mathbf{\Delta} = \left[\begin{array}{cccccc}
\delta_{11} & \delta_{12} & \delta_{13} & \delta_{14} & \delta_{15} & \delta_{16} \\
\delta_{21} & \delta_{22} & \delta_{23} & \delta_{24} & \delta_{25} & \delta_{26} \\
\delta_{31} & \delta_{32} & \delta_{33} & \delta_{34} & \delta_{35} & \delta_{36}
\end{array}\right]
$$

Notice that

$$
\hat{\mathbf{H}}\mathbf{B} = \left[\begin{array}{ccc}
0 & 0 & 0 \\
0 & 0 & 0 \\
1 & b_{12} & b_{13} \\
0 & 0 & 0 \\
0 & 1 & b_{23} \\
0 & 0 & 1
\end{array}\right]\left[\begin{array}{ccc}
1 & b_{12} & b_{13} \\
0 & 1 & b_{23} \\
0 & 0 & 1
\end{array}\right]^{-1} = \left[\begin{array}{ccc}
0 & 0 & 0 \\
0 & 0 & 0 \\
1 & 0 & 0 \\
0 & 0 & 0 \\
0 & 1 & 0 \\
0 & 0 & 1
\end{array}\right]
$$

The effect of postmultiplying matrix $\mathbf{B}$ to matrix $\hat{\mathbf{H}}$ is to eliminate the b_{ij} from the product matrix $\hat{\mathbf{H}}\mathbf{B}$.

Note that if $\mathbf{u}(k)$ is an r-vector the general form of $\mathbf{B}$ matrix is

$$
\mathbf{B} = \left[\begin{array}{cccc}
1 & b_{12} & \cdots & b_{1r} \\
0 & 1 & \cdots & b_{2r} \\
0 & 0 & \cdots & b_{3r} \\
\vdots & \vdots & & \vdots \\
0 & 0 & \cdots & 1
\end{array}\right]^{-1}
\tag{C-24}
$$

where the constants b_{ij}'s are those that will appear in the $n \times r$ matrix $\hat{\mathbf{H}}$. (The elements of $\hat{\mathbf{H}}\mathbf{B}$ are either 0 or 1.)

Example C–1

Consider the system

$$
\mathbf{x}(k + 1) = \mathbf{Gx}(k) + \mathbf{Hu}(k)
$$

where

$$\mathbf{x}(k) = \text{state vector (3-vector)}$$

$$\mathbf{u}(k) = \text{control vector (2-vector)}$$

and

$$\mathbf{G} = \begin{bmatrix} 0 & 1 & 0 \\ 0 & 0 & 1 \\ -0.25 & 0 & 0.5 \end{bmatrix}, \quad \mathbf{H} = \begin{bmatrix} 0 & 1 \\ 0 & 0 \\ 1 & 0 \end{bmatrix}$$

It is desired to determine the state feedback gain matrix $\mathbf{K}$ so that the response to the initial state $\mathbf{x}(0)$ is deadbeat. Note that with state feedback $\mathbf{u}(k) = -\mathbf{K}\mathbf{x}(k)$ the system equation becomes

$$\mathbf{x}(k+1) = (\mathbf{G} - \mathbf{H}\mathbf{K})\mathbf{x}(k) \tag{C–25}$$

We shall first examine the controllability matrix:

$$[\mathbf{H} \vdots \mathbf{GH} \vdots \mathbf{G}^2\mathbf{H}] = [\mathbf{H}_1 \vdots \mathbf{H}_2 \vdots \mathbf{GH}_1 \vdots \mathbf{GH}_2 \vdots \mathbf{G}^2\mathbf{H}_1 \vdots \mathbf{G}^2\mathbf{H}_2]$$

$$= \begin{bmatrix} 0 & 1 & 0 & 0 & 1 & 0 \\ 0 & 0 & 1 & 0 & 0.5 & -0.25 \\ 1 & 0 & 0.5 & -0.25 & 0.25 & -0.125 \end{bmatrix}$$

Clearly, the rank of this controllability matrix is 3. Therefore, arbitrary pole placement is possible. We now choose three linearly independent vectors starting from the left end. These vectors are shown enclosed by dashed lines. (The three linearly independent vectors chosen are $\mathbf{H}_1$, $\mathbf{H}_2$, and $\mathbf{GH}_1$.) Now we rearrange these three vectors according to Equation (C–10) and define matrix $\mathbf{F}$ as follows:

$$\mathbf{F} = [\mathbf{H}_1 \vdots \mathbf{GH}_1 \vdots \mathbf{H}_2]$$

We note that $n_1 = 2$ and $n_2 = 1$.

Rewriting matrix $\mathbf{F}$, we have

$$\mathbf{F} = \begin{bmatrix} 0 & 0 & 1 \\ 0 & 1 & 0 \\ 1 & 0.5 & 0 \end{bmatrix}$$

The inverse of matrix $\mathbf{F}$ becomes

$$\mathbf{F}^{-1} = \begin{bmatrix} 0 & -0.5 & 1 \\ 0 & 1 & 0 \\ 1 & 0 & 0 \end{bmatrix}$$

We now define the η_ith row vector of $\mathbf{F}^{-1}$ as $\mathbf{f}_1$, where $\eta_1 = n_1$ and $\eta_2 = n_1 + n_2$. Since $n_1 = 2$ and $n_2 = 1$, the vectors $\mathbf{f}_1$ and $\mathbf{f}_2$ are the second and third row vectors, respectively. That is,

$$\mathbf{f}_1 = [0 \quad 1 \quad 0]$$

$$\mathbf{f}_2 = [1 \quad 0 \quad 0]$$

Next, define the transformation matrix $\mathbf{T}$ by

$$\mathbf{T} = \begin{bmatrix} \dfrac{\mathbf{S}_1}{\mathbf{S}_2} \end{bmatrix}^{-1}$$

where

$$\mathbf{S}_1 = \begin{bmatrix} \dfrac{\mathbf{f}_1}{\mathbf{f}_1 \mathbf{G}} \end{bmatrix}, \quad \mathbf{S}_2 = \mathbf{f}_2$$

Hence,

$$\mathbf{T} = \begin{bmatrix} 0 & 1 & 0 \\ 0 & 0 & 1 \\ 1 & 0 & 0 \end{bmatrix}^{-1} = \begin{bmatrix} 0 & 0 & 1 \\ 1 & 0 & 0 \\ 0 & 1 & 0 \end{bmatrix}$$

and

$$\mathbf{T}^{-1} = \begin{bmatrix} 0 & 1 & 0 \\ 0 & 0 & 1 \\ 1 & 0 & 0 \end{bmatrix}$$

With this transformation matrix $\mathbf{T}$, we define

$$\mathbf{x}(k) = \mathbf{T}\hat{\mathbf{x}}(k)$$

Then

$$\mathbf{T}^{-1}\mathbf{GT} = \hat{\mathbf{G}}$$

$$= \begin{bmatrix} 0 & 1 & 0 \\ 0 & 0 & 1 \\ 1 & 0 & 0 \end{bmatrix} \begin{bmatrix} 0 & 1 & 0 \\ 0 & 0 & 1 \\ -0.25 & 0 & 0.5 \end{bmatrix} \begin{bmatrix} 0 & 0 & 1 \\ 1 & 0 & 0 \\ 0 & 1 & 0 \end{bmatrix}$$

$$= \begin{bmatrix} 0 & 1 & 0 \\ 0 & 0.5 & -0.25 \\ 1 & 0 & 0 \end{bmatrix}$$

Also,

$$\mathbf{T}^{-1}\mathbf{H} = \hat{\mathbf{H}}$$

$$= \begin{bmatrix} 0 & 1 & 0 \\ 0 & 0 & 1 \\ 1 & 0 & 0 \end{bmatrix} \begin{bmatrix} 0 & 1 \\ 0 & 0 \\ 1 & 0 \end{bmatrix} = \begin{bmatrix} 0 & 0 \\ 1 & 0 \\ 0 & 1 \end{bmatrix}$$

Next, we determine the state feedback gain matrix $\mathbf{K}$, where

$$\mathbf{K} = \mathbf{B}\boldsymbol{\Delta}\mathbf{T}^{-1}$$

From Equation (C–24), matrix $\mathbf{B}$ for the present case is a 2×2 matrix. Noting that $b_{12} = 0$, we have

$$\mathbf{B} = \begin{bmatrix} 1 & b_{12} \\ 0 & 1 \end{bmatrix}^{-1} = \begin{bmatrix} 1 & 0 \\ 0 & 1 \end{bmatrix}$$

For the present case, $\boldsymbol{\Delta}$ is a 2×3 matrix:

$$\boldsymbol{\Delta} = \begin{bmatrix} \delta_{11} & \delta_{12} & \delta_{13} \\ \delta_{21} & \delta_{22} & \delta_{23} \end{bmatrix}$$

Now we determine matrix $\hat{\mathbf{G}} - \hat{\mathbf{H}}\mathbf{B}\boldsymbol{\Delta}$:

$$\hat{\mathbf{G}} - \hat{\mathbf{H}}\mathbf{B}\boldsymbol{\Delta} = \begin{bmatrix} 0 & 1 & 0 \\ 0 & 0.5 & -0.25 \\ 1 & 0 & 0 \end{bmatrix} - \begin{bmatrix} 0 & 0 \\ 1 & 0 \\ 0 & 1 \end{bmatrix} \begin{bmatrix} 1 & 0 \\ 0 & 1 \end{bmatrix} \begin{bmatrix} \delta_{11} & \delta_{12} & \delta_{13} \\ \delta_{21} & \delta_{22} & \delta_{23} \end{bmatrix}$$

$$= \begin{bmatrix} 0 & 1 & 0 \\ 0 & 0.5 & -0.25 \\ 1 & 0 & 0 \end{bmatrix} - \begin{bmatrix} 0 & 0 & 0 \\ \delta_{11} & \delta_{12} & \delta_{13} \\ \delta_{21} & \delta_{22} & \delta_{23} \end{bmatrix}$$

$$= \begin{bmatrix} 0 & 1 & 0 \\ -\delta_{11} & 0.5 - \delta_{12} & -0.25 - \delta_{13} \\ 1 - \delta_{21} & -\delta_{22} & -\delta_{23} \end{bmatrix}$$

The characteristic equation $|z\mathbf{I} - \hat{\mathbf{G}} + \hat{\mathbf{H}}\mathbf{B}\boldsymbol{\Delta}| = 0$ is given as follows:

$$|z\mathbf{I} - \hat{\mathbf{G}} + \hat{\mathbf{H}}\mathbf{B}\boldsymbol{\Delta}| = \begin{vmatrix} z & -1 & 0 \\ \delta_{11} & z - 0.5 + \delta_{12} & 0.25 + \delta_{13} \\ -1 + \delta_{21} & \delta_{22} & z + \delta_{23} \end{vmatrix}$$

$$= 0$$

Since the deadbeat response is desired, the desired characteristic equation is

$$z^3 = 0$$

Note that the choice of the δ's is not unique and matrix $\boldsymbol{\Delta}$ is not unique. Suppose we choose the δ's so that

$$\delta_{11} = 0, \qquad \delta_{12} = 0.5, \qquad \delta_{13} = -0.25$$
$$\delta_{21} = 1, \qquad \delta_{22} = 0, \qquad \delta_{23} = 0$$

Then

$$|z\mathbf{I} - \hat{\mathbf{G}} + \hat{\mathbf{H}}\mathbf{B}\boldsymbol{\Delta}| = \begin{vmatrix} z & -1 & 0 \\ 0 & z & 0 \\ 0 & 0 & z \end{vmatrix} = z^3 = 0$$

and thus

$$\boldsymbol{\Delta} = \begin{bmatrix} 0 & 0.5 & -0.25 \\ 1 & 0 & 0 \end{bmatrix}$$

is acceptable. Then matrix $\mathbf{K}$ is obtained as follows:

$$\mathbf{K} = \mathbf{B}\boldsymbol{\Delta}\mathbf{T}^{-1} = \begin{bmatrix} 1 & 0 \\ 0 & 1 \end{bmatrix} \begin{bmatrix} 0 & 0.5 & -0.25 \\ 1 & 0 & 0 \end{bmatrix} \begin{bmatrix} 0 & 1 & 0 \\ 0 & 0 & 1 \\ 1 & 0 & 0 \end{bmatrix}$$

$$= \begin{bmatrix} -0.25 & 0 & 0.5 \\ 0 & 1 & 0 \end{bmatrix}$$

With this choice of matrix $\mathbf{K}$, $(\hat{\mathbf{G}} - \hat{\mathbf{H}}\mathbf{B}\boldsymbol{\Delta})^k = 0$ for $k \geq n_{\min}$, where

$$n_{\min} = \max{(n_1, n_2)} = \max{(2, 1)} = 2$$

In fact,

$$\hat{\mathbf{G}} - \hat{\mathbf{H}}\mathbf{B}\boldsymbol{\Delta} = \begin{bmatrix} 0 & 1 & 0 \\ 0 & 0 & 0 \\ 0 & 0 & 0 \end{bmatrix}$$

$$(\hat{\mathbf{G}} - \hat{\mathbf{H}}\mathbf{B}\boldsymbol{\Delta})^2 = \begin{bmatrix} 0 & 0 & 0 \\ 0 & 0 & 0 \\ 0 & 0 & 0 \end{bmatrix}$$

Thus,

$$(\hat{\mathbf{G}} - \hat{\mathbf{H}}\mathbf{B}\boldsymbol{\Delta})^k = \mathbf{0}, \qquad k = 2, 3, 4, \ldots$$

Note that

$$\hat{\mathbf{G}} - \hat{\mathbf{H}}\mathbf{B}\boldsymbol{\Delta} = \mathbf{T}^{-1}\mathbf{G}\mathbf{T} - \mathbf{T}^{-1}\mathbf{H}\mathbf{B}\boldsymbol{\Delta} = \mathbf{T}^{-1}\mathbf{G}\mathbf{T} - \mathbf{T}^{-1}\mathbf{H}\mathbf{K}\mathbf{T} = \mathbf{T}^{-1}(\mathbf{G} - \mathbf{H}\mathbf{K})\mathbf{T}$$

Referring to Equation (C-25) and its solution $\mathbf{x}(k) = (\mathbf{G} - \mathbf{H}\mathbf{K})^k\mathbf{x}(0)$, we have $\mathbf{x}(k) = \mathbf{0}$ for $k = 2, 3, 4, \ldots$, since

$$G - HK = T(\hat{G} - \hat{H}B\Delta)T^{-1}$$

$$(G - HK)^2 = T(\hat{G} - \hat{H}B\Delta)T^{-1}T(\hat{G} - \hat{H}B\Delta)T^{-1} = T(\hat{G} - \hat{H}B\Delta)^2T^{-1} = 0$$

and

$$x(k) = (G - HK)^k x(0) = 0, \qquad k = 2, 3, 4 \ldots$$

We have thus designed the state feedback gain matrix K so that the system's response to any initial state $x(0)$ is deadbeat. The state $x(k)$ can be transferred to the origin in at most two sampling periods. [Note that if the control signal $u(k)$ were a scalar then it would take at most three sampling periods, rather than at most two sampling periods, for deadbeat response.]

EXAMPLE PROBLEMS AND SOLUTIONS

Problem C–1

Consider the system given by

$$x(k + 1) = Gx(k) + H_1 u(k)$$

where

$$x(k) = \text{state vector (n-vector)}$$
$$u(k) = \text{control signal (scalar)}$$
$$G = n \times n \text{ matrix}$$
$$H_1 = n \times 1 \text{ matrix}$$

Assume that the system is completely state controllable.

Define

$$[H_1 \vdots GH_1 \vdots \cdots \vdots G^{n-1}H_1]^{-1} = \begin{bmatrix} f_1 \\ f_2 \\ \vdots \\ f_n \end{bmatrix}$$

where the f_i's ($i = 1, 2, \ldots, n$) are row vectors. Define also

$$T_1 = \begin{bmatrix} f_n \\ f_n G \\ \vdots \\ f_n G^{n-1} \end{bmatrix}^{-1}$$

Show that

$$T_1^{-1}GT_1 = \begin{bmatrix} 0 & 1 & 0 & \cdots & 0 \\ 0 & 0 & 1 & \cdots & 0 \\ \vdots & \vdots & \vdots & & \vdots \\ 0 & 0 & 0 & \cdots & 1 \\ -a_n & -a_{n-1} & -a_{n-2} & \cdots & -a_1 \end{bmatrix} \qquad \text{(C–26)}$$

and

$$T_1^{-1} H_1 = \begin{bmatrix} 0 \\ 0 \\ \vdots \\ 0 \\ 1 \end{bmatrix} \qquad \text{(C-27)}$$

where the a_i's are the coefficients appearing in the characteristic polynomial of $\mathbf{G}$, or

$$|z\mathbf{I} - \mathbf{G}| = z^n + a_1 z^{n-1} + \cdots + a_{n-1} z + a_n$$

Solution   We shall prove Equations (C–26) and (C–27) for the case where $n = 3$. (Extension of the derivation to an arbitrary positive integer n is straightforward.) Thus, we shall derive that

$$T_1^{-1} \mathbf{G} T_1 = \begin{bmatrix} 0 & 1 & 0 \\ 0 & 0 & 1 \\ -a_3 & -a_2 & -a_1 \end{bmatrix} \qquad \text{(C-28)}$$

Since

$$T_1^{-1} = \begin{bmatrix} \mathbf{f}_3 \\ \mathbf{f}_3\,\mathbf{G} \\ \mathbf{f}_3\,\mathbf{G}^2 \end{bmatrix}$$

it is possible to rewrite Equation (C–28) as follows:

$$T_1^{-1}\mathbf{G} = \begin{bmatrix} 0 & 1 & 0 \\ 0 & 0 & 1 \\ -a_3 & -a_2 & -a_1 \end{bmatrix}\begin{bmatrix} \mathbf{f}_3 \\ \mathbf{f}_3\,\mathbf{G} \\ \mathbf{f}_3\,\mathbf{G}^2 \end{bmatrix} \qquad \text{(C-29)}$$

Now consider the conjugate transpose of the right-hand side of Equation (C–29). Noting that for physical systems the coefficients $a_1, a_2, \ldots, a_n$ of the characteristic polynomial are real, we have

$$[\mathbf{f}_3^*\,\vdots\,\mathbf{G}^*\mathbf{f}_3^*\,\vdots\,(\mathbf{G}^*)^2\,\mathbf{f}_3^*]\begin{bmatrix} 0 & 0 & -a_3 \\ 1 & 0 & -a_2 \\ 0 & 1 & -a_1 \end{bmatrix} = [\mathbf{G}^*\mathbf{f}_3^*\,\vdots\,(\mathbf{G}^*)^2\,\mathbf{f}_3^*\,\vdots\,-a_3\mathbf{f}_3^* - a_2\,\mathbf{G}^*\mathbf{f}_3^* - a_1(\mathbf{G}^*)^2\mathbf{f}_3^*]$$

Note that $\mathbf{G}^*$ satisfies its own characteristic equation:

$$\phi(\mathbf{G}^*) = (\mathbf{G}^*)^3 + a_1(\mathbf{G}^*)^2 + a_2\,\mathbf{G}^* + a_3\mathbf{I} = \mathbf{0}$$

Hence,

$$-[a_3\mathbf{I} + a_2\,\mathbf{G}^* + a_1(\mathbf{G}^*)^2]\mathbf{f}_3^* = (\mathbf{G}^*)^3\,\mathbf{f}_3^*$$

Consequently,

$$[\mathbf{f}_3^*\,\vdots\,\mathbf{G}^*\mathbf{f}_3^*\,\vdots\,(\mathbf{G}^*)^2\mathbf{f}_3^*]\begin{bmatrix} 0 & 0 & -a_3 \\ 1 & 0 & -a_2 \\ 0 & 1 & -a_1 \end{bmatrix} = [\mathbf{G}^*\mathbf{f}_3^*\,\vdots\,(\mathbf{G}^*)^2\mathbf{f}_3^*\,\vdots\,(\mathbf{G}^*)^3\mathbf{f}_3^*] = \mathbf{G}^*[\mathbf{f}_3^*\,\vdots\,\mathbf{G}^*\mathbf{f}_3^*\,\vdots\,(\mathbf{G}^*)^2\mathbf{f}_3^*]$$

Taking the conjugate transpose of both sides of this last equation, we obtain

$$\begin{bmatrix} 0 & 1 & 0 \\ 0 & 0 & 1 \\ -a_3 & -a_2 & -a_1 \end{bmatrix}\begin{bmatrix} \mathbf{f}_3 \\ \mathbf{f}_3\,\mathbf{G} \\ \mathbf{f}_3\,\mathbf{G}^2 \end{bmatrix} = \begin{bmatrix} \mathbf{f}_3 \\ \mathbf{f}_3\,\mathbf{G} \\ \mathbf{f}_3\,\mathbf{G}^2 \end{bmatrix}\mathbf{G} = T_1^{-1}\mathbf{G}$$

which is Equation (C–29). Thus, we have shown that Equation (C–28) is true, or

$$\mathbf{T}_1^{-1}\mathbf{G}\mathbf{T}_1 = \begin{bmatrix} 0 & 1 & 0 \\ 0 & 0 & 1 \\ -a_3 & -a_2 & -a_1 \end{bmatrix}$$

Next, we shall show that

$$\mathbf{T}_1^{-1}\mathbf{H}_1 = \begin{bmatrix} 0 \\ 0 \\ 1 \end{bmatrix}$$

Since

$$[\mathbf{H}_1 \vdots \mathbf{G}\mathbf{H}_1 \vdots \mathbf{G}^2\mathbf{H}_1]^{-1} = \begin{bmatrix} \mathbf{f}_1 \\ \mathbf{f}_2 \\ \mathbf{f}_3 \end{bmatrix}$$

we obtain

$$\mathbf{I} = \begin{bmatrix} \mathbf{f}_1 \\ \mathbf{f}_2 \\ \mathbf{f}_3 \end{bmatrix} [\mathbf{H}_1 \vdots \mathbf{G}\mathbf{H}_1 \vdots \mathbf{G}^2\mathbf{H}_1]$$

or

$$\begin{bmatrix} 1 & 0 & 0 \\ 0 & 1 & 0 \\ 0 & 0 & 1 \end{bmatrix} = \begin{bmatrix} \mathbf{f}_1\mathbf{H}_1 & \mathbf{f}_1\mathbf{G}\mathbf{H}_1 & \mathbf{f}_1\mathbf{G}^2\mathbf{H}_1 \\ \mathbf{f}_2\mathbf{H}_1 & \mathbf{f}_2\mathbf{G}\mathbf{H}_1 & \mathbf{f}_2\mathbf{G}^2\mathbf{H}_1 \\ \mathbf{f}_3\mathbf{H}_1 & \mathbf{f}_3\mathbf{G}\mathbf{H}_1 & \mathbf{f}_3\mathbf{G}^2\mathbf{H}_1 \end{bmatrix}$$

Hence,

$$\mathbf{f}_3\mathbf{H}_1 = 0, \qquad \mathbf{f}_3\mathbf{G}\mathbf{H}_1 = 0, \qquad \mathbf{f}_3\mathbf{G}^2\mathbf{H}_1 = 1$$

By using these equations, we obtain

$$\mathbf{T}_1^{-1}\mathbf{H}_1 = \begin{bmatrix} \mathbf{f}_3 \\ \mathbf{f}_3\mathbf{G} \\ \mathbf{f}_3\mathbf{G}^2 \end{bmatrix} \mathbf{H}_1 = \begin{bmatrix} \mathbf{f}_3\mathbf{H}_1 \\ \mathbf{f}_3\mathbf{G}\mathbf{H}_1 \\ \mathbf{f}_3\mathbf{G}^2\mathbf{H}_1 \end{bmatrix} = \begin{bmatrix} 0 \\ 0 \\ 1 \end{bmatrix}$$

Note that the extension of the derivations presented here to the case of an arbitrary positive integer n can be made easily.

Problem C–2

Consider the system

$$\mathbf{x}(k+1) = \mathbf{G}\mathbf{x}(k) + \mathbf{H}\mathbf{u}(k)$$

where

$\mathbf{x}(k)$ = state vector (4-vector)

$\mathbf{u}(k)$ = control vector (2-vector)

and

$$\mathbf{G} = \begin{bmatrix} -1 & 1 & 0 & 0 \\ 1 & -2 & 1 & 0 \\ 0 & 1 & -1 & 2 \\ 1 & 0 & 0 & 1 \end{bmatrix}, \qquad \mathbf{H} = [\mathbf{H}_1 \vdots \mathbf{H}_2] = \begin{bmatrix} 1 & 0 \\ 0 & 0 \\ 0 & 0 \\ 0 & 1 \end{bmatrix}$$

Referring to Equation (C–10), obtain matrix **F**. Then, by use of the transformation matrix **T** defined by Equation (C–12), determine matrices $\hat{\mathbf{G}} = \mathbf{T}^{-1}\mathbf{GT}$ and $\hat{\mathbf{H}} = \mathbf{T}^{-1}\mathbf{H}$. Finally, derive Equation (C–14).

Solution   We shall first write the controllability matrix as follows:

$$[\mathbf{H}_1 \vdots \mathbf{H}_2 \vdots \mathbf{GH}_1 \vdots \mathbf{GH}_2 \vdots \mathbf{G}^2\mathbf{H}_1 \vdots \mathbf{G}^2\mathbf{H}_2 \vdots \mathbf{G}^3\mathbf{H}_1 \vdots \mathbf{G}^3\mathbf{H}_2]$$

$$= \begin{bmatrix} 1 & 0 & -1 & 0 & 2 & 0 & -5 & 2 \\ 0 & 0 & 1 & 0 & -3 & 2 & 11 & -4 \\ 0 & 0 & 0 & 2 & 3 & 0 & -6 & 4 \\ 0 & 1 & 1 & 1 & 0 & 1 & 2 & 1 \end{bmatrix}$$

We now choose four linearly independent vectors from this 4×8 matrix, starting from the left end. (These vectors are shown enclosed by dashed lines.) The four linearly independent vectors chosen are $\mathbf{H}_1$, $\mathbf{H}_2$, $\mathbf{GH}_1$, and $\mathbf{GH}_2$. Next, we rearrange these four vectors according to Equation (C–10) and define matrix **F** as follows:

$$\mathbf{F} = [\mathbf{H}_1 \vdots \mathbf{GH}_1 \vdots \mathbf{H}_2 \vdots \mathbf{GH}_2]$$

(Note that in this case $n_1 = 2$ and $n_2 = 2$.) Thus,

$$\mathbf{F} = \begin{bmatrix} 1 & -1 & 0 & 0 \\ 0 & 1 & 0 & 0 \\ 0 & 0 & 0 & 2 \\ 0 & 1 & 1 & 1 \end{bmatrix}$$

The inverse of this matrix is given by

$$\mathbf{F}^{-1} = \begin{bmatrix} 1 & 1 & 0 & 0 \\ 0 & 1 & 0 & 0 \\ 0 & -1 & -0.5 & 1 \\ 0 & 0 & 0.5 & 0 \end{bmatrix}$$

Since in this case $n_1 = 2$ and $n_2 = 2$, we define the second row vector of $\mathbf{F}^{-1}$ as $\mathbf{f}_1$ and the fourth row vector as $\mathbf{f}_2$. Then

$$\mathbf{f}_1 = [0 \quad 1 \quad 0 \quad 0]$$
$$\mathbf{f}_2 = [0 \quad 0 \quad 0.5 \quad 0]$$

The transformation matrix **T** is given by

$$\mathbf{T} = \begin{bmatrix} \mathbf{S}_1 \\ \overline{\mathbf{S}_2} \end{bmatrix}^{-1}$$

where

$$\mathbf{S}_1 = \begin{bmatrix} \mathbf{f}_1 \\ \overline{\mathbf{f}_1\mathbf{G}} \end{bmatrix}, \qquad \mathbf{S}_2 = \begin{bmatrix} \mathbf{f}_2 \\ \overline{\mathbf{f}_2\mathbf{G}} \end{bmatrix}$$

Hence,

$$\mathbf{T} = \begin{bmatrix} \mathbf{f}_1 \\ \mathbf{f}_1\mathbf{G} \\ \mathbf{f}_2 \\ \mathbf{f}_2\mathbf{G} \end{bmatrix}^{-1} = \begin{bmatrix} 0 & 1 & 0 & 0 \\ 1 & -2 & 1 & 0 \\ 0 & 0 & 0.5 & 0 \\ 0 & 0.5 & -0.5 & 1 \end{bmatrix}^{-1} = \begin{bmatrix} 2 & 1 & -2 & 0 \\ 1 & 0 & 0 & 0 \\ 0 & 0 & 2 & 0 \\ -0.5 & 0 & 1 & 1 \end{bmatrix}$$

With this transformation matrix $\mathbf{T}$ we obtain

$$\hat{\mathbf{G}} = \mathbf{T}^{-1}\mathbf{G}\mathbf{T} = \begin{bmatrix} 0 & 1 & 0 & 0 \\ 1 & -2 & 1 & 0 \\ 0 & 0 & 0.5 & 0 \\ 0 & 0.5 & -0.5 & 1 \end{bmatrix} \begin{bmatrix} -1 & 1 & 0 & 0 \\ 1 & -2 & 1 & 0 \\ 0 & 1 & -1 & 2 \\ 1 & 0 & 0 & 1 \end{bmatrix} \begin{bmatrix} 2 & 1 & -2 & 0 \\ 1 & 0 & 0 & 0 \\ 0 & 0 & 2 & 0 \\ -0.5 & 0 & 1 & 1 \end{bmatrix}$$

$$= \begin{bmatrix} 0 & 1 & 0 & 0 \\ -1 & -3 & 2 & 2 \\ 0 & 0 & 0 & 1 \\ 1.5 & 1.5 & -1 & 0 \end{bmatrix}$$

and

$$\hat{\mathbf{H}} = \mathbf{T}^{-1}\mathbf{H} = \begin{bmatrix} 0 & 1 & 0 & 0 \\ 1 & -2 & 1 & 0 \\ 0 & 0 & 0.5 & 0 \\ 0 & 0.5 & -0.5 & 1 \end{bmatrix} \begin{bmatrix} 1 & 0 \\ 0 & 0 \\ 0 & 0 \\ 0 & 1 \end{bmatrix} = \begin{bmatrix} 0 & 0 \\ 1 & 0 \\ 0 & 0 \\ 0 & 1 \end{bmatrix}$$

Notice that, when $n_1 = n_2 = 2$, matrix $\hat{\mathbf{G}}$ has the form given by Equation (C–13) and matrix $\hat{\mathbf{H}}$ has the form given by Equation (C–14), or

$$\hat{\mathbf{G}} = \begin{bmatrix} 0 & 1 & 0 & 0 \\ -a_{11} & -a_{12} & -a_{13} & -a_{14} \\ 0 & 0 & 0 & 1 \\ -a_{21} & -a_{22} & -a_{23} & -a_{24} \end{bmatrix}, \qquad \hat{\mathbf{H}} = \begin{bmatrix} 0 & 0 \\ 1 & b_{12} \\ 0 & 0 \\ 0 & 1 \end{bmatrix}$$

(Note that b_{12} is zero in this case.)

Finally, we shall derive Equation (C–14). Notice that

$$\mathbf{F}^{-1}\mathbf{F} = \begin{bmatrix} \mathbf{m}_1 \\ \mathbf{f}_1 \\ \mathbf{m}_2 \\ \mathbf{f}_2 \end{bmatrix} [\mathbf{H}_1 \quad \mathbf{G}\mathbf{H}_1 \quad \mathbf{H}_2 \quad \mathbf{G}\mathbf{H}_2]$$

$$= \begin{bmatrix} \mathbf{m}_1\mathbf{H}_1 & \mathbf{m}_1\mathbf{G}\mathbf{H}_1 & \mathbf{m}_1\mathbf{H}_2 & \mathbf{m}_1\mathbf{G}\mathbf{H}_2 \\ \mathbf{f}_1\mathbf{H}_1 & \mathbf{f}_1\mathbf{G}\mathbf{H}_1 & \mathbf{f}_1\mathbf{H}_2 & \mathbf{f}_1\mathbf{G}\mathbf{H}_2 \\ \mathbf{m}_2\mathbf{H}_1 & \mathbf{m}_2\mathbf{G}\mathbf{H}_1 & \mathbf{m}_2\mathbf{H}_2 & \mathbf{m}_2\mathbf{G}\mathbf{H}_2 \\ \mathbf{f}_2\mathbf{H}_1 & \mathbf{f}_2\mathbf{G}\mathbf{H}_1 & \mathbf{f}_2\mathbf{H}_2 & \mathbf{f}_2\mathbf{G}\mathbf{H}_2 \end{bmatrix}$$

$$= \begin{bmatrix} 1 & 0 & 0 & 0 \\ 0 & 1 & 0 & 0 \\ 0 & 0 & 1 & 0 \\ 0 & 0 & 0 & 1 \end{bmatrix}$$

where $\mathbf{m}_1$ and $\mathbf{m}_2$ are the first row vector and the third row vector of $\mathbf{F}^{-1}$, respectively. Since $\mathbf{F}^{-1}\mathbf{F}$ is an identity matrix, we have $\mathbf{f}_1\mathbf{H}_1 = 0$, $\mathbf{f}_1\mathbf{H}_2 = 0$, $\mathbf{f}_1\mathbf{G}\mathbf{H}_1 = 1$, $\mathbf{f}_1\mathbf{G}\mathbf{H}_2 = 0$, $\mathbf{f}_2\mathbf{H}_1 = 0$, $\mathbf{f}_2\mathbf{H}_2 = 0$, $\mathbf{f}_2\mathbf{G}\mathbf{H}_1 = 0$, and $\mathbf{f}_2\mathbf{G}\mathbf{H}_2 = 1$. Thus, we have

$$\hat{\mathbf{H}} = \mathbf{T}^{-1}\mathbf{H} = \begin{bmatrix} \mathbf{f}_1 \\ \mathbf{f}_1\mathbf{G} \\ \mathbf{f}_2 \\ \mathbf{f}_2\mathbf{G} \end{bmatrix} [\mathbf{H}_1 \quad \mathbf{H}_2] = \begin{bmatrix} \mathbf{f}_1\mathbf{H}_1 & \mathbf{f}_1\mathbf{H}_2 \\ \mathbf{f}_1\mathbf{G}\mathbf{H}_1 & \mathbf{f}_1\mathbf{G}\mathbf{H}_2 \\ \mathbf{f}_2\mathbf{H}_1 & \mathbf{f}_2\mathbf{H}_2 \\ \mathbf{f}_2\mathbf{G}\mathbf{H}_1 & \mathbf{f}_2\mathbf{G}\mathbf{H}_2 \end{bmatrix} = \begin{bmatrix} 0 & 0 \\ 1 & 0 \\ 0 & 0 \\ 0 & 1 \end{bmatrix}$$

which is Equation (C–14).

Problem C–3

Consider the system defined by Equation (C–8):

$$\hat{\mathbf{x}}(k + 1) = \mathbf{T}^{-1}\mathbf{G}\mathbf{T}\hat{\mathbf{x}}(k) + \mathbf{T}^{-1}\mathbf{H}\mathbf{u}(k) = \hat{\mathbf{G}}\hat{\mathbf{x}}(k) + \hat{\mathbf{H}}\mathbf{u}(k)$$

where the transformation matrix $\mathbf{T}$ is defined by Equation (C–12). Assume that the matrix $\hat{\mathbf{G}}$ is given by Equation (C–15) and the matrix $\hat{\mathbf{H}}$ is given by Equation (C–16). That is,

$$\hat{\mathbf{G}} = \left[\begin{array}{cccc} 0 & 1 & 0 & 0 \\ 0 & 0 & 1 & 0 \\ -a_{11} & -a_{12} & -a_{13} & -a_{14} \\ \hline -a_{21} & -a_{22} & -a_{23} & -a_{24} \end{array}\right], \qquad \hat{\mathbf{H}} = \left[\begin{array}{cc} 0 & 0 \\ 0 & 0 \\ 1 & b_{12} \\ 0 & 1 \end{array}\right]$$

Show that

$$|z\mathbf{I} - \hat{\mathbf{G}}| = \left|\begin{array}{cc} z^3 + a_{13}z^2 + a_{12}z + a_{11} & a_{14} \\ a_{23}z^2 + a_{22}z + a_{21} & z + a_{24} \end{array}\right|$$

and

$$\hat{\mathbf{G}} - \hat{\mathbf{H}}\mathbf{B}\mathbf{\Delta} = \left[\begin{array}{cccc} 0 & 1 & 0 & 0 \\ 0 & 0 & 1 & 0 \\ -a_{11} - \delta_{11} & -a_{12} - \delta_{12} & -a_{13} - \delta_{13} & -a_{14} - \delta_{14} \\ -a_{21} - \delta_{21} & -a_{22} - \delta_{22} & -a_{23} - \delta_{23} & -a_{24} - \delta_{24} \end{array}\right]$$

where

$$\mathbf{B} = \left[\begin{array}{cc} 1 & b_{12} \\ 0 & 1 \end{array}\right]^{-1}, \qquad \mathbf{\Delta} = \left[\begin{array}{cccc} \delta_{11} & \delta_{12} & \delta_{13} & \delta_{14} \\ \delta_{21} & \delta_{22} & \delta_{23} & \delta_{24} \end{array}\right]$$

Show also that if we choose, for example,

$$\mathbf{\Delta} = \left[\begin{array}{cccc} -a_{11} & -a_{12} & -a_{13} & -a_{14} \\ * & * & * & -a_{24} \end{array}\right] \tag{C–30}$$

where the elements shown by asterisks are arbitrary constants, the system will exhibit the deadbeat response to any initial state $\mathbf{x}(0)$; that is,

$$(\hat{\mathbf{G}} - \hat{\mathbf{H}}\mathbf{B}\mathbf{\Delta})^k = \mathbf{0}, \qquad k = 4, 5, 6, \ldots$$

Show also that if we choose

$$\mathbf{\Delta} = \left[\begin{array}{cccc} -a_{11} & -a_{12} & -a_{13} & -a_{14} \\ -a_{21} & -a_{22} & -a_{23} & -a_{24} \end{array}\right] \tag{C–31}$$

then

$$(\hat{\mathbf{G}} - \hat{\mathbf{H}}\mathbf{B}\mathbf{\Delta})^k = \mathbf{0}$$

for $k \geq n_{\min}$, where

$$n_{\min} = \max(n_1, n_2) = \max(3, 1) = 3$$

Solution   For the case where $\hat{\mathbf{G}}$ is as given by Equation (C–15), we have

$$|z\mathbf{I} - \hat{\mathbf{G}}| = \left|\begin{array}{cccc} z & -1 & 0 & 0 \\ 0 & z & -1 & 0 \\ a_{11} & a_{12} & z + a_{13} & a_{14} \\ a_{21} & a_{22} & a_{23} & z + a_{24} \end{array}\right|$$

Expanding this determinant using the Laplace's expansion formula, we obtain

$$|z\mathbf{I} - \hat{\mathbf{G}}| = \begin{vmatrix} z & -1 \\ 0 & z \end{vmatrix} \begin{vmatrix} z + a_{13} & a_{14} \\ a_{23} & z + a_{24} \end{vmatrix} - \begin{vmatrix} z & -1 \\ a_{11} & a_{12} \end{vmatrix} \begin{vmatrix} -1 & 0 \\ a_{23} & z + a_{24} \end{vmatrix}$$

$$+ \begin{vmatrix} z & -1 \\ a_{21} & a_{22} \end{vmatrix} \begin{vmatrix} -1 & 0 \\ z + a_{13} & a_{14} \end{vmatrix}$$

$$= (z + a_{24})(z^3 + a_{13} z^2 + a_{12} z + a_{11}) - a_{14}(a_{23} z^2 + a_{22} z + a_{21})$$

Hence, the determinant $|z\mathbf{I} - \hat{\mathbf{G}}|$ may be written as follows:

$$|z\mathbf{I} - \hat{\mathbf{G}}| = \begin{vmatrix} z^3 + a_{13} z^2 + a_{12} z + a_{11} & a_{14} \\ a_{23} z^2 + a_{22} z + a_{21} & z + a_{24} \end{vmatrix} \qquad \text{(C–32)}$$

Next, compute

$$\hat{\mathbf{H}}\mathbf{B}\boldsymbol{\Delta} = \begin{bmatrix} 0 & 0 \\ 0 & 0 \\ 1 & b_{12} \\ 0 & 1 \end{bmatrix} \begin{bmatrix} 1 & b_{12} \\ 0 & 1 \end{bmatrix}^{-1} \begin{bmatrix} \delta_{11} & \delta_{12} & \delta_{13} & \delta_{14} \\ \delta_{21} & \delta_{22} & \delta_{23} & \delta_{24} \end{bmatrix}$$

$$= \begin{bmatrix} 0 & 0 \\ 0 & 0 \\ 1 & 0 \\ 0 & 1 \end{bmatrix} \begin{bmatrix} \delta_{11} & \delta_{12} & \delta_{13} & \delta_{14} \\ \delta_{21} & \delta_{22} & \delta_{23} & \delta_{24} \end{bmatrix} = \begin{bmatrix} 0 & 0 & 0 & 0 \\ 0 & 0 & 0 & 0 \\ \delta_{11} & \delta_{12} & \delta_{13} & \delta_{14} \\ \delta_{21} & \delta_{22} & \delta_{23} & \delta_{24} \end{bmatrix}$$

(Notice that the effect of postmultiplying matrix $\hat{\mathbf{H}}$ by matrix $\mathbf{B}$ is to eliminate b_{12} from the product matrix $\hat{\mathbf{H}}\mathbf{B}$.) Thus,

$$\hat{\mathbf{G}} - \hat{\mathbf{H}}\mathbf{B}\boldsymbol{\Delta} = \begin{bmatrix} 0 & 1 & 0 & 0 \\ 0 & 0 & 1 & 0 \\ -a_{11} - \delta_{11} & -a_{12} - \delta_{12} & -a_{13} - \delta_{13} & -a_{14} - \delta_{14} \\ -a_{21} - \delta_{21} & -a_{22} - \delta_{22} & -a_{23} - \delta_{23} & -a_{24} - \delta_{24} \end{bmatrix}$$

If we choose $\boldsymbol{\Delta}$ as given by Equation (C–30), then

$$\hat{\mathbf{G}} - \hat{\mathbf{H}}\mathbf{B}\boldsymbol{\Delta} = \begin{bmatrix} 0 & 1 & 0 & 0 \\ 0 & 0 & 1 & 0 \\ 0 & 0 & 0 & 0 \\ * & * & * & 0 \end{bmatrix}$$

where the elements shown by asterisks are arbitrary constants. Notice that

$$(\hat{\mathbf{G}} - \hat{\mathbf{H}}\mathbf{B}\boldsymbol{\Delta})^2 = \begin{bmatrix} 0 & 0 & 1 & 0 \\ 0 & 0 & 0 & 0 \\ 0 & 0 & 0 & 0 \\ 0 & * & * & 0 \end{bmatrix}$$

$$(\hat{\mathbf{G}} - \hat{\mathbf{H}}\mathbf{B}\boldsymbol{\Delta})^3 = \begin{bmatrix} 0 & 0 & 0 & 0 \\ 0 & 0 & 0 & 0 \\ 0 & 0 & 0 & 0 \\ 0 & 0 & * & 0 \end{bmatrix}$$

$$(\hat{\mathbf{G}} - \hat{\mathbf{H}}\mathbf{B}\boldsymbol{\Delta})^4 = \begin{bmatrix} 0 & 0 & 0 & 0 \\ 0 & 0 & 0 & 0 \\ 0 & 0 & 0 & 0 \\ 0 & 0 & 0 & 0 \end{bmatrix}$$

Hence,

$$\mathbf{x}(k) = (\mathbf{G} - \mathbf{HK})^k \mathbf{x}(0) = \mathbf{T}(\hat{\mathbf{G}} - \hat{\mathbf{H}}\mathbf{B}\mathbf{\Delta})^k \mathbf{T}^{-1}\mathbf{x}(0) = \mathbf{0}, \qquad k \geq 4$$

We have thus seen that the deadbeat response is achieved by choosing $\mathbf{\Delta}$ as given by Equation (C–30).

However, if we choose $\mathbf{\Delta}$ as given by Equation (C–31), then the deadbeat response can be achieved in at most three sampling periods, because the asterisk appearing in $(\hat{\mathbf{G}} - \hat{\mathbf{H}}\mathbf{B}\mathbf{\Delta})^3$ becomes zero and

$$\mathbf{x}(k) = \mathbf{T}(\hat{\mathbf{G}} - \hat{\mathbf{H}}\mathbf{B}\mathbf{\Delta})^k \mathbf{T}^{-1}\mathbf{x}(0) = \mathbf{0}, \qquad k \geq n_{min} = 3$$

Problem C–4

Consider the following system:

$$\mathbf{x}(k + 1) = \mathbf{G}\mathbf{x}(k) + \mathbf{H}\mathbf{u}(k)$$

where

$$\mathbf{x}(k) = \text{state vector (4-vector)}$$

$$\mathbf{u}(k) = \text{control vector (2-vector)}$$

and

$$\mathbf{G} = \begin{bmatrix} -1 & 1 & 0 & 0 \\ 1 & -2 & 1 & 0 \\ 0 & 1 & -1 & 2 \\ 1 & 0 & 0 & 1 \end{bmatrix}, \qquad \mathbf{H} = [\mathbf{H}_1 \vdots \mathbf{H}_2] = \begin{bmatrix} 0 & 1 \\ 1 & 0 \\ 0 & 0 \\ 1 & 0 \end{bmatrix}$$

By use of the state feedback control $\mathbf{u}(k) = -\mathbf{K}\mathbf{x}(k)$, we wish to place the closed-loop poles at the following locations:

$$z_1 = 0.5 + j0.5, \qquad z_2 = 0.5 - j0.5$$

$$z_3 = -0.2, \qquad z_4 = -0.8$$

Determine the required state feedback gain matrix $\mathbf{K}$. Then, using the given $\mathbf{G}$ and $\mathbf{H}$ matrices, derive Equation (C–16).

Solution   We shall first examine the controllability matrix:

$$[\mathbf{H} \vdots \mathbf{GH} \vdots \mathbf{G}^2\mathbf{H} \vdots \mathbf{G}^3\mathbf{H}] = [\mathbf{H}_1 \vdots \mathbf{H}_2 \vdots \mathbf{GH}_1 \vdots \mathbf{GH}_2 \vdots \mathbf{G}^2\mathbf{H}_1 \vdots \mathbf{G}^2\mathbf{H}_2 \vdots \mathbf{G}^3\mathbf{H}_1 \vdots \mathbf{G}^3\mathbf{H}_2]$$

$$= \begin{bmatrix} 0 & 1 & 1 & -1 & -3 & 2 & 11 & -5 \\ 1 & 0 & -2 & 1 & 8 & -3 & -22 & 11 \\ 0 & 0 & 3 & 0 & -3 & 3 & 15 & -6 \\ 1 & 0 & 1 & 1 & 2 & 0 & -1 & 2 \end{bmatrix} \qquad \text{(C–33)}$$

The rank of this controllability matrix is 4. Thus, arbitrary pole placement is possible. Four linearly independent vectors are chosen starting from the left end. (These vectors are shown enclosed by dashed lines.) The four linearly independent vectors chosen are $\mathbf{H}_1$, $\mathbf{H}_2$, $\mathbf{GH}_1$, and $\mathbf{G}^2\mathbf{H}_1$. Now we rearrange these four vectors according to Equation (C–10) and define matrix $\mathbf{F}$ as follows:

$$\mathbf{F} = [\mathbf{H}_1 \vdots \mathbf{GH}_1 \vdots \mathbf{G}^2\mathbf{H}_1 \vdots \mathbf{H}_2]$$

We note that $n_1 = 3$ and $n_2 = 1$ in this case. Rewriting matrix $\mathbf{F}$, we have

$$\mathbf{F} = \begin{bmatrix} 0 & 1 & -3 & 1 \\ 1 & -2 & 8 & 0 \\ 0 & 3 & -3 & 0 \\ 1 & 1 & 2 & 0 \end{bmatrix} = \begin{bmatrix} \mathbf{A} & \vdots & \mathbf{B} \\ \mathbf{C} & \vdots & \mathbf{D} \end{bmatrix}$$

Next, we compute $\mathbf{F}^{-1}$. Referring to Appendix A, we have

$$\mathbf{F}^{-1} = \begin{bmatrix} \mathbf{A}^{-1} + \mathbf{A}^{-1}\mathbf{B}(\mathbf{D} - \mathbf{CA}^{-1}\mathbf{B})^{-1}\mathbf{CA}^{-1} & -\mathbf{A}^{-1}\mathbf{B}(\mathbf{D} - \mathbf{CA}^{-1}\mathbf{B})^{-1} \\ -(\mathbf{D} - \mathbf{CA}^{-1}\mathbf{B})^{-1}\mathbf{CA}^{-1} & (\mathbf{D} - \mathbf{CA}^{-1}\mathbf{B})^{-1} \end{bmatrix}$$

$$= \begin{bmatrix} 0 & -1 & -\frac{4}{3} & 2 \\ 0 & \frac{1}{3} & \frac{2}{3} & -\frac{1}{3} \\ 0 & \frac{1}{3} & \frac{1}{3} & -\frac{1}{3} \\ 1 & \frac{2}{3} & \frac{1}{3} & -\frac{2}{3} \end{bmatrix}$$

(The same result can be obtained easily by use of MATLAB.) Since $n_1 = 3$ and $n_2 = 1$, we choose the third row vector as $\mathbf{f}_1$ and the fourth row vector as $\mathbf{f}_2$. (Note that we define the η_ith row vector, where $\eta_i = n_1 + n_2 + \cdots + n_i$, as $\mathbf{f}_i$.) That is,

$$\mathbf{f}_1 = \begin{bmatrix} 0 & \frac{1}{3} & \frac{1}{3} & -\frac{1}{3} \end{bmatrix}$$

$$\mathbf{f}_2 = \begin{bmatrix} 1 & \frac{2}{3} & \frac{1}{3} & -\frac{2}{3} \end{bmatrix}$$

Next, we define the transformation matrix $\mathbf{T}$ by

$$\mathbf{T} = \begin{bmatrix} \mathbf{S}_1 \\ \mathbf{S}_2 \end{bmatrix}^{-1}$$

where

$$\mathbf{S}_1 = \begin{bmatrix} \mathbf{f}_1 \\ \mathbf{f}_1\mathbf{G} \\ \mathbf{f}_1\mathbf{G}^2 \end{bmatrix}, \qquad \mathbf{S}_2 = [\mathbf{f}_2]$$

Hence,

$$\mathbf{T} = \begin{bmatrix} 0 & \frac{1}{3} & \frac{1}{3} & -\frac{1}{3} \\ 0 & -\frac{1}{3} & 0 & \frac{1}{3} \\ 0 & \frac{2}{3} & -\frac{1}{3} & \frac{1}{3} \\ 1 & \frac{2}{3} & \frac{1}{3} & -\frac{2}{3} \end{bmatrix}^{-1} = \begin{bmatrix} -1 & 1 & 0 & 1 \\ 1 & 0 & 1 & 0 \\ 3 & 3 & 0 & 0 \\ 1 & 3 & 1 & 0 \end{bmatrix}$$

With this transformation matrix $\mathbf{T}$, if we define

$$\mathbf{x}(k) = \mathbf{T}\hat{\mathbf{x}}(k)$$

then

$$\hat{\mathbf{G}} = \mathbf{T}^{-1}\mathbf{GT} = \begin{bmatrix} 0 & 1 & 0 & | & 0 \\ 0 & 0 & 1 & | & 0 \\ 0 & 3 & -2 & | & 1 \\ \hline 2 & 0 & 0 & | & -1 \end{bmatrix} \tag{C-34}$$

Also,

$$\hat{\mathbf{H}} = \mathbf{T}^{-1}\mathbf{H} = \begin{bmatrix} 0 & \frac{1}{3} & \frac{1}{3} & -\frac{1}{3} \\ 0 & -\frac{1}{3} & 0 & \frac{1}{3} \\ 0 & \frac{2}{3} & -\frac{1}{3} & \frac{1}{3} \\ 1 & \frac{2}{3} & \frac{1}{3} & -\frac{2}{3} \end{bmatrix}\begin{bmatrix} 0 & 1 \\ 1 & 0 \\ 0 & 0 \\ 1 & 0 \end{bmatrix} = \begin{bmatrix} 0 & 0 \\ 0 & 0 \\ 1 & 0 \\ 0 & 1 \end{bmatrix} \tag{C-35}$$

Now we shall determine the state feedback gain matrix $\mathbf{K}$, where

$$\mathbf{K} = \mathbf{B\Delta T}^{-1}$$

Referring to Equation (C–24) and noting that $b_{12} = 0$ in this case, matrix **B** is a 2×2 matrix given by

$$\mathbf{B} = \begin{bmatrix} 1 & b_{12} \\ 0 & 1 \end{bmatrix}^{-1} = \begin{bmatrix} 1 & 0 \\ 0 & 1 \end{bmatrix} \qquad (C–36)$$

For the present case, $\boldsymbol{\Delta}$ is a 2×4 matrix:

$$\boldsymbol{\Delta} = \begin{bmatrix} \delta_{11} & \delta_{12} & \delta_{13} & \delta_{14} \\ \delta_{21} & \delta_{22} & \delta_{23} & \delta_{24} \end{bmatrix}$$

Hence,

$$\hat{\mathbf{G}} - \hat{\mathbf{H}}\mathbf{B}\boldsymbol{\Delta} = \begin{bmatrix} 0 & 1 & 0 & \vdots & 0 \\ 0 & 0 & 1 & \vdots & 0 \\ -\delta_{11} & 3 - \delta_{12} & -2 - \delta_{13} & \vdots & 1 - \delta_{14} \\ \hline 2 - \delta_{21} & -\delta_{22} & -\delta_{23} & \vdots & -1 - \delta_{24} \end{bmatrix}$$

Referring to Equation (C–32), we have

$$|z\mathbf{I} - \hat{\mathbf{G}} + \hat{\mathbf{H}}\mathbf{B}\boldsymbol{\Delta}| = \begin{vmatrix} z^3 + (2 + \delta_{13})z^2 + (-3 + \delta_{12})z + \delta_{11} & \vdots & -1 + \delta_{14} \\ \hline \delta_{23}z^2 + \delta_{22}z + (-2 + \delta_{21}) & \vdots & z + 1 + \delta_{24} \end{vmatrix} = 0$$

This characteristic equation must be equal to the desired characteristic equation, which is

$$(z - 0.5 - j0.5)(z - 0.5 + j0.5)(z + 0.2)(z + 0.8) = z^4 - 0.34z^2 + 0.34z + 0.08 = 0$$

If we equate the coefficients of the equal powers of z of the two characteristic equations, we will have four equations for the determination of eight δ's. Hence, matrix $\boldsymbol{\Delta}$ is not unique. Suppose we arbitrarily choose

$$\delta_{14} = 0, \qquad \delta_{22} = 0, \qquad \delta_{23} = 0, \qquad \delta_{24} = -1$$

Then

$$|z\mathbf{I} - \hat{\mathbf{G}} + \hat{\mathbf{H}}\mathbf{B}\boldsymbol{\Delta}| = z^4 + (2 + \delta_{13})z^3 + (-3 + \delta_{12})z^2 + \delta_{11}z - 2 + \delta_{21} = 0$$

By equating this characteristic equation with the desired characteristic equation, we have

$$\delta_{11} = 0.34$$
$$\delta_{12} = 2.66$$
$$\delta_{13} = -2$$
$$\delta_{21} = 2.08$$

Thus,

$$\boldsymbol{\Delta} = \begin{bmatrix} 0.34 & 2.66 & -2 & 0 \\ 2.08 & 0 & 0 & -1 \end{bmatrix}$$

Then matrix **K** is obtained as follows:

$$\mathbf{K} = \mathbf{B}\boldsymbol{\Delta}\mathbf{T}^{-1} = \begin{bmatrix} 0 & -2.1067 & 0.7800 & 0.1067 \\ -1 & 0.02667 & 0.3600 & -0.02667 \end{bmatrix}$$

With the matrix **K** thus determined, state feedback control

$$u(k) = -\mathbf{K}\mathbf{x}(k)$$

will place the closed-loop poles at $z_1 = 0.5 + j0.5$, $z_2 = 0.5 - j0.5$, $z_3 = -0.2$, and $z_4 = -0.8$. It is noted that matrix **K** is not unique; there are many other possible matrices for **K**.

Finally, we shall derive Equation (C–16). Notice that

$$\mathbf{F}^{-1}\mathbf{F} = \begin{bmatrix} \mathbf{m}_1 \\ \mathbf{m}_2 \\ \mathbf{f}_1 \\ \mathbf{f}_2 \end{bmatrix} [\mathbf{H}_1 \quad \mathbf{GH}_1 \quad \mathbf{G}^2\mathbf{H}_1 \quad \mathbf{H}_2]$$

$$= \begin{bmatrix} \mathbf{m}_1\mathbf{H}_1 & \mathbf{m}_1\mathbf{GH}_1 & \mathbf{m}_1\mathbf{G}^2\mathbf{H}_1 & \mathbf{m}_1\mathbf{H}_2 \\ \mathbf{m}_2\mathbf{H}_1 & \mathbf{m}_2\mathbf{GH}_1 & \mathbf{m}_2\mathbf{G}^2\mathbf{H}_1 & \mathbf{m}_2\mathbf{H}_2 \\ \mathbf{f}_1\mathbf{H}_1 & \mathbf{f}_1\mathbf{GH}_1 & \mathbf{f}_1\mathbf{G}^2\mathbf{H}_1 & \mathbf{f}_1\mathbf{H}_2 \\ \mathbf{f}_2\mathbf{H}_1 & \mathbf{f}_2\mathbf{GH}_1 & \mathbf{f}_2\mathbf{G}^2\mathbf{H}_1 & \mathbf{f}_2\mathbf{H}_2 \end{bmatrix} = \begin{bmatrix} 1 & 0 & 0 & 0 \\ 0 & 1 & 0 & 0 \\ 0 & 0 & 1 & 0 \\ 0 & 0 & 0 & 1 \end{bmatrix}$$

where $\mathbf{m}_1$ and $\mathbf{m}_2$ are the first row vector and second row vector of $\mathbf{F}^{-1}$, respectively. Since $\mathbf{F}^{-1}\mathbf{F}$ is an identity matrix, $\mathbf{f}_1\mathbf{H}_1 = 0$, $\mathbf{f}_1\mathbf{H}_2 = 0$, $\mathbf{f}_1\mathbf{GH}_1 = 0$, $\mathbf{f}_1\mathbf{G}^2\mathbf{H}_1 = 1$, $\mathbf{f}_2\mathbf{H}_1 = 0$, and $\mathbf{f}_2\mathbf{H}_2 = 1$. From Equation (C–33) we see that $\mathbf{GH}_2$ is linearly dependent on $\mathbf{H}_1$, $\mathbf{H}_2$, and $\mathbf{GH}_1$. Hence, $\mathbf{f}_1\mathbf{GH}_2 = \alpha\mathbf{f}_1\mathbf{H}_1 + \beta\mathbf{f}_1\mathbf{H}_2 + \gamma\mathbf{f}_1\mathbf{GH}_1 = 0$, where α, β, and γ are constants. Note that $\mathbf{f}_1\mathbf{G}^2\mathbf{H}_2$ may or may not be zero. Consequently,

$$\hat{\mathbf{H}} = \mathbf{T}^{-1}\mathbf{H} = \begin{bmatrix} \mathbf{f}_1 \\ \mathbf{f}_1\mathbf{G} \\ \mathbf{f}_1\mathbf{G}^2 \\ \mathbf{f}_2 \end{bmatrix} [\mathbf{H}_1 \quad \mathbf{H}_2] = \begin{bmatrix} \mathbf{f}_1\mathbf{H}_1 & \mathbf{f}_1\mathbf{H}_2 \\ \mathbf{f}_1\mathbf{GH}_1 & \mathbf{f}_1\mathbf{GH}_2 \\ \mathbf{f}_1\mathbf{G}^2\mathbf{H}_1 & \mathbf{f}_1\mathbf{G}^2\mathbf{H}_2 \\ \mathbf{f}_2\mathbf{H}_1 & \mathbf{f}_2\mathbf{H}_2 \end{bmatrix} = \begin{bmatrix} 0 & 0 \\ 0 & 0 \\ 1 & b_{12} \\ 0 & 1 \end{bmatrix}$$

where $b_{12} = \mathbf{f}_1\mathbf{G}^2\mathbf{H}_2$. This last equation is Equation (C–16).

Problem C–5

Referring to Problem C–4, consider the same system. Suppose that we desire the deadbeat response to an arbitrary initial state $\mathbf{x}(0)$. Determine the state feedback gain matrix **K**.

Solution   Referring to Equations (C–34), (C–35), and (C–36), we have

$$\hat{\mathbf{G}} = \begin{bmatrix} 0 & 1 & 0 & 0 \\ 0 & 0 & 1 & 0 \\ 0 & 3 & -2 & 1 \\ 2 & 0 & 0 & -1 \end{bmatrix}, \qquad \hat{\mathbf{H}} = \begin{bmatrix} 0 & 0 \\ 0 & 0 \\ 1 & 0 \\ 0 & 1 \end{bmatrix}$$

$$\mathbf{B} = \begin{bmatrix} 1 & b_{12} \\ 0 & 1 \end{bmatrix}^{-1} = \begin{bmatrix} 1 & 0 \\ 0 & 1 \end{bmatrix}$$

where b_{12} is zero. For the deadbeat response, we choose $\boldsymbol{\Delta}$ as follows:

$$\boldsymbol{\Delta} = \begin{bmatrix} -a_{11} & -a_{12} & -a_{13} & -a_{14} \\ -a_{21} & -a_{22} & -a_{23} & -a_{24} \end{bmatrix} = \begin{bmatrix} 0 & 3 & -2 & 1 \\ 2 & 0 & 0 & -1 \end{bmatrix}$$

where the a_{ij}'s are as defined in Equation (C–15). Then

$$\hat{\mathbf{G}} - \hat{\mathbf{H}}\mathbf{B}\boldsymbol{\Delta} = \begin{bmatrix} 0 & 1 & 0 & 0 \\ 0 & 0 & 1 & 0 \\ 0 & 0 & 0 & 0 \\ 0 & 0 & 0 & 0 \end{bmatrix}$$

and we find

$$(\hat{\mathbf{G}} - \hat{\mathbf{H}}\mathbf{B}\boldsymbol{\Delta})^k = \mathbf{0}, \qquad k = 3, 4, 5, \ldots$$

The deadbeat response is reached in at most three sampling periods. [Note that in this problem $n_1 = 3$ and $n_2 = 1$. Hence, $n_{min} = \max(n_1, n_2) = 3$.] The desired state feedback gain matrix $\mathbf{K}$ is obtained as follows:

$$\mathbf{K} = \mathbf{B}\mathbf{\Delta}\mathbf{T}^{-1}$$

$$= \begin{bmatrix} 1 & 0 \\ 0 & 1 \end{bmatrix} \begin{bmatrix} 0 & 3 & -2 & 1 \\ 2 & 0 & 0 & -1 \end{bmatrix} \begin{bmatrix} 0 & \frac{1}{3} & \frac{1}{3} & -\frac{1}{3} \\ 0 & -\frac{1}{3} & 0 & \frac{1}{3} \\ 0 & \frac{2}{3} & -\frac{1}{3} & \frac{1}{3} \\ 1 & \frac{2}{3} & \frac{1}{3} & -\frac{2}{3} \end{bmatrix}$$

$$= \begin{bmatrix} 1 & -\frac{5}{3} & 1 & -\frac{1}{3} \\ -1 & 0 & \frac{1}{3} & 0 \end{bmatrix}$$

With this matrix $\mathbf{K}$, the state feedback control

$$\mathbf{u}(k) = -\mathbf{K}\mathbf{x}(k)$$

will place the four closed-loop poles at the origin and thus will produce the deadbeat response to any initial state $\mathbf{x}(0)$.

References

A–1. Antoniou, A., *Digital Filters: Analysis and Design*. New York: McGraw-Hill Book Company, 1979.

A–2. Aseltine, J. A., *Transform Method in Linear System Analysis*. New York: McGraw-Hill Book Company, 1958.

A–3. Åström, K. J., and B. Wittenmark, *Computer Controlled Systems: Theory and Design*. Englewood Cliffs, N.J.: Prentice Hall, Inc., 1984.

B–1. Bellman, R., *Introduction to Matrix Analysis*. New York: McGraw-Hill Book Company, 1960.

B–2. Bristol, E. H., "Design and Programming Control Algorithms for DDC Systems," *Control Engineering*, **24**, Jan. 1977, pp. 24–26.

B–3. Butman, S., and R. Sivan (Sussman), "On Cancellations, Controllability and Observability," *IEEE Trans. Automatic Control*, **AC-9** (1964), pp. 317–18.

C–1. Cadzow, J. A., and H. R. Martens, *Discrete-Time and Computer Control Systems*. Englewood Cliffs, N.J.: Prentice Hall, Inc., 1970.

C–2. Chan, S. W., G. C. Goodwin, and K. S. Sin, "Convergence Properties of the Riccati Difference Equation in Optimal Filtering of Nonstabilizable Systems," *IEEE Trans. Automatic Control*, **AC-29** (1984), pp. 110–18.

C–3. Churchill, R. V., and J. W. Brown, *Complex Variables and Applications*, 4th ed., New York: McGraw-Hill Book Company, 1984.

D–1. Dorato, P., and A. H. Levis, "Optimal Linear Regulators: The Discrete-Time Case," *IEEE Trans. Automatic Control*, **AC-16** (1971), pp. 613–20.

E–1. Evans, W. R., "Control System Synthesis by Root Locus Method," *AIEE Trans. Part II*, **69** (1950), pp. 66–69.

F–1. Falb, P. L., and M. Athans, "A Direct Constructive Proof of the Criterion for Complete Controllability of Time-Invariant Linear Systems," *IEEE Trans. Automatic Control*, **AC-9** (1964), pp. 189–90.

F–2. Fortmann, T. E., "A Matrix Inversion Identity," *IEEE Trans. Automatic Control*, **AC-15** (1970), p. 599.

F–3. Franklin, G. F., J. D. Powell, and M. L. Workman, *Digital Control of Dynamic Systems*, 2nd ed., Reading, Mass.: Addison-Wesley Publishing Co., Inc., 1990.

F–4. Freeman, H., *Discrete-Time Systems*. New York: John Wiley & Sons, Inc., 1965.

G–1. Gantmacher, F. R., *Theory of Matrices*, Vols. I and II. New York: Chelsea Publishing Co., Inc., 1959.

G–2. Gopinath, B., "On the Control of Linear Multiple Input–Output Systems," *Bell Syst. Tech. J.*, **50** (1971), pp. 1063–81.

H–1. Hahn, W., *Theory and Application of Liapunov's Direct Method*. Englewood Cliffs, N.J.: Prentice Hall, Inc., 1963.

H–2. Halmos, P. R., *Finite Dimensional Vector Spaces*. Princeton, N.J.: D. Van Nostrand Company, 1958.

I–1. Ichikawa, K., *Theory for Design of Control Systems* (in Japanese). Tokyo: Gijutsu-Shoin, 1989.

J–1. Jerri, A. J., "The Shannon Sampling Theorem—Its Various Extensions and Applications: A Tutorial Review," *Proc. IEEE*, **65** (1977), pp. 1565–95.

J–2. Jury, E. I., "Hidden Oscillations in Sampled-Data Control Systems," *AIEE Trans. Part II*, **75** (1956), pp. 391–95.

J–3. Jury, E. I., *Sampled-Data Control Systems*. New York: John Wiley & Sons, Inc., 1958.

J–4. Jury, E. I., "Sampling Schemes in Sampled-Data Control Systems," *IRE Trans. Automatic Control*, **AC-6** (1961), pp. 88–90.

J–5. Jury, E. I., *Theory and Applications of the z Transform Method*. New York: John Wiley & Sons, Inc., 1964.

J–6. Jury, E. I., "A General *z*-Transform Formula for Sampled-Data Systems," *IEEE Trans. Automatic Control*, **AC-12** (1967), pp. 606–8.

J–7. Jury, E. I., "Sampled-Data Systems, Revisited: Reflections, Recollections, and Reassessments," *ASME J. Dynamic Systems, Measurement, and Control*, **102** (1980), pp. 208–16.

J–8. Jury, E. I., and J. Blanchard, "A Stability Test for Linear Discrete-Time Systems in Table Forms," *Proc. IRE*, **49** (1961), pp. 1947–48.

K–1. Kailath, T., *Linear Systems*. Englewood Cliffs, N.J.: Prentice Hall, Inc., 1980.

K–2. Kailath, T., and P. Frost, "An Innovations Approach to Least-Squares Estimation, Part II: Linear Smoothing in Additive White Noise," *IEEE Trans. Automatic Control*, **AC-13** (1968), pp. 655–60.

K–3. Kalman, R. E., "On the General Theory of Control Systems," *Proc. First Intern. Cong. IFAC*, Moscow, 1960. *Automatic and Remote Control*. London: Butterworth & Co., Ltd., 1961, pp. 481–92.

K–4. Kalman, R. E., and J. E. Bertram, "Control System Analysis and Design via the Second Method of Lyapunov: I. Continuous-Time Systems; II. Discrete-Time Systems," *ASME J. Basic Engineering*, ser. **D**, **82** (1960), pp. 371–93, 394–400.

K–5. Kalman, R. E., Y. C. Ho, and K. S. Narendra, "Controllability of Linear Dynamical Systems," *Contributions to Differential Equations*, **1** (1963), pp. 189–213.

K–6. Kanai, K., and N. Hori, *Introduction to Digital Control Systems* (in Japanese). Tokyo: Maki Shoten, 1992.

K–7. Katz, P., *Digital Control Using Microprocessors*. London: Prentice Hall International, Inc., 1981.

K–8. Kreindler, E., and P. E. Sarachik, "On the Concepts of Controllability and Observability of Linear Systems," *IEEE Trans. Automatic Control*, **AC-9** (1964), pp. 129–36.

K–9. Kuo, B. C., *Digital Control Systems*. New York: Holt, Rinehart and Winston, Inc., 1980.

L–1. LaSalle, J. P., and S. Lefschetz, *Stability by Liapunov's Direct Method with Applications*. New York: Academic Press, Inc., 1961.

L–2. Lee, E. B., and L. Markus, *Foundations of Optimal Control Theory*. New York: John Wiley & Sons, Inc., 1967.

L–3. Leondes, C. T., and M. Novak, "Reduced-Order Observers for Linear Discrete-Time Systems," *IEEE Trans. Automatic Control*, **AC-19** (1974), pp. 42–46.

L–4. Li, Y. T., J. L. Meiry, and R. E. Curry, "On the Ideal Sampler Approximation," *IEEE Trans. Automatic Control*, **AC-17** (1972), pp. 167–68.

L–5. Luenberger, D. G., "Observing the State of a Linear System," *IEEE Trans. Military Electronics*, **MIL-8** (1964), pp. 74–80.

L–6. Luenberger, D. G., "An Introduction to Observers," *IEEE Trans. Automatic Control*, **AC-16** (1971), pp. 596–602.

M–1. Melsa, J. L., and D. G. Schultz, *Linear Control Systems*. New York: McGraw-Hill Book Company, 1969.

M–2. Middleton, R. H., and G. C. Goodwin, *Digital Control and Estimation—A Unified Approach*. Englewood Cliffs, N.J.: Prentice Hall, Inc., 1990.

M–3. Mitra, S. K., and R. J. Sherwood, "Canonic Realizations of Digital Filters Using the Continued Fraction Expansion," *IEEE Trans. Audio and Electroacoustics*, **AU-20** (1972), pp. 185–94.

M–4. Mitra, S. K., and R. J. Sherwood, "Digital Ladder Networks," *IEEE Trans. Audio and Electroacoustics*, **AU-21** (1973), pp. 30–36.

N–1. Neuman, C. P., and C. S. Baradello, "Digital Transfer Functions for Microcomputer Control.," *IEEE Trans. Systems, Man, and Cybernetics*, **SMC-9** (1979), pp. 856–60.

N–2. Noble, B., and J. Daniel, *Applied Linear Algebra*, 2nd ed. Englewood Cliffs, N.J.: Prentice Hall, Inc., 1977.

O–1. Ogata, K., *State Space Analysis of Control Systems*. Englewood Cliffs, N.J.: Prentice Hall, Inc., 1967.

O–2. Ogata, K., *Modern Control Engineering*, 2nd ed. Englewood Cliffs, N.J.: Prentice Hall, Inc., 1990.

O–3. Ogata, K., *System Dynamics*, 2nd ed., Englewood Cliffs, N.J.: Prentice Hall, Inc., 1992.

O–4. Ogata, K., *Solving Control Engineering Problems with MATLAB*. Englewood Cliffs, N.J.: Prentice Hall, Inc., 1994.

O–5. Ogata, K., *Designing Linear Control Systems with MATLAB*. Englewood Cliffs, N.J.: Prentice Hall, Inc., 1994.

P–1. Pappas, T., A. J. Laub, and N. R. Sandell, Jr., "On the Numerical Solution of the Discrete-Time Algebraic Riccati Equation," *IEEE Trans. Automatic Control*, **AC-25** (1980), pp. 631–41.

P–2. Payne, H. J., and L. M. Silverman, "On the Discrete Time Algebraic Riccati Equation," *IEEE Trans. Automatic Control*, **AC-18** (1973), pp. 226–34.

P–3. Phillips, C. L., and H. T. Nagle, Jr., *Digital Control Systems Analysis and Design*. Englewood Cliffs, N.J.: Prentice Hall, Inc., 1984.

R–1. Ragazzini, J. R., and G. F. Franklin, *Sampled-Data Control Systems*. New York: McGraw-Hill Book Company, 1958.

R–2. Ragazzini, J. R., and L. A. Zadeh, "The Analysis of Sampled-Data Systems," *AIEE Trans. Part II*, **71** (1952), pp. 225–34.

S–1. Strang, G., *Linear Algebra and Its Applications*. New York: Academic Press, Inc., 1976.

T–1. Tou, J. T., *Digital and Sampled-Data Control Systems*. New York: McGraw-Hill Book Company, 1959.

T–2. Turnbull, H. W., and A. C. Aitken, *An Introduction to the Theory of Canonical Matrices*. London: Blackie and Son, Ltd., 1932.

V–1. Van Dooren, P., "A Generalized Eigenvalue Approach for Solving Riccati Equations," *SIAM J. Scientific and Statistical Computing*, **2** (1981), pp. 121–35.

W–1. Willems, J. C., and S. K. Mitter, "Controllability, Observability, Pole Allocation, and State Reconstruction," *IEEE Trans. Automatic Control*, **AC-16** (1971), pp. 582–95.

W–2. Wolovich, W. A., *Linear Multivariable Systems*. New York: Springer-Verlag, 1974.

W–3. Wonham, W. M., "On Pole Assignment in Multi-Input Controllable Linear Systems," *IEEE Trans. Automatic Control*, **AC-12** (1967), pp. 660–65.

Z–1. Zadeh, L. A., and C. A. Desoer, *Linear System Theory: The State Space Approach*. New York: McGraw-Hill Book Company, 1963.

Index

The
MATH
WORKS
Inc.

BUSINESS REPLY MAIL

FIRST CLASS PERMIT NO. 82 NATICK, MA

POSTAGE WILL BE PAID BY ADDRESSEE

The MathWorks, Inc.
24 Prime Park Way
Natick, MA 01760-9889